Grundlehren der mathematischen Wissenschaften 347

A Series of Comprehensive Studies in Mathematics

For further volumes:
www.springer.com/series/138

Camille Laurent-Gengoux • Anne Pichereau • Pol Vanhaecke

Poisson Structures

Springer

Camille Laurent-Gengoux
CNRS UMR 7122, Laboratoire de Mathématiques
Université de Lorraine
Metz, France

Anne Pichereau
CNRS UMR 5208, Institut Camille Jordan
Université Jean Monnet
Saint-Etienne, France

Pol Vanhaecke
CNRS UMR 7348, Lab. Mathématiques et Applications
Université de Poitiers
Futuroscope Chasseneuil, France

ISSN 0072-7830 Grundlehren der mathematischen Wissenschaften
ISBN 978-3-642-31089-8 ISBN 978-3-642-31090-4 (eBook)
DOI 10.1007/978-3-642-31090-4
Springer Heidelberg New York Dordrecht London

Library of Congress Control Number: 2012947315

Mathematics Subject Classification: 17B63, 53D17, 17B80, 53D55, 17B62

Printed on acid-free paper

Springer is part of Springer Science+Business Media (www.springer.com)

For Clio

Preface

Poisson structures naturally appear in very different forms and contexts. Symplectic manifolds, Lie algebras, singularity theory, r-matrices, for example, all lead to a certain type of Poisson structures, sharing several features, despite the distances between the mathematics they originate from. This observation motivated us to bring the different worlds in which Poisson structures live together in a single volume, providing a multitude of entrances to the book, and hence to the subject. Thus, the idea of the body of the book, Part II, was born.

It is well known, i.e., it is commonly agreed upon by the experts, that results and techniques which are valid for one type of Poisson structure ought to apply, *mutatis mutandis*, to other types of Poisson structures. When starting to write Part I of the book, we soon realized that not everything could be derived from the general concept of a Poisson algebra, so we faced the challenge of presenting the concepts and the results in detail, for the algebraic, algebraic-geometric and geometric contexts, without copy-pasting large parts of the text two or three times. But as the writing moved on, both the algebraic and geometric contexts kept imposing themselves; finally, each found its proper place, appearing as being complementary and dual to the other, rather than a consequence or rephrasing, one of the other. It added, unexpectedly, an extra dimension to the book.

It was pointed out by one of the referees that the main applications of Poisson structures should be present in a book which has Poisson structures as its main subject. This was quite another challenge, giving rise to the third and final part of the book, Part III, undoubtedly an important addition.

Several years were necessary for this project, a big part was done at Poitiers, when the three of us were appointed to the "Laboratoire de Mathématiques et Applications". We were spending long hours together in what our colleagues called *The Aquarium* (for an explanation of the name, look up "poisson" in a French dictionary). When two of us moved away from Poitiers, we sometimes worked together in other places, which always offered us a pleasant and stimulating working atmosphere. Thus, we are happy to acknowledge the hospitality of our colleagues from the mathematics departments at the universities of Poitiers, Antwerp, Coimbra,

Saint-Etienne and at the CRM in Barcelona and the Max-Planck Institute in Bonn. Each one of us was partially supported by an ANR contract (TcChAm, DPSing and GIMP, respectively), which made it possible to keep working on a blackboard, rather than seeing each other on a computer screen.

We learned Poisson structures from our teachers, friends and collaborators: Mark Adler, Paulo Antunes, Pantelis Damianou, Rui Fernandes, Benoit Fresse, Eva Miranda, Joana Nunes da Costa, Marco Pedroni, Michael Penkava, Claude Roger, Pierre van Moerbeke, Yvette Kosmann-Schwarzbach, Hervé Sabourin, Mathieu Stiénon, Friedrich Wagemann, Alan Weinstein, Ping Xu and Marco Zambon. The multitude of different points of view on the subject, which we learned from them, have given a quite specific flavor to the book. Even if a book cannot replace what one learns through lectures, discussions and collaborations, it is our hope that this book may further transfer what we learned from them and that it invites other researchers to explore the subject of Poisson structures, which turns out to be as diverse and rich as the colorful fauna and flora which inhabit the bottom of our oceans.

April 1, 2012 Camille Laurent-Gengoux, Anne Pichereau and Pol Vanhaecke

Contents

Introduction

Poisson structures appear in a large variety of different contexts, ranging from string theory, classical/quantum mechanics and differential geometry to abstract algebra, algebraic geometry and representation theory. In each one of these contexts, it turns out that the Poisson structure is not a theoretical artifact, but a key element which, unsolicited, comes along with the problem which is investigated and its delicate properties are in basically all cases decisive for the solution to the problem.

Hamiltonian mechanics and integrable systems. A first striking example of this phenomenon appears in classical mechanics. Poisson's classical bracket, which is given for smooth functions F and G on $\mathbb{R}^{2r}$ by

$$\{F,G\} := \sum_{i=1}^{r} \left(\frac{\partial F}{\partial q_i}\frac{\partial G}{\partial p_i} - \frac{\partial G}{\partial q_i}\frac{\partial F}{\partial p_i} \right), \tag{0.1}$$

allows one to write Hamilton's equations of motion

$$\dot{q}_i = \frac{\partial H}{\partial p_i}, \qquad \dot{p}_i = -\frac{\partial H}{\partial q_i}, \qquad i = 1,\ldots,r, \tag{0.2}$$

associated to a Hamiltonian $H = H(q,p)$, in a coordinate-free manner, treating positions q_i and momenta p_i on an equal footing. Indeed, using (0.1), the equations of motion (0.2) can be written as

$$\dot{q}_i = \{q_i, H\}, \qquad \dot{p}_i = \{p_i, H\}, \qquad i = 1,\ldots,r\,. \tag{0.3}$$

For every solution $t \mapsto (q(t),p(t))$ to these differential equations, it follows from (0.1) and (0.3) that

$$\frac{\mathrm{d}}{\mathrm{d}t} F(q(t),p(t)) = \{F,H\}\,(q(t),p(t))\,, \tag{0.4}$$

for all smooth functions F on $\mathbb{R}^{2r}$.

In differential geometrical terms, the Poisson bracket defines an operator which associates to every smooth function H on $\mathbb{R}^{2r}$ a vector field $\mathscr{X}_H$ on $\mathbb{R}^{2r}$. In terms of local coordinates, the vector field $\mathscr{X}_H$ is given by (0.3) and the variation of a smooth function F on $\mathbb{R}^{2r}$ along any integral curve $(q(t), p(t))$ of the vector field $\mathscr{X}_H$ is given by (0.4). In fact, when the function H is the Hamiltonian (the energy) of the system, the integral curves of $\mathscr{X}_H$ describe the time evolution of the system. Thus, the equations of motion of classical mechanics come along with an associated Poisson bracket; for simple systems, such as systems of particles with a classical interaction, the Poisson structure is the above one, but for constrained and reduced systems, the Poisson bracket takes a more complicated form.

The rôle which the Poisson structure plays in solving the problems of classical mechanics was known partly to Poisson, who pointed out that the vanishing of $\{F,H\}$ and $\{G,H\}$ implies the vanishing of $\{\{F,G\},H\}$. In mechanical terms, this means in view of (0.4) that the Poisson bracket of two constants of motion is again a constant of motion. Since constants of motion are a great help in the explicit integration of Hamilton's equations, Poisson's theorem brings to light the Poisson bracket as a tool for integration.

A strong amplification of this fact is the Liouville theorem, which states that r independent constants of motion $H_1 = H, H_2, \ldots, H_r$ of (0.3) suffice for integrating Hamilton's equations by quadratures, under the hypothesis that these functions commute for the Poisson bracket (are in involution). A yet further amplification is the action-angle theorem, which states that in the neighborhood of a compact submanifold, traced out by the flows of the (commuting!) vector fields $\mathscr{X}_{H_i}$, there exist $(S^1)^r \times \mathbb{R}^r$-valued coordinates $(\theta_1, \ldots, \theta_r, \rho_1, \ldots, \rho_r)$ in which the Poisson bracket takes the canonical form (0.1), namely

$$\{F,G\} = \sum_{i=1}^{r} \left(\frac{\partial F}{\partial \theta_i} \frac{\partial G}{\partial \rho_i} - \frac{\partial G}{\partial \theta_i} \frac{\partial F}{\partial \rho_i} \right),$$

and such that the functions $H_1, \ldots, H_r$ (including the Hamiltonian H) depend on the coordinates $\rho_1, \ldots, \rho_r$ only. According to (0.2), Hamilton's equations take in these coordinates the simple form

$$\dot{\theta}_i = \frac{\partial H}{\partial \rho_i}, \qquad \dot{\rho}_i = -\frac{\partial H}{\partial \theta_i} = 0, \qquad i = 1, \ldots, r,$$

which entails that the ρ_i are constant and hence that the θ_i are affine functions of time (since H does not depend on $\theta_1, \ldots, \theta_r$).

Thus, in a nutshell, the equations of motion of classical mechanics come with a Poisson bracket, the Poisson bracket is decisive for their integrability and it permits us to construct coordinates in which the problem (including the Poisson bracket) takes a very simple form, from which the solutions and their characteristics can be read off at once.

Abstraction and generalization. Before giving another example of the key rôle played by Poisson brackets, we explain the algebraic and geometric abstraction of the classical Poisson bracket (0.1). It is well known that Poisson's theorem is explained by the Jacobi identity

$$\{\{F,G\},H\}+\{\{G,H\},F\}+\{\{H,F\},G\}=0\,, \tag{0.5}$$

valid for arbitrary smooth functions F,G and H on $\mathbb{R}^{2r}$. It is however noteworthy to point out that Jacobi discovered this identity only thirty years after Poisson announced his theorem. Taking arbitrary smooth functions $x_{ij}=-x_{ji}$ defined on a non-empty open subset of $\mathbb{R}^d$, with coordinates $x_1,\dots,x_d$, the skew-symmetric bilinear operation $\{\cdot\,,\cdot\}$, defined on smooth functions by

$$\{F,G\}:=\sum_{i,j=1}^{d} x_{ij}\frac{\partial F}{\partial x_i}\frac{\partial G}{\partial x_j}\,, \tag{0.6}$$

satisfies the Jacobi identity if and only if the functions x_{ij} satisfy the identity

$$\sum_{\ell=1}^{d}\left(x_{\ell k}\frac{\partial x_{ij}}{\partial x_\ell}+x_{\ell i}\frac{\partial x_{jk}}{\partial x_\ell}+x_{\ell j}\frac{\partial x_{ki}}{\partial x_\ell}\right)=0\,.$$

This fact was observed by Lie, who also pointed out that under the assumption of constancy of the rank (which he assumes implicitly), there exist local coordinates $q_1,\dots,q_r,p_1,\dots,p_r,z_1,\dots,z_s$, where $d=2r+s$, such that (0.6) takes the form

$$\{F,G\}=\sum_{i=1}^{r}\left(\frac{\partial F}{\partial q_i}\frac{\partial G}{\partial p_i}-\frac{\partial G}{\partial q_i}\frac{\partial F}{\partial p_i}\right)\,,$$

which is formally the same as (0.1), but on a $d=2r+s$ dimensional space, with s of the variables being totally absent from the bracket. An important class of examples of brackets (0.6), satisfying (0.5), is defined on the dual $\mathfrak{g}^*$ of a (finite-dimensional) Lie algebra $\mathfrak{g}$, where the bracket is directly inherited by the Lie bracket. Many important facts about the Poisson bracket on $\mathfrak{g}^*$, such as the relation to the coadjoint representation of the adjoint group $\mathbf{G}$ of $\mathfrak{g}$, were observed by Lie and were rediscovered and further expanded by Berezin [21], Kirillov [105, 106], Kostant [117] and Souriau [185].

It was pointed out by Lichnerowicz in [126] that, in abstract differential geometrical terms, a Poisson structure on a smooth manifold M is a smooth bivector field π on M, satisfying $[\pi,\pi]_S=0$, where $[\cdot\,,\cdot]_S$ stands for the Schouten bracket (a natural extension of the Lie bracket of vector fields to arbitrary multivector fields). It implies that for every open subset U of M, the bilinear operation $\{\cdot\,,\cdot\}_U$, defined for $F,G\in C^\infty(U)$ by

$$\{F,G\}_U\,(m):=\langle \mathrm{d}_m F\wedge \mathrm{d}_m G,\pi_m\rangle\,, \tag{0.7}$$

for all $m\in M$, satisfies the Jacobi identity, hence defines a Lie algebra structure on $C^\infty(U)$.

Moreover, this Lie algebra structure on $C^\infty(U)$ and the associative structure on $C^\infty(U)$ (pointwise multiplication of functions) are compatible in the sense that

$$\{FG,H\}_U = F\{G,H\}_U + G\{F,H\}_U ,$$

for all $F,G,H \in C^\infty(U)$.

It is from here easy to make the algebraic abstraction of the notion of a Poisson structure, leading to the notion of a Poisson algebra. For concreteness, we give the definition in the case of a vector space $\mathscr{A}$ over a field $\mathbb{F}$ of characteristic zero (one may think of $\mathbb{R}$ or $\mathbb{C}$). Two products (bilinear maps from $\mathscr{A}$ to itself) are assumed to be given on $\mathscr{A}$, one denoted by "$\cdot$", which is assumed to be associative and commutative, while the other one, denoted by $\{\cdot,\cdot\}$, is assumed to be a Lie bracket. The triple $(\mathscr{A},\cdot,\{\cdot,\cdot\})$ is called a Poisson algebra if the two products are compatible in the sense that

$$\{F\cdot G,H\} = F\cdot\{G,H\} + G\cdot\{F,H\} , \tag{0.8}$$

for all $F,G,H \in \mathscr{A}$. It is clear that in the case of a Poisson manifold (M,π), for each open subset U of M, the triple $(C^\infty(U),\cdot,\{\cdot,\cdot\}_U)$ is a Poisson algebra, where $\cdot$ stands for the pointwise product of functions and $\{\cdot,\cdot\}_U$ stands for the Lie bracket defined by (0.7).

Deformation theory and quantization. We now come to a second striking example of Poisson structures making an unexpected appearance, and playing next a very definite rôle in the statement of the problem and in its final solution. We give a purely algebraic, but very simple and natural, introduction to the problem. In this approach, one speaks of "deformation theory". A physical interpretation of the problem will be given afterwards; one then speaks of "quantization" or of "deformation quantization".

We start with a commutative associative algebra $\mathscr{A} = (\mathscr{A},\cdot)$. The reader may think of the example of the polynomial algebra $\mathscr{A} = \mathbb{F}[x_1,\dots,x_d]$. We suppose that the product in $\mathscr{A}$ is part of a family of associative (not necessarily commutative) products, say we have a family $\star_t$ of bilinear maps

$$\begin{aligned} \star_t : \mathscr{A}\times\mathscr{A} &\to \mathscr{A} \\ (a,b) &\mapsto a\star_t b \end{aligned} \tag{0.9}$$

such that, for each fixed $t\in\mathbb{F}$, the product $\star_t$ is associative, with $\star_0$ being the original associative commutative product on $\mathscr{A}$, i.e., $(\mathscr{A},\cdot) = (\mathscr{A},\star_0)$. We refer to the family of associative products $\star_t$ on $\mathscr{A}$ as a *deformation* of $\mathscr{A}$. It is assumed that the family depends *nicely* on t, in a way which we do not make precise here. We consider, for every $t\in\mathbb{F}$, the commutator of $\star_t$, which is defined for all $a,b\in\mathscr{A}$ by

$$[a,b]_t := a\star_t b - b\star_t a .$$

Since $\star_t$ is associative, the commutator $[\cdot,\cdot]_t$ satisfies for each $t \in \mathbb{F}$ the following identities, valid for all $a,b,c \in \mathscr{A}$:

$$\begin{aligned} [a \star_t b, c]_t &= a \star_t [b,c]_t + [a,c]_t \star_t b\,, \\ [[a,b]_t, c]_t + [[b,c]_t, a]_t + [[c,a]_t, b]_t &= 0\,. \end{aligned} \tag{0.10}$$

Consider the skew-symmetric bilinear map $\{\cdot,\cdot\} : \mathscr{A} \times \mathscr{A} \to \mathscr{A}$, which is defined for $a,b \in \mathscr{A}$ by

$$\{a,b\} := \frac{\mathrm{d}}{\mathrm{d}t}_{|t=0} [a,b]_t = \frac{\mathrm{d}}{\mathrm{d}t}_{|t=0} (a \star_t b - b \star_t a)\,.$$

Then the first equation in (0.10) implies that $\{\cdot,\cdot\}$ satisfies the Leibniz property (0.8), while the second equation in (0.10) implies that $\{\cdot,\cdot\}$ satisfies the Jacobi identity. Combined, it means that $(\mathscr{A},\cdot,\{\cdot,\cdot\})$ is a Poisson algebra. Thus, a family of deformations of a given commutative associative algebra leads naturally to a Poisson bracket on this algebra! In the particular case of the associative algebras of operators of quantum mechanics, in which t is a multiple of Planck's constant $\hbar$, the operators become in the limit $\hbar \to 0$ commuting classical variables and the above procedure yields a Poisson bracket on the algebra of classical variables.

The main problem of deformation theory is to invert this procedure; in the context of physics, this is called the quantization problem or, more precisely, the problem of quantization by deformation. In its algebraic version, the first question is whether given a Poisson algebra $(\mathscr{A},\cdot,\{\cdot,\cdot\})$ there exists a family of associative products $\star_t$ on $\mathscr{A}$, such that $(\mathscr{A},\cdot) = (\mathscr{A},\star_0)$ and such that $\{\cdot,\cdot\}$, obtained from $\star_t$ by the above procedure, is a Poisson bracket on $\mathscr{A}$. The second question is to classify all such deformations, for a given $(\mathscr{A},\cdot,\{\cdot,\cdot\})$. The physical interpretation of these questions is that, starting from the algebra of classical variables, which inherits from phase space a Poisson bracket, as explained above in the context of mechanical systems, one wishes to (re-)construct the corresponding algebra of quantum mechanical operators.

A mathematical simplification of this problem is to ask whether the above procedure can be inverted formally. Stated in a precise way, this means the following. Let $(\mathscr{A},\cdot,\{\cdot,\cdot\})$ be a Poisson algebra and consider the vector space $\mathscr{A}[[\nu]]$ of formal power series in one variable ν, whose coefficients belong to $\mathscr{A}$. Let $\mu_\star$ denote an associative product on $\mathscr{A}[[\nu]]$, i.e., an $\mathbb{F}[[\nu]]$-bilinear map from $\mathscr{A}[[\nu]]$ to itself, which is associative. Given $a,b \in \mathscr{A}$, we can write

$$\mu_\star(a,b) = \sum_{k\in\mathbb{N}} \mu_k(a,b)\nu^k\,,$$

which defines bilinear maps $\mu_i : \mathscr{A} \times \mathscr{A} \to \mathscr{A}$ for $i \in \mathbb{N}$. By definition, $\mu_\star$ is a formal deformation of $(\mathscr{A},\cdot,\{\cdot,\cdot\})$ if $\mu_0(a,b) = a \cdot b$ and $\mu_1(a,b) - \mu_1(b,a) = \{a,b\}$, for all $a,b \in \mathscr{A}$. The main question, stated formally, is then if every Poisson algebra $(\mathscr{A},\cdot,\{\cdot,\cdot\})$ admits a formal deformation $\mu_\star$.

Consider for example $\mathbb{R}^{2r}$, with coordinates $(x_1,\dots,x_{2r}) = (q_1,\dots,q_r,p_1,\dots,p_r)$. The *Moyal product* is the $\mathbb{R}[[\nu]]$-bilinear product on $C^\infty(\mathbb{R}^{2r})[[\nu]]$, which is given for $F,G \in C^\infty(\mathbb{R}^{2r})$ by

$$\mu_\star(F,G) := \sum_{k\in\mathbb{N}} \sum_{1\leqslant i_1,j_1,\dots,i_k,j_k\leqslant 2r} J_{i_1,j_1}\dots J_{i_k,j_k} \frac{\partial^k F}{\partial x_{i_1}\dots\partial x_{i_r}} \frac{\partial^k G}{\partial x_{j_1}\dots\partial x_{j_r}} \frac{\nu^k}{k!},$$

where J is the skew-symmetric matrix of size $2r$, given by $J := \begin{pmatrix} 0 & \mathbb{1}_r \\ -\mathbb{1}_r & 0 \end{pmatrix}$. The Moyal product, also called the Moyal–Weyl product, defines an associative product on $C^\infty(\mathbb{R}^{2r})[[\nu]]$, whose leading terms satisfy $\mu_0(F,G) = FG$ and

$$\mu_1(F,G) - \mu_1(G,F) = \sum_{j=1}^{r} \left(\frac{\partial F}{\partial q_j}\frac{\partial G}{\partial p_j} - \frac{\partial G}{\partial q_j}\frac{\partial F}{\partial p_j} \right),$$

where the latter right-hand side is the Poisson bracket (0.1). This shows that the classical Poisson bracket admits a formal deformation.

It was shown by De Wilde and Lecomte in [57] that the Poisson algebra of an arbitrary symplectic manifold (regular Poisson manifolds of maximal rank) admits a formal deformation. A geometrical proof of their result, which also works in the case of arbitrary regular Poisson manifolds, was given by Fedosov [73]. Finally, it was Kontsevich [107] who proved, using ideas which come from string theory, that the algebra of functions on any Poisson manifold admits a formal deformation. He proved in fact a quite stronger result, which says that for a given Poisson manifold (M,π), the equivalence classes of formal deformations of the algebra of smooth functions on M are classified by the equivalence classes of formal deformations of the Poisson structure π.

Summarizing, when one deforms a commutative associative algebra, a Poisson structure shows up and it plays a dominant rôle in the entire deformation process.

The examples. The main idea which led us to the writing of this book, and to its actual structure, is that Poisson structures come in big classes (families) where roughly speaking each class has its own tools and is related to a very definite part of mathematics, hence leading to specific questions. As a consequence, the main part of the book (about half of it) is Part II, dedicated to six classes of examples. They make their appearance in six different chapters, namely Chapters 6–11, as follows.

• Constant Poisson structures, such as Poisson's original bracket (0.1), are according to the Darboux theorem the model (normal form) for Poisson structures on real or complex manifolds, in the neighborhood of any point where the rank is locally constant. Regular Poisson manifolds and symplectic manifolds (such as cotangent bundles and Kähler manifolds) are the main examples, which exhibit even in the absence of singularities some of the main phenomena in Poisson geometry, which distinguish it from Riemannian geometry and show its pertinence for classical and quantum mechanics. Constant and regular Poisson structures are studied in Chapter 6.

• Linear Poisson structures are in one-to-one correspondence with Lie algebras and many features of Lie algebras are most naturally approached in terms of the canonical Poisson bracket on the dual of a (finite-dimensional) Lie algebra, the so-called Lie–Poisson structure. The coadjoint orbits, for example, of a Lie algebra are the symplectic leaves of the Lie–Poisson structure, showing on the one hand that they are even dimensional, and on the other hand that they carry a canonical symplectic structure, the Kostant–Kirillov–Souriau symplectic structure. Linear Poisson structures appear also on an arbitrary Poisson manifold (M,π) at any point x where the Poisson structure vanishes: the tangent space T_xM inherits a linear Poisson structure, the linearization of π at x. It leads to the linearization problem, which inquires if π and its linearization at x are isomorphic, at least on a neighborhood of x. Linear Poisson structures and their relation to Lie algebras are the subject of Chapter 7.

• Many Poisson structures of interest are neither constant nor linear (or affine). In fact, while the basic properties of the latter Poisson structures can be traced back to (known) properties of bilinear forms and of Lie algebras, new phenomena appear when considering quadratic Poisson structures (i.e., homogeneous Poisson structures of degree two) and higher degree Poisson structures, which often turn out to be weight homogeneous. A prime example of this is the transverse Poisson structure to an adjoint orbit in a semi-simple Lie algebra. Moreover, this Poisson structure arises in the case of the subregular orbit from a Nambu–Poisson structure, defined by the invariant functions of the Lie algebra. Higher degree Poisson structures are studied in Chapter 8.

• In dimension two, every bivector field is a Poisson structure, yet many questions about Poisson structures on surfaces are non-trivial. For example, their local classification has up to now only been accomplished under quite strong regularity assumptions on the singular locus of the Poisson structure. In dimension three, the Jacobi identity can be stated as an integrability condition of a distribution (or of a one-form), which eventually leads to the symplectic foliation. A surface in $\mathbb{C}^3$ which is defined by the zero locus of a polynomial φ inherits a Poisson structure from the standard Nambu–Poisson structure on $\mathbb{C}^3$, with φ as Casimir. In the case of singular surfaces $\mathbb{C}^2/\mathbf{G}$, where $\mathbf{G}$ is a finite subgroup of $\mathbf{SL}_2(\mathbb{C})$, this Poisson structure coincides with the Poisson structure obtained from the canonical symplectic structure on $\mathbb{C}^2$ by reduction. Thus, rather than being trivial, Poisson structures in dimensions two and three form a rich playground on which a variety of phenomena can be observed. Poisson structures in dimensions two and three are discussed in Chapter 9.

• In a Lie-theoretical context, a linear Poisson structure, different from the canonical Lie–Poisson structure, often pops up. This Poisson structure appears on the Lie algebra $\mathfrak{g}$ at hand, so that the dual $\mathfrak{g}^*$ of $\mathfrak{g}$ comes equipped with a Lie algebra structure, or it appears on $\mathfrak{g}^*$, so that $\mathfrak{g}$ is equipped with a second Lie algebra structure. The underlying operator, which relates either of these Lie structures with the original Lie bracket on $\mathfrak{g}$ is in the first case an element $r \in \mathfrak{g}\otimes\mathfrak{g}$, called an r-matrix, while it is in the second case a (vector space) endomorphism R of $\mathfrak{g}$, called an R-matrix. These structures appeared first in the theory of integrable systems, but are nowadays equally important in the theory of Lie–Poisson groups (see the next item) and of quantum groups. In the context of Lie algebras which come from associative algebras, r-matrices and R-matrices also lead to a quadratic Poisson structure on the Lie algebra, which plays an important rôle in the theory of Lie–Poisson groups; they also lead to a cubic Poisson structure, whose virtue seems at this point still very mysterious. Poisson structures coming from r-matrices or R-matrices are the subject of Chapter 10.

• A Poisson–Lie group is a Lie group $\mathbf{G}$, equipped with a Poisson structure π for which the group multiplication $\mathbf{G} \times \mathbf{G} \to \mathbf{G}$ is a Poisson map. The Poisson structure π leads to a Lie algebra structure on the dual of the Lie algebra $\mathfrak{g}$ of $\mathbf{G}$, making $\mathfrak{g}$ into a Lie bialgebra, a structure generalizing the structure which comes from an r-matrix. Conversely, every finite-dimensional Lie bialgebra is obtained in this way from a Poisson–Lie group, a result which extends Lie's third theorem (every finite-dimensional Lie algebra is the Lie algebra of a Lie group). Poisson–Lie groups have many applications, for example in the description of the Schubert cells of a Lie group. Poisson–Lie groups are discussed in Chapter 11.

Theoretical foundations. The examples (Chapters 6–11) are preceded by a first part (5 chapters) with a systematic exposition of the general theory of Poisson structures. We used two principles in the writing of these chapters. The first one is that we wanted to develop both the algebraic and geometric points of view of the theory. In fact, there has for the last decade been an increasing interest in the algebraic aspects of Poisson structures, while of course geometrical intuition underlies all constructions and the vast majority of examples. In the algebraic context, we have a Poisson algebra, whose underlying space is an $\mathbb{F}$-vector space, where $\mathbb{F}$ is an arbitrary field of characteristic zero. In the geometric context, we have a Poisson manifold, where the manifold is either real or complex. In between the algebraic and geometric contexts, Poisson varieties appear, whose underlying object can both be viewed as a variety or as a finitely generated algebra. Throughout this first part, the reader will experience that although most of the results are formally very similar in the algebraic and in the geometric setting, their concrete implementation (including proofs of the propositions) is quite different and needs to be worked out in detail in both settings.

The second principle which we used was to keep the prerequisites in (commutative and Lie) algebra and in (differential and algebraic) geometry to a minimum. Thus, for example, when we describe Poisson structures geometrically as bivector fields, we explain the notion of a bivector field by first recalling the notion of a vector field, rather than simply saying that a bivector field is a section of the exterior square of the tangent bundle to the manifold. Similarly, while the basic facts about Lie groups and Lie algebras are supposed to be known, the interplay between Lie groups and Lie algebras is recalled in detail, as it is further amplified in the presence of Poisson structures. We added an appendix at the end of the book with some facts on multilinear algebra (tensor products, wedges, algebra and coalgebra structures) and an appendix which recalls the basic facts about differential geometry. The objects and properties which are recalled in these appendices are used throughout the book.

Chapters 1 and 2 contain the basic definitions and constructions. A good familiarity with these chapters should suffice for reading a good part of any other chapter of the book. Chapter 3 deals with multi-derivations and their geometrical analog, multivector fields, and with Kähler forms, which are the algebraic analogs of differential forms. Chapter 4 deals with Poisson cohomology and Chapter 5 is devoted to reduction in the context of Poisson structures. At the end of each of these five chapters, we give a list of exercises, so that the reader can test his understanding of the theory.

Applications. The third part of the book contains the two major applications, which we discussed above: integrable systems are discussed in Chapter 12, while deformation quantization is detailed in Chapter 13. They are however not the only applications which are given in the book. In fact, we end each example chapter (Chapters 6–11) with an application of the class of Poisson structures, studied in that chapter.

What is absent from this book. The main topic about Poisson structures which is absent from this book is what should be called "Poisson geometry". By this we mean the global geometry of Poisson structures, which involves the integration of Poisson brackets, Poisson connections, stability of symplectic leaves and related topics. For this, we refer to [51, 53, 74].

Equally absent from this book are the many generalizations of Poisson structures. Quasi-Poisson structures and Poisson structures with background are not only mathematically speaking very interesting, see [112], they have in addition non-trivial applications in physics, see [119, 162, 176], and are themselves a particular case of Dirac structures [48, 89]. The theory of quantum groups, initiated in the seminal ICM talk [59] by Drinfel'd, has several connections with the topics which are developed in this book, but adding several chapters to the book would not have been sufficient to give a fair account of this subject; we simply refer to the excellent books [40, 103].

Equally absent from this book is the theory of Lie algebroids, introduced by Pradines [171], which are both a particular case and a generalization of Poisson manifolds. In particular, after the pioneering work of [47], the problem of the integration of Poisson manifolds, which claims that symplectic Lie groupoids are to Poisson manifolds what Lie groups are to Lie algebras, has gained a long attention. For a definite solution of this problem, given by Cranic and Fernandes, see [49, 50].

The notes and the references. When writing the history of the theory of Poisson brackets, giving each of the main results in the theory and examples the "right" credit is a challenge which is beyond the scope of this book. We claim no originality in this book, but since none of the proofs which are given are copy-and-paste proofs from existing proofs, we did not think it was necessary to cite the proof in the literature which is closest to the proof that we give. Also, many properties of Poisson structures, and many examples, have been through a long series of transformations and generalizations, think for example of Weinstein's splitting theorem or the Poisson and Poisson–Dirac reduction theorems; structuring the long list of important intermediate results which led to the final results is certainly interesting from the epistemological point of view, but this was not our main focus when writing the book.

However, at the end of each chapter, we have put a section called "Notes", in which we indicate some references, including the main references which we know of, for further reading on what has been treated in the chapter. These notes are also used to situate, a posteriori, the treated subjects in the scientific literature and to trace some of their historical developments. Finally, the notes are also used to indicate some further connections to other fields of mathematics or physics. Since Poisson structures are a very active field of research, it is clear that the list of connections with other fields is not exhaustive.

Part I
Theoretical Background

Chapter 1
Poisson Structures: Basic Definitions

In this chapter, we give the basic definitions of a Poisson algebra, of a Poisson variety, of a Poisson manifold and of a Poisson morphism.

Geometrically speaking, a Poisson structure on a smooth manifold M associates to every smooth function H on M, a vector field $\mathscr{X}_H$ on M. In the context of a mechanical system, this vector field yields the equations of motion, when H is taken as the Hamiltonian. The Poisson bracket is demanded to be a Lie bracket, which amounts to demanding that Poisson's theorem is valid, namely that the Poisson bracket of two constants of motion is itself a constant of motion.

Algebraically speaking, one considers on a (typically infinite-dimensional) vector space $\mathscr{A}$ two different algebra structures: (1) a commutative, associative multiplication, (2) a Lie bracket. It results in the following definition:

Poisson algebra := Comm. assoc. algebra + Lie algebra + Compatibility.

The compatibility between the two algebra structures is precisely what allows us to associate to elements of $\mathscr{A}$, derivations of $\mathscr{A}$, derivations being the algebraic analogs of vector fields.

Poisson algebras and their morphisms are defined in Section 1.1. In Section 1.2, respectively Section 1.3, we introduce Poisson varieties, respectively Poisson manifolds, and their morphisms. Sections 1.2 and 1.3 can be read independently and in either order; the reader is invited to discover, by comparing them, up to which point Poisson varieties and Poisson manifolds can be treated uniformly, and how quickly the techniques and results diverge, past this point. In Section 1.4 we specialize the results of Sections 1.2 and 1.3 to the case of Poisson structures on a finite-dimensional vector space; such a space can indeed be viewed as an affine variety or as a manifold, depending on the choice of algebra of functions on it.

Throughout the present chapter, we fix a ground field $\mathbb{F}$ of characteristic zero, which the reader may think of as being $\mathbb{R}$ or $\mathbb{C}$, especially in the context of varieties.

C. Laurent-Gengoux et al., *Poisson Structures*,
Grundlehren der mathematischen Wissenschaften 347,
DOI 10.1007/978-3-642-31090-4_1, © Springer-Verlag Berlin Heidelberg 2013

1.1 Poisson Algebras and Their Morphisms

In this section we introduce the general notions of a Poisson algebra and of a morphism of Poisson algebras. The Hamiltonian operator, which associates to an element of the Poisson algebra a derivation of the Poisson algebra, will be introduced and its basic algebraic properties will be given.

1.1.1 Poisson Algebras

We start with the algebraic notion of a Poisson algebra over a field $\mathbb{F}$ which, as we have already said, is always assumed to be of characteristic zero.

Definition 1.1. A *Poisson algebra* is an $\mathbb{F}$-vector space $\mathscr{A}$ equipped with two multiplications $(F,G) \mapsto F \cdot G$ and $(F,G) \mapsto \{F,G\}$, such that

(1) $(\mathscr{A},\cdot)$ is a commutative associative algebra over $\mathbb{F}$, with unit 1;
(2) $(\mathscr{A},\{\cdot\,,\cdot\})$ is a Lie algebra over $\mathbb{F}$;
(3) The two multiplications are compatible in the sense that

$$\{F\cdot G,H\} = F\cdot\{G,H\} + G\cdot\{F,H\}\ , \tag{1.1}$$

where F,G and H are arbitrary elements of $\mathscr{A}$.

The Lie bracket $\{\cdot\,,\cdot\}$ is then called a *Poisson bracket*.

Poisson brackets are often constructed on a *given* commutative associative algebra $\mathscr{A} = (\mathscr{A},\cdot)$. The algebra of regular functions on an (affine algebraic) variety is a typical example of this, leading to the notion of Poisson variety (Section 1.2); see Section 1.3 for the case of real or complex Poisson manifolds. On $\mathscr{A}$, one considers then a skew-symmetric bilinear map $\{\cdot\,,\cdot\} : \mathscr{A}\times\mathscr{A}\to\mathscr{A}$, which satisfies condition (1.1). It will be a Poisson bracket on $\mathscr{A}$ as soon as the *Jacobi identity*

$$\{F,\{G,H\}\} + \{G,\{H,F\}\} + \{H,\{F,G\}\} = 0 \tag{1.2}$$

is verified for all triples (F, G, H) of elements of $\mathscr{A}$. In order to shorten the formulas, (1.2) is often written as

$$\{F,\{G,H\}\} + \circlearrowleft(F,G,H) = 0\ ,$$

and similarly for other formulas which involve sums over three terms with cyclically permuted variables. The algebraic terminology which expresses (1.1) is that for every $H\in\mathscr{A}$ the linear map $F\mapsto\{F,H\}$ is a derivation of $\mathscr{A}$. Namely, a linear map $\mathscr{V} : \mathscr{A}\to\mathscr{A}$ is called a *derivation* of $\mathscr{A}$ (with values in $\mathscr{A}$) if

$$\mathscr{V}(FG) = F\mathscr{V}(G) + G\mathscr{V}(F)\ , \tag{1.3}$$

for every $F, G \in \mathscr{A}$. We have written FG for $F \cdot G$, a convenient shorthand which we will use freely in the sequel. We denote by $\mathfrak{X}^1(\mathscr{A})$ the Lie algebra of derivations of $\mathscr{A}$, where the Lie bracket $[\cdot,\cdot]$ on $\mathfrak{X}^1(\mathscr{A})$ is given by the commutator of derivations. A bilinear map $P : \mathscr{A} \times \mathscr{A} \to \mathscr{A}$ is called a *biderivation* of $\mathscr{A}$ (with values in $\mathscr{A}$) if

$$\begin{aligned} P(FG,H) &= F\,P(G,H) + G\,P(F,H)\,, \\ P(H,FG) &= F\,P(H,G) + G\,P(H,F)\,, \end{aligned} \tag{1.4}$$

for every $F, G, H \in \mathscr{A}$. Our biderivations will always be defined by a *skew-symmetric* bilinear map $P : \mathscr{A} \times \mathscr{A} \to \mathscr{A}$; we say then that P is a *skew-symmetric biderivation* of $\mathscr{A}$. For such a biderivation, it is convenient to enclose its arguments in square brackets; thus, when P is a skew-symmetric biderivation, we write the first equation in (1.4) as

$$P[FG,H] = F\,P[G,H] + G\,P[F,H]\,;$$

by the skew-symmetry of P, this formula is of course equivalent with the second equation in (1.4). Square brackets will also be used to enclose the arguments of a general skew-symmetric multi-derivation, a notion which will be introduced in Section 3.1; in particular, we will also use square brackets to enclose the argument of a derivation, writing $\mathscr{V}[F]$ for $\mathscr{V}(F)$ when $\mathscr{V}$ is a derivation of $\mathscr{A}$.

The $\mathbb{F}$-vector space of skew-symmetric biderivations of $\mathscr{A}$ is denoted by $\mathfrak{X}^2(\mathscr{A})$. As recalled in Appendix A, a skew-symmetric bilinear map on $\mathscr{A}$ can be viewed as a linear map on $\wedge^2\mathscr{A} = \mathscr{A} \wedge \mathscr{A}$; we therefore often abbreviate $\psi : \mathscr{A} \times \mathscr{A} \to \mathscr{A}$ to $\psi : \wedge^2\mathscr{A} \to \mathscr{A}$ or $\psi : \mathscr{A} \wedge \mathscr{A} \to \mathscr{A}$. According to Definition 1.1, a Poisson bracket on $\mathscr{A}$ is in particular a skew-symmetric biderivation of $\mathscr{A}$.

1.1.2 Morphisms of Poisson Algebras

We now turn to the notion of a morphism between two Poisson algebras (defined over the same field $\mathbb{F}$). Since a Poisson algebra is an algebra in two different ways, it is natural to demand that a morphism of Poisson algebras be a morphism with respect to both algebra structures.

Definition 1.2. Let $(\mathscr{A}_i, \cdot_i, \{\cdot,\cdot\}_i)$, $i = 1,2$, be two Poisson algebras over $\mathbb{F}$. A linear map $\phi : \mathscr{A}_1 \to \mathscr{A}_2$ which satisfies, for all $F, G \in \mathscr{A}_1$,

(1) $\phi(F \cdot_1 G) = \phi(F) \cdot_2 \phi(G)$;
(2) $\phi(\{F,G\}_1) = \{\phi(F), \phi(G)\}_2$,

is called a *morphism of Poisson algebras*.

It is clear that if $\phi : \mathscr{A}_1 \to \mathscr{A}_2$ is a morphism of Poisson algebras and ϕ is bijective, then $\phi^{-1} : \mathscr{A}_2 \to \mathscr{A}_1$ is also a morphism of Poisson algebras. We say then that ϕ is an *isomorphism of Poisson algebras*.

As we will see in the case of Poisson manifolds and of Poisson varieties, one is often led to morphisms of the associative algebras which underlie the two Poisson algebras. In this case, we only have to consider condition (2) in the above definition, which says that ϕ is a morphism of Lie algebras.

When one deals with subalgebras or ideals of Poisson algebras, it is important to distinguish which multiplication (maybe both) is considered. Our convention is that *subalgebra* and *ideal* refer to the associative multiplication, *Lie subalgebra* and *Lie ideal* refer to the Lie bracket, and *Poisson subalgebra* and *Poisson ideal* refer to both. This means that a vector subspace $\mathscr{B} \subset \mathscr{A}$ is a Poisson subalgebra of $\mathscr{A} = (\mathscr{A}, \cdot, \{\cdot,\cdot\})$ if

$$\mathscr{B} \cdot \mathscr{B} \subset \mathscr{B} \quad \text{and} \quad \{\mathscr{B}, \mathscr{B}\} \subset \mathscr{B}\,, \tag{1.5}$$

while a vector subspace $\mathscr{I} \subset \mathscr{A}$ is a Poisson ideal of $\mathscr{A}$ if

$$\mathscr{I} \cdot \mathscr{A} \subset \mathscr{I} \quad \text{and} \quad \{\mathscr{I}, \mathscr{A}\} \subset \mathscr{I}\,. \tag{1.6}$$

In the first case, $\mathscr{B}$ becomes itself a Poisson algebra, when equipped with the restrictions of both multiplications to $\mathscr{B}$ and the inclusion map $\iota : \mathscr{B} \to \mathscr{A}$ is a morphism of Poisson algebras. In the second case, $\mathscr{A}/\mathscr{I}$ inherits a Poisson bracket from $\mathscr{A}$, such that the canonical projection $p : \mathscr{A} \to \mathscr{A}/\mathscr{I}$ is a morphism of Poisson algebras. We will come back at length to the notions of Poisson subalgebra and Poisson ideal in Chapter 2, when we have developed sufficient concepts and tools to illustrate these notions geometrically.

In view of Definitions 1.1 and 1.2 we get, for a fixed field $\mathbb{F}$, a category whose objects are the Poisson algebras over $\mathbb{F}$ and whose morphisms are the morphisms of Poisson algebras.

1.1.3 Hamiltonian Derivations

The biderivation property of the Poisson bracket leads to a fundamental operation which allows one to associate to elements of $\mathscr{A}$, derivations of $\mathscr{A}$; in the case of a real Poisson manifold M, it allows us to associate to a function on M, a vector field on M (see Section 1.3). This operation, which we introduce now, corresponds in the Hamiltonian formulation of classical mechanics to writing the equations of motion for a given Hamiltonian (see Section 6.1).

Definition 1.3. Let $(\mathscr{A}, \cdot, \{\cdot,\cdot\})$ be a Poisson algebra and let $H \in \mathscr{A}$. The derivation $\mathscr{X}_H := \{\cdot, H\}$ of $\mathscr{A}$ is called a *Hamiltonian derivation* and we call H a *Hamiltonian*, associated to $\mathscr{X}_H$. We write

$$\operatorname{Ham}(\mathscr{A}, \{\cdot,\cdot\}) := \{\mathscr{X}_H \mid H \in \mathscr{A}\}$$

for the $\mathbb{F}$-vector space of Hamiltonian derivations of $\mathscr{A}$, so that we have an $\mathbb{F}$-linear map

$$\begin{aligned} \mathscr{X} : \mathscr{A} &\to \mathrm{Ham}(\mathscr{A},\{\cdot,\cdot\}) \\ H &\mapsto \mathscr{X}_H := \{\cdot, H\}\,. \end{aligned} \tag{1.7}$$

An element $H \in \mathscr{A}$ whose Hamiltonian derivation is zero, $\mathscr{X}_H = 0$, is called a *Casimir* and we denote

$$\mathrm{Cas}\,(\mathscr{A},\{\cdot,\cdot\}) := \{H \in \mathscr{A} \mid \mathscr{X}_H = 0\}$$

for the $\mathbb{F}$-vector space of Casimirs. When no confusion can arise, we write $\mathrm{Ham}(\mathscr{A})$ for $\mathrm{Ham}(\mathscr{A},\{\cdot,\cdot\})$ and $\mathrm{Cas}(\mathscr{A})$ for $\mathrm{Cas}(\mathscr{A},\{\cdot,\cdot\})$. It is clear that $\mathrm{Cas}(\mathscr{A})$ is the center of the Lie algebra $(\mathscr{A},\{\cdot,\cdot\})$.

The defining properties of the Poisson bracket lead to the following proposition.

Proposition 1.4. *Let $(\mathscr{A},\cdot,\{\cdot,\cdot\})$ be a Poisson algebra.*

(1) $\mathrm{Cas}(\mathscr{A})$ *is a subalgebra of $(\mathscr{A},\cdot)$, which contains the image of $\mathbb{F}$ in $\mathscr{A}$, under the natural inclusion $a \mapsto a \cdot 1$;*
(2) *If $\mathscr{A}$ has no zero divisors, then* $\mathrm{Cas}(\mathscr{A})$ *is integrally closed in $\mathscr{A}$;*
(3) $\mathrm{Ham}(\mathscr{A})$ *is* not *an $\mathscr{A}$-module (in general); instead, $\mathscr{X}_{FG} = F\mathscr{X}_G + G\mathscr{X}_F$, for every $F,G \in \mathscr{A}$;*
(4) $\mathrm{Ham}(\mathscr{A})$ *is a* $\mathrm{Cas}(\mathscr{A})$*-module;*
(5) *The map $\mathscr{A} \to \mathfrak{X}^1(\mathscr{A})$ which is defined by $H \mapsto -\mathscr{X}_H$ is a morphism of Lie algebras; as a consequence,* $\mathrm{Ham}(\mathscr{A})$ *is a Lie subalgebra of $\mathfrak{X}^1(\mathscr{A})$;*
(6) *The Lie algebra sequence*

$$0 \longrightarrow \mathrm{Cas}(\mathscr{A}) \longrightarrow \mathscr{A} \xrightarrow{-\mathscr{X}} \mathrm{Ham}(\mathscr{A}) \longrightarrow 0$$

is a short exact sequence.

Proof. Properties *(1)–(4)* follow from the biderivation property (1.1), while properties *(5)* and *(6)* follow from the Jacobi identity (1.2) for $\{\cdot,\cdot\}$. We only show *(2)*.

Let F be integral over $\mathrm{Cas}(\mathscr{A})$, that is $F \in \mathscr{A}$ and there exists a monic polynomial $p(X) \in \mathrm{Cas}(\mathscr{A})[X]$, such that $p(F) = 0$. We need to show that $F \in \mathrm{Cas}(\mathscr{A})$. Let p be the monic polynomial of smallest degree such that $p(F) = 0$. If $\deg(p) = 1$, then $F \in \mathrm{Cas}(\mathscr{A})$ and we are done. Let us suppose therefore that $r := \deg(p) > 1$. For every element $G \in \mathscr{A}$ we have by the derivation property that $0 = \{p(F), G\} = p'(F)\{F,G\}$, where p' denotes the derivative of p. Now $p'(F) \neq 0$, as p'/r would otherwise be a monic polynomial of degree $r-1$, such that $p'(F)/r = 0$. Since $\mathscr{A}$ is without zero divisors, $\{F,G\} = 0$. As G is arbitrary, this shows that F is a Casimir of $\mathscr{A}$. □

1.2 Poisson Varieties

When $\mathscr{A}$ is an algebra of functions on some variety or manifold, the properties of a Poisson bracket on $\mathscr{A}$ get a geometrical meaning, leading to many interesting constructions, which can often be generalized to arbitrary Poisson algebras. In this section, we consider affine varieties, see the next section for the case of (real or complex) manifolds.[1] The reader, unfamiliar with affine varieties, may on his first reading of this section think of the affine variety M as being the affine space $\mathbb{F}^d$, where $d \in \mathbb{N}^*$, equipped with the algebra of polynomial functions $\mathbb{F}[x_1,\dots,x_d]$.

1.2.1 Affine Poisson Varieties and Their Morphisms

We use the convention that an *affine variety* is an irreducible algebraic subset M of an affine space $\mathbb{F}^d$, where *algebraic subset* means that M is the zero locus of a family of polynomials (in d variables). It conforms to the American convention, see e.g. [93, Ch. I.1]; in the French literature, affine varieties are usually not assumed to be irreducible. Given an algebraic subset $M \subset \mathbb{F}^d$, one considers the prime ideal $\mathscr{I}$ of $\mathbb{F}[x_1,\dots,x_d]$ which consists of all polynomial functions, vanishing on M. Then $\mathbb{F}[x_1,\dots,x_d]/\mathscr{I}$ becomes a finitely generated (commutative associative) algebra, which can be considered as an algebra of functions on M, since the evaluation of elements of $\mathbb{F}[x_1,\dots,x_d]/\mathscr{I}$ at points of M is well-defined. This algebra of functions on M is denoted by $\mathscr{F}(M)$ and is called the *affine coordinate ring* of M. For $F \in \mathbb{F}[x_1,\dots,x_d]$ we denote its projection in $\mathscr{F}(M) = \mathbb{F}[x_1,\dots,x_d]/\mathscr{I}$ by $\overline{F}$ or by $F_{|M}$, the latter notation being justified by the fact that $\overline{F}$ can be viewed as the restriction of F to M. Since M is irreducible, $\mathscr{F}(M)$ has no zero divisors. Elements of $\mathscr{F}(M)$ are called *regular functions* on M, although the name *polynomial function* is also used, especially when $M = \mathbb{F}^d$.

Thus, the vector space of regular functions on an affine variety is naturally equipped with a first algebra structure: the structure of a commutative associative algebra. The extra datum of a Poisson bracket on this algebra leads to the notion of an affine Poisson variety.

Definition 1.5. Let M be an affine variety and suppose that $\mathscr{F}(M)$ is equipped with a Lie bracket $\{\cdot,\cdot\} : \mathscr{F}(M) \times \mathscr{F}(M) \to \mathscr{F}(M)$, which makes $(\mathscr{F}(M),\cdot,\{\cdot,\cdot\})$ into a Poisson algebra. Then we say that $(M,\{\cdot,\cdot\})$ is an *affine Poisson variety*, or simply a *Poisson variety*. The Poisson bracket on $\mathscr{F}(M)$ is usually referred to as a *Poisson structure* on M.

In the case of a Poisson structure on $\mathbb{F}^d$, the Poisson bracket of two functions can be computed from the following standard formula; see Section 1.2.2 for the case of a general affine Poisson variety.

Proposition 1.6. *Let* $\{\cdot,\cdot\}$ *be a Poisson bracket on* $\mathscr{A} = \mathbb{F}[x_1,\dots,x_d]$. *For all functions* F *and* G *in* $\mathscr{A}$, *their Poisson bracket is given by*

$$\{F,G\} = \sum_{i,j=1}^{d} \{x_i,x_j\} \frac{\partial F}{\partial x_i} \frac{\partial G}{\partial x_j} . \tag{1.8}$$

[1] As we said in the introduction to this chapter, Sections 1.2 and 1.3 can be read independently.

Proof. The proof is based on the standard argument that two biderivations (or p-derivations, in general) of some commutative associative algebra $\mathscr{A}$ are equal as soon as they agree on a system of generators for $\mathscr{A}$. Since we will use this argument several times, we spell it out in detail. Both sides of (1.8) are bilinear in F and G, so it suffices to show (1.8) when F and G are monomials in $x_1,\dots,x_d$. If F or G is a monomial of total degree 0, then the right-hand side in (1.8) is obviously zero, but also the left-hand side is zero, because constant functions are Casimirs (item *(1)* in Proposition 1.4). Also, the fact that (1.8) holds when F and G are both monomials of degree 1, is clear. Suppose now that (1.8) holds when $\deg(F)+\deg(G)\leqslant n$, for some $n\geqslant 2$; we show that it holds for all F and G such that $\deg(F)+\deg(G)=n+1$. Let F and G be non-constant monomials, such that $\deg(F)+\deg(G)=n+1$. By skew-symmetry, we may assume that $\deg(F)>1$. There exist monomials $F_1,F_2\in\mathscr{A}$, with $\deg(F_1)<\deg(F)$ and $\deg(F_2)<\deg(F)$, such that $F=F_1F_2$. Since $\{\cdot,\cdot\}$ is a biderivation and in view of the recursion hypothesis, we have that

$$\begin{aligned}\{F,G\} &= \{F_1F_2,G\} = F_1\{F_2,G\}+F_2\{F_1,G\}\\ &= F_1\sum_{i,j=1}^{d}\{x_i,x_j\}\frac{\partial F_2}{\partial x_i}\frac{\partial G}{\partial x_j}+F_2\sum_{i,j=1}^{d}\{x_i,x_j\}\frac{\partial F_1}{\partial x_i}\frac{\partial G}{\partial x_j}\\ &= \sum_{i,j=1}^{d}\{x_i,x_j\}\frac{\partial F}{\partial x_i}\frac{\partial G}{\partial x_j}.\end{aligned}$$

This proves (1.8) for arbitrary polynomials F and G. □

We now turn to the notion of a Poisson morphism of Poisson varieties.

Definition 1.7. Let $(M_1,\{\cdot,\cdot\}_1)$ and $(M_2,\{\cdot,\cdot\}_2)$ be two Poisson varieties. A morphism of varieties $\Psi: M_1\to M_2$ is called a *Poisson morphism* or a *Poisson map* if the dual morphism

$$\Psi^*:\mathscr{F}(M_2)\to\mathscr{F}(M_1)$$

is a morphism of Poisson algebras.

Recall (see e.g. [182, Ch. I.2.3]) that for a morphism of varieties (also called a regular map) $\Psi: M_1\to M_2$, the dual morphism Ψ^* is (well-)defined by $\Psi^*(F):=F\circ\Psi$, for all $F\in\mathscr{F}(M_2)$, so that the first condition in Definition 1.2 is automatically satisfied:

$$\Psi^*(FG)=(FG)\circ\Psi=(F\circ\Psi)(G\circ\Psi)=(\Psi^*F)(\Psi^*G)\,.$$

In this case, the second condition, which says that

$$\Psi^*:(\mathscr{F}(M_2),\{\cdot,\cdot\}_2)\to(\mathscr{F}(M_1),\{\cdot,\cdot\}_1)$$

is a morphism of Lie algebras, can also be written as

$$\{F,G\}_2\circ\Psi=\{F\circ\Psi,G\circ\Psi\}_1\,,\qquad\text{for all } F,G\in\mathscr{F}(M_2)\,.$$

When N is a subvariety of M, which amounts to saying that the inclusion map $\iota : N \hookrightarrow M$ is a morphism, then N is said to be a *Poisson subvariety* if N admits a Poisson structure such that ι is a Poisson map. See Section 2.2.

Combining Definitions 1.5 and 1.7 we get, for a fixed $\mathbb{F}$, a category whose objects are (affine) Poisson varieties and whose morphisms are Poisson maps.

1.2.2 The Poisson Matrix

In this section, we show how a Poisson structure on an affine variety can be encoded in a matrix. We first consider the case of a Poisson structure $\{\cdot,\cdot\}$ on the affine space $\mathbb{F}^d$ (with its algebra of regular functions $\mathscr{A} := \mathbb{F}[x_1,\ldots,x_d]$). The d^2 *structure functions* $x_{ij} \in \mathscr{A}$, which are defined for $1 \leqslant i,j \leqslant d$ by $x_{ij} := \{x_i, x_j\}$, satisfy

$$x_{ij} = -x_{ji}, \tag{1.9}$$

$$\sum_{\ell=1}^{d}\left(x_{\ell k}\frac{\partial x_{ij}}{\partial x_\ell} + x_{\ell i}\frac{\partial x_{jk}}{\partial x_\ell} + x_{\ell j}\frac{\partial x_{ki}}{\partial x_\ell}\right) = 0\,, \tag{1.10}$$

for all $1 \leqslant i,j,k \leqslant d$. Formula (1.10) is obtained by writing the Jacobi identity (1.2) for the triple (x_i, x_j, x_k) in the form

$$\{x_{ij}, x_k\} + \{x_{jk}, x_i\} + \{x_{ki}, x_j\} = 0\,,$$

and by using (1.8).

If we view the structure functions x_{ij} as the elements of a (skew-symmetric) $d \times d$ matrix X, then the Poisson bracket may, according to (1.8), also be expressed in a compact form by

$$\{F,G\} = [\mathrm{d}F]^\top X\,[\mathrm{d}G]\,,$$

where $[\mathrm{d}F]$ is the column vector which represents $\mathrm{d}F$, the differential of F, in the natural basis $(\mathrm{d}x_1,\ldots,\mathrm{d}x_d)$, i.e., the i-th component of $[\mathrm{d}F]$ is $\frac{\partial F}{\partial x_i}$. The matrix $X \in \mathrm{Mat}_d(\mathscr{A})$ is called the *Poisson matrix* of $(\mathscr{A},\{\cdot,\cdot\})$.

A natural question is: given a skew-symmetric matrix $X = (x_{ij}) \in \mathrm{Mat}_d(\mathscr{A})$, where $\mathscr{A} = \mathbb{F}[x_1,\ldots,x_d]$, does the corresponding skew-symmetric biderivation, defined for $F, G \in \mathscr{A}$ by

$$\{F,G\} := \sum_{i,j=1}^{d} x_{ij}\frac{\partial F}{\partial x_i}\frac{\partial G}{\partial x_j}$$

define a Poisson structure on $\mathscr{A}$, i.e., does it satisfy the Jacobi identity? For F, G and H in $\mathscr{A}$, the biderivation property and the fact that second order derivatives commute, imply that

$$\{\{F,G\},H\} + \circlearrowleft(F,G,H) = \sum_{i,j,k,\ell=1}^{d} x_{\ell k}\frac{\partial x_{ij}}{\partial x_\ell}\frac{\partial F}{\partial x_i}\frac{\partial G}{\partial x_j}\frac{\partial H}{\partial x_k} + \circlearrowleft(F,G,H) \tag{1.11}$$

$$= \sum_{i,j,k=1}^{d} \sum_{\ell=1}^{d} \left(x_{\ell k} \frac{\partial x_{ij}}{\partial x_\ell} + \circlearrowleft (i,j,k) \right) \frac{\partial F}{\partial x_i} \frac{\partial G}{\partial x_j} \frac{\partial H}{\partial x_k},$$

so that the Jacobi identity is satisfied for all triples of elements of $\mathscr{A} = \mathbb{F}[x_1, \ldots, x_d]$ if and only if (1.10) holds, i.e., if and only if the Jacobi identity holds for every triple (x_i, x_j, x_k), with $1 \leqslant i < j < k \leqslant d$.

Summarizing, we have the following proposition.

Proposition 1.8. *If $X = (x_{ij})$ is a skew-symmetric $d \times d$ matrix, with elements in $\mathscr{A} = \mathbb{F}[x_1, \ldots, x_d]$, then*

$$\{F, G\} := \sum_{i,j=1}^{d} x_{ij} \frac{\partial F}{\partial x_i} \frac{\partial G}{\partial x_j} \tag{1.12}$$

defines a Poisson bracket on $\mathscr{A}$ (with Poisson matrix X) if and only if every triple (x_i, x_j, x_k), with $1 \leqslant i < j < k \leqslant d$, satisfies the Jacobi identity

$$\{\{x_i, x_j\}, x_k\} + \{\{x_j, x_k\}, x_i\} + \{\{x_k, x_i\}, x_j\} = 0,$$

which amounts to the condition

$$\sum_{\ell=1}^{d} \left(x_{\ell k} \frac{\partial x_{ij}}{\partial x_\ell} + x_{\ell i} \frac{\partial x_{jk}}{\partial x_\ell} + x_{\ell j} \frac{\partial x_{ki}}{\partial x_\ell} \right) = 0.$$

An analogous result is valid when $\mathscr{A}$ is the algebra of regular functions on an affine variety,

$$\mathscr{A} := \mathbb{F}[x_1, \ldots, x_d] / \mathscr{I},$$

where $\mathscr{I}$ is a prime ideal of $\mathbb{F}[x_1, \ldots, x_d]$. For $F \in \mathbb{F}[x_1, \ldots, x_d]$, let us denote the coset $F + \mathscr{I}$ by $\overline{F}$. Then $\bar{x}_1, \ldots, \bar{x}_d$ are generators of $\mathscr{A}$ and $\mathscr{I}$ is the set of all relations between these generators. We suppose that we are given a skew-symmetric $d \times d$ matrix X, whose elements belong to $\mathscr{A}$. We can write $X = (\bar{x}_{ij})$, where the elements $x_{ij} \in \mathbb{F}[x_1, \ldots, x_d]$, which we choose such that $x_{ij} = -x_{ji}$, are of course not unique. Let us consider the skew-symmetric biderivation $\{\cdot, \cdot\}_0$ of $\mathbb{F}[x_1, \ldots, x_d]$, defined for $F, G \in \mathbb{F}[x_1, \ldots, x_d]$ by

$$\{F, G\}_0 := \sum_{i,j=1}^{d} x_{ij} \frac{\partial F}{\partial x_i} \frac{\partial G}{\partial x_j}. \tag{1.13}$$

We would like to define a skew-symmetric biderivation $\{\cdot, \cdot\}$ of $\mathscr{A}$ by setting $\{\overline{F}, \overline{G}\} := \overline{\{F, G\}_0}$, for all $F, G \in \mathbb{F}[x_1, \ldots, x_d]$. The problem is that, in general, this is not well-defined. Namely, for every $K \in \mathscr{I}$ one has that $\overline{K} = \overline{0}$, so that a necessary condition for the bracket $\{\cdot, \cdot\}$ on $\mathscr{A}$ to be well-defined, is that for every $K \in \mathscr{I}$ and $1 \leqslant i \leqslant d$,

$$\{x_i, K\}_0 = \sum_{j=1}^{d} x_{ij} \frac{\partial K}{\partial x_j} \in \mathscr{I}. \tag{1.14}$$

Notice that this condition is independent of the chosen elements $x_{ij} \in \mathbb{F}[x_1,\ldots,x_d]$, which represent $\bar{x}_{ij}$.

• When condition (1.14) holds for every $K \in \mathscr{I}$ and for every i with $1 \leqslant i \leqslant d$, then $\{\mathbb{F}[x_1,\ldots,x_d],\mathscr{I}\}_0 \subset \mathscr{I}$, since $\{\cdot,\cdot\}_0$ is a biderivation. It follows that $\{\cdot,\cdot\}$ is well-defined. Also, $\{\cdot,\cdot\}$ is a skew-symmetric biderivation since $\{\cdot,\cdot\}_0$ is one. Since $\{\cdot,\cdot\}$ is well-defined,

$$\{\{\overline{F},\overline{G}\},\overline{H}\} = \overline{\{\{F,G\}_0,H\}_0}\,,$$

for all $F,G,H \in \mathbb{F}[x_1,\ldots,x_d]$, so that $\{\cdot,\cdot\}$ is a Poisson bracket on $\mathscr{A}$ if and only if

$$\{\{F,G\}_0,H\}_0+\{\{G,H\}_0,F\}_0+\{\{H,F\}_0,G\}_0 \in \mathscr{I}\,,$$

for all $F,G,H \in \mathbb{F}[x_1,\ldots,x_d]$. Since $\{\cdot,\cdot\}_0$ is a skew-symmetric biderivation, this amounts to checking that

$$\left\{\{x_i,x_j\}_0,x_k\right\}_0+\left\{\{x_j,x_k\}_0,x_i\right\}_0+\left\{\{x_k,x_i\}_0,x_j\right\}_0 \in \mathscr{I}\,,$$

for every $1 \leqslant i < j < k \leqslant d$. Again, notice that this condition is independent of the chosen elements $x_{ij} \in \mathbb{F}[x_1,\ldots,x_d]$, which represent $\bar{x}_{ij}$.

• Suppose that $\{\cdot,\cdot\}$ is a Poisson bracket on $\mathscr{A} = \mathbb{F}[x_1,\ldots,x_d]/\mathscr{I}$. We can choose elements $x_{ij} \in \mathbb{F}[x_1,\ldots,x_d]$ such that $x_{ji} = -x_{ij}$ for all i and j, and such that $\bar{x}_{ij} = \{\bar{x}_i,\bar{x}_j\}$. Define a skew-symmetric biderivation $\{\cdot,\cdot\}_0$ on $\mathbb{F}[x_1,\ldots,x_d]$ by (1.13). Since $\{\bar{x}_i,\cdot\}$ is a derivation of $\mathscr{A}$, we have that

$$0 = \{\bar{x}_i,\overline{K}\} = \sum_{j=1}^d \{\bar{x}_i,\bar{x}_j\}\overline{\frac{\partial K}{\partial x_j}} = \overline{\sum_{j=1}^d x_{ij}\frac{\partial K}{\partial x_j}} = \overline{\{x_i,K\}_0}\,,$$

for every $K \in \mathscr{I}$, so that condition (1.14) holds. This means that $\{\cdot,\cdot\}'$, defined for all $\overline{F},\overline{G} \in \mathscr{A}$ by $\{\overline{F},\overline{G}\}' := \overline{\{F,G\}_0}$, is a biderivation of $\mathscr{A}$, which coincides with $\{\cdot,\cdot\}$ on a system of generators $(\bar{x}_1,\ldots,\bar{x}_d)$ of $\mathscr{A}$. Thus, $\{\cdot,\cdot\}' = \{\cdot,\cdot\}$ and we can compute the Poisson bracket $\{\overline{F},\overline{G}\}$ by using (1.13), as in the case of a polynomial algebra.

We summarize these results in the following proposition.

Proposition 1.9. *Let M be an affine variety in $\mathbb{F}^d$, with algebra of regular functions*

$$\mathscr{A} = \mathbb{F}[x_1,\ldots,x_d]/\mathscr{I}\,,$$

where $\mathscr{I}$ is the ideal of $\mathbb{F}[x_1,\ldots,x_d]$ of polynomial functions vanishing on M, and let $X = (\bar{x}_{ij})$ be a skew-symmetric matrix with coefficients in $\mathscr{A}$. Picking representatives $x_{ij} \in \mathbb{F}[x_1,\ldots,x_d]$ of $\bar{x}_{ij}$, with $x_{ij} = -x_{ji}$, define a skew-symmetric biderivation $\{\cdot,\cdot\}_0$ of $\mathbb{F}[x_1,\ldots,x_d]$ by

$$\{F,G\}_0 := \sum_{i,j=1}^d x_{ij}\frac{\partial F}{\partial x_i}\frac{\partial G}{\partial x_j}\,. \tag{1.15}$$

If, for every $1 \leqslant j \leqslant d$ and for every $K \in \mathscr{I}$,

$$\sum_{i=1}^{d} \frac{\partial K}{\partial x_i} x_{ij} \in \mathscr{I} , \tag{1.16}$$

and, for every $1 \leqslant i < j < k \leqslant d$,

$$\left\{\{x_i,x_j\}_0,x_k\right\}_0 + \left\{\{x_j,x_k\}_0,x_i\right\}_0 + \left\{\{x_k,x_i\}_0,x_j\right\}_0 \in \mathscr{I} , \tag{1.17}$$

then a Poisson bracket $\{\cdot,\cdot\}$ on $\mathscr{A}$ is defined by setting $\{\overline{F},\overline{G}\} := \overline{\{F,G\}_0}$, for all $\overline{F},\overline{G} \in \mathscr{A}$. Conditions (1.16) and (1.17) are independent of the chosen representatives x_{ij} of the elements of the matrix X. Also, the Poisson bracket $\{\cdot,\cdot\}$ on $\mathscr{A}$ is independent of this choice of representatives.

Conversely, given a Poisson structure $\{\cdot,\cdot\}$ on $\mathscr{A}$, the Poisson bracket $\{\overline{F},\overline{G}\}$ of any two elements $\overline{F}$ and $\overline{G}$ of $\mathscr{A}$ can be computed from (1.15), namely

$$\{\overline{F},\overline{G}\} = \sum_{i,j=1}^{d} \{\bar{x}_i,\bar{x}_j\} \overline{\frac{\partial F}{\partial x_i}}\, \overline{\frac{\partial G}{\partial x_j}} , \tag{1.18}$$

where F and G are arbitrary polynomials which represent $\overline{F}$ and $\overline{G}$.

As in the case of $\mathscr{A} = \mathbb{F}[x_1,\dots,x_d]$, the matrix $X = (\{\bar{x}_i,\bar{x}_j\})$ is called the *Poisson matrix* of $(M,\{\cdot,\cdot\})$, with respect to $\bar{x}_1,\dots,\bar{x}_d$. Notice that, even if the Lie algebra $(\mathscr{F}(M),\{\cdot,\cdot\})$ is in general infinite-dimensional, the Poisson bracket of arbitrary elements of $\mathscr{F}(M)$ is, in the case of an affine variety $M \subset \mathbb{F}^d$, completely determined by the Poisson matrix X, i.e., by the brackets between all pairs of elements of an arbitrary set of generators of $\mathscr{F}(M)$, as is seen from the explicit formula (1.18).

1.2.3 The Rank of a Poisson Structure

In this section, we introduce the rank of a Poisson structure at a point of a Poisson variety. Our definition is based on the fact that for a Poisson variety $(M,\{\cdot,\cdot\})$ the rank of its Poisson matrix at a point does not depend on the chosen generators for the algebra of regular functions on M.

Lemma 1.10. *Let $(M,\{\cdot,\cdot\})$ be an affine Poisson variety and let $m \in M$. The rank of the Poisson matrix $X = (\{\bar{x}_i,\bar{x}_j\})$ evaluated at m, is independent of the chosen generators $\bar{x}_1,\dots,\bar{x}_d$ of $\mathscr{F}(M)$.*

Proof. It suffices to show that if $\bar{x}_1,\dots,\bar{x}_d$ are generators of $\mathscr{F}(M)$ and $\bar{x}_0$ is an arbitrary element of $\mathscr{F}(M)$, then the matrices

$$X(m) := (\{\bar{x}_i,\bar{x}_j\}(m))_{1\leqslant i,j\leqslant d} \quad \text{and} \quad X'(m) := (\{\bar{x}_i,\bar{x}_j\}(m))_{0\leqslant i,j\leqslant d} ,$$

have the same rank. Since $\bar{x}_0$ can be written as a polynomial in the generators $\bar{x}_1,\dots,\bar{x}_d$ of $\mathscr{F}(M)$, say $\bar{x}_0 = F(\bar{x}_1,\dots,\bar{x}_d)$, we have in view of (1.18),

$$\{\bar{x}_i,\bar{x}_0\}(m) = \sum_{j=1}^{d} \{\bar{x}_i,\bar{x}_j\}(m)\frac{\partial F}{\partial x_j}(m) ,$$

and so the zeroth column of $X'(m)$ is a linear combination of the other columns of $X'(m)$, i.e., of the columns of $X(m)$. Thus, $X'(m)$ and $X(m)$ have the same rank. □

Definition 1.11. For a Poisson variety $(M,\{\cdot\,,\cdot\})$ and a point $m\in M$, the rank of the Poisson matrix of $\{\cdot\,,\cdot\}$ with respect to an arbitrary system of generators of $\mathscr{F}(M)$, evaluated at m, is called the *rank* of $\{\cdot\,,\cdot\}$ at m, denoted $\mathrm{Rk}_m\{\cdot\,,\cdot\}$. The maximum $\max_{m\in M}\mathrm{Rk}_m\{\cdot\,,\cdot\}$ is called the *rank* of $\{\cdot\,,\cdot\}$, denoted $\mathrm{Rk}\{\cdot\,,\cdot\}$. A point $m\in M$ is said to be a *regular point* of $(M,\{\cdot\,,\cdot\})$, if $\mathrm{Rk}_m\{\cdot\,,\cdot\} = \mathrm{Rk}\{\cdot\,,\cdot\}$, otherwise it is said to be a *singular point* of $(M,\{\cdot\,,\cdot\})$. The set of singular points of $(M,\{\cdot\,,\cdot\})$ is called the *singular locus* of $(M,\{\cdot\,,\cdot\})$.

A more intrinsic definition of the rank of a Poisson structure can be given in terms of the (Zariski) *cotangent space* to M at m. Recall that this vector space is intrinsically defined as $\mathscr{I}_m/\mathscr{I}_m^2$, where $\mathscr{I}_m$ denotes the ideal of $\mathscr{F}(M)$, consisting of all functions which vanish at m; its dual space is the (Zariski) *tangent space* to M at m. We denote the tangent space to M at m by T_mM, while the cotangent space is denoted by T_m^*M. Taking generators of $\mathscr{F}(M)$ which vanish at m, the ideal $\mathscr{I}_m$ corresponds to elements of $\mathscr{F}(M)$ “without constant term”, so that $\mathscr{F}(M)/\mathscr{I}_m \simeq \mathbb{F}$, in a natural way (i.e., by evaluation at m). A point m of M where the dimension of T_mM attains its minimal value is called a *smooth point* of M. The smooth points of M form a Zariski open subset of M and the dimension of T_mM in a smooth point m of M is called the *dimension* of M, denoted $\dim M$.

By the biderivation property (1.1), we have that $\left\{\mathscr{I}_m^2,\mathscr{I}_m\right\}\subset\mathscr{I}_m$. Therefore, the bracket $\{\cdot\,,\cdot\}$ induces for every $m\in M$ a well-defined skew-symmetric bilinear map:

$$\pi_m : \frac{\mathscr{I}_m}{\mathscr{I}_m^2}\times\frac{\mathscr{I}_m}{\mathscr{I}_m^2}\to\frac{\mathscr{F}(M)}{\mathscr{I}_m}\simeq\mathbb{F} . \tag{1.19}$$

The bilinear map π_m is closely related to the Poisson matrix of $\{\cdot\,,\cdot\}$ at m. Namely, let $\bar{x}_1,\dots,\bar{x}_d$ be generators of $\mathscr{F}(M)$, so that $\bar{x}_1-\bar{x}_1(m),\dots,\bar{x}_d-\bar{x}_d(m)$ generate the ideal $\mathscr{I}_m$. If we express the bilinear map π_m in terms of the – possibly linearly dependent – equivalence classes $[\bar{x}_i-\bar{x}_i(m)]\in\mathscr{I}_m/\mathscr{I}_m^2$, then the matrix which we get has entries

$$\pi_m([\bar{x}_i-\bar{x}_i(m)],[\bar{x}_j-\bar{x}_j(m)]) = \left\{\bar{x}_i-\bar{x}_i(m),\bar{x}_j-\bar{x}_j(m)\right\}(m) = \left\{\bar{x}_i,\bar{x}_j\right\}(m) ,$$

i.e., the resulting matrix is nothing but the Poisson matrix of $\{\cdot\,,\cdot\}$, with respect to $\bar{x}_1,\dots,\bar{x}_d$, evaluated at m. It follows that the rank of $\{\cdot\,,\cdot\}$ at m can also be defined as the rank of the intrinsically defined bilinear form π_m on T_m^*M.

The main properties of the rank are given in the following proposition.

Proposition 1.12. *Let $(M,\{\cdot\,,\cdot\})$ be an affine Poisson variety.*

(1) For every $m \in M$, the rank of $\{\cdot\,,\cdot\}$ at m is even;
(2) The rank of $\{\cdot\,,\cdot\}$ is at most equal to the dimension of M;
(3) For every $s \in \mathbb{N}$, the subset $M_{(s)}$ of M, defined by

$$M_{(s)} := \{m \in M \mid \mathrm{Rk}_m\{\cdot\,,\cdot\} \geqslant 2s\}$$

is open; in particular, the subset of points $m \in M$ such that $\mathrm{Rk}_m\{\cdot\,,\cdot\} = \mathrm{Rk}\{\cdot\,,\cdot\}$ is open and dense in M.

Proof. Writing $\mathscr{F}(M)$ as $\mathbb{F}[x_1,\dots,x_d]/\mathscr{I}$, the Poisson matrix of $\{\cdot\,,\cdot\}$ at $m \in M$ is the skew-symmetric matrix $X(m) = (\{\bar{x}_i,\bar{x}_j\}(m))_{1\leqslant i,j\leqslant d}$, whose rank is even. This shows *(1)*. Consider the open subset $R_s \subset \mathfrak{gl}_d$ of all $d \times d$ matrices of rank greater than or equal to $2s$. Since $M_{(s)}$ is the inverse image of R_s by the continuous map

$$\begin{aligned} X : M &\to \mathfrak{gl}_d \\ m &\mapsto \left(\{\bar{x}_i,\bar{x}_j\}(m)\right)_{1\leqslant i,j\leqslant d} \end{aligned} \tag{1.20}$$

it is open; since the topology which is considered here is the Zariski topology, these open subsets are dense as soon as they are non-empty. This yields the proof of *(3)*. Finally, let m be an arbitrary point of M and denote the rank of $\{\cdot\,,\cdot\}$ at m by $2r$. Consider a point m' of $M_{(r)}$, which is a smooth point of M; such a point exists because $M_{(r)}$ and the set of smooth points of M are both dense subsets of M. At such a point m', the dimension of the (co)tangent space coincides with the dimension of M, so that

$$\mathrm{Rk}_m\{\cdot\,,\cdot\} \leqslant \mathrm{Rk}_{m'}\{\cdot\,,\cdot\} \leqslant \dim\left(\mathscr{I}_{m'}/\mathscr{I}_{m'}^2\right) = \dim M\,.$$

Since, in this formula, $m \in M$ is arbitrary, $\mathrm{Rk}\{\cdot\,,\cdot\} \leqslant \dim M$, which is the content of *(2)*. □

A more geometric interpretation of the rank will be given in the next section, when we consider Poisson structures on manifolds.

1.3 Poisson Manifolds

In this section, we introduce the notion of a Poisson structure on a (real or complex) manifold. Our definitions and constructions are more geometric than in the previous section. A geometric approach to Poisson structures is in the case of manifolds not only more natural, but in the case of complex manifolds it is even obligatory, because a complex manifold cannot be replaced by its algebra of (global) functions, as the latter may consist of constant functions only. Thus we will use the geometric analog of a skew-symmetric biderivation, which is a bivector field. We will at the beginning of this section only shortly recall the basic terminology and notations from differential geometry which we will use, referring the reader who needs more details to Appendix B at the end of the book.

1.3.1 Bivector Fields on Manifolds

We consider both real manifolds and complex manifolds, since many constructions in Poisson geometry apply in the same way if all manifolds are considered real or if they are all considered complex; we will therefore make simple statements such as "Let M and N be two manifolds and let $\Psi : M \to N$ be a map", which the reader may specialize to "Let M and N be two differentiable manifolds and let $\Psi : M \to N$ be a smooth map" or to "Let M and N be two complex manifolds and let $\Psi : M \to N$ be a holomorphic map". Similarly, $\dim M$ stands for the real or complex *dimension*, according to the context, $\mathscr{F}(M)$ stands for the appropriate algebra of functions, $\mathbb{F}$ stands for $\mathbb{R}$ or $\mathbb{C}$, and so on. In both cases, the coordinates, real or complex, will be denoted by $x = (x_1, \ldots, x_d)$, where $d = \dim M$.

For $m \in M$ the *tangent space* to M at m is denoted by $T_m M$. Elements of $T_m M$ are by definition pointwise derivation at m, i.e., they are linear forms on the vector space of all function germs at m, satisfying, for all functions F and G, defined on a neighborhood of m in M,

$$\delta_m(F_m G_m) = F(m)\,\delta_m G_m + G(m)\,\delta_m F_m\ ;$$

in this formula, F_m stands for the germ of F at m. The dual space to $T_m M$ is the *cotangent space*, denoted $T_m^* M$. The canonical pairing between $T_m M$ and $T_m^* M$ is denoted by $\langle \cdot, \cdot \rangle$. A map $\Psi : M \to N$ between manifolds leads for every $m \in M$ to a linear map $T_m \Psi : T_m M \to T_{\Psi(m)} N$, called the *tangent map* of Ψ at m. Upon identifying the tangent spaces to $\mathbb{F}$ with $\mathbb{F}$, the tangent map leads to the *differential* of a function; specifically, if F is a function on M, then we view the differential of F at m, denoted $\mathrm{d}_m F$ as a linear function on $T_m M$, that is, as an element of $T_m^* M$. We denote by $\mathfrak{X}^1(M)$ the $\mathscr{F}(M)$-module of vector fields on M and for $\mathscr{V} \in \mathfrak{X}^1(M)$ and $m \in M$ we denote by $\mathscr{V}_m$ the value of $\mathscr{V}$ at m (which is an element of $T_m M$). We often use (and define) vector fields on M through their action on (local) functions: when U is a non-empty open subset of M and $\mathscr{V}$ is a vector field on M (or on U) then $\mathscr{V}[F]$ denotes the function on U, defined by

$$\begin{aligned} \mathscr{V}[F] : U &\to \mathbb{F} \\ m &\mapsto \mathscr{V}_m F_m\ . \end{aligned} \tag{1.21}$$

We also write $\mathscr{V}_m[F]$ for $\mathscr{V}_m F_m$, so that

$$\mathscr{V}_m[F] = \mathscr{V}[F](m) = \langle \mathrm{d}_m F, \mathscr{V}_m \rangle\ . \tag{1.22}$$

Notice that we use square brackets to denote the action of a vector field on a function, as we did in the case of a derivation; this notation will be extended to bivector fields in this section, and to multivector fields in Chapter 3.

We now come to the definition of a bivector field on a manifold. To do this, we first introduce the notion of a pointwise biderivation.

Definition 1.13. Let M be a manifold and let $m \in M$. A bilinear map

$$B_m : \frac{\mathscr{F}_m(M)}{\sim} \times \frac{\mathscr{F}_m(M)}{\sim} \to \mathbb{F}$$

is a *pointwise biderivation* of $\mathscr{F}(M)$ at m, if for all functions F, G and H, defined in a neighborhood of m in M,

$$B_m(F_m G_m, H_m) = F(m) B_m(G_m, H_m) + G(m) B_m(F_m, H_m) ,$$
$$B_m(H_m, F_m G_m) = F(m) B_m(H_m, G_m) + G(m) B_m(H_m, F_m) .$$

We will usually consider pointwise *skew-symmetric* biderivations. Given two pointwise derivations δ_m and ε_m at m, we construct a pointwise skew-symmetric biderivation at m by

$$(\delta_m \wedge \varepsilon_m)(F_m, G_m) := \begin{vmatrix} \delta_m F_m & \delta_m G_m \\ \varepsilon_m F_m & \varepsilon_m G_m \end{vmatrix} ,$$

for all functions F and G, defined in a neighborhood of m. It follows easily from the Hadamard lemma (Lemma B.2) that the basic pointwise skew-symmetric biderivations $\left(\frac{\partial}{\partial x_i}\right)_m \wedge \left(\frac{\partial}{\partial x_j}\right)_m$ span the vector space of all pointwise skew-symmetric biderivations at m; therefore, we denote the latter space by $\wedge^2 T_m M$.

The passage from pointwise skew-symmetric biderivations to bivector fields is similar to the passage from pointwise derivations to vector fields, detailed in Appendix B. Namely, we consider a map P which assigns to every point m of M an element P_m of $\wedge^2 T_m M$. We say that P is a (smooth) *bivector field* on M, if for every open subset U of M, and for all functions $F, G \in \mathscr{F}(U)$, one has that $P[F,G] \in \mathscr{F}(U)$, where $P[F,G]$ is the function on U, which is defined by

$$P[F,G](m) := P_m(F_m, G_m) ,$$

for every $m \in U$. We also write $P_m[F,G]$ for $P_m(F_m, G_m)$. Clearly, P is a (smooth) bivector field as soon as $P[x_i, x_j] \in \mathscr{F}(U)$ for some collection of coordinate charts (U,x) which cover M, and for all i,j, with $1 \leqslant i < j \leqslant d$. Notice that, as we did for vector fields, we enclose the arguments of a bivector field in square brackets. Just like vector fields, bivector fields can be restricted to open subsets; the same notation will be used for P and each of its restrictions to open subsets. The coordinate expression for a bivector field on a coordinate chart (U,x) takes the following form:[2]

$$P = \sum_{1\leqslant i<j\leqslant d} P[x_i,x_j]\frac{\partial}{\partial x_i}\wedge\frac{\partial}{\partial x_j}\,. \tag{1.23}$$

In differential geometric terms, one often refers to bivector fields as being (skew-symmetric) tensors (of type (2, 0)), which refers to the fact that they are $\mathscr{F}(M)$-linear objects. In view of the biderivation property for a bivector field, this may seem absurd, but the point is that, as a tensor, a bivector field acts on a pair of differential one-forms, rather than on a pair of functions. To explain this, notice that the value of a bivector field P on two elements F and G of $\mathscr{F}(M)$, at $m\in M$, depends on the differentials d_mF and d_mG only. Indeed, (1.23) yields

$$P[F,G](m) = \sum_{i,j=1}^{d} P[x_i,x_j](m)\frac{\partial F}{\partial x_i}(m)\,\frac{\partial G}{\partial x_j}(m)\,, \tag{1.24}$$

where $\frac{\partial F}{\partial x_i}(m)$ is the i-th component of d_mF, and similarly for $\frac{\partial G}{\partial x_j}(m)$: in view of (1.22),

$$\left\langle \mathrm{d}_mF, \left(\frac{\partial}{\partial x_i}\right)_m\right\rangle = \frac{\partial}{\partial x_i}[F](m) = \frac{\partial F}{\partial x_i}(m)\,.$$

The fact that $P[F,G](m)$ depends on d_mF and d_mG only, implies that we can define a skew-symmetric $\mathscr{F}(M)$-bilinear map from $\Omega^1(M)$, the $\mathscr{F}(M)$-module of differential one-forms on M, to $\mathscr{F}(M)$: simply put

$$\hat{P}(\mathrm{d}F,\mathrm{d}G) := P[F,G]\,.$$

We will in the sequel usually make no distinction between the bivector field P and the (2,0)-tensor $\hat{P}$, and denote both by the same letter P. Note also that the tensorial interpretation, or equivalently (1.23), implies that we may specialize P to points $m\in M$, giving a skew-symmetric, $\mathbb{F}$-bilinear form

$$P_m : T^*_mM \times T^*_mM \to \mathbb{F}\,, \tag{1.25}$$

which is the differential geometric analog of (1.19). As in the algebraic case, P_m totally determines the bivector field P at m.

Just like for vector fields, we can also define the *pushforward* Ψ_*P of a bivector field P by a diffeomorphism $\Psi : M\to N$ between two manifolds. The bivector field Ψ_*P on N is defined by setting[3]

$$(\Psi_*P)_{\Psi(m)} := \wedge^2(T_m\Psi)P\,,$$

for all $m\in M$.

[2] See (B.8) for the coordinate expression of a vector field, of which (1.23) is a direct generalization.

[3] For a linear map $\phi : V\to W$ between two vector spaces, the linear map $\wedge^2\phi : \wedge^2V\to\wedge^2W$ is defined by $(\wedge^2\phi)(v\wedge w) := \phi(v)\wedge\phi(w)$, for all $v,w\in V$. See Appendix A for details and generalizations.

Remark 1.14. Vector fields on a manifold M can be defined as sections of its *tangent bundle*, TM, the vector bundle over M, whose fiber over $m \in M$ is the vector space T_mM. Similarly, bivector fields on M can be defined as (smooth or holomorphic) sections of $\wedge^2 TM \to M$, where the latter vector bundle has as fiber over $m \in M$ the vector space $\wedge^2 T_mM$.

1.3.2 Poisson Manifolds and Poisson Maps

We are now ready to define the notion of a Poisson structure on a (real or complex) manifold M. The idea is that a bivector field π on M leads for every open subset U of M to a skew-symmetric product on $\mathscr{F}(U)$, which is demanded to make $\mathscr{F}(U)$ into a Poisson algebra.

Definition 1.15. Let π be a bivector field on a manifold M. We say that π is a *Poisson structure* on M if for every open subset U of M, the restriction of π to U makes $\mathscr{F}(U)$ into a Poisson algebra. We then call (M,π) a *Poisson manifold*.

The Poisson structure π of a Poisson manifold (M,π) will often be denoted by $\{\cdot,\cdot\}$, in particular the Poisson bracket $\pi[F,G]$ of two functions F,G, defined on an open subset U of M, will usually be denoted by $\{F,G\}$. The notation $\{F,G\}$ for $\pi[F,G]$ will be referred to as the *bracket notation*.

Let (U,x) be a coordinate neighborhood of a d-dimensional Poisson manifold (M,π). On U, the bivector field π can, according to (1.23), be written as

$$\pi = \sum_{1\leqslant i<j\leqslant d} \{x_i,x_j\} \frac{\partial}{\partial x_i} \wedge \frac{\partial}{\partial x_j}\,, \tag{1.26}$$

so that the matrix $X := \left(\{x_i,x_j\}\right)_{1\leqslant i,j\leqslant d}$ encodes the restriction of π to U. This matrix X, whose elements belong to $\mathscr{F}(U)$, is called the *Poisson matrix* of π with respect to the coordinates $x_1,\dots,x_d$. It is the differential geometric analog of the Poisson matrix, introduced in Section 1.2.2 in the context of affine Poisson varieties. Evaluating the Poisson matrix X at $m \in M$, we get the matrix of the skew-symmetric bilinear map $\pi_m : T^*_mM \times T^*_mM \to \mathbb{F}$, defined by the Poisson structure at m.

Proposition 1.16. *Let $\pi = \{\cdot,\cdot\}$ be a bivector field on a manifold M of dimension d. Then π is a Poisson structure on M if and only if one of the following equivalent conditions holds.*

(i) For every open subset U of M and for all functions $F,G,H \in \mathscr{F}(U)$, the Jacobi identity holds:

$$\{\{F,G\},H\} + \{\{G,H\},F\} + \{\{H,F\},G\} = 0\,; \tag{1.27}$$

(ii) For some collection of coordinate charts (U,x) of M which cover M and for all functions $F,G,H \in \mathscr{F}(U)$, the Jacobi identity (1.27) holds;

(iii) *For some collection of coordinate charts (U,x) of M which cover M and for all i,j,k with $1 \leqslant i < j < k \leqslant d$ the following equality holds:*

$$\sum_{\ell=1}^{d} \left(x_{\ell k} \frac{\partial x_{ij}}{\partial x_\ell} + x_{\ell i} \frac{\partial x_{jk}}{\partial x_\ell} + x_{\ell j} \frac{\partial x_{ki}}{\partial x_\ell} \right) = 0 , \tag{1.28}$$

where $x_{ij} := \{x_i, x_j\}$, for $1 \leqslant i,j \leqslant d$;

(iv) *For some collection of coordinate charts (U,x) of M which cover M and for all i,j,k with $1 \leqslant i < j < k \leqslant d$, the Jacobi identity holds:*

$$\{\{x_i,x_j\},x_k\} + \{\{x_j,x_k\},x_i\} + \{\{x_k,x_i\},x_j\} = 0 . \tag{1.29}$$

Proof. Clearly, *(i)* holds if and only if π makes every $\mathscr{F}(U)$ into a Poisson algebra, hence π is a Poisson structure if and only if *(i)* holds. *(ii)* is a consequence of *(i)*, by specialization. The equivalence of *(ii)* and *(iii)* follows from the fact that

$$\begin{aligned}
&\{\{F,G\},H\} + \circlearrowleft (F,G,H) \\
&= \sum_{i,j,k,\ell=1}^{d} x_{\ell k} \frac{\partial x_{ij}}{\partial x_\ell} \frac{\partial F}{\partial x_i} \frac{\partial G}{\partial x_j} \frac{\partial H}{\partial x_k} + \circlearrowleft (F,G,H) \\
&= \sum_{i,j,k=1}^{d} \sum_{\ell=1}^{d} \left(x_{\ell k} \frac{\partial x_{ij}}{\partial x_\ell} + \circlearrowleft (i,j,k) \right) \frac{\partial F}{\partial x_i} \frac{\partial G}{\partial x_j} \frac{\partial H}{\partial x_k} ,
\end{aligned} \tag{1.30}$$

where we used (1.26) to compute the Poisson brackets. Notice that this computation shows that the value of (1.30) at a point $m \in U$ depends only on F, G and H in a neighborhood of m, in other words on the germs F_m, G_m and H_m. Therefore, if the Jacobi identity holds on a coordinate neighborhood (U,x), then it will hold on every open subset $V \subset U$, which leads to the implication *(ii)* $\Rightarrow$ *(i)*. The equivalence of *(iii)* and *(iv)* is clear because (1.28) is just (1.29) written out explicitly, using (1.26). □

Remark 1.17. In the case of a real manifold, every germ is the germ of a globally defined function, hence it suffices to verify the Jacobi identity for all triples of functions on M. For complex manifolds, this is not sufficient, see Remark B.4.

We now turn to the notion of a Poisson map between two Poisson manifolds.

Definition 1.18. A map $\Psi : M \to N$ between Poisson manifolds (M,π) and (N,π') is called a *Poisson map* if for all open subsets $U \subset M$ and $V \subset N$, with $\Psi(U) \subset V$, the induced map $\Psi^* : \mathscr{F}(V) \to \mathscr{F}(U)$, which is defined for all $F \in \mathscr{F}(V)$ by $\Psi^*(F) := F \circ \Psi$, is a morphism of Poisson algebras, i.e., for all functions $F, G \in \mathscr{F}(V)$,

$$\{F \circ \Psi, G \circ \Psi\} = \{F,G\}' \circ \Psi , \tag{1.31}$$

where $\{\cdot\,,\cdot\} := \pi$ and $\{\cdot\,,\cdot\}' := \pi'$.

The condition that $\Psi : M \to N$ be a Poisson map can be written directly in terms of the bivector field, without reference to local functions on N, as shown in the following proposition.

Proposition 1.19. *Let $\Psi : M \to N$ be a map between two Poisson manifolds (M,π) and (N,π'). Then Ψ is a Poisson map if and only if*

$$\wedge^2(T\Psi)\pi = \pi' , \tag{1.32}$$

i.e., $\wedge^2(T_m\Psi)\pi_m = \pi'_{\Psi(m)}$ for every $m \in M$.

Proof. Let F and G be functions, defined on a neighborhood of $\Psi(m)$ in N, and let (U,x) be a coordinate chart for M around m, so that π can be written on U as $\pi = \sum_{i<j} x_{ij}\frac{\partial}{\partial x_i}\wedge\frac{\partial}{\partial x_j}$.

$$\begin{aligned}
\wedge^2(T_m\Psi)\pi_m[F,G] &= \sum_{1\leqslant i<j\leqslant d} \wedge^2(T_m\Psi)\; x_{ij}(m)\left(\frac{\partial}{\partial x_i}\right)_m \wedge \left(\frac{\partial}{\partial x_j}\right)_m [F,G] \\
&= \sum_{1\leqslant i<j\leqslant d} x_{ij}(m)\, T_m\Psi\left(\frac{\partial}{\partial x_i}\right)_m \wedge T_m\Psi\left(\frac{\partial}{\partial x_j}\right)_m [F,G] \\
&= \sum_{1\leqslant i<j\leqslant d} x_{ij}(m)\left(\frac{\partial}{\partial x_i}\right)_m \wedge \left(\frac{\partial}{\partial x_j}\right)_m [F\circ\Psi, G\circ\Psi] \\
&= \{F\circ\Psi, G\circ\Psi\}(m) ,
\end{aligned}$$

where we have set, as usual, $\{\cdot\,,\cdot\} := \pi$. Setting $\{\cdot\,,\cdot\}' := \pi'$ we clearly have that

$$\pi'_{\Psi(m)}[F,G] = \{F,G\}'(\Psi(m)) ,$$

so that $\wedge^2(T_m\Psi)\pi_m = \pi'_{\Psi(m)}$ if and only if $\{F\circ\Psi, G\circ\Psi\}(m) = \{F,G\}'(\Psi(m))$ for all functions F and G, defined in a neighborhood of $\Psi(m)$. This proves that Ψ is a Poisson map if and only if $\wedge^2(T\Psi)\pi = \pi'$. □

In the context of Poisson manifolds, one calls the Hamiltonian derivation $\mathscr{X}_H$, associated to a function $H \in \mathscr{F}(M)$, the *Hamiltonian vector field*, associated to H. One says then that H is a *Hamiltonian function*, corresponding to this vector field. More generally, a vector field $\mathscr{V}$ is called a *locally Hamiltonian vector field*, if every point $m \in M$ belongs to a neighborhood U on which $\mathscr{V}$ is Hamiltonian, i.e., there exists $H \in \mathscr{F}(U)$ such that $\mathscr{V} = \mathscr{X}_H$ on U; such a function is called a *local Hamiltonian* of $\mathscr{V}$. According to (1.26), the (local) Hamiltonian vector field $\mathscr{X}_H$, associated to $H \in \mathscr{F}(U)$, is given in a coordinate chart (U,x) by

$$\mathscr{X}_H = \sum_{i,j=1}^{d} \{x_i, x_j\}\frac{\partial H}{\partial x_j}\frac{\partial}{\partial x_i} . \tag{1.33}$$

It follows that the first order differential equation, which describes the integral curves of $\mathscr{X}_H$ in a coordinate chart (U,x), is given by

$$\frac{\mathrm{d}x_i}{\mathrm{d}t} = \{x_i, H\} = \sum_{j=1}^{d} \{x_i, x_j\} \frac{\partial H}{\partial x_j}, \qquad i = 1, \dots, d. \tag{1.34}$$

Besides the $\mathrm{Cas}(M)$-module $\mathrm{Ham}(M)$ of Hamiltonian vector fields on M, we will also consider the vector space $\mathrm{Ham}_m(M)$, which consists of all tangent vectors of the form $(\mathscr{X}_H)_m$, where H is an arbitrary function, defined on some neighborhood of m. Elements of $\mathrm{Ham}_m(M)$ are called *Hamiltonian vectors* at m.

Definition 1.20. For a Poisson manifold (M, π) and a point $m \in M$ the integer $\mathrm{Rk}\, \pi_m$, which is also the rank of every Poisson matrix of π at m, is called the *rank* of π at m, denoted $\mathrm{Rk}_m \pi$. One says that (M, π) is a *regular Poisson manifold* when its rank is constant (independent of $m \in M$), and that the rank is *locally constant* at m, when it is constant on some neighborhood of m in M. The rank of π is said to be *maximal* at $m \in M$, when it coincides with the dimension of M. The maximum $\max_{m \in M} \mathrm{Rk}_m \pi$ is called the *rank* of π, denoted $\mathrm{Rk}\, \pi$. A point $m \in M$ is said to be a *regular point* of (M, π), if $\mathrm{Rk}_m \pi = \mathrm{Rk}\, \pi$, otherwise it is said to be a *singular point* of $(M, \{\cdot, \cdot\})$. The set of singular points of (M, π) is called the *singular locus* of (M, π).

The main properties of the rank are given in the following proposition.

Proposition 1.21. *Let (M, π) be a Poisson manifold and let $m \in M$.*

(1) $\mathrm{Rk}_m \pi$ *is even and is equal to* $\dim \mathrm{Ham}_m(M)$*;*
(2) *For every $s \in \mathbb{N}$, the subset $M_{(s)}$ of M, defined by*

$$M_{(s)} := \{m \in M \mid \mathrm{Rk}_m \pi \geqslant 2s\}$$

is open (and dense in the complex case); in particular, the subset of points $m \in M$ such that $\mathrm{Rk}_m \pi = \mathrm{Rk}\, \pi$ is open.

Proof. The rank of π at m is the rank of the bilinear map π_m, so it is the dimension of the image of the linear map $T^*_m M \to T_m M$, defined by $\xi \mapsto \pi_m(\xi, \cdot)$. In turn, the latter equals the dimension of $\mathrm{Ham}_m(M)$, since every $\xi \in T^*_m M$ can be written as $\mathrm{d}_m F$, for some function F, defined in a neighborhood of m. Also, since π_m is skew-symmetric, its rank is even. This shows *(1)*. The set R_s of $d \times d$ matrices of rank greater than or equal to $2s$ is an open subset of $\mathfrak{gl}_d$. Let m be a point in $M_{(s)}$ and let $(x_1, \dots, x_d)$ be local coordinates on a neighborhood U of m. The restriction to U of $M_{(s)}$ is open since it is the inverse image of R_s by the continuous map $X : U \to \mathfrak{gl}_d$ defined by $m \mapsto \left(\{x_i, x_j\}(m)\right)_{1 \leqslant i,j \leqslant d}$. This yields the proof of *(2)*. □

In the real case, the rank of a Poisson structure does not necessarily attain its maximum on a *dense* open subset. This is shown in the following example.

Example 1.22. Let φ be a smooth function on $\mathbb{R}^2$ which is positive on the interior of the unit disk and which is zero elsewhere. Then the rank of the Poisson structure on $\mathbb{R}^2$, which is defined by the bivector field $\varphi\, \partial/\partial x \wedge \partial/\partial y$, is of rank 2 only on the inside of the disk.

To finish this section, we show that the flow of a (locally) Hamiltonian vector field leaves the Poisson structure invariant.

Proposition 1.23. *Let (M,π) be a Poisson manifold. The Lie derivative of π with respect to every (locally) Hamiltonian vector field is zero. As a consequence, the flow of each (locally) Hamiltonian vector field preserves the Poisson structure.*

Proof. Let U be an arbitrary open subset of M and let $F,G \in \mathscr{F}(U)$. We need to show that the bivector field $\mathscr{L}_{\mathscr{V}}\pi$, which is the Lie derivative of π with respect to $\mathscr{V}$, vanishes on the pair (F,G). To do this, we use the classical formula for the Lie derivative of a tensor; in the case of a bivector field P, this formula is given by

$$\mathscr{L}_{\mathscr{V}}P[F,G] := \mathscr{V}[P[F,G]] - P[\mathscr{V}[F],G] - P[F,\mathscr{V}[G]]\,, \qquad (1.35)$$

for all $F,G \in \mathscr{F}(U)$ (see Section 3.3.4 for generalizations). For $P := \pi = \{\cdot\,,\cdot\}$ and $\mathscr{V} := \mathscr{X}_H$, with $H \in \mathscr{F}(U)$, this yields

$$\begin{aligned}\mathscr{L}_{\mathscr{X}_H}\pi[F,G] &= \mathscr{X}_H[\{F,G\}] - \{\mathscr{X}_H[F],G\} - \{F,\mathscr{X}_H[G]\}\\ &= \{\{F,G\},H\} - \{\{F,H\},G\} - \{F,\{G,H\}\}\,,\end{aligned}$$

whose vanishing is equivalent to the Jacobi identity (1.2). □

A vector field $\mathscr{V}$ is called a *Poisson vector field* if taking the Lie derivative with respect to $\mathscr{V}$ kills the Poisson structure, i.e., if $\mathscr{L}_{\mathscr{V}}\pi = 0$. Proposition 1.23 implies that all Hamiltonian vector fields are Poisson vector fields. This property will be reformulated in cohomological terms in Section 4.1.

Remark 1.24. For a Poisson algebra $(\mathscr{A},\cdot,\pi)$ one says that a derivation $\mathscr{V}$ of $(\mathscr{A},\cdot)$ is a *Poisson derivation* if the Lie derivative of π with respect to $\mathscr{V}$ vanishes. One has as in Proposition 1.23 that all Hamiltonian derivations are Poisson derivations; there is, however, no interpretation of Poisson derivations in terms of flows, since the notion of a flow does not make sense for derivations of algebras, in general.

1.3.3 Local Structure: Weinstein's Splitting Theorem

This section deals with Weinstein's splitting theorem, which states that, in the neighborhood of a point where the rank of the Poisson structure is $2r$, the Poisson manifold is a product of a symplectic manifold of dimension $2r$, and a Poisson manifold which has rank zero at the origin. It implies the Darboux theorem, which states that all symplectic manifolds of the same dimension are locally isomorphic, "there are no local invariants in symplectic geometry", as well as a generalization to regular Poisson manifolds.

Theorem 1.25 (Weinstein's splitting theorem). *Let (M,π) be a (real or complex) Poisson manifold, let $m \in M$ be an arbitrary point and denote the rank* *of π at m by $2r$. There exists a coordinate neighborhood U of m with coordinates $q_1,\dots,q_r,p_1,\dots,p_r,z_1,\dots,z_s$, centered at m, such that, on U,*

$$\pi = \sum_{i=1}^{r} \frac{\partial}{\partial q_i} \wedge \frac{\partial}{\partial p_i} + \sum_{1\leqslant k<\ell\leqslant s} \phi_{k\ell}(z) \frac{\partial}{\partial z_k} \wedge \frac{\partial}{\partial z_\ell}\,, \tag{1.36}$$

where the functions $\phi_{k\ell}$ are (smooth or holomorphic) functions, which depend on $z=(z_1,\dots,z_s)$ only, and which vanish when $z=0$. Such local coordinates $q_1,\dots,q_r,p_1,\dots,p_r,z_1,\dots,z_s$ are called splitting coordinates, *centered at m.*

Proof. We use induction on r. For the case $r=0$, it is clear that for every Poisson manifold (M,π) and for every point m such that the rank of π at m is zero, an arbitrary system of local coordinates $(z_1,\dots,z_d)$, centered at m, works ($d:=\dim M$). Let $r\in\mathbb{N}^*$ and assume that Theorem 1.25 holds true for every Poisson manifold, at every point where the rank is $2(r-1)$. Let (M,π) be a Poisson manifold and let $m\in M$ be a point for which $\mathrm{Rk}_m\,\pi = 2r$. We will show that the theorem holds for (M,π) at m.

Since $\mathrm{Rk}_m\,\pi>0$, there exists a function p on a neighborhood of m, whose Hamiltonian vector field $\mathscr{X}_p$ does not vanish at $m\in M$; we may suppose that $p(m)=0$. Since $\mathscr{X}_p(m)\neq 0$, there exists by the straightening theorem (Theorem B.7) a system of coordinates $(q,y'_2,\dots,y'_d)$ on a neighborhood U' of m, centered at m, such that $\mathscr{X}_p=\partial/\partial q$. Writing $\{\cdot,\cdot\}=\pi$, it follows that

$$\{q,p\}=\mathscr{X}_p[q]=\frac{\partial q}{\partial q}=1\,,\quad [\mathscr{X}_q,\mathscr{X}_p]=\mathscr{X}_{\{p,q\}}=-\mathscr{X}_1=0 \quad\text{and}\quad \mathscr{X}_p[y'_i]=0\,,$$

for $i=2,\dots,d$, on U'. Writing $\mathscr{X}_q$ in terms of these coordinates,

$$\mathscr{X}_q=\xi_1\frac{\partial}{\partial q}+\sum_{i=2}^{d}\xi_i\frac{\partial}{\partial y'_i}\,,$$

we have that $\xi_1=\mathscr{X}_q[q]=\{q,q\}=0$, and all coefficients $\xi_2,\dots,\xi_d$ are independent of q, since $\mathscr{X}_q$ and $\mathscr{X}_p=\partial/\partial q$ commute. At m, we have

$$-1=\{p,q\}(m)=\mathscr{X}_q[p](m)=\sum_{i=2}^{d}\xi_i(m)\frac{\partial p}{\partial y'_i}(m)\,,$$

so that the vector field $\mathscr{X}_q=\sum_{i=2}^{d}\xi_i\partial/\partial y'_i$ is independent of q and does not vanish at m. Applying the straightening theorem once more, we may introduce a system of coordinates $(q,p',y_3,\dots,y_d)$ on a neighborhood of m, centered at m, where $p',y_3,\dots,y_d$ depend on $y'_2,\dots,y'_d$ only, with

$$\frac{\partial}{\partial p'}=-\sum_{i=2}^{d}\xi_i\frac{\partial}{\partial y'_i}=-\mathscr{X}_q\,.$$

Substituting p' by p we consider $(q, p, y_3, \ldots, y_d)$ which is also a system of coordinates on a neighborhood U of m, since

$$\frac{\partial p}{\partial p'} = -\mathscr{X}_q[p] = \{q, p\} = 1\,,$$

in a neighborhood of m. Since $p', y_3, \ldots, y_d$ depend on $y_2', \ldots, y_d'$ only, $\partial/\partial q$ has the same meaning in both coordinate systems, so that the Poisson brackets take in the new coordinates the following form,

$$\begin{aligned}\{q, p\} &= 1\,,\\ \{q, y_i\} &= -\mathscr{X}_q[y_i] = \partial y_i/\partial p' = 0\,,\\ \{p, y_i\} &= -\mathscr{X}_p[y_i] = -\partial y_i/\partial q = 0\,,\end{aligned}$$

for $i = 3, \ldots, d$, and we conclude that the Poisson structure π is given, in terms of the coordinates $q, p, y_3, \ldots, y_d$, by

$$\pi = \frac{\partial}{\partial q} \wedge \frac{\partial}{\partial p} + \sum_{3 \leqslant k < \ell \leqslant d} \{y_k, y_\ell\} \frac{\partial}{\partial y_k} \wedge \frac{\partial}{\partial y_\ell} \tag{1.37}$$

To show that $\{y_k, y_\ell\}$ is independent of p and q, for all values of k, ℓ, we just have to check that $\{\{y_k, y_\ell\}, p\} = \{\{y_k, y_\ell\}, q\} = 0$, an easy consequence of the Jacobi identity for π. The Jacobi identity also yields that the second term in (1.37) defines a Poisson structure π' on a neighborhood V of the origin o of $\mathbb{F}^{d-2}$. The Poisson matrix of π with respect to $q, p, y_3, \ldots, y_d$ has a diagonal block form, where the lower block is the Poisson matrix of π' with respect to $y_3, \ldots, y_d$. Thus, π' has rank $2(r-1)$ at o, and by the induction hypothesis there exist local coordinates $q_2, \ldots, q_r, p_2, \ldots, p_r, z_1, \ldots, z_s$ centered at o, such that π' takes on V the following form:

$$\pi' = \sum_{i=2}^{r} \frac{\partial}{\partial q_i} \wedge \frac{\partial}{\partial p_i} + \sum_{1 \leqslant k < \ell \leqslant s} \phi_{k\ell}(z) \frac{\partial}{\partial z_k} \wedge \frac{\partial}{\partial z_\ell}$$

In terms of the system of coordinates $(q_1, q_2, \ldots, q_r, p_1, p_2, \ldots, p_r, z_1, \ldots, z_s)$, which is centered at m, π takes the required form (1.36), where we have set $q_1 := q$ and $p_1 := p$. □

For a given point m in a Poisson manifold M, splitting coordinates are not unique. We will see however in Section 5.3.3 that the Poisson structure, which is defined in a neighborhood of $z = 0$ in $\mathbb{F}^s$ by the second term in (1.36), is unique, up to isomorphism.

In terms of the splitting coordinates $q_1, \ldots, q_r, p_1, \ldots, p_r, z_1, \ldots, z_s$, the Poisson structure (1.36) has the block form

$$\begin{pmatrix} 0 & \mathbb{1}_r & 0 \\ -\mathbb{1}_r & 0 & 0 \\ 0 & 0 & \Phi \end{pmatrix}$$

where $\Phi \in \mathrm{Mat}_s(\mathscr{F}(M))$ is given by $\Phi_{ij} := \{z_i, z_j\}$ and $\mathbb{1}_r$ denotes the identity matrix of size r. It follows that $\Phi = 0$ in a neighborhood of m, when the rank of π is locally constant ($= 2r$) at m. This leads to the following strengthening of the splitting theorem for points at which the rank is locally constant.

Theorem 1.26 (Darboux theorem). *Let (M,π) be a (real or complex) Poisson manifold of dimension d, and suppose that m is a point where the rank of π is locally constant and equal to $2r$. There exists a coordinate neighborhood U of m with coordinates $(q_1,\dots,q_r,\, p_1,\dots,p_r, z_1,\dots,z_s)$ such that, on U,*

$$\pi = \sum_{i=1}^{r} \frac{\partial}{\partial q_i} \wedge \frac{\partial}{\partial p_i}\,. \tag{1.38}$$

Moreover, π is locally of the form (1.38), in terms of arbitrary splitting coordinates on M. Such coordinates are called Darboux coordinates.

Corollary 1.27. *The only local invariant of a regular (smooth or complex) d-dimensional Poisson manifold (M,π) is its rank* $\mathrm{Rk}\,\pi$.

To close this section we state a generalization of Weinstein's splitting theorem which we will use in Chapter 12. It provides a set of Darboux coordinates for the Poisson structure π, which contains a given set $p_1,\dots,p_r$ of functions *in involution* (i.e., functions which pairwise commute for the Poisson structure), whose Hamiltonian vector fields are assumed to be independent at a point $m \in M$. See [120] for a proof and see Section 12.3 below for an application.

Theorem 1.28. *Let m be a point of a Poisson manifold (M,π) of dimension d. Let $p_1,\dots,p_r$ be r functions in involution, defined on a neighborhood of m, which vanish at m and whose Hamiltonian vector fields are linearly independent at m. There exist, on a neighborhood U of m, functions $q_1,\dots,q_r,z_1,\dots,z_{d-2r}$, such that*

(1) The d functions $(q_1,\dots,q_r,p_1,\dots,p_r,z_1,\dots,z_{d-2r})$ form a system of coordinates on U, centered at m;

(2) The Poisson structure π is given on U by

$$\pi = \sum_{i=1}^{r} \frac{\partial}{\partial q_i} \wedge \frac{\partial}{\partial p_i} + \sum_{i,j=1}^{d-2r} g_{ij}(z) \frac{\partial}{\partial z_i} \wedge \frac{\partial}{\partial z_j}\,, \tag{1.39}$$

where each function $g_{ij}(z)$ is a smooth function on U and is independent of $q_1,\dots,q_r,p_1,\dots,p_r$.

The rank of π at m is $2r$ if and only if all the functions $g_{ij}(z)$ vanish for $z = 0$.

1.3.4 Global Structure: The Symplectic Foliation

We now come to the global structure of a Poisson manifold. As we will explain in this section, a Poisson manifold naturally decomposes into immersed submanifolds, where each submanifold inherits a Poisson structure from the ambient Poisson structure; since this induced Poisson structure is of maximal rank, it is symplectic,[4] so one usually refers to this decomposition as the *symplectic foliation* of the Poisson manifold. We combine two approaches, which are both due to Weinstein [199], to obtain the symplectic foliation; both have their advantage, a fact that we exploit to get at the decomposition in a natural way and to prove its properties with minimal effort. For another approach, using the theory of generalized distributions, we refer to [125]. Throughout this section (M,π) is an arbitrary Poisson manifold.

The idea which underlies the decomposition is based on the following two observations. First, recall from Proposition 1.23 that the flow of every Hamiltonian vector field preserves the Poisson structure, hence its rank at each point. Second, the rank of the Poisson structure at a point is precisely the dimension of the space of Hamiltonian vector fields, at that point. Thus, the flow of all Hamiltonian vector fields, starting from a given point $m \in M$ where the rank of the Poisson structure is $2r$, should trace out locally a submanifold of dimension $2r$, on which the Poisson structure has constant rank $2r$. This leads, for $m \in M$, to the following definition of a subset of M,

$$\mathscr{S}_m(M) := \left\{m' \in M \mid \exists \text{ a piecewise Hamiltonian path in } M \text{ from } m \text{ to } m'\right\} .$$

By a *Hamiltonian path* from m to m' we mean a curve γ, defined on an open neighborhood[5] of $[0,1]$, with $\gamma(0) = m$ and $\gamma(1) = m'$, which is an integral curve of a Hamiltonian vector field $\mathscr{X}_F$, where F is a function, defined on an open neighborhood of $\gamma([0,1])$. More generally, when points $m_0,\dots,m_N$ in M are such that there exists a Hamiltonian path from m_{i-1} to m_i, for $i = 1,\dots,N$, we say that there exists a *piecewise Hamiltonian path* from m_0 to m_N.

It is clear that the subsets $\mathscr{S}_m(M)$ define a well-defined global decomposition of M: the subsets $\mathscr{S}_m(M)$, with $m \in M$ fill up M, and two subsets $\mathscr{S}_m(M)$ and $\mathscr{S}_{m'}(M)$ are disjoint, unless they coincide. Notice however that some care has to be taken when passing to open subsets: if U is an open subset of M, then $\mathscr{S}_m(U)$ is, in general, not the intersection of $\mathscr{S}_m(M)$ and U: the subset $\mathscr{S}_m(U)$ is by definition path-connected, while $\mathscr{S}_m(M) \cap U$ does not have to be connected, even if U is connected.

The differential geometric properties of $\mathscr{S}_m(M)$ are best understood by relating $\mathscr{S}_m(M)$ to an immersed submanifold of M, passing through m, which appears naturally in the splitting theorem (Theorem 1.25). Recall that a subset N of a manifold M is called an *immersed submanifold* if the inclusion map $\imath : N \to M$ is an immersion, i.e., for every $m \in N$ the tangent map $T_m\imath : T_mN \to T_mM$ is injective. We are viewing here N as a manifold in its own right, i.e., with a topology, which is not necessarily the induced topology from M (one speaks of an *embedded submanifold*

[4] Symplectic manifolds will be introduced and discussed in Chapter 6. For now, the reader may think of a symplectic manifold as being a Poisson manifold (M,π) whose rank $\mathrm{Rk}_m\,\pi$ at each point is equal to the dimension of M.

[5] The neighborhood is taken in $\mathbb{R}$ when M is a real manifold and is taken in $\mathbb{C}$ when M is a complex manifold.

when the manifold topology on N is the induced topology). However, locally, with respect to the manifold topology on N, every point in N has a neighborhood, which sits in M in the same way an open subset of $\mathbb{F}^{d'}$ sits in $\mathbb{F}^d$ ($d' = \dim N \leqslant d = \dim M$). To construct this immersed submanifold of M, through m, we choose on a neighborhood U of m, splitting coordinates (at m) which we denote by $x = (q_1,\dots,q_r, p_1,\dots,p_r,z_1,\dots,z_s)$ and we define

$$\mathscr{S}'_m(U,x) := \{m' \in U \mid z_1(m') = \cdots = z_s(m') = 0\} .$$

It is clear that $\mathscr{S}'_m(U,x)$ is an embedded submanifold of M of dimension $2r$, and that $(q_1,\dots,q_r, p_1,\dots,p_r)$ (restricted to $\mathscr{S}'_m(U,x)$) forms a system of coordinates on $\mathscr{S}'_m(U,x)$. Since the Poisson structure has on U the form

$$\pi = \sum_{i=1}^{r} \frac{\partial}{\partial q_i} \wedge \frac{\partial}{\partial p_i} + \sum_{1\leqslant k<\ell\leqslant s} \phi_{k\ell}(z) \frac{\partial}{\partial z_k} \wedge \frac{\partial}{\partial z_\ell} , \tag{1.40}$$

where the functions $\phi_{k\ell}(z)$ vanish for $z = 0$, it follows that at points $m' \in \mathscr{S}'_m(U,x)$, the Poisson structure takes the form $\sum_{i=1}^{r} \frac{\partial}{\partial q_i} \wedge \frac{\partial}{\partial p_i}$, which defines a Poisson structure on $\mathscr{S}'_m(U,x)$, which we denote by $\pi_{\mathscr{S}}$. The map $\imath : (\mathscr{S}'_m(U,x), \pi_{\mathscr{S}}) \to (M,\pi)$ is a Poisson map, since for all functions $F, G \in \mathscr{F}(U)$, and for every $m' \in \mathscr{S}'_m(U,x)$, we have that

$$\begin{aligned} \{F,G\}(\imath(m')) = \{F,G\}(m') &= \sum_{i=1}^{r} \left(\frac{\partial F}{\partial q_i}(m') \frac{\partial G}{\partial p_i}(m') - \frac{\partial F}{\partial p_i}(m') \frac{\partial G}{\partial q_i}(m') \right) \\ &= \{F \circ \imath, G \circ \imath\}_{\mathscr{S}}(m') . \end{aligned}$$

In the language of Section 2.2, $(\mathscr{S}'_m(U,x), \pi_{\mathscr{S}})$ is a Poisson submanifold of (M,π). Notice that $\pi_{\mathscr{S}}$ has constant rank $2r = \dim \mathscr{S}'_m(U,x)$. In the following proposition we show that, for each $m \in M$, the subsets $\mathscr{S}_m(U)$ and $\mathscr{S}'_m(U,x)$ coincide, for small enough U.

Proposition 1.29. *Let m be a point in a Poisson manifold (M,π) and let $x = (q_1,\dots,q_r, p_1,\dots,p_r,z_1,\dots,z_s)$ be splitting coordinates at m, on a coordinate neighborhood U, with $x(U)$ convex. Then $\mathscr{S}_m(U) = \mathscr{S}'_m(U,x)$.*

Proof. For $F \in \mathscr{F}(U)$ and $m' \in \mathscr{S}'_m(U,x)$ we have, in view of (1.40), that

$$\mathscr{X}_F[z_i](m') = \{z_i, F\}(m') = \sum_{\ell=1}^{s} \phi_{i\ell}(m') \frac{\partial F}{\partial z_\ell}(m') = 0 ,$$

since the functions $\phi_{i\ell}$ vanish at all points m' of $\mathscr{S}'_m(U,x)$. It follows that all Hamiltonian vector fields of U are tangent to $\mathscr{S}'_m(U,x)$ at points $m' \in \mathscr{S}'_m(U,x)$. Thus, a (piecewise) Hamiltonian path in U which starts at m must stay in $\mathscr{S}'_m(U,x)$. It follows that $\mathscr{S}_m(U) \subset \mathscr{S}'_m(U,x)$. In order to show the other inclusion, suppose that $m' \in \mathscr{S}'_m(U,x)$. We show that there exists a Hamiltonian path from m to m' in U. Consider the (linear) function H on U, given by

$$H := \sum_{i=1}^{r} (q_i(m')p_i - p_i(m')q_i)\,,$$

whose (constant) Hamiltonian vector field $\mathscr{X}_H$ is given by

$$\mathscr{X}_H : \begin{cases} \dfrac{\mathrm{d}q_i}{\mathrm{d}t} = \dfrac{\partial H}{\partial p_i} = q_i(m')\,, \\ \dfrac{\mathrm{d}p_i}{\mathrm{d}t} = -\dfrac{\partial H}{\partial q_i} = p_i(m')\,. \end{cases}$$

Consider the integral curve of $\mathscr{X}_H$, defined for t in a small neighborhood of $[0,1]$ (in $\mathbb{R}$ or $\mathbb{C}$) by

$$\gamma : t \mapsto (q_1(m')t, \ldots, q_r(m')t, p_1(m')t, \ldots, p_r(m')t, 0, \ldots, 0)\,.$$

The image of γ is contained in U, because $x(U) \subset \mathbb{F}^{2r+s}$ is a convex open subset. Since γ is a Hamiltonian path from $m = \gamma(0)$ to $m' = \gamma(1)$ in U, it follows that $m' \in \mathscr{S}_m(U)$. $\square$

We use Proposition 1.29 to build a topology and a differential structure on each one of the subsets $\mathscr{S}_m(M)$. A subset $V \subset \mathscr{S}_m(M)$ is by definition an open subset if and only if for all splitting coordinates (U,x) at $m' \in V$, with $x(U)$ convex, $V \cap U$ is an open subset of $\mathscr{S}'_{m'}(U,x) = \mathscr{S}_{m'}(U)$. Similarly, an algebra of functions is defined on $\mathscr{S}_m(M)$ by saying that $F : \mathscr{S}_m(M) \to \mathbb{F}$ is smooth (holomorphic in the complex case), when its restriction to every such $\mathscr{S}'_{m'}(U,x) = \mathscr{S}_{m'}(U)$ is smooth (or holomorphic). Since the topology and the differential structure on these subsets $\mathscr{S}_{m'}(U)$ is intrinsically defined, as embedded submanifolds of M, these definitions yield a well-defined manifold structure on $\mathscr{S}_m(M)$. In general, the topology which we defined on $\mathscr{S}_m(M)$ is different from (finer than) the induced topology which $\mathscr{S}_m(M)$ inherits as a subset of M: it is an immersed submanifold, which is not necessarily an embedded submanifold. The Poisson structure on the submanifolds $\mathscr{S}'_{m'}(U,x)$ yields a well-defined Poisson structure of maximal rank (symplectic structure) on $\mathscr{S}_m(M)$. The results of the present section, combined, prove the following fundamental theorem.

Theorem 1.30. *Every Poisson manifold (M,π) is the disjoint union of immersed submanifolds, whose tangent spaces are spanned by the Hamiltonian vector fields of (M,π). The Poisson structure, restricted to each of these submanifolds yields a Poisson structure of maximal rank (symplectic structure). This decomposition is called the* symplectic foliation *of M and the immersed submanifolds are called the* symplectic leaves *of M. For $m \in M$ the symplectic leaf which contains m is given by*

$$\mathscr{S}_m(M) = \{m' \in M \mid \exists \text{ a piecewise Hamiltonian path in } M \text{ from } m \text{ to } m'\}\,.$$

The first part of the theorem can also be restated by saying that the (singular) distribution, defined by the Hamiltonian vector fields, is *integrable* (admits an integral manifold through each point), see [125, Appendix 3]. The theorem leads to the following description of the symplectic leaves of a Poisson manifold.

Proposition 1.31. *Let (M,π) be a Poisson manifold of dimension d and rank 2r.*

(1) A function F on M is a Casimir function if and only if F is constant on each symplectic leaf;

(2) The (non-empty open subsets of the) symplectic leaves are the smallest embedded manifolds of M which are Poisson submanifolds (i.e., such that the inclusion map is a Poisson map);

(3) If $\mathscr{Y}$ is a Poisson vector field on M which is tangent to every symplectic leaf of M, then $\mathscr{Y}$ is Hamiltonian in the neighborhood of every point $m \in M$ where the rank of π is 2r.

Proof. The fact that the manifold $\mathscr{S}_m(M)$ admits locally (in the sense of the topology on $\mathscr{S}_m(M)$ which we have constructed) the description as a submanifold $\mathscr{S}'_m(U,x)$, leads at once to the proof of *(1)* and *(2)*. We proceed to the proof of *(3)*. If the rank of π at m is $2r$, so that m is a regular point of π, then there exists local coordinates $q_1,\ldots,q_r,p_1,\ldots,p_r,z_1,\ldots,z_{d-2r}$ in a contractible neighborhood U of m with respect to which the Poisson structure π is given by:

$$\pi = \sum_{i=1}^{r} \frac{\partial}{\partial q_i} \wedge \frac{\partial}{\partial p_i}. \tag{1.41}$$

The vector fields $\frac{\partial}{\partial q_1},\frac{\partial}{\partial p_1},\ldots,\frac{\partial}{\partial q_r},\frac{\partial}{\partial p_r}$ span the symplectic leaves of π on U. Therefore, every vector field $\mathscr{Y}$, which is tangent to the symplectic leaves of π, is of the form

$$\mathscr{Y} = \sum_{i=1}^{r} F_i \frac{\partial}{\partial p_i} + \sum_{i=1}^{r} G_i \frac{\partial}{\partial q_i} \tag{1.42}$$

for some smooth or holomorphic functions $F_1,\ldots,F_r,G_1,\ldots,G_r$, defined on U. Suppose now that $\mathscr{Y}$ is a Poisson vector field, so that $\mathscr{L}_{\mathscr{Y}}\pi = 0$. Then one computes easily from (1.35), (1.41) and (1.42) that

$$\frac{\partial F_i}{\partial q_j} = \frac{\partial F_j}{\partial q_i}, \quad \frac{\partial G_i}{\partial p_j} = \frac{\partial G_j}{\partial p_i} \quad \text{and} \quad \frac{\partial F_i}{\partial p_j} = -\frac{\partial G_i}{\partial q_j},$$

for $i,j = 1,\ldots,r$. By the classical Poincaré lemma, there exists a function H, defined on the contractible subset U, which satisfies

$$F_i = -\frac{\partial H}{\partial q_i} \quad \text{and} \quad G_i = \frac{\partial H}{\partial p_i},$$

for $i = 1,\ldots,r$. Hence,

$$\mathscr{X}_H = \sum_{i=1}^{r} \frac{\partial H}{\partial q_i} \mathscr{X}_{q_i} + \sum_{i=1}^{r} \frac{\partial H}{\partial p_i} \mathscr{X}_{p_i} + \sum_{k=1}^{d-2r} \frac{\partial H}{\partial z_k} \mathscr{X}_{z_k} = \mathscr{Y},$$

which shows that $\mathscr{Y}$ is a Hamiltonian vector field on U. □

Locally, the regular leaves (i.e., the leaves which pass through a point where the rank is maximal, hence locally constant) are given as the level sets of the (local) Casimir functions. This is shown in the following proposition.

Proposition 1.32. *Let (M,π) be a Poisson manifold of dimension d, let U be a non-empty open subset of M and let $F_1,\dots,F_s \in \mathscr{F}(U)$, satisfying:*

(1) The rank of π is constant on U and is equal to $d-s$;
(2) The functions $F_1,\dots,F_s$ are Casimirs of the restriction of π to U;
(3) For every point m of U, the differentials $\mathrm{d}_m F_1,\dots,\mathrm{d}_m F_s$ are independent.

Then the symplectic foliation of the restriction of π to U coincides with the foliation which is defined on U by the map $F := (F_1,\dots,F_s) : M \to \mathbb{F}^s$.

Proof. Under the stated assumptions, the restriction of π to U is a regular Poisson structure, hence the symplectic foliation is a regular foliation of U, where every leaf has dimension $d-s$. Since $F := (F_1,\dots,F_s)$ has constant, maximal rank, F also defines a foliation $\mathscr{F}$ on U, whose leaves are $(d-s)$-dimensional. In view of *(1)* in Proposition 1.31, F is constant on every symplectic leaf in U, hence every symplectic leaf is contained in a leaf of $\mathscr{F}$. Since all symplectic leaves and all leaves of $\mathscr{F}$ have the same dimension $(d-s)$, it follows that both foliations coincide. □

In good cases most or all of the symplectic leaves are level sets of the Casimir functions, but this is not true in general.

Example 1.33. Consider the Poisson structure $\partial/\partial x \wedge \partial/\partial y$ on $\mathbb{R}^3$, where x,y,z are the natural coordinates on $\mathbb{R}^3$. Since translations over a constant vector are obviously Poisson maps, it descends to a Poisson structure on every torus $\mathbb{R}^3/\Lambda$, where Λ is a lattice in $\mathbb{R}^3$. Unless Λ is very special, all symplectic leaves are dense on the torus, hence they cannot be the level sets of a (Casimir) function.

Proposition 1.34. *Let π_1 and π_2 be two Poisson structures on a manifold M. Suppose that both Poisson structures define the same symplectic foliation on M and that for every symplectic leaf $\mathscr{S}$ (of both structures), the Poisson (symplectic) structure induced on $\mathscr{S}$ by π_1 is the same as the Poisson structure induced on $\mathscr{S}$ by π_2. Then π_1 and π_2 are equal.*

Proof. Let $\pi_1 = \{\cdot\,,\cdot\}_1$ and $\pi_2 = \{\cdot\,,\cdot\}_2$ be two Poisson structures on M which determine the same symplectic foliation on M. Let $m \in M$ and let $\mathscr{S}$ be the leaf which passes through m; we denote by $\imath_{\mathscr{S}}$ the inclusion map $\imath_{\mathscr{S}} : \mathscr{S} \hookrightarrow M$. It is assumed that the induced Poisson structures $\{\cdot\,,\cdot\}_{i,\mathscr{S}}$ on $\mathscr{S}$ coincide, so that $\{F\circ\imath_{\mathscr{S}}, G\circ\imath_{\mathscr{S}}\}_{1,\mathscr{S}} = \{F\circ\imath_{\mathscr{S}}, G\circ\imath_{\mathscr{S}}\}_{2,\mathscr{S}}$ for every $F,G \in \mathscr{F}(M)$. Since $\imath_{\mathscr{S}}$ is a Poisson map, when M is equipped with π_1, or with π_2, we have for every $m \in \mathscr{S}$,

$$\{F,G\}_1(m) = \{F\circ\imath_{\mathscr{S}}, G\circ\imath_{\mathscr{S}}\}_{1,\mathscr{S}}(m) = \{F\circ\imath_{\mathscr{S}}, G\circ\imath_{\mathscr{S}}\}_{2,\mathscr{S}}(m) = \{F,G\}_2(m)\,.$$

This applies to every point m of M, so that $\{F,G\}_1(m) = \{F,G\}_2(m)$, for all $m \in M$, which was to be shown. □

1.4 Poisson Structures on a Vector Space

On a finite-dimensional $\mathbb{F}$-vector space V, one can consider its algebra of polynomial functions, which is canonically isomorphic to the symmetric algebra SV^*, but when $\mathbb{F} = \mathbb{R}$ one may also consider the algebra of smooth functions on V, while when $\mathbb{F} = \mathbb{C}$ one may consider the algebra of holomorphic functions on V. The difference corresponds to viewing V as an affine variety or as a real/complex manifold. What makes a vector space V special, from either point of view, is that it admits a basis, say $(e_1, \ldots, e_d)$, and hence a dual basis, consisting of linear forms $x_i : V \to \mathbb{F}$, for $i = 1, \ldots, d$.

(1) From the affine variety point of view, $SV^* = \mathbb{F}[x_1, \ldots, x_d]$, so every function can be written *uniquely* as a polynomial in $x_1, \ldots, x_d$. More generally, every derivation $\mathscr{V}$ of SV^* can be uniquely written as $\mathscr{V} = \sum_{i=1}^{d} \mathscr{V}_i \frac{\partial}{\partial x_i}$, where $\mathscr{V}_i := \mathscr{V}[x_i]$, and similarly for biderivations, such as Poisson structures.

(2) From the manifold point of view, $x_1, \ldots, x_d$ are global coordinates on V, i.e., V is covered by a single coordinate chart. Functions, vector fields, Poisson structures, and so on, admit a *global* coordinate representation in terms of these global coordinates.

Many results about Poisson structures on a vector space take formally the same form when the Poisson structures are considered on the algebra of polynomial functions or on the algebra of smooth/holomorphic functions. Therefore we will often refer to $\mathscr{F}(V)$ as *the* algebra of functions on V, independently of whether the algebra of polynomial, smooth or holomorphic functions is considered; however, when $\mathbb{F}$ is different from $\mathbb{R}$ and $\mathbb{C}$, then $\mathscr{F}(V)$ always stands for the algebra of polynomial functions on V. To stress that we consider a Poisson bracket on the algebra of polynomial functions on V, we often use the term *polynomial Poisson structure*.

Vector fields on a real or complex vector space V can be identified with derivations of the algebra of smooth or holomorphic functions on V, so we will simply write $\mathfrak{X}^1(V)$ for the Lie algebra of vector fields on V; even when $\mathbb{F}$ is different from $\mathbb{R}$ and $\mathbb{C}$, we refer to elements of $\mathfrak{X}^1(V)$ as vector fields on V. Similarly, we refer to elements of $\mathfrak{X}^2(V)$ as bivector fields on V. Vector fields and bivector fields admit in this general context the following description.

Proposition 1.35. *Let V be a vector space and let $\mathscr{F}(V)$ be its algebra of functions. Let $\mathscr{V}$ be a vector field on V and let P be a bivector field on V. In terms of linear coordinates $x_1, \ldots, x_d$ for V, the elements $\mathscr{V}$ and P can be written (uniquely) as*

$$\mathscr{V} = \sum_{i=1}^{d} \mathscr{V}_i \frac{\partial}{\partial x_i}, \qquad P = \sum_{1 \leqslant i < j \leqslant d} P_{ij} \frac{\partial}{\partial x_i} \wedge \frac{\partial}{\partial x_j},$$

where $\mathscr{V}_i \in \mathscr{F}(V)$ and $P_{ij} \in \mathscr{F}(V)$, for all $1 \leqslant i, j \leqslant d$. In particular, every vector field on V is determined by its values on a system of linear coordinates on V, and every bivector field on V is determined by its values on all pairs of functions, taken from a system of linear coordinates on V.

Suppose now that we have a bivector field $\pi = \{\cdot,\cdot\}$ on V and that we wish to express that π is a Poisson structure. When $\mathscr{F}(V)$ is the algebra of polynomial functions on V, then π is a Poisson structure if and only if π satisfies the Jacobi identity,

$$\{F,\{G,H\}\}+\{G,\{H,F\}\}+\{H,\{F,G\}\}=0\,, \tag{1.43}$$

for all $F,G,H \in \mathscr{F}(V)$; however, when $\mathscr{F}(V)$ is the algebra of smooth of holomorphic functions on V, precisely the same condition expresses that π is a Poisson structure, because V can be covered with a single coordinate chart. We formulate this result in the following proposition, which is a transcription and a specialization of Propositions 1.8 and 1.16 to the case of a vector space V.

Proposition 1.36. *Let V be a finite-dimensional vector space, with algebra of functions $\mathscr{F}(V)$ and let $(x_1,\dots,x_d)$ be a system of linear coordinates on V. Given a skew-symmetric matrix $(x_{ij})_{1\leqslant i,j\leqslant d}$ of elements of $\mathscr{F}(V)$, consider the skew-symmetric biderivation π of $\mathscr{F}(V)$, which is defined, for $F,G \in \mathscr{F}(V)$, by*

$$\{F,G\} := \sum_{i,j=1}^{d} x_{ij}\frac{\partial F}{\partial x_i}\frac{\partial G}{\partial x_j}\,,$$

which is the unique skew-symmetric biderivation of $\mathscr{F}(V)$ such that $\left\{x_i,x_j\right\}=x_{ij}$, for all $1\leqslant i,j\leqslant d$. The following conditions are equivalent.

(i) *π is a Poisson structure on V;*
(ii) *π satisfies the Jacobi identity: for all $F,G,H \in \mathscr{F}(V)$,*

$$\{F,\{G,H\}\}+\{G,\{H,F\}\}+\{H,\{F,G\}\}=0\,;$$

(iii) *For all i,j,k, with $1\leqslant i<j<k\leqslant d$,*

$$\left\{\left\{x_i,x_j\right\},x_k\right\}+\left\{\left\{x_j,x_k\right\},x_i\right\}+\left\{\left\{x_k,x_i\right\},x_j\right\}=0\,;$$

(iv) *For all i,j,k, with $1\leqslant i<j<k\leqslant d$,*

$$\sum_{\ell=1}^{d}\left(x_{\ell k}\frac{\partial x_{ij}}{\partial x_\ell}+x_{\ell i}\frac{\partial x_{jk}}{\partial x_\ell}+x_{\ell j}\frac{\partial x_{ki}}{\partial x_\ell}\right)=0\,.$$

In this case, the skew-symmetric matrix $(x_{ij})_{1\leqslant i,j\leqslant d}$ is the Poisson matrix *of π with respect to the coordinates $x_1,\dots,x_d$.*

At every point $m \in V$, the tangent space T_mV is in a canonical way isomorphic to V, and similarly the cotangent space T_m^*V is canonically isomorphic to V^*. Therefore, one usually thinks of the Poisson bivector at m, which is a skew-symmetric bilinear map $\pi_m : T_m^*V \times T_m^*V \to \mathbb{F}$, as a skew-symmetric bilinear map $V^* \times V^* \to \mathbb{F}$, as an element of $V \wedge V$, or as a linear map $V^* \to V$. The rank of the latter linear map is the rank of the Poisson structure at m.

It is clear that the results in this section are also valid when the vector space V is replaced by a non-empty open subset of V.

1.5 Exercises

1. Suppose that $\mathscr{V}$ is a derivation of $\mathscr{A} := \mathbb{F}[x_1,\dots,x_d]$. Show that, for every $F \in \mathscr{A}$,

$$\mathscr{V}[F] = \sum_{i=1}^{d} \frac{\partial F}{\partial x_i} \mathscr{V}[x_i] \; .$$

Deduce from it an alternative proof of Proposition 1.6.

2. Two elements F, G of a Poisson algebra $(\mathscr{A}, \cdot, \pi)$ are said to be in *involution* if $\{F,G\} = 0$. Prove *Poisson's theorem*: if F and G are in involution with a third element $H \in \mathscr{A}$, then $\{F,G\}$ is in involution with H. The original motivation of Poisson's theorem is that, since functions which are in involution with the Hamiltonian are constants of motion, taking the Poisson bracket of two constants of motion produces a (a priori new) constant of motion.

3. Let π be a polynomial Poisson structure on $\mathbb{F}^d$, with Poisson matrix X, and suppose that X vanishes at the origin $o \in \mathbb{F}^d$ (i.e., all entries x_{ij} of X vanish at o). Let Y be the matrix whose entry y_{ij} is the linear part of x_{ij}, for $1 \leqslant i,j \leqslant d$. Show that Y is a Poisson matrix.

4. Let (M_1, π_1) and (M_2, π_2) be two affine Poisson varieties, or two Poisson manifolds. Let $\Psi : M_1 \to M_2$ be a Poisson map. Show that

$$\mathrm{Rk}_m M_1 \geqslant \mathrm{Rk}_{\Psi(m)} M_2 \; ,$$

for every $m \in M_1$.

5. Given a skew-symmetric biderivation $\{\cdot\,,\cdot\}$ on a commutative associative algebra $\mathscr{A}$, define its *Jacobiator* $\mathscr{J}$ as the skew-symmetric tri-linear map $\mathscr{A}^3 \to \mathscr{A}$, given by

$$\mathscr{J}(F,G,H) := \{F,\{G,H\}\} + \{G,\{H,F\}\} + \{H,\{F,G\}\} \; .$$

Show that $\mathscr{J}$ is a derivation in each of its arguments (as in (1.4); in the language of Section 3.1, it is a triderivation) and derive from it an alternative proof of Proposition 1.8.

6. Consider on $\mathbb{F}^2$ the bivector field π, defined by $\{x,y\} := x^2$, and let $\mathscr{Y}$ denote the vector field on $\mathbb{F}^2$, defined by $\mathscr{Y}[x] := 0$ and $\mathscr{Y}[y] := x$. Show that $\mathscr{Y}$ is a Poisson vector field, but that there exists no neighborhood of $o = (0,0)$ on which $\mathscr{Y}$ is a Hamiltonian vector field (see item *(3)* of Proposition 1.31).

7. The following is an open problem: given a Poisson algebra $(\mathscr{A}, \cdot, \{\cdot\,,\cdot\})$ of finite type, do there exist generators $\bar{x}_1, \dots, \bar{x}_d$ of $\mathscr{A}$ and representatives $x_{ij} = -x_{ji} \in \mathbb{F}[x_1,\dots,x_d]$ of $\{\bar{x}_i, \bar{x}_j\}$ such that

$$\sum_{\ell=1}^{d} \left(x_{\ell k} \frac{\partial x_{ij}}{\partial x_\ell} + x_{\ell i} \frac{\partial x_{jk}}{\partial x_\ell} + x_{\ell j} \frac{\partial x_{ki}}{\partial x_\ell} \right) = 0 \; ,$$

for every $1 \leqslant i,j,k \leqslant d$; in other words, such that the matrix $X := (x_{ij})_{1\leqslant i,j\leqslant d}$ is a Poisson matrix?

1.6 Notes

In the mathematics and physics literature, Poisson structures are most often considered in the case of smooth manifolds. In this context, Poisson structures are usually defined as sections of the exterior square of the tangent bundle of the manifold (see Vaisman [194]), as Lie algebra structures on the algebra of functions on the manifold (see Cannas da Silva–Weinstein [34] or Dufour–Zung [63]) or as sheafs of Poisson algebras on the manifold (see Przybylski [172]). Abstract Poisson algebras are considered in Bhaskara–Viswanath [23]; see also Huebschmann [96]. For more information on the general theory of real manifolds, we refer to Warner [198] or to Spivak [187]; for complex manifolds, see Wells [203]. For a gentle introduction to algebraic geometry, in particular the link between commutative associative algebras and affine varieties, see Perrin [165] or Shafarevich [182].

The main result in this chapter is Weinstein's splitting theorem (Theorem 1.25), from which we derived the symplectic foliation. This general theorem admits further generalizations: for example, an equivariant version is given in Dufour–Zung [63], while a generalization to arbitrary Lie algebroids is given in Fernandes [75]. See Laurent–Miranda–Vanhaecke [120] for the Carathéodory–Jacobi–Lie theorem for Poisson manifolds, which was stated without proof in Theorem 1.28 above; for regular Poisson manifolds, this theorem goes back to Lie and Engel [128].

Chapter 2
Poisson Structures: Basic Constructions

In this chapter we give a few basic, general constructions which allow one to build new Poisson structures from given ones. These constructions are fundamental and will be used throughout the book. More advanced constructions, which all fall under the general concept of reduction, will be discussed in detail in Chapter 5, while several constructions which are specific to a particular class of examples will be given in Part II.

As in the previous chapter, all Poisson algebras will be defined over an arbitrary field $\mathbb{F}$ of characteristic zero, our Poisson varieties will be affine varieties over $\mathbb{F}$ and our Poisson manifolds will be either real smooth manifolds ($\mathbb{F} = \mathbb{R}$) or complex manifolds ($\mathbb{F} = \mathbb{C}$). We will adhere to our habit of reformulating all our algebraic constructions in geometrical terms, or, conversely, present the geometrical construction in general algebraic terms. We stress that although the geometrical and algebraic constructions are based on the same idea, their concrete implementation is usually quite different.

Section 2.1 deals with the tensor product of Poisson algebras, which geometrically corresponds to the construction of a Poisson structure on the product of two Poisson manifolds. We investigate in Section 2.2 the notion of a Poisson ideal, whose geometrical counterpart is that of a Poisson submanifold. A Poisson structure cannot be restricted to any submanifold; a necessary and sufficient conditions for this is that all Hamiltonian vector fields be tangent to the submanifold. In Section 2.3, we show on the one hand how real and complex Poisson structures are related, and on the other hand how complex algebraic and holomorphic Poisson structures are related. In Section 2.4, we assemble a few constructions which are of a different nature, namely changing the base field, localization and germification.

Unless otherwise stated, $\mathbb{F}$ stands throughout this chapter for an arbitrary field of characteristic zero.

C. Laurent-Gengoux et al., *Poisson Structures*,
Grundlehren der mathematischen Wissenschaften 347,
DOI 10.1007/978-3-642-31090-4_2, © Springer-Verlag Berlin Heidelberg 2013

2.1 The Tensor Product of Poisson Algebras and the Product of Poisson Manifolds

The product of two Poisson algebras/varieties/manifolds is in a natural way a Poisson algebra/variety/manifold. We first give the algebraic construction and we make precise what we mean by "a natural way" (Proposition 2.1); it yields by a direct translation the corresponding construction for Poisson varieties, where we give an explicit formula for the product structure in terms of the Poisson matrix (Proposition 2.4). Finally, we give a purely geometric construction, in terms of bivector fields, for the product of two Poisson manifolds (Proposition 2.5).

2.1.1 The Tensor Product of Poisson Algebras

Let $(\mathscr{A}_1, \cdot_1)$ and $(\mathscr{A}_2, \cdot_2)$ be two commutative associative algebras over $\mathbb{F}$, with unit. Their tensor product $\mathscr{A}_1 \otimes \mathscr{A}_2$ is itself a commutative associative algebra with unit, with respect to the $\mathbb{F}$-bilinear product, which is defined for $F_1, G_1 \in \mathscr{A}_1$ and for $F_2, G_2 \in \mathscr{A}_2$ by:

$$(F_1 \otimes F_2) \cdot (G_1 \otimes G_2) := (F_1 G_1) \otimes (F_2 G_2) . \tag{2.1}$$

In this formula, and in the formulas which follow, we use the obvious abbreviations $F_1 G_1$ for $F_1 \cdot_1 G_1$ and similarly for $F_2 G_2$. The algebras $\mathscr{A}_1$ and $\mathscr{A}_2$ are naturally identified with subalgebras of $\mathscr{A}_1 \otimes \mathscr{A}_2$ via the inclusion maps $\jmath_i : \mathscr{A}_i \to \mathscr{A}_1 \otimes \mathscr{A}_2$ $(i = 1, 2)$, given by $\jmath_1(F) = F \otimes 1$ for every $F \in \mathscr{A}_1$, and similarly $\jmath_2(F) = 1 \otimes F$ for every $F \in \mathscr{A}_2$. Also, to a derivation $\mathscr{V}$ of $\mathscr{A}_1$ we can associate in a natural way a derivation $\widehat{\mathscr{V}}$ of $\mathscr{A}_1 \otimes \mathscr{A}_2$ (with values in $\mathscr{A}_1 \otimes \mathscr{A}_2$), such that $\widehat{\mathscr{V}}[\jmath_1(F)] = \jmath_1(\mathscr{V}[F])$, for every $F \in \mathscr{A}_1$. It is defined by

$$\begin{aligned} \widehat{\mathscr{V}} : \mathscr{A}_1 \otimes \mathscr{A}_2 &\to \mathscr{A}_1 \otimes \mathscr{A}_2 \\ F_1 \otimes F_2 &\mapsto \mathscr{V}[F_1] \otimes F_2 . \end{aligned} \tag{2.2}$$

Notice that $\widehat{\mathscr{V}}[\jmath_2(F)] = 0$, for every $F \in \mathscr{A}_2$. To see that $\widehat{\mathscr{V}}$ is indeed a derivation of $\mathscr{A}_1 \otimes \mathscr{A}_2$, use the fact that $\mathscr{V}$ is a derivation of $\mathscr{A}_1$, combined with (2.1). Similarly, to a derivation $\mathscr{W}$ of $\mathscr{A}_2$, we can associate in a natural way a derivation $\widehat{\mathscr{W}}$ of $\mathscr{A}_1 \otimes \mathscr{A}_2$, such that $\widehat{\mathscr{W}}[\jmath_2(F)] = \jmath_2(\widehat{\mathscr{W}}[F])$, for every $F \in \mathscr{A}_2$.

By extending this construction to the case of skew-symmetric biderivations, we can construct a skew-symmetric biderivation of $\mathscr{A}_1 \otimes \mathscr{A}_2$ (with values in $\mathscr{A}_1 \otimes \mathscr{A}_2$), starting from a skew-symmetric biderivation of $\mathscr{A}_1$ (or of $\mathscr{A}_2$). We use this in the following proposition to construct a Poisson structure on $\mathscr{A}_1 \otimes \mathscr{A}_2$, starting from a Poisson structure on $\mathscr{A}_1$ and a Poisson structure on $\mathscr{A}_2$.

Proposition 2.1. *Let* $(\mathscr{A}_i, \cdot_i, \{\cdot\,,\cdot\}_i)$ *be two Poisson algebras over* $\mathbb{F}$, *where* $i = 1,2$. *The tensor product* $\mathscr{A}_1 \otimes \mathscr{A}_2$ *admits a unique Poisson bracket* $\{\cdot\,,\cdot\}$, *making the canonical inclusions* $\jmath_i : \mathscr{A}_i \to \mathscr{A}_1 \otimes \mathscr{A}_2$ *into Poisson morphisms with Poisson-commuting images,* $\{\jmath_1(\mathscr{A}_1), \jmath_2(\mathscr{A}_2)\} = 0$. *For* $F_i, G_i \in \mathscr{A}_i$, *this Poisson bracket is given by*

$$\{F_1 \otimes F_2, G_1 \otimes G_2\} = \{F_1, G_1\}_1 \otimes (F_2G_2) + (F_1G_1) \otimes \{F_2, G_2\}_2 \,. \tag{2.3}$$

For every Hamiltonian derivation $\mathscr{X}_H$ *of* $\mathscr{A}_1$, *with* $H \in \mathscr{A}_1$, $\widehat{\mathscr{X}_H}$ *is a Hamiltonian derivation of* $\mathscr{A}_1 \otimes \mathscr{A}_2$, *more precisely,*

$$\widehat{\mathscr{X}_H} = \mathscr{X}_{H\otimes 1} \,,$$

and similarly for the Hamiltonian derivations of $\mathscr{A}_2$.

We call $(\mathscr{A}_1 \otimes \mathscr{A}_2, \cdot, \{\cdot\,,\cdot\})$ the *Poisson tensor product* of $\mathscr{A}_1$ and $\mathscr{A}_2$, while $\{\cdot\,,\cdot\}$ is called the *product Poisson bracket*.

Proof. Let $\pi_1 : \mathscr{A}_1 \times \mathscr{A}_1 \to \mathscr{A}_1$ be a skew-symmetric biderivation of $\mathscr{A}_1$. Then it can be verified, as in the case of derivations, that the skew-symmetric bilinear map $\widehat{\pi}_1$, defined by

$$\begin{aligned} \widehat{\pi}_1 : (\mathscr{A}_1 \otimes \mathscr{A}_2) \times (\mathscr{A}_1 \otimes \mathscr{A}_2) &\to \mathscr{A}_1 \otimes \mathscr{A}_2 \\ (F_1 \otimes F_2, G_1 \otimes G_2) &\mapsto \pi_1[F_1, G_1] \otimes (F_2G_2) \end{aligned} \tag{2.4}$$

is a biderivation of $\mathscr{A}_1 \otimes \mathscr{A}_2$. If π_1 satisfies moreover the Jacobi identity, so that π_1 is a Poisson bracket on $\mathscr{A}_1$, then $\widehat{\pi}_1$ is a Poisson bracket on $\mathscr{A}_1 \otimes \mathscr{A}_2$, since

$$\widehat{\pi}_1\left[\widehat{\pi}_1[F_1 \otimes F_2, G_1 \otimes G_2], H_1 \otimes H_2\right] = \pi_1\left[\pi_1[F_1, G_1], H_1\right] \otimes (F_2G_2H_2) \,.$$

Notice also that $\widehat{\pi}_1$ has the following properties:

$$\begin{aligned} \widehat{\pi}_1[F_1 \otimes 1, G_1 \otimes 1] &= \pi_1[F_1, G_1] \otimes 1 \,, \\ \widehat{\pi}_1[F_1 \otimes 1, 1 \otimes G_2] &= 0 \,, \\ \widehat{\pi}_1[1 \otimes F_2, 1 \otimes G_2] &= 0 \,, \end{aligned}$$

for every $F_i, G_i \in \mathscr{A}_i$, $(i = 1,2)$. The first formula tells us that $\jmath_1 : \mathscr{A}_1 \to \mathscr{A}_1 \otimes \mathscr{A}_2$ is a morphism of Poisson algebras, when we take $\widehat{\pi}_1$ as a Poisson bracket on $\mathscr{A}_1 \otimes \mathscr{A}_2$; according to the second formula, $\jmath_1$ and $\jmath_2$ have Poisson-commuting images; however, the third formula indicates that $\jmath_2 : \mathscr{A}_2 \to \mathscr{A}_1 \otimes \mathscr{A}_2$ is only a morphism of Poisson algebras when $\mathscr{A}_2$ is equipped with the trivial Poisson bracket.

However, letting π_2 denote a Poisson bracket on $\mathscr{A}_2$ and taking the sum $\pi := \widehat{\pi}_1 + \widehat{\pi}_2$, we obtain a skew-symmetric biderivation of $\mathscr{A}_1 \otimes \mathscr{A}_2$, and the above formulas lead to

$$\begin{aligned}
\pi[F_1 \otimes 1, G_1 \otimes 1] &= \pi_1[F_1, G_1] \otimes 1\,, \\
\pi[F_1 \otimes 1, 1 \otimes G_2] &= 0\,, \\
\pi[1 \otimes F_2, 1 \otimes G_2] &= 1 \otimes \pi_2[F_2, G_2]\,.
\end{aligned}$$

By the biderivation property, these formulas can be summarized in the single formula

$$\pi[F_1 \otimes F_2, G_1 \otimes G_2] = \pi_1[F_1, G_1] \otimes (F_2 G_2) + (F_1 G_1) \otimes \pi_2[F_2, G_2] \tag{2.5}$$

for every $F_i, G_i \in \mathscr{A}_i$, $(i = 1, 2)$. It follows that π satisfies the Jacobi identity, because (2.5), combined with the fact that π is a biderivation, implies that

$$\begin{aligned}
&\pi[\pi[F_1 \otimes F_2, G_1 \otimes G_2], H_1 \otimes H_2] + \circlearrowleft (F, G, H) \\
&= \pi_1[\pi_1[F_1, G_1], H_1] \otimes (F_2 G_2 H_2) + (F_1 G_1 H_1) \otimes \pi_2[\pi_2[F_2, G_2], H_2] + \circlearrowleft (F, G, H)\,.
\end{aligned}$$

Therefore, π is a Poisson bracket on $\mathscr{A}_1 \otimes \mathscr{A}_2$, both maps $\jmath_i : \mathscr{A}_i \to \mathscr{A}_1 \otimes \mathscr{A}_2$ are morphisms of Poisson algebras and they have Poisson-commuting images. Notice that in order to establish formula (2.5) we have used only that π is a biderivation of $\mathscr{A}_1 \otimes \mathscr{A}_2$, such that canonical inclusions $\jmath_i : \mathscr{A}_i \to \mathscr{A}_1 \otimes \mathscr{A}_2$ are Lie algebra homomorphisms and have Poisson-commuting images, which yields uniqueness of a Poisson bracket on $\mathscr{A}_1 \otimes \mathscr{A}_2$ with the stated properties, and it provides us with an explicit formula for it. In the bracket notation, (2.5) is (2.3).

Finally, let us show that the linear map $\mathscr{V} \mapsto \widehat{\mathscr{V}}$ sends Hamiltonian derivations of $\mathscr{A}_1$ to Hamiltonian derivations of $\mathscr{A}_1 \otimes \mathscr{A}_2$. To do this, it suffices to show that for every $H \in \mathscr{A}_1$, the derivation $\widehat{\mathscr{X}_H}$ is the Hamiltonian derivation of $\mathscr{A}_1 \otimes \mathscr{A}_2$, associated to $H \otimes 1$. Let $F_1 \in \mathscr{A}_1$ and $F_2 \in \mathscr{A}_2$ be arbitrary elements. According to (2.2) and (2.5)

$$\begin{aligned}
\widehat{\mathscr{X}_H}[F_1 \otimes F_2] &= \mathscr{X}_H[F_1] \otimes F_2 = \pi_1[F_1, H] \otimes F_2 = \pi[F_1 \otimes F_2, H \otimes 1] \\
&= \mathscr{X}_{H \otimes 1}[F_1 \otimes F_2]\,,
\end{aligned}$$

so that $\widehat{\mathscr{X}_H} = \mathscr{X}_{H \otimes 1}$, as was to be shown. □

Given two affine varieties M_1 and M_2, their product $M_1 \times M_2$ is an affine variety, with algebra of functions $\mathscr{F}(M_1 \times M_2) \simeq \mathscr{F}(M_1) \otimes \mathscr{F}(M_2)$. Under the latter isomorphism, the natural projection maps $p_i : M_1 \times M_2 \to M_i$, with $i = 1, 2$, are dual to the inclusion maps $\jmath_i : \mathscr{F}(M_i) \to \mathscr{F}(M_1) \otimes \mathscr{F}(M_2)$; specifically, the following triangle of algebra homomorphisms is commutative:

$$\begin{array}{ccc}
\mathscr{F}(M_i) & \xrightarrow{\quad p_i^* \quad} & \mathscr{F}(M_1 \times M_2) \\
& {}_{\jmath_i}\searrow \qquad \nearrow_{\simeq} & \\
& \mathscr{F}(M_1) \otimes \mathscr{F}(M_2) &
\end{array}$$

Translated in the language of varieties, Proposition 2.1 yields the following result.

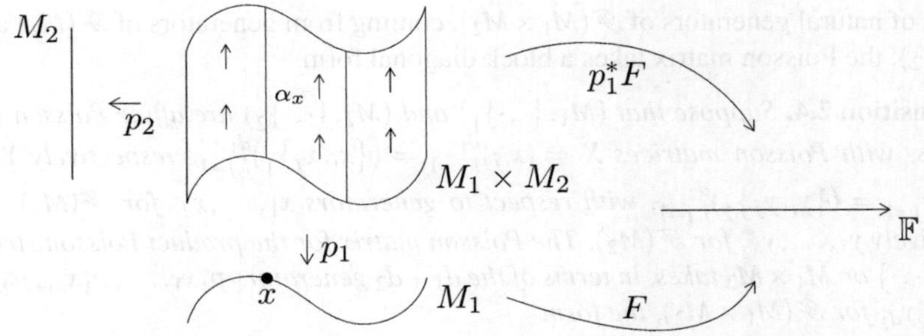

Fig. 2.1 On the product of two affine varieties M_1 and M_2, functions which are constant on the fibers of the canonical projection map $p_1 : M_1 \times M_2 \to M_1$ are of the form $p_1^*F = F \circ p_1$, for some $F \in \mathscr{F}(M_1)$. Every vector field on $M_1 \times M_2$ which annihilates all these functions is tangent to the fibers of p_1.

Proposition 2.2. *Let (M_i, π_i) be two affine Poisson varieties, where $i = 1, 2$. The product variety $M_1 \times M_2$ has a unique Poisson structure $\pi = \{\cdot,\cdot\}$ such that the projection maps $p_i : M_1 \times M_2 \to M_i$ $(i = 1, 2)$ are Poisson maps and such that*

$$\{p_1^*F_1, p_2^*F_2\} = 0\,, \tag{2.6}$$

for all functions $F_1 \in \mathscr{F}(M_1)$ and $F_2 \in \mathscr{F}(M_2)$. Each Hamiltonian vector field on M_1 (or M_2) extends to a Hamiltonian vector field on $M_1 \times M_2$, tangent to the fibers of p_2 (or p_1).

The Poisson variety $(M_1 \times M_2, \pi)$ is called the *product Poisson variety* of (M_1, π_1) and (M_2, π_2), while π is called its *product Poisson structure.*

Remark 2.3. For $i = 1, 2$, the fibers of the projection maps $p_i : M_1 \times M_2 \to M_i$ have the following properties (see Fig. 2.1):

1. A function $F \in \mathscr{F}(M_1 \times M_2)$ is of the form $p_i^*F_i$ for some $F_i \in \mathscr{F}(M_i)$ if and only if it is constant on all the fibers of p_i;
2. A vector field $\mathscr{V}$ on $M_1 \times M_2$ satisfies $\mathscr{V}[p_i^*F_i] = 0$ for every function $F_i \in \mathscr{F}(M_i)$ if and only if it is tangent to all the fibers of p_i.

It follows that the condition given by (2.6) can be rephrased in either of the following equivalent ways.

(i) Every function which is constant on all the fibers of p_1 is in involution with every function which is constant on all the fibers of p_2;
(ii) The Hamiltonian vector field of every function which is constant on all the fibers of p_1 is tangent to all the fibers of p_2;
(iii) The Hamiltonian vector field of every function which is constant on all the fibers of p_2 is tangent to all the fibers of p_1.

To finish this section, we describe the Poisson matrix of the product Poisson structure on $M_1 \times M_2$, when M_1 and M_2 are affine Poisson varieties. We show that, in terms of natural generators of $\mathscr{F}(M_1 \times M_2)$, coming from generators of $\mathscr{F}(M_1)$ and $\mathscr{F}(M_2)$, the Poisson matrix takes a block diagonal form.

Proposition 2.4. *Suppose that $(M_1, \{\cdot,\cdot\}_1)$ and $(M_2, \{\cdot,\cdot\}_2)$ are affine Poisson varieties, with Poisson matrices $X = (x_{ij})_{i,j=1}^{d_1} = (\{x_i, x_j\}_1)_{i,j=1}^{d_1}$, respectively $Y = (y_{ij})_{i,j=1}^{d_2} = (\{y_i, y_j\}_2)_{i,j=1}^{d_2}$, with respect to generators $x_1, \dots, x_{d_1}$ for $\mathscr{F}(M_1)$, respectively $y_1, \dots, y_{d_2}$ for $\mathscr{F}(M_2)$. The Poisson matrix for the product Poisson structure $\{\cdot,\cdot\}$ on $M_1 \times M_2$ takes, in terms of the $d_1 + d_2$ generators $p_1^* x_1, \dots, p_1^* x_{d_1}, p_2^* y_1, \dots, p_2^* y_{d_2}$ for $\mathscr{F}(M_1 \times M_2)$, the form*

$$\begin{pmatrix} \left(p_1^* x_{ij}\right)_{i,j=1}^{d_1} & 0 \\ 0 & \left(p_2^* y_{ij}\right)_{i,j=1}^{d_2} \end{pmatrix}. \tag{2.7}$$

Proof. Since p_1 is a Poisson map we have that

$$\{p_1^* x_i, p_1^* x_j\} = p_1^* \{x_i, x_j\}_1 = p_1^* x_{ij} ,$$

which yields the first diagonal block of the Poisson matrix of $\{\cdot,\cdot\}$. Similarly, the fact that p_2 is a Poisson map yields the second diagonal block. The other entries $\{p_1^* x_i, p_2^* y_j\}$ of the Poisson matrix are zero, as follows from (2.6). □

By a slight abuse of notation, one usually simply writes x_i for $p_1^* x_i$ and y_j for $p_2^* y_j$. Then the Poisson matrix (2.7) takes the simple form $\begin{pmatrix} X & 0 \\ 0 & Y \end{pmatrix}$ and the product Poisson structure $\pi = \{\cdot,\cdot\}$ can be explicitly written in the following form

$$\pi = \sum_{1 \leqslant i < j \leqslant d_1} x_{ij} \frac{\partial}{\partial x_i} \wedge \frac{\partial}{\partial x_j} + \sum_{1 \leqslant i < j \leqslant d_2} y_{ij} \frac{\partial}{\partial y_i} \wedge \frac{\partial}{\partial y_j} .$$

2.1.2 The Product of Poisson Manifolds

The product of two manifolds M_1 and M_2 has a natural structure of a manifold, whose construction we briefly recall. First, as a topological space, $M_1 \times M_2$ is endowed with the product topology. Second, an atlas of $M_1 \times M_2$ is constructed from an atlas of M_1 and of M_2: let $(U_i, x^{(i)})_{i \in I}$ be coordinate charts covering M_1 and let $(V_j, y^{(j)})_{j \in J}$ be coordinate charts covering M_2, then $(U_i \times V_j)_{(i,j) \in I \times J}$ is an open covering of $M_1 \times M_2$ and the homeomorphisms

$$(x^{(i)}, y^{(j)}) : U_i \times V_j \longrightarrow x^{(i)}(U_i) \times y^{(j)}(V_j) \subset \mathbb{F}^{d_1 + d_2}$$

define compatible coordinate charts, hence endow $M_1 \times M_2$ with a manifold structure; in this formula, d_1 and d_2 are the dimensions of M_1 and M_2 respectively; as before, $\mathbb{F} = \mathbb{R}$ or $\mathbb{F} = \mathbb{C}$, depending on whether the manifolds which are considered are real or complex.

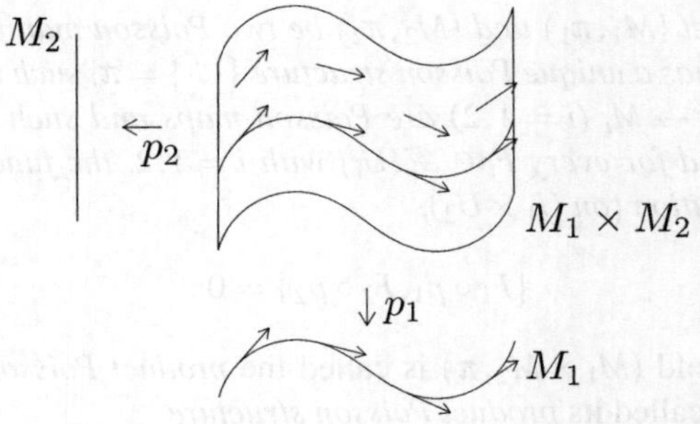

Fig. 2.2 A vector field on M_1 can be naturally lifted to a vector field on $M_1 \times M_2$, tangent to the fibers of $p_2 : M_1 \times M_2 \to M_2$.

As in the case of varieties, we denote by $p_i : M_1 \times M_2 \to M_i$ the canonical projection maps ($i = 1,2$). For $m_1 \in M_1$ and $m_2 \in M_2$, we will also use the inclusion maps $\imath_{m_2} : M_1 \hookrightarrow M_1 \times M_2$ and $\imath'_{m_1} : M_2 \hookrightarrow M_1 \times M_2$, defined by $\imath_{m_2}(x) := (x, m_2)$ and $\imath'_{m_1}(y) := (m_1, y)$, for $x \in M_1$ and $y \in M_2$.

To a vector field on an open subset of M_1 (or of M_2), we can associate a vector field on an open subset of $M_1 \times M_2$ by using the tangent maps $T\imath_{m_2}$ (respectively $T\imath_{m_2}$). Namely, to a vector field $\mathscr{V} \in \mathfrak{X}^1(U_1)$, where U_1 is an open subset of M_1, we associate the vector field $\widehat{\mathscr{V}} \in \mathfrak{X}^1(U_1 \times M_2)$, whose value at $m = (m_1, m_2) \in U_1 \times M_2$ is by definition

$$\widehat{\mathscr{V}}_m := T_{m_1}\imath_{m_2}(\mathscr{V}_{m_1}) \in T_m(M_1 \times M_2) .$$

This means that for every $F \in \mathscr{F}(U_1)$ and $G \in \mathscr{F}(U_2)$, where U_2 is an open subset of M_2,

$$\widehat{\mathscr{V}}[F \circ p_1] = \mathscr{V}[F] \circ p_1 \qquad \text{and} \qquad \widehat{\mathscr{V}}[G \circ p_2] = 0 , \tag{2.8}$$

since for all $m = (m_1, m_2) \in U_1 \times U_2$,

$$\begin{aligned} \widehat{\mathscr{V}}[F \circ p_1](m) &= \widehat{\mathscr{V}}_m(F \circ p_1) = T_{m_1}\imath_{m_2}(\mathscr{V}_{m_1})(F \circ p_1) \\ &= \mathscr{V}_{m_1}(F \circ p_1 \circ \imath_{m_2}) = \mathscr{V}_{m_1}(F) = \mathscr{V}[F](p_1(m)) , \end{aligned}$$

and similarly for the second equality in (2.8). These equalities can be used to write $\widehat{\mathscr{V}}$ in terms of local coordinates, showing in particular that $\widehat{\mathscr{V}}$ is a vector field (i.e., it is smooth/holomorphic). Similarly, every vector field $\mathscr{W}$ on an open subset U_2 of M_2 leads to a vector field $\widehat{\mathscr{W}}$ on $M_1 \times U_2$. See Fig. 2.2.

By extending this construction to the case of bivector fields, we can construct a bivector field on open subsets of $M_1 \times M_2$, starting from a bivector field defined on open subsets of M_1 (or M_2). We will use this in the following proposition to construct a Poisson structure on $M_1 \times M_2$, starting from a Poisson structure on M_1 and a Poisson structure on M_2.

Proposition 2.5. *Let (M_1,π_1) and (M_2,π_2) be two Poisson manifolds. The product manifold $M_1 \times M_2$ has a unique Poisson structure $\{\cdot\,,\cdot\} = \pi$, such that the projection maps $p_i : M_1 \times M_2 \to M_i$ $(i = 1,2)$ are Poisson maps and such that, for all open subsets $U_i \subset M_i$ and for every $F_i \in \mathscr{F}(U_i)$ with $i = 1,2$, the functions $F_1 \circ p_1$ and $F_2 \circ p_2$ are in involution (on $U_1 \times U_2$),*

$$\{F_1 \circ p_1, F_2 \circ p_2\} = 0\,. \tag{2.9}$$

The Poisson manifold $(M_1 \times M_2, \pi)$ is called the *product Poisson manifold* of M_1 and M_2, while π is called its *product Poisson structure*.

Proof. We have seen how we can associate in a natural way to a vector field on M_1 a vector field on $M_1 \times M_2$. In order to generalize this to bivector fields, we use that every $\mathbb{F}$-linear map $\phi : V \to W$ between vector spaces V and W, leads to a linear map $\wedge^2\phi : \wedge^2 V \to \wedge^2 W$. As recalled in Appendix A (Section A.2) it is defined for $v_1, v_2 \in V$ by

$$\wedge^2\phi\,(v_1 \wedge v_2) := \phi(v_1) \wedge \phi(v_2)\,.$$

Applied to the linear map

$$T_{m_1}\imath_{m_2} : T_{m_1}M_1 \to T_m(M_1 \times M_2)$$

we can define a bivector field $\widehat{\pi_1}$ on $M_1 \times M_2$ by its value at $m = (m_1, m_2)$:

$$\widehat{\pi_1}_m := \wedge^2(T_{m_1}\imath_{m_2})(\pi_1)_{m_1} \in \wedge^2 T_m(M_1 \times M_2)\,.$$

For open subsets $U_1 \subset M_1$ and $U_2 \subset M_2$ and for functions $F_1, G_1 \in \mathscr{F}(U_1)$ and $F_2, G_2 \in \mathscr{F}(U_2)$ we have, as in (2.8), that

$$\begin{aligned}
\widehat{\pi_1}\,[F_1 \circ p_1, G_1 \circ p_1] &= \pi_1[F_1, G_1] \circ p_1\,,\\
\widehat{\pi_1}\,[F_1 \circ p_1, G_2 \circ p_2] &= 0\,,\\
\widehat{\pi_1}\,[F_2 \circ p_2, G_2 \circ p_2] &= 0\,.
\end{aligned}$$

It follows that the bivector field $\pi := \widehat{\pi_1} + \widehat{\pi_2}$, where $\widehat{\pi_2}$ is constructed in the same way as $\widehat{\pi_1}$, but using $T_{m_2}\imath'_{m_1}$ instead of $T_{m_1}\imath_{m_2}$, has the following properties:

$$\begin{aligned}
\pi\,[F_1 \circ p_1, G_1 \circ p_1] &= \pi_1[F_1, G_1] \circ p_1\,,\\
\pi\,[F_1 \circ p_1, G_2 \circ p_2] &= 0\,,\\
\pi\,[F_2 \circ p_2, G_2 \circ p_2] &= \pi_2[F_2, G_2] \circ p_2\,.
\end{aligned} \tag{2.10}$$

Since every point in $M_1 \times M_2$ admits local coordinates of the form $x_i \circ p_1$ and $y_j \circ p_2$, where x_i and y_j are coordinate functions on M_1, respectively on M_2, it follows from (2.10) that every point admits local coordinates in which π satisfies the Jacobi identity. In view of *(iv)* in Proposition 1.16, this implies that π is a Poisson structure.

Moreover, the first and third equations in (2.10) express that p_1 and p_2 are Poisson maps, while the second one says that π satisfies (2.9).

If a Poisson structure on $M_1 \times M_2$ satisfies the requirements of Proposition 2.5, then it satisfies the equations (2.10), from which one can obtain explicit formulas for it, in local coordinates, which provides uniqueness of such a Poisson structure. In order to show how these explicit formulas are obtained from (2.10), let (U,x) and (V,y) be coordinate charts of M_1 and M_2, where $x = (x_1,\ldots,x_{d_1})$ and $y = (y_1,\ldots,y_{d_2})$. By a slight abuse of notation, we write the corresponding coordinates on $U \times V$ as $x_1,\ldots,x_{d_1},y_1,\ldots,y_{d_2}$. According to (1.26) and (2.10), π is given in terms of these coordinates by

$$\begin{aligned}\pi &= \sum_{1\leqslant i<j\leqslant d_1} \pi[x_i,x_j]\frac{\partial}{\partial x_i}\wedge\frac{\partial}{\partial x_j} + \sum_{1\leqslant i<j\leqslant d_2} \pi[y_i,y_j]\frac{\partial}{\partial y_i}\wedge\frac{\partial}{\partial y_j}\\ &= \sum_{1\leqslant i<j\leqslant d_1} \pi_1[x_i,x_j]\circ p_1\frac{\partial}{\partial x_i}\wedge\frac{\partial}{\partial x_j} + \sum_{1\leqslant i<j\leqslant d_2} \pi_2[y_i,y_j]\circ p_2\frac{\partial}{\partial y_i}\wedge\frac{\partial}{\partial y_j},\end{aligned}$$

which is written, in the bracket notation, as

$$\{\cdot\,,\cdot\} = \sum_{1\leqslant i<j\leqslant d_1} \left\{x_i,x_j\right\}_1\frac{\partial}{\partial x_i}\wedge\frac{\partial}{\partial x_j} + \sum_{1\leqslant i<j\leqslant d_2} \left\{y_i,y_j\right\}_2\frac{\partial}{\partial y_i}\wedge\frac{\partial}{\partial y_j},$$

where $\left\{x_i,x_j\right\}_1$ and $\left\{y_i,y_j\right\}_2$ are taken as functions on $U \times V$. □

Remark 2.6. Note that for all open subsets $U_1 \subset M_1$ and $U_2 \subset M_2$, the restriction of the Poisson structure $\{\cdot\,,\cdot\}$ on $M_1 \times M_2$ to $\mathscr{F}(U_1)\otimes\mathscr{F}(U_2)\subset\mathscr{F}(U_1\times U_2)$ is precisely equal to the product Poisson bracket, introduced in the previous section, of the Poisson algebras $\mathscr{F}(U_1)$ and $\mathscr{F}(U_2)$.

Remark 2.7. Remark 2.3 applies also to manifolds, upon replacing the functions and vector fields which appear in that remark by arbitrary *local* functions and vector fields, i.e., functions and vector fields which are defined in the neighborhood of a point.

To finish this section, we give a useful formula for computing the product Poisson bracket. As above, let (M_1,π_1) and (M_2,π_2) be Poisson manifolds and let π denote the product Poisson structure on $M_1 \times M_2$. Let F,G be functions on a non-empty open subset U of $M_1 \times M_2$ and let $(m_1,m_2)\in U$. By definition of π, we have that

$$\begin{aligned}\pi_{(m_1,m_2)} &= \widehat{\pi_1}_{(m_1,m_2)} + \widehat{\pi_2}_{(m_1,m_2)}\\ &= \wedge^2(T_{m_1}\imath_{m_2})(\pi_1)_{m_1} + \wedge^2(T_{m_2}\imath'_{m_1})(\pi_2)_{m_2}\,,\end{aligned}\tag{2.11}$$

so that

$$\pi_{(m_1,m_2)}(F,G) = (\pi_1)_{m_1}(F\circ\imath_{m_2},G\circ\imath_{m_2}) + (\pi_2)_{m_2}(F\circ\imath'_{m_1},G\circ\imath'_{m_1})\,,$$

which yields, in the bracket notation, the following formula for the product bracket

$$\{F,G\}(m_1,m_2) = \{F\circ\iota_{m_2}, G\circ\iota_{m_2}\}_1(m_1) + \{F\circ\iota'_{m_1}, G\circ\iota'_{m_1}\}_2(m_2), \qquad (2.12)$$

valid for all $F,G \in \mathscr{F}(U)$, where U is an open subset of $M_1 \times M_2$, which contains (m_1,m_2).

2.2 Poisson Ideals and Poisson Submanifolds

The simplest sub-objects of Poisson manifolds are Poisson submanifolds, which are submanifolds, equipped with a Poisson structure, with respect to which the inclusion map is a Poisson map. For a submanifold, several conditions will be given, which are equivalent to the fact that it is a Poisson submanifold, including the well known condition that all Hamiltonian vector fields (of locally defined functions) be tangent to the submanifold. The algebraic condition, which corresponds to it, is that the ideal of functions which vanish on the submanifold, is a Poisson ideal. In general algebraic terms, for an ideal $\mathscr{I}$ of a Poisson algebra $(\mathscr{A},\cdot,\{\cdot,\cdot\})$, one has the following fact: $\mathscr{A}/\mathscr{I}$ has a Poisson bracket which makes the canonical projection into a morphism of Poisson algebras if and only if $\mathscr{I}$ is a Poisson ideal; if such a Poisson bracket on $\mathscr{A}/\mathscr{I}$ exists, then it is unique.

We give in Section 2.2.1 the algebraic construction, and we reformulate the result in geometrical terms for subvarieties of affine Poisson varieties. In Section 2.2.2, we give a purely geometrical construction for (immersed or embedded) submanifolds of a Poisson manifold. For the latter, we also relate the Poisson submanifold condition to the intersection of the submanifold with the symplectic leaves.

2.2.1 Poisson Ideals and Poisson Subvarieties

Let $(\mathscr{A},\cdot)$ be a commutative associative algebra and let $\mathscr{I}$ be an ideal of $\mathscr{A}$. The quotient $\mathscr{A}/\mathscr{I}$ inherits from $\mathscr{A}$ the structure of an associative algebra. We denote the quotient map by $p:\mathscr{A}\to\mathscr{A}/\mathscr{I}$ and we use, for $F\in\mathscr{A}$, the convenient notation $\overline{F} := p(F)$. For a derivation $\mathscr{V}$ of $\mathscr{A}$, the following conditions are equivalent:

(i) There exists a derivation $\mathscr{V}_0$ of $\mathscr{A}/\mathscr{I}$ such that $\overline{\mathscr{V}[F]} = \mathscr{V}_0[\overline{F}]$, for every $F\in\mathscr{A}$;

(ii) $\mathscr{V}$ maps $\mathscr{I}$ to itself: for every $F\in\mathscr{I}$, $\mathscr{V}[F]\in\mathscr{I}$.

The proof (i) $\Rightarrow$ (ii) is immediate; for the other implication, we define $\mathscr{V}_0$ on elements $\overline{F}\in\mathscr{A}/\mathscr{I}$ by $\mathscr{V}_0[\overline{F}] := \overline{\mathscr{V}[F]}$, which is well-defined in view of (ii), and which is then easily shown to be a derivation of $\mathscr{A}/\mathscr{I}$.

Similarly, if $\mathscr{A}$ comes equipped with a Poisson bracket, i.e., $(\mathscr{A},\cdot,\{\cdot,\cdot\})$ is a Poisson algebra, it is a natural question to ask what are the conditions on the ideal $\mathscr{I}$ which imply that $\mathscr{A}/\mathscr{I}$ carries a Poisson bracket $\{\cdot,\cdot\}_0$, which comes from

$\{\cdot,\cdot\}$, i.e., such that p is a Poisson morphism. The fact that p is surjective implies that, if such a bracket exists, then it is unique.

Suppose that such a Poisson bracket $\{\cdot,\cdot\}_0$ exists. The morphism property

$$\{\overline{F},\overline{G}\}_0 = \overline{\{F,G\}} \tag{2.13}$$

implies that $\overline{\{F,G\}} = 0$ for every $F \in \mathscr{I}$ and $G \in \mathscr{A}$, since $\overline{F} = 0$ if $F \in \mathscr{I}$. This means that $\mathscr{I}$ is a Lie ideal of $(\mathscr{A}, \{\cdot,\cdot\})$, i.e., $\{\mathscr{I},\mathscr{A}\} \subset \mathscr{I}$. Thus $\mathscr{I}$ is an ideal for both multiplications of the Poisson algebra: one calls such an ideal a *Poisson ideal*.

Conversely, suppose that $\mathscr{I}$ is a Poisson ideal of $\mathscr{A}$. Then (2.13) can be taken as the definition of a bilinear map $\{\cdot,\cdot\}_0$ from $\mathscr{A}/\mathscr{I}$ into itself. Since it is well-defined,

$$\{\{\overline{F},\overline{G}\}_0,\overline{H}\}_0 = \overline{\{\{F,G\},H\}},$$

for all elements F, G and H of $\mathscr{A}$, so that $\{\cdot,\cdot\}_0$ defines a Lie bracket on $\mathscr{A}/\mathscr{I}$. Since p is a morphism for both algebra structures on $\mathscr{A}$, $\{\cdot,\cdot\}_0$ is also a biderivation. Thus, $\{\cdot,\cdot\}_0$ is a Poisson bracket on $\mathscr{A}$ which makes p into a Poisson morphism.

The condition $\{F,G\} \in \mathscr{I}$ for all $F \in \mathscr{I}$, $G \in \mathscr{A}$, rewritten as $\mathscr{X}_G(F) \in \mathscr{I}$ means that every Hamiltonian derivation of $\mathscr{A}$ maps $\mathscr{I}$ to itself, and rewritten as $\mathscr{X}_F(G) \in \mathscr{I}$ means that every Hamiltonian derivation of $F \in \mathscr{I}$ maps $\mathscr{A}$ to $\mathscr{I}$.

Summarizing, the following proposition holds.

Proposition 2.8. *Let $(\mathscr{A},\cdot,\{\cdot,\cdot\})$ be a Poisson algebra and let $\mathscr{I}$ be an ideal of $(\mathscr{A},\cdot)$. The following conditions are equivalent:*

(i) There exists a Poisson bracket $\{\cdot,\cdot\}_0$ on $\mathscr{A}/\mathscr{I}$ such that $p : \mathscr{A} \to \mathscr{A}/\mathscr{I}$ is a Poisson morphism;
(ii) $\mathscr{I}$ is a Poisson ideal of $(\mathscr{A},\cdot,\{\cdot,\cdot\})$;
(iii) For every $F \in \mathscr{A}$, the Hamiltonian derivation $\mathscr{X}_F$ preserves $\mathscr{I}$;
(iv) For every $F \in \mathscr{I}$, the Hamiltonian derivation $\mathscr{X}_F$ is $\mathscr{I}$-valued.

Moreover, if one of these conditions holds, then the Poisson bracket $\{\cdot,\cdot\}_0$ on $\mathscr{A}/\mathscr{I}$, satisfying (i), is unique and it is defined by

$$\{\overline{F},\overline{G}\}_0 := \overline{\{F,G\}},$$

for all $F, G \in \mathscr{A}$.

We now turn to the case of affine Poisson varieties, which will bring us to the notion of a Poisson subvariety. We first recall that if N and M are affine varieties which belong to the same affine space $\mathbb{F}^d$, then N is said to be an *affine subvariety* of M if $N \subset M$. Then N is the zero locus of a prime ideal $\mathscr{I}_N$ in $\mathscr{F}(M)$ and $\mathscr{F}(N)$ is in a natural way isomorphic to $\mathscr{F}(M)/\mathscr{I}_N$, with which it will in the sequel be identified. We denote by $\kappa : \mathscr{F}(M) \to \mathscr{F}(N)$ the canonical projection, which we think of as a restriction map.

Translating the equivalent algebraic conditions, given at the beginning of this section, to the case of varieties, we have that for a derivation $\mathscr{V}$ of $\mathscr{F}(M)$, the following conditions are equivalent:

(i) There exists a derivation $\mathscr{V}_0$ of $\mathscr{F}(N)$ such that $\kappa(\mathscr{V}[F]) = \mathscr{V}_0[\kappa(F)]$, for every $F \in \mathscr{F}(M)$, which can be expressed as the commutativity of the following diagram:

$$\begin{array}{ccc} \mathscr{F}(M) & \xrightarrow{\ \mathscr{V}\ } & \mathscr{F}(M) \\ \Big\downarrow{\scriptstyle \kappa} & & \Big\downarrow{\scriptstyle \kappa} \\ \mathscr{F}(N) & \xrightarrow[\ \mathscr{V}_0\ ]{} & \mathscr{F}(N) \end{array}$$

(ii) $\mathscr{V}$ maps the prime ideal $\mathscr{I}_N$ of all functions which vanish on N to itself: for every $F \in \mathscr{I}_N$, $\mathscr{V}[F] \in \mathscr{I}_N$.

As we will see in the next section, these two equivalent conditions on $\mathscr{V}$ correspond in the context of manifolds to the fact that $\mathscr{V}$, viewed as a vector field on M, is tangent to N, and $\mathscr{V}_0$ is the restriction of $\mathscr{V}$ to N. In the case of varieties, we therefore also think of $\mathscr{V}_0$ as being the restriction of $\mathscr{V}$ to N and, by a slight abuse of language, we say that $\mathscr{V}$ is *tangent* to N (at points of N), if $\mathscr{V}$ satisfies $\mathscr{V}[\mathscr{I}_N] \subset \mathscr{I}_N$.

Definition 2.9. Let $(M, \{\cdot\,,\cdot\})$ be a Poisson variety and let $N \subset M$ be a subvariety. We say that N is a *Poisson subvariety* of M if there exists a Poisson structure on N which turns the inclusion map $\imath : N \hookrightarrow M$ into a Poisson map.

This means that, if N is a subvariety of M, and $\mathscr{I}_N \subset \mathscr{F}(M)$ denotes the prime ideal of all functions on M which vanish on N, then N is a Poisson subvariety of $(M, \{\cdot\,,\cdot\})$ if and only if $\mathscr{F}(M)/\mathscr{I}_N$ has a Poisson bracket such that the quotient map $\kappa : \mathscr{F}(M) \to \mathscr{F}(M)/\mathscr{I}_N$ is a morphism of Poisson algebras. It leads, for affine Poisson varieties, to the following geometric analog of Proposition 2.8.

Proposition 2.10. *Let $(M, \{\cdot\,,\cdot\})$ be an affine Poisson variety and let N be a subvariety of M, defined by a prime ideal $\mathscr{I}_N \subset \mathscr{F}(M)$. The following are equivalent:*

(i) N is a Poisson subvariety of $(M, \{\cdot\,,\cdot\})$;
(ii) $\mathscr{I}_N$ is a Poisson ideal of $\mathscr{F}(M)$;
(iii) For every $F \in \mathscr{F}(M)$, the Hamiltonian vector field $\mathscr{X}_F$ is tangent to N;
(iv) For every $F \in \mathscr{F}(M)$ which vanishes on N, the restriction of the Hamiltonian vector field $\mathscr{X}_F$ to N is zero.

Moreover, if one of these conditions holds, then the Poisson structure on N which makes the inclusion map into a Poisson map is unique.

2.2.2 Poisson Submanifolds

Another approach is needed in the case of submanifolds of a Poisson manifold, as submanifolds are, in general, not defined by ideals. Moreover, important examples of submanifolds in Poisson geometry are not embedded submanifolds, but immersed submanifolds; even locally they are, in general, not defined by an ideal. Think, for example, of the leaves of the symplectic foliation of a Poisson manifold (see Section 1.3.4).

First, let us recall the basic vocabulary. Suppose that $\imath : N_0 \to M$ is a map, where both M and N_0 are manifolds. If for every point $n_0 \in N_0$ the tangent map $T_{n_0}\imath : T_{n_0}N_0 \to T_{\imath(n_0)}M$ is injective, then $\imath$ is said to be an *immersion*. If the immersion is injective one says that $(N_0, \imath)$ is an *immersed submanifold*, or *submanifold* for short; one often refers to the subset $N := \imath(N_0)$ of M as being an immersed submanifold, keeping in mind that N comes with its own differential structure and topology. Since $T_{n_0}\imath$ is injective we can, by the implicit function theorem, find a coordinate chart $(V, y = (y_1, \dots, y_s))$ of N_0, containing n_0, and a coordinate chart $(U, x = (x_1, \dots, x_d))$ of M, containing $\imath(n_0)$, with $\imath(V) \subset U$, and such that $\imath$ is given in terms of these coordinates by

$$(y_1, \dots, y_s) \mapsto (x_1, \dots, x_d) = (y_1, \dots, y_s, 0, \dots, 0),$$

where s is the dimension of N_0 and d is the dimension of M. The coordinate chart (U, x) for M is then said to be *adapted* to N at $\imath(n_0)$.

In the special case in which the topology on N which comes from N_0 coincides with the induced topology from M, i.e., $\imath$ is a homeomorphism between N_0 and N (equipped with the induced topology from M), one says that N is an *embedded submanifold*. Open subsets of N are then restrictions to N of open subsets of M and every coordinate chart for N, can be extended to a coordinate chart of M, adapted to N.

When $N = \imath(N_0)$ is a submanifold of M, then for $n := \imath(n_0) \in N$, the subspace $T_{n_0}\imath(T_{n_0}N_0)$ of T_nM is denoted by T_nN and is called the *tangent space* to N at n. A vector field $\mathscr{V}$, defined on an open subset $U \subset M$, is said to be *tangent* to N if $\mathscr{V}_n \in T_nN$ for every $n \in U \cap N$.

Definition 2.11. Let (M, π) be a Poisson manifold. An (immersed or embedded) submanifold $(N_0, \imath)$ of M is called a *Poisson submanifold* if there exists on N_0 a Poisson structure π_N such that the inclusion map $\imath$ is a Poisson map. Viewing $N := \imath(N_0)$ as a submanifold of M, we refer to π_N as a Poisson structure on N and we say that N is an (immersed or embedded) Poisson submanifold of M.

We give in the following proposition a characterization of Poisson submanifolds in terms of (locally defined) Hamiltonian vector fields, as well as in terms of the symplectic leaves of the ambient Poisson manifold.

Proposition 2.12. *Let (M, π) be a Poisson manifold. For an (immersed or embedded) submanifold $N = \imath(N_0)$ of M, the following conditions are equivalent:*

(i) *N is a Poisson submanifold of (M,π);*
(ii) *For every open subset $U \subset M$ and for every $F \in \mathscr{F}(U)$, the Hamiltonian vector field $\mathscr{X}_F$ is tangent to N;*
(iii) *For every $n \in N$, the intersection of N with the symplectic leaf $\mathscr{S}_n(M)$ contains a neighborhood of n in $\mathscr{S}_n(M)$;*
(iv) *For every $n \in N$, the bivector π_n belongs to $\wedge^2 T_nN$.*

Moreover, if one of these conditions holds, then the Poisson structure on N, which makes the inclusion map into a Poisson map, is unique.

Proof. (i) $\Longrightarrow$ (ii). Let $n := \imath(n_0) \in N$, where $n_0 \in N_0$ is an arbitrary point. Let (U,x) be a coordinate chart of M, adapted to N at n. In terms of the coordinates $x = (x_1,\dots,x_d)$, where d is the dimension of M, the Hamiltonian vector field of $F \in \mathscr{F}(U)$ is given, according to (1.33), by

$$\mathscr{X}_F = \sum_{i=1}^{s}\{x_i,F\}\frac{\partial}{\partial x_i} + \sum_{j=s+1}^{d}\{x_j,F\}\frac{\partial}{\partial x_j},$$

where s is the dimension of N_0 and $\{\cdot\,,\cdot\} := \pi$. Let us denote the Poisson structure on N by $\{\cdot\,,\cdot\}_0 = \pi_0$. Since $\imath$ is a Poisson morphism, $\{x_j,F\}(n) = \{x_j\circ\imath, F\circ\imath\}_0(n_0)$, which is zero when $s+1 \leqslant j \leqslant d$, since $x_j\circ\imath = 0$, for such j. It follows that, for $n \in N\cap U$,

$$(\mathscr{X}_F)_n = \sum_{i=1}^{s}\{x_i,F\}(n)\left(\frac{\partial}{\partial x_i}\right)_n = \sum_{i=1}^{s}\{x_i,F\}(n)\; T_{n_0}\imath\left(\frac{\partial}{\partial x_i}\right)_{n_0},$$

so that $(\mathscr{X}_F)_{\imath(n_0)} \in T_{n_0}\imath\,(T_{n_0}N_0) = T_nN$, for all $n_0 \in N_0$, which means that $\mathscr{X}_F$ is tangent to N. This shows that all (locally defined) Hamiltonian vector fields are tangent to N.

(ii) $\Longrightarrow$ (iii). Let $n \in N$ and consider the symplectic leaf $\mathscr{S}_n(M)$ which contains n (see Section 1.3.4). Since $\imath$ is an immersion, there exists an open subset V of N_0 which contains n_0, where $n = \imath(n_0)$, and an open subset U of M, such that $\imath(V)$ is a closed, embedded submanifold of U. We also consider the symplectic leaf $\mathscr{S}_n(U)$, which is an open subset of $\mathscr{S}_n(M)$ (recall that it is, in general, strictly smaller than $\mathscr{S}_n(M)\cap U$). We claim that $\mathscr{S}_n(U) \subset \imath(V)$. To show this, it suffices to show that every Hamiltonian path in U which starts in $\imath(V)$, stays in $\imath(V)$. Consider a Hamiltonian path $\gamma : I \to U$, where I is an open neighborhood of 0 in $\mathbb{F}$ (which is $\mathbb{R}$ or $\mathbb{C}$), with $\gamma(0) \in \imath(V)$. It is the integral curve of a Hamiltonian vector field $\mathscr{X}_F$, with $F \in \mathscr{F}(U')$, which is tangent to $\imath(V)$, at all points of $\imath(V)$, where U' is an open subset of U, containing $\gamma(I)$. Suppose that $\gamma(I)$ is not entirely contained in $\imath(V)$. Consider

$$I' := \{t \in I \mid \gamma(t) \in \imath(V)\},$$

which is an open subset of I, since γ is an integral curve of a vector field which is tangent to $\imath(V)$. Let I'_0 denote the connected component of I' which contains 0. Let t' be an arbitrary element of the (topological) boundary of I'_0 and let $n' := \gamma(t')$. Then

$n' \in \imath(V)$, since $\imath(V)$ is a closed subset of U. Since γ is an integral curve, passing through n', of a vector field which is tangent to $\imath(V)$, we have that $\gamma(t'') \subset \imath(V)$ for t'' in a neighborhood of t', which contradicts the fact that t' belongs to the boundary of I_0'. Therefore, all Hamiltonian paths $\gamma : I \to U$, defined on a neighborhood I of 0, with $\gamma(0) \in \imath(V)$, stay in $\imath(V)$ and $\imath(V) \supset \mathscr{S}_n(U)$. Since the latter is an open subset of $\mathscr{S}_n(M)$, property *(iii)* follows.

(iii) $\Longrightarrow$ *(i)*. Let $n_0 \in N_0$ be an arbitrary point and let $n := \imath(n_0)$. Since $\imath$ is an immersion,

$$T_{n_0}\imath : T_{n_0}N_0 \to T_nN \subset T_nM$$

is an isomorphism, while *(iii)* implies that $T_n\mathscr{S}_n(M) \subset T_nN$. It follows that there exists a unique bivector $(\pi_N)_{n_0} \in \wedge^2 T_{n_0}N_0$ such that

$$\wedge^2 (T_{n_0}\imath)(\pi_N)_{n_0} = \pi_n \ , \tag{2.14}$$

as the latter belongs to $\wedge^2 T_n\mathscr{S}_n(M)$. We claim that the map

$$\begin{aligned} \pi_N : N_0 &\to \wedge^2 TN_0 \\ n_0 &\mapsto (\pi_N)_{n_0} \end{aligned} \tag{2.15}$$

is a Poisson structure on N_0, such that $\imath$ is a Poisson map. To see that π_N is a (smooth/holomorphic) bivector field on N_0, let n_0 be an arbitrary point and let F_0, G_0 be arbitrary functions, defined on a neighborhood of n_0 in N_0. There exists a neighborhood U of $n := \imath(n_0)$, and functions $F, G \in \mathscr{F}(U)$, such that $F \circ \imath = F_0$ and $G \circ \imath = G_0$ on some neighborhood V of n_0 in N_0. For every $n' = \imath(n_0') \in U \cap \imath(V)$ we have that

$$\begin{aligned} \pi_N[F_0, G_0](n_0') &= \pi_N[F \circ \imath, G \circ \imath](n_0') \\ &= (\pi_N)_{n_0'}\left(\mathrm{d}_{n_0'}(F \circ \imath), \mathrm{d}_{n_0'}(G \circ \imath)\right) \\ &= \pi_{n'}(\mathrm{d}_{n'}F, \mathrm{d}_{n'}G) = \pi[F, G](\imath(n_0')) \ , \end{aligned}$$

where we have used the definition (2.14) of π_N to obtain the last line. Since $n_0' \mapsto \pi[F, G](\imath(n_0'))$ is clearly a smooth/holomorphic function on $V \subset N_0$, it follows that π_N is a (smooth/holomorphic) bivector field on N_0. The same computation, with three functions F_0, G_0 and H_0, defined on a neighborhood of n_0 in N_0, yields

$$\pi_N\left[\pi_N\left[F_0, G_0\right], H_0\right](n_0') = \pi\left[\pi\left[F, G\right], H\right](\imath(n_0')) \ ,$$

for all n_0' in a neighborhood V of n_0 in N_0. It leads to the Jacobi identity for π_N, so that π_N is a Poisson structure on N_0.

According to Proposition 1.19, $\imath$ is a Poisson map; it is clear that π_N is the unique Poisson structure on N_0, such that $\imath$ is a Poisson map.

(ii) $\Longleftrightarrow$ *(iv)*. As in the first step of the proof, let (U, x) be a coordinate chart of M, adapted to N at $n \in N$. On U we can write π as

$$\pi = \sum_{1 \leqslant i < j \leqslant s} \alpha_{ij} \frac{\partial}{\partial x_i} \wedge \frac{\partial}{\partial x_j} + \sum_{i=1}^{d} \sum_{j=s+1}^{d} \alpha_{ij} \frac{\partial}{\partial x_i} \wedge \frac{\partial}{\partial x_j},$$

where all the coefficients α_{ij} are functions, defined on U, with $\alpha_{ji} = -\alpha_{ij}$. If $\pi_n \in \wedge^2 T_n N$, then $\alpha_{ij}(n) = 0$ for all i, j with $1 \leqslant i \leqslant d$ and $s+1 \leqslant j \leqslant d$, which is equivalent to saying that every Hamiltonian vector field on U is tangent to N at n. This shows that *(ii)* and *(iv)* are equivalent. □

2.3 Real and Holomorphic Poisson Structures

This section is devoted to three constructions which are intimately related to the real or holomorphic nature of Poisson structures.

2.3.1 *Real Poisson Structures Associated to Holomorphic Poisson Structures*

Our aim is to construct from a holomorphic Poisson structure on a complex manifold, real Poisson structures on the underlying real manifold. Let M be a complex manifold of (complex) dimension $d \in \mathbb{N}^*$, whose underlying real manifold will be denoted by $M_{\mathbb{R}}$. We fix some notations for the different algebras of functions which will be considered on M or $M_{\mathbb{R}}$:

(a) $\mathscr{H}(M)$ denotes the $\mathbb{C}$-algebra of all holomorphic functions on M,
(b) $C^\infty(M_{\mathbb{R}}, \mathbb{C})$ denotes the $\mathbb{C}$-algebra of all $\mathbb{C}$-valued $\mathbb{R}$-differentiable functions on $M_{\mathbb{R}}$,
(c) $C^\infty(M_{\mathbb{R}})$ denotes the $\mathbb{R}$-algebra of all $\mathbb{R}$-valued $\mathbb{R}$-differentiable functions on $M_{\mathbb{R}}$,
(d) $\overline{\mathscr{H}(M)}$ denotes the $\mathbb{C}$-algebra of all anti-holomorphic functions on M, i.e. the functions of the form $\overline{F}$, with $F \in \mathscr{H}(M)$.

We also use analogous notations for the functions defined on an open subset U of M (or $U_{\mathbb{R}}$ of $M_{\mathbb{R}}$). We then have

$$\mathscr{H}(U) \hookrightarrow C^\infty(U_{\mathbb{R}}, \mathbb{C}) \simeq C^\infty(U_{\mathbb{R}}) + \sqrt{-1}\, C^\infty(U_{\mathbb{R}}), \tag{2.16}$$

where the latter isomorphism consists of writing a $\mathbb{C}$-valued function as the sum of its real and its imaginary parts. It is clear that every coordinate chart (U, z) of M, with $z = (z_1, \ldots, z_d)$ leads to a coordinate chart $(U_{\mathbb{R}}, Z)$ of $M_{\mathbb{R}}$, with $Z = (x_1, \ldots, x_d, y_1, \ldots, y_d)$, where $x_k, y_k \in C^\infty(U_{\mathbb{R}})$ are the real smooth functions obtained by writing $z_k = x_k + \sqrt{-1}\, y_k$ under the decomposition (2.16).

Let $\mathscr{V}$ be a holomorphic vector field on M. According to (B.8), in a coordinate chart (U, z), we can write

$$\mathscr{V} = \sum_{k=1}^{d} \mathscr{V}[z_k]\frac{\partial}{\partial z_k} = \frac{1}{2}\sum_{k=1}^{d} \mathscr{V}[z_k]\left(\frac{\partial}{\partial x_k} - \sqrt{-1}\,\frac{\partial}{\partial y_k}\right). \tag{2.17}$$

Decomposing $\mathscr{V}[z_k]$ into its real and imaginary parts, leads to the definition of two derivations of $C^\infty(U_\mathbb{R})$, the real and imaginary parts of $\mathscr{V}$, denoted by $\Re(\mathscr{V})$ and $\Im(\mathscr{V})$, by putting:

$$\Re(\mathscr{V}) := \frac{1}{2}\sum_{k=1}^{d}\left(\Re(\mathscr{V}[z_k])\frac{\partial}{\partial x_k} + \Im(\mathscr{V}[z_k])\frac{\partial}{\partial y_k}\right), \tag{2.18}$$

$$\Im(\mathscr{V}) := \frac{1}{2}\sum_{k=1}^{d}\left(\Im(\mathscr{V}[z_k])\frac{\partial}{\partial x_k} - \Re(\mathscr{V}[z_k])\frac{\partial}{\partial y_k}\right). \tag{2.19}$$

These two derivations of the $\mathbb{R}$-algebra $C^\infty(U_\mathbb{R})$ extend uniquely to the $\mathbb{C}$-algebra $C^\infty(U_\mathbb{R},\mathbb{C}) \simeq C^\infty(U_\mathbb{R}) + \sqrt{-1}C^\infty(U_\mathbb{R})$, by $\mathbb{C}$-linearity; we will denote these extensions still by $\Re(\mathscr{V})$ and $\Im(\mathscr{V})$. Writing a holomorphic function F on U as $F = G + \sqrt{-1}H$ and using the Cauchy–Riemann equations

$$\frac{\partial G}{\partial x_k} = \frac{\partial H}{\partial y_k}, \quad \frac{\partial G}{\partial y_k} = -\frac{\partial H}{\partial x_k},$$

one verifies by a direct computation that, for all $F \in \mathscr{H}(U)$,

$$\Re(\mathscr{V})[F] = \frac{1}{2}\mathscr{V}[F], \quad \Re(\mathscr{V})[\overline{F}] = \frac{1}{2}\overline{\mathscr{V}[F]}. \tag{2.20}$$

These equations characterize $\Re(\mathscr{V})$ since (2.18) can be derived from it. Similarly, the imaginary part $\Im(\mathscr{V})$ is characterized by the fact that

$$\Im(\mathscr{V})[F] = -\frac{\sqrt{-1}}{2}\mathscr{V}[F], \quad \Im(\mathscr{V})[\overline{F}] = \frac{\sqrt{-1}}{2}\overline{\mathscr{V}[F]},$$

for all $F \in \mathscr{H}(U)$. These characterizations imply that $\Re(\mathscr{V})$ and $\Im(\mathscr{V})$ are independent of the choice of the coordinates and therefore, define (smooth) vector fields on $M_\mathbb{R}$.

An analogous construction works in the case of bivector fields. Indeed, let P be a bivector field of M. According to (1.23), we can write P in a coordinate chart (U,z) as

$$\begin{aligned} P &= \sum_{1\leqslant k<\ell\leqslant d} P[z_k,z_\ell]\frac{\partial}{\partial z_k}\wedge\frac{\partial}{\partial z_\ell} \\ &= \frac{1}{4}\sum_{1\leqslant k<\ell\leqslant d} P[z_k,z_\ell]\left(\frac{\partial}{\partial x_k} - \sqrt{-1}\frac{\partial}{\partial y_k}\right)\wedge\left(\frac{\partial}{\partial x_\ell} - \sqrt{-1}\frac{\partial}{\partial y_\ell}\right). \end{aligned}$$

As above, decomposing $P[z_k,z_\ell]$ into its real and imaginary parts, leads to the definition of two skew-symmetric biderivations $\Re(P)$ and $\Im(P)$ of $C^\infty(U_\mathbb{R})$, by putting,

$$\mathfrak{R}(P) := \frac{1}{4}\sum_{1\leqslant k<\ell\leqslant d} \mathfrak{R}(P[z_k,z_\ell])\left(\frac{\partial}{\partial x_k}\wedge\frac{\partial}{\partial x_\ell}-\frac{\partial}{\partial y_k}\wedge\frac{\partial}{\partial y_\ell}\right)$$
$$+\frac{1}{4}\sum_{1\leqslant k<\ell\leqslant d} \mathfrak{I}(P[z_k,z_\ell])\left(\frac{\partial}{\partial x_k}\wedge\frac{\partial}{\partial y_\ell}+\frac{\partial}{\partial y_k}\wedge\frac{\partial}{\partial x_\ell}\right),$$

and similarly for $\mathfrak{I}(P)$. By definition, $\mathfrak{R}(P)$ and $\mathfrak{I}(P)$ are skew-symmetric biderivations of $C^\infty(U_\mathbb{R})$ and they uniquely extend to skew-symmetric biderivations of $C^\infty(U_\mathbb{R},\mathbb{C})$, by $\mathbb{C}$-linearity. Then, $\mathfrak{R}(P)$ is, as in (2.20), characterized by the following properties: for all $F,G\in\mathscr{H}(U)$,

$$\begin{aligned}
\mathfrak{R}(P)[F,G] &= \frac{1}{2}P[F,G]\,,\\
\mathfrak{R}(P)[\overline{F},G] &= 0\,,\\
\mathfrak{R}(P)[\overline{F},\overline{G}] &= \frac{1}{2}\overline{P[F,G]}\,.
\end{aligned}\tag{2.21}$$

There is an analogous characterization for the imaginary part $\mathfrak{I}(P)$. Because of these characterizations, we can conclude, as in the case of vector fields, that $\mathfrak{R}(P)$ and $\mathfrak{I}(P)$ are well-defined bivector fields of $M_\mathbb{R}$. We show in the next proposition that, if the bivector field P has the additional property of being a Poisson structure, then $\mathfrak{R}(P)$ is also Poisson structure. One shows similarly that $\mathfrak{I}(P)$ is a Poisson structure.

Proposition 2.13. *If (M,π) is a complex Poisson manifold of dimension d, then the bivector field $\mathfrak{R}(\pi)$ is a Poisson structure on $M_\mathbb{R}$. In local coordinates, $\mathfrak{R}(\pi)$ is given by:*

$$\mathfrak{R}(\pi) = \frac{1}{4}\sum_{1\leqslant k<\ell\leqslant d} \mathfrak{R}(\pi_{k\ell})\left(\frac{\partial}{\partial x_k}\wedge\frac{\partial}{\partial x_\ell}-\frac{\partial}{\partial y_k}\wedge\frac{\partial}{\partial y_\ell}\right)$$
$$+\frac{1}{4}\sum_{1\leqslant k<\ell\leqslant d} \mathfrak{I}(\pi_{k\ell})\left(\frac{\partial}{\partial x_k}\wedge\frac{\partial}{\partial y_\ell}+\frac{\partial}{\partial y_k}\wedge\frac{\partial}{\partial x_\ell}\right)$$

where $z_k = x_k+\sqrt{-1}y_k$ and $\pi_{k\ell} := \pi(z_k,z_\ell)$, for $1\leqslant k,\ell\leqslant d$.

Proof. According to the characterization of $\mathfrak{R}(\pi)$, given above, and since π is a holomorphic Poisson bracket, we have, for every coordinate chart (U,z) and for all $F,G,H\in\mathscr{H}(U)$,

$$\begin{aligned}
\mathfrak{R}(\pi)[\mathfrak{R}(\pi)[F,G],H] &= \frac{1}{4}\{\{F,G\},H\}\,,\\
\mathfrak{R}(\pi)[\mathfrak{R}(\pi)[\overline{F},G],H] &= 0\,,\\
\mathfrak{R}(\pi)[\mathfrak{R}(\pi)[\overline{F},\overline{G}],H] &= 0\,,\\
\mathfrak{R}(\pi)[\mathfrak{R}(\pi)[\overline{F},\overline{G}],\overline{H}] &= \frac{1}{4}\overline{\{\{F,G\},H\}}\,,
\end{aligned}$$

where we have written $\{\cdot,\cdot\}$ for π. Since π satisfies the Jacobi identity, these equations imply that $\Re(\pi)$ satisfies the Jacobi identity, when applied to arbitrary triples of holomorphic or anti-holomorphic functions on U. In particular, for every coordinate chart $(U_\mathbb{R},Z)$, the Jacobi identity will be satisfied by $\Re(\pi)$ for the coordinate functions x_k and y_ℓ, because they can be written as $\mathbb{C}$-linear combinations of the holomorphic and anti-holomorphic functions z_k and $\bar{z}_k$. Then, according to Proposition 1.16, $\Re(\pi)$ is a bivector field of $M_\mathbb{R}$, satisfying the Jacobi identity, i.e., it is a Poisson structure on $M_\mathbb{R}$. □

2.3.2 Holomorphic Poisson Structures on Smooth Affine Poisson Varieties

Let N be an s-dimensional complex affine variety, $N\subset\mathbb{C}^d$. It is the zero locus of a prime ideal $\mathscr{I}$ of $\mathbb{C}[x_1,\dots,x_d]$ and the algebra of regular functions on N is related to $\mathscr{I}$ by $\mathscr{F}(N)=\mathbb{C}[x_1,\dots,x_d]/\mathscr{I}$ (see Section 1.2.1). Let $H_1,\dots,H_t$ be a system of generators of $\mathscr{I}$. The set of points n of N where the rank of the Jacobian matrix

$$\left(\frac{\partial H_i}{\partial x_j}(n)\right)_{\substack{1\leqslant i\leqslant t\\ 1\leqslant j\leqslant d}}$$

equals $d-s$ is a (Zariski) open subset of N, called the *smooth part* of N, and denoted N^{sm}. The points of N^{sm} are called *smooth points* of N. By the implicit function theorem, N admits in the neighborhood of every smooth point the structure of a complex manifold, which is the unique structure for which the functions $x_{1|_N},\dots,x_{d|_N}$ are holomorphic. In particular, N^{sm} is a complex manifold. For every point n in N^{sm} we can choose s functions among $x_{1|_N},\dots,x_{d|_N}$, which constitute local coordinates of N in a neighborhood of n in N^{sm} and the other functions $x_{i|_N}$ are then expressible as holomorphic functions in terms of these local coordinates.

Suppose now that N comes equipped with a Poisson structure, i.e., $(N,\{\cdot,\cdot\})$ is a (complex) affine Poisson variety, $N\subset\mathbb{C}^d$. Since $\{\cdot,\cdot\}$ associates to two regular functions[1] on N a regular function on N, we may choose, for every i,j, with $1\leqslant i,j\leqslant d$, a representative x_{ij} of the Poisson bracket $\left\{x_{i|_N},x_{j|_N}\right\}$, i.e., an element of $\mathbb{C}[x_1,\dots,x_d]$ such that $x_{ij|_N}=\left\{x_{i|_N},x_{j|_N}\right\}$. Since all x_{ij} are polynomials, the skew-symmetric biderivation

$$\sum_{1\leqslant i<j\leqslant d} x_{ij}\frac{\partial}{\partial x_i}\wedge\frac{\partial}{\partial x_j}$$

defines a holomorphic bivector field on $\mathbb{C}^d$. We use it to construct a holomorphic bivector field on N^{sm}. Let n be an arbitrary point of N^{sm} and let F,G be holomorphic

---

[1] Recall that a regular function on N is, by definition, the restriction to N of a polynomial function on $\mathbb{C}^d$.

functions, defined on a neighborhood W of n in $N^{\rm sm}$. We extend F and G to functions $\tilde{F}$ and $\tilde{G}$, defined on a neighborhood U of n in M. For n' in a neighborhood of V in $N^{\rm sm}$, contained in $W \cap U$, we define,

$$\{F,G\}_{\mathscr{H}}(n') := \sum_{1 \leqslant i < j \leqslant d} x_{ij}(n') \frac{\partial \tilde{F}}{\partial x_i}(n') \frac{\partial \tilde{G}}{\partial x_j}(n').$$

It defines the unique bivector field on V such that $\left\{x_{i|_N}, x_{j|_N}\right\}_{\mathscr{H}} = \left\{x_{i|_N}, x_{j|_N}\right\}$ on V, for all $1 \leqslant i, j \leqslant d$, hence it is a well-defined holomorphic bivector field on $N^{\rm sm}$. Since $\{\cdot,\cdot\}$ satisfies the Jacobi identity for all triples $\left(x_{i|_N}, x_{j|_N}, x_{k|_N}\right)$, with $1 \leqslant i < j < k \leqslant d$, the same holds true for $\{\cdot,\cdot\}_{\mathscr{H}}$, which means that the Jacobi identity holds for $\{\cdot,\cdot\}_{\mathscr{H}}$, since in a neighborhood of every point, a subset of $x_1,\dots,x_d$ yields a system of local coordinates. This shows that $\{\cdot,\cdot\}$ leads to a holomorphic Poisson structure $\{\cdot,\cdot\}_{\mathscr{H}}$ on $N^{\rm sm}$. In particular, if N is a smooth Poisson variety, so that $N^{\rm sm} = N$, the above construction yields a holomorphic Poisson structure on N, starting from the Poisson bracket on the algebra of regular functions on N.

2.4 Other Constructions

We finish this chapter with a few isolated constructions, which are of a different nature.

2.4.1 Field Extension

If $\mathscr{A}$ is a commutative associative $\mathbb{F}$-algebra and if $\hat{\mathbb{F}}$ is a field extension of $\mathbb{F}$ (think of $\mathbb{C}$, which is a field extension of $\mathbb{R}$), then the tensor product $\hat{\mathbb{F}} \otimes_{\mathbb{F}} \mathscr{A}$ carries a natural structure of a commutative associative $\hat{\mathbb{F}}$-algebra, simply by putting, for all $a,b \in \hat{\mathbb{F}}$ and for all $F,G \in \mathscr{A}$,

$$\begin{aligned} a(b \otimes F) &:= (ab) \otimes F\,, \\ (a \otimes F)(b \otimes G) &:= (ab) \otimes (FG)\,. \end{aligned} \tag{2.22}$$

This very general algebraic procedure is known as "extension of scalars". In fact, an $\mathbb{F}$-Lie algebra structure $[\cdot,\cdot]$ on $\mathscr{A}$ leads, in a similar way, to an $\hat{\mathbb{F}}$-Lie algebra structure on $\hat{\mathbb{F}} \otimes_{\mathbb{F}} \mathscr{A}$: simply replace the second line in (2.22) by

$$[a \otimes F, b \otimes G] := (ab) \otimes [F,G]\,. \tag{2.23}$$

Combining (2.22) and (2.23) shows that every $\mathbb{F}$-Poisson algebra $(\mathscr{A}, \cdot, \{\cdot,\cdot\})$, leads in a natural way to an $\hat{\mathbb{F}}$-Poisson algebra $(\hat{\mathbb{F}} \otimes_{\mathbb{F}} \mathscr{A}, \cdot, \{\cdot,\cdot\})$. For example, it allows one to pass from a Poisson bracket on an $\mathbb{F}$-algebra $\mathscr{A}$ to a Poisson bracket on $\bar{\mathbb{F}}\otimes_{\mathbb{F}}\mathscr{A}$, where $\bar{\mathbb{F}}$ is the algebraic closure of $\mathbb{F}$.

2.4.2 Localization

A classical construction in algebra and in algebraic geometry is to localize a commutative associative algebra $\mathscr{A}$ with respect to a multiplicative system S in $\mathscr{A}$ (*multiplicative* means that S, which is a subset of $\mathscr{A}\setminus\{0\}$, is closed under multiplication). When $\mathscr{A}$ is without zero divisors, which we assume in the sequel of this section, the localization $\mathscr{A}/S$ may be defined as the smallest algebra which contains the elements of $\mathscr{A}$ and the inverses of elements of S. Formally this means that one forms the quotient of $\mathscr{A}\times S$ by the equivalence relation $\sim$, defined by

$$(F,H)\sim(F',H')\Leftrightarrow FH'=F'H\,. \tag{2.24}$$

The class of (F,H) is denoted by F/H or by $\frac{F}{H}$ and $\mathscr{A}/S$ is made into an algebra by defining the sum and product of such elements in the way one adds and multiplies fractions. We think of $\mathscr{A}$ as a subalgebra of $\mathscr{A}/S$ by using the inclusion $F\mapsto F/1$, where the latter element is also simply written as F.

A derivation $\mathscr{V}$ of $\mathscr{A}$ extends to a unique derivation $\mathscr{V}_S$ of $\mathscr{A}/S$. To show this, we first prove that a derivation $\mathscr{V}_S$ of $\mathscr{A}/S$ is entirely determined by its values $\mathscr{V}_S[F]=\mathscr{V}[F]$ on elements $F=F/1$ of $\mathscr{A}$. For $F\in\mathscr{A}$ and $H\in S$, applying the derivation $\mathscr{V}_S$ to the equality $F=\frac{F}{H}H$, we find the familiar formula from calculus,

$$\mathscr{V}_S\left[\frac{F}{H}\right]=\frac{H\mathscr{V}[F]-F\mathscr{V}[H]}{H^2}, \tag{2.25}$$

showing that $\mathscr{V}_S$ is indeed determined by its values on $\mathscr{A}$. In order to show that every derivation $\mathscr{V}$ of $\mathscr{A}$ extends to a derivation of $\mathscr{A}/S$ one takes (2.25) as a definition of $\mathscr{V}_S$. One needs to check that (2.25) is well-defined: to do this, suppose that $F_1H_2=F_2H_1$, with $F_i\in\mathscr{A}$ and $H_i\in S$. First, notice that the equality $F_1H_2=F_2H_1$ implies that $\mathscr{V}[F_1H_2-F_2H_1]=0$, which we write as

$$\frac{\mathscr{V}[F_1]}{H_1}-\frac{\mathscr{V}[F_2]}{H_2}=\frac{F_2\mathscr{V}[H_1]-F_1\mathscr{V}[H_2]}{H_1H_2}\,. \tag{2.26}$$

The definition (2.25) of $\mathscr{V}_S$, combined with (2.26) and with the equality $F_1H_2=F_2H_1$, yields

$$\begin{aligned}
&\mathscr{V}_S[F_1/H_1] - \mathscr{V}_S[F_2/H_2] \\
&\quad = -F_1\mathscr{V}[H_1]/H_1^2 + \mathscr{V}[F_1]/H_1 + F_2\mathscr{V}[H_2]/H_2^2 - \mathscr{V}[F_2]/H_2 \\
&\quad = -F_1\mathscr{V}[H_1]/H_1^2 + F_2\mathscr{V}[H_2]/H_2^2 + F_2\mathscr{V}[H_1]/(H_1H_2) - F_1\mathscr{V}[H_2]/(H_1H_2) \\
&\quad = (F_2H_1 - F_1H_2)\mathscr{V}[H_1]/(H_1^2H_2) + (F_2H_1 - F_1H_2)\mathscr{V}[H_2]/(H_1H_2^2) \\
&\quad = 0\,.
\end{aligned}$$

This shows that $\mathscr{V}_S$ is well-defined. It is easy to verify that (2.25) and the fact that $\mathscr{V}$ is a derivation, imply that, for all $F, F' \in \mathscr{A}$ and $H, H' \in S$, one has that

$$\mathscr{V}_S\left[\frac{F}{H}\frac{F'}{H'}\right] = \frac{F}{H}\,\mathscr{V}_S\left[\frac{F'}{H'}\right] + \frac{F'}{H'}\,\mathscr{V}_S\left[\frac{F}{H}\right], \tag{2.27}$$

which says that $\mathscr{V}_S$ is a derivation of $\mathscr{A}/S$.

In the same way, every skew-symmetric biderivation P of $\mathscr{A}$ extends uniquely to a skew-symmetric biderivation P_S of $\mathscr{A}/S$. As above, one obtains that the basic formulas which allow one to compute P_S, are given as follows: if $F, G \in \mathscr{A}/1 \simeq \mathscr{A}$ and $H, K \in S$, then $P_S[F,G] = P[F,G]$, and

$$\begin{aligned}
P_S\left[\frac{1}{H}, G\right] &= -P_S\left[G, \frac{1}{H}\right] = -\frac{P[H,G]}{H^2}\,, \\
P_S\left[\frac{1}{H}, \frac{1}{K}\right] &= \frac{P[H,K]}{H^2K^2}\,.
\end{aligned} \tag{2.28}$$

As above, it is easily checked that P_S is a skew-symmetric biderivation of $\mathscr{A}/S$. If P is a Poisson bracket on $\mathscr{A}$, then P_S is a Poisson bracket on $\mathscr{A}/S$. That is the content of the following proposition.

Proposition 2.14. *Let $(\mathscr{A}, \cdot, \{\cdot,\cdot\})$ be a Poisson algebra without zero divisors and let S be a multiplicative system of $\mathscr{A}$. The localization algebra $\mathscr{A}/S$ carries a unique Poisson structure $\{\cdot,\cdot\}_S$ which makes the natural inclusion $\mathscr{A} \hookrightarrow \mathscr{A}/S$ into a morphism of Poisson algebras.*

Proof. If a Poisson structure $\{\cdot,\cdot\}_S$ on A/S, for which $\mathscr{A} \hookrightarrow \mathscr{A}/S$ is a Poisson morphism, exists, then $\{\cdot,\cdot\}_S$ is the extension of $\{\cdot,\cdot\}$ to $\mathscr{A}/S$, as constructed above. Therefore, we only need to prove the Jacobi identity for $\{\cdot,\cdot\}_S$. Since $\mathscr{A}/S$ is generated by all elements F, where $F \in \mathscr{A}$ and $1/H$, where $H \in S$, we only need to verify the Jacobi identity for triples consisting of such elements. Let us do this for a triple $(F, 1/H, 1/K)$, where $F \in \mathscr{A}$ and $H, K \in S$. From (2.28) we compute,

$$\begin{aligned}
&\{\{1/H, 1/K\}_S, F\}_S + \{\{1/K, F\}_S, 1/H\}_S + \{\{F, 1/H\}_S, 1/K\}_S \\
&\quad = \{\{H,K\}/(HK)^2, F\}_S - \{\{K,F\}/K^2, 1/H\}_S - \{\{F,H\}/H^2, 1/K\}_S \\
&\quad = (\{\{H,K\},F\} + \circlearrowleft(H,K,F))/(HK)^2 - 2\{K,H\}\{K,F\}/(H^2K^3) \\
&\qquad - 2\{H,K\}\{HK,F\}/(HK)^3 - 2\{F,H\}\{H,K\}/(H^3K^2)
\end{aligned}$$

$$= \frac{\{\{H,K\},F\} + \circlearrowleft(H,K,F)}{(HK)^2},$$

which is zero, in view of the Jacobi identity for $\{\cdot,\cdot\}$. □

In the case of an affine variety M there are two important cases of localization. The first one consists in taking for S the set S_m of all functions $F \in \mathscr{F}(M)$ which do not vanish at a given point m. Then the localization $\mathscr{F}(M)/S_m$ is the algebra of rational functions F/G, with $F,G \in \mathscr{F}(M)$ and $G(m) \neq 0$. By the above, every Poisson structure π on M will induce a Poisson structure π_{S_m} on this algebra of functions. Of course, the algebra of all rational functions on M also inherits a Poisson structure from M, leading to a Poisson algebra which contains all $(\mathscr{F}(M)/S_m, \pi_{S_m})$ as Poisson subalgebras.

The second special case of localization consists in taking a non-constant function $G \in \mathscr{A}$ and considering for S the set $S_G := \{G^n \mid n \in \mathbb{N}\}$. Then $\mathscr{F}(M)/S_G$ is the algebra of rational functions F/G^n, with $F \in \mathscr{F}(M)$ and $n \in \mathbb{N}$. Picking an indeterminate ξ we have an algebra isomorphism

$$\rho : \frac{\mathscr{F}(M)[\xi]}{\langle G\xi - 1\rangle} \to \mathscr{F}(M)/S_G$$
$$\sum_{i=0}^{n} F_i\xi^i \mapsto \sum_{i=0}^{n} F_i/G^i.$$

It follows that the localization $\mathscr{F}(M)/S_G$ is the algebra of regular functions on an affine variety, which we may think of as being obtained by removing the zero locus $G = 0$ from M. Such an open subset of M is sometimes called a *principal open subset*. By the above, every Poisson structure on M will lead to a Poisson structure on this open subset, making it into an affine Poisson variety.

2.4.3 Germification

Recall that a function germ at a point m of a manifold M is an equivalence class of functions, defined on a neighborhood of m, where the equivalence is defined by $F \sim G$ if there exists an open neighborhood U of m in M, such that $F_{|U} = G_{|U}$. Recall also that we denote the germ of F at m by F_m. Suppose that M (or at least a neighborhood of m in M) is equipped with a Poisson structure and let F_m and G_m be two germs at m, represented by functions F and G, which are both defined on a neighborhood U of m. For every open subset U' which is contained in U, the restriction of $\pi[F,G]$ to U' depends on the restrictions of F and G to U only, hence the germ $(\pi[F,G])_m$ of $\pi[F,G]$ at m is well-defined. Thus, a Poisson structure π, defined on a neighborhood of $m \in M$ leads to a Poisson bracket π_m on the algebra of germs of functions at m. We may now define an equivalence relation on the set of all Poisson structures, defined in a neighborhood of m by saying that $\pi \sim_m \pi'$ if

$\pi_m = \pi'_m$. It amounts to saying that if F and G are functions which are defined in a neighborhood of m in M, then $\pi[F,G]$ and $\pi'[F,G]$ agree on some neighborhood of m in M. A *germ* of a Poisson structure at m is then an equivalence class of $\sim_m$; by a slight abuse of notation we denote the germ of π at m by π_m. In bracket notation, the germ π_m is given by

$$\{F_m, G_m\}_m = \sum_{i,j=1}^{d} (x_{ij})_m \left(\frac{\partial F}{\partial x_i}\right)_m \left(\frac{\partial G}{\partial x_j}\right)_m , \tag{2.29}$$

where F and G are arbitrary representatives of F_m and G_m; also $x_1, \ldots, x_d$ are local coordinates, in a neighborhood of m and $x_{ij} := \{x_i, x_j\}$. Evaluating $\{F_m, G_m\}_m$ at m yields a skew-symmetric bilinear map $T^*_m M \times T^*_m M \to \mathbb{F}$, which is precisely the pointwise biderivation at m, associated to the Poisson structure, which we denoted earlier by π_m. Although we have used the same notation π_m for the germ of the Poisson structure at m and for the Poisson bivector at m, one should not forget that the germ at m contains more information than the Poisson bivector at m.

2.5 Exercises

1. Determine the algebra of Casimirs of the Poisson tensor product of two Poisson algebras.

2. Let $\mathscr{A}$ and $\mathscr{B}$ be two Poisson algebras and let $\rho : \mathscr{A} \to \mathscr{B} \otimes \mathscr{A}$ be a Poisson morphism, where the right-hand side is equipped with the product bracket. Prove that

$$\mathscr{A}^{\mathscr{B}} := \{F \in \mathscr{A} \mid \rho(F) = 1 \otimes F\}$$

is a Poisson subalgebra of $\mathscr{A}$.

3. Prove Proposition 2.14 in the case in which $\mathscr{A}$ may have zero divisors (in this case, the definition (2.24) of the equivalence relation $\sim$ is replaced by $(a,s) \sim (a',s') \Leftrightarrow \exists u \in S$ such that $u(as' - a's) = 0$).

4. Reprove Proposition 2.14 by first showing that a biderivation of $\mathscr{A}/S$ vanishes, as soon as it vanishes on $\mathscr{A}$.

5. Let (M_1, π_1) and (M_2, π_2) be two Poisson manifolds. Let $\Psi : M_1 \to M_2$ be a Poisson map. Show that if $\Psi(M_1)$ is a submanifold of M_2, then it is a Poisson submanifold.

6. Let (M, π) be a complex Poisson manifold. Show that every linear combination of the Poisson structures $\Re(\pi)$ and $\Im(\pi)$, defined in Section 2.3.1, is a Poisson structure on $M_{\mathbb{R}}$.

7. Let π be a Poisson structure on an open subset U of $\mathbb{R}^d$, where $d \geqslant 2$, and let F be a smooth function on U. Show that if π has rank at most 2 at every point of U, then $F\pi$ is a Poisson structure on U.

8. We assume in this exercise that the reader is familiar with the notion of symplectic manifold (see Section 6.3). Let (M,ω) be a symplectic manifold, and denote by π the canonical Poisson structure, associated to ω. Prove that the open subsets of M are the only Poisson submanifolds of (M,π).

9. The purpose of this exercise is to establish the existence of non-trivial Poisson structures on every (smooth real) manifold of dimension at least 2. Throughout the exercise we consider $\mathbb{R}^d$ with its algebra of smooth functions.

a. Show that, if $d \geqslant 2$, then there exists a Poisson structure on $\mathbb{R}^d$ which has constant rank 2 in a neighborhood of 0 and which has rank zero outside some compact subset of $\mathbb{R}^d$;
b. Show that, if $d \geqslant 4$, then there exists a Poisson structure on $\mathbb{R}^d$ which has constant rank 4 in a neighborhood of 0 and which has rank at most 2 outside some compact subset of $\mathbb{R}^d$. Using Exercise 6, conclude that there exists such a Poisson structure, which vanishes outside some compact subset of $\mathbb{R}^d$;
c. Generalizing parts a. and b., show that for every $s \in \mathbb{N}$, such that $2s \leqslant d$, there exists a Poisson structure on $\mathbb{R}^d$ which has constant rank $2s$ in a neighborhood of 0 and which vanishes outside some compact subset of $\mathbb{R}^d$;
d. Using a coordinate chart, conclude that every manifold of dimension $d \geqslant 2$, admits a non-trivial Poisson structure; it can be chosen such that its rank, at a given point, equals a given even integer $2s \leqslant d$;
e. The following is an open problem: show that for every Poisson structure π on $\mathbb{R}^d$ there exists a Poisson structure on $\mathbb{R}^d$ which coincides with π in a neighborhood of the origin of 0 and which vanishes outside some compact subset of $\mathbb{R}^d$.

2.6 Notes

The constructions which were given in this chapter are so basic that it is difficult to give the original references. Most of the geometrical constructions are at least implicit in Lichnerowicz's seminal paper [126], in which Poisson manifolds (of constant rank) were introduced. The algebraic formulation of these constructions follows easily. For the link between real and holomorphic Poisson structures, explained in Section 2.3.1, see Laurent–Stiénon–Xu [121].

Chapter 3
Multi-Derivations and Kähler Forms

The Jacobi identity, which is a key element in the definition of a Poisson bracket on an algebra $\mathscr{A}$ (or a Poisson structure on a manifold M), can be naturally formulated in terms of the Schouten bracket on the space $\mathfrak{X}^\bullet(\mathscr{A})$ of all skew-symmetric multi-derivations of $\mathscr{A}$ (or on the space $\mathfrak{X}^\bullet(M)$ of all multivector fields on M). In a sense, the Schouten bracket is dual to the differential on the space of Kähler forms $\Omega^\bullet(\mathscr{A})$ (or on $\Omega^\bullet(M)$, the space of differential forms on M). See Table 3.1 for an overview of the algebraic structures which will be explained in detail in this chapter, with special emphasis on their connections with Poisson structures.

Table 3.1 A summary of the operations which will be considered in this chapter. In the table, $P \in \mathfrak{X}^p$, where $p > 0$.

	$\mathfrak{X}^\bullet(\mathscr{A})$, $\mathfrak{X}^\bullet(M)$	$\Omega^\bullet(\mathscr{A})$, $\Omega^\bullet(M)$
Algebraic	multi-derivation	Kähler form
Geometric	multivector field	differential form
Wedge product	$\wedge : \mathfrak{X}^\bullet \times \mathfrak{X}^\bullet \to \mathfrak{X}^\bullet$	$\wedge : \Omega^\bullet \times \Omega^\bullet \to \Omega^\bullet$
Differential	–	$\mathrm{d} : \Omega^\bullet \to \Omega^{\bullet+1}$
Schouten bracket	$[\cdot,\cdot]_S : \mathfrak{X}^\bullet \times \mathfrak{X}^\bullet \to \mathfrak{X}^{\bullet-1}$	–
Lie derivative	$\mathscr{L}_P : \mathfrak{X}^\bullet \to \mathfrak{X}^{\bullet+p-1}$	$\mathscr{L}_P : \Omega^\bullet \to \Omega^{\bullet-p+1}$
Internal product	–	$\iota_P : \Omega^\bullet \to \Omega^{\bullet-p}$

Since we are using mostly the algebraic properties of these operations, our constructions will be mainly algebraic, but they will always be reformulated in geometrical terms. Multi-derivations are introduced in Section 3.1 and Kähler forms in Section 3.2. The Schouten bracket and the closely related notion of (generalized) Lie derivative will be introduced in Section 3.3.

Unless otherwise stated, $\mathbb{F}$ stands throughout this chapter for an arbitrary field of characteristic zero.

C. Laurent-Gengoux et al., *Poisson Structures*,
Grundlehren der mathematischen Wissenschaften 347,
DOI 10.1007/978-3-642-31090-4_3, © Springer-Verlag Berlin Heidelberg 2013

3.1 Multi-Derivations and Multivector Fields

In this section, we generalize the notion of a derivation and of a skew-symmetric biderivation, defined in Section 1.1.1, to the notion of a skew-symmetric multi-derivation. We assemble all multi-derivations in one graded vector space, which becomes a graded algebra, upon introducing the wedge product. Multivector fields, which are the geometrical analogs of skew-symmetric multi-derivations, will be discussed in Section 3.1.3.

3.1.1 Multi-Derivations

Let $\mathscr{A}$ be a commutative associative $\mathbb{F}$-algebra and let $p \geqslant 1$. A skew-symmetric, p-linear map $P \in \mathrm{Hom}_{\mathbb{F}}(\wedge^p \mathscr{A}, \mathscr{A})$ is called a skew-symmetric *p-derivation* of $\mathscr{A}$ (with values in $\mathscr{A}$), if P is a derivation in each of its arguments. By skew-symmetry, this is equivalent to demanding that P be a derivation in its first argument, i.e.,

$$P(FG, F_2, \ldots, F_p) = F\,P(G, F_2, \ldots, F_p) + GP(F, F_2, \ldots, F_p)\,, \tag{3.1}$$

for arbitrary elements $F, G, F_2, \ldots, F_p$ of $\mathscr{A}$. We keep using our convention, introduced in Chapter 1, to enclose the list of arguments of a skew-symmetric multi-derivation in square brackets; in this notation, (3.1) becomes

$$P[FG, F_2, \ldots, F_p] = F\,P[G, F_2, \ldots, F_p] + GP[F, F_2, \ldots, F_p]\,. \tag{3.2}$$

In view of the derivation property (3.2), two skew-symmetric p-derivations of $\mathscr{A}$ are equal as soon as they are equal for all p-tuples, taken from a system of generators of $\mathscr{A}$; in particular, a p-derivation of $\mathscr{A}$ vanishes as soon as it vanishes on all p-tuples, taken from a system of generators of $\mathscr{A}$ (a detailed argument of this fact was given in the case of derivations in the proof of Proposition 1.6). The vector space of all skew-symmetric p-derivations of $\mathscr{A}$ is denoted by $\mathfrak{X}^p(\mathscr{A})$ and we introduce the graded vector space

$$\mathfrak{X}^\bullet(\mathscr{A}) := \bigoplus_{p\in\mathbb{N}} \mathfrak{X}^p(\mathscr{A}) \subset \bigoplus_{p\in\mathbb{N}} \mathrm{Hom}_{\mathbb{F}}(\wedge^p \mathscr{A}, \mathscr{A})\,,$$

whose elements are called skew-symmetric *multi-derivations*. By convention, the first term in this sum, $\mathfrak{X}^0(\mathscr{A})$, is just $\mathscr{A}$, and $\mathfrak{X}^p(\mathscr{A}) := \{0\}$ for $p < 0$. Every $\mathfrak{X}^p(\mathscr{A})$ has a natural $\mathscr{A}$-module structure, where for $F \in \mathscr{A}$ and $P \in \mathfrak{X}^p(\mathscr{A})$, the product $FP \in \mathfrak{X}^p(\mathscr{A})$ is defined by setting

$$(FP)[F_1, \ldots, F_p] := F\,P[F_1, \ldots, F_p]\,,$$

for all $F_1, \ldots, F_p \in \mathscr{A}$. Thus, $\mathfrak{X}^\bullet(\mathscr{A})$ is actually a graded $\mathscr{A}$-module.

We often think of $\mathscr{A}$ as being the algebra of regular functions on an affine variety M; then P is an operator which associates to p regular functions on M a regular function on M.

3.1.2 Basic Operations on Multi-Derivations

There are two natural composition laws on the graded $\mathscr{A}$-module $\mathfrak{X}^\bullet(\mathscr{A})$ of skew-symmetric multi-derivations of a commutative associative $\mathbb{F}$-algebra $\mathscr{A}$. The first one, introduced here, is the wedge product. The second one is a graded Lie bracket, the Schouten bracket, which will be introduced in Section 3.3.

We first recall the notion of a shuffle. For $p,q \in \mathbb{N}$, a (p,q)-*shuffle* is a permutation σ of the set $\{1,\dots,p+q\}$, such that $\sigma(1) < \cdots < \sigma(p)$ and $\sigma(p+1) < \cdots < \sigma(p+q)$. The set of all (p,q)-shuffles is denoted by $S_{p,q}$. For a shuffle $\sigma \in S_{p,q}$, we denote the signature of σ (as a permutation) by $\operatorname{sgn}(\sigma)$. It is also convenient to define $S_{p,-1} := \emptyset$ and $S_{-1,q} := \emptyset$, for $p,q \in \mathbb{N}$.

For $P \in \mathfrak{X}^p(\mathscr{A})$ and $Q \in \mathfrak{X}^q(\mathscr{A})$, their *wedge product* $P \wedge Q \in \mathfrak{X}^{p+q}(\mathscr{A})$ is the skew-symmetric $(p+q)$-derivation of $\mathscr{A}$, defined by

$$
\begin{aligned}
&(P \wedge Q)[F_1,\dots,F_{p+q}] := \\
&\qquad \sum_{\sigma \in S_{p,q}} \operatorname{sgn}(\sigma)\, P[F_{\sigma(1)},\dots,F_{\sigma(p)}]\; Q[F_{\sigma(p+1)},\dots,F_{\sigma(p+q)}]\,,
\end{aligned}
\tag{3.3}
$$

for all $F_1,\dots,F_{p+q} \in \mathscr{A}$. Notice that, in particular, $F \wedge P = P \wedge F = FP$ when $F \in \mathfrak{X}^0(\mathscr{A}) = \mathscr{A}$. Stated briefly, the wedge product is the skew-symmetrization of the tensor product (of skew-symmetric multi-linear maps). A priori, we have only that $P \wedge Q \in \operatorname{Hom}(\wedge^{p+q}\mathscr{A},\mathscr{A})$, but the reader will easily check that indeed $P \wedge Q \in \mathfrak{X}^{p+q}(\mathscr{A})$, by showing that $P \wedge Q$ verifies the derivation property (3.2). The wedge product has the following properties.

Proposition 3.1. *Let $\mathscr{A}$ be a commutative associative algebra. The wedge product*

$$\bigwedge : \mathfrak{X}^\bullet(\mathscr{A}) \times \mathfrak{X}^\bullet(\mathscr{A}) \to \mathfrak{X}^\bullet(\mathscr{A})$$

makes $\mathfrak{X}^\bullet(\mathscr{A})$ into an associative graded $\mathscr{A}$-algebra. Moreover, it is graded commutative: for $P \in \mathfrak{X}^p(\mathscr{A})$ and $Q \in \mathfrak{X}^q(\mathscr{A})$, one has

$$P \wedge Q = (-1)^{pq} Q \wedge P\,. \tag{3.4}$$

Example 3.2. If $\mathscr{V}$ and $\mathscr{W}$ are derivations, then $\mathscr{V} \wedge \mathscr{W}$ is the skew-symmetric biderivation, given by

$$(\mathscr{V} \wedge \mathscr{W})[F,G] = \mathscr{V}[F]\,\mathscr{W}[G] - \mathscr{V}[G]\,\mathscr{W}[F]\,,$$

for $F,G \in \mathscr{A}$. More generally, if $\mathscr{V}_1,\dots,\mathscr{V}_k$ are derivations of $\mathscr{A}$ and $F_1,\dots,F_k$ are elements of $\mathscr{A}$, then

$$(\mathscr{V}_1 \wedge \cdots \wedge \mathscr{V}_k)[F_1,\dots,F_k] = \begin{vmatrix} \mathscr{V}_1[F_1] & \cdots & \mathscr{V}_1[F_k] \\ \vdots & & \vdots \\ \mathscr{V}_k[F_1] & \cdots & \mathscr{V}_k[F_k] \end{vmatrix} .$$

Each element F of $\mathscr{A}$ defines a graded $\mathscr{A}$-linear map

$$\imath_F : \mathfrak{X}^\bullet(\mathscr{A}) \to \mathfrak{X}^{\bullet-1}(\mathscr{A}) ,$$

of degree -1, by taking F as the first element on which the multi-derivation is evaluated: for $P \in \mathfrak{X}^p(\mathscr{A})$, the skew-symmetric $(p-1)$-derivation $\imath_F P$ is defined by

$$\imath_F P[F_2,\dots,F_p] := P[F,F_2\dots,F_p] , \tag{3.5}$$

for all $F_2,\dots,F_p \in \mathscr{A}$. For a derivation $\mathscr{V} \in \mathfrak{X}^1(\mathscr{A})$, this yields $\imath_F \mathscr{V} = \mathscr{V}[F]$; for a function G (element of $\mathfrak{X}^0(\mathscr{A})$), the above definition should be understood as $\imath_F G := 0$, since $\mathfrak{X}^p(\mathscr{A}) = \{0\}$ for $p<0$. For every $F \in \mathscr{A}$, the graded linear map $\imath_F$ is a graded derivation[1] of degree -1 of $\mathfrak{X}^\bullet(\mathscr{A})$, i.e.,

$$\imath_F(P \wedge Q) = \imath_F P \wedge Q + (-1)^p P \wedge \imath_F Q . \tag{3.6}$$

This follows immediately from (3.3), upon using that for every shuffle $\sigma \in S_{p,q}$, one has that either $\sigma(1)=1$ or $\sigma(p+1)=1$. Also, each derivation $\mathscr{V} \in \mathfrak{X}^1(\mathscr{A})$ defines a graded $\mathbb{F}$-linear map of degree 0,

$$\mathscr{L}_\mathscr{V} : \mathfrak{X}^\bullet(\mathscr{A}) \to \mathfrak{X}^\bullet(\mathscr{A}) ,$$

which is called the *Lie derivative* with respect to $\mathscr{V}$. For $P \in \mathfrak{X}^p(\mathscr{A})$, its Lie derivative $\mathscr{L}_\mathscr{V} P \in \mathfrak{X}^p(\mathscr{A})$ is defined as follows:

$$\mathscr{L}_\mathscr{V} P[F_1,\dots,F_p] := \mathscr{V}[P[F_1,\dots,F_p]] - \sum_{i=1}^p P[F_1,\dots,\mathscr{V}[F_i],\dots,F_p] , \tag{3.7}$$

for all $F_1,\dots,F_p \in \mathscr{A}$. It is easy to check that indeed $\mathscr{L}_\mathscr{V} P \in \mathfrak{X}^p(\mathscr{A})$ by showing that $\mathscr{L}_\mathscr{V} P$, as defined in (3.7), satisfies (3.2). Notice that, in particular, $\mathscr{L}_\mathscr{V} F = \mathscr{V}[F]$, for $F \in \mathscr{A}$, and $\mathscr{L}_\mathscr{V} \mathscr{W} = [\mathscr{V},\mathscr{W}]$, for all $\mathscr{V},\mathscr{W} \in \mathfrak{X}^1(\mathscr{A})$; for a skew-symmetric biderivation P, (3.7) is precisely (1.35). It follows at once from the definition that the Lie derivative is a graded derivation of degree 0 of $\mathfrak{X}^\bullet(\mathscr{A})$:

$$\mathscr{L}_\mathscr{V}(P \wedge Q) = \mathscr{L}_\mathscr{V} P \wedge Q + P \wedge \mathscr{L}_\mathscr{V} Q , \tag{3.8}$$

for all $P,Q \in \mathfrak{X}^\bullet(A)$.

[1] See Appendix A (in particular, Section A.5) for the basic definitions and properties of graded derivations and coderivations.

3.1.3 Multivector Fields

We have already pointed out in Section B.2 that, in the context of manifolds, the analog of a derivation, respectively skew-symmetric biderivation, is a vector field, respectively a bivector field. Similarly, multivector fields are the geometrical analogs of skew-symmetric multi-derivations. Since the basic definitions and properties of multivector fields are a direct generalization of what we have seen for bivector fields in Section B.2, the explanations and justifications which are given in this section are kept to a minimum.

Let M be a manifold (a real manifold when $\mathbb{F}=\mathbb{R}$, a complex manifold when $\mathbb{F}=\mathbb{C}$) and let $p\in\mathbb{N}^*$ and $m\in M$. Generalizing Definition 1.13, we call a skew-symmetric p-linear map

$$\Psi_m:\left(\frac{\mathscr{F}_m(M)}{\sim}\right)^p\to\mathbb{F}$$

a *skew-symmetric pointwise p-derivation* at m, if for all functions $F,G,H,\dots,K$, defined in a neighborhood of m in M,

$$\begin{aligned}&\Psi_m(F_mG_m,H_m,\dots,K_m)\\&\quad=F(m)\,\Psi_m(G_m,H_m,\dots,K_m)+G(m)\,\Psi_m(F_m,H_m,\dots,K_m)\,.\end{aligned}$$

The vector space of all pointwise, skew-symmetric p-derivations at m is denoted by $\wedge^pT_mM$, since, in a coordinate chart (U,x) around m, a basis of this vector space is given by the pointwise p-derivations,

$$\left(\frac{\partial}{\partial x_{i_1}}\right)_m\wedge\left(\frac{\partial}{\partial x_{i_2}}\right)_m\wedge\cdots\wedge\left(\frac{\partial}{\partial x_{i_p}}\right)_m,$$

where $1\leqslant i_1<i_2<\cdots<i_p\leqslant\dim M$; in this formula, the wedge product is the wedge product of (linear) maps, as in the previous section (see for example Eq. 3.3). A map, which assigns to every point $m\in M$ an element P_m of $\wedge^pT_mM$ is called a p-vector field on M if for every open subset $U\subset M$ and for all functions $F_1,\dots,F_p\in\mathscr{F}(U)$, one has that $P[F_1,\dots,F_p]\in\mathscr{F}(U)$, where $P[F_1,\dots,F_p]$ is the function on U, defined for all $m\in U$ by $P[F_1,\dots,F_p](m):=P_m((F_1)_m,\dots,(F_p)_m)$, which we also write as $P_m[F_1,\dots,F_p]$. Thus, on every coordinate chart (U,x), the p-vector field P becomes a skew-symmetric p-derivation of the algebra of functions $\mathscr{F}(U)$ and P admits the local coordinate expression

$$P=\sum_{1\leqslant i_1<\cdots<i_p\leqslant d}P[x_{i_1},\dots,x_{i_p}]\frac{\partial}{\partial x_{i_1}}\wedge\cdots\wedge\frac{\partial}{\partial x_{i_p}},\tag{3.9}$$

where $d:=\dim M$. It follows that a p-vector field is a tensorial object: for $m\in M$, the value of $P[F_1,\dots,F_p](m)$ depends on the differentials $\mathrm{d}_mF_1,\dots,\mathrm{d}_mF_p$ only. This is a useful fact for constructing multivector fields, since it implies that it is sufficient to define a multivector field for all local functions, defined on open subsets which cover the manifold M.

We denote the $\mathscr{F}(M)$-module of p-vector fields on M by $\mathfrak{X}^p(M)$ and we define the graded $\mathscr{F}(M)$-module $\mathfrak{X}^\bullet(M) := \oplus_{p\in\mathbb{N}}\mathfrak{X}^p(M)$, which is the geometrical analog of $\mathfrak{X}^\bullet(\mathscr{A})$. The operations which we have introduced in Section 3.1.2 for skew-symmetric p-derivations, namely the wedge product $\wedge$, the contraction $\imath_F$ by a function F and the Lie derivative $\mathscr{L}_{\mathscr{V}}$, are introduced in the same way for p-vector fields, upon replacing the functions on which they act by local functions. Clearly, all properties of these operations can be repeated word-for-word, as they are purely algebraic. However, in a geometrical approach, the formula (3.7) for the Lie derivative of a p-vector field should be viewed as a property of the Lie derivative, not as its definition. To give the geometrical definition of the Lie derivative, let us first recall that if $\mathscr{V}$ and $\mathscr{W}$ are vector fields on a manifold M, then their commutator $[\mathscr{V},\mathscr{W}]$ is the vector field on M, whose value at $m \in M$ is given by

$$[\mathscr{V},\mathscr{W}]_m := (\mathscr{L}_{\mathscr{V}}\mathscr{W})_m := \frac{\mathrm{d}}{\mathrm{d}t}_{|t=0} T_{\Phi_{-t}(m)}\Phi_t\left(\mathscr{W}_{\Phi_{-t}(m)}\right), \tag{3.10}$$

where Φ_t is the local flow of $\mathscr{V}$. Notice that the tangent map $T_{\Phi_{-t}(m)}\Phi_t$ of the map Φ_t at $\Phi_{-t}(m)$ sends tangent vectors at $\Phi_{-t}(m)$ (here, the vector $\mathscr{W}_{\Phi_{-t}(m)}$) to tangent vectors at m, so taking the derivative in (3.10) amounts to differentiating a curve in T_mM. More generally, for P a p-vector field on a manifold M and $\mathscr{V}$ a vector field on M, the *Lie derivative* of P with respect to $\mathscr{V}$ is defined as the p-vector field, whose value at $m \in M$ is given by

$$(\mathscr{L}_{\mathscr{V}}P)_m := \frac{\mathrm{d}}{\mathrm{d}t}_{|t=0} \wedge^p T_{\Phi_{-t}(m)}\Phi_t\left(P_{\Phi_{-t}(m)}\right). \tag{3.11}$$

One immediate consequence of this definition is that $\mathscr{L}_{\mathscr{V}}P = 0$ if and only if P is preserved by the local flow of the vector field $\mathscr{V}$. The graded derivation property (3.8) of the Lie derivative, and hence also (3.7), follow easily from (3.11).

3.2 Kähler Forms and Differential Forms

In this section, we introduce the objects which are, in a sense, dual to skew-symmetric multi-derivations of a (commutative associative) algebra $\mathscr{A}$: the Kähler forms of $\mathscr{A}$. In the case of a manifold M, the dual objects to multivector fields are differential forms. The modules of Kähler forms and of differential forms will be described in this section, together with their algebraic structure.

3.2.1 Kähler Differentials

Let $\mathscr{A}$ be a commutative associative $\mathbb{F}$-algebra. The module of *Kähler differentials* of $\mathscr{A}$ (over $\mathbb{F}$) is the $\mathscr{A}$-module $\Omega^1(\mathscr{A})$, which is the free $\mathscr{A}$-module generated by the set $\{\mathrm{d}(F) \mid F \in \mathscr{A}\}$, modulo the submodule, generated by all elements of either one of the following forms:

$$\begin{array}{l} \mathrm{d}(F+G) - \mathrm{d}(F) - \mathrm{d}(G)\,, \\ \mathrm{d}(FG) - F\,\mathrm{d}(G) - G\,\mathrm{d}(F)\,, \\ \mathrm{d}(a)\,, \end{array}$$

where $a \in \mathbb{F}$ and $F, G \in \mathscr{A}$. We write $\mathrm{d}F$ rather than $\mathrm{d}(F)$ and we consider d as a map $\mathscr{A} \to \Omega^1(\mathscr{A})$, defined by $F \mapsto \mathrm{d}F$. The pair $(\Omega^1(\mathscr{A}), \mathrm{d})$ satisfies the following universal property: for every derivation $\mathscr{V} : \mathscr{A} \to \mathscr{A}$, there exists a unique $\mathscr{A}$-linear map $\hat{\mathscr{V}} : \Omega^1(\mathscr{A}) \to \mathscr{A}$, such that $\hat{\mathscr{V}} \circ \mathrm{d} = \mathscr{V}$. In other words, there exists a unique arrow $\hat{\mathscr{V}}$ which makes the following diagram commutative:

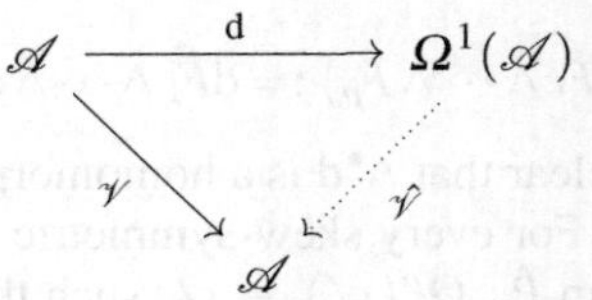

In order to prove the universal property, notice that the commutativity of the diagram amounts to saying that $\hat{\mathscr{V}}(\mathrm{d}F) = \mathscr{V}[F]$, for all $F \in \mathscr{A}$. It is then clear that $\hat{\mathscr{V}}$ should be defined as the (unique) $\mathscr{A}$-linear map, such that

$$\hat{\mathscr{V}}(G\mathrm{d}F) = G\mathscr{V}[F]\,, \tag{3.12}$$

for all $F, G \in \mathscr{A}$. It is indeed easily verified that $\hat{\mathscr{V}}$, given by (3.12), is well-defined; for example, if $F, G \in \mathscr{A}$, then

$$\hat{\mathscr{V}}\big(\mathrm{d}(FG) - F\mathrm{d}G - G\mathrm{d}F\big) = \mathscr{V}[FG] - F\mathscr{V}[G] - G\mathscr{V}[F] = 0\,,$$

since $\mathscr{V}$ is a derivation of $\mathscr{A}$. Given an arbitrary $\mathscr{A}$-linear map $\hat{\mathscr{V}} : \Omega^1(\mathscr{A}) \to \mathscr{A}$, it is also clear that $\hat{\mathscr{V}} \circ \mathrm{d} : \mathscr{A} \to \mathscr{A}$ is a derivation of $\mathscr{A}$. This fact, combined with the universal property, implies the existence of a natural isomorphism (of $\mathscr{A}$-modules)

$$\mathfrak{X}^1(\mathscr{A}) \simeq \mathrm{Hom}_{\mathscr{A}}(\Omega^1(\mathscr{A}), \mathscr{A})\,, \tag{3.13}$$

which allows us to think of derivations of $\mathscr{A}$ as $\mathscr{A}$-linear maps on some $\mathscr{A}$-module, intimately related to $\mathscr{A}$.

3.2.2 Kähler Forms

In order to generalize (3.13) to skew-symmetric p-derivations, we introduce, for $p \in \mathbb{N}$, the $\mathscr{A}$-module $\Omega^p(\mathscr{A}) := \wedge^p \Omega^1(\mathscr{A})$, where $\wedge$ is the wedge product over $\mathscr{A}$; for $p = 0$, this should be read as $\Omega^0(\mathscr{A}) := \mathscr{A}$. Together, the $\mathscr{A}$-modules $\Omega^p(\mathscr{A})$ form the graded $\mathscr{A}$-module

$$\Omega^\bullet(\mathscr{A}) := \bigoplus_{p \in \mathbb{N}} \Omega^p(\mathscr{A}) .$$

The elements of $\Omega^p(\mathscr{A})$, with $p > 0$, are called *Kähler p-forms*, or simply *Kähler forms*. As a vector space, respectively as an $\mathscr{A}$-module, $\Omega^p(\mathscr{A})$ is generated by elements of the form $G\mathrm{d}F_1 \wedge \cdots \wedge \mathrm{d}F_p$, respectively of the form $\mathrm{d}F_1 \wedge \cdots \wedge \mathrm{d}F_p$, where $G, F_1, \ldots, F_p \in \mathscr{A}$. Being, by definition, an exterior algebra, it is an associative, graded commutative $\mathscr{A}$-algebra with product $\wedge$. The $\mathbb{F}$-linear map $\mathrm{d} : \mathscr{A} \to \Omega^1(\mathscr{A})$ extends by functoriality of $\wedge$ to an $\mathbb{F}$-linear map $\wedge^\bullet \mathrm{d} : \wedge^\bullet \mathscr{A} \to \Omega^\bullet(\mathscr{A})$. Explicitly, it is given by

$$\wedge^\bullet \mathrm{d}(F_1 \wedge \cdots \wedge F_p) := \mathrm{d}F_1 \wedge \cdots \wedge \mathrm{d}F_p , \tag{3.14}$$

where $F_1, \ldots, F_p \in \mathscr{A}$. It is clear that $\wedge^\bullet \mathrm{d}$ is a homomorphism between the algebras $(\wedge^\bullet \mathscr{A}, \wedge)$ and $(\Omega^\bullet(\mathscr{A}), \wedge)$. For every skew-symmetric p-derivation P of $\mathscr{A}$, there exists a unique $\mathscr{A}$-linear map $\hat{P} : \Omega^p(\mathscr{A}) \to \mathscr{A}$, such that the following triangle is commutative:

$$\begin{array}{ccc} \wedge^p \mathscr{A} & \xrightarrow{\ \wedge^p \mathrm{d}\ } & \wedge^p \Omega^1(\mathscr{A}) = \Omega^p(\mathscr{A}) \\ & {\scriptstyle P} \searrow \qquad \swarrow {\scriptstyle \hat{P}} & \\ & \mathscr{A} & \end{array}$$

The definition of $\hat{P}$ for arbitrary p is the natural generalization of the definition (3.12) of $\hat{\mathscr{V}}$, which corresponds to the case of $p = 1$. It is given by

$$\hat{P}(G\mathrm{d}F_1 \wedge \cdots \wedge \mathrm{d}F_p) := G P[F_1, \ldots, F_p] , \tag{3.15}$$

for all $G, F_1, \ldots, F_p \in \mathscr{A}$. It leads to a natural isomorphism (of $\mathscr{A}$-modules):

$$\mathfrak{X}^p(\mathscr{A}) \simeq \mathrm{Hom}_{\mathscr{A}}(\Omega^p(\mathscr{A}), \mathscr{A}) . \tag{3.16}$$

For some of the computations which we will do later, it is useful to view $\hat{P}$ as an $\mathscr{A}$-linear map $\Omega^\bullet(\mathscr{A}) \to \mathscr{A}$, so we extend $\hat{P}$, defined in (3.15), trivially to Kähler q-forms, with $q \neq p$:

$$\hat{P}(G\mathrm{d}F_1 \wedge \cdots \wedge \mathrm{d}F_q) := 0, \text{ if } q \neq p , \tag{3.17}$$

for all $G, F_1, \ldots, F_q \in \mathscr{A}$. With this notation, the map $P \mapsto \hat{P}$ yields a natural isomorphism

$$\mathfrak{X}^\bullet(\mathscr{A}) \to \bigoplus_{p\in\mathbb{N}} \mathrm{Hom}_{\mathscr{A}}(\Omega^p(\mathscr{A}),\mathscr{A}) .$$

Remark 3.3. For $p \in \mathbb{N}$, the natural pairing

$$\begin{aligned} \Psi : \mathfrak{X}^p(\mathscr{A}) \otimes_{\mathscr{A}} \Omega^p(\mathscr{A}) &\to \mathscr{A} \\ P \otimes \omega &\mapsto \langle \omega, P \rangle := \hat{P}(\omega) \end{aligned} \tag{3.18}$$

induces two $\mathscr{A}$-linear maps

$$\begin{aligned} \Psi_1 &: \mathfrak{X}^p(\mathscr{A}) \to \mathrm{Hom}_{\mathscr{A}}(\Omega^p(\mathscr{A}),\mathscr{A}) , \\ \Psi_2 &: \Omega^p(\mathscr{A}) \to \mathrm{Hom}_{\mathscr{A}}(\mathfrak{X}^p(\mathscr{A}),\mathscr{A}) . \end{aligned}$$

As we have seen in (3.16), Ψ_1 is an isomorphism. In general, Ψ_2 is neither injective nor surjective; however, if $\mathscr{A}$ is the algebra $C^\infty(M)$ of smooth functions on a real manifold M, the map Ψ_2 is also an isomorphism (see Section 3.2.5).

3.2.3 Algebra and Coalgebra Structure

We now consider the algebra and coalgebra structures on $\Omega^\bullet(\mathscr{A})$, which are closely related to the corresponding structures on $\wedge^\bullet\mathscr{A}$, where $\mathscr{A}$ is as before a commutative associative $\mathbb{F}$-algebra. We assume that the reader is familiar with the facts, recalled in Appendix A (in particular, in Sections A.3 and A.4), that $(\wedge^\bullet\mathscr{A},\wedge)$ is an associative, graded commutative $\mathbb{F}$-algebra, and that $(\wedge^\bullet\mathscr{A},\Delta)$ is a coassociative $\mathbb{F}$-coalgebra. Since $\Omega^\bullet(\mathscr{A})$ is an exterior algebra (over $\mathscr{A}$), we have, as we mentioned in the previous section, a wedge product on $\Omega^\bullet(\mathscr{A})$, also denoted by $\wedge$, which makes $(\Omega^\bullet(\mathscr{A}),\wedge)$ into an associative, graded commutative $\mathscr{A}$-algebra. Also, it leads to an $\mathscr{A}$-linear coassociative coproduct

$$\Delta : \Omega^\bullet(\mathscr{A}) \to \Omega^\bullet(\mathscr{A}) \otimes_{\mathscr{A}} \Omega^\bullet(\mathscr{A}) ,$$

which is defined in the same way as in the case of $\wedge^\bullet\mathscr{A}$ (see (A.17)). Namely, $\Delta_{p,q}$ is given, for $p,q \in \mathbb{N}$, and $F_1,\dots,F_{p+q} \in \mathscr{A}$ by

$$\begin{aligned} &\Delta_{p,q}(\mathrm{d}F_1 \wedge \cdots \wedge \mathrm{d}F_{p+q}) \\ &\quad := \sum_{\sigma\in S_{p,q}} \mathrm{sgn}(\sigma)(\mathrm{d}F_{\sigma(1)} \wedge \cdots \wedge \mathrm{d}F_{\sigma(p)}) \otimes (\mathrm{d}F_{\sigma(p+1)} \wedge \cdots \wedge \mathrm{d}F_{\sigma(p+q)}) . \end{aligned} \tag{3.19}$$

Recall that coassociativity means that

$$(\mathbb{1}_{\Omega^\bullet(\mathscr{A})} \otimes \Delta) \circ \Delta = (\Delta \otimes \mathbb{1}_{\Omega^\bullet(\mathscr{A})}) \circ \Delta . \tag{3.20}$$

It is clear, from its definition (3.14), that $\wedge^\bullet\mathrm{d}$ is a homomorphism of the $\mathbb{F}$-coalgebras $(\wedge^\bullet\mathscr{A},\Delta)$ and $(\Omega^\bullet(\mathscr{A}),\Delta)$.

A useful application of the coproduct Δ of $\Omega^\bullet(\mathscr{A})$, is that the wedge product (3.3) of two multi-derivations $P, Q \in \mathfrak{X}^\bullet(\mathscr{A})$ can be expressed in terms of Δ, as follows:

$$\widehat{P \wedge Q} = \nu \circ (\hat{P} \otimes \hat{Q}) \circ \Delta \,, \tag{3.21}$$

where ν is the canonical isomorphism

$$\begin{aligned} \nu : \mathscr{A} \otimes_{\mathscr{A}} \Omega^\bullet(\mathscr{A}) &\to \Omega^\bullet(\mathscr{A}) \\ F \otimes \omega \quad &\mapsto F\,\omega \,. \end{aligned} \tag{3.22}$$

We used ν in (3.21) only for the case of $\omega \in \Omega^0(\mathscr{A}) = \mathscr{A}$, but we will need its general form later on; in particular we will need the following properties of ν:

$$\begin{aligned} \hat{P} \circ \nu &= \nu \circ (\mathbb{1}_{\mathscr{A}} \otimes \hat{P}) \,, \\ \Delta \circ \nu &= (\nu \otimes \mathbb{1}_{\Omega^\bullet(\mathscr{A})}) \circ (\mathbb{1}_{\mathscr{A}} \otimes \Delta) \,, \end{aligned} \tag{3.23}$$

where $P \in \mathfrak{X}^\bullet(\mathscr{A})$. The proof of these properties follows at once from the definition (3.22) of ν.

3.2.4 The de Rham Differential and Cohomology

The differential $\mathrm{d} : \mathscr{A} \to \Omega^1(\mathscr{A})$ extends to a well-defined graded $\mathbb{F}$-linear map

$$\mathrm{d} : \Omega^\bullet(\mathscr{A}) \to \Omega^{\bullet+1}(\mathscr{A}) \,,$$

by putting

$$\mathrm{d}(G\,\mathrm{d}F_1 \wedge \cdots \wedge \mathrm{d}F_p) := \mathrm{d}G \wedge \mathrm{d}F_1 \wedge \cdots \wedge \mathrm{d}F_p \,, \tag{3.24}$$

for all $G, F_1, \ldots, F_p \in \mathscr{A}$, where $p \in \mathbb{N}$. It is called the *(algebraic) de Rham differential*. It is a graded derivation, of degree 1, of $(\Omega^\bullet(\mathscr{A}), \wedge)$, as

$$\mathrm{d}(\omega \wedge \eta) = \mathrm{d}\omega \wedge \eta + (-1)^p \omega \wedge \mathrm{d}\eta \,, \tag{3.25}$$

for $\omega \in \Omega^p(\mathscr{A})$ and $\eta \in \Omega^\bullet(\mathscr{A})$, an easy consequence of (3.24). Since $\mathrm{d}a = 0$ for every $a \in \mathbb{F}$, it follows from (3.24) that $\mathrm{d} \circ \mathrm{d} = 0$, so that d is a coboundary operator and the sequence

$$0 \longrightarrow \Omega^0(\mathscr{A}) \xrightarrow{\ \mathrm{d}\ } \Omega^1(\mathscr{A}) \xrightarrow{\ \mathrm{d}\ } \Omega^2(\mathscr{A}) \xrightarrow{\ \mathrm{d}\ } \cdots$$

is a complex, called the *(algebraic) de Rham complex*. A Kähler p-form ω for which $\mathrm{d}\omega = 0$ is called *closed*, while it is called *exact* if it is contained in the image of d. By the above, exact Kähler forms are closed, which leads, for $p \in \mathbb{N}$, to the cohomology space[2]

$$H^p_{dR}(\mathscr{A}) := \frac{\mathrm{Ker}(\mathrm{d} : \Omega^p(\mathscr{A}) \to \Omega^{p+1}(\mathscr{A}))}{\mathrm{Im}(\mathrm{d} : \Omega^{p-1}(\mathscr{A}) \to \Omega^p(\mathscr{A}))},$$

where it is understood that $H^0_{dR}(\mathscr{A}) := \mathrm{Ker}(\mathrm{d} : \mathscr{A} \to \Omega^1(\mathscr{A})) = \mathbb{F}$. The $\mathbb{F}$-vector space $H^p_{dR}(\mathscr{A})$ is called the *p-th de Rham cohomology space* and the graded vector space

$$H^\bullet_{dR}(\mathscr{A}) := \bigoplus_{p\in\mathbb{N}} H^p_{dR}(\mathscr{A})$$

is called the *(algebraic) de Rham cohomology* of $\mathscr{A}$.

The derivation property (3.25) of d implies the following two facts: (1) the wedge product of two closed Kähler forms is closed; (2) the wedge product of an exact Kähler form with a closed Kähler form is exact. Combining these two properties shows that there is an induced product

$$\begin{aligned} \wedge : H^\bullet_{dR}(\mathscr{A}) \times H^\bullet_{dR}(\mathscr{A}) &\to H^\bullet_{dR}(\mathscr{A}) \\ ([\omega],[\eta]) &\mapsto [\omega \wedge \eta] \end{aligned}$$

where $[\omega]$ denotes the de Rham cohomology class of a closed Kähler form ω. It follows that $(H^\bullet_{dR}(\mathscr{A}), \wedge)$ is, just like $(\Omega^\bullet(\mathscr{A}), \wedge)$, an associative, graded commutative $\mathbb{F}$-algebra.

To close this section, we wish to point out that $\Omega^\bullet$ leads to a covariant functor from the category of commutative associative $\mathbb{F}$-algebras to the category of graded $\mathbb{F}$-vector spaces (associative, graded commutative $\mathbb{F}$-algebras, if you wish). Namely, let $\phi : \mathscr{A} \to \mathscr{B}$ be an algebra homomorphism, where $\mathscr{A}$ and $\mathscr{B}$ are commutative associative algebras. For every $p \in \mathbb{N}$, the induced linear map $\Omega^p(\phi) : \Omega^p(\mathscr{A}) \to \Omega^p(\mathscr{B})$ is defined as follows:

$$\begin{aligned} \Omega^p(\phi) : \quad \Omega^p(\mathscr{A}) &\to \Omega^p(\mathscr{B}) \\ G\,\mathrm{d}F_1 \wedge \cdots \wedge \mathrm{d}F_p &\mapsto \phi(G)\,\mathrm{d}\phi(F_1) \wedge \cdots \wedge \mathrm{d}\phi(F_p) . \end{aligned}$$

From this definition, it is easily verified that $\Omega^\bullet$ is indeed a covariant functor. Since $\Omega^\bullet(\phi)$ and d commute, $\Omega^\bullet(\phi)$ induces a graded linear map of degree 0:

$$H^\bullet_{dR}(\phi) : H^\bullet_{dR}(\mathscr{A}) \to H^\bullet_{dR}(\mathscr{B}) .$$

It follows that $H^\bullet_{dR}$ is also a (covariant) functor.

[2] Notice that it is only a vector space and not an $\mathscr{A}$-module, as the differential d is only $\mathbb{F}$-linear.

3.2.5 Differential Forms

There are several equivalent ways of defining the notion of a differential form on a manifold. The most economical definition of a differential p-form on a *real* manifold is that it is an $\mathscr{F}(M)$-linear map from $\mathfrak{X}^p(M)$ to $\mathscr{F}(M)$. In other words, it is a skew-symmetric, $\mathscr{F}(M)$-p-linear map from $\mathfrak{X}^1(M)$ to $\mathscr{F}(M)$. This definition does not work in the case of a complex manifold since, as we have already said, the space of holomorphic functions on M may consist of constant functions only. To remedy this, one considers local functions and local vector fields. Namely, given a manifold M, a map which assigns to every $m \in M$ an element $\omega_m \in \wedge^p T_m^* M$, is called a *differential p-form* if for every open subset $U \subset M$ and for all vector fields $\mathscr{V}_1, \dots, \mathscr{V}_p \in \mathfrak{X}^1(U)$, one has that $\omega(\mathscr{V}_1, \dots, \mathscr{V}_p) \in \mathscr{F}(U)$, where $\omega(\mathscr{V}_1, \dots, \mathscr{V}_p)$ is the function on U, defined by

$$\omega(\mathscr{V}_1, \dots, \mathscr{V}_p)(m) := \omega_m((\mathscr{V}_1)_m, \dots, (\mathscr{V}_p)_m) ,$$

for all $m \in U$. Thus, on every coordinate chart (U, x), the differential p-form ω becomes a skew-symmetric $\mathscr{F}(U)$-p-linear map from $\mathscr{F}(U)$ to itself. It follows that ω admits the local coordinate expression

$$\omega = \sum_{1 \leqslant i_1 < \dots < i_p \leqslant d} \omega\left(\frac{\partial}{\partial x_{i_1}}, \dots, \frac{\partial}{\partial x_{i_p}}\right) \mathrm{d}x_{i_1} \wedge \dots \wedge \mathrm{d}x_{i_p} , \tag{3.26}$$

where d denotes the dimension of M. Just like a p-vector field, a differential p-form is a tensorial object: for $m \in M$, the value of $\omega(\mathscr{V}_1, \dots, \mathscr{V}_p)(m)$ depends only on the values at m of the vector fields $\mathscr{V}_1, \dots, \mathscr{V}_p$, which is a useful fact for constructing differential p-forms.

The $\mathscr{F}(M)$-module of differential p-forms on M is denoted by $\Omega^p(M)$ and we define the graded $\mathscr{F}(M)$-module

$$\Omega^\bullet(M) := \bigoplus_{p \in \mathbb{N}} \Omega^p(M) ,$$

which is the geometrical analog of the $\mathscr{A}$-module $\Omega^\bullet(\mathscr{A})$, which was defined in Section 3.2.2. If U is the domain of a coordinate chart on M, then $\Omega^p(U) = \wedge^p \Omega^1(U)$, just like in the case of Kähler differentials, although $\Omega^p(M) \neq \wedge^p \Omega^1(M)$, in general. The fact that, locally, $\Omega^p(M)$ is an exterior algebra, leads to a product, called the *wedge product* of differential forms. It makes $\Omega^\bullet(M)$ into an associative, graded commutative algebra. The differential of a function, which we introduced in Section 1.3.1, is naturally viewed as a linear map $\mathrm{d} : \Omega^0(M) \to \Omega^1(M)$. It extends to a unique graded derivation of $\Omega^\bullet(M)$ of degree 1, which means that

$$\mathrm{d}(\omega \wedge \eta) = \mathrm{d}\omega \wedge \eta + (-1)^p \omega \wedge \mathrm{d}\eta , \tag{3.27}$$

for $\omega \in \Omega^p(M)$ and $\eta \in \Omega^\bullet(M)$, just like in (3.25). In a coordinate chart (U, x), it leads to the classical formula for the differential of a differential p-form, namely,

$$\mathrm{d}\left(\sum_{1\leqslant i_1<\cdots<i_p\leqslant d}\omega_{i_1,\ldots,i_p}\,\mathrm{d}x_{i_1}\wedge\cdots\wedge\mathrm{d}x_{i_p}\right)$$
$$=\sum_{1\leqslant i_1<\cdots<i_p\leqslant d}\sum_{j=1}^{d}\frac{\partial\omega_{i_1,\ldots,i_p}}{\partial x_j}\,\mathrm{d}x_j\wedge\mathrm{d}x_{i_1}\wedge\cdots\wedge\mathrm{d}x_{i_p},$$

and d is called the *de Rham differential*. It has formally the same properties as the algebraic de Rham differential, in particular it leads to a complex, known as the *de Rham complex*, whose cohomology

$$H^\bullet_{dR}(M):=\bigoplus_{p\in\mathbb{N}}H^p_{dR}(M):=\bigoplus_{p\in\mathbb{N}}\frac{\mathrm{Ker}(\mathrm{d}:\Omega^p(M)\to\Omega^{p+1}(M))}{\mathrm{Im}(\mathrm{d}:\Omega^{p-1}(M)\to\Omega^p(M))},$$

is called the *de Rham cohomology* of M. In the manifold case, the properties of the de Rham differential are usually derived from the following explicit formula: $\mathrm{d}\omega$ is given, for $\omega\in\Omega^p(M)$, by

$$\mathrm{d}\omega(\mathscr{V}_0,\mathscr{V}_1,\ldots,\mathscr{V}_p)=\sum_{i=0}^{p}(-1)^i\mathscr{V}_i\left[\omega\left(\mathscr{V}_0,\mathscr{V}_1,\ldots,\widehat{\mathscr{V}_i},\ldots,\mathscr{V}_p\right)\right] \tag{3.28}$$
$$+\sum_{0\leqslant i<j\leqslant p}(-1)^{i+j}\omega\left([\mathscr{V}_i,\mathscr{V}_j],\mathscr{V}_0,\ldots,\widehat{\mathscr{V}_i},\ldots,\widehat{\mathscr{V}_j},\ldots,\mathscr{V}_p\right),$$

where $\mathscr{V}_0,\ldots,\mathscr{V}_p$ are arbitrary local vector fields on M.

3.3 The Schouten Bracket

In this section we introduce a few extra natural operations on the graded algebras of skew-symmetric multi-derivations and of Kähler forms on a commutative associative algebra $\mathscr{A}$. The first one is an action of $\mathfrak{X}^\bullet(\mathscr{A})$ on $\Omega^\bullet(\mathscr{A})$, which generalizes the natural pairing between multi-derivations and Kähler forms, which we encountered in Remark 3.3. The second operation is the Schouten bracket which extends the classical commutator of derivations to the case of skew-symmetric multi-derivations. It will lead at the end of the section to the so-called Gerstenhaber algebra structure on $\mathfrak{X}^\bullet(\mathscr{A})$ and to two generalizations of the Lie derivative, acting in one case on the algebra of multi-derivations, in the other case on the algebra of Kähler forms, where the derivative is in both cases taken with respect to a multi-derivation. In the geometrical setting, this yields for every manifold M on the one hand an action of the graded algebra of multivector fields $\mathfrak{X}^\bullet(M)$ on the graded algebra of differential forms $\Omega^\bullet(M)$ and on the other hand a graded Lie bracket, the Schouten bracket, on $\mathfrak{X}^\bullet(M)$.

3.3.1 Internal Products

The universal property of the graded $\mathscr{A}$-module $\Omega^\bullet(\mathscr{A})$ of Kähler differentials on a commutative associative algebra $\mathscr{A}$ leads to a natural action of $\mathfrak{X}^\bullet(\mathscr{A})$ on $\Omega^\bullet(\mathscr{A})$, often referred to as the *internal product*. To introduce it, we associate to $P \in \mathfrak{X}^p(\mathscr{A})$, the graded $\mathscr{A}$-linear map of degree $-p$,

$$\imath_P : \Omega^\bullet(\mathscr{A}) \to \Omega^{\bullet-p}(\mathscr{A}) ,$$

which is defined for $\omega = \mathrm{d}F_1 \wedge \cdots \wedge \mathrm{d}F_k \in \Omega^k(\mathscr{A})$, by

$$\imath_P(\mathrm{d}F_1 \wedge \cdots \wedge \mathrm{d}F_k) := \sum_{\sigma \in S_{p,k-p}} \operatorname{sgn}(\sigma) P[F_{\sigma(1)}, \dots, F_{\sigma(p)}]\, \mathrm{d}F_{\sigma(p+1)} \wedge \cdots \wedge \mathrm{d}F_{\sigma(k)} \tag{3.29}$$

when $k \geqslant p$, and otherwise $\imath_P\omega := 0$; for $F \in \mathfrak{X}^0(\mathscr{A}) = \mathscr{A}$, (3.29) has to be understood as $\imath_F\omega := F\omega$. The reader may wish to write the definition of $\imath_P$ differently, using

$$P[F_{\sigma(1)}, \dots, F_{\sigma(p)}] = \hat{P}\left(\mathrm{d}F_{\sigma(1)} \wedge \cdots \wedge \mathrm{d}F_{\sigma(p)}\right) .$$

With this notation, it is obvious that the restriction of $\imath_P$ to $\Omega^p(\mathscr{A})$ is just $\hat{P}$, which was defined in (3.15); the restriction to the spaces $\Omega^q(\mathscr{A})$, with $q \neq p$, is however different, since $\hat{P}$ acts trivially on them (see (3.17)). It also follows that $\imath_P$ can be written, for every $P \in \mathfrak{X}^\bullet(\mathscr{A})$, in terms of the coproduct Δ, defined in (3.19), as

$$\imath_P = \nu \circ (\hat{P} \otimes \mathbb{1}_{\Omega^\bullet(\mathscr{A})}) \circ \Delta . \tag{3.30}$$

As a consequence, $\imath_P$ is the adjoint of taking the wedge product with P: for $P \in \mathfrak{X}^p(\mathscr{A})$ and $Q \in \mathfrak{X}^q(\mathscr{A})$ and $\omega \in \Omega^{p+q}(\mathscr{A})$, we have in terms of the notations in (3.18),

$$\langle \imath_P\omega, Q\rangle = \langle \omega, P \wedge Q\rangle . \tag{3.31}$$

Indeed, (3.30), (3.23) and (3.21) imply that, on $\Omega^{p+q}(\mathscr{A})$,

$$\begin{aligned}
\hat{Q} \circ \imath_P &= \hat{Q} \circ \nu \circ (\hat{P} \otimes \mathbb{1}_{\Omega^\bullet(\mathscr{A})}) \circ \Delta \\
&= \nu \circ (\mathbb{1}_{\mathscr{A}} \otimes \hat{Q}) \circ (\hat{P} \otimes \mathbb{1}_{\Omega^\bullet(\mathscr{A})}) \circ \Delta \\
&= \nu \circ (\hat{P} \otimes \hat{Q}) \circ \Delta \\
&= \widehat{P \wedge Q} ,
\end{aligned}$$

which yields (3.31), when applied to $\omega \in \Omega^{p+q}(\mathscr{A})$. Equation (3.31) explains the notation $\imath_P$: as in the case of $\imath_F$ ($F \in \mathscr{A}$), defined in (3.5), $\imath_P$ provides ω with the p-derivation P as its first p arguments. Yet, when considering $\imath_F G$ for $F, G \in \mathscr{A}$ one should decide from the context whether G is interpreted as a 0-derivation or as a 0-form, since in the first case $\imath_F G = 0$ (as defined in (3.5)), while in the second case $\imath_F G = FG$ (as defined in (3.29)).

In the following proposition, we establish the main properties of $\imath$.

Proposition 3.4. *Let $\mathscr{A}$ be a commutative associative algebra. The following identities hold, for $P \in \mathfrak{X}^p(\mathscr{A})$ and $Q \in \mathfrak{X}^q(\mathscr{A})$:*

(1) $\imath_P \circ \imath_Q = \imath_{Q \wedge P}$;
(2) $[\imath_P, \imath_Q] = \imath_P \circ \imath_Q - (-1)^{pq} \imath_Q \circ \imath_P = 0$;
(3) *For every $F \in \mathscr{A}$ and $\omega \in \Omega^\bullet(\mathscr{A})$:*

$$\imath_P(\mathrm{d}F \wedge \omega) = \imath_{(\imath_F P)} \omega + (-1)^p \mathrm{d}F \wedge \imath_P \omega \,. \tag{3.32}$$

Proof. We give three different proofs of *(1)*. The first one is a quick proof, but it is not valid for general algebras, because we assume that $\mathrm{Hom}_{\mathscr{A}}(\mathfrak{X}^p(\mathscr{A}), \mathscr{A}) \simeq \Omega^p(\mathscr{A})$, for all $p \in \mathbb{N}$ (see Remark 3.3 at the end of Section 3.2.2). Applying (3.31) three times, we find for every $\omega \in \Omega^k(\mathscr{A})$ and for homogeneous elements P, Q and R of $\mathfrak{X}^\bullet(\mathscr{A})$, whose degrees sum up to k,

$$\begin{aligned} \langle \imath_{Q \wedge P} \omega, R \rangle &= \langle \omega, (Q \wedge P) \wedge R \rangle = \langle \omega, Q \wedge (P \wedge R) \rangle \\ &= \langle \imath_Q \omega, P \wedge R \rangle = \langle \imath_P(\imath_Q \omega), R \rangle \,. \end{aligned} \tag{3.33}$$

Under the above assumption, the fact that (3.33) is valid for all R, implies that $\imath_{Q \wedge P} \omega = \imath_P(\imath_Q \omega)$, for all $\omega \in \Omega^\bullet(\mathscr{A})$, which leads to *(1)*. Our second proof is based on the fact that Δ is coassociative. To simplify the notation, we will write $\mathbb{1}$ for $\mathbb{1}_{\Omega^\bullet(\mathscr{A})}$; using (3.30), the properties (3.23) of ν, the coassociativity (3.20) of Δ and (3.21) (in that order), we find

$$\begin{aligned} \imath_P \circ \imath_Q &= \nu \circ (\hat{P} \otimes \mathbb{1}) \circ \Delta \circ \nu \circ (\hat{Q} \otimes \mathbb{1}) \circ \Delta \\ &= \nu \circ (\hat{P} \otimes \mathbb{1}) \circ (\nu \otimes \mathbb{1}) \circ (\mathbb{1}_{\mathscr{A}} \otimes \Delta) \circ (\hat{Q} \otimes \mathbb{1}) \circ \Delta \\ &= \nu \circ (\nu \otimes \mathbb{1}) \circ (\mathbb{1}_{\mathscr{A}} \otimes \hat{P} \otimes \mathbb{1}) \circ (\hat{Q} \otimes \mathbb{1} \otimes \mathbb{1}) \circ (\mathbb{1} \otimes \Delta) \circ \Delta \\ &= \nu \circ (\nu \otimes \mathbb{1}) \circ (\hat{Q} \otimes \hat{P} \otimes \mathbb{1}) \circ (\Delta \otimes \mathbb{1}) \circ \Delta \\ &= \nu \circ ((\nu \circ (\hat{Q} \otimes \hat{P}) \circ \Delta) \otimes \mathbb{1}) \circ \Delta \\ &= \nu \circ \left(\widehat{Q \wedge P} \otimes \mathbb{1} \right) \circ \Delta \\ &= \imath_{Q \wedge P} \,. \end{aligned}$$

This terminates the second proof. For the third proof, we show by direct computation that

$$(\imath_P \circ \imath_Q)(\mathrm{d}F_1 \wedge \cdots \wedge \mathrm{d}F_k) = \imath_{Q \wedge P}(\mathrm{d}F_1 \wedge \cdots \wedge \mathrm{d}F_k) \tag{3.34}$$

for all $F_1, \ldots, F_k$. Since $\imath_P$ is $\mathscr{A}$-linear for all $P \in \mathfrak{X}^p(\mathscr{A})$, this will provide a third proof of *(1)*. Since $\imath_P \omega = 0$ for $\omega \in \Omega^q(\mathscr{A})$ when $q < p$, where $P \in \mathfrak{X}^p(\mathscr{A})$, it suffices to prove (3.34) when $k \geqslant p + q$. On the one hand,

$$\begin{aligned} &\imath_P \circ \imath_Q (\mathrm{d}F_1 \wedge \cdots \wedge \mathrm{d}F_k) \\ &= \imath_P \left(\sum_{\sigma \in S_{q,k-q}} \mathrm{sgn}(\sigma) Q[F_{\sigma(1)}, \ldots, F_{\sigma(q)}] \mathrm{d}F_{\sigma(q+1)} \wedge \cdots \wedge \mathrm{d}F_{\sigma(k)} \right) \end{aligned}$$

$$= \sum_{\tau \in S_{q,p,k-q-p}} \operatorname{sgn}(\tau) Q[F_{\tau(1)}, \dots, F_{\tau(q)}] P[F_{\tau(q+1)}, \dots, F_{\tau(q+p)}] \\ \mathrm{d}F_{\tau(q+p+1)} \wedge \dots \wedge \mathrm{d}F_{\tau(k)} ,$$

where $S_{q,p,k-p-q}$ is the set of all permutations σ of the set $\{1,\dots,k\}$ which are increasing on the intervals $1,\dots,q$ and $q+1,\dots,q+p$ and $q+p+1,\dots,k$. On the other hand, (3.29) and (3.3) lead to

$$\begin{aligned} &\imath_{Q\wedge P}(\mathrm{d}F_1 \wedge \dots \wedge \mathrm{d}F_k) \\ &= \sum_{\rho \in S_{q+p,k-q-p}} \operatorname{sgn}(\rho)(Q \wedge P)[F_{\rho(1)}, \dots, F_{\rho(q+p)}] \mathrm{d}F_{\rho(q+p+1)} \wedge \dots \wedge \mathrm{d}F_{\rho(k)} \\ &= \sum_{\tau \in S_{q,p,k-q-p}} \operatorname{sgn}(\tau) Q[F_{\tau(1)}, \dots, F_{\tau(q)}] P[F_{\tau(q+1)}, \dots, F_{\tau(q+p)}] \\ &\qquad \mathrm{d}F_{\tau(q+p+1)} \wedge \dots \wedge \mathrm{d}F_{\tau(k)} , \end{aligned}$$

which yields the third proof. The point of this computation is that an element of $S_{q,p,k-p-q}$ can be uniquely obtained from a shuffle $\sigma \in S_{q,k-q}$, followed by a shuffle of the set $\sigma(q+1),\dots,\sigma(k)$, or from a shuffle $\rho \in S_{q+p,k-p-q}$, followed by a shuffle of the set $\rho(1),\dots,\rho(q+p)$.

The proof of *(2)* is immediate from *(1)*, since the graded commutator is given by $[\imath_P, \imath_Q] = \imath_P \circ \imath_Q - (-1)^{pq} \imath_Q \circ \imath_P = \imath_{Q\wedge P} - (-1)^{pq} \imath_{P\wedge Q}$, and since $P \wedge Q = (-1)^{pq} Q \wedge P$ (graded commutativity, see (3.4)).

We now prove *(3)*. Since, for every $Q \in \mathfrak{X}^\bullet(\mathscr{A})$, the map $\imath_Q$ is $\mathscr{A}$-linear and the wedge product is also $\mathscr{A}$-bilinear, we can suppose that ω is of the form $\omega = \mathrm{d}F_1 \wedge \dots \wedge \mathrm{d}F_k$, with $k \in \mathbb{N}^*$, $k \geqslant p-1$ and $F_1,\dots,F_k \in \mathscr{A}$. In this case,

$$\begin{aligned} &\imath_P(\mathrm{d}F \wedge \mathrm{d}F_1 \wedge \dots \wedge \mathrm{d}F_k) \\ &= \sum_{\sigma \in S_{p-1,k-p+1}} \operatorname{sgn}(\sigma) P[F, F_{\sigma(1)}, \dots, F_{\sigma(p-1)}] \, \mathrm{d}F_{\sigma(p)} \wedge \dots \wedge \mathrm{d}F_{\sigma(k)} \\ &\quad + \sum_{\tau \in S_{p,k-p}} (-1)^p \operatorname{sgn}(\tau) P[F_{\tau(1)}, \dots, F_{\tau(p)}] \, \mathrm{d}F \wedge \mathrm{d}F_{\tau(p+1)} \wedge \dots \wedge \mathrm{d}F_{\tau(k)} , \end{aligned}$$

since every $(p, k-p+1)$-shuffle of $\{0,\dots,k\}$ has 0 as the first element in its first part or in its second part. The first sum is exactly $\imath_{(\imath_F P)}(\mathrm{d}F_1 \wedge \dots \wedge \mathrm{d}F_k) = \imath_{(\imath_F P)}\omega$, while the second sum is given by $(-1)^p \mathrm{d}F \wedge \imath_P(\mathrm{d}F_1 \wedge \dots \wedge \mathrm{d}F_k) = (-1)^p \mathrm{d}F \wedge \imath_P \omega$. This yields a proof of Eq. (3.32). □

For $P \in \mathfrak{X}^p(\mathscr{A})$, the operation $\imath_P : \Omega^\bullet(\mathscr{A}) \to \Omega^{\bullet-p}(\mathscr{A})$ is closely related to the natural extension of $P : \wedge^p \mathscr{A} \to \mathscr{A}$ to a coderivation

$$\tilde{P} : \wedge^\bullet \mathscr{A} \to \wedge^{\bullet-p+1}(\mathscr{A})$$

of degree $p-1$ of $(\wedge^\bullet \mathscr{A}, \Delta)$. It is defined by

$$\tilde{P}(F_1 \wedge \cdots \wedge F_k) := \sum_{\tau \in S_{p,k-p}} \operatorname{sgn}(\tau) P[F_{\tau(1)}, \ldots, F_{\tau(p)}] \wedge F_{\tau(p+1)} \wedge \cdots \wedge F_{\tau(k)}\,, \quad (3.35)$$

for $k \geqslant p$ and $\tilde{P}(F_1 \wedge \cdots \wedge F_k) := 0$ otherwise (see Example A.4 in Appendix A). Comparing this formula with (3.29), it is clear that $\mathrm{d} \circ \iota_P$ and $\tilde{P}$ correspond under the map $\wedge^\bullet \mathrm{d} : \wedge^\bullet \mathscr{A} \to \Omega^\bullet(\mathscr{A})$, i.e., the following diagram is commutative:

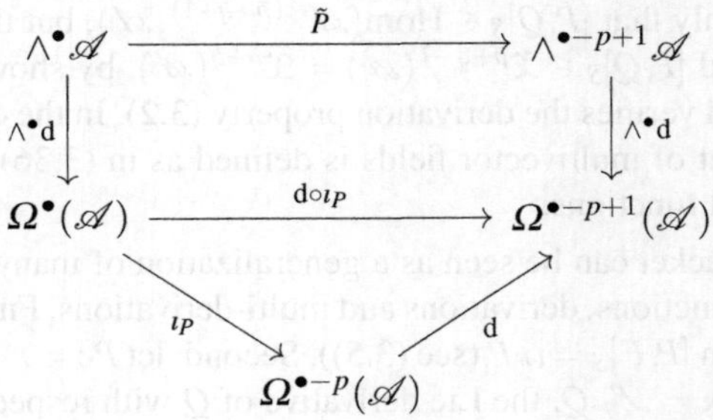

As before, the algebraic constructions in this section, in particular the internal product, can easily be transcribed for the case of a manifolds. For a manifold M, the internal product is an action of $\mathfrak{X}^\bullet(M)$ on $\Omega^\bullet(M)$, where we recall that $\mathfrak{X}^\bullet(M)$ is the $\mathscr{F}(M)$-module of multivector fields on M and $\Omega^\bullet(M)$ is the $\mathscr{F}(M)$-module of differential forms on M. This action is defined as in (3.29), the functions $F_1, \ldots, F_k$ being now arbitrary *local* functions on M. The formulas in Proposition 3.4 remain valid, with $P, Q \in \mathfrak{X}^\bullet(M)$, with $\omega \in \Omega^\bullet(M)$ and $F \in \mathscr{F}(M)$.

3.3.2 The Schouten Bracket

We now introduce the *Schouten bracket*, which is a product of skew-symmetric multi-derivations, based on the operation of composition. It is a family of maps

$$[\cdot,\cdot]_S : \mathfrak{X}^p(\mathscr{A}) \times \mathfrak{X}^q(\mathscr{A}) \to \mathfrak{X}^{p+q-1}(\mathscr{A})\,,$$

for $p, q \in \mathbb{N}$, which therefore only respects the grading up to a shift over 1, suggesting to shift the degree of all multi-derivations by 1. Namely, we define $\overline{\mathfrak{X}}^{p-1}(\mathscr{A}) := \mathfrak{X}^p(\mathscr{A})$, for $p \in \mathbb{N}$, and we call $\bar{p} := p - 1$ the *shifted degree* of an element P of $\mathfrak{X}^p(\mathscr{A}) = \overline{\mathfrak{X}}^{\bar{p}}(\mathscr{A})$. Thus, elements of $\mathscr{A}$ have shifted degree -1, derivations of $\mathscr{A}$ have shifted degree 0, and so on. Notice that $p + q - 1 = \bar{p} + \bar{q}$, so that the Schouten bracket is a family of maps

$$[\cdot,\cdot]_S : \overline{\mathfrak{X}}^{\bar{p}}(\mathscr{A}) \times \overline{\mathfrak{X}}^{\bar{q}}(\mathscr{A}) \to \overline{\mathfrak{X}}^{\bar{p}+\bar{q}}(\mathscr{A})\,,$$

for $\bar{p}, \bar{q} \in \mathbb{N} \cup \{-1\}$, hence respects the (new) grading. It is defined, for $P \in \overline{\mathfrak{X}}^{\bar{p}}(\mathscr{A})$ and $Q \in \overline{\mathfrak{X}}^{\bar{q}}(\mathscr{A})$, and for $F_1, \ldots, F_{\bar{p}+\bar{q}+1} \in \mathscr{A}$, by

$$
\begin{aligned}
&[P,Q]_S[F_1,\ldots,F_{\bar p+\bar q+1}] \\
&\quad = \sum_{\sigma\in S_{q,\bar p}} \operatorname{sgn}(\sigma) P\left[Q[F_{\sigma(1)},\ldots,F_{\sigma(q)}],F_{\sigma(q+1)},\ldots,F_{\sigma(q+\bar p)}\right] \\
&\qquad -(-1)^{\bar p\bar q}\sum_{\sigma\in S_{p,\bar q}} \operatorname{sgn}(\sigma) Q\left[P[F_{\sigma(1)},\ldots,F_{\sigma(p)}],F_{\sigma(p+1)},\ldots,F_{\sigma(p+\bar q)}\right] .
\end{aligned}
\tag{3.36}
$$

A priori, we have only that $[P,Q]_S \in \operatorname{Hom}(\mathscr{A}^{\otimes(\bar p+\bar q+1)},\mathscr{A})$, but the reader will easily check that indeed $[P,Q]_S \in \mathfrak{X}^{\bar p+\bar q+1}(\mathscr{A}) = \overline{\mathfrak{X}}^{\bar p+\bar q}(\mathscr{A})$, by showing that $[P,Q]_S$ is skew-symmetric and verifies the derivation property (3.2). In the case of a manifold, the Schouten bracket of multivector fields is defined as in (3.36), but replacing the functions F_i by local functions.

The Schouten bracket can be seen as a generalization of many classical elementary operations on functions, derivations and multi-derivations. First, let $Q := F \in \mathscr{A}$ and $P \in \mathfrak{X}^p(\mathscr{A})$, then $[P,F]_S = \imath_F P$ (see (3.5)). Second, let $P := \mathscr{V} \in \mathfrak{X}^1(\mathscr{A})$ and $Q \in \mathfrak{X}^q(\mathscr{A})$, then $[\mathscr{V},Q]_S = \mathscr{L}_{\mathscr{V}} Q$, the Lie derivative of Q with respect to $\mathscr{V}$ (see (3.7)). Third, the Schouten bracket of two skew-symmetric biderivations $P,Q \in \mathfrak{X}^2(\mathscr{A})$ is given by

$$
[P,Q]_S[F_1,F_2,F_3] = P[Q[F_1,F_2],F_3] + Q[P[F_1,F_2],F_3] + \circlearrowleft (F_1,F_2,F_3) , \tag{3.37}
$$

where $F_1,F_2,F_3 \in \mathscr{A}$. This leads to the following result.

Proposition 3.5. *Let $\mathscr{A}$ be a commutative associative algebra. If P is a skew-symmetric biderivation of $\mathscr{A}$, i.e., $P \in \mathfrak{X}^2(\mathscr{A})$, then P defines a Poisson bracket on $\mathscr{A}$ if and only if $[P,P]_S = 0$.*

Similarly, in the case of a manifold M, a bivector field P on M defines a Poisson structure on M if and only if $[P,P]_S = 0$.

We say that $P,Q \in \mathfrak{X}^2(\mathscr{A})$ are *compatible* if $[P,Q]_S = 0$. It follows that two Poisson brackets P and Q are compatible if and only if their sum (equivalently, an arbitrary linear combination with non-zero coefficients) is a Poisson bracket.

In the following proposition, we express the Schouten bracket in terms of the de Rham differential.

Proposition 3.6. *Let $\mathscr{A}$ be a commutative associative algebra. If P and Q are two skew-symmetric multi-derivations of $\mathscr{A}$, then*

$$
[[\imath_P,\mathrm{d}],\imath_Q] = \imath_{[P,Q]_S} . \tag{3.38}
$$

This formula is called Cartan's formula.

Proof. We prove Cartan's formula for elements $P \in \mathfrak{X}^p(\mathscr{A})$ and $Q \in \mathfrak{X}^q(\mathscr{A})$. Let us first show that the linear map $[[\imath_P,\mathrm{d}],\imath_Q]$ is $\mathscr{A}$-linear. We take $F \in \mathscr{A}$ and $\omega \in \Omega^\bullet(\mathscr{A})$. Then, using (3.32),

$$
[\imath_P,\mathrm{d}](F\omega) = \imath_P(\mathrm{d}F\wedge\omega) + F\imath_P\mathrm{d}\omega - (-1)^p\mathrm{d}(F\imath_P\omega)
$$

$$= \imath_{(\imath_F P)}\omega + F\imath_P \mathrm{d}\omega - (-1)^p F \mathrm{d}\imath_P\omega$$
$$= F[\imath_P, \mathrm{d}]\omega + \imath_{(\imath_F P)}\omega\,,$$

so that

$$\begin{aligned}
&[[\imath_P, \mathrm{d}], \imath_Q](F\omega)\\
&= [\imath_P, \mathrm{d}](F\imath_Q\omega) - (-1)^{\bar{p}q}\imath_Q[\imath_P, \mathrm{d}](F\omega)\\
&= F[\imath_P, \mathrm{d}](\imath_Q\omega) + \imath_{(\imath_F P)}\imath_Q\omega - (-1)^{\bar{p}q}F\imath_Q[\imath_P, \mathrm{d}]\omega - (-1)^{\bar{p}q}\imath_Q\imath_{(\imath_F P)}\omega\\
&= F[[\imath_P, \mathrm{d}], \imath_Q]\omega\,,
\end{aligned}$$

where we have used in the last step that $\left[\imath_{(\imath_F P)}, \imath_Q\right] = 0$ (Proposition 3.4). This shows that

$$[[\imath_P, \mathrm{d}]\,, \imath_Q] : \Omega^\bullet(\mathscr{A}) \to \Omega^{\bullet - p - q + 1}(\mathscr{A})$$

is $\mathscr{A}$-linear. It follows that, in order to establish Cartan's formula (3.38), we only need to show that

$$\imath_P \mathrm{d}\imath_Q\omega - (-1)^p \mathrm{d}\imath_P\imath_Q\omega - (-1)^{\bar{p}\bar{q}}\imath_Q \mathrm{d}\imath_P\omega = \imath_{[P,Q]_S}\omega\,, \tag{3.39}$$

for ω of the form $\omega = \mathrm{d}F_1 \wedge \cdots \wedge \mathrm{d}F_k$, where $F_1, \ldots, F_k \in \mathscr{A}$ and where $k \in \mathbb{N}$.

We do this by induction on k, starting with $k = 0$, so that we take $\omega := 1 \in \mathscr{A}$. Notice that both sides in (3.39) have degree $k - p - q + 1$, so that when $k = 0$, both sides of (3.39) are zero as soon as $p + q > 1$. If $p + q = 1$, we may suppose that $p = 1$ and $q = 0$ (the case $p = 0$ and $q = 1$ is obtained by symmetry): then $P := \mathscr{V} \in \mathfrak{X}^1(\mathscr{A})$ and $Q := F \in \mathscr{A}$ and the right-hand side of (3.39) (with $\omega = 1$) is given by $\imath_{[\mathscr{V},F]_S}1 = \imath_{\mathscr{V}[F]}1 = \mathscr{V}[F]$, while the left-hand side is given by $\imath_{\mathscr{V}}\mathrm{d}F + \mathrm{d}\imath_{\mathscr{V}}F - \imath_F\mathrm{d}\imath_{\mathscr{V}}1 = \imath_{\mathscr{V}}\mathrm{d}F = \mathscr{V}[F]$, as $\imath_{\mathscr{V}}F = 0$ for degree reasons. If $p + q = 0$, then $p = q = 0$ so that $P := F$ and $Q := G$, with $F, G \in \mathscr{A}$. Then $[F, G]_S = 0$, while the left-hand side of (3.39) reduces to $\imath_F\mathrm{d}G - \mathrm{d}(FG) + \imath_G\mathrm{d}F = F\mathrm{d}G - \mathrm{d}(FG) + G\mathrm{d}F = 0$. This shows that (3.39) holds for ω of degree 0, for all P and Q.

We now assume that (3.39) holds for $\omega \in \Omega^k(\mathscr{A})$ and we show that it holds for $\mathrm{d}F \wedge \omega$, where $F \in \mathscr{A}$ is arbitrary. On the one hand, (3.32) and the induction hypothesis imply that

$$\begin{aligned}
&\imath_{[P,Q]_S}(\mathrm{d}F \wedge \omega)\\
&= \imath_{(\imath_F[P,Q]_S)}\omega - (-1)^{p+q}\mathrm{d}F \wedge \imath_{[P,Q]_S}\omega\\
&= \imath_{(\imath_F[P,Q]_S)}\omega - (-1)^{p+q}\mathrm{d}F \wedge ([[\imath_P, \mathrm{d}]\,, \imath_Q]\,\omega)\\
&= (-1)^{\bar{q}}\imath_{[\imath_F P,Q]_S}\omega + \imath_{[P,\imath_F Q]_S}\omega - (-1)^{p+q}\mathrm{d}F \wedge ([[\imath_P, \mathrm{d}]\,, \imath_Q]\,\omega)\,,
\end{aligned} \tag{3.40}$$

where we have used

$$\imath_F[P,Q]_S = (-1)^{\bar{q}}\,[\imath_F P, Q]_S + [P, \imath_F Q]_S$$

to obtain the last line, which is similar to (3.6), and is a direct consequence of the definition of $[P,Q]_S$. Equation (3.40) has to be compared to $[[\iota_P,\mathrm{d}]\,,\iota_Q]\,(\mathrm{d}F\wedge\omega)$, i.e., to

$$(\iota_P \mathrm{d}\iota_Q - (-1)^p \mathrm{d}\iota_{Q\wedge P} - (-1)^{\bar{p}\bar{q}}\iota_Q \mathrm{d}\iota_P)(\mathrm{d}F\wedge\omega)\,. \tag{3.41}$$

Using (3.32) twice and (3.25) once, we find

$$\begin{aligned}\iota_P \mathrm{d}\iota_Q(\mathrm{d}F\wedge\omega) &= \iota_P \mathrm{d}\iota_{(\iota_F Q)}\omega - (-1)^q \iota_P(\mathrm{d}F\wedge \mathrm{d}\iota_Q\omega)\\ &= \iota_P \mathrm{d}\iota_{(\iota_F Q)}\omega - (-1)^q \iota_{(\iota_F P)}\mathrm{d}\iota_Q\omega - (-1)^{p+q}\mathrm{d}F\wedge\iota_P \mathrm{d}\iota_Q\omega\,.\end{aligned}$$

Similarly, we compute, using (3.32) and (3.6)

$$\begin{aligned}\mathrm{d}\iota_{Q\wedge P}(\mathrm{d}F\wedge\omega) &= \mathrm{d}\iota_{(\iota_F(Q\wedge P))}\omega + (-1)^{q+\bar{p}}\mathrm{d}F\wedge \mathrm{d}\iota_{Q\wedge P}\omega\\ &= \mathrm{d}\iota_{(\iota_F Q\wedge P)}\omega + (-1)^q \mathrm{d}\iota_{(Q\wedge\iota_F P)}\omega + (-1)^{q+\bar{p}}\mathrm{d}F\wedge \mathrm{d}\iota_{Q\wedge P}\omega\,.\end{aligned}$$

If we substitute these results in (3.41), then we find nine terms which can, by using the induction hypothesis twice, be written as follows:

$$\begin{aligned}(\iota_P \mathrm{d}\iota_{(\iota_F Q)} - (-1)^p \mathrm{d}\iota_{(\iota_F Q\wedge P)} - (-1)^{\bar{p}q}\iota_{(\iota_F Q)}\mathrm{d}\iota_P)\,\omega &= \iota_{[P,\iota_F Q]_S}\,\omega\\ (-1)^{\bar{q}}(\iota_{(\iota_F P)}\mathrm{d}\iota_Q + (-1)^p \mathrm{d}\iota_{(Q\wedge\iota_F P)} - (-1)^{p\bar{q}}\iota_Q \mathrm{d}\iota_{(\iota_F P)})\,\omega &= (-1)^{\bar{q}}\iota_{[\iota_F P,Q]_S}\,\omega\\ \mathrm{d}F\wedge(-(-1)^{p+q}\iota_P \mathrm{d}\iota_Q + (-1)^q \mathrm{d}\iota_{Q\wedge P} - (-1)^{pq}\iota_Q \mathrm{d}\iota_P)\,\omega &=\\ &\quad -(-1)^{p+q}\mathrm{d}F\wedge([[\iota_P,\mathrm{d}]\,,\iota_Q]\,\omega)\,.\end{aligned}$$

This yields precisely the three terms in (3.40). □

Cartan's formula leads to a simple proof of the main properties of the Schouten bracket, as given in the following proposition.

Proposition 3.7. *Let $\mathscr{A}$ be a commutative associative algebra. The Schouten bracket*

$$[\cdot\,,\cdot]_S : \overline{\mathfrak{X}}^\bullet(\mathscr{A})\times\overline{\mathfrak{X}}^\bullet(\mathscr{A})\to\overline{\mathfrak{X}}^\bullet(\mathscr{A})\,,$$

defines a graded Lie algebra structure on $\overline{\mathfrak{X}}^\bullet(\mathscr{A})$, meaning that, for all homogeneous elements P,Q and R in $\overline{\mathfrak{X}}^\bullet(\mathscr{A})$, of respective shifted degree $\bar{p}$, $\bar{q}$ and $\bar{r}$,

(1) $[P,Q]_S\in\overline{\mathfrak{X}}^{\bar{p}+\bar{q}}(\mathscr{A})$;
(2) $[P,Q]_S = -(-1)^{\bar{p}\bar{q}}[Q,P]_S$, *(graded skew-symmetry);*
(3) $(-1)^{\bar{p}\bar{r}}[P,[Q,R]_S]_S + \circlearrowleft(P,Q,R) = 0$ *(graded Jacobi identity).*

Moreover, the associative product $\wedge$ and the Lie bracket $[\cdot\,,\cdot]_S$ are compatible in the sense that $[\cdot,P]_S$ is a graded derivation of $(\mathfrak{X}^\bullet(\mathscr{A}),\wedge)$ of degree $\bar{p}$:

$$[Q\wedge R,P]_S = [Q,P]_S\wedge R + (-1)^{\bar{p}q}Q\wedge[R,P]_S\,, \tag{3.42}$$

(graded Leibniz identity),

with P,Q and R as above. In view of the graded skew-symmetry of $[\cdot\,,\cdot]_S$, one also has, for such elements,

$$[P, Q \wedge R]_S = Q \wedge [P,R]_S + (-1)^{\bar{p}r}[P,Q]_S \wedge R\,. \tag{3.43}$$

Proof. *(1)* and *(2)* follow at once from the explicit formula for the Schouten bracket (Eq. (3.36)). Notice that the graded skew-symmetry also follows from Cartan's formula (3.38). Indeed, the graded Jacobi identity for the graded commutator $[\cdot\,,\cdot]$,

$$(-1)^{pq}[[\iota_P, \mathrm{d}], \iota_Q] + (-1)^p[[\mathrm{d}, \iota_Q], \iota_P] + (-1)^q[[\iota_Q, \iota_P], \mathrm{d}] = 0\,,$$

leads, since $[\iota_Q, \iota_P] = 0$, to

$$\iota_{[Q,P]_S} = [[\iota_Q, \mathrm{d}], \iota_P] = -(-1)^{\bar{p}\bar{q}}[[\iota_P, \mathrm{d}], \iota_Q] = -(-1)^{\bar{p}\bar{q}}\iota_{[P,Q]_S}$$

which proves that $[Q,P]_S = -(-1)^{\bar{p}\bar{q}}[P,Q]_S$, since $P \mapsto \iota_P$ is injective. We next prove the graded Jacobi identity for $[\cdot\,,\cdot]_S$. Let $P \in \mathfrak{X}^p(\mathscr{A})$ and $Q \in \mathfrak{X}^q(\mathscr{A})$. We associate to these multi-derivations, the coderivations $\tilde{P}$ (of degree $\bar{p}$) and $\tilde{Q}$ (of degree $\bar{q}$) of the coalgebra $(\wedge^\bullet \mathscr{A}, \Delta)$, as in (3.35). We claim that

$$[\tilde{P}, \tilde{Q}] = \widetilde{[P,Q]_S}\,. \tag{3.44}$$

Since both members are coderivations of $\wedge^\bullet \mathscr{A}$ of degree $p+q-2$, it suffices to verify that they agree in degree $k := p+q-1$ (see Example A.4 in Appendix A). For $F_1, \dots, F_k \in \mathscr{A}$, we have that

$$\begin{aligned}
&(\tilde{P} \circ \tilde{Q})(F_1, \dots, F_k) \\
&= \tilde{P}\left(\sum_{\sigma \in S_{q,\bar{p}}} \mathrm{sgn}(\sigma) Q[F_{\sigma(1)}, \dots, F_{\sigma(q)}] \wedge F_{\sigma(q+1)} \wedge \dots \wedge F_{\sigma(k)}\right) \\
&= \sum_{\sigma \in S_{q,\bar{p}}} \mathrm{sgn}(\sigma) P\left[Q[F_{\sigma(1)}, \dots, F_{\sigma(q)}], F_{\sigma(q+1)}, \dots, F_{\sigma(k)}\right],
\end{aligned}$$

which leads, in view of (3.36), to

$$[\tilde{P}, \tilde{Q}](F_1, \dots, F_k) = [P,Q]_S[F_1, \dots, F_k] = \widetilde{[P,Q]_S}(F_1, \dots, F_k)\,.$$

This shows (3.44). As a consequence, for all skew-symmetric multi-derivations P, Q and R,

$$([P,[Q,R]_S]_S)^\sim = \left[\tilde{P}, \widetilde{[Q,R]_S}\right] = [\tilde{P}, [\tilde{Q}, \tilde{R}]]$$

and the graded Jacobi identity for the Schouten bracket follows from the graded Jacobi identity for the commutator of coderivations.

The Leibniz property of the Schouten bracket follows easily from Cartan's formula. Indeed, for P, Q and R as in (3.43), Cartan's formula and *(1)* of Proposition 3.4 imply that

$$\iota_{[P, Q \wedge R]_S} = [[\iota_P, \mathrm{d}], \iota_{Q \wedge R}] = [[\iota_P, \mathrm{d}], \iota_R \circ \iota_Q]$$

$$= [[\iota_P, \mathrm{d}], \iota_R] \circ \iota_Q + (-1)^{r\bar{p}} \iota_R \circ [[\iota_P, \mathrm{d}], \iota_Q]$$
$$= \iota_{Q\wedge[P,R]_S} + (-1)^{r\bar{p}} \iota_{[P,Q]_S\wedge R} \,,$$

which yields (3.43), and hence also (3.42). We have used the formula[3]

$$[\psi, \phi_1 \circ \phi_2] = [\psi, \phi_1] \circ \phi_2 + (-1)^{d_1 d} \phi_1 \circ [\psi, \phi_2] \,, \tag{3.45}$$

which is valid for graded linear maps ϕ_1, ϕ_2 and ψ, of respective degrees d_1, d_2 and d. □

We have equipped $\mathfrak{X}^\bullet(\mathscr{A})$ with two graded algebra structures: the wedge product $\wedge$, which is associative and graded commutative, and the Schouten bracket $[\cdot\,,\cdot]_S$, which is a graded Lie bracket (with respect to a different grading). Moreover, we have seen that these two products are, in a sense, compatible. Formalizing these properties leads to the notion of a Gerstenhaber algebra.

Definition 3.8. Let $V^\bullet = \oplus_{p\in\mathbb{Z}} V_p$ be a graded $\mathbb{F}$-vector space and denote by $\bar{p} := p-1$ the shifted degree of $P \in V_p$. Suppose that V is equipped with two multiplications $\wedge$ and $[\cdot\,,\cdot]$, such that

(1) $(V^\bullet, \wedge)$ is an associative, graded commutative algebra over $\mathbb{F}$;
(2) $(V^\bullet, [\cdot\,,\cdot])$ is a graded Lie algebra over $\mathbb{F}$, with respect to the shifted grading;
(3) The two algebra structures are compatible in the sense that

$$[Q \wedge R, P] = [Q, P] \wedge R + (-1)^{\bar{p}q} Q \wedge [R, P] \,,$$

where P, Q and R are arbitrary homogeneous elements of $V^\bullet$, of degree p, q and r.

Then $(V^\bullet, \wedge, [\cdot\,,\cdot])$ is called a *Gerstenhaber algebra* (over $\mathbb{F}$).

Comparing the three items in Definition 3.8 to the ones in Definition 1.1, one sees that the notion of a Gerstenhaber algebra is a natural graded analog of the notion of a Poisson algebra.

Proposition 3.7 can be reformulated by saying that for every commutative associative $\mathbb{F}$-algebra $\mathscr{A}$, the graded $\mathbb{F}$-vector space $\mathfrak{X}^\bullet(\mathscr{A})$, equipped with the wedge product and the Schouten bracket, is a Gerstenhaber algebra (over $\mathbb{F}$). In the case of a manifold M, the graded $\mathbb{F}$-vector space $\mathfrak{X}^\bullet(M)$ of multivector fields on M, equipped with the wedge product and the Schouten bracket, is a Gerstenhaber algebra.

3.3.3 The Algebraic Schouten Bracket

For any Lie algebra $(\mathfrak{g}, [\cdot\,,\cdot])$ there is a graded Lie algebra structure on $\wedge^\bullet \mathfrak{g}$, which is the natural extension of the Lie bracket $[\cdot\,,\cdot]$ on $\mathfrak{g}$. It is very similar to the Schouten

[3] The formula is a graded version of the well-known matrix identity $[x, yz] = [x, y]z + y[x, z]$; also, it is half of the Jacobi identity (A.13).

bracket on multi-derivations, in particular it only becomes a graded Lie algebra structure when we shift the grading: for $p \in \mathbb{N}^*$, we let $\overline{\wedge}^{p-1}\mathfrak{g} := \wedge^p\mathfrak{g}$ and we let $\bar{p} := p-1$. The definition and main properties of the graded bracket on the graded vector space $\overline{\wedge}^\bullet\mathfrak{g} := \oplus_{k\in\mathbb{N}}\overline{\wedge}^k\mathfrak{g}$ are given in the following proposition.

Proposition 3.9. *Let* $(\mathfrak{g}, [\cdot\,,\cdot])$ *be a Lie algebra and let* $[\![\cdot\,,\cdot]\!]$ *denote the bracket on* $\overline{\wedge}^\bullet\mathfrak{g}$, *defined for all pairs of monomials by*

$$\begin{aligned}&[\![x_1\wedge\cdots\wedge x_p, y_1\wedge\cdots\wedge y_q]\!] &(3.46)\\ &:= (-1)^{\bar{p}\bar{q}}\sum_{i=1}^{p}\sum_{j=1}^{q}(-1)^{i+j}\,[x_i,y_j]\wedge x_1\wedge\cdots\wedge\widehat{x_i}\wedge\cdots\wedge x_p\wedge y_1\wedge\cdots\wedge\widehat{y_j}\wedge\cdots\wedge y_q\,.\end{aligned}$$

Then $(\wedge^\bullet\mathfrak{g}, \wedge, [\![\cdot\,,\cdot]\!])$ *is a Gerstenhaber algebra. The graded Lie bracket* $[\![\cdot\,,\cdot]\!]$ *on* $\wedge^\bullet\mathfrak{g}$ *is called the* algebraic Schouten bracket.

Proof. The proof of each one of the properties of $(\wedge^\bullet\mathfrak{g}, \wedge, [\![\cdot\,,\cdot]\!])$ is either immediate from the definitions or follows from a direct computation. Using the graded Leibniz identity

$$[\![Y\wedge Z, X]\!] = [\![Y,X]\!]\wedge Z + (-1)^{\bar{p}q}\,Y\wedge[\![Z,X]\!]\,, \tag{3.47}$$

valid for all $X\in\overline{\wedge}^{\bar{p}}\mathfrak{g}$, $Y\in\overline{\wedge}^{q}\mathfrak{g}$ and $Z\in\overline{\wedge}^\bullet\mathfrak{g}$, the graded Jacobi identity can be proven easily by recursion, since for three elements of shifted degree zero, it is just the Jacobi identity in $\mathfrak{g}$, while if at least one element is a monomial of higher shifted degree, say X, it can be written as $X = X'\wedge x$, with $x\in\mathfrak{g}$ and $X'\in\wedge^\bullet\mathfrak{g}$, of lower degree. For a quick proof of (3.47), one clearly only needs to check that the signs in the right-hand side of (3.47) are correct, which follows at once from the graded skew-symmetry of $[\![\cdot\,,\cdot]\!]$. □

3.3.4 The (Generalized) Lie Derivative

We have already encountered the Lie derivative of a multi-derivation with respect to a derivation in Section 3.1.2; it is the algebraic analog of the Lie derivative of a multivector field with respect to a vector field, as briefly discussed in Section 3.1.3. As we have shown in Section 3.3.2, the Lie derivative can be written in terms of the Schouten bracket, namely $\mathscr{L}_\mathscr{V}Q = [\mathscr{V}, Q]_S$, for $\mathscr{V}\in\mathfrak{X}^1(\mathscr{A})$ and $Q\in\mathfrak{X}^q(\mathscr{A})$ (as before, $\mathscr{A}$ is a commutative associative algebra). Since $[\cdot\,,\cdot]_S$ is a graded derivation of degree one in its first argument (see (3.42)), taking the Schouten bracket with a given p-derivation P, is a natural generalization of the Lie derivative, called *generalized Lie derivative*, or simply *Lie derivative*. Explicitly, for $P\in\mathfrak{X}^p(\mathscr{A})$, the generalized Lie derivative $\mathscr{L}_P$ is given by the graded linear map (of degree $\bar{p} = p-1$)

$$\begin{aligned}\mathscr{L}_P : \overline{\mathfrak{X}}^\bullet(\mathscr{A}) &\to \overline{\mathfrak{X}}^{\bullet+\bar{p}}(\mathscr{A})\\ Q &\mapsto [P,Q]_S\,.\end{aligned} \tag{3.48}$$

The properties of the Schouten bracket, given in Proposition 3.7, lead at once to the following properties for this generalized Lie derivative.

Proposition 3.10. *Let $\mathscr{A}$ be a commutative associative algebra and let P, Q and R be homogeneous elements of $\mathfrak{X}^\bullet(\mathscr{A})$ of shifted degrees $\bar{p}, \bar{q}$ and $\bar{r}$. Then*

(1) $\mathscr{L}_P Q \in \mathfrak{X}^{\bar{p}+\bar{q}}(\mathscr{A})$;
(2) $\mathscr{L}_P(Q \wedge R) = Q \wedge \mathscr{L}_P R + (-1)^{\bar{p}r}(\mathscr{L}_P Q) \wedge R$;
(3) $\mathscr{L}_P [Q,R]_S = [\mathscr{L}_P Q, R]_S + (-1)^{\bar{p}\bar{q}} [Q, \mathscr{L}_P R]_S$.

In the geometrical context, the generalized Lie derivative becomes a Lie derivative with respect to an arbitrary p-vector field P; it associates to a q-vector field a $(q+p-1)$-vector field, i.e., it raises the degree of a multivector field by $\bar{p} = p-1$.

Since the calculus of differential forms has been more popularized than the calculus of multivector fields, the notion of the Lie derivative of a differential form (rather than of a multivector field) with respect to a vector field is better known. We recall this definition and generalize it to the Lie derivative of a differential form with respect to a multivector field. We start from the geometrical definition, which is similar to the geometrical definition (3.11). Namely, let $\mathscr{V}$ be a vector field on a manifold M and let ω be a differential p-form on M. Then $\mathscr{L}_\mathscr{V} \omega$ is the differential p-form on M, whose value at $m \in M$ is given by

$$(\mathscr{L}_\mathscr{V} \omega)_m = \frac{\mathrm{d}}{\mathrm{d}t}_{|t=0} \Phi_t^* \left(\omega_{\Phi_t(m)}\right) ,$$

where Φ_t is the local flow of $\mathscr{V}$ and Φ_t^* the pull-back of ω by Φ_t, so that $\Phi_t^*(\omega_{\Phi_t(m)})$ is a differential p-form on $T_m M$ (depending on t). The classical[4] Cartan formula says that

$$\mathscr{L}_\mathscr{V} = \imath_\mathscr{V} \circ \mathrm{d} + \mathrm{d} \circ \imath_\mathscr{V} , \tag{3.49}$$

a formula which we can take as a definition of the Lie derivative in the algebraic setting. Writing (3.49) as $\mathscr{L}_\mathscr{V} = [\imath_\mathscr{V}, \mathrm{d}]$, where we recall that $[\cdot,\cdot]$ denotes the graded commutator (see (A.12)), leads to the following natural generalization of the Lie derivative to an action of a multivector field (or multi-derivation) on a differential form (Kähler form), namely we define the *Lie derivative* with respect to a skew-symmetric multi-derivation $P \in \mathfrak{X}^p(\mathscr{A})$ as the graded $\mathbb{F}$-linear map $\Omega^\bullet(\mathscr{A}) \to \Omega^{\bullet-\bar{p}}(\mathscr{A})$ of degree $-\bar{p}$, given by

$$\mathscr{L}_P := [\imath_P, \mathrm{d}] ,$$

where we recall that $\imath_P$ was defined in (3.29). Explicitly, this means that for $\omega \in \Omega^\bullet(\mathscr{A})$,

$$\mathscr{L}_P \omega := \imath_P(\mathrm{d}\omega) - (-1)^p \mathrm{d}(\imath_P \omega) . \tag{3.50}$$

Comparing the Definitions (3.48) and (3.50) of the generalized Lie derivative, one notices that there is some ambiguity in the definition of the generalized Lie

---

[4] The *classical* Cartan formula should not be confused with Cartan's formula (3.38), which relates the Schouten bracket to the de Rham differential.

derivative of a function with respect to a p-vector field, when $p \geqslant 2$. In fact, a function $F \in \mathscr{A}$ can be viewed as an element of $\mathfrak{X}^0(\mathscr{A})$ or as an element of $\Omega^0(\mathscr{A})$. According to (3.48), $\mathscr{L}_P F = \iota_F(P) \in \mathfrak{X}^{p-1}(\mathscr{A})$, which is a priori different from zero; however, according to (3.50), $\mathscr{L}_P(F) \in \Omega^{1-p}(\mathscr{A}) = \{0\}$. Therefore, when applying the generalized Lie derivative to functions, one should decide from the context and/or from the problem at hand, whether Definition (3.48) is used, or Definition (3.50).

The main properties of the Lie derivative $\mathscr{L}_P : \Omega^\bullet(\mathscr{A}) \to \Omega^{\bullet-\bar{p}}(\mathscr{A})$ are summarized in the following proposition.

Proposition 3.11. *Let $\mathscr{A}$ be a commutative associative algebra. For $P \in \mathfrak{X}^p(\mathscr{A})$ and $Q \in \mathfrak{X}^q(\mathscr{A})$, the following formulas hold:*

(1) $[\mathscr{L}_P, \mathrm{d}] = 0$;
(2) $[\mathscr{L}_P, \iota_Q] = \iota_{[P,Q]_S}$;
(3) $\mathscr{L}_{[P,Q]_S} = [\mathscr{L}_P, \mathscr{L}_Q]$;
(4) $\mathscr{L}_{P\wedge Q} = \iota_Q \circ \mathscr{L}_P + (-1)^p \mathscr{L}_Q \circ \iota_P$.

Proof. Since $\mathrm{d}\circ\mathrm{d} = 0$, we have that $[\mathscr{L}_P, \mathrm{d}] = -(-1)^p \mathrm{d}\circ\iota_P\circ\mathrm{d} + (-1)^p \mathrm{d}\circ\iota_P\circ\mathrm{d} = 0$, which leads to *(1)*, while *(2)* is precisely Cartan's formula (3.38). Combined with *(1)* and *(2)*, the graded Jacobi identity for $[\cdot,\cdot]$ yields

$$[\mathscr{L}_P, \mathscr{L}_Q] = [\mathscr{L}_P, [\iota_Q, \mathrm{d}]] = [[\mathscr{L}_P, \iota_Q], \mathrm{d}] = \left[\iota_{[P,Q]_S}, \mathrm{d}\right] = \mathscr{L}_{[P,Q]_S} ,$$

which proves *(3)*. Finally, using $\iota_{P\wedge Q} = \iota_Q \circ \iota_P$ (see Proposition 3.4) and (3.45),

$$\begin{aligned} \mathscr{L}_{P\wedge Q} &= [\iota_{P\wedge Q}, \mathrm{d}] = [\iota_Q \circ \iota_P, \mathrm{d}] \\ &= \iota_Q \circ [\iota_P, \mathrm{d}] + (-1)^p [\iota_Q, \mathrm{d}] \circ \iota_P \\ &= \iota_Q \circ \mathscr{L}_P + (-1)^p \mathscr{L}_Q \circ \iota_P , \end{aligned}$$

which proves *(4)*. □

3.4 Exercises

1. Prove the explicit formula (3.28) for a Kähler k-form ω on an arbitrary commutative associative algebra $\mathscr{A}$; the arguments $\mathscr{V}_i$ are in this context arbitrary derivations of $\mathscr{A}$.

2. Give an alternative proof of the graded Jacobi identity for the Schouten bracket by using Cartan's formula (3.38).

3. Let $\mathscr{A}$ be an arbitrary commutative associative algebra. Prove the following formulas, for $F \in \mathscr{A}$ and $\omega \in \Omega^\bullet(\mathscr{A})$:

$$\begin{aligned} \mathscr{L}_{FP}\omega &= F\,\mathscr{L}_P\omega - (-1)^p \mathrm{d}F \wedge \iota_P\omega , \\ \mathscr{L}_P(F\omega) &= F\,\mathscr{L}_P\omega + \iota_{(\iota_F P)}\omega . \end{aligned}$$

These formulas show that $\mathscr{L}_P\omega$ is neither $\mathscr{A}$-linear in P nor in ω.

4. Let $\mathscr{A}$ be a commutative associative algebra and let $\mathscr{B}$ be an $\mathscr{A}$-module. A linear map $\phi : \mathscr{A} \to \mathscr{B}$ is called a *derivation* of $\mathscr{A}$ with values in $\mathscr{B}$ if for all $F_1, F_2 \in \mathscr{A}$, one has $\phi(F_1F_2) = F_1\phi(F_2) + F_2\phi(F_1)$.

a. Interpret the Hamiltonian operator $\mathscr{X}$, associated to a Poisson bracket on $\mathscr{A}$ (see Definition 1.3), as a derivation of $\mathscr{A}$ with values in $\mathfrak{X}^1(\mathscr{A})$;
b. For $\mathscr{A} = \mathscr{F}(M)$ an algebra of functions on a variety or manifold, interpret pointwise derivations at a point $m \in M$ as derivations with values in $\mathbb{F}$, where $\mathbb{F}$ is viewed as an $\mathscr{A}$-module, via $F \cdot a := F(m)a$, for $F \in \mathscr{A}$ and $a \in \mathbb{F}$;
c. Interpret the de Rham differential d as a derivation of $\mathscr{A}$ with values in $\Omega^1(\mathscr{A})$.

5. Suppose that $\mathscr{A}$ is a commutative associative algebra, that $\{\cdot,\cdot\}$ is a Poisson bracket on $\mathscr{A}$ and that $\mathscr{V}$ is a derivation of $\mathscr{A}$. Show that $\{\cdot,\cdot\}$ and the Lie derivative of $\{\cdot,\cdot\}$ with respect to $\mathscr{V}$ are compatible (see Section 3.3).

6. Let $\mathscr{A} := \mathbb{F}[x,y]/\langle x^2+y^2\rangle$. Show that $\mathfrak{X}^2(\mathscr{A}) = 0$ and that $\Omega^2(\mathscr{A}) \neq 0$.

7. The purpose of this exercise is to give an example of an algebra $\mathscr{A}$, which admits skew-symmetric biderivations which cannot be written as the sum of wedge products of derivations of $\mathscr{A}$. In formulas, this means that $\mathfrak{X}^2(\mathscr{A}) \neq \wedge^2\mathfrak{X}^1(\mathscr{A})$. Consider the quotient algebra

$$\mathscr{A} := \frac{\mathbb{F}[x,y,z]}{\langle yx, yz, y^2\rangle}\ .$$

a. Show that $P := y\dfrac{\partial}{\partial y} \wedge \dfrac{\partial}{\partial z}$ induces a non-zero skew-symmetric biderivation $\bar{P}$ of $\mathscr{A}$;
b. Show that $\bar{P}$ does not belong to the exterior algebra $\wedge^2\mathfrak{X}^1(\mathscr{A})$.

8. Let $\mathscr{A}$ be a commutative associative algebra, and let $P \in \mathfrak{X}^2(\mathscr{A})$ be a skew-symmetric biderivation of $\mathscr{A}$. Let $\mathscr{A}[\nu]$ denote the vector space of all polynomials in ν with coefficients in $\mathscr{A}$. Consider on $\mathscr{A}[\nu]/\nu^2$ the product, defined for elements $F_1 + \nu F_2$ and $G_1 + \nu G_2$ of $\mathscr{A}[\nu]/\nu^2$ by

$$(F_1+\nu F_2)\star(G_1+\nu G_2) := F_1G_1 + \nu(F_1G_2 + F_2G_1 + P[F_1,G_1])\ .$$

a. Show that $\star$ is $\mathbb{F}[\nu]$-bilinear and associative;
b. Show that, if $\star$ can be extended to an associative $\mathbb{F}[\nu]$-bilinear product on $\mathscr{A}[\nu]/\nu^3$, then P satisfies the Jacobi identity.

3.5 Notes

The calculus of differential forms on manifolds is explained in detail in all basic books on differential geometry, see for example Spivak [186, 187] or Warner [198].

For the algebraic analog, which is the calculus of Kähler forms, we refer to Hartshorne [93].

Multivector fields are rarely explicitly treated in books on differential geometry; even if those fields themselves are particular cases of tensor fields, everywhere treated in detail, their algebraic structure does not derive from the general tensor formalism. The calculus of multivector fields is presented in detail in Vaisman [194] and in Bhaskara–Viswanath [23]. Cannas da Silva–Weinstein [34] concentrate on the structure of the space of multivector fields as a Gerstenhaber algebra, which they present as a particular case of the Gerstenhaber algebra of a Lie algebroid.

The Schouten bracket, also called the Schouten–Nijenhuis bracket, was introduced by Schouten in [177, 178]; a generalization of the Schouten bracket, which also generalizes the Fröhlicher–Nijenhuis bracket on vector valued differential forms, is given in Vinogradov [197].

Chapter 4
Poisson (Co)Homology

A Poisson bracket on a commutative associative algebra $\mathscr{A}$, or a Poisson structure on a manifold M, leads in a natural way to cohomology spaces, derived from the multi-derivations of $\mathscr{A}$ (multivector fields on M), and to homology spaces, derived from the Kähler differentials of $\mathscr{A}$ (differential forms on M). These spaces give information on the derivations, normal forms, deformations and several invariants of the Poisson structure. In some specific, but important, cases they are related to classically known cohomology spaces, as de Rham cohomology or Lie algebra cohomology. In general, Poisson cohomology is finer, but is also more difficult to compute. See Table 4.1 for a concise overview of the spaces and the (co)boundary operations which will be introduced in this chapter.

Table 4.1 A summary of the notations which we use for Lie and Poisson homology and cohomology. $(\mathscr{A},\cdot,\pi)$ is an arbitrary Poisson algebra over $\mathbb{F}$, and $\mathfrak{g}$ is a Lie algebra with a representation space V.

	k-(co)chains	(co)boundary	(co)homology spaces
Lie cohomology	$\mathrm{Hom}(\wedge^k\mathfrak{g},V)$	δ_L^k	$H_L^k(\mathfrak{g};V)$
Poisson cohomology	$\mathfrak{X}^k(\mathscr{A})$	δ_π^k	$H_\pi^k(\mathscr{A})$
Lie homology	$V\otimes\wedge^k\mathfrak{g}$	∂_k^L	$H_k^L(\mathfrak{g};V)$
Poisson homology	$\Omega^k(\mathscr{A})$	∂_k^π	$H_k^\pi(\mathscr{A})$

In Sections 4.1 and 4.2, we construct the various complexes which lead to the homologies and cohomologies which we will consider. In Section 4.3 we describe a few natural operations in Poisson cohomology and homology and we show in Section 4.4 that the Poisson cohomology and homology spaces of a Poisson manifold are, under certain conditions, isomorphic to each other.

Throughout this chapter, $\mathbb{F}$ is an arbitrary field of characteristic zero.

C. Laurent-Gengoux et al., *Poisson Structures*,
Grundlehren der mathematischen Wissenschaften 347,
DOI 10.1007/978-3-642-31090-4_4,

4.1 Lie Algebra and Poisson Cohomology

Before constructing the Poisson cohomology complex, and deriving the cohomology spaces from it, we recall Lie algebra cohomology, which is formally very similar to Poisson cohomology, and more widely known (see [156]). The basic notions which we recall will also be useful when we compare Poisson and Lie algebra cohomology in the case of the canonical Poisson structure, defined on the dual of a Lie algebra (see Chapter 7).

4.1.1 Lie Algebra Cohomology

Let $(\mathfrak{g},[\cdot\,,\cdot])$ be a Lie algebra over $\mathbb{F}$. We recall that a *representation* of $\mathfrak{g}$ is a Lie algebra homomorphism $\rho:\mathfrak{g}\to\mathrm{End}(V)$, where V is an $\mathbb{F}$-vector space, called the *representation space* and the Lie algebra structure on $\mathrm{End}(V)$ is given by the commutator: $\phi\circ\psi-\psi\circ\phi$, for $\phi,\psi\in\mathrm{End}(V)$. The vector space V, together with the representation ρ, is called a $\mathfrak{g}$*-module*. We write ρ_x for $\rho(x)$, where $x\in\mathfrak{g}$. We fix a representation (ρ,V) of $\mathfrak{g}$ and we introduce, for $k\in\mathbb{N}$, the $\mathbb{F}$-vector space

$$C^k(\mathfrak{g};V):=\mathrm{Hom}(\wedge^k\mathfrak{g},V)$$

of skew-symmetric k-linear maps from $\mathfrak{g}$ to V. Elements of $C^k(\mathfrak{g};V)$ are called *V-valued k-cochains* of $\mathfrak{g}$. The *Lie coboundary map* $\delta_L^k:C^k(\mathfrak{g};V)\to C^{k+1}(\mathfrak{g};V)$ is given, for $c\in C^k(\mathfrak{g};V)$ and $x_0,\dots,x_k\in\mathfrak{g}$, by

$$\begin{aligned}\delta_L^k(c)(x_0,\dots,x_k)&=\sum_{i=0}^{k}(-1)^i\rho_{x_i}c(x_0,\dots,\widehat{x_i},\dots,x_k)\\&\quad+\sum_{0\leqslant i<j\leqslant k}(-1)^{i+j}c\Big([x_i,x_j],x_0,\dots,\widehat{x_i},\dots,\widehat{x_j},\dots,x_k\Big),\end{aligned}\tag{4.1}$$

where the symbol $\widehat{\ }$ over an argument means that we omit that argument. Elements of $\mathrm{Ker}\,\delta_L^k$ are called *Lie k-cocycles*, while elements in $\mathrm{Im}\,\delta_L^{k-1}$ are called *Lie k-coboundaries*. The Jacobi identity for $[\cdot\,,\cdot]$ and the fact that ρ is a Lie algebra homomorphism permits one to show that $\delta_L^k\circ\delta_L^{k-1}=0$, for every $k\in\mathbb{N}^*$. In words: every coboundary is a cocycle. The converse is not true in general and the cohomology spaces precisely measure the obstruction. Namely, for $k\in\mathbb{N}^*$, we define

$$H_L^k(\mathfrak{g};V):=\mathrm{Ker}\,\delta_L^k/\,\mathrm{Im}\,\delta_L^{k-1}\ ,$$

and $H_L^0(\mathfrak{g};V):=\mathrm{Ker}\,\delta_L^0$. Putting these spaces together, we obtain a graded $\mathbb{F}$-vector space

$$H_L^\bullet(\mathfrak{g};V):=\bigoplus_{k\in\mathbb{N}}H_L^k(\mathfrak{g};V)\,,$$

which is called the *V-valued Lie algebra cohomology of* $\mathfrak{g}$.

For example, for $k=0$, the coboundary operator is given by $\delta_L^0(v)(x) = \rho_x v$, for $v \in V \simeq C^0(\mathfrak{g};V)$ and $x \in \mathfrak{g}$. It follows that $H_L^0(\mathfrak{g};V) = \operatorname{Ker}\delta_L^0$ is given by

$$H_L^0(\mathfrak{g};V) = \bigcap_{x\in\mathfrak{g}} \operatorname{Ker}\rho_x ,$$

the subspace of all *invariant elements* in V, also denoted by $V^{\mathfrak{g}}$. The following lemma is fundamental in Lie theory; we will use it several times to illustrate the phenomena which we encounter. For a proof, see [102].

Lemma 4.1 (Whitehead's lemma). *Let $\mathfrak{g}$ be a semi-simple Lie algebra and let $\rho : \mathfrak{g} \to \operatorname{End}(V)$ be a representation of $\mathfrak{g}$. If V is finite-dimensional, then $H_L^1(\mathfrak{g};V) = 0$ and $H_L^2(\mathfrak{g};V) = 0$.*

The following two examples of representations (ρ, V) are of particular interest and will be our main cases.

Example 4.2. Let $\rho : \mathfrak{g} \to \mathbb{F}$ be the trivial representation: $\rho_x = 0$ for every $x \in \mathfrak{g}$. Then

$$C^k(\mathfrak{g};\mathbb{F}) = \operatorname{Hom}(\wedge^k\mathfrak{g},\mathbb{F})$$

and (4.1) becomes:

$$\delta_L^k(c)(x_0,\ldots,x_k) = \sum_{0\leqslant i<j\leqslant k} (-1)^{i+j}\, c\Big([x_i,x_j],x_0,\ldots,\widehat{x_i},\ldots,\widehat{x_j},\ldots,x_k\Big). \tag{4.2}$$

In this case, the Lie algebra cohomology of $\mathfrak{g}$ is called the *trivial Lie algebra cohomology* of $\mathfrak{g}$ and is simply denoted by $H_L^\bullet(\mathfrak{g})$. It follows from (4.2) that $\operatorname{Im}\delta_L^0 = \{0\}$ and that $\operatorname{Ker}\delta_L^1 = \{c \in \mathfrak{g}^* \mid \forall x,y \in \mathfrak{g},\ c([x,y]) = 0\}$, so that

$$H_L^1(\mathfrak{g}) \simeq (\mathfrak{g}/[\mathfrak{g},\mathfrak{g}])^* .$$

For a semi-simple Lie algebra $\mathfrak{g}$, for example, it is well known that $[\mathfrak{g},\mathfrak{g}] = \mathfrak{g}$ (see [190, Ch. 20]). Therefore, $H_L^1(\mathfrak{g}) = 0$ when $\mathfrak{g}$ is semi-simple, which is a special case of Whitehead's lemma (Lemma 4.1).

Example 4.3. Let $\rho := \operatorname{ad}$, the adjoint representation of $\mathfrak{g}$ on itself, which is given by $\rho_x = \operatorname{ad}_x := [x,\cdot]$, for $x \in \mathfrak{g}$. Then

$$C^k(\mathfrak{g};\mathfrak{g}) = \operatorname{Hom}(\wedge^k\mathfrak{g},\mathfrak{g})$$

and (4.1) becomes:

$$\begin{aligned}\delta_L^k(c)(x_0,\ldots,x_k) = &\sum_{i=0}^{k}(-1)^i\,[x_i, c(x_0,\ldots,\widehat{x_i},\ldots,x_k)]\\ &+ \sum_{0\leqslant i<j\leqslant k}(-1)^{i+j}\, c\big([x_i,x_j],x_0,\ldots,\widehat{x_i},\ldots,\widehat{x_j},\ldots,x_k\big)\end{aligned} \tag{4.3}$$

In this case, the Lie algebra cohomology of $\mathfrak{g}$ is called the *Chevalley–Eilenberg cohomology of* $\mathfrak{g}$. We denote it by $H_{CE}^\bullet(\mathfrak{g})$. Let us give an interpretation of the space $H^1_{CE}(\mathfrak{g})$. The 1-cocycles are the derivations of $\mathfrak{g}$, since

$$\delta^1_L(c)(x,y) = [x,c(y)] + [c(x),y] - c([x,y])$$

for $x,y \in \mathfrak{g}$ and $c \in \operatorname{Hom}(\mathfrak{g},\mathfrak{g})$. Since $\delta^0_L(x) = -\operatorname{ad}_x$, for $x \in \mathfrak{g} \simeq C^0(\mathfrak{g};\mathfrak{g})$, the space of 1-coboundaries $\operatorname{Im}\delta^0_L$ consists of all derivations of $\mathfrak{g}$ of the form ad_x, where $x \in \mathfrak{g}$; these derivations are called *inner derivations*. It follows that $H^1_{CE}(\mathfrak{g})$ is the quotient of the space of all derivations of $\mathfrak{g}$, by the space of all inner derivations of $\mathfrak{g}$. The elements of this quotient are called the *exterior derivations of* $\mathfrak{g}$.

The spaces $H^2_{CE}(\mathfrak{g})$ and $H^3_{CE}(\mathfrak{g})$ play an important rôle in the theory of formal deformations of Lie algebras. This fact is quite similar (by forgetting the multi-derivation aspect) to what happens in the case of Poisson algebras. Therefore, we will not make it explicit in the Lie algebra case, but in the Poisson algebra case, see Section 13.2.

4.1.2 Poisson Cohomology

We now define Poisson cohomology for a Poisson algebra $(\mathscr{A},\cdot,\{\cdot\,,\cdot\})$; the Poisson bracket on $\mathscr{A}$ will also be denoted by π, so that $\pi(F,G) = \{F,G\}$, for $F,G \in \mathscr{A}$. For $k \in \mathbb{N}$, the space of k-cochains of the Poisson cohomology complex is by definition $\mathfrak{X}^k(\mathscr{A})$, the $\mathbb{F}$-vector space of skew-symmetric k-derivations of $\mathscr{A}$ (see Section 3.1.1). The *Poisson coboundary operator* is the Poisson analog of the Chevalley–Eilenberg coboundary operator (4.3), where $\operatorname{Hom}(\wedge^k\mathfrak{g},\mathfrak{g})$ has been replaced by $\mathfrak{X}^k(\mathscr{A})$, and where the Lie bracket $[\cdot\,,\cdot]$ is the Poisson bracket $\{\cdot\,,\cdot\}$. Namely, the graded $\mathbb{F}$-linear map (of degree 1) $\delta_\pi : \mathfrak{X}^\bullet(\mathscr{A}) \to \mathfrak{X}^{\bullet+1}(\mathscr{A})$ is defined, for $Q \in \mathfrak{X}^q(\mathscr{A})$, where $q \in \mathbb{N}$, by

$$\begin{aligned}\delta^q_\pi(Q)[F_0,\dots,F_q] &:= \sum_{i=0}^{q}(-1)^i\Big\{F_i, Q[F_0,\dots,\widehat{F_i},\dots,F_q]\Big\} \qquad (4.4)\\ &+ \sum_{0\leqslant i<j\leqslant q}(-1)^{i+j}Q\Big[\{F_i,F_j\},F_0,\dots,\widehat{F_i},\dots,\widehat{F_j},\dots,F_q\Big],\end{aligned}$$

for all $F_0,\dots,F_q \in \mathscr{A}$. In order to establish that $\delta^q_\pi(Q)$ is indeed a skew-symmetric multi-derivation of $\mathscr{A}$ and that $\delta^{q+1}_\pi \circ \delta^q_\pi = 0$, for $q \in \mathbb{N}$, we relate the coboundary operator to the Schouten bracket. In fact, the explicit formula (3.36) for the Schouten bracket $[\cdot\,,\cdot]_S$, implies that

$$\delta^q_\pi(Q) = -[Q,\pi]_S\,, \qquad (4.5)$$

for $Q \in \mathfrak{X}^q(\mathscr{A})$. Thus, $\delta^q_\pi(Q)$ is the Schouten bracket of two skew-symmetric multi-derivations, so it is itself a skew-symmetric multi-derivation. In order to see that $\delta^{q+1}_\pi \circ \delta^q_\pi = 0$, we write down the graded Jacobi identity of the Schouten bracket (Proposition 3.7) for the triple (Q,π,π),

$$(-1)^{q-1}[Q,[\pi,\pi]_S]_S - [\pi,[Q,\pi]_S]_S + (-1)^{q-1}[\pi,[\pi,Q]_S]_S = 0\,, \tag{4.6}$$

where we have used that the shifted degree $\bar{q}$ of a q-derivation is $q-1$, in particular the shifted degree of π is 1. Since $[\pi,\pi]_S = 0$ (because π is a Poisson bracket) and since $[Q,\pi]_S = -(-1)^{\bar{q}}[\pi,Q]_S$ (graded skew-symmetry of $[\cdot,\cdot]_S$), the identity (4.6) reduces to $[\pi,[\pi,Q]_S]_S = 0$, for every $Q \in \mathfrak{X}^q(\mathscr{A})$, showing that $\delta_\pi^{q+1} \circ \delta_\pi^q = 0$, for all $q \in \mathbb{N}$. Thus, we have a complex, the *Poisson cohomology complex* of $\mathscr{A}$,

$$\cdots \longrightarrow \mathfrak{X}^{q-1}(\mathscr{A}) \xrightarrow{\delta_\pi^{q-1}} \mathfrak{X}^q(\mathscr{A}) \xrightarrow{\delta_\pi^q} \mathfrak{X}^{q+1}(\mathscr{A}) \xrightarrow{\delta_\pi^{q+1}} \cdots$$

The elements of $\operatorname{Ker}\delta_\pi^q$ are called *Poisson q-cocycles*, while the elements of $\operatorname{Im}\delta_\pi^{q-1}$ are called *Poisson q-coboundaries*. Elements of the *q-th Poisson cohomology space* are Poisson q-cocycles modulo Poisson q-coboundaries,

$$H_\pi^q(\mathscr{A}) := \operatorname{Ker}\delta_\pi^q / \operatorname{Im}\delta_\pi^{q-1}\,,$$

for $q \in \mathbb{N}^*$ and $H_\pi^0(\mathscr{A}) := \operatorname{Ker}\delta_\pi^0$. The graded vector space

$$H_\pi^\bullet(\mathscr{A}) := \bigoplus_{q\in\mathbb{N}} H_\pi^q(\mathscr{A})\,,$$

is called the *Poisson cohomology* of $\mathscr{A}$.

In the case of a Poisson manifold (M,π), one replaces in the above construction the multi-derivations of $\mathscr{A}$ by the multivector fields on M, leading to the following complex,

$$\cdots \longrightarrow \mathfrak{X}^{q-1}(M) \xrightarrow{\delta_\pi^{q-1}} \mathfrak{X}^q(M) \xrightarrow{\delta_\pi^q} \mathfrak{X}^{q+1}(M) \xrightarrow{\delta_\pi^{q+1}} \cdots$$

where the coboundary operator δ_π is defined by (4.5), for $Q \in \mathfrak{X}^q(M)$ (a q-vector field on M). This complex is called the *Poisson cohomology complex* of M. The q-th Poisson cohomology space of (M,π), respectively the Poisson cohomology of (M,π), is denoted by $H_\pi^q(M)$ respectively $H_\pi^\bullet(M)$.

Remark 4.4. Let $(\mathscr{A},\cdot,\{\cdot,\cdot\})$ be a Poisson algebra over $\mathbb{F}$. Forgetting the associative product, $(\mathscr{A},\{\cdot,\cdot\})$ is simply a Lie algebra and it makes sense to consider $C^\bullet(\mathscr{A},\mathscr{A})$, with the Chevalley–Eilenberg coboundary operator (4.3). This yields precisely (4.4), so that the Poisson coboundary operator of $(\mathscr{A},\cdot,\{\cdot,\cdot\})$ is the Chevalley–Eilenberg coboundary operator, restricted to the multi-derivations of $(\mathscr{A},\cdot)$.

For small q, the Poisson coboundary operator δ_π^q has a natural interpretation, which yields a natural interpretation for the Poisson cohomology space $H_\pi^q(\mathscr{A})$. One easily reads off from (4.4) that for $F \in \mathfrak{X}^0(\mathscr{A}) = \mathscr{A}$ and for $\mathscr{V} \in \mathfrak{X}^1(\mathscr{A})$,

$$\delta_\pi^0(F) = \mathscr{X}_F\,, \qquad \delta_\pi^1(\mathscr{V}) = -\mathscr{L}_\mathscr{V}\pi\,, \tag{4.7}$$

where we recall that $\mathscr{X}_F$ is the Hamiltonian derivation associated to $F \in \mathscr{A}$ (see Definition 1.3), and $\mathscr{L}_{\mathscr{V}}$ denotes the Lie derivative with respect to $\mathscr{V}$ (see (3.7)). Also, for $Q \in \mathfrak{X}^2(\mathscr{A})$ and $F, G, H \in \mathscr{A}$,

$$\delta^2_\pi(Q)[F,G,H] = \{F, Q[G,H]\} + Q[F,\{G,H\}] + \circlearrowleft(F,G,H) \,.$$

It follows from (4.7) that Poisson 0-cocycles are Casimirs,

$$H^0_\pi(\mathscr{A}) = \mathrm{Cas}(\mathscr{A}) \,,$$

Poisson 1-cocycles are Poisson derivations and Poisson 1-coboundaries are Hamiltonian derivations. Denoting the space of all Poisson derivations of $\mathscr{A}$ by $\mathscr{P}(\mathscr{A})$, it follows that

$$H^1_\pi(\mathscr{A}) = \frac{\mathscr{P}(\mathscr{A})}{\mathrm{Ham}(\mathscr{A})} \,.$$

Poisson 2-cocycles $Q \in \mathfrak{X}^2(\mathscr{A})$ are skew-symmetric biderivations which satisfy $[\pi, Q]_S = 0$, i.e., they are the ones which are compatible with $\{\cdot\,,\cdot\}$, see Section 3.3.2; Poisson 2-coboundaries are biderivations, obtained as a Lie derivative of the Poisson structure. It follows that

$$H^2_\pi(\mathscr{A}) = \frac{\text{skew-symmetric biderivations compatible with } \{\cdot\,,\cdot\}}{\text{Lie derivatives of } \{\cdot\,,\cdot\}} \,.$$

As we will see in Section 13.2, $H^2_\pi(\mathscr{A})$ and $H^3_\pi(\mathscr{A})$ show up naturally in deformation theory.

There is a natural morphism from de Rham cohomology to Poisson cohomology, which we explain here in the algebraic context; this morphism is defined in the same way in the case of a smooth manifold. We first construct an $\mathscr{A}$-linear map

$$\begin{aligned} \pi^\sharp : \Omega^1(\mathscr{A}) &\to \mathfrak{X}^1(\mathscr{A}) \\ G\,\mathrm{d}F &\mapsto G\mathscr{X}_F \,. \end{aligned} \tag{4.8}$$

It is clear that this map is well-defined; for example, we have for all $F, G \in \mathscr{A}$ that

$$\pi^\sharp(\mathrm{d}(FG) - F\mathrm{d}G - G\mathrm{d}F) = \mathscr{X}_{FG} - F\mathscr{X}_G - G\mathscr{X}_F = 0 \,,$$

where we used *(3)* of Proposition 1.4 in the last step. The map $\pi^\sharp$ extends naturally to an $\mathscr{A}$-linear map

$$\wedge^\bullet \pi^\sharp : \Omega^\bullet(\mathscr{A}) \to \wedge^\bullet \mathfrak{X}^1(\mathscr{A}) \to \mathfrak{X}^\bullet(\mathscr{A}) \,,$$

which is explicitly given by

$$\wedge^q \pi^\sharp(F\mathrm{d}G_1 \wedge \cdots \wedge \mathrm{d}G_q) = F\mathscr{X}_{G_1} \wedge \cdots \wedge \mathscr{X}_{G_q} \,,$$

for all $F, G_1, \ldots, G_q \in \mathscr{A}$. It leads to a natural morphism between the de Rham cohomology and the Poisson cohomology of $(\mathscr{A}, \pi)$.

Proposition 4.5. *Let $(\mathscr{A}, \cdot, \{\cdot,\cdot\})$ be a Poisson algebra. The maps $\wedge^q \pi^\sharp : \Omega^q(\mathscr{A}) \to \mathfrak{X}^q(\mathscr{A})$ define a chain map from $(\Omega^\bullet(\mathscr{A}), \mathrm{d})$ to $(\mathfrak{X}^\bullet(\mathscr{A}), \delta_\pi)$, i.e., for every $q \in \mathbb{N}$, the following diagram is commutative:*

$$\begin{array}{ccc} \Omega^q(\mathscr{A}) & \xrightarrow{\ \mathrm{d}\ } & \Omega^{q+1}(\mathscr{A}) \\ \Big\downarrow{\scriptstyle \wedge^q \pi^\sharp} & & \Big\downarrow{\scriptstyle \wedge^{q+1} \pi^\sharp} \\ \mathfrak{X}^q(\mathscr{A}) & \xrightarrow{\ \delta_\pi^q\ } & \mathfrak{X}^{q+1}(\mathscr{A}) \end{array}$$

It leads for every $q \in \mathbb{N}$ to an $\mathbb{F}$-linear map $H^q_{dR}(\mathscr{A}) \to H^q_\pi(\mathscr{A})$.

Proof. Let $\omega := G \mathrm{d}F_1 \wedge \cdots \wedge \mathrm{d}F_q \in \Omega^q(\mathscr{A})$. On the one hand,

$$\wedge^{q+1} \pi^\sharp(\mathrm{d}\omega) = \wedge^{q+1} \pi^\sharp(\mathrm{d}G \wedge \mathrm{d}F_1 \wedge \cdots \wedge \mathrm{d}F_q) = \mathscr{X}_G \wedge \mathscr{X}_{F_1} \wedge \cdots \wedge \mathscr{X}_{F_q}.$$

On the other hand, (4.5) and (3.42) imply that

$$\begin{aligned} \delta_\pi^q(\wedge^q \pi^\sharp(\omega)) &= \delta_\pi^q(G \mathscr{X}_{F_1} \wedge \cdots \wedge \mathscr{X}_{F_q}) = -\left[G \mathscr{X}_{F_1} \wedge \cdots \wedge \mathscr{X}_{F_q}, \pi\right]_S \\ &= -[G, \pi]_S \wedge \mathscr{X}_{F_1} \wedge \cdots \wedge \mathscr{X}_{F_q} = \mathscr{X}_G \wedge \mathscr{X}_{F_1} \wedge \cdots \wedge \mathscr{X}_{F_q}, \end{aligned}$$

where we also used that each Hamiltonian vector field $\mathscr{X}_{F_i}$ is a Poisson 1-cocycle, $[\mathscr{X}_{F_i}, \pi]_S = 0$. This leads to the commutativity of the above diagram. It follows that $\wedge^q \pi^\sharp$ sends q-cocycles to q-cocycles and sends q-coboundaries to q-coboundaries. Thus, it induces a map in cohomology. □

Remark 4.6. Contrary to what we will see for Poisson homology in Section 4.2, Poisson cohomology is not a functor: a homomorphism $\varphi : \mathscr{A} \to \mathscr{B}$ between two Poisson algebras does not lead in general to a homomorphism between the spaces of all multi-derivations of $\mathscr{A}$ and $\mathscr{B}$, not even between their corresponding Poisson cohomology spaces. In the case of manifolds, for example, it is well known that vector fields cannot be transferred, in either direction, by a smooth map, which is not a diffeomorphism.

4.2 Lie Algebra and Poisson Homology

A Poisson bracket on a commutative associative algebra $\mathscr{A}$ also leads to homology spaces. In special cases they are isomorphic to the Poisson cohomology spaces, but in general they define new invariants for a Poisson algebra. The Poisson homology spaces have fewer immediate applications than the Poisson cohomology spaces, but have the advantage of being sometimes easier to compute and have better functorial properties, as we will see. As in the previous section, we start with the Lie algebra case.

4.2.1 Lie Algebra Homology

The basic setup of Lie algebra homology is the same as in the case of Lie algebra cohomology: $(\mathfrak{g},[\cdot\,,\cdot])$ is a Lie algebra and (ρ,V) is a representation of $\mathfrak{g}$. For $k\in\mathbb{N}$, we define

$$C_k(\mathfrak{g};V) := V\otimes\wedge^k\mathfrak{g}\,,$$

whose elements are called *V-valued k-chains* of $\mathfrak{g}$. Note that $C_0(\mathfrak{g},V)\simeq V$, since $\wedge^0\mathfrak{g}=\mathbb{F}$. The *Lie boundary operator* $\partial_k^L : C_k(\mathfrak{g};V)\to C_{k-1}(\mathfrak{g};V)$ is the linear map given, for all $v\in V$ and $x_1,\dots,x_k\in\mathfrak{g}$, by:

$$\begin{aligned}
&\partial_k^L\left(v\otimes(x_1\wedge\dots\wedge x_k)\right)\\
&\quad=\sum_{i=1}^k(-1)^i\rho_{x_i}v\otimes\left(x_1\wedge\dots\wedge\widehat{x_i}\wedge\dots\wedge x_k\right)\\
&\qquad+\sum_{1\leqslant i<j\leqslant k}(-1)^{i+j}v\otimes\left([x_i,x_j]\wedge x_1\wedge\dots\wedge\widehat{x_i}\wedge\dots\wedge\widehat{x_j}\wedge\dots\wedge x_k\right)
\end{aligned}$$

and $\partial_0^L:=0$. We have that $\partial_{k-1}^L\circ\partial_k^L=0$ as a consequence of the Jacobi identity for $[\cdot,\cdot]$ and the Lie homomorphism property for ρ. The operators ∂_k^L lead to a homology, called the *V-valued Lie algebra homology of* $\mathfrak{g}$. Namely, the k-th V-valued Lie algebra homology space of $\mathfrak{g}$ is denoted and defined by:

$$H_k^L(\mathfrak{g};V) := \operatorname{Ker}\partial_k^L/\operatorname{Im}\partial_{k+1}^L\,.$$

We denote by $H_\bullet^L(\mathfrak{g};V)$ the graded vector space

$$H_\bullet^L(\mathfrak{g};V) := \bigoplus_{k\in\mathbb{N}}H_k^L(\mathfrak{g};V)\,.$$

For $v\in V$ and $x\in\mathfrak{g}$, we have that

$$\partial_1^L(v\otimes x)=-\rho_x v\,.$$

Therefore, the subspace $\operatorname{Im}\partial_1^L$ of V is denoted by $\mathfrak{g}V$ and $H_0^L(\mathfrak{g};V)=V/\mathfrak{g}V$, since $\partial_0^L=0$. When $V=\mathfrak{g}$ and $\rho=\operatorname{ad}$, as in Example 4.3, the boundary operator takes the following form

$$\begin{aligned}\partial_k^L\left(x_0\otimes(x_1\wedge\cdots\wedge x_k)\right)&\\ =\sum_{i=1}^{k}(-1)^i\,[x_i,x_0]\otimes\Big(x_1\wedge\cdots\wedge\widehat{x_i}\wedge\cdots\wedge x_k\Big)&\\ +\sum_{1\leqslant i<j\leqslant k}(-1)^{i+j}\,x_0\otimes\Big([x_i,x_j]\wedge x_1\wedge\cdots\wedge\widehat{x_i}\wedge\cdots\wedge\widehat{x_j}\wedge\cdots\wedge x_k\Big)&.\end{aligned}\tag{4.9}$$

In this case, $\mathfrak{g}V=[\mathfrak{g},\mathfrak{g}]$, so that $H_0^{CE}(\mathfrak{g}):=H_0^L(\mathfrak{g};\mathfrak{g})=\mathfrak{g}/[\mathfrak{g},\mathfrak{g}]$. This space also appears when one computes $H_1^L(\mathfrak{g};\mathbb{F})$, where $\rho:\mathfrak{g}\to\mathbb{F}$ is the trivial representation, as in Example 4.2. Indeed, in this case,

$$\begin{aligned}\operatorname{Ker}\partial_1^L&=C_1(\mathfrak{g};\mathbb{F})=\{a\otimes x\mid a\in\mathbb{F}\text{ and }x\in\mathfrak{g}\}\simeq\mathfrak{g}\,,\\ \operatorname{Im}\partial_2^L&=\{a\otimes[x,y]\mid a\in\mathbb{F}\text{ and }x,y\in\mathfrak{g}\}\simeq[\mathfrak{g},\mathfrak{g}]\,.\end{aligned}$$

This leads to $H_1^L(\mathfrak{g}):=H_1^L(\mathfrak{g};\mathbb{F})\simeq\mathfrak{g}/[\mathfrak{g},\mathfrak{g}]$.

4.2.2 Poisson Homology

A Poisson bracket $\pi=\{\cdot,\cdot\}$ on a commutative associative algebra $\mathscr{A}$ leads to a homology, called its *Poisson homology*. The k-chains which define the Poisson homology complex are the Kähler differentials on $\mathscr{A}$ (the differential forms when $\mathscr{A}=\mathscr{F}(M)$, the algebra of smooth functions on a manifold M). In analogy with (4.9), the *Poisson boundary operator* $\partial^\pi:\Omega^\bullet(\mathscr{A})\to\Omega^{\bullet-1}(\mathscr{A})$ is given by

$$\begin{aligned}\partial_k^\pi(F_0\,\mathrm{d}F_1\wedge\cdots\wedge\mathrm{d}F_k)&\\ =\sum_{i=1}^{k}(-1)^i\{F_i,F_0\}\,\mathrm{d}F_1\wedge\cdots\wedge\widehat{\mathrm{d}F_i}\wedge\cdots\wedge\mathrm{d}F_k&\\ +\sum_{1\leqslant i<j\leqslant k}(-1)^{i+j}F_0\,\mathrm{d}\{F_i,F_j\}\wedge\mathrm{d}F_1\wedge\cdots\wedge\widehat{\mathrm{d}F_i}\wedge\cdots\wedge\widehat{\mathrm{d}F_j}\wedge\cdots\wedge\mathrm{d}F_k&\,,\end{aligned}$$

where $F_0,\ldots,F_k\in\mathscr{A}$. For $k=0$, this formula should be read as $\partial_0^\pi(F_0):=0$. The main properties of ∂^π follow from the fact that $\partial^\pi=\mathscr{L}_\pi=[\imath_\pi,\mathrm{d}]$, the Lie derivative with respect to π, as stated in the following proposition.

Proposition 4.7. *Let $(\mathscr{A},\cdot,\pi)$ be a Poisson algebra. Then the linear map ∂^π, defined above, is given by the Lie derivative*

$$\partial^\pi=\mathscr{L}_\pi=[\imath_\pi,\mathrm{d}]=\imath_\pi\circ\mathrm{d}-\mathrm{d}\circ\imath_\pi\,.$$

It commutes (in the graded sense) with d *and with $\imath_\pi$,*

$$[\partial^\pi,\mathrm{d}]=\partial^\pi\circ\mathrm{d}+\mathrm{d}\circ\partial^\pi=0\,,\tag{4.10}$$

$$[\partial^\pi,\imath_\pi]=\partial^\pi\circ\imath_\pi-\imath_\pi\circ\partial^\pi=0\,,\tag{4.11}$$

and satisfies $\partial^\pi\circ\partial^\pi=0$ (i.e., it is a boundary operator).

Proof. For the biderivation $\pi = \{\cdot,\cdot\}$, the internal product $\imath_\pi$ is, according to (3.29), explicitly given by

$$\begin{aligned}\imath_\pi(F_0\mathrm{d}F_1\wedge\cdots\wedge\mathrm{d}F_k) \\ = \sum_{1\leqslant i<j\leqslant k}(-1)^{i+j+1}F_0\{F_i,F_j\}\,\mathrm{d}F_1\wedge\cdots\wedge\widehat{\mathrm{d}F_i}\cdots\widehat{\mathrm{d}F_j}\wedge\cdots\wedge\mathrm{d}F_k\,,\end{aligned}$$

where $F_0,\ldots,F_k\in\mathscr{A}$, so that

$$\begin{aligned}\imath_\pi\circ\mathrm{d}(F_0\mathrm{d}F_1\wedge\cdots\wedge\mathrm{d}F_k) \\ = \sum_{0\leqslant i<j\leqslant k}(-1)^{i+j+1}\{F_i,F_j\}\,\mathrm{d}F_0\wedge\cdots\wedge\widehat{\mathrm{d}F_i}\ldots\widehat{\mathrm{d}F_j}\wedge\cdots\wedge\mathrm{d}F_k\,,\end{aligned}$$

and

$$\begin{aligned}\mathrm{d}\circ\imath_\pi(F_0\mathrm{d}F_1\wedge\cdots\wedge\mathrm{d}F_k) \\ = \sum_{1\leqslant i<j\leqslant k}(-1)^{i+j+1}\{F_i,F_j\}\,\mathrm{d}F_0\wedge\cdots\wedge\widehat{\mathrm{d}F_i}\ldots\widehat{\mathrm{d}F_j}\wedge\cdots\wedge\mathrm{d}F_k\,, \\ +\sum_{1\leqslant i<j\leqslant k}(-1)^{i+j+1}F_0\,\mathrm{d}\{F_i,F_j\}\wedge\mathrm{d}F_1\wedge\cdots\wedge\widehat{\mathrm{d}F_i}\ldots\widehat{\mathrm{d}F_j}\wedge\cdots\wedge\mathrm{d}F_k\,.\end{aligned}$$

Taking the difference of these two formulas leads at once to $\partial^\pi = [\imath_\pi,\mathrm{d}] = \mathscr{L}_\pi$. The properties of the Lie derivative (Proposition 3.11) imply that $\partial^\pi\circ\mathrm{d}+\mathrm{d}\circ\partial^\pi = 0$, and that

$$[\partial^\pi,\imath_\pi] = [\mathscr{L}_\pi,\imath_\pi] = \imath_{[\pi,\pi]_S} = 0\,,$$

where we used in the last equality that π is a Poisson structure. This yields (4.10) and (4.11). Similarly, since $\mathscr{L}_\pi$ has degree -1,

$$2\,\mathscr{L}_\pi\circ\mathscr{L}_\pi = [\mathscr{L}_\pi,\mathscr{L}_\pi] = \mathscr{L}_{[\pi,\pi]_S} = 0\,,$$

which shows that $\partial^\pi = \mathscr{L}_\pi$ is a boundary operator. □

According to the proposition, we have a complex

$$\cdots\longrightarrow\Omega^{k+1}(\mathscr{A})\xrightarrow{\partial^\pi_{k+1}}\Omega^k(\mathscr{A})\xrightarrow{\partial^\pi_k}\Omega^{k-1}(\mathscr{A})\xrightarrow{\partial^\pi_{k-1}}\cdots$$

whose homology is called the *Poisson homology* of $(\mathscr{A},\cdot,\pi)$. Namely, we define the *k-th Poisson homology space*

$$H^\pi_k(\mathscr{A}) := \operatorname{Ker}\partial^\pi_k/\operatorname{Im}\partial^\pi_{k+1}\,,$$

and we call elements of $\operatorname{Ker}\partial^\pi_k$ *Poisson k-cycles* and elements of $\operatorname{Im}\partial^\pi_{k+1}$ *Poisson k-boundaries*. Putting the Poisson homology spaces together leads to the graded vector space $H^\pi_\bullet(\mathscr{A}) := \oplus_{k\in\mathbb{N}}H^\pi_k(\mathscr{A})$. In the case of a Poisson manifold (M,π),

Poisson homology is defined similarly, using the differential forms on M, rather than the Kähler forms on $\mathscr{A}$. It is denoted by $H^{\pi}_{\bullet}(M)$.

For example, $H^{\pi}_0(\mathscr{A})$ is the abelianization of $\mathscr{A}$: since $\operatorname{Ker}\partial^{\pi}_0 = \mathscr{A}$ while $\operatorname{Im}\partial^{\pi}_1 = \pi(\mathscr{A},\mathscr{A}) = \{\mathscr{A},\mathscr{A}\}$, we have that

$$H^{\pi}_0(\mathscr{A}) = \frac{\mathscr{A}}{\{\mathscr{A},\mathscr{A}\}} .$$

For the higher Poisson homology spaces, no simple interpretation is known.

Remark 4.8. Unlike Poisson cohomology, Poisson homology is a (covariant) functor. Indeed, every algebra homomorphism $\varphi : (\mathscr{A},\cdot) \to (\mathscr{B},\cdot)$ extends to a degree zero map $\Omega^{\bullet}(\varphi) : \Omega^{\bullet}(\mathscr{A}) \to \Omega^{\bullet}(\mathscr{B})$, which commutes with d. For $k \in \mathbb{N}$, it is given by $\Omega^k(\varphi)(F_0 \mathrm{d}F_1 \wedge \cdots \wedge \mathrm{d}F_k) = \varphi(F_0)\mathrm{d}\varphi(F_1) \wedge \cdots \wedge \mathrm{d}\varphi(F_k)$, where $F_0, \ldots, F_k \in \mathscr{A}$. If $\mathscr{A}$ and $\mathscr{B}$ are equipped with Poisson brackets $\pi_{\mathscr{A}}$ and $\pi_{\mathscr{B}}$, and φ is a morphism of Poisson algebras, then $\Omega^{\bullet}(\varphi) \circ \imath_{\pi_{\mathscr{A}}} = \imath_{\pi_{\mathscr{B}}} \circ \Omega^{\bullet}(\varphi)$, so that $\Omega^{\bullet}(\varphi) \circ \partial^{\pi_{\mathscr{A}}} = \partial^{\pi_{\mathscr{B}}} \circ \Omega^{\bullet}(\varphi)$. Thus, there is an induced map $H^{\pi}_{\bullet}(\varphi) : H^{\pi}_{\bullet}(\mathscr{A}) \to H^{\pi}_{\bullet}(\mathscr{B})$, which is a homomorphism of graded vector spaces. Clearly, H^{π} has all properties which are required for a (covariant) functor.

4.3 Operations in Homology and Cohomology

We have introduced on the graded vector space $\mathfrak{X}^{\bullet}(\mathscr{A})$ (where $\mathscr{A}$ is a commutative associative algebra) two multiplications, the wedge product $\wedge$, which is associative, graded commutative and the Schouten bracket $[\cdot\,,\cdot]_S$, which is a graded Lie bracket (with a grading, shifted by 1). As we discussed in Section 3.3.2, these two products make $(\mathfrak{X}^{\bullet}(\mathscr{A}), \wedge, [\cdot\,,\cdot]_S)$ into a Gerstenhaber algebra. Moreover, we have two natural operations $\imath$ (the internal product) and $\mathscr{L}$ (the Lie derivative) of $\mathfrak{X}^{\bullet}(\mathscr{A})$ on the module of Kähler differentials $\Omega^{\bullet}(\mathscr{A})$. We will show in this section that these four operations induce similar operations, with the same algebraic properties, in (co)homology. The reader will have no difficulty in transcribing these operations (and their properties) for the case of (real or complex) manifolds.

We first define the wedge product and the Schouten bracket in cohomology. The graded Leibniz identity and the graded Jacobi identity for the Schouten bracket,

$$[Q \wedge R, P]_S = [Q,P]_S \wedge R + (-1)^{\bar{p}q} Q \wedge [R,P]_S ,$$
$$(-1)^{\bar{p}\bar{r}}[P,[Q,R]_S]_S + (-1)^{\bar{q}\bar{p}}[Q,[R,P]_S]_S + (-1)^{\bar{r}\bar{q}}[R,[P,Q]_S]_S = 0 ,$$

become, for $P = \pi$, $Q \in \mathfrak{X}^q(\mathscr{A})$ and $R \in \mathfrak{X}^r(\mathscr{A})$, with $\delta_\pi = -[\cdot,\pi]_S$,

$$\delta_\pi(Q \wedge R) = \delta_\pi(Q) \wedge R + (-1)^q Q \wedge \delta_\pi(R) ,$$
$$\delta_\pi([Q,R]_S) = [Q,\delta_\pi(R)]_S + (-1)^{\bar{r}} [\delta_\pi(Q),R]_S .$$

First, it follows from these formulas that if Q and R are Poisson cocycles, then $Q \wedge R$ and $[Q,R]_S$ are Poisson cocycles. This means that both operations can be restricted to cocycles. Second, it follows from the same formulas that either product of a Poisson cocycle with a Poisson coboundary is a Poisson coboundary. This means that, for Poisson cocycles Q and R, the cohomology class of $Q \wedge R$ and the cohomology class of $[Q,R]_S$ are independent of the representatives Q and R of their cohomology classes. Thus, we have two well-defined products in Poisson cohomology, which we also denote by $\wedge$ and $[\cdot,\cdot]_S$. Obviously, their algebraic properties are the same as on $\mathfrak{X}^\bullet(\mathscr{A})$ and we obtain:

Proposition 4.9. *For every Poisson algebra* $(\mathscr{A},\cdot,\pi)$, $(H^\bullet_\pi(\mathscr{A}),\wedge,[\cdot,\cdot]_S)$ *is a Gerstenhaber algebra.*

We now consider the action of Poisson cohomology on Poisson homology, which comes from the internal product and from the Lie derivative. Remembering that the Poisson boundary operator ∂^π is the Lie derivative $\mathscr{L}_\pi$, we take from Proposition 3.11 the formulas which compute the graded commutator of a Lie derivative with an internal product, or with another Lie derivative, namely,

$$[\mathscr{L}_P,\imath_Q] = \imath_{[P,Q]_S} \qquad \text{and} \qquad [\mathscr{L}_P,\mathscr{L}_Q] = \mathscr{L}_{[P,Q]_S}\,,$$

which can be written, for $P = \pi$ and $\omega \in \Omega^\bullet(\mathscr{A})$, as

$$\begin{aligned}
(-1)^q\partial^\pi(\imath_Q\omega) &= \imath_Q\,\partial^\pi(\omega) - \imath_{\delta_\pi(Q)}\,\omega\,,\\
(-1)^{\bar{q}}\partial^\pi(\mathscr{L}_Q\omega) &= \mathscr{L}_Q(\partial^\pi(\omega)) + \mathscr{L}_{\delta_\pi(Q)}\,\omega\,.
\end{aligned}$$

It follows from these formulas, as above, that $\imath$ and $\mathscr{L}$ induce indeed actions of the Poisson cohomology of $\mathscr{A}$ on the Poisson homology of $\mathscr{A}$.

4.4 The Modular Class of a Poisson Manifold

The modular class of a Poisson manifold M is a Poisson cohomology class (element of $H^1_\pi(M)$) which is canonically associated to the Poisson structure. We will not give the description in the most general case and restrict ourselves to the case of an orientable (real) Poisson manifold. The definitions and arguments remain valid in the case of $\mathbb{C}^d$, equipped with its algebra of holomorphic functions, but some minor adaptations are needed for the affine space $\mathbb{F}^d$, equipped with its algebra of polynomial functions: see Remark 4.13 below. Our exposition is based on the following:

Question: "Let M be a Poisson manifold. Does there exist a volume form λ on M, preserved by all Hamiltonian flows?"

We call a Poisson manifold, which admits such a volume form, a *unimodular Poisson manifold*; the Poisson structure is then called a *unimodular Poisson structure.*

For a symplectic[1] manifold (M,ω), the answer to the above question is yes. Indeed, since the symplectic structure ω is preserved by all Hamiltonian flows, the same is true for the Liouville form ω^d on M, where $\dim M = 2d$. Thus, every symplectic manifold is a unimodular Poisson manifold. Many properties of symplectic manifolds generalize to the case of unimodular Poisson manifolds. For example, for unimodular Poisson manifolds the Poisson cohomology and the Poisson homology spaces are isomorphic to each other (Theorem 4.18).

4.4.1 The Modular Vector Field

We first list a few basic facts about volume forms on a real orientable manifold M. Suppose that λ is an arbitrary *volume form* on M. Recall that this means that λ is a differential d-form on M, where $d := \dim M$, vanishing at no point of M, and that such a form exists on a manifold if and only if the manifold is orientable; moreover, two such forms differ only by a factor, which is a nowhere vanishing smooth function on M. We have that $\mathrm{d}\lambda = 0$ and that $\mathrm{d}F \wedge \lambda = 0$, for every $F \in \mathscr{F}(M)$. Combined with (3.32), this yields

$$\imath_{(\imath_F P)}\lambda = -(-1)^p \mathrm{d}F \wedge \imath_P \lambda \,, \tag{4.12}$$

for all $P \in \mathfrak{X}^p(M)$. In particular,

$$\mathscr{V}[F]\lambda = \mathrm{d}F \wedge \imath_{\mathscr{V}}\lambda \,, \tag{4.13}$$

for every vector field $\mathscr{V} \in \mathfrak{X}^1(M)$. Also, if π is a Poisson structure on M, then (4.12) specializes to

$$\imath_{\mathscr{X}_F}\lambda = \mathrm{d}F \wedge \imath_\pi \lambda \,. \tag{4.14}$$

The Lie derivative of λ with respect to some vector field is again a top form, hence it is a multiple of λ. For $F \in \mathscr{F}(M)$, let us denote by $\mathscr{D}_F \in \mathscr{F}(M)$ the smooth function on M defined by $\mathscr{L}_{\mathscr{X}_F}\lambda = \mathscr{D}_F\lambda$. We show that the map $F \mapsto \mathscr{D}_F$ is a derivation. Indeed, for all $F, G \in \mathscr{F}(M)$, we have that

$$\begin{aligned}
\mathscr{D}_{FG}\lambda &= \mathscr{L}_{\mathscr{X}_{FG}}\lambda = \mathscr{L}_{G\mathscr{X}_F + F\mathscr{X}_G}\lambda \\
&= G\mathscr{L}_{\mathscr{X}_F}\lambda + \{G,F\}\lambda + F\mathscr{L}_{\mathscr{X}_G}\lambda + \{F,G\}\lambda \\
&= (G\mathscr{D}_F + F\mathscr{D}_G)\lambda \,,
\end{aligned}$$

where the identity

$$\mathscr{L}_{H\mathscr{V}}\lambda = H\mathscr{L}_{\mathscr{V}}\lambda + \mathscr{V}[H]\lambda \tag{4.15}$$

for all $H \in \mathscr{F}(M)$, $\mathscr{V} \in \mathfrak{X}^1(M)$ has been used (to prove (4.15), write $\mathscr{L}_{\mathscr{V}} = \imath_{\mathscr{V}} \circ \mathrm{d} + \mathrm{d} \circ \imath_{\mathscr{V}}$ and use (4.13)). Since $\mathscr{D} : F \mapsto \mathscr{D}_F$ is a derivation of $\mathscr{F}(M)$, it corresponds to a vector field.

[1] See Section 6.3.

Definition 4.10. Let (M,π) be a real Poisson manifold, equipped with a volume form λ. The vector field Φ_λ, defined by

$$\mathscr{L}_{\mathscr{X}_F}\lambda = \Phi_\lambda[F]\lambda\,, \tag{4.16}$$

for all $F \in \mathscr{F}(M)$, is called the *modular vector field* of π with respect to λ.

When the chosen volume form is clear from the context, we simply write Φ for Φ_λ. Notice that the modular vector field Φ can also be defined as the unique vector field on M, satisfying

$$\mathscr{L}_\pi\lambda = \imath_\Phi\lambda\,. \tag{4.17}$$

This is seen by writing $\Phi[F]\lambda$ in two different ways. On the one hand, (4.13) tells us that $\Phi[F]\lambda = \mathrm{d}F\wedge\imath_\Phi\lambda$, for all $F\in\mathscr{F}(M)$. On the other hand, (4.16), (4.14) and the formula $\mathscr{L}_\pi = \imath_\pi\circ\mathrm{d} - \mathrm{d}\circ\imath_\pi$ imply that

$$\Phi[F]\lambda = \mathscr{L}_{\mathscr{X}_F}\lambda = \mathrm{d}(\imath_{\mathscr{X}_F}\lambda) = \mathrm{d}(\mathrm{d}F\wedge\imath_\pi\lambda) = -\mathrm{d}F\wedge\mathrm{d}\imath_\pi\lambda = \mathrm{d}F\wedge\mathscr{L}_\pi\lambda\,,$$

for all $F\in\mathscr{F}(M)$. It follows that $\mathrm{d}F\wedge\imath_\Phi\lambda = \mathrm{d}F\wedge\mathscr{L}_\pi\lambda$ for every $F\in\mathscr{F}(M)$, which leads to (4.17).

In view of Definition 4.10, the question which we posed at the beginning of this section now becomes:

Question: "Let M be a Poisson manifold. Does there exist a volume form λ on M, with respect to which the modular vector field Φ is zero?"

By giving a cohomological interpretation of the modular vector field, we will be able to answer the question in the next section.

4.4.2 The Modular Class

The modular vector field which we constructed in the previous section depends on the choice of volume form on M. We will now show that when the volume form is changed, the modular vector field will be modified by a Hamiltonian vector field. We also show that the modular vector field is a Poisson vector field, so that it defines a Poisson cohomology class which, by the above, is independent of the choice of volume form.

Proposition 4.11. *Let $(M,\{\cdot,\cdot\})$ be a real Poisson manifold, which is orientable. For λ a volume form on M, the modular vector field Φ has the following properties:*

(1) Φ is a Poisson vector field, i.e., it is a Poisson 1*-cocycle;*
(2) The Poisson cohomology class of Φ is independent of the chosen volume form λ;
(3) If Φ is a Hamiltonian vector field, i.e., if the Poisson cohomology class of Φ is zero, then there exists a volume form λ', with respect to which the modular vector field is zero (and therefore λ' is preserved by all the Hamiltonian flows).

Proof. In order to prove *(1)* we need to verify that $\Phi[\{F,G\}] = \{\Phi[F],G\} + \{F,\Phi[G]\}$, for all $F,G \in \mathscr{F}(M)$. The basic properties of the Poisson bracket and the Lie derivative yield:

$$\begin{aligned}\Phi[\{F,G\}]\,\lambda &= \mathscr{L}_{\mathscr{X}_{\{F,G\}}}\lambda = \mathscr{L}_{[\mathscr{X}_G,\mathscr{X}_F]}\lambda = \left[\mathscr{L}_{\mathscr{X}_G},\mathscr{L}_{\mathscr{X}_F}\right]\lambda \\ &= \mathscr{L}_{\mathscr{X}_G}(\Phi[F]\lambda) - \mathscr{L}_{\mathscr{X}_F}(\Phi[G]\lambda) \\ &= (\{\Phi[F],G\} + \{F,\Phi[G]\})\,\lambda\,,\end{aligned}$$

where we have used in the last step that

$$\Phi[F]\mathscr{L}_{\mathscr{X}_G}\lambda = \Phi[F]\Phi[G]\lambda = \Phi[G]\mathscr{L}_{\mathscr{X}_F}\lambda\,.$$

This proves *(1)*. Let us consider now another volume form μ on M and let us denote by Φ_λ and Φ_μ the modular class of $\{\cdot\,,\cdot\}$ with respect to λ, respectively μ. There exists a nowhere vanishing function $\rho \in \mathscr{F}(M)$ such that $\mu = \rho\,\lambda$. Using that λ and μ are both top forms and (4.13), we compute, for an arbitrary $F \in \mathscr{F}(M)$,

$$\mathscr{L}_{\mathscr{X}_F}(\rho\lambda) = \mathscr{X}_F[\rho]\lambda + \rho\mathscr{L}_{\mathscr{X}_F}\lambda = -\mathscr{X}_\rho[F]\lambda + \rho\,\Phi_\lambda[F]\lambda\,,$$

to conclude that

$$\Phi_\mu[F] = \Phi_\lambda[F] - \frac{1}{\rho}\mathscr{X}_\rho[F]\,,$$

for all $F \in \mathscr{F}(M)$, and hence that

$$\Phi_\mu = \Phi_\lambda - \frac{1}{\rho}\mathscr{X}_\rho = \Phi_\lambda - \mathscr{X}_{\ln|\rho|}\,.$$

This shows that the vector fields Φ_λ and Φ_μ differ by a Hamiltonian vector field, i.e., a Poisson 1-coboundary. The cohomology class of the modular form is therefore independent of the chosen volume form, which proves *(2)*. The previous calculation shows that if this cohomology class is trivial, i.e., $\Phi = \mathscr{X}_F$ for some $F \in \mathscr{F}(M)$, then $\mu := \exp(F)\lambda$ is a volume form whose modular vector field is zero. □

Summarizing, we have a complete answer to our question:

Answer: "The Poisson manifold (M,π) admits a volume form, preserved by all Hamiltonian flows if and only if its modular class is zero."

It leads to the following definition.

Definition 4.12. Let $(M,\{\cdot\,,\cdot\})$ be a real Poisson manifold, which is assumed orientable. The cohomology class of the modular vector field (with respect to an arbitrary volume form on M) is called the *modular class* of the Poisson manifold. If this cohomology class is trivial, then we say that $(M,\{\cdot\,,\cdot\})$ is a *unimodular Poisson manifold*. A volume form on M, whose modular vector field is zero, is called an *invariant density*.

Notice that if one multiplies an invariant density by a nowhere vanishing Casimir function, then one obtains another invariant density; up to this ambiguity, the invariant density on a unimodular Poisson manifold is unique.

Remark 4.13. In the case of a finite-dimensional vector space V, a natural volume form is given, in terms of linear coordinates $x_1,\ldots,x_d$ on V, by $\lambda_0 := \mathrm{d}x_1 \wedge \cdots \wedge \mathrm{d}x_d$, and every top form on V is of the form $F\lambda_0$, where F is an arbitrary function on V. Up to a non-zero constant, λ_0 is the only translation invariant volume form on V. If $\mathbb{F}$ is assumed to be algebraically closed, then λ_0 is, again up to a non-zero constant, the unique volume form on V, because every non-constant polynomial F has a non-empty zero locus, on which the top form $F\lambda_0$ fails to be a volume form. Notice that, since the modular vector field does not change when the volume form is multiplied by a constant, the modular vector field of a Poisson structure is independent of the chosen volume form, when only translation invariant volume forms on V are considered.

We show in the following proposition that the obstruction to the existence of an invariant density is intimately related to the existence of points where the rank of the Poisson structure drops: in the neighborhood of each regular point, the modular vector field is Hamiltonian, hence an invariant density exists on this neighborhood.

Proposition 4.14. *Let $(M,\{\cdot,\cdot\})$ be a real orientable Poisson manifold and let λ be a volume form on M.*

(1) The modular vector field Φ is a Hamiltonian vector field in a neighborhood of every point at which the rank is locally constant. In particular, it is tangent to all symplectic leaves of maximal dimension;

(2) The modular class is preserved under Poisson diffeomorphisms $M \to M$.

Proof. Let $m \in M$ be a point where the rank of $\{\cdot,\cdot\}$ is locally constant. According to the Darboux theorem (Theorem 1.26), there exist splitting coordinates $q_1,\ldots,q_r$, $p_1,\ldots,p_r,z_1,\ldots,z_s$ on a neighborhood U of m, such that

$$\{\cdot,\cdot\} = \sum_{i=1}^{r} \frac{\partial}{\partial q_i} \wedge \frac{\partial}{\partial p_i},$$

on U, where $\mathrm{Rk}_m\{\cdot,\cdot\} = 2r$. We take on U as volume form

$$\lambda := \bigwedge_{i=1}^{r} (\mathrm{d}q_i \wedge \mathrm{d}p_i) \bigwedge_{j=1}^{s} \mathrm{d}z_j\,,$$

and we denote

$$\lambda_k := \bigwedge_{i\neq k} (\mathrm{d}q_i \wedge \mathrm{d}p_i) \bigwedge_{j=1}^{s} \mathrm{d}z_j\,,$$

for $k = 1,\ldots,r$. For such k and for $F \in \mathscr{F}(M)$ we have

$$\begin{aligned}\mathscr{L}_{\mathscr{X}_F}(\mathrm{d}q_k\wedge\mathrm{d}p_k) &= \left(\mathscr{L}_{\mathscr{X}_F}\mathrm{d}q_k\right)\wedge\mathrm{d}p_k+\mathrm{d}q_k\wedge\left(\mathscr{L}_{\mathscr{X}_F}\mathrm{d}p_k\right)\\ &= \mathrm{d}\left(\frac{\partial F}{\partial p_k}\right)\wedge\mathrm{d}p_k-\mathrm{d}q_k\wedge\mathrm{d}\left(\frac{\partial F}{\partial q_k}\right).\end{aligned}$$

Therefore, for all $k=1,\dots,r$,

$$\lambda_k\wedge\mathscr{L}_{\mathscr{X}_F}(\mathrm{d}q_k\wedge\mathrm{d}p_k)=\lambda\left(\frac{\partial^2 F}{\partial q_k\partial p_k}-\frac{\partial^2 F}{\partial p_k\partial q_k}\right)=0\,,$$

from which it follows that $\mathscr{L}_{\mathscr{X}_F}\lambda=0$, i.e., the modular vector field Φ is zero on U. Then *(2)* of Proposition 4.11 implies that all modular vector fields are Hamiltonian on U, hence they are tangent on U to all symplectic leaves. The second property is obvious, since the modular class depends on $\{\cdot,\cdot\}$ only. □

Example 4.15. If m is a point where the rank of the Poisson structure is not locally constant, then it may happen that there does not exist a neighborhood of m on which the modular vector field is Hamiltonian. In fact, the modular vector field may be non-vanishing at a point where the rank of the Poisson structure vanishes. Consider for example the Poisson structure $\{\cdot,\cdot\}=x\frac{\partial}{\partial x}\wedge\frac{\partial}{\partial y}$ on $\mathbb{R}^2$, equipped with the volume form $\mathrm{d}x\wedge\mathrm{d}y$. The rank at the origin is zero, while $\Phi=\frac{\partial}{\partial y}$. In particular, the modular class of this Poisson structure is not zero.

4.4.3 The Divergence of the Poisson Bracket

We now show that the modular vector field is (up to a sign) the divergence of the Poisson structure (with respect to λ). The *divergence* operator (with respect to λ) is the graded $\mathbb{F}$-linear map (of degree -1),

$$\mathrm{Div}:\mathfrak{X}^\bullet(M)\to\mathfrak{X}^{\bullet-1}(M)\,,$$

which makes the following diagram commute

$$\begin{array}{ccc}\mathfrak{X}^\bullet(M) & \xrightarrow{\;\star\;} & \Omega^{d-\bullet}(M)\\ \Big\downarrow{\scriptstyle\mathrm{Div}} & & \Big\downarrow{\scriptstyle\mathrm{d}}\\ \mathfrak{X}^{\bullet-1}(M) & \xrightarrow[\;\star\;]{} & \Omega^{d-\bullet+1}(M)\end{array}\qquad(4.18)$$

where $d:=\dim M$. The diagram involves the star operator $\star$, which is the family of $\mathscr{F}(M)$-linear isomorphisms $\star:\mathfrak{X}^k(M)\to\Omega^{d-k}(M)$, defined for $Q\in\mathfrak{X}^k(M)$ by

$$\star\,Q:=\imath_Q\lambda\,.\qquad(4.19)$$

Since λ is a top form, we have for every vector field $\mathscr{V}$ on M that

$$\mathscr{L}_{\mathscr{V}}\lambda = \mathrm{d}\iota_{\mathscr{V}}\lambda = \mathrm{d}\star\mathscr{V} = \star\mathrm{Div}(\mathscr{V}) = \mathrm{Div}(\mathscr{V})\,\lambda\,. \tag{4.20}$$

For example, taking the standard volume form on $\mathbb{R}^d$, i.e., $\lambda = \mathrm{d}x_1\wedge\cdots\wedge\mathrm{d}x_d$ the star of a vector field $\mathscr{V} = \sum_{i=1}^d F_i\partial/\partial x_i$, respectively of a bivector field $P = \sum_{i<j} P_{ij}\partial/\partial x_i\wedge\partial/\partial x_j$ is given by

$$\star\mathscr{V} = \sum_{i=1}^{d}(-1)^{i-1}F_i\,\mathrm{d}x_1\wedge\cdots\wedge\widehat{\mathrm{d}x_i}\wedge\cdots\wedge\mathrm{d}x_d\,,$$

$$\star P = \sum_{1\leqslant i<j\leqslant d}(-1)^{i+j-1}P_{ij}\,\mathrm{d}x_1\wedge\cdots\wedge\widehat{\mathrm{d}x_i}\wedge\cdots\wedge\widehat{\mathrm{d}x_j}\wedge\cdots\wedge\mathrm{d}x_d\,,$$

so that $\mathrm{Div}(\mathscr{V})$ is given by the classical formula

$$\mathrm{Div}(\mathscr{V}) = \sum_{i=1}^{d}\frac{\partial F_i}{\partial x_i} \tag{4.21}$$

and $\mathrm{Div}(P)$ is given by

$$\mathrm{Div}(P) = \sum_{1\leqslant i<j\leqslant d}\left(\frac{\partial P_{ij}}{\partial x_j}\frac{\partial}{\partial x_i} - \frac{\partial P_{ij}}{\partial x_i}\frac{\partial}{\partial x_j}\right). \tag{4.22}$$

In general, the divergence of a bivector field, written in terms of vector fields, can be computed using the following proposition.

Proposition 4.16. *Let λ be a volume form on an orientable manifold M and let $\mathscr{V}$ and $\mathscr{W}$ be two vector fields on M. The divergence of $\mathscr{V}\wedge\mathscr{W}$ with respect to λ is given by*

$$\mathrm{Div}(\mathscr{V}\wedge\mathscr{W}) = \mathrm{Div}(\mathscr{W})\mathscr{V} - \mathrm{Div}(\mathscr{V})\mathscr{W} - [\mathscr{V},\mathscr{W}]\,. \tag{4.23}$$

Proof. According to the definition of the (generalized) Lie derivative (3.50) and its properties *(4)* and *(2)* in Proposition 3.11, we have that

$$\mathrm{d}\iota_{\mathscr{V}\wedge\mathscr{W}}\lambda = -\mathscr{L}_{\mathscr{V}\wedge\mathscr{W}}\lambda = \mathscr{L}_{\mathscr{W}}\iota_{\mathscr{V}}\lambda - \iota_{\mathscr{W}}\mathscr{L}_{\mathscr{V}}\lambda = \iota_{\mathscr{V}}\mathscr{L}_{\mathscr{W}}\lambda - \iota_{\mathscr{W}}\mathscr{L}_{\mathscr{V}}\lambda - \iota_{[\mathscr{V},\mathscr{W}]}\lambda\,.$$

According to the definition of the divergence, this leads to

$$\begin{aligned}\mathrm{Div}(\mathscr{V}\wedge\mathscr{W}) &= \star^{-1}\mathrm{d}\iota_{\mathscr{V}\wedge\mathscr{W}}\lambda\\ &= \star^{-1}\iota_{\mathscr{V}}\mathscr{L}_{\mathscr{W}}\lambda - \star^{-1}\iota_{\mathscr{W}}\mathscr{L}_{\mathscr{V}}\lambda - \star^{-1}\iota_{[\mathscr{V},\mathscr{W}]}\lambda\\ &= \mathrm{Div}(\mathscr{W})\mathscr{V} - \mathrm{Div}(\mathscr{V})\mathscr{W} - [\mathscr{V},\mathscr{W}]\,,\end{aligned}$$

where we used in the last step that $\iota_{\mathscr{V}}\mathscr{L}_{\mathscr{W}}\lambda = \iota_{\mathscr{V}}(\mathrm{Div}(\mathscr{W})\lambda) = \iota_{\mathrm{Div}(\mathscr{W})\mathscr{V}}\lambda = \star(\mathrm{Div}(\mathscr{W})\mathscr{V})$ and that $\star[\mathscr{V},\mathscr{W}] = \iota_{[\mathscr{V},\mathscr{W}]}\lambda$. □

We now show that the modular vector field is, up to a sign, the divergence of the Poisson structure.

Proposition 4.17. *Let (M,π) be a real Poisson manifold, which is assumed orientable. For every volume form λ on M,*

$$\Phi = -\operatorname{Div}(\pi)\ . \tag{4.24}$$

In particular, the divergence of the modular vector field of π is zero.

Proof. Let λ be an arbitrary volume form on M and let $F \in \mathscr{F}(M)$. The definitions of Φ and of $\star$, combined with (4.17), imply that

$$\star\operatorname{Div}(\pi) = \mathrm{d}\star\pi = \mathrm{d}(\imath_\pi\lambda) = -\mathscr{L}_\pi\lambda = -\imath_\Phi\lambda = -\star\Phi\ . \tag{4.25}$$

Since $\star$ is an isomorphism, this shows that the modular vector field is minus the divergence of π. Since $\operatorname{Div}\circ\operatorname{Div} = \star^{-1}\circ\mathrm{d}\circ\mathrm{d}\circ\star = 0$, the modular vector field is divergence-free. □

According to (4.25), the modular vector field can also be defined by the formula

$$\mathrm{d}\star\pi = -\star\Phi\ ;$$

forgetting the stars, the modular vector field is the differential of the Poisson bracket. The string (4.25) also contains the identity $\mathscr{L}_\pi\lambda = \star\Phi$, which amounts to

$$\partial^\pi(\lambda) = \star\Phi\ . \tag{4.26}$$

Thus, $\star\Phi$ is always a Poisson 1-boundary, while Φ is a Poisson 1-coboundary if and only if the underlying Poisson manifold is unimodular.

4.4.4 Unimodular Poisson Manifolds

We now prove that, for unimodular Poisson manifolds, the Poisson homology and cohomology spaces are isomorphic. Let (M,π) be a d-dimensional Poisson manifold. We first show how the Poisson boundary and coboundary operators are related via the modular vector field Φ. To do this, we specialize Cartan's formula (3.38)

$$[[\imath_P,\mathrm{d}],\imath_Q] = \imath_{[P,Q]_S}\ ,$$

valid for $P \in \mathfrak{X}^p(M)$ and $Q \in \mathfrak{X}^q(M)$, to the case $P = \pi$, giving

$$\partial^\pi\circ\imath_Q - (-1)^q\imath_Q\circ\partial^\pi = (-1)^{q-1}\imath_{\delta_\pi(Q)}\ . \tag{4.27}$$

If we apply (4.27) to λ and use (4.26), then we find

$$\partial^\pi(\star Q) - (-1)^q\imath_Q\circ\imath_\Phi\lambda = (-1)^{q-1}\star\delta_\pi(Q)\ ,$$

which we also write, using $\imath_Q\circ\imath_P = \imath_{P\wedge Q}$, valid for $P,Q \in \mathfrak{X}^\bullet(M)$ (see Proposition 3.4), as follows,

$$\star(\Phi\wedge Q)=\star\delta_\pi(Q)+(-1)^q\partial^\pi(\star Q)\,. \tag{4.28}$$

The latter formula says that the modular vector field Φ measures the non-commutativity of the following diagram

$$\begin{array}{ccc} \mathfrak{X}^\bullet(M) & \xrightarrow{\ \star\ } & \Omega^{d-\bullet}(M) \\ \delta_\pi\big\downarrow & \overset{\star\Phi\wedge}{\searrow} & \big\downarrow\partial^\pi \\ \mathfrak{X}^{\bullet+1}(M) & \xrightarrow[\ \star\ ]{} & \Omega^{d-1-\bullet}(M) \end{array} \tag{4.29}$$

The reader is invited to compare the diagrams (4.29) and (4.18), keeping in mind that both depend on the volume form λ, but only the second one depends on the Poisson structure π. Equation (4.28) leads at once to the following duality theorem.

Proposition 4.18. *For every unimodular Poisson manifold $(M,\{\cdot\,,\cdot\})$ of dimension d, the Poisson cohomology spaces and Poisson homology spaces are isomorphic:*

$$H^k_\pi(M)\simeq H^\pi_{d-k}(M), \qquad 0\leqslant k\leqslant d\,. \tag{4.30}$$

Proof. Let λ be an invariant density on M, so that $\Phi=0$. Then (4.28) implies that the isomorphism $\star:\mathfrak{X}^\bullet(M)\to\Omega^{d-\bullet}(M)$ is a bijection between the Poisson cocycles (respectively coboundaries) of M and the Poisson cycles (respectively boundaries) of M. It follows that $\star$ induces isomorphisms, as indicated in (4.30). □

Notice that the isomorphism in (4.30) is not canonical, since there is no canonical invariant density (in general; symplectic manifolds form an important exception).

Remark 4.19. The proposition is also valid for $\mathbb{F}^d$, equipped with a Poisson structure, in case it is a unimodular Poisson variety (see Remark 4.13).

4.5 Exercises

1. Show that $\wedge$ does not induce a product in Poisson homology (Hint: compute $\partial^\pi(\mathrm{d}F\wedge\mathrm{d}G)$ for $F,G\in\mathscr{A}$).

2. Show that the Poisson cohomology spaces are modules over the algebra of Casimirs.

3. Find a formula for the Poisson cohomology of the tensor product of two Poisson algebras.

4. Prove property *(2)* of Proposition 4.11 by using the star operator (4.19).

5. Suppose that $\mathfrak{g}$ is a Lie algebra, equipped with a symmetric bilinear form $\langle\cdot\,|\,\cdot\rangle$ which is ad-invariant, i.e., $\langle[x,y]\,|\,z\rangle=\langle x\,|\,[y,z]\rangle$ for all x, y, $z\in\mathfrak{g}$. Consider the element $C \in \operatorname{Hom}(\wedge^3 \mathfrak{g}, \mathbb{F})$, defined by $C(x,y,z) := \langle [x,y] \mid z \rangle$. Show that C is a 3-cocycle in the trivial Lie algebra cohomology.

6. Consider the Poisson structure π on $\mathbb{F}^2$ given by

$$\pi := x\frac{\partial}{\partial x} \wedge \frac{\partial}{\partial y} \;.$$

a. Show that the modular class of π is not trivial;
b. Compute the Poisson homology and the Poisson cohomology of π;
c. Conclude from part b. that the conclusion of Proposition 4.18 does not hold for the Poisson manifold $(\mathbb{F}^2, \pi)$, i.e., that the vector spaces $H^k_\pi(\mathbb{F}^2)$ and $H^\pi_{2-k}(\mathbb{F}^2)$ are not isomorphic, for $k = 0, 1, 2$.

7. Let $(\mathscr{A}, \cdot, \{\cdot\,,\cdot\})$ be a Poisson algebra.

a. Show that the skew-symmetric biderivation $\{\cdot\,,\cdot\}$ is a Poisson 2-cocycle for the Poisson cohomology of $\mathscr{A}$;
b. Assume that $\mathscr{A}$ is a polynomial algebra, $\mathscr{A} = \mathbb{F}[x_1, \dots, x_d]$, and that the polynomials $\{x_i, x_j\}$, for $i, j = 1, \dots, d$, are all homogeneous of the same degree k. Show that if $k \neq 2$, then the skew-symmetric biderivation $\{\cdot\,,\cdot\}$ is a coboundary (Hint: compute $\delta^1_\pi(\mathscr{E})$ where $\mathscr{E} := \sum_{i=1}^d x_i \frac{\partial}{\partial x_i}$);
c. Assume now that $\mathscr{A} := \mathbb{C}[x,y,z]$, and that

$$\pi := x^2\frac{\partial}{\partial y} \wedge \frac{\partial}{\partial z} + y^2\frac{\partial}{\partial z} \wedge \frac{\partial}{\partial x} + z^2\frac{\partial}{\partial x} \wedge \frac{\partial}{\partial y} \;.$$

Show that π is a Poisson bracket on $\mathscr{A} := \mathbb{C}[x,y,z]$ which is not a coboundary (Hint: show first that if there exists a derivation $\mathscr{V} = F_1\frac{\partial}{\partial x} + F_2\frac{\partial}{\partial y} + F_3\frac{\partial}{\partial z} \in \mathfrak{X}^1(\mathscr{A})$ satisfying $\delta^1_\pi(\mathscr{V}) = \pi$, then there exists such a derivation where F_1, F_2 and F_3 are linear functions).

4.6 Notes

Poisson cohomology was first introduced in Lichnerowicz's paper [127], where it was shown that the Poisson cohomology of a symplectic manifold is canonically isomorphic to its de Rham cohomology. See Bhaskara–Viswanath [23] and Huebschmann [96] for the algebraic cohomology of a general Poisson algebra. The relation between Poisson cohomology and singularities is studied by Monnier [152] and Roger–Vanhaecke [174] in the two-dimensional case and by Pichereau [166] in the three-dimensional case. The definition of Poisson homology goes back to Gelfand–Dorfman [81], Koszul [118] and Brylinski [28]. The modular vector field of a Poisson manifold was first introduced by Koszul [118]. It was studied for general Poisson manifolds and extended to Lie algebroids by Weinstein [201], Evans–Lu–Weinstein [71] and Kosmann–Weinstein [116]. See Fernandes–Crainic [51, 52] for applications of Poisson cohomology to the linearization and stability of Poisson structures; for applications of Poisson cohomology to deformation theory, see Chapter 13 below.

Chapter 5
Reduction

In Chapter 2, we have given a few basic constructions which allow one to build new Poisson structures from given ones. In the present chapter, we explain a few more advanced constructions, which all fall under the general concept of reduction. Roughly speaking, reduction means that the object under study (here a Poisson structure), is replaced by an object of the same type, but on a manifold of smaller dimension, in classical terms "with fewer degrees of freedom". These constructions are based on geometrical considerations, yet their algebraic formulation highlights some key elements of the reduction mechanism. We therefore present all constructions in both algebraic and geometrical terms.

We present two constructions for reducing Poisson structures. Geometrically speaking, the first one, called Poisson reduction, deals with Poisson structures on quotients of Poisson manifolds, or, more generally on quotients of (coisotropic) submanifolds of Poisson manifolds. Poisson reduction is presented in Section 5.2. The second construction, called Poisson–Dirac reduction, concerns Poisson structures on submanifolds, which are not necessarily Poisson submanifolds; a typical example is a submanifold which is transverse to a symplectic leaf, leading to the notion of transverse Poisson structure. Poisson–Dirac reduction is discussed in Section 5.3.

As an application of these constructions, we consider in Section 5.4 Poisson varieties and Poisson manifolds on which a group acts; we show how a Poisson structure is inherited on the quotient space, the subspace of fixed points of the action, and on the reduced space of the so-called momentum map. For the convenience of the reader, and in order to fix both our notations and conventions, the basic facts on Lie groups, and their relation with Lie algebras are recalled in the first section of this chapter.

As in the previous chapters, all Poisson algebras are defined over an arbitrary field $\mathbb{F}$ of characteristic zero, all Poisson varieties are affine varieties over $\mathbb{F}$ and the Poisson manifolds are real or complex manifolds.

C. Laurent-Gengoux et al., *Poisson Structures*,
Grundlehren der mathematischen Wissenschaften 347,
DOI 10.1007/978-3-642-31090-4_5, © Springer-Verlag Berlin Heidelberg 2013

5.1 Lie Groups and (Their) Lie Algebras

We recall in this section the natural correspondence between Lie groups and Lie algebras, we give the basic definitions and properties of group actions and we explain construction of invariant (multi-)vector fields on a Lie group.

5.1.1 Lie Groups and Lie Algebras

A *Lie group* is a set $\mathbf{G}$ endowed with two structures, a group structure and a manifold structure, which are required to be compatible, in the sense that the product map (the group structure)

$$\mu : \mathbf{G}\times\mathbf{G}\to\mathbf{G} : (g,h)\mapsto gh$$

is assumed to be smooth (or holomorphic). Then the inverse map

$$\mathrm{inv} : \mathbf{G}\to\mathbf{G} : g\mapsto g^{-1}$$

is also smooth (or holomorphic). The product of two elements g,h of $\mathbf{G}$ is denoted by $\mu(g,h)$ or simply by gh, and the unit of $\mathbf{G}$ is denoted by e; as a manifold, $\mathbf{G}$ may be either real or complex. The identity map on $\mathbf{G}$ is denoted by $\mathbb{1}_{\mathbf{G}}$. Given two Lie groups $\mathbf{G}$ and $\mathbf{H}$, a *Lie group morphism* from $\mathbf{G}$ to $\mathbf{H}$ is a group homomorphism $\Phi : \mathbf{G}\to\mathbf{H}$, which is a smooth (or holomorphic) map. If $\mathbf{G}$ is contained in $\mathbf{H}$ and the inclusion map is a Lie group morphism, then $\mathbf{G}$ is said to be a *Lie subgroup* of $\mathbf{H}$. The multiplication in $\mathbf{G}$ leads to three, a priori different, sets of diffeomorphisms of $\mathbf{G}$: given $g\in\mathbf{G}$, we define *left translation*, *right translation* and *conjugation* by $g\in\mathbf{G}$ respectively as follows:

$$\begin{aligned} L_g &: \mathbf{G}\to\mathbf{G} : h\mapsto \quad gh\,,\\ R_g &: \mathbf{G}\to\mathbf{G} : h\mapsto \quad hg\,,\\ C_g &: \mathbf{G}\to\mathbf{G} : h\mapsto ghg^{-1}\,. \end{aligned}$$

The tangent maps of μ and inv can be expressed in terms of these diffeomorphisms: since $\mu(g,h)=L_g(h)=R_h(g)$ for all $g,h\in\mathbf{G}$, the tangent map of μ at (g,h) is given by

$$T_{(g,h)}\mu\,(v,w) = T_gR_h(v)+T_hL_g(w) \tag{5.1}$$

for all $(v,w)\in T_{(g,h)}(\mathbf{G}\times\mathbf{G})\simeq T_g\mathbf{G}\times T_h\mathbf{G}$. Differentiating the identities $L_g\circ L_{g^{-1}} = \mathbb{1}_{\mathbf{G}}$ at e and $\mu(g,\mathrm{inv}(g))=e$ at $g\in\mathbf{G}$, leads, upon using (5.1), to the following expression for the tangent map of inv at $g\in\mathbf{G}$,

$$T_g\,\mathrm{inv} = -T_eL_{g^{-1}}\circ T_gR_{g^{-1}}\,. \tag{5.2}$$

Notice that, in particular, $T_e \operatorname{inv} = -\mathbb{1}_{T_e\mathbf{G}}$. Similarly, one computes from $C_g = L_g \circ R_{g^{-1}}$, valid for $g \in \mathbf{G}$, that the tangent map of the conjugation map C_g is given, at the unit e, by

$$T_e C_g = T_{g^{-1}} L_g \circ T_e R_{g^{-1}} \,. \tag{5.3}$$

Since for a fixed $g \in \mathbf{G}$, left translation by g is a diffeomorphism of $\mathbf{G}$, which sends e to g, we can identify the tangent spaces $T_e\mathbf{G}$ and $T_g\mathbf{G}$ by using the linear map $T_e L_g$. For a fixed $x \in T_e\mathbf{G}$, this allows us to define a vector field $\overleftarrow{x}$ on $\mathbf{G}$ by setting, for all $g \in \mathbf{G}$:

$$\overleftarrow{x}_g := T_e L_g(x) \,.$$

The same can be done using right translation by g, yielding a vector field $\overrightarrow{x}$ on $\mathbf{G}$, defined for all $g \in \mathbf{G}$ by

$$\overrightarrow{x}_g := T_e R_g(x) \,.$$

It is clear that the vector fields $\overleftarrow{x}$ and $\overrightarrow{x}$ are determined by their value, x, at e. This fact is closely related to an invariance property which we now define. A vector field $\mathscr{V}$ on $\mathbf{G}$ is said to be *left-invariant* (respectively *right-invariant*) if $T_h L_g(\mathscr{V}_h) = \mathscr{V}_{gh}$, for all $g, h \in \mathbf{G}$ (respectively, $T_g R_h(\mathscr{V}_g) = \mathscr{V}_{gh}$, for all $g, h \in \mathbf{G}$). Each one of the vector fields $\overleftarrow{x}$, where $x \in T_e\mathbf{G}$, is left-invariant, since

$$T_h L_g(\overleftarrow{x}_h) = T_h L_g(T_e L_h(x)) = T_e(L_g \circ L_h)(x) = T_e L_{gh}(x) = \overleftarrow{x}_{gh} \,,$$

for all $g, h \in \mathbf{G}$. Similarly, each of the vector fields $\overrightarrow{x}$ is right-invariant. It follows that there is a natural identification between $T_e\mathbf{G}$ and the vector space of all left-invariant vector fields on $\mathbf{G}$ (or the vector space of all right-invariant vector fields on $\mathbf{G}$). Since the commutator of two left-invariant vector fields on $\mathbf{G}$ is a left-invariant vector field on $\mathbf{G}$, the tangent space $T_e\mathbf{G}$ at the unit of a Lie group $\mathbf{G}$ has a natural Lie algebra structure. It leads to the following basic correspondence between Lie groups and Lie algebras (for a proof, see [66]).

Theorem 5.1 (Lie's theorem). *Let* $\mathbf{G}$ *and* $\mathbf{H}$ *be Lie groups, whose units are denoted by* e *and* e' *respectively, and let* $\Phi : \mathbf{G} \to \mathbf{H}$ *be a Lie group homomorphism.*

(1) The tangent space $T_e\mathbf{G}$ *is equipped with a natural Lie bracket* $[\cdot\,,\cdot]$*;*
(2) The linear map $T_e\Phi : T_e\mathbf{G} \to T_{e'}\mathbf{H}$ *is a morphism of Lie algebras.*

The Lie algebra $(T_e\mathbf{G}, [\cdot\,,\cdot])$ *is called the* Lie algebra of $\mathbf{G}$ *and is denoted by* $(\mathfrak{g}, [\cdot\,,\cdot])$ *(or* $\mathfrak{g}$*, for short). Denoting* $\mathfrak{h} := T_{e'}\mathbf{H}$*, the linear map* $T_e\Phi : \mathfrak{g} \to \mathfrak{h}$ *is called the* Lie algebra homomorphism associated to Φ.

Conversely, let $\mathfrak{g}$ *and* $\mathfrak{h}$ *be finite-dimensional Lie algebras and let* $\phi : \mathfrak{g} \to \mathfrak{h}$ *be a Lie algebra homomorphism.*

(3) There exists a connected and simply connected Lie group, whose Lie algebra is isomorphic to $\mathfrak{g}$*; up to isomorphism, such a Lie group is unique;*
(4) Let $\mathbf{G}$ *be a connected and simply connected Lie group with Lie algebra* $\mathfrak{g}$*, and let* $\mathbf{H}$ *be a Lie group with Lie algebra* $\mathfrak{h}$*. There exists a unique Lie group homomorphism* $\Phi : \mathbf{G} \to \mathbf{H}$*, such that* ϕ *is the Lie algebra homomorphism associated to* Φ*;*

(5) *Let* **G** *be a connected and simply connected Lie group with Lie algebra* $\mathfrak{g}$, *and let* $\mathfrak{h}$ *be a Lie subalgebra of* $\mathfrak{g}$. *There exists a unique connected Lie subgroup* **H** *of* **G** *with Lie algebra* $\mathfrak{h}$.

In a different vein, a Lie group and its Lie algebra are related as follows: there exists a diffeomorphism exp between a neighborhood of the origin o in $\mathfrak{g}$ and a neighborhood of the unit e in **G**, having the fundamental property that for every $x \in \mathfrak{g}$, the integral curve of $\overleftarrow{x}$ which starts from o is given (for small $|t|$) by $t \mapsto \exp(tx)$. In fact, one possible construction of the diffeomorphism exp is by integrating, for x in a neighborhood of o in $\mathfrak{g}$, the left-invariant vector fields $\overleftarrow{x}$. We refer to exp as the *exponential map*.

We end this section with a few comments on algebraic groups, which for us will always be complex (defined over $\mathbb{C}$). By definition, a *(complex) algebraic group* is an algebraic variety with a group structure, such that the product map and the inverse map are morphisms (of algebraic varieties). Much of what we discussed in this section carries over to algebraic groups, for example there is canonically associated to every algebraic group **G** a Lie algebra, whose underlying vector space $\mathfrak{g}$ is the (Zariski) tangent space to **G** at the unit e of **G**. However, there is in general no algebraic analog of the exponential map and parts *(3)* and *(4)* of Lie's theorem do not hold in general. When dealing with quotients of algebraic varieties by algebraic group actions, we will often assume that the algebraic group is *reductive*, which means that all its finite-dimensional representations are completely reducible; as we will explain at the end of Section 5.1.2, such quotients are rather well-behaved.

5.1.2 Lie Group Actions

Given a Lie group **G** and a manifold M, a group action $\chi : \mathbf{G} \times M \to M$ is called a *Lie group action*, if χ is smooth (or holomorphic); since the only actions of Lie groups on manifolds which are considered here are Lie group actions, we will simply call them *group actions*. When the action is clear, we write gm for $\chi(g,m)$, where $(g,m) \in \mathbf{G} \times M$. Clearly, a group action of **G** on M induces a group action of **G** on $\mathscr{F}(M)$, the algebra of smooth (or holomorphic) functions on M, by defining for $g \in \mathbf{G}$ and $F \in \mathscr{F}(M)$

$$(g \cdot F)(m) := F(g^{-1}m) , \tag{5.4}$$

for all $m \in M$. The virtue of the latter group action is that it is a representation of **G** (on $\mathscr{F}(M)$); a *representation* of a Lie group **G** on a vector space V is a group action $\chi : \mathbf{G} \times V \to V$ with the property that for every $g \in \mathbf{G}$ the map $\chi_g : V \to V$, defined for all $x \in \mathfrak{g}$ by $\chi_g(x) := \chi(g,x)$, is a linear map (an endomorphism of V). The action of an algebraic group on an algebraic variety is defined in an analogous way, demanding now that the map χ is a morphism of algebraic varieties.

The main group actions which we will consider in this chapter are the group actions of **G** on itself, given by

$$
\begin{aligned}
L &: \mathbf{G}\times\mathbf{G}\to\mathbf{G} : (g,h)\mapsto gh\,,\\
R &: \mathbf{G}\times\mathbf{G}\to\mathbf{G} : (g,h)\mapsto hg^{-1}\\
C &: \mathbf{G}\times\mathbf{G}\to\mathbf{G} : (g,h)\mapsto ghg^{-1}\,.
\end{aligned}
$$

These actions are called *left translation*, *right translation* and *conjugation* (in that order). Notice that taking the inverse of g in the definition of R is essential to make it into an action (otherwise it would be a so-called *right action*); we stress that it amounts to $R(g,h) = R_{g^{-1}}(h)$ for all $g,h \in \mathbf{G}$.

For a fixed group action χ of $\mathbf{G}$ on a manifold M, the *fundamental vector field* associated to $x \in T_e\mathbf{G}$, denoted $\underline{x}$, is the vector field on M, whose value at $m \in M$ is given by

$$
\underline{x}_m[F] := \frac{\mathrm{d}}{\mathrm{d}t}_{|t=0} F(\chi(\exp(-tx),m))\,, \tag{5.5}
$$

for all functions F, defined on a neighborhood of m in M. Writing $g = \exp(tx)$, one sees that Eq. (5.5) is the infinitesimal version of (5.4). Applied to the case of right translation (recall that $R(g,h) = R_{g^{-1}}h = hg^{-1}$), this yields for all functions F, defined in a neighborhood of g in $\mathbf{G}$,

$$
\begin{aligned}
\underline{x}_g[F] &= \frac{\mathrm{d}}{\mathrm{d}t}_{|t=0} F(R(\exp(-tx),g)) = \frac{\mathrm{d}}{\mathrm{d}t}_{|t=0} F(g\exp(tx))\\
&= \frac{\mathrm{d}}{\mathrm{d}t}_{|t=0} ((F\circ L_g)\exp(tx)) = x[F\circ L_g]\\
&= T_eL_g(x)[F] = \overleftarrow{x}_g[F]\,.
\end{aligned}
$$

We conclude that $\underline{x}_g = \overleftarrow{x}_g$, for all $g \in \mathbf{G}$, i.e., $\underline{x} = \overleftarrow{x}$. In words: the *left*-invariant vector field on $\mathbf{G}$, whose value at e is x, is the fundamental vector field of *right* translation, corresponding to x. Similarly, the fundamental vector field of left translation on $\mathbf{G}$, corresponding to $x \in \mathfrak{g}$, is given by $-\overrightarrow{x}$. Also, since $C_g = R_{g^{-1}} \circ L_g$, the fundamental vector field of conjugation on $\mathbf{G}$, corresponding to $x \in \mathfrak{g}$, is given by $\overleftarrow{x} - \overrightarrow{x}$. The identity

$$
\mathrm{inv}\circ R_g = L_{g^{-1}}\circ \mathrm{inv}
$$

implies that a vector field $\mathscr{V}$ on $\mathbf{G}$ is right-invariant if and only if the vector field on $\mathbf{G}$, defined[1] by $\mathrm{inv}_*\mathscr{V}$, is left-invariant. It follows, using $T_e\,\mathrm{inv} = -\mathbb{1}_{T_e\mathbf{G}}$, that $\mathrm{inv}_*\overleftarrow{x} = -\overrightarrow{x}$, for every $x \in \mathfrak{g}$.

Given a group action of a Lie group $\mathbf{G}$ on a manifold M, the map $\mathfrak{g} \to \mathfrak{X}^1(M)$, which associates, as above, to $x \in \mathfrak{g}$ the fundamental vector field $\underline{x}$ on M, admits as a natural generalization a map $\wedge^\bullet\mathfrak{g} \to \mathfrak{X}^\bullet(M)$, which associates to a multivector X of $\mathfrak{g}$ a multivector field $\underline{X}$ on M. For $X \in \wedge^p\mathfrak{g}$ of the form $X = x_1\wedge\cdots\wedge x_p$, the p-vector field $\underline{X}$ is defined as

[1] Recall that for a diffeomorphism $\Phi : M \to N$ between manifolds and for $\mathscr{V}$ a vector field on M, we denote by $\Phi_*\mathscr{V}$ the vector field on N, which is defined by $(\Phi_*\mathscr{V})_{\Phi(m)} := T_m\Phi(\mathscr{V}_m)$, for all $m \in M$, and similarly for multivector fields on M.

$$\underline{X} := \underline{x_1} \wedge \cdots \wedge \underline{x_p}$$

and is called the *fundamental p-vector field*, corresponding to X. In the case of right and left translations on a Lie group $\mathbf{G}$, we write $\overleftarrow{X}$, respectively $(-1)^p\overrightarrow{X}$ for the fundamental multivector field, associated to $X \in \wedge^p\mathfrak{g}$. Equivalently, for $X \in \wedge^p\mathfrak{g}$, the vector fields $\overleftarrow{X}$ and $\overrightarrow{X}$ can be defined by $\overleftarrow{X}_g := \wedge^p(T_eL_g)X$ and $\overrightarrow{X}_g := \wedge^p(T_eR_g)X$, for all $g \in \mathbf{G}$. As above, it follows from this definition that $\overleftarrow{X}$ is a left-invariant multivector field and that $\overrightarrow{X}$ is a right-invariant multivector field.

We show in the following proposition that the map, which sends $X \in \wedge^\bullet\mathfrak{g}$ to $\overleftarrow{X}$, is a homomorphism of Gerstenhaber algebras.[2]

Proposition 5.2. *Let* $\mathbf{G}$ *be a Lie group with Lie algebra* $\mathfrak{g}$. *For all* $X,Y \in \wedge^\bullet\mathfrak{g}$, *we have*

$$\overleftarrow{X \wedge Y} = \overleftarrow{X} \wedge \overleftarrow{Y}\,, \qquad \overleftarrow{[\![X,Y]\!]} = \left[\overleftarrow{X}, \overleftarrow{Y}\right]_S, \tag{5.6}$$

so the map $X \mapsto \overleftarrow{X}$ *preserves the wedge product and the Schouten bracket. Up to a sign, the same is true for the map* $X \mapsto \overrightarrow{X}$: *for all* $X,Y \in \wedge^\bullet\mathfrak{g}$, *we have*

$$\overrightarrow{X \wedge Y} = \overrightarrow{X} \wedge \overrightarrow{Y}\,, \qquad \overrightarrow{[\![X,Y]\!]} = -\left[\overrightarrow{X}, \overrightarrow{Y}\right]_S. \tag{5.7}$$

Moreover,

$$\left[\overleftarrow{X}, \overrightarrow{Y}\right]_S = 0\,, \tag{5.8}$$

for all $X,Y \in \wedge^\bullet\mathfrak{g}$.

Proof. Since every element of $\wedge^\bullet\mathfrak{g}$ is a sum of elements of the form $x_1 \wedge \cdots \wedge x_p$, where $1 \leqslant p \leqslant \dim\mathfrak{g}$, and $x_1,\ldots,x_p \in \mathfrak{g}$, it suffices to prove (5.6)–(5.8) for X and Y of this form. Then the first equation in (5.6) follows from the associativity of $\wedge$ on $\wedge^\bullet\mathfrak{g}$ and on $\mathfrak{X}^\bullet(\mathbf{G})$. The second equation in (5.6) clearly holds when X and Y belong to $\mathfrak{g}$; upon using the graded Leibniz identities (3.42) and (3.47) of the Schouten brackets $[\![\cdot,\cdot]\!]$ on $\wedge^\bullet\mathfrak{g}$ and $[\cdot,\cdot]_S$ on $\mathfrak{X}^\bullet(\mathbf{G})$, it follows that this formula holds for general $X,Y \in \wedge^\bullet\mathfrak{g}^*$. Since $\mathrm{inv}_*\,\overleftarrow{x} = -\overrightarrow{x}$, for all $x \in \mathfrak{g}$, we have $\mathrm{inv}_*\,\overleftarrow{X} = (-1)^p\overrightarrow{X}$, for all $X \in \wedge^p\mathfrak{g}$; therefore, applying inv_* to both equations in (5.6), we find the equations in (5.7). We now prove (5.8). In view of the graded Leibniz rule for the Schouten bracket, it suffices to prove (5.8) when X and Y are elements of $\mathfrak{g}$, which we write as x and y. The vector fields $\overleftarrow{x}$ and $\overrightarrow{y}$ are the fundamental vector fields of the actions of right, respectively left, translation of $\mathbf{G}$ on itself. These actions commute, hence the flows of $\overleftarrow{x}$ and $\overrightarrow{y}$ also commute. The commutator $[\overleftarrow{x}, \overrightarrow{y}]$ is therefore zero. □

For algebraic groups there is, in general, no exponential map, but for linear actions on vector spaces (representations), such as the actions which we will consider in the next section, taking the differential of the action, viewed as a map $\chi : \mathbf{G} \to \mathrm{End}(V)$, yields a linear vector field on V, which is the analog (up to a sign) of the fundamental vector field, defined in (5.5).

---

[2] Recall that $(\wedge^\bullet\mathfrak{g}, \wedge, [\![\cdot,\cdot]\!])$ is a Gerstenhaber algebra, just like $(\mathfrak{X}^\bullet(M), \wedge, [\cdot,\cdot]_S)$, where M is an arbitrary manifold (see Proposition 3.9).

To finish this section, we recall some rather general conditions under which the quotient space $M/\mathbf{G}$ of a group action $\mathbf{G}\times M\to M$ is of the same type as M, i.e., is an affine variety or a manifold.

(1) Suppose that M is an affine variety on which a reductive algebraic group $\mathbf{G}$ acts. Then the algebra of $\mathbf{G}$-invariant functions $\mathscr{F}(M)^{\mathbf{G}}$ is finitely generated, hence is the algebra of regular functions on an affine variety, which can be identified with the orbit space $M/\mathbf{G}$. Since, in particular, every finite group $\mathbf{G}$ is reductive, the quotient of an affine variety by a finite group is an affine variety.

(2) Suppose that M is a manifold and that $\mathbf{G}$ is a Lie group, which acts properly and locally freely on M. Then the orbit space $M/\mathbf{G}$ has a unique manifold structure for which the canonical map $p:M\to M/\mathbf{G}$ is smooth. Interesting particular cases include free actions of a finite group and locally free actions of a compact group on a manifold. We recall that an action $\chi:\mathbf{G}\times M\to M$ is called a *proper action*, if the map

$$\begin{aligned}\chi\times p_2:\mathbf{G}\times M&\to M\times M\\(g,m)&\mapsto(gm,m)\end{aligned}\tag{5.9}$$

is a proper map, i.e., the inverse image of every compact subset of $M\times M$ is a compact subset of $\mathbf{G}\times M$. Also, χ is called a *locally free action* if, for every $m\in M$, there exists a neighborhood U of the identity in $\mathbf{G}$ such that the restriction to U of the map $g\mapsto gm$ is injective.

5.1.3 Adjoint and Coadjoint Action

We next consider the adjoint action of a Lie group $\mathbf{G}$ on its Lie algebra $\mathfrak{g}$, and more generally on the space $\wedge^\bullet\mathfrak{g}$; we will also consider its infinitesimal version, which is a Lie algebra action of $\mathfrak{g}$ on itself, and more generally on $\wedge^\bullet\mathfrak{g}$. Dually, we will also consider the corresponding group action of $\mathbf{G}$ and the corresponding Lie algebra action of $\mathfrak{g}$ on its dual vector space $\mathfrak{g}^*$ and on $\wedge^\bullet\mathfrak{g}^*$. The formulas are most naturally written in terms on the natural pairing $\langle\cdot,\cdot\rangle$ between $\wedge^p\mathfrak{g}^*$ and $\wedge^p\mathfrak{g}$, given for all $\xi_1,\dots,\xi_p\in\mathfrak{g}^*$ and for all $x_1,\dots,x_p\in\mathfrak{g}$ by

$$\langle\xi_1\wedge\cdots\wedge\xi_p,x_1,\dots,x_p\rangle:=\begin{vmatrix}\xi_1[x_1]&\cdots&\xi_1[x_p]\\\vdots&&\vdots\\\xi_p[x_1]&\cdots&\xi_p[x_p]\end{vmatrix}.\tag{5.10}$$

For $g\in\mathbf{G}$, the conjugation map $C_g:\mathbf{G}\to\mathbf{G}$ is a Lie group automorphism of $\mathbf{G}$. According to Theorem 5.1, the tangent map to C_g at e is a Lie algebra automorphism of $\mathfrak{g}$. It is customary to denote this map by Ad_g. Thus, $\mathrm{Ad}_g:=T_eC_g:\mathfrak{g}\to\mathfrak{g}$ and

$$\mathrm{Ad}_g([x,y])=[\mathrm{Ad}_g x,\mathrm{Ad}_g y]\ ,$$

for all $x,y \in \mathfrak{g}$. For $g,h \in \mathbf{G}$, we have that

$$\mathrm{Ad}_{gh} = T_e C_{gh} = T_e(C_g \circ C_h) = (T_e C_g) \circ (T_e C_h) = \mathrm{Ad}_g \circ \mathrm{Ad}_h ,$$

so that the assignment $(g,x) \mapsto \mathrm{Ad}_g x$ defines a group action of $\mathbf{G}$ on $\mathfrak{g}$. It is called the *adjoint action* of $\mathbf{G}$ on $\mathfrak{g}$, denoted Ad. Since the maps Ad_g are linear maps, Ad is actually a representation of $\mathbf{G}$, the *adjoint representation* of $\mathbf{G}$ on $\mathfrak{g}$. The transpose map of $\mathrm{Ad}_{g^{-1}}$ is denoted by Ad^*_g: for all $\xi \in \mathfrak{g}^*$ and for all $x \in \mathfrak{g}$,

$$\left\langle \mathrm{Ad}^*_g \xi, x \right\rangle = \left\langle \xi, \mathrm{Ad}_{g^{-1}} x \right\rangle .$$

The resulting map $(g,\xi) \mapsto \mathrm{Ad}^*_g \xi$ is a group action of $\mathbf{G}$ on $\mathfrak{g}^*$, called the *coadjoint action* of $\mathbf{G}$ on $\mathfrak{g}^*$, denoted Ad^*. As in the case of the adjoint action, *Ad** is a representation of $\mathbf{G}$, the *coadjoint representation* of $\mathbf{G}$ on $\mathfrak{g}$.

These adjoint and coadjoint actions extend to group actions of $\mathbf{G}$ on $\wedge^\bullet \mathfrak{g}$ and on $\wedge^\bullet \mathfrak{g}^*$, also denoted by Ad and Ad^*, upon setting

$$\mathrm{Ad}_g(x_1 \wedge \cdots \wedge x_p) := \mathrm{Ad}_g x_1 \wedge \cdots \wedge \mathrm{Ad}_g x_p , \tag{5.11}$$

$$\mathrm{Ad}^*_g(\xi_1 \wedge \cdots \wedge \xi_p) := \mathrm{Ad}^*_g \xi_1 \wedge \cdots \wedge \mathrm{Ad}^*_g \xi_p , \tag{5.12}$$

for all $g \in \mathbf{G}$, all $p \in \mathbb{N}^*$, all $x_1,\ldots,x_p \in \mathfrak{g}$ and all $\xi_1,\ldots,\xi_p \in \mathfrak{g}^*$. It follows from these definitions that

$$\left\langle \mathrm{Ad}^*_g \Omega, X \right\rangle = \left\langle \Omega, \mathrm{Ad}_{g^{-1}} X \right\rangle ,$$

for all $g \in \mathbf{G}$, all $\Omega \in \wedge^p \mathfrak{g}^*$ and all $X \in \wedge^p \mathfrak{g}$.

The tangent map of $\mathrm{Ad} : \mathbf{G} \to \mathrm{End}(\mathfrak{g})$ at e is the linear map

$$\mathrm{ad} := T_e \mathrm{Ad} : \mathfrak{g} \to \mathrm{End}(\mathfrak{g}) ,$$

called the *adjoint representation* of $\mathfrak{g}$. Writing ad_x for the image of $x \in \mathfrak{g}$ by ad, it is a fundamental fact that the endomorphism ad_x is given by $\mathrm{ad}_x(y) = [x,y]$, for all $y \in \mathfrak{g}$. It follows that ad is indeed a Lie algebra representation: $\mathrm{ad}_{[x,y]} = \mathrm{ad}_x \circ \mathrm{ad}_y - \mathrm{ad}_y \circ \mathrm{ad}_x$, for all $x,y \in \mathfrak{g}$. More generally, taking the tangent map of $\mathrm{Ad} : \mathbf{G} \to \mathrm{End}(\wedge^\bullet \mathfrak{g})$ at e we obtain the *adjoint representation* of $\mathfrak{g}$ on $\wedge^\bullet \mathfrak{g}$, still denoted by ad. Equation (5.11) implies that

$$\mathrm{ad}_x(x_1 \wedge \cdots \wedge x_p) = \sum_{i=1}^{p} x_1 \wedge \cdots \wedge [x,x_i] \wedge \cdots \wedge x_p ,$$

for all $x_1,\ldots,x_p \in \mathfrak{g}$ hence that $\mathrm{ad}_x X = [\![x,X]\!]$, for all $X \in \wedge^\bullet \mathfrak{g}$. In view of the graded Jacobi identity for the algebraic Schouten bracket $[\![\cdot\,,\cdot]\!]$, we have for all $x,y \in \mathfrak{g}$ and for all $X \in \wedge^\bullet \mathfrak{g}$ that

$$\mathrm{ad}_{[x,y]} X = [\![[x,y]\,,X]\!] = [\![x,[\![y,X]\!]]\!] - [\![y,[\![x,X]\!]]\!] = \mathrm{ad}_x(\mathrm{ad}_y X) - \mathrm{ad}_y(\mathrm{ad}_x X) ,$$

so that $\mathrm{ad}_{[x,y]} = \mathrm{ad}_x \circ \mathrm{ad}_y - \mathrm{ad}_y \circ \mathrm{ad}_x$, which shows that ad is indeed a Lie algebra representation of $\mathfrak{g}$ (on $\wedge^\bullet \mathfrak{g}$). Similarly, we obtain the coadjoint representation by taking the tangent map of $\mathrm{Ad}^* : \mathbf{G} \to \mathrm{End}(\wedge^\bullet \mathfrak{g}^*)$ at e, which yields $\mathrm{ad}^* : \mathfrak{g} \to \mathrm{End}(\wedge^\bullet \mathfrak{g}^*)$. As above, ad^* is for all $x \in \mathfrak{g}$ a derivation of $\wedge^\bullet \mathfrak{g}^*$ and ad_x^* is related to ad_x by

$$\langle \mathrm{ad}_x^* \Omega, X \rangle = -\langle \Omega, \mathrm{ad}_x X \rangle = -\langle \Omega, [\![x, X]\!] \rangle , \tag{5.13}$$

for all $\Omega \in \wedge^p \mathfrak{g}^*$ and all $X \in \wedge^p \mathfrak{g}$.

To finish this section, we shortly discuss invariant multivectors. An element $X \in \wedge^\bullet \mathfrak{g}$ is said to be Ad-*invariant* (respectively ad-invariant) if $\mathrm{Ad}_g X = X$, for all $g \in \mathbf{G}$ (respectively $\mathrm{ad}_x X = 0$, for all $x \in \mathfrak{g}$). The following lemma relates Ad and ad-invariance and gives a useful characterization in terms of the associated infinitesimal multivector fields.

Proposition 5.3. *Let* $\mathbf{G}$ *be a Lie group with Lie algebra* $\mathfrak{g}$*, and let* $X \in \wedge^\bullet \mathfrak{g}$.

(1) *If* X *is* Ad-*invariant, then* X *is* ad-*invariant;*
(2) *If* $\mathbf{G}$ *is connected, then* X *is* Ad-*invariant if and only if* X *is* ad-*invariant;*
(3) X *is* Ad-*invariant if and only if* $\overleftarrow{X} = \overrightarrow{X}$.

Proof. Let $X \in \wedge^\bullet \mathfrak{g}$. If X is Ad-invariant, then the map $\mathbf{G} \to \wedge^\bullet \mathfrak{g}$, defined for $g \in \mathbf{G}$ by $g \mapsto \mathrm{Ad}_g X$ is a constant map, hence its tangent map is zero, $\mathrm{ad}_x X = T_e \mathrm{Ad}(x)(X) = 0$, for all $x \in \mathfrak{g}$. Conversely, the latter map is locally a constant map on $\mathbf{G}$ when $\mathrm{ad}_x X = 0$ for all $x \in \mathfrak{g}$; if $\mathbf{G}$ is connected, this implies that $g \mapsto \mathrm{Ad}_g X$ is a constant map, hence that X is Ad-invariant. This proves *(1)* and *(2)*. For $X \in \wedge^\bullet \mathfrak{g}$, say $X \in \wedge^p \mathfrak{g}$, and for $g \in \mathbf{G}$, the equality $\overleftarrow{X}_g = \overrightarrow{X}_g$, which can also be written as

$$\wedge^p (T_e L_g) X = \wedge^p (T_e R_g) X ,$$

is equivalent to $\wedge^p \mathrm{Ad}_g X = X$, because

$$\wedge^p \mathrm{Ad}_g = \wedge^p (T_e C_g) = \wedge^p (T_g R_{g^{-1}}) \circ \wedge^p (T_e L_g) = \wedge^p (T_e R_g)^{-1} \circ \wedge^p (T_e L_g) .$$

Therefore, $\overleftarrow{X}_g = \overrightarrow{X}_g$ for all $g \in \mathbf{G}$ if and only if X is Ad-invariant, which is the content of *(3)*. □

5.1.4 Invariant Bilinear Forms and Invariant Functions

Let $\mathbf{G}$ be a Lie group and let $\mathfrak{g}$ be its Lie algebra. A bilinear form $\langle \cdot \,|\, \cdot \rangle$ on $\mathfrak{g}$ is called Ad-*invariant*, if for all $g \in \mathbf{G}$ and for all $x, y \in \mathfrak{g}$,

$$\langle \mathrm{Ad}_g x \,|\, \mathrm{Ad}_g y \rangle = \langle x \,|\, y \rangle .$$

If $\langle \cdot \,|\, \cdot \rangle$ is Ad-invariant, then

$$\langle [z, x] \,|\, y \rangle + \langle x \,|\, [z, y] \rangle = 0 ,$$

for all $x,y,z \in \mathfrak{g}$; a bilinear form on $\mathfrak{g}$ satisfying the latter condition is called *ad-invariant*. When $\mathbf{G}$ is connected, every bilinear form which is ad-invariant is Ad-invariant, so the two notions of invariance need not be distinguished.

We often use a non-degenerate symmetric bilinear form to identify $\mathfrak{g}$ with its dual vector space $\mathfrak{g}^*$. *Non-degeneracy* of a bilinear form $\langle \cdot | \cdot \rangle$ means that the linear map $\chi : \mathfrak{g} \to \mathfrak{g}^*$, which sends $x \in \mathfrak{g}$ to the linear form $\langle x | \cdot \rangle : y \mapsto \langle x | y \rangle$, is an isomorphism. Its inverse will be denoted by ψ. Explicitly, χ is given by

$$\langle \chi(x), y \rangle = \langle x | y \rangle = \langle \chi(y), x \rangle \ ,$$

for all $x,y \in \mathfrak{g}$. If $\langle \cdot | \cdot \rangle$ is moreover Ad-invariant, then for every $g \in \mathbf{G}$ and for all $x,y \in \mathfrak{g}$ it follows from the definitions that

$$\begin{aligned} \langle \mathrm{Ad}^*_g(\chi(x)), y \rangle &= \left\langle \chi(x), \mathrm{Ad}_{g^{-1}} y \right\rangle = \left\langle x | \, \mathrm{Ad}_{g^{-1}} y \right\rangle \\ &= \langle \mathrm{Ad}_g x | y \rangle = \langle \chi(\mathrm{Ad}_g x), y \rangle \ , \end{aligned}$$

so that the following diagram is commutative:

$$\begin{array}{ccc} \mathfrak{g} & \xrightarrow{\ \chi\ } & \mathfrak{g}^* \\ {\scriptstyle \mathrm{Ad}_g}\downarrow & & \downarrow{\scriptstyle \mathrm{Ad}^*_g} \\ \mathfrak{g} & \xrightarrow[\ \chi\ ]{} & \mathfrak{g}^* \end{array} \tag{5.14}$$

A Lie algebra $\mathfrak{g}$, equipped with a non-degenerate symmetric bilinear form $\langle \cdot | \cdot \rangle$, which is ad-invariant is called a *quadratic Lie algebra*.

A function F on $\mathfrak{g}$ is called Ad-*invariant* if it is invariant for the adjoint action of $\mathbf{G}$ on $\mathfrak{g}$, i.e., if $F(\mathrm{Ad}_g x) = F(x)$ for all $g \in \mathbf{G}$ and $x \in \mathfrak{g}$. Similarly, a function F on $\mathfrak{g}^*$ is called Ad^*-*invariant* if it is invariant for the coadjoint action of $\mathbf{G}$ on $\mathfrak{g}^*$. The Ad^*-invariance of a function F on $\mathfrak{g}^*$ is characterized infinitesimally by

$$\mathrm{ad}^*_{\mathrm{d}_\xi F} \xi = 0 \ , \tag{5.15}$$

for all $\xi \in \mathfrak{g}^*$. Indeed, for $z \in \mathfrak{g}$ and $\xi \in \mathfrak{g}^*$,

$$\left\langle \mathrm{ad}^*_{\mathrm{d}_\xi F} \xi, z \right\rangle = \langle \xi, [z, \mathrm{d}_\xi F] \rangle = -\langle \mathrm{d}_\xi F, \mathrm{ad}^*_z \xi \rangle = 0 \ ,$$

where we used in the last step that $\frac{\mathrm{d}}{\mathrm{d}t}_{|t=0} F(\mathrm{Ad}^*_{\exp tz} \xi) = 0$, itself a direct consequence of the Ad^*-invariance of F. If $F : \mathfrak{g}^* \to \mathbb{F}$ is an Ad^*-invariant function and if the bilinear form $\langle \cdot | \cdot \rangle$ is Ad-invariant, then the commutativity of (5.14) leads for every $g \in \mathbf{G}$ to

$$F \circ \chi \circ \mathrm{Ad}_g = F \circ \mathrm{Ad}^*_g \circ \chi = F \circ \chi \ ,$$

so that $F \circ \chi$ is Ad-invariant. Similarly, $\langle \cdot | \cdot \rangle$ is Ad-invariant and $F \circ \chi$ is Ad-invariant, then F is Ad^*-invariant. Thus, χ establishes a one-to-one correspondence between Ad^*-invariant functions on $\mathfrak{g}^*$ and Ad-invariant functions on $\mathfrak{g}$. For a function $F : \mathfrak{g} \to \mathbb{F}$, the Ad-invariance of $\langle \cdot \mid \cdot \rangle$ also leads to

$$\psi\left(\mathrm{ad}^*_{\mathrm{d}_\xi (F\circ\psi)}\, \xi\right) = \left[\nabla_{\psi(\xi)} F, \psi(\xi)\right] , \tag{5.16}$$

where $\nabla_x F := \psi(\mathrm{d}_x F)$, for $x \in \mathfrak{g}$. It follows that a function F on $\mathfrak{g}$ is Ad-invariant if and only if $[\nabla_x F, x] = 0$ for all $x \in \mathfrak{g}$. The proof of (5.16) follows at once from the definitions of ψ, ad, ∇ and the ad-invariance of $\langle \cdot \mid \cdot \rangle$.

5.2 Poisson Reduction

In many geometrical situations of interest, one deals with a quotient of a submanifold of a Poisson manifold, and while the submanifold itself does not inherit a Poisson structure from its ambient space (i.e., it is not a Poisson submanifold, as in Section 2.2), the quotient *does* inherit a Poisson structure. The construction of such a Poisson structure is called *Poisson reduction*, while the Poisson structure itself is called a *reduced Poisson structure*. We give in this section necessary and sufficient conditions for Poisson reduction: we formulate them first algebraically (Section 5.2.1) and then for manifolds (Section 5.2.2). The geometrical picture to keep in mind, reading this section, consists of the following manifolds and maps:

$$\begin{array}{c} N \subset M \\ p \big\downarrow \\ P \end{array} \tag{5.17}$$

where $M = (M, \pi)$ is a Poisson manifold, N is a submanifold of M and p is a submersion. The example to keep in mind is that of a Poisson manifold on which a group $\mathbf{G}$ acts, a submanifold N which is $\mathbf{G}$-invariant and the quotient $P = N/\mathbf{G}$. We will come back to this example in Section 5.4.2.

5.2.1 Algebraic Poisson Reduction

We first describe Poisson reduction in purely algebraic terms. Dualizing the diagram (5.17), we consider a Poisson algebra $(\mathscr{A}, \cdot, \{\cdot,\cdot\})$, an ideal $\mathscr{I}$ of $(\mathscr{A}, \cdot)$, the canonical map $\kappa : \mathscr{A} \to \mathscr{A}/\mathscr{I}$ and a subalgebra $\mathscr{B}$ of $\mathscr{A}/\mathscr{I}$, as in the following diagram:

$$\begin{array}{c} \mathscr{A} \\ \Big\downarrow \kappa \\ \mathscr{B} \subset \mathscr{A}/\mathscr{I} \end{array} \tag{5.18}$$

We would like to have a Poisson structure $\{\cdot,\cdot\}_{\mathscr{B}}$ on $\mathscr{B}$, such that $\{F,G\}_{\mathscr{B}} = \kappa\left(\{\tilde{F},\tilde{G}\}\right)$, for every $F,G \in \mathscr{B}$, where $\tilde{F}$ and $\tilde{G}$ are arbitrary representatives in $\mathscr{A}$ of F and G.

Definition 5.4. Let $(\mathscr{A},\cdot,\{\cdot,\cdot\})$ be a Poisson algebra, let $\mathscr{I}$ be an ideal of $(\mathscr{A},\cdot)$ and let $\mathscr{B}$ be a subalgebra of $\mathscr{A}/\mathscr{I}$. We say that the triple $(\mathscr{A},\mathscr{I},\mathscr{B})$ is *Poisson reducible* if there exists a Poisson bracket $\{\cdot,\cdot\}_{\mathscr{B}}$ on $\mathscr{B}$, such that

$$\{F,G\}_{\mathscr{B}} = \kappa\left(\{\tilde{F},\tilde{G}\}\right) , \tag{5.19}$$

for all $F,G \in \mathscr{B}$, where $\tilde{F},\tilde{G}$ are arbitrary representatives of F and G in $\mathscr{A}$. The Poisson bracket $\{\cdot,\cdot\}_{\mathscr{B}}$ on $\mathscr{B}$ is called the *reduced Poisson bracket*.

We give in the following proposition a necessary and sufficient condition for a triple $(\mathscr{A},\mathscr{I},\mathscr{B})$ to be Poisson reducible. For the ideal $\mathscr{I}$ of $(\mathscr{A},\cdot)$ we denote its *normalizer* in $(\mathscr{A},\{\cdot,\cdot\})$ by $\mathscr{N}(\mathscr{I})$,

$$\mathscr{N}(\mathscr{I}) = \{F \in \mathscr{A} \mid \{\mathscr{I},F\} \subset \mathscr{I}\} . \tag{5.20}$$

Clearly, $\mathscr{N}(\mathscr{I})$ is a Poisson subalgebra of $\mathscr{A}$: since $\{\cdot,\cdot\}$ is a biderivation, $\mathscr{N}(\mathscr{I})$ is a subalgebra of $(\mathscr{A},\cdot)$, while the Jacobi identity for $\{\cdot,\cdot\}$ implies that $\mathscr{N}(\mathscr{I})$ is a Lie subalgebra of $(\mathscr{A},\{\cdot,\cdot\})$.

Proposition 5.5. *Let $(\mathscr{A},\cdot,\{\cdot,\cdot\})$ be a Poisson algebra, let $\mathscr{I}$ be an ideal of $(\mathscr{A},\cdot)$ and let $\mathscr{B}$ be a subalgebra of $\mathscr{A}/\mathscr{I}$. Then $(\mathscr{A},\mathscr{I},\mathscr{B})$ is Poisson reducible if and only if the following conditions are satisfied:*

(1) $\kappa^{-1}(\mathscr{B}) \subset \mathscr{N}(\mathscr{I})$;
(2) $\kappa^{-1}(\mathscr{B})$ *is a Lie subalgebra of* $(\mathscr{A},\{\cdot,\cdot\})$.

For $F,G \in \mathscr{B}$, their reduced Poisson bracket $\{F,G\}_{\mathscr{B}}$ is then given by

$$\{F,G\}_{\mathscr{B}} = \kappa\left(\{\tilde{F},\tilde{G}\}\right) , \tag{5.21}$$

where $\tilde{F}$ and $\tilde{G}$ are arbitrary representatives in $\mathscr{A}$ of F and G.

Proof. Suppose first that conditions *(1)* and *(2)* are satisfied. We wish to define a Poisson bracket on $\mathscr{B}$ by using Eq. (5.21). Let $F,G \in \mathscr{B}$ and let $\tilde{F},\tilde{G}$ be representatives in $\mathscr{A}$ of F and G; both $\tilde{F}$ and $\tilde{G}$ are determined uniquely up to an element of $\mathscr{I}$. According to condition *(1)*, $\tilde{F}$ and $\tilde{G}$ belong to $\mathscr{N}(\mathscr{I})$, so their bracket $\{\tilde{F},\tilde{G}\}$ is, modulo $\mathscr{I}$, independent of the chosen representatives $\tilde{F}$ and $\tilde{G}$ for F and G. Condition *(2)* implies that $\{\tilde{F},\tilde{G}\} \in \kappa^{-1}(\mathscr{B})$, so that $\kappa\left(\{\tilde{F},\tilde{G}\}\right) \in \mathscr{B}$. It follows that (5.21) defines a map $\{\cdot,\cdot\}_{\mathscr{B}} : \mathscr{B} \times \mathscr{B} \to \mathscr{B}$. Since κ is an algebra homomorphism, the fact that $\{\cdot,\cdot\}$ is a skew-symmetric biderivation of $\mathscr{A}$ implies that $\{\cdot,\cdot\}_{\mathscr{B}}$ is a skew-symmetric biderivation of $\mathscr{B}$. For the Jacobi identity, observe that, for all $F,G,H \in \mathscr{B}$, (5.21) yields that

$$\{\{F,G\}_{\mathscr{B}},H\}_{\mathscr{B}} = \kappa\left(\{\{\tilde{F},\tilde{G}\},\tilde{H}\}\right)$$

where $\tilde{F},\tilde{G}$ and $\tilde{H}$ are arbitrary representatives of F,G and H. The Jacobi identity for $\{\cdot,\cdot\}_{\mathscr{B}}$ therefore follows from the Jacobi identity for the bracket $\{\cdot,\cdot\}$. We conclude that $(\mathscr{A},\mathscr{I},\mathscr{B})$ is Poisson reducible, in particular the reduced Poisson bracket on $\mathscr{B}$ is given by (5.21).

Suppose now that $(\mathscr{A},\mathscr{I},\mathscr{B})$ is Poisson reducible, so that $\mathscr{B}$ admits a Poisson bracket $\{\cdot,\cdot\}_{\mathscr{B}}$ which satisfies (5.19). Since $\tilde{F}$ and $\tilde{G}$ are arbitrary elements of $\kappa^{-1}(\mathscr{B})$, the latter formula implies that $\kappa\left(\{\kappa^{-1}(\mathscr{B}),\kappa^{-1}(\mathscr{B})\}\right) \subset \mathscr{B}$, which yields condition *(2)*. For arbitrary elements $\tilde{F} \in \kappa^{-1}(\mathscr{B})$ and $\tilde{G} \in \mathscr{I}$ we have from (5.19), since $\kappa(\tilde{G}) = 0$, that $\kappa\left(\{\tilde{F},\tilde{G}\}\right) = 0$, so that $\left\{\kappa^{-1}(\mathscr{B}),\mathscr{I}\right\} \subset \mathscr{I}$, which yields condition *(1)*. □

Notice that since $\kappa^{-1}(\mathscr{B})$ is a subalgebra of $(\mathscr{A},\cdot)$ and $\mathscr{N}(\mathscr{I})$ is a Poisson subalgebra of $(\mathscr{A},\cdot,\{\cdot,\cdot\})$, the two conditions *(1)* and *(2)* in Proposition 5.5 can be summarized in the single condition:

$$\kappa^{-1}(\mathscr{B}) \text{ is a Poisson subalgebra of } \mathscr{N}(\mathscr{I}).$$

Moreover, viewing the canonical map κ as a map $\kappa : \kappa^{-1}(\mathscr{B}) \to \mathscr{B}$, the formula (5.21) for the reduced Poisson bracket can be simply stated as κ being a morphism of Poisson algebras, where $\kappa^{-1}(\mathscr{B})$, which is a Poisson subalgebra of $\mathscr{N}(\mathscr{I})$, and hence of $\mathscr{A}$, is equipped with the Poisson bracket which it inherits from $\mathscr{A}$.

We next consider Poisson reduction in the setting of affine varieties. We do this by replacing each of the algebras in (5.18) by an affine variety, keeping in mind that under the basic correspondence between algebras of finite type and affine varieties, surjective algebra homomorphisms are replaced by injective maps between varieties (with the direction of the arrow being reversed) while injective algebra homomorphisms are replaced by dominant maps between varieties, i.e., maps whose image is (Zariski) dense in the target variety. We are thus led to consider the following diagram,

$$\begin{array}{c} N \subset M \\ p \big\downarrow \\ P \end{array} \qquad (5.22)$$

where M,N and P are affine varieties and p is a dominant map.

Definition 5.6. Let N be a subvariety of an affine Poisson variety $(M,\{\cdot,\cdot\})$ and let $p : N \to P$ be a dominant map, where P is also an affine variety. We say that the triple (M,N,P) is *Poisson reducible* if there exists a Poisson structure $\{\cdot,\cdot\}_P$ on P, such that, for every $n \in N$,

$$\{F,G\}_P(p(n)) = \{\tilde{F},\tilde{G}\}(n),$$

for all $F,G \in \mathscr{F}(P)$ and for all extensions $\tilde{F},\tilde{G} \in \mathscr{F}(M)$ of $F \circ p$ and $G \circ p$. The Poisson structure $\{\cdot,\cdot\}_P$ is called a *reduced Poisson structure*.

It is clear from the definitions that a triple of affine varieties (M,N,P) is Poisson reducible if and only if the triple of algebras $(\mathscr{A},\mathscr{I},\mathscr{B}) := (\mathscr{F}(M),\mathscr{I}_N,\mathscr{F}(P))$ is Poisson reducible, where $\mathscr{I}_N$ denotes the ideal of N. The geometric interpretation of the normalizer $\mathscr{N}(\mathscr{I}_N)$ of $\mathscr{I}_N$, defined in (5.20), is that

$$\mathscr{N}(\mathscr{I}_N) = \{F \in \mathscr{F}(M) \mid \mathscr{X}_F \text{ is tangent to } N \text{ at every point of } N\}\ . \qquad (5.23)$$

Also, $\kappa^{-1}(\mathscr{B}) = \kappa^{-1}(\mathscr{F}(P))$ consists of all functions on M, whose restriction to N is of the form $H \circ p$, for some function $H \in \mathscr{F}(P)$. It follows that Proposition 5.5 leads for affine varieties to the following result.

Proposition 5.7. *Let $(M,\{\cdot,\cdot\})$ be an affine Poisson variety, let N be a subvariety of M and let P be an affine variety. Suppose that $p : N \to P$ is a dominant map, as in diagram (5.22). Then (M,N,P) is Poisson reducible if and only if the following conditions are satisfied:*

(1) All Hamiltonian vector fields, which are associated to functions, whose restriction to N is of the form $H \circ p$, for some function $H \in \mathscr{F}(P)$, are tangent to N;

(2) For every two functions $F,G \in \mathscr{F}(P)$ and for all extensions $\tilde{F},\tilde{G} \in \mathscr{F}(M)$ of $F \circ p$ and $G \circ p$ to M, there exists a function $H \in \mathscr{F}(P)$, such that $\{\tilde{F},\tilde{G}\}(n) = H(p(n))$, for every $n \in N$.

In this case, the Poisson bracket of $F,G \in \mathscr{F}(P)$ at $p(n)$, where $n \in N$, is given by

$$\{F,G\}_P(p(n)) = \{\tilde{F},\tilde{G}\}(n),$$

where $\tilde{F}$ and $\tilde{G}$ are arbitrary extensions to M, of $F \circ p$ and $G \circ p$.

There are two extreme cases of Proposition 5.7. The first one occurs when $N = P$. Comparing Definitions 2.9 and 5.6, one has that (M,N,N) is defined to be Poisson reducible if and only if N is a Poisson subvariety of M; Proposition 5.7 says that this is the case precisely when all Hamiltonian vector fields are tangent to N. Thus the equivalence $(i) \Leftrightarrow (iii)$ in Proposition 2.10 is a particular case of Proposition 5.7. For the second extreme case of Proposition 5.7, suppose now that $N = M$. Then Proposition 5.7 says that (M,M,P) is Poisson reducible if and only if the Poisson bracket of every pair of functions which are of the form $H \circ p$, with $H \in \mathscr{F}(P)$, is also of that form. We will see an important particular case of this in Section 5.4, where these functions are the invariant functions for some group action. Another particular case of Proposition 5.7 is given in the following corollary.

Corollary 5.8. *Let $(M,\{\cdot,\cdot\})$ be an affine Poisson variety, let N be a subvariety of M and let P be an affine variety. Suppose that $p : N \to P$ is a dominant map and that for a function $\tilde{F} \in \mathscr{F}(M)$ the following conditions are equivalent:*

(i) *The restriction of $\tilde{F}$ to N is of the form $H \circ p$, for some $H \in \mathscr{F}(P)$;*
(ii) *The Hamiltonian vector field $\mathscr{X}_{\tilde{F}}$ is tangent to N.*

Then (M,N,P) is Poisson reducible.

Proof. If *(i)* and *(ii)* are equivalent, then condition *(1)* in Proposition 5.7 is clearly satisfied. For condition *(2)*, let $F,G \in \mathscr{F}(P)$ and let $\tilde{F},\tilde{G} \in \mathscr{F}(M)$ be extensions of $F \circ p$ and $G \circ p$. According to *(i)* $\Rightarrow$ *(ii)*, $\mathscr{X}_{\tilde{F}}[\mathscr{I}_N] \subset \mathscr{I}_N$ and $\mathscr{X}_{\tilde{G}}[\mathscr{I}_N] \subset \mathscr{I}_N$, where, as before, we denote the ideal of all functions on M which vanish on N by $\mathscr{I}_N$. Using the Jacobi identity for $\{\cdot,\cdot\}$ in the form of Proposition 1.4 *(5)*, it follows that

$$\mathscr{X}_{\{\tilde{F},\tilde{G}\}}[\mathscr{I}_N] \subset \mathscr{X}_{\tilde{F}}\left[\{\tilde{G},\mathscr{I}_N\}\right] + \mathscr{X}_{\tilde{G}}\left[\{\tilde{F},\mathscr{I}_N\}\right] \subset \mathscr{X}_{\tilde{F}}[\mathscr{I}_N] + \mathscr{X}_{\tilde{G}}[\mathscr{I}_N] \subset \mathscr{I}_N\,,$$

which implies, in view of *(ii)* $\Rightarrow$ *(i)*, that the restriction of $\{\tilde{F},\tilde{G}\}$ to N is of the form $H \circ p$, for some $H \in \mathscr{F}(P)$. □

The sufficient condition for Poisson reducibility, given in Corollary 5.8, is not a necessary one. For example, let P and Q be affine Poisson varieties and consider the triple $(P \times Q, P \times Q, P)$, where $P \times Q$ is equipped with the product Poisson structure. If Q is of positive dimension (i.e., is not a point), then conditions *(i)* and *(ii)* in the corollary are not equivalent. However, the triple is reducible, since condition *(1)* in Proposition 5.7 is trivially satisfied, while condition *(2)* is a consequence of the fact that the projection map $p_2 : P \times Q \to Q$ is a Poisson map (see Proposition 2.2).

5.2.2 Poisson Reduction for Poisson Manifolds

We now turn to the case of Poisson manifolds. The basic setup and the basic definition are very similar to the case of affine Poisson varieties (Section 5.2.1). First, recall that, by definition, a map $p : N \to P$ between manifolds is a *submersion* if the tangent map $T_n p : T_n N \to T_{p(n)} P$ is surjective for every $n \in N$; in particular, every submersion is an open map (it maps open subsets of N to open subsets of P).

Definition 5.9. Let $(M, \{\cdot,\cdot\})$ be a Poisson manifold, N an (immersed or embedded) submanifold of M, P a manifold and $p : N \to P$ a surjective submersion, as in the following diagram:

$$\begin{array}{c} N \subset M \\ p \big\downarrow \\ P \end{array} \tag{5.24}$$

We say that the triple (M,N,P) is *Poisson reducible* if there exists a Poisson structure $\pi_P = \{\cdot,\cdot\}_P$ on P, such that for all open subsets $V \subset N$ and $U \subset M$, with $V \subset U \cap N$, and for all functions $F,G \in \mathscr{F}(p(V))$, one has

$$\{F,G\}_P\,(p(n)) = \{\tilde{F},\tilde{G}\}(n)\,, \tag{5.25}$$

at all points n of V, where $\tilde{F},\tilde{G} \in \mathscr{F}(U)$ are arbitrary extensions of $F \circ p_{|V}$ and $G \circ p_{|V}$. The Poisson structure π_P on P is called the *reduced Poisson structure*.

Remark 5.10. Recall from Section 2.2.2 that an immersed submanifold comes with its own topology and differential structure, so the open subset V of N need not be of the form $U \cap N$ for some open subset U of M. However, when N is an embedded submanifold of M, it carries the induced topology from M and the condition in the above definition simplifies to: for every open subset $U \subset M$ and for all functions $F,G \in \mathscr{F}(p(U \cap N))$, one has (5.25), for all $n \in U \cap N$, where $\tilde{F},\tilde{G} \in \mathscr{F}(U)$ are arbitrary extensions of $F \circ p_{|U\cap N}$ and $G \circ p_{|U\cap N}$.

Proposition 5.11. *Let $(M,\{\cdot,\cdot\})$ be a Poisson manifold, let N be an (immersed or embedded) submanifold of M and let $p : N \to P$ be a surjective submersion onto a manifold P. Then (M,N,P) is Poisson reducible if and only if for all open subsets $V \subset N$ and $U \subset M$, with $V \subset U \cap N$, the following conditions are satisfied:*

(1) For every function $\tilde{F} \in \mathscr{F}(U)$, whose restriction to V is constant on the fibers of p, the Hamiltonian vector field $\mathscr{X}_{\tilde{F}}$ is tangent to N at every point of V;
(2) For every pair of functions $\tilde{F},\tilde{G} \in \mathscr{F}(U)$ whose restrictions to V are constant on the fibers of p, their Poisson bracket $\{\tilde{F},\tilde{G}\}$, restricted to V, is constant on the fibers of p.

In this case, for every open subset $V \subset N$ and for all $n \in V$, the Poisson bracket of $F,G \in \mathscr{F}(p(V))$ at $p(n) \in P$ is given by

$$\{F,G\}_P(p(n)) = \{\tilde{F},\tilde{G}\}(n)\,, \tag{5.26}$$

where $\tilde{F}$ and $\tilde{G}$ are arbitrary extensions to a neighborhood of n in M, of $F \circ p_{|V}$ and $G \circ p_{|V}$. In particular, the Hamiltonian vector field associated to F is given, at $p(n)$, by

$$(\mathscr{X}_F)_{p(n)} = \mathrm{d}_n p\,(\mathscr{X}_{\tilde{F}})_n\,. \tag{5.27}$$

Proof. We first point out that for an open subset $V \subset N$ and a function $F \in \mathscr{F}(V)$ the following conditions are equivalent:

(i) F is of the form $H \circ p$ for some function $H \in \mathscr{F}(p(V))$;
(ii) F is constant on all the fibers of p.

The only thing which is not obvious in this equivalence is the fact that the function H, constructed in (ii) $\Rightarrow$ (i), is smooth/holomorphic; it is a consequence of the implicit function theorem, applied to the surjective submersion p.

Suppose that (M,N,P) is Poisson reducible. Let $V \subset N$ and $U \subset M$ be open subsets, with $V \subset U \cap N$, and suppose that $\tilde{F} \in \mathscr{F}(U)$ is constant on the fibers of p, when restricted to V. Then $\tilde{F}_{|V} = F \circ p_{|V}$ for some function $F \in \mathscr{F}(p(V))$. For an arbitrary $n \in V$ and an arbitrary function $\tilde{G}$, defined on a neighborhood U' of n in M, and vanishing on some neighborhood of n in N, we have in view of (5.25) that $\{\tilde{F},\tilde{G}\}(n) = \{F,G\}_P(p(n)) = 0$, since $G = 0$, in a neighborhood of $p(n)$. Therefore, $\mathscr{X}_{\tilde{F}}$ is tangent to N at n, leading to condition *(1)*. Condition *(2)* follows also from (5.25), because the left-hand side of (5.25) depends only on $p(n)$, i.e., on the fiber which contains n, rather than on n itself.

Suppose now that conditions *(1)* and *(2)* hold. For every $w \in P$, we define a skew-symmetric biderivation $(\pi_P)_w$ at w by: for all germs F_w, G_w at w,

$$(\pi_P)_w[F_w, G_w] := \pi_n[\tilde{F}_n, \tilde{G}_n] = \left\{\tilde{F}, \tilde{G}\right\}(n)\,, \tag{5.28}$$

where n is an arbitrary point of N for which $p(n) = w$ and $\tilde{F}_n, \tilde{G}_n$ are germs at n of arbitrary functions $\tilde{F}, \tilde{G}$, defined on a neighborhood of n in M, and satisfying $\tilde{F}_{|V} = F \circ p_{|V}$ and $\tilde{G}_{|V} = G \circ p_{|V}$ for some neighborhood V of n in N. The right-hand side in (5.28) is independent of the choice of n (with $p(n) = w$) because, according to *(2)*, the value of $\left\{\tilde{F}, \tilde{G}\right\}$ at points n of V depends on $p(n)$ only. Similarly, it is independent of the chosen $\tilde{F}$ and $\tilde{G}$ because, according to *(1)*, if $\tilde{G}$ is such that $\tilde{G}_{|V} = 0$ and $\tilde{F}$ is such that $\tilde{F}_{|V} = F \circ p_{|V}$, then $\left\{\tilde{F}, \tilde{G}\right\}(n) = 0$. Thus π_P is well-defined at every point $w \in P$ and it inherits from π the biderivation property at each point. We need to show that the bivector field π_P, defined on P by $w \mapsto (\pi_P)_w$ is smooth/holomorphic. Let $w \in P$, let $n \in p^{-1}(w)$ and let V be a neighborhood of n in N, having the property that every function in $\mathscr{F}(V)$ extends to a function in $\mathscr{F}(U)$, where U is a neighborhood of V in M. Such neighborhoods V and U exist because N is an immersed submanifold of M. For every pair of functions $F, G \in \mathscr{F}(p(V))$, we then have, using the equivalence (i)$\Leftrightarrow$(ii), given at the beginning of the proof, that

$$\{F, G\}_P \circ p = \left\{\tilde{F}, \tilde{G}\right\}_{|V} = \tilde{H} \circ p\,,$$

for some function $\tilde{H} \in \mathscr{F}(p(V))$, in particular $\{F, G\}_P \in \mathscr{F}(p(V))$. This shows that π_P is a smooth/holomorphic bivector field in a neighborhood of w; since w was arbitrary, π_P is a smooth/holomorphic bivector field on P. In the bracket notation, it is given by (5.26), for arbitrary points $p(n)$ in P and for arbitrary functions F and G, defined in a neighborhood of $p(n)$ in P. Notice that (5.26) implies that

$$\{\{F, G\}_P, H\}_P\,(p(n)) = \left\{\left\{\tilde{F}, \tilde{G}\right\}, \tilde{H}\right\}(n)\,, \tag{5.29}$$

for arbitrary functions F, G, H, defined in a neighborhood of $p(n)$, and $\tilde{F}, \tilde{G}, \tilde{H}$ extensions as before. It follows from (5.29) that the Jacobi identity for $\{\cdot,\cdot\}$ implies the Jacobi identity for $\{\cdot,\cdot\}_P$. Also, (5.27) is an immediate consequence of (5.26). □

According to *(1)* in Proposition 5.11, if (M, N, P) is Poisson reducible and $\tilde{F}$ is a function, defined on an open subset of M, whose restriction to an open subset V of N vanishes, then the Hamiltonian vector field $\mathscr{X}_{\tilde{F}}$ is tangent to N at every point of V. This motivates the following definition, from which it follows that if (M, N, P) is Poisson reducible, then N is a coisotropic submanifold of the Poisson manifold (M, π).

Definition 5.12. A submanifold N of a Poisson manifold (M, π) is said to be a *coisotropic submanifold* of M if for every open subset V of N and for every function $\tilde{F}$, defined on a neighborhood of V in M, and such that $\tilde{F}_{|N\cap V} = 0$, the Hamiltonian vector field $\mathscr{X}_{\tilde{F}}$ is tangent to N, at every point of V.

The notion of a coisotropic submanifold generalizes the notion of a Poisson submanifold, since the Hamiltonian vector fields, associated to (local) functions which vanish on a Poisson submanifold, are not only tangent to the Poisson submanifold, but they actually vanish on it.

The condition that a submanifold N of a Poisson manifold is coisotropic can be stated pointwise. To do this, we introduce, for every $m \in M$, the linear map $\pi_m^\sharp : T_m^*M \to T_mM$, which corresponds[3] to the bivector $\pi_m \in \wedge^2 T_mM$, and for $n \in N$, the subspace $T_n^\perp N$ of T_n^*M, which consists of the annihilator of T_nN in T_n^*M,

$$T_n^\perp N := \{\xi \in T_n^*M \mid \forall v \in T_nN, \, \xi(v) = 0\} \ .$$

By the implicit function theorem, every element of $T_n^\perp N$ can be realized as the differential at n of some function $\tilde{F}$, defined on a neighborhood of n in M, and whose restriction to N vanishes. It leads to the following two pointwise characterizations of a coisotropic submanifold.

Lemma 5.13. *Let N be a submanifold of a Poisson manifold (M,π). The following conditions on N are equivalent.*

(i) *N is a coisotropic submanifold;*
(ii) *For every $n \in N$, $\pi_n(T_n^\perp N, T_n^\perp N) = \{0\}$;*
(iii) *For every $n \in N$, $\pi_n^\sharp(T_n^\perp N) \subset T_nN$.*

Similarly, the stronger condition that all Hamiltonian vector fields $\mathscr{X}_{\tilde{F}}$ are tangent to the fibers $p_n := p^{-1}(p(n))$ of the surjective map $p : N \to P$ can be stated as follows: for every $n \in N$,

$$\pi_n^\sharp(T_n^\perp N) \subset T_n p_n \ .$$

In the following proposition we show that for a triple (M,N,P) as above, if all the tangent spaces $T_n p_n$ are spanned by Hamiltonian vector fields which come from functions which vanish on N, then (M,N,P) is Poisson reducible.

Proposition 5.14. *Let (M,π) be a Poisson manifold, N an (immersed or embedded) submanifold of M and let $p : N \to P$ be a surjective submersion with connected fibers. If*

$$\pi_n^\sharp\left(T_n^\perp N\right) = T_n p_n \tag{5.30}$$

for every $n \in N$, then N is a coisotropic submanifold of M and (M,N,P) is Poisson reducible.

---

[3] Our sign convention is that $\pi_m^\sharp(\mathrm{d}_m H)[F] = \pi_m[H_m, F_m] = \{H,F\}(m)$, for all functions F and H, defined on a neighborhood of m.

Proof. Notice first that in view of *(iii)* ⇒ *(i)* in Lemma 5.13, N is a coisotropic submanifold of M. To show that (M,N,P) is Poisson reducible, we verify both conditions *(1)* and *(2)* in Proposition 5.11.

(1) The skew-symmetry of π_n, combined with (5.30), yields that $\pi_n^\sharp(T_n^\perp p_n) \subset T_nN$, for every $n \in N$. It implies that the Hamiltonian vector field of every function, defined on a neighborhood of n in M, whose restriction to N is constant on the fibers of p, is tangent to N, which is condition *(1)*.

(2) Let $U \subset M$ and $V \subset N$ be non-empty open subsets, with $V \subset U \cap N$. Let $\tilde{F}, \tilde{G} \in \mathscr{F}(U)$ be two functions whose restrictions to V are constant on the fibers of p. We show that the restriction of $\left\{\tilde{F}, \tilde{G}\right\}$ to V is constant on the fibers of p. Let $n \in V$ and $v \in T_n p_n$ be arbitrary. Then v can be extended, on a neighborhood of n in M, to a Hamiltonian vector field, which is tangent to the fibers of p. Indeed, by (5.30), there exists a function $\tilde{H}$ on a neighborhood $U' \subset U$ of n in M, whose restriction to $V' := V \cap U'$ is zero, with $(\mathscr{X}_{\tilde{H}})_n = -\pi_n^\sharp(\mathrm{d}_n\tilde{H}) = v$. Since $\tilde{H}$ is zero on V', its Hamiltonian vector field $\mathscr{X}_{\tilde{H}}$ is indeed, again by (5.30), tangent to the fibers of p at every point of V'. It follows that the function $\mathscr{X}_{\tilde{H}}[\tilde{F}]$ is zero on V', because $\tilde{F}_{|V}$ is constant on the fibers of p. By (5.30), $\left\{\mathscr{X}_{\tilde{H}}[\tilde{F}], \tilde{G}\right\}$ vanishes on N (recall that $\tilde{G}$ is also constant on the fibers of p). Similarly, $\left\{\mathscr{X}_{\tilde{H}}[\tilde{G}], \tilde{F}\right\}(n) = 0$. Using the Jacobi identity, it follows that

$$v\left[\{\tilde{F},\tilde{G}\}\right] = \mathscr{X}_{\tilde{H}}\left[\{\tilde{F},\tilde{G}\}\right](n) = \left\{\mathscr{X}_{\tilde{H}}[\tilde{F}],\tilde{G}\right\}(n) + \left\{\tilde{F},\mathscr{X}_{\tilde{H}}[\tilde{G}]\right\}(n) = 0\,.$$

Since $n \in N$ and $v \in T_n p_n$ are arbitrary, the restriction of $\left\{\tilde{F},\tilde{G}\right\}$ to each connected component of $p_n \cap V$ is constant. Notice that, since $p_n \cap V$ is not necessarily connected, we cannot conclude from it that the restriction of $\left\{\tilde{F},\tilde{G}\right\}$ to every fiber of p is constant. However, there is a function H on $p^{-1}(p(V))$, which agrees with $\left\{\tilde{F},\tilde{G}\right\}$ at every point of V and which has the same property of being constant, when restricted to each fiber p_n, where $n \in V$; since each fiber p_n is connected, we can conclude that $\left\{\tilde{F},\tilde{G}\right\}$ is constant, when restricted to the fibers of p. In order to construct H, let n be an arbitrary point of $p^{-1}(p(V))$. Since $\tilde{F}$ is constant on the fibers of p, its restriction to V is of the form $F \circ p_{|V}$ for some function F. On a small neighborhood U'' of n in M we can construct an extension $\tilde{F}_0$ of $F \circ p_{|V''}$, where V'' is a neighborhood of n in N, with $V'' \subset U'' \cap N$. Similarly, we construct $\tilde{G}_0$ from $\tilde{G}$. It follows from (5.30) that $H := \left\{\tilde{F}_0, \tilde{G}_0\right\}$ is at points of V'' independent of the chosen extensions, hence leads to a well-defined function on $p^{-1}(p(V))$, which coincides with $\left\{\tilde{F},\tilde{G}\right\}$ on V. Moreover, it follows as earlier in this proof that H is constant on the fibers of p. This proves the existence of H, with the announced properties, hence achieves the proof. □

Remark 5.15. Let N be a coisotropic submanifold of a Poisson manifold (M,π). Assume that the dimension of $\pi_n^\sharp(T_n^\perp N)$ is independent of $n \in N$. According to Lemma 5.13, a distribution is defined on N by assigning to every $n \in N$ the subspace $\pi_n^\sharp(T_n^\perp N)$ of T_nM. This distribution is differentiable, because it is locally defined by the Hamiltonian vector fields which correspond to local functions on M

which vanish on N. For functions F, G, defined on an open subset U of M, and vanishing on N, their commutator $[\mathscr{X}_F, \mathscr{X}_G] = -\mathscr{X}_{\{F,G\}}$ is the Hamiltonian vector field of the function $\{F, G\}$, which vanishes on N; indeed, $\{F, G\} = \mathscr{X}_G[F]$ vanishes on N because F is constant on N (in fact, it vanishes on N) and $\mathscr{X}_G$ is tangent to N (since N is coisotropic). Thus, $[\mathscr{X}_F, \mathscr{X}_G]$ belongs to the distribution and the latter is integrable in the sense of Frobenius. If the space P of leaves of this distribution is a manifold, then the conditions of Proposition 5.14 hold, so that (M, N, P) is Poisson reducible and P inherits a Poisson structure from (M, π).

Remark 5.16. It is instructive to write down the conditions of Poisson reducibility in terms of a well-chosen system of local coordinates for the ambient manifold. It leads to an expression for the Poisson matrix of the reduced Poisson structure in terms of the Poisson matrix of the initial Poisson structure.

Let (M, N, P) be three manifolds with $N \subset M$ a submanifold of M, and $p : N \to P$ a surjective submersion. For simplicity, we assume that N is an embedded submanifold. We define s, t and u by

$$s := \dim P, \qquad s + t := \dim N, \qquad s + t + u := \dim M\,.$$

Let n be an arbitrary point of N and let $(x_1, \ldots, x_s)$ be an arbitrary coordinate system, defined on a neighborhood of $p(n)$ in P. For $i = 1, \ldots, s$, choose an extension $\tilde{x}_i$ of $x_i \circ p$ to a neighborhood of n in M. Then there exist on a neighborhood of n in M functions $y_1, \ldots, y_t$ and $z_1, \ldots, z_u$, such that $(\tilde{x}_1, \ldots, \tilde{x}_s, y_1, \ldots, y_t, z_1, \ldots, z_u)$ is a coordinate system for M in a neighborhood U of n, and such that $N \cap U$ is given by the equations $z_1 = \cdots = z_u = 0$.

Let π be a Poisson structure on M. The Poisson matrix X of π with respect to the coordinates $\tilde{x}_1, \ldots, \tilde{x}_s, y_1, \ldots, y_t, z_1, \ldots, z_u$ (in that order) is given by the block matrix

$$X = \begin{pmatrix} \tilde{A} & B & C \\ -B^\top & D & E \\ -C^\top & -E^\top & F \end{pmatrix}$$

where $\tilde{A}$, D and F are square matrices of size s, t and u respectively. Each of the blocks $\tilde{A}$, B, $C, \ldots$ is a matrix-valued function on U, whose restriction to $U \cap N$ is denoted by $\tilde{A}_{|N}$, $B_{|N}$, $C_{|N}, \ldots$ We summarize the conditions related to Poisson reducibility, expressed in terms of the Poisson matrix X, in Table 5.1.

The proof of each one of the lines of the table is a direct consequence of the definitions. To start with, $N \cap U$ is coisotropic in $U \subset M$ if and only if every Hamiltonian vector field $\mathscr{X}_G$, with G a function on U which vanishes on $N \cap U$, is tangent to N at every point of $N \cap U$. Since $N \cap U$ is the zero locus of the local coordinates $z_1, \ldots, z_s$, this happens if and only if the Hamiltonian vector fields $\mathscr{X}_{z_i}$ are tangent to $N \cap U$ at every point of $N \cap U$ for all $i = 1, \ldots, u$, i.e. if and only if $\mathscr{X}_{z_i}[z_j](n') = 0$ for all $i, j = 1, \ldots, s$ and $n' \in N \cap U$. This is equivalent to saying that $F_{|_N} = 0$.

Let us explain the second line in the table. Condition (1) in Proposition 5.11 demands that the Hamiltonian vector field $\mathscr{X}_{\tilde{G}}$ is tangent to N for every function $\tilde{G}$ whose restriction to N is of the form $G \circ p$ for some function G defined in a

Table 5.1 A summary of the conditions related to Poisson reducibility, expressed in terms of the Poisson matrix X.

Condition	Poisson matrix
N is a coisotropic submanifold	$F_{\|N} = 0$
Condition (1) in Proposition 5.11	$C_{\|N} = F_{\|N} = 0$
Condition (2) in Proposition 5.11	$\tilde{A}_{\|N}$ depends only on $\tilde{x}_1,\ldots,\tilde{x}_s$
(M,N,P) is Poisson reducible	$C_{\|N} = F_{\|N} = 0$ $\tilde{A}_{\|N}$ depends only on $\tilde{x}_1,\ldots,\tilde{x}_s$
Condition (5.30) in Proposition 5.14	$C_{\|N} = F_{\|N} = 0$ $\mathrm{Rk}(E(n')) = t$ for every $n' \in N \cap U$

neighborhood of $p(n)$ in P. It implies that the Hamiltonian vector fields associated to the functions $\tilde{x}_1,\ldots,\tilde{x}_s,z_1,\ldots,z_u$ are tangent to N, hence that $C_{|N} = F_{|N} = 0$. The inverse implication follows from the fact that if the restriction of a function G to N is constant on the fibers of p, then this restriction is independent of the coordinates $y_1,\ldots,y_t$.

For the third line, condition (2) in Proposition 5.11 is tantamount to saying that the functions $\{\tilde{x}_i,\tilde{x}_j\}$, restricted to $U \cap N$ depend on the functions $\tilde{x}_1,\ldots,\tilde{x}_s$ only; in terms of the Poisson matrix X, this condition amounts to saying that $\tilde{A}_{|N}$ depends only on $\tilde{x}_1,\ldots,\tilde{x}_s$.

The fourth line follows from the second and third lines, because (M,N,P) being Poisson reducible means precisely that conditions (1) and (2) in Proposition 5.11 are satisfied.

For the last line in the table, condition (5.30) in Proposition 5.14, which states that $\pi_n^\sharp\left(T_n^\perp N\right) = T_n p_n$ for every $n \in N$, implies that condition (1) in Proposition 5.11 is satisfied, so that $C_{|N} = F_{|N} = 0$. It also implies that, at every point $n' \in N \cap U$, the u columns of the matrix E, which represent the Hamiltonian vector fields of the functions $z_1,\ldots,z_u$, generate a space of dimension t, namely the tangent space of the fiber of p through n'.

When (M,N,P) is Poisson reducible, $\tilde{A}_{|N}$ depends only on the variables $\tilde{x}_1,\ldots,\tilde{x}_s$, so that there exists a matrix-valued function A, such that $A \circ p = \tilde{A}_{|U}$. In view of (5.26), the Poisson matrix of the reduced Poisson structure on P, expressed in terms of the local coordinates $x_1,\ldots,x_s$, is precisely the matrix A.

5.3 Poisson–Dirac Reduction

We saw in Section 2.2 that if $(\mathscr{A},\cdot,\{\cdot,\cdot\})$ is a Poisson algebra and $\mathscr{I}$ is a Poisson ideal of $(\mathscr{A},\cdot)$, then $\mathscr{A}/\mathscr{I}$ inherits a Poisson algebra structure from $\mathscr{A}$. As we will show in this section, $\mathscr{A}/\mathscr{I}$ may inherit a Poisson structure from $\mathscr{A}$ under weaker conditions; we then say that the Poisson structure on $\mathscr{A}/\mathscr{I}$ is obtained by *Poisson–Dirac reduction*. For Poisson manifolds it amounts to obtaining a Poisson structure on a submanifold, which is not a Poisson submanifold.

5.3.1 Algebraic Poisson–Dirac Reduction

Let $(\mathscr{A},\cdot,\{\cdot,\cdot\})$ be a Poisson algebra and let $\mathscr{I}$ be an ideal of $(\mathscr{A},\cdot)$. Consider the composition of algebra homomorphisms

$$\mathscr{N}(\mathscr{I}) \hookrightarrow \mathscr{A} \xrightarrow{\ \kappa\ } \frac{\mathscr{A}}{\mathscr{I}} \tag{5.31}$$

where we recall from Section 5.2.1 that $\mathscr{N}(\mathscr{I})$ is the normalizer of $\mathscr{I}$,

$$\mathscr{N}(\mathscr{I}) := \{F \in \mathscr{A} \mid \{F,\mathscr{I}\} \subset \mathscr{I}\}\,,$$

which is a Poisson subalgebra of $\mathscr{A}$. The kernel of the composite map in (5.31) is the Poisson ideal $\mathscr{N}(\mathscr{I})\cap\mathscr{I}$ of $\mathscr{N}(\mathscr{I})$. It follows that $\mathscr{N}(\mathscr{I})/(\mathscr{N}(\mathscr{I})\cap\mathscr{I})$ is itself a Poisson algebra. Notice that it is isomorphic to a subalgebra of $\mathscr{A}/\mathscr{I}$, although the latter is not a Poisson algebra, in general. It leads to the following definition.

Definition 5.17. Let $(\mathscr{A},\cdot,\{\cdot,\cdot\})$ be a Poisson algebra. An ideal $\mathscr{I}$ of $(\mathscr{A},\cdot)$ is said to be a *Poisson–Dirac ideal* if one of the following equivalent conditions holds:

(i) The natural inclusion of algebras $\mathscr{N}(\mathscr{I})/(\mathscr{N}(\mathscr{I})\cap\mathscr{I}) \hookrightarrow \mathscr{A}/\mathscr{I}$ is an isomorphism;

(ii) The composition of algebra homomorphisms $\mathscr{N}(\mathscr{I}) \hookrightarrow \mathscr{A} \to \mathscr{A}/\mathscr{I}$ is surjective;

(iii) $\mathscr{I} + \mathscr{N}(\mathscr{I}) = \mathscr{A}$.

The three conditions in the definition are clearly equivalent. Condition (i) implies that $\mathscr{A}/\mathscr{I}$ is a Poisson algebra, isomorphic to $\mathscr{N}(\mathscr{I})/(\mathscr{N}(\mathscr{I})\cap\mathscr{I})$, as stated in the following proposition.

Proposition 5.18. *Let $\mathscr{I}$ be a Poisson–Dirac ideal of a Poisson algebra $(\mathscr{A},\cdot,\{\cdot,\cdot\})$. Then $\mathscr{A}/\mathscr{I}$ has a unique Poisson bracket which makes the surjective morphism $\mathscr{N}(\mathscr{I}) \hookrightarrow \mathscr{A} \to \mathscr{A}/\mathscr{I}$ into a morphism of Poisson algebras. This bracket is said to be obtained by* Poisson–Dirac reduction.

When $\mathscr{I}$ is a Poisson–Dirac ideal, the different Poisson algebras which appear in Poisson–Dirac reduction can be summarized in the following commutative diagram,

$$
\begin{array}{ccc}
\mathscr{N}(\mathscr{I}) & \hookrightarrow & \mathscr{A} \\
\big\downarrow & \searrow & \big\downarrow \kappa \\
\dfrac{\mathscr{N}(\mathscr{I})}{\mathscr{N}(\mathscr{I})\cap\mathscr{I}} & \xrightarrow{\simeq} & \mathscr{A}/\mathscr{I}
\end{array}
\tag{5.32}
$$

where all maps are morphisms of Poisson algebras, except (in general) for the canonical projection $\kappa : \mathscr{A} \to \mathscr{A}/\mathscr{I}$.

We now consider Poisson–Dirac reduction for affine Poisson varieties. Suppose that $N \subset M$ is a subvariety of an affine Poisson variety $(M, \{\cdot,\cdot\})$. We denote the ideal of N (in $\mathscr{F}(M)$) by $\mathscr{I}_N$. Recall from (5.23) that $\mathscr{N}(\mathscr{I}_N)$ consists of all functions on M whose Hamiltonian vector field is tangent to N. Translating (ii) of Definition 5.17 into geometrical terms, leads to the following definition.

Definition 5.19. Let $(M, \{\cdot,\cdot\})$ be an affine Poisson variety. A subvariety $N \subset M$ is said to be a *Poisson–Dirac subvariety* if every function on N is the restriction of a function on M whose Hamiltonian vector field is tangent to N.

Thus, a subvariety N is, by definition, a Poisson–Dirac subvariety if and only if its ideal $\mathscr{I}_N$ is a Poisson–Dirac ideal. In view of Definition 5.17 this leads to other, algebraic, characterizations of Poisson–Dirac subvarieties and Proposition 5.18 shows that a Poisson–Dirac subvariety inherits a Poisson structure from its ambient Poisson variety. That is the content of the following proposition.

Proposition 5.20. *Let N be a Poisson–Dirac subvariety of an affine Poisson variety $(M, \{\cdot,\cdot\})$. There exists a unique Poisson structure $\{\cdot,\cdot\}_N$ on N such that, for every $F, G \in \mathscr{F}(N)$, their Poisson bracket is given by*

$$
\{F, G\}_N = \{\tilde{F}, \tilde{G}\}\big|_N \tag{5.33}
$$

where $\tilde{F}, \tilde{G}$ are arbitrary extensions of F, G to M, whose Hamiltonian vector fields $\mathscr{X}_{\tilde{F}}$ and $\mathscr{X}_{\tilde{G}}$ are tangent to N at all points of N. In particular, every Hamiltonian vector field on $(N, \{\cdot,\cdot\}_N)$ is the restriction, to N, of a Hamiltonian vector field on $(M, \{\cdot,\cdot\})$. The Poisson structure $\{\cdot,\cdot\}_N$ is called a reduced Poisson structure.

Remark 5.21. Suppose that N is a Poisson–Dirac subvariety of an affine Poisson variety $(M, \{\cdot,\cdot\})$. Then the subalgebra $\mathscr{N}(\mathscr{I}_N)$ of $\mathscr{F}(M)$ may not be finitely generated, and therefore may not be the algebra of functions on some affine variety. However, when $\mathscr{N}(\mathscr{I}_N)$ is finitely generated, then the commutative diagram (5.32) leads to the following commutative diagram of affine varieties

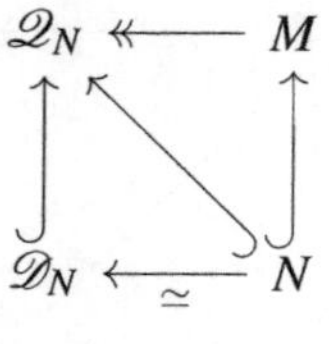

where $\mathscr{Q}_N$, respectively $\mathscr{D}_N$, is the affine variety, whose algebra of regular functions is $\mathscr{N}(\mathscr{I}_N)$, respectively $\mathscr{N}(\mathscr{I}_N)/(\mathscr{N}(\mathscr{I}_N)\cap\mathscr{I}_N)$. All maps in this diagram are Poisson maps, except (in general) for the inclusion $N \hookrightarrow M$.

We have seen in Section 2.2 that if N is a Poisson subvariety of a Poisson variety $(M,\{\cdot,\cdot\})$, then for every function F on M, which vanishes on N, the Hamiltonian vector field $\mathscr{X}_F$ vanishes at every point of N. For Poisson–Dirac subvarieties a similar, but weaker, property holds.

Proposition 5.22. *Let $(M,\{\cdot,\cdot\})$ be an affine Poisson variety and let N be a Poisson–Dirac subvariety, whose ideal is denoted by $\mathscr{I}_N$. Let $F\in\mathscr{I}_N$. If the Hamiltonian vector field $\mathscr{X}_F$ is tangent to N at some point $n\in N$, then it vanishes at that point, $(\mathscr{X}_F)_n=0$. In other words, $\mathrm{Ham}_n(\mathscr{I}_N)\cap T_nN=\{0\}$ for every $n\in N$, where $\mathrm{Ham}_n(\mathscr{I}_N)=\{(\mathscr{X}_F)_n \mid F\in\mathscr{I}_N\}$ is the vector space of Hamiltonian vectors at n, which come from elements of $\mathscr{I}_N$.*

Proof. Let $F\in\mathscr{I}_N$ and let $n\in N$. Suppose that $\mathscr{X}_F$ is tangent to N at n, which means that $\mathscr{X}_F[G](n)=0$, for every $G\in\mathscr{I}_N$. We wish to show that $(\mathscr{X}_F)_n=0$. If $H\in\mathscr{F}(M)$, then H can be written as H_1+H_2 with $H_1\in\mathscr{I}_N$ and $H_2\in\mathscr{N}(\mathscr{I}_N)$, since $\mathscr{I}_N$ is a Poisson–Dirac ideal (condition (iii) in Definition 5.17). By the above, $\mathscr{X}_F[H_1](n)=0$, while $\mathscr{X}_F[H_2](n)=\{H_2,F\}(n)=0$, since H_2 belongs to the normalizer of $\mathscr{I}_N$. Therefore, $\mathscr{X}_F[H](n)=0$ for every $H\in\mathscr{F}(M)$, which means that $(\mathscr{X}_F)_n=0$. □

To finish this section, we write a Poisson matrix for the reduced Poisson structures, obtained by Poisson–Dirac reduction. Assume that N is a Poisson–Dirac subvariety of an affine Poisson variety $(M,\{\cdot,\cdot\})$. As before, the ideal of N is denoted by $\mathscr{I}_N$ and its normalizer by $\mathscr{N}(\mathscr{I}_N)$. There exist generators $F_1,\dots,F_s,G_1,\dots,G_t$ of $\mathscr{F}(M)$ with every F_i belonging to $\mathscr{N}(\mathscr{I}_N)$ and every G_i belonging to $\mathscr{I}_N$; to construct such generators one can for example start from generators $H_1,\dots,H_s$ for $\mathscr{F}(M)=\mathscr{N}(\mathscr{I}_N)+\mathscr{I}_N$ and decompose every H_i as F_i+G_i, with $F_i\in\mathscr{N}(\mathscr{I}_N)$ and $G_i\in\mathscr{I}_N$. In terms of the generators $F_1,\dots,F_s,G_1,\dots,G_t$, the Poisson matrix of $(M,\{\cdot,\cdot\})$ is a 2×2 block matrix

$$\begin{pmatrix} A & B \\ -B^\top & D \end{pmatrix}$$

where

$$A_{ij}=\{F_i,F_j\}\,,\; B_{ik}=\{F_i,G_k\}\,,\; D_{k\ell}=\{G_k,G_\ell\}\,,$$

for $1\leqslant i,j\leqslant s$ and $1\leqslant k,\ell\leqslant t$. Since N is a Poisson–Dirac subvariety, the functions $F_1,\dots,F_s$, restricted to N, generate $\mathscr{F}(N)$. Also, the Poisson matrix of $(N,\{\cdot,\cdot\}_N)$, in terms of these generators, is given by the matrix $\overline{A}$, with entries $\overline{A}_{ij} = \{F_i, F_j\}_{|N}$, where $1 \leqslant i, j \leqslant s$; indeed, the Hamiltonian vector fields of $F_1, \ldots, F_s$ are tangent to N at every point of N, hence we can use these functions in (5.33) to compute the Poisson brackets $\left\{F_{i|N}, F_{j|N}\right\}_N$. Note also that all entries of B belong to $\mathscr{I}_N$, hence vanish on N; see Proposition 5.25 below for a converse in the differential-geometrical context.

5.3.2 Poisson–Dirac Reduction for Poisson Manifolds

Let N be an (immersed or embedded) submanifold of a manifold M. Given a point n in N and a function F, defined on a neighborhood W of n in N, a *local extension* of F at n is a triple $(\tilde{F}, U, V)$ where

(1) $U \subset M$ is a neighborhood of n in M;
(2) $V \subset (W \cap U) \subset N$ is a neighborhood of n in N;
(3) $\tilde{F} \in \mathscr{F}(U)$ is a function such that $\tilde{F}_{|V} = F_{|V}$.

The relation between the different subsets and maps is represented in the following diagram:

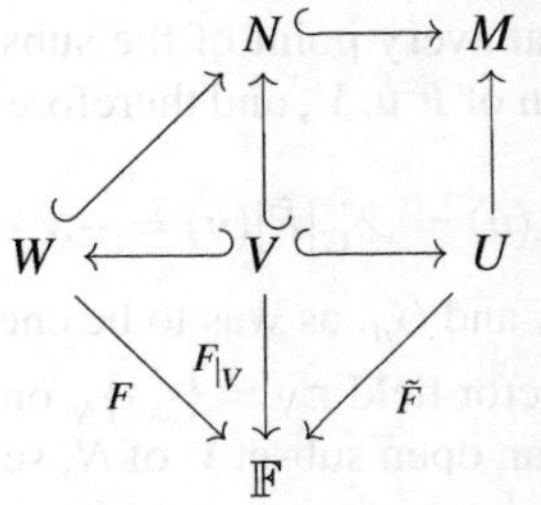

Every function F, defined on an open subset of N, admits a local extension at every point of its domain of definition.

Definition 5.23. Let $(M, \{\cdot,\cdot\})$ be a Poisson manifold and let N be an (immersed or embedded) submanifold of M. We say that N is a *Poisson–Dirac submanifold* of M, if for every $n \in N$, every function F, defined in a neighborhood of n in N, admits a local extension $(\tilde{F}, U, V)$ at n, such that $\mathscr{X}_{\tilde{F}}$ is tangent to N at every point of V.

Note that, in particular, a Poisson submanifold N of a Poisson manifold $(M, \{\cdot,\cdot\})$ is a Poisson–Dirac submanifold, since *all* Hamiltonian vector fields are tangent to N.

We show in the following proposition that every Poisson–Dirac submanifold inherits a Poisson structure from its ambient Poisson manifold.

Proposition 5.24. *Let $(M, \{\cdot,\cdot\})$ be a Poisson manifold and let N be a Poisson–Dirac submanifold of M. There exists on N a unique Poisson structure $\{\cdot,\cdot\}_N$ such that, for all open subsets $V \subset N$ and $U \subset M$ with $V \subset U$, for every function $\tilde{F}, \tilde{G} \in$*

$\mathscr{F}(U)$ whose Hamiltonian vector fields are tangent to N at every point of V, and for every $n \in V$, we have

$$\{F,G\}_N(n) = \{\tilde{F},\tilde{G}\}(n)\,, \tag{5.34}$$

where $F,G \in \mathscr{F}(V)$ are the restrictions of $\tilde{F},\tilde{G}$ to V. In particular, locally, every Hamiltonian vector field on N is the restriction to N of a Hamiltonian vector field on M.

Proof. For every $n \in N$, we define a skew-symmetric pointwise biderivation $(\pi_N)_n$ by its values on germs F_n, G_n at n,

$$(\pi_N)_n(F_n,G_n) := \{\tilde{F},\tilde{G}\}(n)\,, \tag{5.35}$$

where the right-hand side is constructed as follows:

- Choose functions F and G, defined in a neighborhood of n in N, with germs F_n, G_n at n;
- Choose local extensions $(\tilde{F},U,V)$ and $(\tilde{G},U,V)$ of F and G at n, such that the Hamiltonian vector fields $\mathscr{X}_{\tilde{F}}$ and $\mathscr{X}_{\tilde{G}}$ are tangent to N, at every point of V.

We have to check that this definition makes sense: a priori, the right-hand side of (5.35) depends on the choice of representatives F and G of the germs F_n and G_n, and on the choice of local extensions $(\tilde{F},U,V)$ and $(\tilde{G},U,V)$. The Hamiltonian vector field $\mathscr{X}_{\tilde{G}}$ is tangent to N at every point of the subset V of N, so that $\mathscr{X}_{\tilde{G}}[\tilde{F}](n)$ depends only on the restriction of $\tilde{F}$ to V, and therefore depends only on the germ F_n of F at n. Hence

$$\{\tilde{F},\tilde{G}\}(n) = \mathscr{X}_{\tilde{G}}[\tilde{F}](n) = -\mathscr{X}_{\tilde{F}}[\tilde{G}](n)$$

depends only on the germs F_n and G_n, as was to be checked.

We may now define a bivector field $\pi_N = \{\cdot,\cdot\}_N$ on N by its value on every pair of functions F,G, defined on an open subset V of N, setting

$$\{F,G\}_N(n) := (\pi_N)_n(F_n,G_n) = \{\tilde{F},\tilde{G}\}(n)\,, \tag{5.36}$$

for every $n \in V$. To check that the bivector field is smooth/holomorphic, notice that the above extensions $(\tilde{F},U,V)$ and $(\tilde{G},U,V)$ are local extensions of F and G at every $n' \in V$, with Hamiltonian vector fields tangent to N at every point of V, so that (5.36) is not only valid for n, but for every $n' \in V$. Since the right-hand side of (5.36) is a smooth or holomorphic function of n, and since F and G are arbitrary functions, we can conclude that the bivector field π_N is also smooth or holomorphic (on V, hence on N).

Let, as above, $\tilde{F},\tilde{G},\tilde{H}$ be three functions, defined on an open subset U of M, assume that their Hamiltonian vector fields are tangent to N at every point of an open subset V of N, and let F,G,H denote their restrictions to V. The Hamiltonian vector field of $\{\tilde{F},\tilde{G}\}$ is tangent to N, at every point of V, since $\mathscr{X}_{\{\tilde{F},\tilde{G}\}} = -[\mathscr{X}_{\tilde{F}},\mathscr{X}_{\tilde{G}}]$ (item *(5)* in Proposition 1.4). As a consequence, $\left(\{\tilde{F},\tilde{G}\},U,V\right)$ is a local extension of $\{F,G\}_N$ at n, whose Hamiltonian vector field is tangent to N at every point of V, hence

$$\{\{F,G\}_N,H\}_N(n) = \{\{\tilde{F},\tilde{G}\},\tilde{H}\}(n), \tag{5.37}$$

for every $n \in V$, so that the Jacobi identity for the bracket $\{\cdot,\cdot\}_N$ follows from the Jacobi identity for $\{\cdot,\cdot\}$. □

In order to give a geometrical characterization of Poisson–Dirac submanifolds (Proposition 5.25 below), we express the Poisson bivector on a Poisson manifold in terms of local coordinates which are adapted to the Poisson–Dirac submanifold. To do this, let N be a Poisson–Dirac submanifold of a Poisson manifold (M,π), and let $n \in N$. We denote the dimension of M by d and the dimension of N by s. There exists a coordinate chart (U,x) of M, adapted to N, and centered at n, such that the vector fields $\mathscr{X}_{x_1},\dots,\mathscr{X}_{x_s}$ are tangent to N, at every point of an open subset V of $U \cap N$, which contains n. In order to construct such local coordinates, one may start from an arbitrary coordinate chart (U,y) of M, adapted to N at n, and centered at n; since N is a Poisson–Dirac submanifold, each of the functions $y_{1|N},\dots,y_{s|N}$ can be extended in a neighborhood of n, to produce the above functions $x_1,\dots,x_s$, with the stated property, and supplemented with the remaining functions $x_{s+1} := y_{s+1},\dots,x_d := y_d$ they form the required coordinate chart, after possibly shrinking the neighborhoods U and V of n.

Since the vector fields $\mathscr{X}_{x_1},\dots,\mathscr{X}_{x_s}$ are tangent to N, at every point of V, all the functions $\{x_i,x_j\} = \pi[x_i,x_j] = -\mathscr{X}_{x_i}[x_j]$ vanish on V, for $1 \leqslant i \leqslant s < j \leqslant d$. For every point $n' \in V$, the Poisson structure π, at n', is therefore given by:

$$\sum_{1\leqslant i<j\leqslant s} x_{ij}(n')\left(\frac{\partial}{\partial x_i}\right)_{n'} \wedge \left(\frac{\partial}{\partial x_j}\right)_{n'} + \sum_{s<i<j\leqslant d} x_{ij}(n')\left(\frac{\partial}{\partial x_i}\right)_{n'} \wedge \left(\frac{\partial}{\partial x_j}\right)_{n'} \tag{5.38}$$

where $x_{ij}(n') := \{x_i,x_j\}(n')$, for $1 \leqslant i < j \leqslant d$, and where the first sum is the reduced Poisson structure at n. Equation (5.38) implies that, in well-chosen local coordinates, the Poisson matrix of π at points of N takes a block-diagonal form. We show in the following proposition that this fact characterizes Poisson–Dirac submanifolds.

Proposition 5.25. *Let (M,π) be a Poisson manifold of dimension d and let N be a submanifold of M. Then N is a Poisson–Dirac submanifold of M if and only if there exists for each $n \in N$ a coordinate chart (U,x) of M, adapted at N, and centered at n with the following property: if the Poisson matrix $X := (\{x_i,x_j\})_{1\leqslant i,j\leqslant d}$ of π in terms of these coordinates is written in block form*

$$X = \begin{pmatrix} A & B \\ -B^\top & D \end{pmatrix},$$

where A has size $s := \dim N$, then there exists a neighborhood V of n in $U \cap N$, such that $B(n') = 0$ for every $n' \in V$. Moreover, when these equivalent conditions are satisfied, then the Poisson matrix of the reduced Poisson structure π_N on N is given, in terms of the above coordinates $x_{1|V},\dots,x_{s|V}$, on a neighborhood of n in N, by $A_{|V} = \left(\{x_i,x_j\}_{|V}\right)_{1\leqslant i,j\leqslant s}$.

Proof. We start with the stated characterization of a Poisson–Dirac submanifold. By the above, every Poisson–Dirac submanifold admits indeed such an adapted coordinate chart, for each $n \in N$. For the inverse implication, let $n \in N$ and assume that a coordinate chart (U,x) of M is given, adapted at N, and centered at n. Let V denote the corresponding open subset of N which contains n, and on which $x_{1|_V}, \ldots, x_{s|_V}$ are local coordinates for N. Suppose that the Poisson matrix of $\{\cdot,\cdot\}$ in terms of the coordinate chart (U,x) takes a block-diagonal form at every point n' in a neighborhood of n in N, as in (5.38). Let F be a function, defined in a neighborhood of n in N and let $\tilde{F} := F \circ p$, where $p : U \to V$ is the projection map, associated to (U,x). Then $(\tilde{F}, U, V)$ is an extension of F at n. Since the coordinate expression of $\tilde{F}$ on (U,x) depends on the first s variables $x_1, \ldots, x_s$ only, the Hamiltonian vector field $\mathscr{X}_{\tilde{F}}$, associated to $\tilde{F}$, takes at points n' in a neighborhood of n in N the form

$$(\mathscr{X}_{\tilde{F}})_{n'} = \sum_{i=1}^{s} x_{ij}(n') \frac{\partial \tilde{F}}{\partial x_j}(n') \left(\frac{\partial}{\partial x_i}\right)_{n'},$$

so it is tangent to N at n'. According to Definition 5.23, this shows that N is a Poisson–Dirac submanifold.

We can use these coordinate charts to compute the Poisson matrix of the reduced Poisson structure on N. Indeed, on a neighborhood V of n in N, the functions $x_{1|_V}, \ldots, x_{s|_V}$ are local coordinates on N and their Poisson brackets $\left\{x_{i|_V}, x_{j|_V}\right\}_N$ can, according to Proposition 5.24, be computed as $\left\{x_i, x_j\right\}\big|_V$. It follows that, in terms of these coordinates, the Poisson matrix of the reduced Poisson structure is the matrix A, restricted to V. □

We give in the following proposition a description of the symplectic leaves of a Poisson–Dirac submanifold.

Proposition 5.26. *Let (N, π_N) be a Poisson–Dirac submanifold of a Poisson manifold (M, π) and let $n \in N$. Denoting by $\mathscr{S}_n(N)$, respectively $\mathscr{S}_n(M)$, the symplectic leaf of (N, π_N), respectively of (M, π), which contains n, the symplectic leaf $\mathscr{S}_n(N)$ is the connected component of $N \cap \mathscr{S}_n(M)$ which contains n.*

Proof. Let $n \in N$ and let $n' \in \mathscr{S}_n(N)$. By the definition of the symplectic leaf $\mathscr{S}_n(N)$, passing through n, this means that there exists a piecewise Hamiltonian path γ for π_N with endpoints n and n' (recall from Section 1.3.4 that a Hamiltonian path is by definition an integral curve of a Hamiltonian vector field). Since N is a Poisson–Dirac submanifold of M, every function F on N admits at every point n' in its domain of definition a local extension $(\tilde{F}, U, V)$, whose Hamiltonian vector field $\mathscr{X}_{\tilde{F}}$ is equal to $\mathscr{X}_F$ on V. Since the image of γ is compact, it can be covered by finitely many such local extensions, and γ is therefore a piecewise Hamiltonian path for π. It follows that

$$\mathscr{S}_n(N) \subset \mathscr{S}_n(M) \cap N. \tag{5.39}$$

Let $\mathscr{C}$ be the connected component of $\mathscr{S}_n(M) \cap N$ containing $n \in N$. Since $\mathscr{S}_n(N)$ is a connected set, (5.39) implies the stronger inclusion $\mathscr{S}_n(N) \subset \mathscr{C}$. We need to prove that $\mathscr{S}_n(N) = \mathscr{C}$.

To do this, let (U,x) be local coordinates in a neighborhood of $n \in N$, which are adapted to the Poisson–Dirac submanifold, as in (5.38). For every function $\tilde{F} \in \mathscr{F}(U)$, the Hamiltonian vector field $\mathscr{X}_{\tilde{F}}$ of $\tilde{F}$, with respect to $\{\cdot,\cdot\}$, and the Hamiltonian vector field $\mathscr{X}_F$ of its restriction $F = \tilde{F}_{|V} \in \mathscr{F}(V)$, with respect to $\{\cdot,\cdot\}_N$, are related at $n' \in V$ by

$$(\mathscr{X}_{\tilde{F}})_{n'} = (\mathscr{X}_F)_{n'} + \sum_{s<i<j\leqslant d} x_{ij}(n')\frac{\partial F}{\partial x_i}(n')\left(\frac{\partial}{\partial x_j}\right)_{n'} .$$

In particular, the tangent vector $(\mathscr{X}_{\tilde{F}})_{n'}$ in $T_{n'}M$ belongs to $T_{n'}N$ if and only if it is equal to $(\mathscr{X}_F)_{n'}$. Since, by Theorem 1.30, the tangent space of a symplectic leaf at a given point n' is the space of all Hamiltonian vectors $(\mathscr{X}_F)_{n'}$ at n', an element in $T_{n'}\mathscr{S}_n(M)$ belongs to $T_{n'}N$ if and only if it belongs to $T_{n'}\mathscr{S}_n(N)$. In short, we have, for all $n' \in \mathscr{S}_n(N)$,

$$T_{n'}\mathscr{S}_n(N) = T_{n'}\mathscr{S}_n(M) \cap T_{n'}N . \tag{5.40}$$

In general geometrical terms, (5.39) and (5.40) say that we have three submanifolds M_0, M_1 and M_2 of M, such that

$$M_0 \subset M_1 \cap M_2 \quad \text{and} \quad T_{m_0}M_0 = T_{m_0}M_1 \cap T_{m_0}M_2 ,$$

for all $m_0 \in M_0$. According to the implicit function theorem, there exist in this case neighborhoods U_0, U_1 and U_2 of m_0 in M_0, M_1 and M_2 respectively such that $U_0 = U_1 \cap U_2$. In particular, there is a neighborhood of m_0 in $M_1 \cap M_2$, which is contained in M_0. Applied to (5.39) and (5.40), this means that $\mathscr{S}_n(N)$ is an open subset in $\mathscr{C}$. Let us show that $\mathscr{S}_n(N)$ is also a closed subset of $\mathscr{C}$. Let n' be a point in the closure of $\mathscr{S}_n(N)$ in $\mathscr{C}$. There is, for the above reasons, an open subset W of $\mathscr{C}$ contained in $\mathscr{S}_{n'}(N)$. Since W has a non-empty intersection with $\mathscr{S}_n(N)$, the symplectic leaves $\mathscr{S}_n(N)$ and $\mathscr{S}_{n'}(N)$ coincide, so that in particular $n' \in \mathscr{S}_n(N)$, i.e., $\mathscr{S}_n(N)$ is a closed subset of $\mathscr{C}$. In conclusion, $\mathscr{S}_n(N)$ is, for every $n \in N$, an open and closed subset of $\mathscr{C}$, the connected component of $\mathscr{S}_n(M) \cap N$, which contains n. □

We now explain how to compute the Poisson matrix of a Poisson structure obtained by Poisson–Dirac reduction out of the Poisson matrix of the Poisson structure on the ambient space. According to (5.34), if $\tilde{x}_1,\dots,\tilde{x}_s$ are functions whose Hamiltonian vector fields are tangent to N, at every point in a neighborhood V of n in N, and such that their restrictions $x_1,\dots,x_s$ to V are local coordinates for N, then the matrix $\left(\{\tilde{x}_i,\tilde{x}_j\}_{|V}\right)_{1\leqslant i,j\leqslant s}$ is the Poisson matrix of $\{\cdot,\cdot\}_N$ with respect to the coordinates $x_1,\dots,x_s$. We show in the following proposition how to construct such functions and we derive from it an explicit formula for the latter Poisson structure.

Proposition 5.27. *Let (M,π) be a Poisson manifold and let N be an s-dimensional submanifold of M. Suppose that (U,x) is a coordinate chart of M, adapted to N and centered at some point $n \in N$. Let V denote an open subset of N, containing n, having the property that $(x_{1|V},\dots,x_{s|V})$ is a system of local coordinates on V. Let X denote the Poisson matrix $(\{x_i,x_j\})_{1\leqslant i,j\leqslant d}$ of π with respect to these coordinates.*

We write X in the block form

$$X = \begin{pmatrix} A & B \\ -B^\top & D \end{pmatrix}$$

where A and D are square matrices, of size s, respectively $d-s$, while B has size $s \times (d-s)$. If the matrix $D(m)$ is invertible for every $m \in U$, then

(1) V is a Poisson–Dirac submanifold of (M,π);
(2) The Poisson matrix X_N of the reduced Poisson structure on V is given, at $n' \in V$, by

$$X_N(n') = A(n') + B(n')D(n')^{-1}B(n')^\top .$$

Proof. We write the above coordinates $x_1,\ldots,x_d$ as $x_1,\ldots,x_s,y_1,\ldots,y_t$, where $t := d-s$, and we consider on U alternative coordinates $(x',y') = (x'_1,\ldots,x'_s,y'_1,\ldots,y'_t)$, defined by

$$\begin{pmatrix} x' \\ y' \end{pmatrix} = \begin{pmatrix} \mathbb{1}_s & -C \\ 0 & \mathbb{1}_t \end{pmatrix} \begin{pmatrix} x \\ y \end{pmatrix},$$

where C is a matrix which we will determine. These alternative coordinates are also adapted to N, since $x'_i - x_i$ vanishes on V, for $i = 1,\ldots,s$. We determine C by demanding that the s Hamiltonian vector fields $\mathscr{X}_{x'_i}$ are tangent to N at every point of V. This means that $\left\langle \mathrm{d}_{n'}y_j, \left(\mathscr{X}_{x'_i}\right)_{n'} \right\rangle = 0$ for all $1 \leqslant i \leqslant s$, $1 \leqslant j \leqslant t$ and $n' \in V$, where $\langle\cdot,\cdot\rangle$ stands for the canonical pairing between $T^*_{n'}M$ and $T_{n'}M$. Using the fact that, for all $F,G \in \mathscr{F}(V)$, we have that $(\mathscr{X}_{FG})_{n'} = F(n')(\mathscr{X}_G)_{n'}$ at every point n' of N where the function G vanishes, we find that

$$\begin{aligned}
\left\langle \mathrm{d}_{n'}y_j, \left(\mathscr{X}_{x'_i}\right)_{n'} \right\rangle &= \langle \mathrm{d}_{n'}y_j, (\mathscr{X}_{x_i})_{n'} \rangle - \sum_{k=1}^t C_{ik}(n') \langle \mathrm{d}_{n'}y_j, (\mathscr{X}_{y_k})_{n'} \rangle \\
&= \{y_j,x_i\}(n') - \sum_{k=1}^t C_{ik}(n') \{y_j,y_k\}(n') \\
&= (-B+CD)_{ij}(n'),
\end{aligned}$$

which is zero, precisely if we choose $C := BD^{-1}$. According to Proposition 5.25, V is a Poisson–Dirac submanifold of M and the Poisson matrix of the reduced Poisson structure on $V \subset N$ is given by the matrix $\left(\left\{x'_i,x'_j\right\}(n')\right)_{1\leqslant i,j\leqslant s}$, whose entries we compute as above,

$$\begin{aligned}
\left\{x'_i,x'_j\right\}(n') &= -\mathscr{X}_{x'_i}[x'_j](n') = -\mathscr{X}_{x'_i}[x_j](n') = \left\{ x_i - \sum_{k=1}^t C_{ik}y_k, x_j \right\}(n') \\
&= A_{ij}(n') + \sum_{k=1}^t C_{ik}(n')B^\top_{kj}(n') = \left(A+CB^\top\right)_{ij}(n') .
\end{aligned}$$

Since $C = BD^{-1}$, the Poisson matrix of the reduced structure is given on V by $A + BD^{-1}B^{\top}$. □

5.3.3 The Transverse Poisson Structure

Let (M,π) be a Poisson manifold and let $m \in M$ be a point where the rank of π is $2r$. Suppose that $q_1,\dots,q_r,p_1,\dots,p_r,z_1,\dots,z_s$ are splitting coordinates, defined on a neighborhood U of m in M, and centered at m. According to Weinstein's splitting theorem (Theorem 1.25), this means that, on U,

$$\pi = \sum_{i=1}^{r} \frac{\partial}{\partial q_i} \wedge \frac{\partial}{\partial p_i} + \sum_{1 \leqslant k < \ell \leqslant s} \phi_{k\ell}(z) \frac{\partial}{\partial z_k} \wedge \frac{\partial}{\partial z_\ell},$$

where the functions $\phi_{k\ell}$ are (smooth or holomorphic) functions, which depend on $z = (z_1,\dots,z_s)$ only, and which vanish when $z = 0$. In view of Definition 5.23, the embedded submanifold N of M, defined by

$$N := \{m' \in U \mid q_1(m') = p_1(m') = \cdots = q_r(m') = p_r(m') = 0\}$$

is a Poisson–Dirac submanifold of M. According to Proposition 5.20, the reduced Poisson structure π_N on N is given by

$$\pi_N = \sum_{1 \leqslant k < \ell \leqslant s} \phi_{k\ell}(z) \frac{\partial}{\partial z_k} \wedge \frac{\partial}{\partial z_\ell},$$

with the understanding that $(z_1,\dots,z_s)$ stands for the restriction of $(z_1,\dots,z_s)$ to N. The submanifold N of M which appears in the latter example is transverse at m to the symplectic leaf $\mathscr{S}_m$ through m, since $T_mN \oplus T_m\mathscr{S}_m = T_mM$. Recall that two submanifolds N and O of a manifold M are said to be *transverse* at a point $m \in M$ if $m \in N \cap O$ and $T_mN + T_mO = T_mM$; when N and O have complementary dimension in M, this is equivalent to $T_mN \oplus T_mO_m = T_mM$. The following theorem states that, in a neighborhood of m in M, every submanifold of M of dimension $\dim N$, which is transverse to $\mathscr{S}_m$ at m, is in a neighborhood of m a Poisson–Dirac submanifold of M and is isomorphic, as a Poisson manifold, to a neighborhood of m in N.

Theorem 5.28. *Let (M,π) be a Poisson manifold and let $m \in M$. The rank of π at m is denoted by $2r$. Suppose that $q_1,\dots,q_r,p_1,\dots,p_r,z_1,\dots,z_s$ are splitting coordinates, defined on a neighborhood U of m in M, and centered at m. We denote by N_0 the Poisson–Dirac submanifold of M, defined by*

$$N_0 := \{m' \in U \mid q_1(m') = p_1(m') = \cdots = q_r(m') = p_r(m') = 0\}.$$

Let $N_1 \subset M$ be an arbitrary s-dimensional submanifold of M, transverse at m to the symplectic leaf $\mathscr{S}_m$ of π which contains m. Then there exists a neighborhood V_0 of m in N_0 and a neighborhood V_1 of m in N_1, such that

(1) V_1 *is a Poisson–Dirac submanifold of* M;

(2) *There exists a Poisson diffeomorphism between* V_0 *and* V_1*, where* V_0 *and* V_1 *are both equipped with their reduced Poisson structure as Poisson–Dirac submanifolds of* M.

Proof. Let $q_1,\dots,q_r,p_1,\dots,p_r,z_1,\dots,z_s$ be splitting coordinates, centered at m, defined on a neighborhood U of m in M. On U, the Poisson structure π is given by

$$\pi = \sum_{i=1}^{r} \frac{\partial}{\partial q_i} \wedge \frac{\partial}{\partial p_i} + \pi' , \tag{5.41}$$

where $\pi' = \sum_{1\leqslant k<\ell\leqslant s} \phi_{k\ell}(z) \frac{\partial}{\partial z_k} \wedge \frac{\partial}{\partial z_\ell}$ is a Poisson structure on U, whose structure functions $\phi_{k\ell}$ depend only on the variables $z = (z_1,\dots,z_s)$ and vanish for $z = 0$. These coordinates are adapted to the symplectic leaf $\mathscr{S}_m$, since $\mathscr{S}_m$ is defined on U by $z = 0$. Consider an arbitrary s-dimensional submanifold N_1 of M, transverse at m to the symplectic leaf $\mathscr{S}_m$ of π, and of dimension $s = \dim N_0$. Since N_1 is of codimension $2r$ in M, N_1 is in a neighborhood of m in M given as the intersection of $2r$ hypersurfaces

$$H_1(q,p,z) = \cdots = H_{2r}(q,p,z) = 0 . \tag{5.42}$$

Since N_1 is transverse to $\mathscr{S}_m$ at m, these hypersurfaces are transverse to $\mathscr{S}_m$ at m and their tangent spaces have T_mN_1 as their common intersection. By the implicit function theorem, applied to (5.42), there exist functions $F_1,\dots,F_r,G_1,\dots,G_r$, such that N_1 is, in a neighborhood of m, given by

$$q_i = F_i(z_1,\dots,z_s) , \qquad p_i = G_i(z_1,\dots,z_s) ,$$

for $i = 1,\dots,r$. Letting

$$q_i' := q_i - F_i(z_1,\dots,z_s) , \qquad p_i' := p_i - G_i(z_1,\dots,z_s) ,$$

it follows that $(q_1',\dots,q_r',p_1',\dots,p_r',z_1,\dots,z_s)$ is in a neighborhood $U' \subset U$ of m a coordinate system for M, adapted to both submanifolds N_1 and $\mathscr{S}_m$. Since π' vanishes at m,

$$\left\{q_i',p_j'\right\}(m) = \delta_{i,j}, \qquad \left\{q_i',q_j'\right\}(m) = \left\{p_i',p_j'\right\}(m) = 0 ,$$

where $\{\cdot,\cdot\} := \pi$ and $\delta_{i,j}$ stands for the Kronecker delta. As a consequence, the matrix

$$\begin{pmatrix} \left(\left\{q_i',q_j'\right\}(m')\right)_{i,j=1}^{r} & \left(\left\{q_i',p_j'\right\}(m')\right)_{i,j=1}^{r} \\ \left(\left\{p_i',q_j'\right\}(m')\right)_{i,j=1}^{r} & \left(\left\{p_i',p_j'\right\}(m')\right)_{i,j=1}^{r} \end{pmatrix}$$

is invertible for every m' in a neighborhood of m in M. It proves, in view of Proposition 5.27, that some neighborhood V_1 of m in N_1 is a Poisson–Dirac submanifold of M, which is the content of *(1)*.

The proof of *(2)* uses some of the objects which were constructed in the first part of the proof, namely the open neighborhood U' of m in M and the functions $F_1, G_1, \dots, F_r, G_r$ which are defined on U'. Consider the product manifold $U' \times \mathbb{F}$, which we equip with its product Poisson structure (the Poisson structure on $\mathbb{F}$ is trivial because $\mathbb{F}$ is one-dimensional). The splitting coordinates which we have constructed on U', plus the natural coordinate on $\mathbb{F}$ which we denote by τ, yield coordinates on $U' \times \mathbb{F}$. We define on $U' \times \mathbb{F}$ the following $2r$ functions

$$Q_i := q_i - \tau F_i(z_1, \dots, z_s) \text{ and } P_i := p_i - \tau G_i(z_1, \dots, z_s) \,,$$

where $i = 1, \dots, r$. For $\tau \in \mathbb{F}$, their common zero locus is denoted by N'_τ,

$$N'_\tau := \left\{ m' \in U' \mid Q_i(m', \tau) = P_i(m', \tau) = 0 \text{ for } 1 \leqslant i \leqslant r \right\}.$$

Letting τ vary between 0 and 1 allows us to relate the Poisson–Dirac submanifolds N_0 and N_1; precisely, $N'_0 = (N_0 \cap U') \times \{0\}$ and $N'_1 = (N_1 \cap U') \times \{1\}$, so that N'_0 and N'_1 can be identified as Poisson manifolds with open neighborhoods of m in N_0 and in N_1 respectively. We construct a Poisson vector field $\mathscr{V}$ on $U' \times \mathbb{F}$, whose time 1 flow yields a diffeomorphism between N'_0 and N'_1, at least in a neighborhood of $m \times \{0\}$; under the above identification, this diffeomorphism is a Poisson diffeomorphism between a neighborhood of m in N_0 and a neighborhood of m in N_1, equipped with their reduced structure, because the reduced Poisson structure on a Poisson–Dirac submanifold (of U', or equivalently of $U' \times \mathbb{F}$) is uniquely defined. The construction of the Poisson vector field $\mathscr{V}$ with the stated property will therefore prove *(2)*.

The vector field $\mathscr{V}$ is chosen of the form $\mathscr{V} = \frac{\partial}{\partial \tau} + \mathscr{X}_H$, where H is a function which will be specified below. This choice ensures that $\mathscr{V}$ is a Poisson vector field on $U' \times \mathbb{F}$, since $\mathscr{X}_H$ is a Hamiltonian vector field on U'. The Hamiltonian H is chosen as

$$H := \sum_{j=1}^{r} A_j(\tau, z_1, \dots, z_s)\, Q_j + \sum_{j=1}^{r} B_j(\tau, z_1, \dots, z_s)\, P_j \,, \tag{5.43}$$

where the functions A_j and B_j are solutions to the linear system

$$\begin{cases} \sum_{j=1}^{r} A_j \{Q_i, Q_j\} + \sum_{j=1}^{r} B_j \{Q_i, P_j\} = F_i \,, \\ \sum_{j=1}^{r} A_j \{P_i, Q_j\} + \sum_{j=1}^{r} B_j \{P_i, P_j\} = G_i \,, \end{cases} \qquad i = 1, \dots, r \,. \tag{5.44}$$

Since

$$\{Q_i, Q_j\}(m, \tau) = \{P_i, P_j\}(m, \tau) = 0 \,, \qquad \{Q_i, P_j\}(m, \tau) = \delta_{i,j} \,,$$

for all $1 \leqslant i, j \leqslant r$ and for all $\tau \in \mathbb{F}$, this linear system has a unique solution, which is defined on a neighborhood $U'' \times W$ of $m \times [0, 1]$ in $U' \times \mathbb{F}$, and which is smooth. Notice that since all coefficients in (5.44) are independent of the coordinates p_i and q_i, the functions A_j and B_j depend only on $z_1, \ldots, z_s$ and τ. Moreover, these functions vanish at $(m,0)$ for all τ, since the functions F_i and G_i vanish at m. The integral curve of $\mathscr{V}$, starting from $(m,0)$ is therefore defined for all $t \in W$, being given by $t \mapsto (m,t)$. It follows that by eventually shrinking U'' to a smaller neighborhood of m and W to a smaller neighborhood of $[0,1]$, we may assume that the flow Φ_t of $\mathscr{V}$ is well-defined on $U'' \times \{0\}$ for all $t \in W$.

It remains to be shown that the time 1 flow Φ_1 of $\mathscr{V}$ sends $U'' \times \{0\}$ to a neighborhood of $(m,1)$ in N_1. Consider the submanifold $\mathscr{N}$ of $U'' \times W$, defined by the common zeros of the functions Q_i and P_i,

$$\mathscr{N} := \{(m',\tau) \in U'' \times W \mid Q_i(m',\tau) = P_i(m',\tau) = 0 \text{ for } 1 \leqslant i \leqslant r\} .$$

The vector field $\mathscr{V}$ is tangent to $\mathscr{N}$, since,

$$\mathscr{V}[Q_i](m'',\tau) = \frac{\partial Q_i}{\partial \tau}(m'',\tau) + \mathscr{X}_H[Q_i](m'',\tau) = F_i(m'') - F_i(m'') = 0 ,$$

and similarly $\mathscr{V}[P_i](m'',\tau) = 0$, for all $(m'',\tau) \in \mathscr{N}$ and all $i = 1, \ldots, r$. The time 1 flow of $\mathscr{V}$ maps $U'' \times \{0\}$ to a neighborhood of $(m,1)$ in $U' \times \{1\}$. Since it preserves $\mathscr{N}$, it maps $(N_0 \cap U'') \times \{0\}$ to a neighborhood of $(m,1)$ in $N_1 \times \{1\}$, as was to be shown. □

It follows from Theorem 5.28 that all s-dimensional submanifolds which are transverse to the symplectic leaf which contains m, are isomorphic as Poisson manifolds in a neighborhood of m, when they are equipped with their reduced Poisson structure.

We show in the following proposition that all s-dimensional submanifolds which are transverse to a given symplectic leaf $\mathscr{S}$ of M are isomorphic as Poisson manifolds in a neighborhood of their intersection point with $\mathscr{S}$, when they are equipped with their reduced Poisson structure.

Proposition 5.29. *Let (M,π) be a Poisson manifold and let $\mathscr{S}$ be a symplectic leaf of (M,π). We denote the dimension of $\mathscr{S}$ by $2r$ and we let $s := \dim M - 2r$. Assume that we are given:*

(1) Two points m, m' which belong to $\mathscr{S}$;
(2) Two s-dimensional submanifolds N and N' of M, which are transverse to $\mathscr{S}$ at m and at m' respectively.

Then there exist neighborhoods V and V' of m and m' in N and N' respectively, such that:

(1) V and V' are Poisson–Dirac submanifolds of (M,π);
(2) Equipped with their reduced Poisson structure, V and V' are isomorphic as Poisson manifolds.

Proof. Let m and m' be two points of M which belong to the same symplectic leaf $\mathscr{S}$ of (M,π). By definition, such points can be joined by a piecewise Hamiltonian path. It is therefore sufficient to prove the proposition for points m and m' which can be joined by a Hamiltonian path. Assume therefore that U is an open subset of M, that H is a function on U and that the flow Φ_t of $\mathscr{X}_H$ is defined for t in a neighborhood of $[0,1]$ and for all points of U, with in particular $\Phi_1(m) = m'$. Since Φ_t is the flow of a Hamiltonian vector field, Φ_1 is a Poisson diffeomorphism between a neighborhood U of m and a neighborhood $\Phi_1(U)$ of $\Phi_1(m) = m'$. Since Φ_1 preserves the Poisson structure, it sends the symplectic leaf $\mathscr{S}$ through m to the symplectic leaf $\mathscr{S}$ through $\Phi_1(m)$; moreover, since Φ_1 is a diffeomorphism, it sends every submanifold, transverse to $\mathscr{S}$ at m, to a submanifold transverse to $\mathscr{S}$ at m', and Φ_1 realizes a Poisson diffeomorphism between them, when both are equipped with the Poisson structure which they inherit from (M,π) as Poisson–Dirac submanifolds. Since we know from item *(2)* of Theorem 5.28 that the transverse Poisson structure is independent of the chosen transversal through a given point, this shows in addition that the transverse Poisson structure is independent of the chosen point on a given symplectic leaf of (M,π). □

Theorem 5.28 and Proposition 5.29 motivate and justify the following definition.

Definition 5.30. Let $\mathscr{S}$ be a symplectic leaf of a Poisson manifold (M,π). Let N be an arbitrary submanifold of M, transverse to $\mathscr{S}$ at some point $m \in \mathscr{S}$, and let π' denote the reduced Poisson structure, which is defined on a neighborhood of m in N. The germ of π' at m is called the *transverse Poisson structure* to $\mathscr{S}$.

We have seen two particular situations in which one constructs a (natural) representative of the transverse Poisson structure to a symplectic leaf $\mathscr{S}$ in a Poisson manifold (M,π). The first one is Weinstein's splitting theorem (Theorem 1.25), which allows us to write π in terms of splitting coordinates on a neighborhood of a point $m \in \mathscr{S}$ as

$$\pi = \sum_{i=1}^{r} \frac{\partial}{\partial q_i} \wedge \frac{\partial}{\partial p_i} + \sum_{1 \leqslant k < \ell \leqslant s} \phi_{k\ell}(z) \frac{\partial}{\partial z_k} \wedge \frac{\partial}{\partial z_\ell}, \tag{5.45}$$

where the functions $\phi_{k\ell}$ are (smooth or holomorphic) functions, which depend on $z = (z_1, \dots, z_s)$ only, and which vanish when $z = 0$; in this case, the second sum in (5.45) is a Poisson structure, which is a representative for the transverse Poisson structure to the symplectic leaf $\mathscr{S}$, while the first sum in (5.45) yields the canonical Poisson structure on the symplectic leaf $\mathscr{S}$. This shows that the decomposition of Poisson structures, which appears in Weinstein's splitting theorem, is unique. The second situation is when one selects a submanifold N of M, which is transverse to the symplectic leaf $\mathscr{S}$; in this case, Theorem 5.28 says that N is in a neighborhood V of the intersection point $\{m\} := \mathscr{S} \cap N$ a Poisson–Dirac submanifold of (M,π) and the reduced Poisson structure on V is a representative for the transverse Poisson structure to the symplectic leaf $\mathscr{S}$.

5.4 Poisson Structures and Group Actions

Poisson manifolds often come with symmetry groups which lead, in favorable cases, to new Poisson manifolds. If a Lie group $\mathbf{G}$ acts on a Poisson manifold, preserving the Poisson structure, then there are three natural constructions which lead to new Poisson manifolds:

- By Poisson reduction, the quotient $M/\mathbf{G}$, assumed to be a manifold, inherits a Poisson structure;
- By Poisson–Dirac reduction, the set of fixed points of the action, assumed to be a manifold, inherits a Poisson structure;
- By Poisson reduction, the action yields a Poisson structure on certain quotients of the fibers of the momentum map, assuming that the latter map exists.

Similarly, such new Poisson structures arise in the case of an affine Poisson variety, when an algebraic group is acting. As indicated in the three items above, this section is an application of the previous sections, in the particular case of a Poisson manifold (or affine Poisson variety) upon which a group is acting.

5.4.1 Poisson Actions

There are two natural types of actions of a group on a Poisson manifold or on a Poisson variety, as given in the following two definitions.

Definition 5.31. Let $\mathbf{G}$ be a group and let (M,π) be an affine Poisson variety or a Poisson manifold. We say that an action $\chi : \mathbf{G}\times M \to M$ *preserves* the Poisson structure when, for every $g\in\mathbf{G}$, the map $\chi_g : M\to M$, defined for all $m\in M$ by $\chi_g(m) := \chi(g,m)$, is a Poisson map.

Definition 5.32. Let $\mathbf{G}$ be a group acting on M, where M and $\mathbf{G}$ are either one of the following:

- $(M,\{\cdot\,,\cdot\})$ and $\mathbf{G}=(\mathbf{G},\{\cdot\,,\cdot\}_{\mathbf{G}})$ are affine Poisson varieties and $\mathbf{G}$ is an affine algebraic group;
- $(M,\{\cdot\,,\cdot\})$ and $\mathbf{G}=(\mathbf{G},\{\cdot\,,\cdot\}_{\mathbf{G}})$ are Poisson manifolds and $\mathbf{G}$ is a Lie group.

Then we say that the action is a *Poisson action* when the action map $\chi : \mathbf{G}\times M\to M$ is a Poisson map, where $\mathbf{G}\times M$ is equipped with the product Poisson structure.

When $\mathbf{G}\times M\to M$ is a Poisson action and the Poisson structure on $\mathbf{G}$ is trivial, then the product bracket $\{\cdot\,,\cdot\}_{\mathbf{G}\times M}$ at $(g,m)\in\mathbf{G}\times M$ reduces, in view of (2.12), to

$$\{F\circ\chi, G\circ\chi\}_{\mathbf{G}\times M}(g,m) = \left\{F\circ\chi_g, G\circ\chi_g\right\}(m)\,,$$

so that a Poisson action is in this case precisely an action preserving the Poisson structure. In general, however, a Poisson action does not preserve the Poisson structure.

5.4.2 Poisson Actions and Quotient Spaces

In this section, we show that a Poisson action $\mathbf{G}\times M\to M$ leads to a Poisson structure on the quotient space $M/\mathbf{G}$ and we give conditions on a $\mathbf{G}$-invariant subvariety N of M, which lead to a Poisson structure on $N/\mathbf{G}$. The following proposition gives sufficient conditions for $M/\mathbf{G}$ to inherit a Poisson structure from M and $\mathbf{G}$.

Proposition 5.33. *Let $\chi:\mathbf{G}\times M\to M$ be a group action, where either one of the following is satisfied:*

(1) $(M,\{\cdot,\cdot\})$ and $(\mathbf{G},\{\cdot,\cdot\}_{\mathbf{G}})$ are affine Poisson varieties, with $\mathbf{G}$ being a reductive algebraic group;
(2) $(M,\{\cdot,\cdot\})$ and $(\mathbf{G},\{\cdot,\cdot\}_{\mathbf{G}})$ are Poisson manifolds, with $\mathbf{G}$ being a Lie group, and the action is proper and locally free.

If χ is a Poisson action, then $M/\mathbf{G}$ carries a unique Poisson structure such that the canonical projection $p:M\to M/\mathbf{G}$ is a Poisson map.

Proof. In the case of an affine Poisson variety, $M/\mathbf{G}$ is an affine variety. According to Proposition 5.7, in which *(1)* is automatically satisfied, $(M,M,M/\mathbf{G})$ is Poisson reducible as soon as the $\mathbf{G}$-invariant functions on M form a Lie subalgebra of $\mathscr{F}(M)$, in formulas

$$\left\{\mathscr{F}(M)^{\mathbf{G}},\mathscr{F}(M)^{\mathbf{G}}\right\}\subset\mathscr{F}(M)^{\mathbf{G}}. \tag{5.46}$$

We show that (5.46) is a consequence of the fact that χ is a Poisson action. To do this, first notice that a function $F\in\mathscr{F}(M)$ belongs to $\mathscr{F}(M)^{\mathbf{G}}$ if and only if $\chi^*F=p_2^*F$ where $p_2:\mathbf{G}\times M\to M$ is the projection onto the second component. Thus, if $F,G\in\mathscr{F}(M)^{\mathbf{G}}$, then the fact that χ is a Poisson map implies that

$$\chi^*\{F,G\}=\{\chi^*F,\chi^*G\}_{\mathbf{G}\times M}=\{p_2^*F,p_2^*G\}_{\mathbf{G}\times M}=p_2^*\{F,G\}, \tag{5.47}$$

where we used in the last step that p_2 is a Poisson map (see Proposition 2.2). This shows, in view of the above characterization of $\mathbf{G}$-invariant functions, that $\{F,G\}\in\mathscr{F}(M)^{\mathbf{G}}$. Hence, $\mathscr{F}(M)^{\mathbf{G}}$ is indeed a Lie subalgebra of $(M,\{\cdot,\cdot\})$, which achieves the proof for the case of affine Poisson varieties.

In the manifold case, $M/\mathbf{G}$ is a manifold since the action is proper and locally free. Functions which are defined on open subsets of M are constant on the fibers of $p:M\to M/\mathbf{G}$ if and only if they are $\mathbf{G}$-invariant. Therefore, $(M,M,M/\mathbf{G})$ is, in view of Proposition 5.11, Poisson reducible if and only if the Poisson bracket of every pair of $\mathbf{G}$-invariant functions, defined on an arbitrary open subset of M, is $\mathbf{G}$-invariant. This is checked precisely as in (5.47), taking for F and G arbitrary functions, which are defined on an open subset of M. □

In applications, one often considers the quotient of a $\mathbf{G}$-invariant[4] subvariety or submanifold N of M and while N does not inherit a Poisson structure from M, the

[4] Saying that N is $\mathbf{G}$-*invariant* means that $\chi(g,n)\in N$ for every $(g,n)\in\mathbf{G}\times N$.

quotient $N/\mathbf{G}$ does. This is the content of the following generalization of Proposition 5.33, which is proved in exactly the same way as a direct application of Proposition 5.7 in the case of an affine Poisson variety, and of Proposition 5.11 in the case of a Poisson manifold.

Proposition 5.34. *Let $\chi : \mathbf{G} \times M \to M$ be a group action and let N be a $\mathbf{G}$-invariant subset of M; denote by $p : N \to N/\mathbf{G}$ the quotient map. Suppose that either one of the following is satisfied:*

(1) $(M, \{\cdot,\cdot\})$ and $(\mathbf{G}, \{\cdot,\cdot\}_{\mathbf{G}})$ are affine Poisson varieties, with $\mathbf{G}$ being a reductive algebraic group, and N is an affine subvariety of M;
(2) $(M, \{\cdot,\cdot\})$ and $(\mathbf{G}, \{\cdot,\cdot\}_{\mathbf{G}})$ are Poisson manifolds, with $\mathbf{G}$ being a Lie group, N is a submanifold of M and the action of $\mathbf{G}$ on N is proper and locally free.

Suppose that, in addition, the following conditions hold:

(1) χ is a Poisson action;
(2) For every function F, whose restriction to N is $\mathbf{G}$-invariant, the Hamiltonian vector field $\mathscr{X}_F$ is tangent to N; it suffices that this property holds for $F \in \mathscr{F}(M)$ in the affine variety case, but it is required to hold for F defined on arbitrary open subsets in the manifold case.

Then $N/\mathbf{G}$ carries a unique Poisson structure $\{\cdot,\cdot\}_{N/\mathbf{G}}$ such that for every open subset V of $N/\mathbf{G}$ and for every pair of functions $F, G \in \mathscr{F}(V)$, and for all $n \in p^{-1}(V)$,

$$\left\{\tilde{F}, \tilde{G}\right\}(n) = \{F, G\}_{N/\mathbf{G}}\,(p(n))\,,$$

where $\tilde{F}$ and $\tilde{G}$ are arbitrary extensions of $F \circ p$ and $G \circ p$ to an arbitrary open subset of M.

5.4.3 Fixed Point Sets as Poisson–Dirac Submanifolds

In this section we show that, under some general assumptions, if a group action on a Poisson variety or manifold leaves the Poisson structure invariant, then the fixed point set of the action is a Poisson–Dirac subvariety or submanifold. We do this in the case of a finite or linear algebraic group, acting on an affine variety; see the end of this section for the case of Lie groups acting on manifolds.

Let $\mathbf{G}$ be a finite group or linear algebraic group, acting on an affine variety M. We write both $\chi_g(m)$ and gm for the action of $g \in \mathbf{G}$ on $m \in M$, while the induced action on $F \in \mathscr{F}(M)$ is denoted by $\psi_g(F)$, as in (5.4). We denote the algebra of invariant functions by $\mathscr{F}(M)^{\mathbf{G}}$,

$$\mathscr{F}(M)^{\mathbf{G}} = \left\{F \in \mathscr{F}(M) \mid \psi_g(F) = F \text{ for all } g \in \mathbf{G}\right\}\,,$$

and we denote the fixed point set of χ by $M^{\mathbf{G}}$,

$$M^{\mathbf{G}} := \{m \in M \mid gm = m \text{ for all } g \in \mathbf{G}\} .$$

In order to describe the algebraic structure of $M^{\mathbf{G}}$, consider for a fixed $g \in \mathbf{G}$ the regular map

$$\begin{aligned} L_g : M &\to M \times M \\ m &\mapsto (m, gm) . \end{aligned} \tag{5.48}$$

Then $M^{\mathbf{G}} = \bigcap_{g \in \mathbf{G}} L_g^{-1}(\Delta)$, where Δ is the diagonal of $M \times M$,

$$\Delta := \{(m,m) \mid m \in M\} .$$

Thus, the group action leads to a natural ideal $\mathscr{I}(M^{\mathbf{G}})$ of $\mathscr{F}(M)$, consisting of functions which vanish on $M^{\mathbf{G}}$, namely the ideal of $\mathscr{F}(M)$, generated by the functions $L_g^*(\mathscr{I}_\Delta)$, where $\mathscr{I}_\Delta$ is the ideal of all regular functions which vanish on Δ; for $i = 1,2$, if we denote by $p_i : M \times M \to M$ the canonical projection on the i-th component, then $\mathscr{I}_\Delta$ is the ideal of $\mathscr{F}(M \times M)$, generated by all functions $p_1^*(F) - p_2^*(F)$, where $F \in \mathscr{F}(M)$. It follows that the ideal $\mathscr{I}(M^{\mathbf{G}})$ is generated by the set of all functions of the form $F - \psi_g(F)$, where $g \in \mathbf{G}$ and $F \in \mathscr{F}(M)$. Notice that although the algebraic subset which is defined by (the common zeros of the elements of) $\mathscr{I}(M^{\mathbf{G}})$ coincides with $M^{\mathbf{G}}$, the latter ideal is in general neither prime nor reduced.

Proposition 5.35. *Let* $\mathbf{G}$ *be a finite group or an algebraic group, acting on an affine Poisson variety* $(M, \{\cdot,\cdot\})$. *If the group action preserves the Poisson structure, then* $\mathscr{F}(M)^{\mathbf{G}}$ *is contained in the normalizer* $\mathscr{N}(\mathscr{I}(M^{\mathbf{G}}))$ *of* $\mathscr{I}(M^{\mathbf{G}})$. *As a consequence, if*

$$\mathscr{F}(M)^{\mathbf{G}} + \mathscr{I}(M^{\mathbf{G}}) = \mathscr{F}(M) , \tag{5.49}$$

then $\mathscr{I}(M^{\mathbf{G}})$ *is a Poisson–Dirac ideal of* $\mathscr{F}(M)$.

Proof. Let $F \in \mathscr{F}(M)^{\mathbf{G}}$ be a $\mathbf{G}$-invariant function. For all $G \in \mathscr{F}(M)$ and all $g \in \mathbf{G}$, we have

$$\{F, G - \psi_g(G)\} = \{F,G\} - \{\psi_g(F), \psi_g(G)\} = \{F,G\} - \psi_g(\{F,G\}) ,$$

where we used in the last step that every map ψ_g is a morphism of Poisson algebras. Since $\mathscr{I}(M^{\mathbf{G}})$ is generated by all functions of the form $G - \psi_g(G)$, with $g \in \mathbf{G}$ and $G \in \mathscr{F}(M)$, the inclusion $\{\mathscr{F}(M)^{\mathbf{G}}, \mathscr{I}(M^{\mathbf{G}})\} \subset \mathscr{I}(M^{\mathbf{G}})$ follows, i.e., $\mathscr{F}(M)^{\mathbf{G}}$ is contained in the normalizer of $\mathscr{I}(M^{\mathbf{G}})$. In view of *(iii)* in Definition 5.17, $\mathscr{I}(M^{\mathbf{G}})$ will be a Poisson–Dirac ideal of $\mathscr{F}(M)$ if $\mathscr{F}(M)^{\mathbf{G}} + \mathscr{I}(M^{\mathbf{G}}) = \mathscr{F}(M)$. □

We consider two particular cases of the above proposition, which both lead to the result that the fixed point set of the group action is a Poisson–Dirac subvariety. We first consider the simpler case in which $\mathbf{G}$ is a finite group.

Proposition 5.36. *Let* $\mathbf{G}$ *be a finite group, acting on an affine Poisson variety* $(M, \{\cdot,\cdot\})$. *If the action preserves the Poisson structure, then the fixed point set* $M^{\mathbf{G}}$ *is a (not necessarily irreducible) Poisson–Dirac subvariety of* $(M, \{\cdot,\cdot\})$. *For* $F, G \in \mathscr{F}(M^{\mathbf{G}})$, *their reduced bracket is given, at* $n \in M^{\mathbf{G}}$, *by*

$$\{F,G\}_{M^{\mathbf{G}}}(n) = \frac{1}{|\mathbf{G}|^2} \sum_{g_1,g_2 \in \mathbf{G}} \{\psi_{g_1}(\tilde{F}), \psi_{g_2}(\tilde{G})\}(n) = \frac{1}{|\mathbf{G}|} \sum_{g \in \mathbf{G}} \{\psi_g(\tilde{F}), \tilde{G}\}(n)$$

where $\tilde{F}, \tilde{G} \in \mathscr{F}(M)$ *are arbitrary extensions of* F, G.

Proof. By averaging the action over the group, one constructs a linear projection $\rho : \mathscr{F}(M) \to \mathscr{F}(M)^{\mathbf{G}}$; explicitly, $\rho(F)$ is given, for $F \in \mathscr{F}(M)$, by

$$\rho(F) := \frac{1}{|\mathbf{G}|} \sum_{g \in \mathbf{G}} \psi_g(F) \,, \tag{5.50}$$

where $|\mathbf{G}|$ stands for the order of the group $\mathbf{G}$. Since

$$F - \rho(F) = \frac{1}{|\mathbf{G}|} \sum_{g \in \mathbf{G}} (F - \psi_g(F)),$$

it is an element of $\mathscr{I}(M^{\mathbf{G}})$, which shows that $\mathscr{F}(M)^{\mathbf{G}} + \mathscr{I}(M^{\mathbf{G}}) = \mathscr{F}(M)$. In view of Proposition 5.35, it follows that $\mathscr{I}(M^{\mathbf{G}})$ is a Poisson–Dirac ideal of $\mathscr{F}(M)$. Since $\mathbf{G}$ is finite, the (dual) action of $\mathbf{G}$ on $\mathscr{F}(M)$ is completely reducible, so that the ideal $\mathscr{I}(M^{\mathbf{G}})$ is the ideal of *all* regular functions, vanishing on $M^{\mathbf{G}}$, so that $M^{\mathbf{G}}$ is a (not necessarily irreducible) Poisson–Dirac subvariety. According to (5.33), the Poisson–Dirac bracket of two functions $F, G \in \mathscr{F}(M^{\mathbf{G}})$ is given, at $n \in M^{\mathbf{G}}$, by $\{F,G\}_{M^{\mathbf{G}}}(n) = \{\hat{F}, \hat{G}\}(n)$, where $\hat{F}$ and $\hat{G}$ are arbitrary $\mathbf{G}$-invariant extensions of F and G to M. Let $\tilde{F}$ and $\tilde{G}$ be arbitrary extensions of F and G to M. Clearly, the $\mathbf{G}$-invariant functions $\rho(\tilde{F})$ and $\rho(\tilde{G})$ coincide with F and G on $M^{\mathbf{G}}$, so we can take $\hat{F} := \rho(\tilde{F})$ and $\hat{G} := \rho(\tilde{G})$, to compute

$$\{F,G\}_{M^{\mathbf{G}}}(n) = \{\rho(\tilde{F}), \rho(\tilde{G})\}(n) = \frac{1}{|\mathbf{G}|^2} \sum_{g_1,g_2 \in \mathbf{G}} \{\psi_{g_1}(\tilde{F}), \psi_{g_2}(\tilde{G})\}(n) \,,$$

for all $n \in M^{\mathbf{G}}$. Since $\tilde{G} - \rho(\tilde{G}) \in \mathscr{I}(M^{\mathbf{G}})$, the latter line can also be written as

$$\{F,G\}_{M^{\mathbf{G}}}(n) = \{\rho(\tilde{F}), \tilde{G}\}(n) = \frac{1}{|\mathbf{G}|} \sum_{g \in \mathbf{G}} \{\psi_g(\tilde{F}), \tilde{G}\}(n) \,,$$

for all $n \in M^{\mathbf{G}}$. □

As a general principle in invariant theory, results which hold for the action of a finite group, acting on an affine variety, can be extended to the action of a reductive group, acting on an affine variety, by replacing the above averaging operator with the Reynolds operator (see [190]). It leads to the following result.

Proposition 5.37. *Let* $\mathbf{G}$ *be a reductive group, acting on an affine Poisson variety* (M, π). *If the action preserves the Poisson structure, then the fixed point set* $M^{\mathbf{G}}$ *is a (not necessarily irreducible) Poisson–Dirac subvariety of* (M, π).

For a differential geometric analog of the proposition, see [54, Th. 2.1].

5.4.4 Reduction with Respect to a Momentum Map

Let **G** be a Lie group, with Lie algebra $\mathfrak{g}$. Suppose that **G** is acting on a Poisson manifold (M,π). It is natural to demand that the fundamental vector fields $\underline{x}$, with $x \in \mathfrak{g}$, are Hamiltonian. It leads to the following definition.

Definition 5.38. Let **G** be a Lie group acting on a Poisson manifold (M,π). The action is said to be a *Hamiltonian action* if there exists a Lie algebra homomorphism

$$\begin{aligned} \tilde{\mu} : \mathfrak{g} &\to \mathscr{F}(M) \\ x &\to \tilde{\mu}_x \end{aligned} \tag{5.51}$$

such that the following diagram is commutative:

$$\begin{array}{ccc} \mathfrak{g} & \longrightarrow & \mathfrak{X}^1(M) \\ & \overset{\tilde{\mu}}{\searrow} & \uparrow -\mathscr{X} \\ & & \mathscr{F}(M) \end{array} \tag{5.52}$$

where the horizontal arrow is the map $x \mapsto \underline{x}$, and where we recall that $-\mathscr{X}$ is a morphism of Lie algebras (see *(5)* of Proposition 1.4 and (1.7)). We call $\tilde{\mu}$ a *co-momentum map* of the action. The dual map,

$$\begin{aligned} \mu : M &\to \mathfrak{g}^* \\ m &\mapsto \mu(m)\,, \end{aligned} \tag{5.53}$$

where $\mu(m) \in \mathfrak{g}^*$ is the linear map $x \mapsto \tilde{\mu}_x(m)$, is said to be a *momentum map* of the action.

We show in the following proposition that, under some assumptions, the orbit space $\mu^{-1}(0)/\mathbf{G}$ inherits a Poisson structure from M.

Proposition 5.39. *Let* **G** *be a connected Lie group and let* (M,π) *be a Poisson manifold. We assume that* **G** *acts locally freely and properly on M and that the action is Hamiltonian. Let* μ *denote a momentum map of the action. If* 0 *is a regular value of* μ*, then*

(1) $\mu^{-1}(0)$ *is an embedded submanifold of M, invariant under the action of* **G***;*
(2) $\mu^{-1}(0)/\mathbf{G}$ *is a manifold;*
(3) The triple $(M,\mu^{-1}(0),\mu^{-1}(0)/\mathbf{G})$ *is Poisson reducible.*

The different spaces involved in the theorem are represented in the following diagram:

$$\begin{array}{ccc} \mu^{-1}(0) & \hookrightarrow & M \\ \downarrow p & & \\ \mu^{-1}(0)/\mathbf{G} & & \end{array} \tag{5.54}$$

Proof. Let us denote $N := \mu^{-1}(0)$. It is an (embedded) submanifold of M since 0 is a regular value of μ. We claim that, for every $x \in \mathfrak{g}$, the vector field $\underline{x}$ is tangent to N. Indeed, N is precisely the zero locus of the vector space of functions $\{\tilde{\mu}_y \mid y \in \mathfrak{g}\}$,

$$N = \bigcap_{y \in \mathfrak{g}} \{m \in M \mid \tilde{\mu}_y(m) = 0\} \ ,$$

and this vector space is invariant for all vector fields $\underline{x}$, since,

$$\underline{x}[\tilde{\mu}_y] = -\mathscr{X}_{\tilde{\mu}_x}[\tilde{\mu}_y] = \{\tilde{\mu}_x, \tilde{\mu}_y\} = \tilde{\mu}_{[x,y]} \ ,$$

where we have used in the last step that $\tilde{\mu}$ is a Lie algebra homomorphism. As a corollary, since $\mathbf{G}$ is connected, N is $\mathbf{G}$-invariant. The quotient space $N/\mathbf{G}$ is a manifold because the action of $\mathbf{G}$ on M is locally free and proper. This proves *(1)* and *(2)*.

In order to prove *(3)*, we check that the assumptions of Proposition 5.14 are satisfied. The fibers of the canonical projection $p : N \to N/\mathbf{G}$ are connected, since $\mathbf{G}$ is connected. Therefore, we only need to verify that (5.30) holds, i.e., that

$$\pi_n^\sharp(T_n^\perp N) = T_n p_n \tag{5.55}$$

for every $n \in N$, where we recall that p_n is the fiber of p which passes through n. Since $0 \in \mathfrak{g}^*$ is a regular value for the momentum map μ, the space $T_n^\perp N$ is, for every $n \in N$, given by

$$T_n^\perp N = \{\mathrm{d}_n \tilde{\mu}_x \mid x \in \mathfrak{g}\} \ .$$

Since $\pi_n^\sharp(\mathrm{d}_n \tilde{\mu}_x) = -\left(\mathscr{X}_{\tilde{\mu}_x}\right)_n = \underline{x}_n$, this means that

$$\pi_n^\sharp(T_n^\perp N) = \{\underline{x}_n \mid x \in \mathfrak{g}\} = T_n p_n \ ,$$

which proves (5.55), and hence that $(M, \mu^{-1}(0), \mu^{-1}(0)/\mathbf{G})$ is Poisson reducible. □

Remark 5.40. Let $\mathbf{G}$ be a reductive algebraic group, acting on an affine variety M. Then one can define as in the case of a Lie group, acting on a manifold, the fundamental vector field $\underline{x}$ for all $x \in \mathfrak{g}$, the Lie algebra of $\mathbf{G}$. If M is equipped with a Poisson structure, one can define the notion of a momentum map as in Definition 5.38. The algebraic version of Proposition 5.39 takes then the following form: let $\mathbf{G}$ be a reductive algebraic group, acting on an affine Poisson variety (M, π). If the action is Hamiltonian with momentum map μ, then $\mu^{-1}(0)$ is $\mathbf{G}$-invariant and $\mu^{-1}(0)/\mathbf{G}$ is a Poisson variety with respect to a Poisson structure obtained by Poisson reduction from the diagram (5.54). In order to prove this, one uses Proposition 5.7 instead of Proposition 5.14.

Remark 5.41. Also, one can generalize Proposition 5.39 and obtain a Poisson structure on $\mu^{-1}(\xi)/\mathbf{G}_\xi$ for every regular value $\xi \in \mathfrak{g}^*$ of μ, where $\mathbf{G}_\xi$ is the subgroup of $\mathbf{G}$, defined in terms of the adjoint action $\mathrm{Ad} : \mathbf{G} \times \mathfrak{g} \to \mathfrak{g}$, by $\mathbf{G}_\xi := \{g \in \mathbf{G} \mid \xi \circ \mathrm{Ad}_g = \xi\}$. See [144] for this and for other generalizations.

5.5 Exercises

1. Let $\mathbf{G}$ be a Lie group, with Lie algebra $\mathfrak{g}$. For $r \in \wedge^2\mathfrak{g}$, show that the following are equivalent:

(i) $\overrightarrow{r}$ is a Poisson structure on $\mathbf{G}$;
(ii) $\overleftarrow{r}$ is a Poisson structure on $\mathbf{G}$;
(iii) $[\![r,r]\!] = 0$.

Show that, when these equivalent conditions are satisfied, the Poisson structures $\overrightarrow{r}$ and $\overleftarrow{r}$ are compatible. For r of the form $r = x \wedge y$, where $x, y \in \mathfrak{g}$, show that these conditions are satisfied if and only if x and y span a Lie subalgebra of $\mathfrak{g}$.

2. Let M be a manifold on which a Lie group $\mathbf{G}$ acts. For $x \in \mathfrak{g}$, the Lie algebra of $\mathbf{G}$, denote by $\underline{x}$ the fundamental vector field, associated to x. Let $r = \sum_{i=1}^k x_i \wedge y_i \in \wedge^2\mathfrak{g}$, satisfying $[\![r,r]\!] = 0$ and consider the bivector field $\underline{r} := \sum_{i=1}^k \underline{x_i} \wedge \underline{y_i}$ on M.

a. Show that $\underline{r}$ is a Poisson structure on M;
b. Prove that the action of $(\mathbf{G}, \overrightarrow{r} - \overleftarrow{r})$ on $(M, \underline{r})$ is a Poisson action;
c. Suppose that M is a vector space and the action of $\mathbf{G}$ on M is a representation. Show that $\underline{r}$ is a quadratic Poisson structure (see Section 8.2).

3. Let M_1 and M_2 be Poisson manifolds and endow $M_1 \times M_2$ with the product Poisson bracket. Show that for every $m_1 \in M_1$ (respectively $m_2 \in M_2$), the submanifold $\{m_1\} \times M_2$ (respectively $M_1 \times \{m_2\}$) is a Poisson–Dirac submanifold of $M_1 \times M_2$.

4. Let M be a Poisson manifold, and let N be a Poisson–Dirac submanifold of M. Suppose that $\mathscr{V}$ is a Poisson vector field on M, which is tangent to N at every point of N. Show that the restriction of $\mathscr{V}$ to N is a Poisson vector field.

5. Let P and M be Poisson manifolds and let $\phi : P \to M$ be a submersive Poisson map. Let N be a submanifold of M.

a. Show that if N is a coisotropic submanifold of M, then $\phi^{-1}(N)$ is a coisotropic submanifold of P;
b. Show, by giving a counterexample, that if N is a Poisson–Dirac submanifold of M, then $\phi^{-1}(N)$ need not be a Poisson–Dirac submanifold of P.

6. We assume in this exercise that the reader is familiar with the notion of symplectic manifold (see Section 6.3). Let (M,ω) be a symplectic manifold, and denote by π the canonical Poisson structure, associated to ω. Let N be a submanifold of M. Show that the following are equivalent:

(i) N is a Poisson–Dirac submanifold of (M,π);
(ii) The restriction of ω to N is non-degenerate.

Conclude that for symplectic manifolds the notions of symplectic submanifold and Poisson–Dirac submanifold coincide.

7. We assume in this exercise that the reader is familiar with the Lie–Poisson structure on the dual of a Lie algebra (see Chapter 7). Let $\mathfrak{g}$ be a finite-dimensional Lie algebra and let $\mathfrak{g}=\mathfrak{g}_\xi\oplus\mathfrak{m}$ be a vector space decomposition of $\mathfrak{g}$, where $\mathfrak{g}_\xi:=\{x\in\mathfrak{g}\mid \mathrm{ad}^*_x\xi=0\}$ denotes the centralizer of an element $\xi\in\mathfrak{g}^*$. We consider the affine subspace $N:=\xi+\mathfrak{m}^\perp$ of $\mathfrak{g}^*$, which is isomorphic to $\mathfrak{g}^*_\xi$, via the affine map χ, which is defined, for all $\eta\in N$, by $\chi(\eta):=(\eta-\xi)_{|\mathfrak{g}_\xi}$.

a. Show that, if $\mathfrak{m}$ satisfies $[\mathfrak{g}_\xi,\mathfrak{m}]\subset\mathfrak{m}$, then N is a Poisson–Dirac submanifold of $\mathfrak{g}^*$, equipped with its Lie–Poisson structure, and the reduced Poisson structure on N is isomorphic to the Lie–Poisson structure on $\mathfrak{g}^*_\xi$, via χ.

b. As an example, consider $\mathfrak{g}=\mathfrak{gl}_d(\mathbb{F})$, let $x\in\mathfrak{g}$ and let $\xi\in\mathfrak{g}^*$ be the linear map, defined by $\xi(y):=\mathrm{Trace}(xy)$, for all $y\in\mathfrak{g}$. Show that if x is diagonalizable, then there exists a complementary subspace $\mathfrak{m}$ of $\mathfrak{g}_\xi$ in $\mathfrak{g}$, such that $[\mathfrak{g}_\xi,\mathfrak{m}]\subset\mathfrak{m}$.

8. Let $(M,\{\cdot,\cdot\})$ be a Poisson manifold and let N be a submanifold of M. We endow $\mathscr{F}(M)[\nu]/\nu^2$ with the structure of an $\mathbb{F}[\nu]$-algebra, defined by

$$F\star_\nu G:=FG+\nu\{F,G\}$$

for all $F,G\in\mathscr{F}(M)$ (see Exercise 8 of Chapter 3, where this product is proven to be associative). Suppose that

$$\Phi:\mathscr{F}(M)[\nu]/\nu^2\times\mathscr{F}(N)[\nu]/\nu^2\to\mathscr{F}(N)[\nu]/\nu^2$$

defines the structure of a left module over the $\mathbb{F}[\nu]$-algebra $(\mathscr{F}(M)[\nu]/\nu^2,\star_\nu)$ on $\mathscr{F}(N)[\nu]/\nu^2$, satisfying $\Phi(F,H)=F_{|N}H \pmod{\nu}$ for every $F\in\mathscr{F}(M)$ and for every $H\in\mathscr{F}(N)$. Let $\sigma:\mathscr{F}(M)\times\mathscr{F}(N)\to\mathscr{F}(N)$ be the bilinear map, given for all $F\in\mathscr{F}(M)$ and $H\in\mathscr{F}(N)$ by

$$\sigma(F,H)=\left.\frac{\Phi(F,H)-F_{|N}H}{\nu}\right|_{\nu=0}.$$

a. Show that σ satisfies the identity

$$\{F,G\}_{|N}H=\sigma(F,G_{|N}H)-\sigma(FG,H)+F_{|N}\sigma(G,H) \tag{5.56}$$

for all $F,G\in\mathscr{F}(M)$ and $H\in\mathscr{F}(N)$;

b. Deduce from it that N is a coisotropic submanifold of M.

5.6 Notes

Lie groups and Lie algebras are standard tools in geometry and in physics; in turn, the study of their structure and their representations is strongly inspired by ideas which come from geometry and physics. A quick introduction to Lie groups is given in Warner [198]; see Bump [29] for a more extensive treatment. For the theory of Lie algebras, we refer to Humphreys [99]. A great introduction to both Lie groups and Lie algebras and their representation theory is given in Fulton–Harris [80].

The starting point of reduction theory comes from the fact that integrals of motion permit to eliminate degrees of freedom in Hamiltonian systems. One may cite here Noether's theorem, which is the predecessor of the notion of momentum map and the theory of Dirac constraints, which eventually led to the notion of Poisson–Dirac reduction. The modern, geometrical development of these ideas first appears in the seminal papers by Śniatycki–Tulczyjew [184], Meyer [148] and Marsden–Weinstein [146]. See Kosmann–Schwarzbach [114], Weinstein [199] and the introduction of Ortega–Ratiu [160] for more details on the history of the subject; in the latter introduction one also finds a detailed list of topics in the theory of moment maps and reduction.

In our presentation of reduction, we have separated the Poisson reduction of coisotropic submanifolds from the theory of reduction of Poisson–Dirac submanifolds, which are both particular cases of the so-called Marsden–Ratiu reduction procedure [144]. The notion of momentum map, presented here, has also been generalized in many different ways, see for example Alekseev–Malkin–Meinrenken [9] for Lie group valued momentum mappings and Lu [133] for Poisson–Lie group valued momentum mappings. These generalizations and their associated reduction can be nicely understood with the help of Lie groupoids, see Xu [208].

We have not discussed the geometry of the image of a momentum map (which is, under some assumptions, a convex polytope). See Guillemin [91] and the references therein.

Part II
Examples

Chapter 6
Constant Poisson Structures, Regular and Symplectic Manifolds

Poisson's original bracket on $\mathbb{R}^{2r}$, given for smooth functions F and G by

$$\{F,G\} := \sum_{i=1}^{r} \left(\frac{\partial F}{\partial q_i}\frac{\partial G}{\partial p_i} - \frac{\partial G}{\partial q_i}\frac{\partial F}{\partial p_i} \right), \tag{6.1}$$

is of fundamental importance in classical mechanics, in quantum mechanics and in other areas of mathematical physics. In fact, locally, the Poisson structure of every symplectic manifold of dimension $2r$ is of this form, and symplectic manifolds appear as phase spaces of many mechanical systems. The rank of the Poisson structure defined by (6.1), is maximal at every point; if we replace maximality of the rank by constancy of the rank (i.e., the rank is independent of the point) we arrive at the notion of a regular Poisson structure. Locally the Poisson structure of a regular Poisson manifold is also given by (6.1), but $2r$ denotes now the rank of the Poisson structure, and not the dimension d of the manifold, which may be larger: there are $d-2r$ local coordinates which are absent in the formula for the bracket. Notice that the Poisson bracket (6.1), applied to a pair of coordinate functions, always yields a constant function: this will be the starting point of the algebraic definition of a *constant* Poisson structure. In fact, the Darboux theorem (Theorem 1.26) shows that regularity of the Poisson structure corresponds locally exactly to constancy of the Poisson structure (in an appropriate coordinate chart).

Constant Poisson structures are introduced in Section 6.1, and they will be related to regular Poisson manifolds in Section 6.2. The particular class of symplectic manifolds will be discussed in Section 6.3. Entire (excellent!) books have been written on symplectic manifolds, so we will restrict ourselves here to the most common classes of examples of symplectic and regular Poisson manifolds, and our focus will be on their properties from the Poisson point of view.

Unless stated otherwise, $\mathbb{F}$ is an arbitrary field of characteristic zero.

C. Laurent-Gengoux et al., *Poisson Structures*,
Grundlehren der mathematischen Wissenschaften 347,
DOI 10.1007/978-3-642-31090-4_6, © Springer-Verlag Berlin Heidelberg 2013

6.1 Constant Poisson Structures

Poisson's original bracket on $\mathbb{R}^{2r}$, defined by

$$\{F,G\} := \sum_{i=1}^{r} \left(\frac{\partial F}{\partial q_i}\frac{\partial G}{\partial p_i} - \frac{\partial G}{\partial q_i}\frac{\partial F}{\partial p_i} \right), \tag{6.2}$$

has the property that the Poisson bracket $\{F,G\}$ of two linear functions F and G is a constant function on $\mathbb{R}^{2r}$. This property leads to the following definition, which will cover a first class of basic examples of Poisson structures.

Definition 6.1. A Poisson structure π on a finite-dimensional vector space V is called a *constant Poisson structure*, if for each pair of linear functions F and G on V, their Poisson bracket $\pi[F,G]$, also denoted by $\{F,G\}$, is a constant function on V.

Let $X = (x_{ij}) \in \mathrm{Mat}_d(\mathbb{F})$ be an arbitrary skew-symmetric matrix, where $d \in \mathbb{N}^*$. We show that X defines a Poisson structure on $\mathbb{F}^d$, i.e., a Poisson bracket on $\mathscr{F}(\mathbb{F}^d)$, where we recall that, for a finite-dimensional $\mathbb{F}$-vector space V, the notation $\mathscr{F}(V)$ stands for the algebra of polynomial functions on V, in general, but may also be chosen as the algebra of smooth functions on V, when $\mathbb{F} = \mathbb{R}$, or as the algebra of holomorphic functions on V, when $\mathbb{F} = \mathbb{C}$. In order to construct this Poisson structure, consider the skew-symmetric biderivation of $\mathscr{F}(\mathbb{F}^d)$, defined for all $F,G \in \mathscr{F}(\mathbb{F}^d)$, by

$$\{F,G\} := \sum_{i,j=1}^{d} x_{ij} \frac{\partial F}{\partial x_i}\frac{\partial G}{\partial x_j}, \tag{6.3}$$

where $x_1,\dots,x_d$ are the standard coordinates on $\mathbb{F}^d$. Since each of the brackets $\{x_i,x_j\} = x_{ij}$ is a constant function, we have that $\{\{x_i,x_j\},x_k\} = 0$, where $1 \leqslant i,j,k \leqslant d$, so that the Jacobi identity holds for every triple of coordinate functions (x_i,x_j,x_k) for $\mathscr{F}(\mathbb{F}^d)$. It implies that the Jacobi identity holds for all triples of functions in $\mathscr{F}(\mathbb{F}^d)$ (see Proposition 1.36), so that (6.3) defines a Poisson structure on $\mathbb{F}^d$, which we denote by π. Clearly, π is a constant Poisson structure and every constant Poisson structure on $\mathbb{F}^d$ is of this form. Since every d-dimensional vector space V can be identified with $\mathbb{F}^d$, by choosing a basis of V, the matrix X defines a Poisson structure on every d-dimensional vector space V, equipped with a basis (equivalently, with a system of linear coordinates). Thus, the following proposition holds.

Proposition 6.2. *Let V be a finite-dimensional vector space, with a fixed system of linear coordinates $(x_1,\dots,x_d)$. There is a one-to-one correspondence between skew-symmetric matrices $X \in \mathrm{Mat}_d(\mathbb{F})$ and constant Poisson structures on V. It is defined by assigning to $X = (x_{ij}) \in \mathrm{Mat}_d(\mathbb{F})$ the Poisson structure, defined for all $F,G \in \mathscr{F}(V)$ by*

$$\{F,G\} := \sum_{i,j=1}^{d} x_{ij} \frac{\partial F}{\partial x_i}\frac{\partial G}{\partial x_j}. \tag{6.4}$$

Example 6.3. Poisson's original bracket (6.2) corresponds to the skew-symmetric matrix $\begin{pmatrix} 0 & \mathbb{1}_r \\ -\mathbb{1}_r & 0 \end{pmatrix}$, where the standard coordinates on $\mathbb{R}^{2r}$ are ordered as follows: $q_1,\dots,q_r,p_1,\dots,p_r$.

The matrix X in Proposition 6.2 is the Poisson matrix of π with respect to the coordinates $x_1,\dots,x_d$ (see Section 1.2.2). Since the entries of X are constant, the rank of π is constant. According to the Darboux theorem (Theorem 1.26), if the rank of a Poisson structure on a manifold is locally constant at a point m, then there exist local coordinates around m in which the Poisson structure takes a canonical form (so-called Darboux coordinates). We show that in the case of a constant Poisson bracket on a d-dimensional vector space V, linear coordinates which are *global* Darboux coordinates can be constructed explicitly. Namely, since X is skew-symmetric, there exists an invertible matrix $A \in \mathrm{Mat}_d(\mathbb{F})$ such that

$$A^\top X A = \begin{pmatrix} 0 & \mathbb{1}_r & 0 \\ -\mathbb{1}_r & 0 & 0 \\ 0 & 0 & 0 \end{pmatrix},$$

where $2r$ is the rank of X and hence of the Poisson bracket π (at every point). It means that, in terms of new linear coordinates on V, defined in terms of $x_1,\dots,x_d$ by

$$(q_1,\dots,q_r,p_1,\dots,p_r,z_1,\dots,z_{d-2r}) := (x_1,\dots,x_d)A$$

one obtains the following brackets

$$\{q_i,p_j\} = \delta_{i,j}\,, \qquad \{q_i,q_j\} = \{p_i,p_j\} = \{q_i,z_k\} = \{p_i,z_k\} = \{z_k,z_\ell\} = 0\,,$$

where $1 \leqslant i,j \leqslant r$ and $1 \leqslant k,\ell \leqslant d-2r$. As a consequence, the d coordinates $q_1,\dots,q_r,p_1,\dots,p_r,z_1,\dots,z_{d-2r}$ are Darboux coordinates for π. Thus, by simple linear algebra, we can construct (global) Darboux coordinates for every constant Poisson structure on V. In terms of these Darboux coordinates, the Hamiltonian vector field $\mathscr{X}_H$ which corresponds to $H \in \mathscr{F}(V)$ takes the well known form

$$\dot{q}_i = \frac{\partial H}{\partial p_i}\,, \qquad \dot{p}_i = -\frac{\partial H}{\partial q_i}\,, \qquad \dot{z}_k = 0\,,$$

where $\dot{q}_i$ is a commonly used shorthand for $\mathscr{X}_H[q_i] = \{q_i,H\}$, and similarly for the other variables.

In order to obtain a more intrinsic formulation of the notion of a constant Poisson structure on a finite-dimensional vector space V, we consider the vector space $\wedge^2 V = V \wedge V$, whose elements are called *bivectors* of V. A bivector b of V is often viewed as a linear map, or as a bilinear form, on V^*. This is done as follows: writing $b = \sum_{i=1}^s v_i \wedge w_i$, where $v_i, w_i \in V$, we define[1] the linear map $b^\sharp : V^* \to V$, by

[1] In terms of internal products, as recalled in Appendix A, $b^\sharp(\xi) = \imath_\xi b$, which shows that $b^\sharp$, as given by (6.5), is well-defined.

$$b^\sharp(\xi) = \sum_{i=1}^{s} (\langle \xi, v_i \rangle \, w_i - \langle \xi, w_i \rangle \, v_i) \,, \tag{6.5}$$

for every $\xi \in V^*$, and we define a skew-symmetric bilinear form $\tilde{b}$ on V^*, by

$$\tilde{b}(\xi,\eta) := \left\langle \eta, b^\sharp(\xi) \right\rangle = \sum_{i=1}^{s} (\langle \xi, v_i \rangle \langle \eta, w_i \rangle - \langle \eta, v_i \rangle \langle \xi, w_i \rangle) \,, \tag{6.6}$$

for all $\xi, \eta \in V^*$, where $\langle \cdot, \cdot \rangle$ denotes the canonical pairing between V^* and V. Both $b^\sharp$ and $\tilde{b}$ are skew-symmetric, which means for the former that $\langle \eta, b^\sharp(\xi) \rangle = -\langle \xi, b^\sharp(\eta) \rangle$, for all $\xi, \eta \in V^*$. In terms of a basis $(e_1, \ldots, e_d)$ of V and the dual basis $(e_1^*, \ldots, e_d^*)$ of V^*, the bivector b can be written as $b = \sum_{1 \leqslant i < j \leqslant d} \tilde{b}(e_i^*, e_j^*) \, e_i \wedge e_j$, so that b can be recovered from $\tilde{b}$ (and hence from $b^\sharp$).

To a bivector $b \in V \wedge V$ one naturally associates a bivector field on V, by identifying all cotangent spaces to V with V^*, which can be done thanks to the affine structure of V. If $F \in \mathscr{F}(V)$ and $m \in V$ then we can view $\mathrm{d}_m F$ as an element of V^*, hence we can define a bivector field on V, by setting, for $F, G \in \mathscr{F}(V)$ and $m \in V$,

$$\{F,G\}(m) := \tilde{b}\,(\mathrm{d}_m F, \mathrm{d}_m G) \,, \tag{6.7}$$

where $\tilde{b} : V^* \times V^* \to \mathbb{F}$ is the bilinear form, associated to b, as in (6.6). Equivalently, we can view $\tilde{b}$ under the same identification as a bilinear form $\tilde{b}_m$ on $T_m V^*$, for any $m \in V$, and the right-hand side of (6.7) can be written as $\tilde{b}_m\,(\mathrm{d}_m F, \mathrm{d}_m G)$, where the pairing now takes place between $\wedge^2 T_m V^* \simeq (\wedge^2 T_m V)^*$ and $\wedge^2 T_m V$, rather than between $\wedge^2 V^* \simeq (\wedge^2 V)^*$ and $\wedge^2 V$.

For linear functions F, G on V, we have that $\{F,G\}$ is independent of m, i.e., is constant; first, this implies that the bivector field defined by (6.7) satisfies the Jacobi identity, so that it is a Poisson structure, and second that it is a *constant* Poisson structure. For $H \in \mathscr{F}(M)$, the Hamiltonian vector field $\mathscr{X}_H$ at m is given, as an element of V, by

$$(\mathscr{X}_H)_m = -\, b^\sharp(\mathrm{d}_m H) \,. \tag{6.8}$$

Writing b as $b = \sum_{i=1}^{s} v_i \wedge w_i$, where v_i, $w_i \in V$, the formula for the Poisson bracket can also be written as

$$\{F,G\} = \sum_{i=1}^{s} (v_i[F] w_i[G] - v_i[G] w_i[F]) \,. \tag{6.9}$$

The square brackets in this formula have the following meaning: for $v \in V$, $v[\cdot]$ is the unique derivation of $\mathscr{F}(V)$ such that $v[\xi] = \langle \xi, v \rangle$ for all linear functions ξ on V. Geometrically, we think of $v[\cdot]$ as the directional derivative, defined by a vector v. Equation (6.9) is a coordinate-free analog of (6.4).

The fact that every bivector on V defines a Poisson structure on V, motivates the following definition.

Definition 6.4. A *Poisson vector space* (V,b) is a finite-dimensional vector space V, equipped with a bivector $b \in V \wedge V$.

By the above, we have the following abstract version and amplification of Proposition 6.2.

Proposition 6.5. *For every finite-dimensional vector space V, there is a one-to-one correspondence between bivectors on V (making V into a Poisson vector space) and constant Poisson structures on V. If $b \in \wedge^2 V$ is a bivector of rank $2r$, then the Poisson structure $\pi = \{\cdot,\cdot\}$ on V which corresponds to b, has rank $2r$ and V admits a system of linear coordinates $(q_1,\dots,q_r,p_1,\dots,p_r,z_1,\dots,z_s)$ in which the Poisson structure is given by*

$$\{F,G\} = \sum_{i=1}^{r} \left(\frac{\partial F}{\partial q_i}\frac{\partial G}{\partial p_i} - \frac{\partial G}{\partial q_i}\frac{\partial F}{\partial p_i} \right),$$

for all $F,G \in \mathscr{F}(V)$.

6.2 Regular Poisson Manifolds

In the case of a Poisson manifold (M,π), there is no notion of linear functions on M, so we cannot use Definition 6.1 as it stands, to define constant Poisson structures on manifolds. Two natural generalizations of the latter definition are the following.

Definition 6.6. A Poisson structure π on a connected manifold M is said to be a *constant Poisson structure* if for every point $m \in M$, there exists a coordinate chart (U,x), centered at m, such that all Poisson brackets $\{x_i,x_j\}$ are constant on U.

Definition 6.7. A Poisson structure on a manifold M is said to be a *regular Poisson structure* if it has the same rank at every point $m \in M$. We say then that (M,π) is a *regular Poisson manifold*.

Comparing the definitions, it is clear that every constant Poisson structure on a connected manifold is regular: taking around an arbitrary point $m \in M$ a coordinate chart (U,x) such that all Poisson brackets $\{x_i,x_j\}$ are constant on U, we see that the rank of π is constant on U, since it is given by the rank of the Poisson matrix of π with respect to x, whose entries are the Poisson brackets $\{x_i,x_j\}$; since M is connected, it follows that the rank of π takes the same value at every point of M, i.e., π is a regular Poisson structure. It is a fundamental fact that the converse is also true, leading to the following proposition.

Proposition 6.8. *Let (M,π) be a Poisson manifold, which is assumed to be connected. Then (M,π) is a regular Poisson manifold if and only if π is a constant Poisson structure.*

Proof. Let (M,π) be a regular Poisson manifold and let $m \in M$. We need to show that there exists a coordinate chart (U,x), centered at m, such that all Poisson brackets $\{x_i,x_j\}$ are constant (functions) on U. Since, by assumption, the rank of π is constant at m, this is an immediate consequence of the Darboux theorem, which we proved in Section 1.3.3 (Theorem 1.26). According to this theorem, if the rank of a Poisson structure is constant (say, equal to $2r$) in the neighborhood of a point $m \in M$, then there exists in a neighborhood of m a coordinate chart (U,x), with $x = (q_1,\ldots,q_r,p_1,\ldots,p_r,z_1,\ldots,z_{d-2r})$, centered at m, in which the Poisson structure takes the canonical form

$$\{q_i,q_j\} = \{p_i,p_j\} = \{q_i,z_k\} = \{p_i,z_k\} = \{z_k,z_\ell\} = 0\,, \qquad \{q_i,p_j\} = \delta_{i,j}\,,$$

where $1 \leqslant i,j \leqslant r$ and $1 \leqslant k,\ell \leqslant d-2r$. In particular, all these Poisson brackets are constant. The Darboux coordinates $q_1,\ldots,q_r,p_1,\ldots,p_r,z_1,\ldots,z_{d-2r}$ are often referred to as *canonical coordinates*, especially in the physics literature. □

The study of regular Poisson manifolds is in general much simpler than the study of arbitrary Poisson manifolds. Being locally the product of a symplectic manifold and $\mathbb{R}^s$ or $\mathbb{C}^s$ with the zero Poisson structure, their (local) study can often be reduced to the case of symplectic manifolds. But sometimes, it is the regularity itself which can be exploited, as we show in the following example.

Example 6.9. According to Proposition 1.21, the rank of a Poisson manifold (M,π) at a point m is equal to the dimension of the vector space of Hamiltonian vector fields at that point, $\mathrm{Rk}_m M = \dim \mathrm{Ham}_m(M)$. Therefore, if the rank is constant, the Hamiltonian vector fields define a distribution of rank $\mathrm{Rk} M$. It is differentiable, because it is defined by smooth vector fields. According to Proposition 1.4, $[\mathscr{X}_F,\mathscr{X}_G] = -\mathscr{X}_{\{F,G\}}$, so the distribution is integrable in the sense of Frobenius, so through every point there passes a unique integral manifold of the distribution, by Frobenius' theorem. By definition, all Hamiltonian vector fields of (M,π) are tangent to this integral manifold, at each of its points. By Proposition 2.10, this integral manifold carries a (unique) Poisson structure of maximal rank. In this way we have recovered, by simple means, the symplectic foliation[2] of a Poisson manifold, in the regular case.

6.3 Symplectic Manifolds

The most important class of regular Poisson manifolds consists of the symplectic manifolds (real or complex). We first consider the particular case of a symplectic vector space.

[2] In this case it is indeed a foliation, and not a singular foliation.

6.3.1 Symplectic Vector Spaces

The skew-symmetric variant of an inner product space is a symplectic vector space. Namely, let V be a finite-dimensional $\mathbb{F}$-vector space and let $\omega \in \mathrm{Hom}(V \wedge V, \mathbb{F})$ be a skew-symmetric bilinear form on V. We say that ω is a *symplectic structure* if ω has maximal rank, $\mathrm{Rk}\,\omega = \dim V$. Then (V, ω) is called a *symplectic vector space*. If (V, ω) is a symplectic vector space, then V is even-dimensional, since the rank of every skew-symmetric bilinear form is even. Since the rank of a symplectic structure ω is maximal, the linear map $\omega^\flat$, which is by definition ω, viewed[3] as an element of $\mathrm{Hom}(V, V^*)$, is invertible. Its inverse $(\omega^\flat)^{-1} \in \mathrm{Hom}(V^*, V)$ is skew-symmetric, hence is of the form $-\pi^\sharp$ for some (unique) bivector $\pi \in \wedge^2 V$. It follows that every symplectic vector space (V, ω) is in a natural way a Poisson vector space, and hence, according to Proposition 6.5, V is equipped with a constant Poisson structure π, called the *canonical Poisson structure*, associated to ω. Explicitly, the Poisson bracket of $F, G \in \mathscr{F}(M)$ is computed at $m \in M$ from $-\pi^\sharp = (\omega^\flat)^{-1}$ using (6.7) and (6.6), namely

$$\begin{aligned} \{F,G\}(m) &= \pi(\mathrm{d}_m F, \mathrm{d}_m G) = \langle \mathrm{d}_m G, \pi^\sharp(\mathrm{d}_m F)\rangle \\ &= -\langle \mathrm{d}_m G, (\omega^\flat)^{-1}(\mathrm{d}_m F)\rangle \\ &= \omega((\omega^\flat)^{-1}(\mathrm{d}_m F), (\omega^\flat)^{-1}(\mathrm{d}_m G)) \\ &= \omega((\mathscr{X}_F)_m, (\mathscr{X}_G)_m)\,, \end{aligned} \tag{6.10}$$

where we used in the last line that $(\mathscr{X}_F)_m = -\pi^\sharp(\mathrm{d}_m F)$ (see (6.8)). The latter definition of $(\mathscr{X}_F)_m$ is, in view of the definition of π, equivalent to saying that $(\mathscr{X}_F)_m$ is the unique element of $T_m V \simeq V$ such that

$$\omega((\mathscr{X}_F)_m, \cdot) = \mathrm{d}_m F\,. \tag{6.11}$$

When ω is viewed as a (constant) differential two-form on V, the formula for the Poisson bracket is also written as

$$\{F,G\} = \omega(\mathscr{X}_F, \mathscr{X}_G)\,, \tag{6.12}$$

and Eq. (6.11) for the Hamiltonian vector field is shortened to

$$\omega(\mathscr{X}_F, \cdot) = \mathrm{d}F\,.$$

The (constant) rank of π equals the rank of $\omega^\flat$, which is $\dim V$.

The above construction is easily reversed: if π is a constant Poisson structure on $\mathscr{F}(V)$ of maximal rank, then it is the canonical Poisson structure of a symplectic

[3] This can be done in two ways, which differ by a sign. Our convention is that $\omega^\flat(v) = \omega(v, \cdot)$, for $v \in V$.

vector space (V,ω), where $\omega \in \mathrm{Hom}(V \wedge V, \mathbb{F})$. We leave this as an exercise to the reader; a globalization of this statement will be worked out in the next section.

6.3.2 Symplectic Manifolds

Let M be a real or complex manifold ($\mathbb{F} = \mathbb{R}$ or $\mathbb{F} = \mathbb{C}$). A (smooth or holomorphic) differential two-form ω on M is called an *almost symplectic structure* if its restriction to each tangent space T_mM is non-degenerate, i.e., ω_m is a symplectic structure on T_mM, for every $m \in M$, as defined in Section 6.3.1. Then (M,ω) is called an *almost symplectic manifold*. The globalization of (6.11) and (6.10) is naturally given by

$$\begin{aligned} (\mathscr{X}_F)_m &:= \left(\omega_m^\flat\right)^{-1} \mathrm{d}_m F \in T_mM \,, \\ \{F,G\}(m) &:= \omega_m((\mathscr{X}_F)_m, (\mathscr{X}_G)_m) \,, \end{aligned} \tag{6.13}$$

for $m \in M$ and $F,G \in \mathscr{F}(M)$. In a compact form, these formulas are usually written as

$$\begin{aligned} \omega(\mathscr{X}_F, \cdot) &= \mathrm{d}F \,, \\ \{F,G\} &= \omega(\mathscr{X}_F, \mathscr{X}_G) \,. \end{aligned} \tag{6.14}$$

As in the case of symplectic vector spaces, these definitions imply that

$$\{F,G\} = \mathrm{d}F(\mathscr{X}_G) = \mathscr{X}_G[F] \,, \tag{6.15}$$

so that $\{\cdot,\cdot\}$ defines a bivector field on V and $\mathscr{X}_F$, as defined by (6.14), is the Hamiltonian vector field of F with respect to this bivector field. However, the Jacobi identity does not need to hold. It holds if and only if ω is closed, $\mathrm{d}\omega = 0$, in which case ω is called a *symplectic structure* and (M,ω) is called a *symplectic manifold*. This equivalence is shown in the following proposition.

Proposition 6.10. *Let (M,ω) be an almost symplectic manifold. Then π, as defined by (6.14), is a Poisson structure on M if and only if (M,ω) is a symplectic manifold. In particular, every symplectic manifold (M,ω) carries a canonical Poisson structure π, which makes (M,π) into a Poisson manifold; it is a regular Poisson structure whose rank is equal to the dimension of M.*

Proof. Let us first assume that ω is closed. Notice that one important consequence of this assumption is that ω is conserved by the flow of every Hamiltonian vector field. To see this, use Cartan's formula (3.50) for the Lie derivative $\mathscr{L}_\mathscr{V}$,

$$\mathscr{L}_\mathscr{V} = \imath_\mathscr{V}\mathrm{d} + \mathrm{d}\imath_\mathscr{V} \tag{6.16}$$

when $\mathscr{V}$ is a (locally) Hamiltonian vector field, $\mathscr{V} = \mathscr{X}_F$, and use that $\imath_{\mathscr{X}_F}\omega = \mathrm{d}F$ and ω are closed; it yields $\mathscr{L}_{\mathscr{X}_F}\omega = 0$, for every function F, defined on an open subset of M. It implies that the following formula,

$$\imath_{[\mathscr{V}_1,\mathscr{V}_2]}\omega = \mathscr{L}_{\mathscr{V}_1}\imath_{\mathscr{V}_2}\omega - \imath_{\mathscr{V}_2}\mathscr{L}_{\mathscr{V}_1}\omega\,,$$

which is a special case of *(2)* in Proposition 3.11, simplifies for Hamiltonian vector fields to

$$\imath_{[\mathscr{X}_F,\mathscr{X}_G]}\omega = \mathscr{L}_{\mathscr{X}_F}\imath_{\mathscr{X}_G}\omega = \mathrm{d}\imath_{\mathscr{X}_F}\imath_{\mathscr{X}_G}\omega = -\mathrm{d}(\omega(\mathscr{X}_F,\mathscr{X}_G)) = -\mathrm{d}\{F,G\}\,, \quad (6.17)$$

where F and G are functions, defined on an open subset U of M. By the definition of $\mathscr{X}$, Eq. (6.17) means that $\imath_{[\mathscr{X}_F,\mathscr{X}_G]}\omega = -\imath_{\mathscr{X}_{\{F,G\}}}\omega$. Non-degeneracy of ω allows us to conclude that $[\mathscr{X}_F,\mathscr{X}_G] = -\mathscr{X}_{\{F,G\}}$. Applying this equation to an arbitrary function $H \in \mathscr{F}(U)$, we find

$$\mathscr{X}_F\mathscr{X}_G[H] - \mathscr{X}_G\mathscr{X}_F[H] + \mathscr{X}_{\{F,G\}}[H] = 0\,,$$

which is precisely the Jacobi identity, as follows from (6.15). Thus, we have shown that if ω is closed, so that (M,π) is a symplectic manifold, then π is a Poisson structure on M.

Assume now that ω is an almost symplectic structure, and that the bivector field π, defined by (6.13), satisfies the Jacobi identity. Using the explicit formula (3.28) for computing the de Rham differential, we obtain, for a triple of *Hamiltonian* vector fields as above,

$$\begin{aligned}\mathrm{d}\omega(\mathscr{X}_F,\mathscr{X}_G,\mathscr{X}_H) &= \mathscr{X}_F[\omega(\mathscr{X}_G,\mathscr{X}_H)] + \omega(\mathscr{X}_F,[\mathscr{X}_G,\mathscr{X}_H]) + \circlearrowleft(F,G,H)\\ &= 2(\{\{F,G\},H\} + \{\{G,H\},F\} + \{\{H,F\},G\})\\ &= 0\,.\end{aligned}$$

We have used that

$$\mathscr{X}_F[\omega(\mathscr{X}_G,\mathscr{X}_H)] = \{\{G,H\},F\} = \omega(\mathscr{X}_F,[\mathscr{X}_G,\mathscr{X}_H])\,,$$

as follows from (6.14), (6.15) and item *(5)* in Proposition 1.4. Since ω_m is non-degenerate for every $m \in M$, we have that $T_mM = \mathrm{Ham}_m(\pi)$ for every $m \in M$; the above computation therefore shows that $\mathrm{d}\omega$ vanishes on every triple of vector fields, i.e., that ω is closed, hence that (M,ω) is a symplectic manifold. □

Proposition 6.11. *Suppose that (M,π) is a regular Poisson manifold, whose rank is equal to the dimension of M. There exists on M a (unique) symplectic structure ω, whose canonical Poisson structure is π.*

Proof. We consider for $m \in M$ the linear map $\pi_m^\sharp : T^*M \to TM$, associated to π. By assumption, π_m is of maximal rank, hence $\pi_m^\sharp$ is an isomorphism for every $m \in M$. It follows that we can use π_m to associate to each differential one-form on M, a vector field on M: given $\eta \in \Omega^1(M)$, the corresponding vector field $\mathscr{V}$ is defined by its value at $m \in M$ as $\mathscr{V}_m := \pi_m^\sharp(\eta_m)$. One often uses the compact[4] notation $\mathscr{V} := \pi^\sharp(\eta)$.

4 Some authors, such as [125] go one step further and write $^\sharp\eta$ for $\pi^\sharp(\eta)$.

In particular, for local functions F and G on M, one has $\pi^\sharp(F\mathrm{d}G) = -F\mathscr{X}_G$. Similarly, one uses the inverse of $\pi_m^\sharp$ to associate to each vector field on M a differential one-form on M; in the compact notation, the differential one-form associated to $\mathscr{V} \in \mathfrak{X}^1(M)$ is $\left(\pi^\sharp\right)^{-1}(\mathscr{V})$. We use $\pi_m^\sharp$ to define a non-degenerate differential two-form ω on M: for vector fields $\mathscr{V}_1$ and $\mathscr{V}_2$, let $\omega(\mathscr{V}_1,\mathscr{V}_2) := \left\langle -\left(\pi^\sharp\right)^{-1}(\mathscr{V}_1), \mathscr{V}_2\right\rangle$. Then

$$\omega(\mathscr{X}_F,\mathscr{X}_G) := -\left\langle \left(\pi^\sharp\right)^{-1}(\mathscr{X}_F), \mathscr{X}_G\right\rangle = \langle \mathrm{d}F, \mathscr{X}_G\rangle = \{F,G\}\ ,$$

for $F,G \in \mathscr{F}(M)$. It follows that (M,ω) is an almost symplectic manifold whose associated bivector field is π. Since π is a Poisson structure, Proposition 6.10 implies that (M,ω) is actually a symplectic manifold. □

The main properties of symplectic manifolds are listed in the following proposition.

Proposition 6.12. *Let (M,ω) be a symplectic manifold of dimension d and let π denote its canonical Poisson structure.*

(1) The dimension d of M is even;
(2) M is an orientable manifold;
(3) (M,π) is a unimodular Poisson manifold;
(4) For every $k \in \mathbb{N}^*$, with $0 \leqslant k \leqslant d$, the de Rham cohomology space $H^k_{dR}(M)$, the Poisson cohomology space $H^k_\pi(M)$ and the Poisson homology space $H^\pi_{d-k}(M)$ are isomorphic,

$$H^k_{dR}(M) \simeq H^k_\pi(M) \simeq H^\pi_{d-k}(M)\ .$$

Proof. Since every symplectic vector space is even-dimensional, every tangent space of a symplectic manifold is even-dimensional, hence also the symplectic manifold itself. In order to prove *(2)*, we assume that M is a *real* manifold (complex manifolds are always orientable). Since ω is non-degenerate (at every point) the differential d-form $\Lambda := \omega^{d/2}$ is nowhere vanishing, hence it is a volume form and M is orientable. The volume form Λ is called the *Liouville volume form* of (M,ω). It is preserved by the flow of every Hamiltonian vector field $\mathscr{X}_F$, since

$$\mathscr{L}_{\mathscr{X}_F}\Lambda = \mathscr{L}_{\mathscr{X}_F}\omega^{d/2} = \sum_{i=1}^{d/2}\omega^{i-1}\wedge\mathscr{L}_{\mathscr{X}_F}\omega\wedge\omega^{d/2-i} = 0\ ,$$

where we have used in the last step that $\mathscr{L}_{\mathscr{X}_F}\omega = 0$ (see the proof of Proposition 6.10). Thus, (M,π) is unimodular; see Section 4.4.4, where it is shown that for a unimodular Poisson manifold, the Poisson cohomology spaces $H^k_\pi(M)$ and the Poisson homology spaces $H^\pi_{d-k}(M)$ (with $0 \leqslant k \leqslant d$) are isomorphic vector spaces, which proves part of *(4)*.

In the case of a symplectic manifold we show that, in addition, the de Rham and Poisson cohomology spaces are isomorphic. To do this, we consider the isomorphisms $\pi_m^\sharp : T_m^*M \to T_mM$, which we used in the proof of Proposition 6.11. Notice that at every point m, the inverse of $\pi_m^\sharp$ is just $\omega_m^\flat$; for a local section $\mathscr{V}$ of TM (i.e., vector field), the inverse of $\pi^\sharp$ is given by $\left(\pi^\sharp\right)^{-1}(\mathscr{V}) = \omega(\mathscr{V},\cdot)$. The natural extension of $\pi^\sharp$ to $\Omega^\bullet(M)$ yields, for every k with $0 \leqslant k \leqslant d$, a map

$$\wedge^k\pi^\sharp : \Omega^k(M) \to \wedge^k\mathfrak{X}^1(M) \to \mathfrak{X}^k(M),$$

which renders the following diagram commutative:

$$\begin{array}{ccc}
\Omega^k(M) & \xrightarrow{\ \mathrm{d}\ } & \Omega^{k+1}(M) \\
\Big\downarrow {\scriptstyle \wedge^k\pi^\sharp} & & \Big\downarrow {\scriptstyle \wedge^{k+1}\pi^\sharp} \\
\mathfrak{X}^k(M) & \xrightarrow{\ -\delta_\pi^k\ } & \mathfrak{X}^{k+1}(M)
\end{array}$$

Indeed, for local functions $F, G_1, \ldots, G_k$ we have on the one hand that

$$\begin{aligned}\left(\wedge^{k+1}\pi^\sharp\right) \mathrm{d}(F\,\mathrm{d}G_1 \wedge \cdots \wedge \mathrm{d}G_k) &= \pi^\sharp(\mathrm{d}F) \wedge \pi^\sharp(\mathrm{d}G_1) \wedge \cdots \wedge \pi^\sharp(\mathrm{d}G_k) \\ &= (-1)^{k+1}\mathscr{X}_F \wedge \mathscr{X}_{G_1} \wedge \cdots \wedge \mathscr{X}_{G_k},\end{aligned}$$

while on the other hand

$$\begin{aligned}\delta_\pi^k\left(\wedge^k\pi^\sharp(F\,\mathrm{d}G_1 \wedge \cdots \wedge \mathrm{d}G_k)\right) &= \delta_\pi^k(F\,\mathscr{X}_{G_1} \wedge \cdots \wedge \mathscr{X}_{G_k}) \\ &= (-1)^k\,\mathscr{X}_F \wedge \mathscr{X}_{G_1} \wedge \cdots \wedge \mathscr{X}_{G_k},\end{aligned}$$

where we have used in the last step that $\mathscr{X}_{G_1}, \ldots, \mathscr{X}_{G_k}$ are Poisson 1-cocycles. The commutativity of the diagram implies that the Poisson complex $(\mathfrak{X}^\bullet(M), [\cdot,\cdot]_S)$ and the de Rham complex $(\Omega^\bullet(M), \mathrm{d})$ are isomorphic. In particular, they have the same cohomology. □

Remark 6.13. A map $\Psi : M_1 \to M_2$ between two symplectic manifolds (M_1, ω_1) and (M_2, ω_2) is called a *symplectic map* if $\Psi^*\omega_2 = \omega_1$. Considering the Liouville volume form on M_1, it is clear that $\dim M_1 \leqslant \dim M_2$ and that Ψ is an immersion. In particular, for every $m \in M_1$,

$$\mathrm{Rk}_{\Psi(m)}\,\pi_2 = \dim M_2 \geqslant \dim M_1 = \mathrm{Rk}_m\,\pi_1 .$$

The resulting inequality $\mathrm{Rk}_{\Psi(m)}\,\pi_2 \geqslant \mathrm{Rk}_m\,\pi_1$, is in sharp contrast with the inequality $\mathrm{Rk}_{\Psi(m)}\,\pi_2 \leqslant \mathrm{Rk}_m\,\pi_1$, which is valid for a Poisson map Ψ (see Exercise 4 of Chapter 1). Thus, a symplectic map $\Psi : M_1 \to M_2$ between two symplectic manifolds (M_1, ω_1) and (M_2, ω_2) is not necessarily a Poisson map, as a map $\Psi : (M_1, \pi_1) \to (M_2, \pi_2)$, where π_i is the canonical Poisson structure, associated to ω_i, for $i = 1, 2$. It follows that the map, which associates to a symplectic manifold the corresponding Poisson manifold, is not functorial.

6.3.3 Symplectic Reduction

In this section, we specialize Poisson reduction, in particular Proposition 5.14, to the case of a symplectic manifold. The main extra result is that the reduced Poisson manifold is also a symplectic manifold.

Proposition 6.14. *Let (M,ω) be a symplectic manifold, N an (immersed or embedded) submanifold of M with inclusion map $\iota : N \hookrightarrow M$, and let $p : N \to P$ be a surjective submersion with connected fibers. If*

$$\pi_n^\sharp\left(T_n^\perp N\right) = T_n p_n \tag{6.18}$$

for every $n \in N$, then N is a coisotropic submanifold of M, the triple (M,N,P) is Poisson reducible and the reduced Poisson structure on P is regular and of maximal rank, i.e., it is the canonical Poisson structure, associated to a symplectic structure ω_P on P. Also, $\iota^*\omega = p^*\omega_P$.

Proof. In view of Proposition 5.14, we know that (M,N,P) is Poisson reducible, hence that P inherits a Poisson structure π_P from (M,ω). In order to show that π_P is of maximal rank (equal to the dimension of P) at every point of P, let $n \in N$ and suppose that F is a function, defined in a neighborhood of $p(n)$ in P, such that $(\mathscr{X}_F)_{p(n)} = 0$. Let $\tilde{F}$ be an extension of $F \circ p$ to a neighborhood of n in M. According to (5.27), $\mathrm{d}_n p\,(\mathscr{X}_{\tilde{F}})_n = 0$, so that

$$(\mathscr{X}_{\tilde{F}})_n \in T_n p_n = \pi_n^\sharp\left(T_n^\perp N\right) .$$

Since M is a symplectic manifold, this means that $\mathrm{d}_n\tilde{F} \in T_n^\perp N$, i.e., $\mathrm{d}_n\tilde{F}_{|T_nN} = 0$, so that $\mathrm{d}_n(F\circ p) = \mathrm{d}_n(\tilde{F}_{|N}) = 0$. Since $\mathrm{d}_n p$ is surjective (as p is a submersion), it follows that $\mathrm{d}_{p(n)}F = 0$. This proves that π_P is of maximal rank at $p(n)$. Since p is surjective, this shows that π_P is of maximal rank at every point of P; it is therefore the canonical Poisson structure, associated to a symplectic structure on P, which we denote by ω_P.

We show that $\iota^*\omega = p^*\omega_P$. To do this, let $n \in N$ and let v,w be arbitrary elements of T_nN, which we identify with elements of T_nM. Since $\pi_n^\sharp$ is invertible, (6.18) implies that $\pi_n^\sharp\left(T_n^\perp p_n\right) = T_nN$, so that we can write v as $(\mathscr{X}_{\tilde{F}})_n$, where $\tilde{F}$ is defined in a neighborhood of n in M and is constant on the fibers of p, so its restriction to N is of the form $\tilde{F}_{|N} = F\circ p$, where F is a function, defined on a neighborhood of $p(n)$ in P. On the one hand,

$$(\iota^*\omega)_n(v,w) = \omega_n((\mathscr{X}_{\tilde{F}})_n, w) = \left\langle \mathrm{d}_n\tilde{F}, w\right\rangle = \left\langle p^*\mathrm{d}_{p(n)}F, w\right\rangle, \tag{6.19}$$

where we have used in the last step that w is tangent to N. On the other hand, using (5.27) it follows that

$$\begin{aligned}(p^*\omega_P)_n(v,w) &= (\omega_P)_{p(n)}(T_n p(\mathscr{X}_{\tilde{F}})_n, T_n p(w)) \\ &= (\omega_P)_{p(n)}((\mathscr{X}_F)_{p(n)}, T_n p(w)) = \langle \mathrm{d}_{p(n)}F, T_n p(w)\rangle \,. \end{aligned} \tag{6.20}$$

Comparing (6.19) and (6.20) we see that $(\iota^*\omega)_n = (p^*\omega_P)_n$ for every $n \in N$, which we needed to show. □

Remark 6.15. The fact that the reduced Poisson structure on P is of maximal rank (symplectic) can also be read off from its Poisson matrix. Recall from Remark 5.16 that in well-chosen coordinates on an open subset U of M, centered at a point $n \in N$, the Poisson matrix of (M,π) is given by

$$X = \begin{pmatrix} \tilde{A} & B & C \\ -B^\top & D & E \\ -C^\top & -E^\top & F \end{pmatrix}$$

and that the condition (6.18) implies that $C_{|N} = F_{|N} = 0$ (see the last line of Table 5.1). Recall also that the Poisson matrix of the reduced Poisson structure π_P is given by the matrix A, defined by $\tilde{A}_{|N} = A \circ p$. Notice that, in the present case, E is a square matrix, because the dimension t of the fibers of $p : N \to P$ is equal to the codimension of N in M, a consequence of condition (6.18), combined with the fact that π is of maximal rank. For every $n' \in N \cap U$,

$$\det X(n') = (\det E(n'))^2 \det \tilde{A}(n') = (\det E(n'))^2 \det A(p(n'))\,,$$

so that $\det A(p(n')) \neq 0$. This gives an alternative proof of the fact that A, and hence π_P, is of maximal rank at every point of P. Notice that the argument also reproves (in the symplectic case) that $\mathrm{Rk}(E(n')) = t$ for every $n' \in N \cap U$ (see again the last line of Table 5.1).

Combining the above proposition with Proposition 5.39 leads to the following result.

Corollary 6.16. *Let* $\mathbf{G}$ *be a connected Lie group and let* (M,ω) *be a symplectic manifold. We assume that* $\mathbf{G}$ *acts locally freely and properly on* M *and that the action is Hamiltonian. If* 0 *is a regular value of its momentum map* μ*, then*

(1) $\mu^{-1}(0)$ *is an embedded submanifold of* M*, invariant under the action of* $\mathbf{G}$*;*
(2) $\mu^{-1}(0)/\mathbf{G}$ *is a manifold;*
(3) The triple $(M, \mu^{-1}(0), \mu^{-1}(0)/\mathbf{G})$ *is Poisson reducible.*

The reduced Poisson structure on $\mu^{-1}(0)/\mathbf{G}$ *is regular and of maximal rank, i.e., it is the canonical Poisson structure, associated to a symplectic structure* ω_0 *on* $\mu^{-1}(0)/\mathbf{G}$*. Also,* $\iota^*\omega = p^*\omega_0$*, where* $p : \mu^{-1}(0) \to \mu^{-1}(0)/\mathbf{G}$ *is the quotient map.*

6.3.4 Quotients by Finite Groups of Symplectomorphisms

Let (V,ω) be a symplectic $\mathbb{C}$-vector space of dimension $2d$. We denote by $\mathbf{SP}(V)$ the group of all linear symplectomorphisms of V, i.e., the group of all endomorphisms of V which preserve ω. Recall from Section 6.3.1 that the symplectic two-form ω leads to a (constant) Poisson structure of rank $2d$ on V, which we called the canonical Poisson of (V,ω). Let $\mathbf{G}$ be a finite subgroup of $\mathbf{SP}(V)$. The purpose of the present section is to study the singular affine variety $V/\mathbf{G}$, both as an affine variety and as a Poisson variety. The case where the dimension of V is equal to 2 will be studied in more detail in Section 9.2.4.

Let us specialize to the present setting what has been said at the very end of Section 5.1.2 about the quotient of an affine variety by a finite group. Since $\mathbf{G}$ is a finite subgroup of $\mathbf{SP}(V)$, it acts on V, and the quotient space $V/\mathbf{G}$ is an affine variety, whose algebra of regular functions $\mathscr{F}(V/\mathbf{G})$ is naturally identified with the algebra $\mathscr{F}(V)^{\mathbf{G}}$ of all $\mathbf{G}$-invariant polynomials on V. The canonical projection $p:V\to V/\mathbf{G}$ corresponds, dually, to the inclusion of algebras $\mathscr{F}(V/\mathbf{G})\simeq\mathscr{F}(V)^{\mathbf{G}}\hookrightarrow\mathscr{F}(V)$. Moreover, since every linear symplectomorphism of (V,ω) preserves the canonical Poisson structure of (V,ω), the finite subgroup $\mathbf{G}\subset\mathbf{SP}(V)$ acts by Poisson isomorphisms. As a consequence, according to the first item of Proposition 5.33, $V/\mathbf{G}$ carries a unique Poisson structure with respect to which $p:V\to V/\mathbf{G}$ is a Poisson map. This justifies the following definition.

Definition 6.17. Let (V,ω) be a symplectic $\mathbb{C}$-vector space and let $\mathbf{G}$ be a finite subgroup of $\mathbf{SP}(V)$. The unique Poisson structure $\{\cdot,\cdot\}_{V/\mathbf{G}}$ on $V/\mathbf{G}$ with respect to which the natural projection map $p:V\to V/\mathbf{G}$ is a Poisson map, is called the *quotient Poisson structure* associated to $\mathbf{G}$. The pair $(V/\mathbf{G},\{\cdot,\cdot\}_{V/\mathbf{G}})$ is called the *quotient Poisson variety* associated to $\mathbf{G}$.

Since the quotient space $V/\mathbf{G}$ is, in general, singular, it makes no sense to ask whether it is a symplectic manifold. However, the second item of the next proposition implies that $V/\mathbf{G}$ is as symplectic as it can be, i.e., $\{\cdot,\cdot\}_{V/\mathbf{G}}$ is of maximal rank at every non-singular point of $V/\mathbf{G}$.

Proposition 6.18. *Let (V,ω) be a symplectic $\mathbb{C}$-vector space of dimension $2d$ and let $\mathbf{G}\neq\{e\}$ be a finite subgroup of $\mathbf{SP}(V)$, the group of all symplectomorphisms of V. Denote by p the natural projection map $p:V\to V/\mathbf{G}$ and by U the (non-empty) Zariski open subset of all $x\in V$ for which the stabilizer $\mathbf{G}_x$ is trivial.*

(1) A point in $V/\mathbf{G}$ is a non-singular point if and only if it belongs to $p(U)$;
(2) The rank of $\{\cdot,\cdot\}_{V/\mathbf{G}}$ is equal to $2d$ at every non-singular point of $V/\mathbf{G}$.

As a consequence, the restriction of $\{\cdot,\cdot\}_{V/\mathbf{G}}$ to $p(U)$ is a holomorphic Poisson structure, which is the canonical Poisson structure associated to a symplectic holomorphic two-form on $p(U)$.

Proof. It is a classical result (valid for the action of an arbitrary finite group on a vector space) that, for every $x\in V$, the point $p(x)$ is a non-singular point of $V/\mathbf{G}$

if and only if its stabilizer $\mathbf{G}_x$ is a *complex reflection group*, i.e., a group generated by elements for which the space of fixed points is a hyperplane (see [42]). But the set of fixed points of a linear symplectomorphism cannot be a hyperplane. As a consequence, in the present case, the point $p(x)$ is non-singular if and only if $\mathbf{G}_x = \{e\}$. This proves *(1)*.

The open subset $U \subset V$ is a $\mathbf{G}$-invariant subset, the action of $\mathbf{G}$ on U is free since the stabilizer of every point of U is by definition trivial and, according to Section 2.3.2, U is equipped with a holomorphic Poisson structure. This implies on the one hand that the projection map $p : U \to U/\mathbf{G} = p(U)$ is a local biholomorphism (when both sets are considered as holomorphic manifolds), and on the other hand, since p is moreover a Poisson map between Poisson manifolds, that p preserves the rank of the Poisson structure at each point, so that the rank of $\{\cdot,\cdot\}_{V/\mathbf{G}}$ is $2d$ at every point of $p(U)$. Therefore, $\{\cdot,\cdot\}_{V/\mathbf{G}}$ restricts to a regular holomorphic Poisson structure of maximal rank on $p(U)$. According to Proposition 6.11, it is the canonical Poisson structure, associated to a symplectic structure on the holomorphic manifold $p(U)$. This completes the proof of *(2)*. □

Remark 6.19. For every $\lambda \in \mathbb{C}^*$, the map $x \mapsto \lambda^{-1}x$ leads to a regular map of $V/\mathbf{G}$ to itself. Since $x \mapsto \lambda^{-1}x$ multiplies the canonical Poisson structure on V by λ^2, this induced automorphism also transforms the quotient Poisson structure $\{\cdot,\cdot\}_{V/\mathbf{G}}$ into $\lambda^2\{\cdot,\cdot\}_{V/\mathbf{G}}$. As a consequence, for every non-zero $\mu \in \mathbb{C}^*$, the Poisson varieties $(V/\mathbf{G}, \{\cdot,\cdot\}_{V/\mathbf{G}})$ and $(V/\mathbf{G}, \mu\{\cdot,\cdot\}_{V/\mathbf{G}})$ are isomorphic as Poisson varieties.

The case where the dimension of V is 2 will be studied in more detail in Section 9.2.4. We give here an immediate corollary of Proposition 6.18, which will be useful then.

Corollary 6.20. *Let V be a symplectic $\mathbb{C}$-vector space of dimension 2 and let $\mathbf{G} \neq \{e\}$ be a finite subgroup of $\mathbf{SP}(V)$. Then the image of the origin o of V through the canonical projection $p : V \to V/\mathbf{G}$ is the only point of $V/\mathbf{G}$ which is singular. It is also the only point where the rank of the quotient Poisson structure is zero.*

Proof. For a vector space V of dimension 2, the group $\mathbf{SP}(V)$ coincides with the group $\mathbf{SL}(V)$ of all linear maps of determinant $+1$. Every element of a finite subgroup $\mathbf{G}$ of $\mathbf{SL}(V)$ is of finite order, hence is diagonalizable. Clearly, a diagonalizable element of $\mathbf{SL}(V)$, different from the unit, admits $o \in V$ as its unique fixed point, hence $U = V \setminus \{o\}$. Clearly, the rank of $\{\cdot,\cdot\}_{V/\mathbf{G}}$ at $p(o)$ is zero. According to Proposition 6.18, the rank of $\{\cdot,\cdot\}_{V/\mathbf{G}}$ is 2 at every other point of $V/\mathbf{G}$. □

6.3.5 Example 1: Cotangent Bundles

It is shown in Section 1.3.4 that every Poisson manifold admits a (singular) foliation whose leaves carry a symplectic structure. Thus, the number of examples of symplectic manifolds is abundant. There are however two large classes of symplectic manifolds which need special attention: cotangent bundles and Kähler manifolds. The first class, considered in this section, is of fundamental importance in classical mechanics, see e.g. [1, 15] and [125].

The idea that cotangent bundles are symplectic manifolds comes from classical mechanics: local coordinates on M and the corresponding momenta, which are linear coordinates on the fibers of $\rho : T^*M \to M$ yield canonical coordinates on phase space. To make this precise, let us first show how to build natural local coordinates on T^*M from local coordinates on M. Suppose that $x_1, \dots, x_d$ are coordinates on $U \subset M$; it yields half of a system of coordinates on $T^*U \subset T^*M$ by letting $q_i := x_i \circ \rho$. For the other half, let p_i denote the function on T^*U, which is defined by

$$p_i(\xi_m) := \left\langle \xi_m, \left(\frac{\partial}{\partial x_i}\right)_m \right\rangle, \tag{6.21}$$

where $\xi_m \in T^*U$ is in the fiber over m, i.e., $\rho(\xi_m) = m$. The functions p_i and x_i are dual in the following sense: for $m \in U$ and $1 \leqslant i, j \leqslant d$,

$$p_i(\mathrm{d}_m x_j) = \left\langle \mathrm{d}_m x_j, \left(\frac{\partial}{\partial x_i}\right)_m \right\rangle = \delta_{i,j}\,.$$

On T^*M there is a natural differential one-form, called the *Liouville form*, which can in terms of the above local coordinates be defined as

$$\theta := \sum_{i=1}^{d} p_i \mathrm{d}q_i\,.$$

More intrinsically, $\theta \in \Omega^1(T^*M)$ is defined by

$$\langle \theta, \mathscr{V} \rangle (\xi_m) := \langle \xi_m, T_{\xi_m}\rho\,(\mathscr{V}_{\xi_m}) \rangle\,, \tag{6.22}$$

where $\xi_m \in T^*M$ and $\mathscr{V} \in \mathfrak{X}^1(T^*M)$. The canonical pairing in this formula is between $T^*_m M$ and $T_m M$: indeed, $T_{\xi_m}\rho : T_{\xi_m}(T^*M) \to T_m M$. To check that θ, as defined by (6.22), is given in terms of the above coordinates q_i, p_i by $\sum_{i=1}^d p_i \mathrm{d}q_i$, take an arbitrary vector field $\mathscr{V} = \sum_{i=1}^d (\beta_i \partial/\partial q_i + \gamma_i \partial/\partial p_i)$ on T^*M and compare

$$\langle \theta, \mathscr{V} \rangle (\xi_m) = \langle \xi_m, T_{\xi_m}\rho\,(\mathscr{V}_{\xi_m}) \rangle = \left\langle \xi_m, \sum_{i=1}^{d} \beta_i(\xi_m) \left(\frac{\partial}{\partial x_i}\right)_m \right\rangle,$$

which was computed from (6.22), using $q_i = x_i \circ \rho$, with

$$\begin{aligned}\left\langle \sum_{i=1}^{d} p_i \mathrm{d}q_i, \mathscr{V} \right\rangle (\xi_m) &= \left\langle \sum_{i=1}^{d} p_i \mathrm{d}q_i, \sum_{j=1}^{d} \left(\beta_j \frac{\partial}{\partial q_j} + \gamma_j \frac{\partial}{\partial p_j}\right) \right\rangle (\xi_m) \\ &= \sum_{i=1}^{d} p_i(\xi_m)\beta_i(\xi_m)\end{aligned}$$

$$= \sum_{i=1}^{d} \beta_i(\xi_m) \left\langle \xi_m, \left(\frac{\partial}{\partial x_i}\right)_m \right\rangle ,$$

which was computed from (6.21).

The canonical symplectic structure on T^*M is by definition the closed differential two-form $\omega := \mathrm{d}\theta$, which is locally given by $\sum_{i=1}^{d} \mathrm{d}p_i \wedge \mathrm{d}q_i$, so that the coordinates q_i and p_i are canonical coordinates. Notice that the differential two-form ω is actually exact, making T^*M into an *exact symplectic manifold*.

To a function $F \in \mathscr{F}(M)$, respectively to a vector field $\mathscr{V} \in \mathfrak{X}^1(M)$, one can associate a function $\tilde{F}$, respectively $\tilde{\mathscr{V}}$, on T^*M by

$$\tilde{F} := F \circ \rho, \qquad \tilde{\mathscr{V}}(\xi_m) := \langle \xi_m, \mathscr{V}(m) \rangle$$

for every $\xi_m \in T^*_m M$. The reader will easily verify that the canonical Poisson bracket of ω satisfies the following formulas: if $F_1, F_2 \in \mathscr{F}(M)$ and $\mathscr{V}_1, \mathscr{V}_2 \in \mathfrak{X}^1(M)$, then

$$\left\{\tilde{F}_1, \tilde{F}_2\right\} = 0, \quad \left\{\tilde{\mathscr{V}}_1, \tilde{\mathscr{V}}_2\right\} = \widetilde{[\mathscr{V}_1, \mathscr{V}_2]} \quad \text{and} \quad \left\{\tilde{\mathscr{V}}_1, \tilde{F}_1\right\} = \widetilde{\mathscr{V}_1[F_1]} .$$

In fact, these formulas can be used to define ω (see [3, Ch. 2]).

6.3.6 Example 2: Kähler Manifolds

We now turn to a second class of examples, which is important from the point of view of complex geometry. Let M be a complex manifold and let us denote the underlying real manifold by $M_{\mathbb{R}}$. Suppose that $\mathrm{d}s^2$ is a *Hermitian metric* on M. This means that, in terms of a system of local (holomorphic) coordinates $z_1, \ldots, z_d$,

$$\mathrm{d}s^2 = \sum_{1 \leqslant i,j \leqslant d} h_{ij}\, \mathrm{d}z_i \otimes \mathrm{d}\bar{z}_j ,$$

where (h_{ij}) is a positive definite Hermitian matrix. The imaginary part (up to a sign) of such a metric is a non-degenerate differential 2-form on $M_{\mathbb{R}}$, given by

$$\omega := \frac{\sqrt{-1}}{2} \sum_{1 \leqslant i,j \leqslant d} h_{ij}\, \mathrm{d}z_i \wedge \mathrm{d}\bar{z}_j .$$

One says that the metric $\mathrm{d}s^2$ is a *Kähler metric* if ω is closed, i.e., if ω is a symplectic structure. Then $(M_{\mathbb{R}}, \omega)$ is called a *Kähler manifold* and ω is called the *associated differential 2-form* of the metric. Notice that the metric can be recovered from its associated differential 2-form.

Two facts underlie the importance of Kähler manifolds. The first one is that the complex projective space $\mathbb{P}^N$ admits a Kähler metric. The second fact is that every submanifold of a Kähler manifold is itself a Kähler manifold. This means that every non-singular complex projective variety has a Kähler structure.

6.4 Notes

The study of symplectic manifolds, initiated by its importance in classical mechanics, has become a subject of its own. Many books have symplectic manifolds as their central theme, yet there are significant variations in how much the theory is connected to classical, and eventually quantum mechanics, as well as in the advanced topics which are treated. For an excellent introduction, see Cannas da Silva [33], where the example of Kähler manifolds is treated in detail. For an introduction to the subject, largely motivated by classical mechanics, see Arnold's classical book [15]. Another classical book is the mechanics book [1] by Abraham and Marsden, where symplectic geometry is used as a tool, rather than a subject of itself; in it, the utility of symplectic structures is plain. The books [125] by Libermann–Marle and [92] by Guillemin–Sternberg, two books of opposite style, are great complements to the above references.

All the above books treat additional topics. Group actions and the momentum map are studied in varying detail in all of them; for this, the books of Audin [18] and Ortega–Ratiu [160] should also be consulted. Symplectic vector bundles and contact manifolds are treated in Libermann–Marle [125], symplectic homogeneous spaces are an additional topic in Guillemin–Sternberg [92], the Hamilton–Jacobi method and perturbation theory are discussed in both Abraham–Marsden [1] and Arnold [15].

Chapter 7
Linear Poisson Structures and Lie Algebras

Together with symplectic manifolds, considered in the previous chapter, Lie algebras provide the first examples of Poisson manifolds. Namely, the dual $\mathfrak{g}^*$ of a finite-dimensional Lie algebra $\mathfrak{g}$ admits a natural Poisson structure, called its Lie–Poisson structure, which provides new insights and technical tools in the study of Lie algebras. For example, the coadjoint orbits in $\mathfrak{g}^*$ are precisely the symplectic leaves of the Lie–Poisson structure, showing in particular that coadjoint orbits are even-dimensional.

Lie–Poisson structures are linear Poisson structures, in the sense that they are Poisson structures on a vector space V, for which the Poisson bracket of every pair of *linear* functions on V (elements of V^*) is a *linear* function on V. Also, every linear Poisson structure (on a finite-dimensional vector space) is a Lie–Poisson structure and we have a functorial equivalence between linear Poisson structures and Lie algebra structures.

In many cases, one considers the Poisson structure on the Lie algebra $\mathfrak{g}$ itself, rather than on $\mathfrak{g}^*$. This is usually done by identifying $\mathfrak{g}$ with $\mathfrak{g}^*$, using a non-degenerate Ad-invariant symmetric bilinear form on $\mathfrak{g}$, which exists for example for every semisimple Lie algebra. The coadjoint orbits are then identified with the adjoint orbits and the Hamiltonian vector fields on $\mathfrak{g}$ take a natural form, a so-called Lax form.

The Lie–Poisson structure on $\mathfrak{g}^*$ is introduced in Section 7.1, while the induced Poisson structure on $\mathfrak{g}$ is discussed in Section 7.2. The main properties of the Lie–Poisson structure are given in Section 7.3. A variant of Lie–Poisson structures, namely affine (= linear + constant) Poisson structures and their Lie theoretical interpretation is discussed in Section 7.4. We finish this chapter with a short introduction to the linearization of Poisson structures (in the neighborhood of a point where the rank is zero).

Unless otherwise stated, $\mathbb{F}$ denotes an arbitrary field of characteristic zero.

C. Laurent-Gengoux et al., *Poisson Structures*,
Grundlehren der mathematischen Wissenschaften 347,
DOI 10.1007/978-3-642-31090-4_7,

7.1 The Lie–Poisson Structure on $\mathfrak{g}^*$

Suppose that $\mathfrak{g}$ is a finite-dimensional Lie algebra over $\mathbb{F}$, with Lie bracket $[\cdot,\cdot]$. We denote the dual vector space to $\mathfrak{g}$ by $\mathfrak{g}^*$. We associate to each element[1] e of $\mathfrak{g}$, a linear function $e^* : \mathfrak{g}^* \to \mathbb{F}$, defined by

$$\begin{aligned} e^* : \mathfrak{g}^* &\to \mathbb{F} \\ \xi &\mapsto \langle \xi, e\rangle := \xi(e)\ . \end{aligned}$$

In words, e^* is "evaluation at e". Taking a basis $(e_1,\dots,e_d)$ of $\mathfrak{g}$, we obtain a system of linear coordinates $(x_1,\dots,x_d)$ on $\mathfrak{g}^*$ by letting $x_i := e_i^*$. We use it to build the skew-symmetric matrix $X = (x_{ij})_{1\leqslant i,j\leqslant d}$, whose elements are defined by $x_{ij} := [e_i,e_j]^*$; each entry of X is a function on $\mathfrak{g}$, which can be expressed as a linear combination of the coordinates $x_1,\dots,x_d$. Consider the skew-symmetric biderivation $\{\cdot,\cdot\}$ of $\mathbb{F}[x_1,\dots,x_d]$, defined by setting, for $F,G\in\mathbb{F}[x_1,\dots,x_d]$,

$$\{F,G\} := \sum_{i,j=1}^{d} x_{ij}\frac{\partial F}{\partial x_i}\frac{\partial G}{\partial x_j}\ . \tag{7.1}$$

It is the unique skew-symmetric biderivation of $\mathbb{F}[x_1,\dots,x_d]$ such that $\left\{e_i^*,e_j^*\right\} = [e_i,e_j]^*$, for all $1\leqslant i<j\leqslant d$. By bilinearity of $\{\cdot,\cdot\}$ and $[\cdot,\cdot]$,

$$\{e^*,f^*\} = [e,f]^*\ , \tag{7.2}$$

for all $e,f\in\mathfrak{g}$. On the one hand, this implies that Definition (7.1) is independent of the chosen basis $(e_1,\dots,e_d)$. On the other hand, it yields, for all i,j,k with $1\leqslant i,j,k\leqslant d$,

$$\left\{\left\{x_i,x_j\right\},x_k\right\} = [[e_i,e_j],e_k]^*\ ,$$

so that the Jacobi identity for $[\cdot,\cdot]$ implies that $\left\{\left\{x_i,x_j\right\},x_k\right\} + \circlearrowleft(i,j,k) = 0$ for all $1\leqslant i<j<k\leqslant d$. According to Proposition 1.8, this shows that $\{\cdot,\cdot\}$ is a Poisson structure on $\mathbb{F}[x_1,\dots,x_d]$. Similarly, according to *(iii)* in Proposition 1.36, Eq. (7.1) also defines a Poisson structure on the algebra of smooth functions on $\mathfrak{g}^*$, when $\mathbb{F}=\mathbb{R}$, or on the algebra of holomorphic functions on $\mathfrak{g}^*$, when $\mathbb{F}=\mathbb{C}$; in either case, it is the unique Poisson structure which satisfies (7.2). Thus, $\mathscr{F}(\mathfrak{g}^*)$ inherits a Poisson structure from the Lie bracket on $\mathfrak{g}$, irrespective of whether we take $\mathscr{F}(\mathfrak{g}^*)$ to be the algebra of polynomial, smooth, or holomorphic functions on $\mathfrak{g}^*$.

An intrinsic formula for $\{F,G\}$ can be written down in terms of the differentials of F and G. Since the differential of $F\in\mathscr{F}(\mathfrak{g}^*)$ at ξ is a linear map $\mathrm{d}_\xi F : T_\xi\mathfrak{g}^*\to\mathbb{F}$, and since $T_\xi\mathfrak{g}^*$ is naturally isomorphic with $\mathfrak{g}^*$, we can think of $\mathrm{d}_\xi F$ as being an element of $(\mathfrak{g}^*)^*$, i.e., as an element of $\mathfrak{g}$, since $\mathfrak{g}$ and its bidual are canonically isomorphic (recall that $\mathfrak{g}$ is finite-dimensional; the canonical isomorphism $\mathfrak{g}\to(\mathfrak{g}^*)^*$

[1] We use in this section letters e and f to denote elements of a Lie algebra $\mathfrak{g}$, to reserve the letters x,y,z for elements of the bidual $(\mathfrak{g}^*)^*$. After identification of $\mathfrak{g}$ with its bidual, the letters x,y,z become elements of $\mathfrak{g}$, just like everywhere else in the book.

is given, for $e \in \mathfrak{g}$, by $e \mapsto e^*$). Thus, we can identify $\mathrm{d}_\xi F$ with an element of $\mathfrak{g}$, which we will also denote by $\mathrm{d}_\xi F$. With this notation, the differential of $F \in \mathscr{F}(\mathfrak{g}^*)$ at $\xi \in \mathfrak{g}^*$, is given by

$$\mathrm{d}_\xi F = \sum_{i=1}^{d} \frac{\partial F}{\partial x_i}(\xi)\, e_i\,, \tag{7.3}$$

and (7.1), evaluated at $\xi \in \mathfrak{g}^*$, can be written in the compact, coordinate-free form,

$$\{F,G\}(\xi) = \langle \xi, [\mathrm{d}_\xi F, \mathrm{d}_\xi G] \rangle\ . \tag{7.4}$$

Definition 7.1. Let $(\mathfrak{g}, [\cdot\,,\cdot])$ be a finite-dimensional Lie algebra and let $\mathscr{F}(\mathfrak{g}^*)$ denote the algebra of polynomial, smooth or holomorphic functions on $\mathfrak{g}^*$. The Poisson structure on $\mathfrak{g}^*$, defined for $F,G \in \mathscr{F}(\mathfrak{g}^*)$, at $\xi \in \mathfrak{g}^*$ by

$$\{F,G\}(\xi) := \langle \xi, [\mathrm{d}_\xi F, \mathrm{d}_\xi G] \rangle \tag{7.5}$$

is called the *Lie–Poisson structure on* $\mathfrak{g}^*$, or the *canonical Poisson structure on* $\mathfrak{g}^*$. As we will show in Proposition 7.3 below, Lie–Poisson structures are in one-to-one correspondence with linear Poisson structures, where the latter are defined as follows.

Definition 7.2. A Poisson structure $\{\cdot\,,\cdot\}$ on a finite-dimensional vector space V is called a *linear Poisson structure* if for each pair of linear functions ξ and η on V, their Poisson bracket $\{\xi, \eta\}$ is a linear function on V.

Proposition 7.3. *For every finite-dimensional vector space V, there is a natural one-to-one correspondence between linear Poisson structures on V^* and Lie algebra structures on V.*

Proof. As we have explained, when $V = \mathfrak{g}$ is a finite-dimensional Lie algebra, its dual vector space $\mathfrak{g}^*$ carries a Poisson structure. If x,y are linear functions on $\mathfrak{g}^*$, then they are of the form $x = e^*$ and $y = f^*$, where $e, f \in \mathfrak{g}$, and it follows from (7.2) that their Poisson bracket is a linear function on $\mathfrak{g}^*$. This shows that the Lie–Poisson structure on $\mathfrak{g}^*$ is a linear Poisson structure. On the other hand, a linear Poisson structure $\{\cdot\,,\cdot\}$ can be restricted to linear functions on V^* and the restriction is a Lie bracket on $(V^*)^* \simeq V$. The Lie–Poisson structure on V^*, associated to this Lie bracket, is the original Poisson bracket $\{\cdot\,,\cdot\}$, since both Poisson brackets coincide for all linear functions on V^*. □

Remark 7.4. For $i = 1,2$, let $(\mathfrak{g}_i, [\cdot\,,\cdot]_i)$ be a Lie algebra whose Lie–Poisson structure (on $\mathfrak{g}_i^*$) is denoted by $\{\cdot\,,\cdot\}_i$. If $\phi : (\mathfrak{g}_1, [\cdot\,,\cdot]_1) \to (\mathfrak{g}_2, [\cdot\,,\cdot]_2)$ is a Lie algebra homomorphism, then the transpose map $\phi^\top : (\mathfrak{g}_2^*, \{\cdot\,,\cdot\}_2) \to (\mathfrak{g}_1^*, \{\cdot\,,\cdot\}_1)$ is a Poisson map. It is easily checked that this leads to a contravariant functor from the category of finite-dimensional Lie algebras to the category of Poisson manifolds (affine Poisson varieties if $\mathbb{F}$ is different from $\mathbb{R}$ and $\mathbb{C}$).

It is often convenient to represent the Lie–Poisson structure by its Poisson matrix (see Section 1.2.2). To do this, consider, as above, a basis $(e_1, \dots, e_d)$ of $\mathfrak{g}$, which leads to a system of linear coordinates $(x_1,\dots,x_d)$ on $\mathfrak{g}^*$. Let us denote by c_{ij}^k the *structure constants* of the Lie algebra $\mathfrak{g}$ in terms of the above basis:

$$[e_i,e_j] = \sum_{k=1}^{d} c_{ij}^k e_k, \qquad 1 \leqslant i,j \leqslant d\,.$$

Then, the entries of the Poisson matrix $(\{x_i,x_j\})_{1\leqslant i,j\leqslant d}$ are given by

$$\{x_i,x_j\} = [e_i,e_j]^* = \sum_{k=1}^{d} c_{ij}^k x_k\,.$$

In this notation, Eq. (7.1) for the Poisson structure takes the form

$$\{F,G\} := \sum_{i,j,k=1}^{d} c_{ij}^k x_k \frac{\partial F}{\partial x_i}\frac{\partial G}{\partial x_j}\,.$$

In tensorial terms, the algebra of polynomial functions on $\mathfrak{g}^*$ is canonically isomorphic to $S\mathfrak{g}$, the symmetric algebra of $\mathfrak{g}$, and the construction of the Lie–Poisson structure on $S\mathfrak{g}$ can be given without passing via $\mathfrak{g}^*$, which has the advantage that its construction does not require $\mathfrak{g}$ to be finite-dimensional. Explicitly, a typical monomial $z_1z_2\dots z_p \in S\mathfrak{g}$ is viewed as a polynomial function on $\mathfrak{g}^*$ by setting, for $\xi \in \mathfrak{g}^*$,

$$(z_1z_2\dots z_p)(\xi) := \xi(z_1)\,\xi(z_2)\dots\xi(z_p) = \langle \xi,z_1\rangle\,\langle \xi,z_2\rangle\cdots\langle \xi,z_p\rangle\,. \tag{7.6}$$

For monomials of degree 1, the Poisson bracket $\{\cdot,\cdot\}$ on $S\mathfrak{g}$ is just the usual Lie bracket, $\{z,z'\} := [z,z']$. For monomials of higher degree, one uses the biderivation property, which gives

$$\{z_1\dots z_p, z_1'\dots z_q'\} = \sum_{k=1}^{p}\sum_{\ell=1}^{q} z_1\dots\widehat{z_k}\dots z_p z_1'\dots\widehat{z_\ell'}\dots z_q'\,[z_k,z_\ell']\,.$$

From this formula, one easily gives a direct proof of the Jacobi identity for the bracket $\{\cdot,\cdot\}$ on $S\mathfrak{g}$.

7.2 The Lie–Poisson Structure on $\mathfrak{g}$

Upon choosing an isomorphism between the finite-dimensional Lie algebra $\mathfrak{g}$ and its dual $\mathfrak{g}^*$, the Lie–Poisson structure on $\mathfrak{g}^*$ can be transferred to a linear Poisson structure on $\mathfrak{g}$. This is usually done by choosing a non-degenerate symmetric bilinear form $\langle\cdot\,|\,\cdot\rangle$ on $\mathfrak{g}$. *Non-degeneracy* of the bilinear form $\langle\cdot\,|\,\cdot\rangle$ means that the linear map $\chi:\mathfrak{g}\to\mathfrak{g}^*$, which sends $x\in\mathfrak{g}$ to the linear form $\langle x\,|\,\cdot\rangle : y\mapsto\langle x\,|\,y\rangle$, is an isomorphism. Explicitly, χ is given by

$$\langle \chi(x), y \rangle = \langle x \,|\, y \rangle = \langle \chi(y), x \rangle , \tag{7.7}$$

for all $x, y \in \mathfrak{g}$. The inverse map to χ is denoted by $\psi : \mathfrak{g}^* \to \mathfrak{g}$. As in (7.7),

$$\langle \xi, \psi(\eta) \rangle = \langle \psi(\xi) \,|\, \psi(\eta) \rangle = \langle \eta, \psi(\xi) \rangle , \tag{7.8}$$

for all $\xi, \eta \in \mathfrak{g}^*$. To a function on $\mathfrak{g}^*$, we can associate a function on $\mathfrak{g}$ and vice versa, simply by composing it (on the right) with χ or with ψ. Similarly, every Poisson structure on $\mathfrak{g}^*$ can be transferred to $\mathfrak{g}$ and vice versa. In particular, the Lie–Poisson structure on $\mathfrak{g}^*$, introduced in Section 7.1, leads to a unique Poisson structure $\{\cdot,\cdot\}_\mathfrak{g}$ on $\mathfrak{g}$ with respect to which χ is an isomorphism of Poisson manifolds. We call $\{\cdot,\cdot\}_\mathfrak{g}$ the *Lie–Poisson structure on* $\mathfrak{g}$ with respect to $\langle \cdot \,|\, \cdot \rangle$.

Explicitly, let $F, G \in \mathscr{F}(\mathfrak{g})$ and compute their Lie–Poisson bracket $\{F, G\}_\mathfrak{g}$ (with respect to $\langle \cdot \,|\, \cdot \rangle$) at a point $x \in \mathfrak{g}$ by using (7.5) as follows.

$$\begin{aligned} \{F, G\}_\mathfrak{g}(x) &= \{F \circ \psi, G \circ \psi\}(\chi(x)) \\ &= \left\langle \chi(x), \left[\mathrm{d}_{\chi(x)}(F \circ \psi), \mathrm{d}_{\chi(x)}(G \circ \psi) \right] \right\rangle \\ &= \langle x \,|\, [\psi(\mathrm{d}_x F), \psi(\mathrm{d}_x G)] \rangle . \end{aligned} \tag{7.9}$$

We have used that $\mathrm{d}_{\chi(x)}(F \circ \psi)$, viewed as an element of $\mathfrak{g}$, is precisely $\psi(\mathrm{d}_x F)$. To check the latter, first use the chain rule and the fact that ψ is a linear map to find that

$$\mathrm{d}_{\chi(x)}(F \circ \psi) = \mathrm{d}_x F \circ \psi ,$$

so that, for every $\xi \in \mathfrak{g}^*$,

$$\left\langle \mathrm{d}_{\chi(x)}(F \circ \psi), \xi \right\rangle = \langle \mathrm{d}_x F, \psi(\xi) \rangle = \langle \xi, \psi(\mathrm{d}_x F) \rangle ,$$

where we have used (7.8) in the last step. Equation (7.9) is usually written in the following form,

$$\{F, G\}_\mathfrak{g}(x) = \langle x \,|\, [\nabla_x F, \nabla_x G] \rangle , \tag{7.10}$$

where $\nabla_x F$, the *gradient* of F at x (with respect to $\langle \cdot \,|\, \cdot \rangle$) is defined, for $F \in \mathscr{F}(\mathfrak{g})$ and $x \in \mathfrak{g}$ by

$$\langle \nabla_x F \,|\, y \rangle = \langle \mathrm{d}_x F, y \rangle , \tag{7.11}$$

for every $y \in \mathfrak{g}$, which is equivalent to saying that $\nabla_x F = \psi(\mathrm{d}_x F)$. Since F is a function on a vector space, (7.11) can also be written in the following form,

$$\langle \nabla_x F \,|\, y \rangle = \frac{\mathrm{d}}{\mathrm{d}t}_{|t=0} F(x + ty) , \tag{7.12}$$

which is the most useful form for explicit computation.

The Lie–Poisson structure on $\mathfrak{g}$ is often restricted to a subspace of $\mathfrak{g}$ which is defined by a Lie ideal of $\mathfrak{g}$. If A is a subset of $\mathfrak{g}$, the *orthogonal* to A, with respect to $\langle \cdot \,|\, \cdot \rangle$, is the subspace $A^\perp \subset \mathfrak{g}$, defined as

$$A^{\perp} := \{x \in \mathfrak{g} \mid \forall y \in A,\ \langle x \mid y \rangle = 0\} \ . \tag{7.13}$$

Proposition 7.5. *Let $(\mathfrak{g}, [\cdot\,,\cdot])$ be a finite-dimensional Lie algebra, equipped with a non-degenerate, symmetric bilinear form $\langle \cdot \mid \cdot \rangle$, and suppose that $\mathfrak{h} \subset \mathfrak{g}$ is a Lie ideal of $\mathfrak{g}$. Then its orthogonal, $\mathfrak{h}^{\perp}$, is a Poisson submanifold of $\mathfrak{g}$. For $F, G \in \mathscr{F}(\mathfrak{h}^{\perp})$, their Poisson bracket $\{F, G\}_{\mathfrak{h}^{\perp}}$ is given, at $x \in \mathfrak{h}^{\perp}$, by*

$$\{F, G\}_{\mathfrak{h}^{\perp}}(x) = \left\{\tilde{F}, \tilde{G}\right\}_{\mathfrak{g}}(x) \ , \tag{7.14}$$

where $\tilde{F}$ and $\tilde{G}$ are arbitrary extensions of F and G to a neighborhood of x in $\mathfrak{g}$.

Proof. We first show that if $\mathfrak{h}$ is a Lie ideal of $\mathfrak{g}$, then $\mathfrak{h}^{\perp}$ is a Poisson submanifold of $\mathfrak{g}$. Let $x \in \mathfrak{h}^{\perp}$ and let G be an arbitrary function, defined in a neighborhood of x in $\mathfrak{g}$. According to Proposition 2.12, we need to show that $(\mathscr{X}_G)_x \in T_x\mathfrak{h}^{\perp} \simeq \mathfrak{h}^{\perp}$. To do this, let F be an arbitrary function, defined in a neighborhood of x in $\mathfrak{g}$, which is constant on $\mathfrak{h}^{\perp}$. For such a function, we have that $\nabla_x F \in \mathfrak{h}$; indeed, $F(x+ty) = F(x)$ for every $y \in \mathfrak{h}^{\perp}$ and every $t \in \mathbb{F}$, so that

$$\langle \nabla_x F \mid y \rangle = \frac{\mathrm{d}}{\mathrm{d}t}_{|t=0} F(x+ty) = 0 \ ,$$

for every $y \in \mathfrak{h}^{\perp}$. Since $\langle \cdot \mid \cdot \rangle$ is non-degenerate, we may conclude that $\nabla_x F \in \mathfrak{h}$. Since $\mathfrak{h}$ is a Lie ideal of $\mathfrak{g}$, we have for every $x \in \mathfrak{h}^{\perp}$ that

$$(\mathscr{X}_G)_x[F] = \{F, G\}_{\mathfrak{g}}(x) = \langle x \mid [\nabla_x F, \nabla_x G] \rangle = 0 \ .$$

This shows that $\mathscr{X}_G$ is tangent to $\mathfrak{h}^{\perp}$ at every point $x \in \mathfrak{h}^{\perp}$. Proposition 2.12 also leads to the formula (7.14) for the Poisson structure $\{\cdot\,,\cdot\}_{\mathfrak{h}^{\perp}}$ on $\mathfrak{h}^{\perp}$. □

We now show that the Hamiltonian vector fields $\mathscr{X}_H$ on $(\mathfrak{g}, \{\cdot\,,\cdot\})$ can be written down in a simple form, if $\langle \cdot \mid \cdot \rangle$ is ad-invariant (i.e., when $\mathfrak{g}$ is a quadratic Lie algebra). Recall from Section 5.1.4 that this means that $\langle x \mid [y,z] \rangle = \langle [x,y] \mid z \rangle$ for all $x, y, z \in \mathfrak{g}$. Then we have that (7.10) can be written, for $H = G$, as

$$\{F, H\}_{\mathfrak{g}}(x) = \langle \nabla_x F \mid [\nabla_x H, x] \rangle = \langle \mathrm{d}_x F, [\nabla_x H, x] \rangle \ .$$

Comparing this to

$$\{F, H\}_{\mathfrak{g}}(x) = (\mathscr{X}_H)_x[F] = \langle \mathrm{d}_x F, (\mathscr{X}_H)_x \rangle \ ,$$

we find that $\dot{x} := (\mathscr{X}_H)_x$, the Hamiltonian vector field $\mathscr{X}_H$ at x, is given by the *Lax equation*

$$\dot{x} = [\nabla_x H, x] \ . \tag{7.15}$$

Lax equations are very important in the theory of integrable systems, as one easily shows that their flows are isospectral, hence admit many constants of motion (see Section 12.2.5 below).

Example 7.6. The basic example of a finite-dimensional quadratic Lie algebra is given by the $\mathbb{F}$-vector space $\mathfrak{g} := \mathfrak{gl}_d(\mathbb{F})$ of $d \times d$ matrices with coefficients in $\mathbb{F}$. The Lie bracket on $\mathfrak{g}$ is the commutator $[x,y] := xy - yx$. If E_{ij} denotes the $d \times d$ matrix with a 1 at position (i,j) and zeros elsewhere, then $[E_{ij}, E_{k\ell}] = \delta_{j,k}E_{i\ell} - \delta_{i,\ell}E_{kj}$, for all $1 \leqslant i,j,k,\ell \leqslant d$. A natural symmetric bilinear form on $\mathfrak{g}$ is given by $\langle x \,|\, y\rangle := \text{Trace}(xy)$. It is non-degenerate, thanks to the orthogonality relations $\langle E_{ij} \,|\, E_{k\ell}\rangle = \delta_{j,k}\delta_{i,\ell}$. Moreover, ad-invariance follows at once from the basic property $\text{Trace}(xy) = \text{Trace}(yx)$ of the trace form. Thus, $\mathfrak{g}$ is a quadratic Lie algebra. Let us denote by ξ_{ij} the linear function on $\mathfrak{g}$ which picks the entry at position (i,j) of a matrix: $\xi_{ij}(x) := x_{ij} = \langle E_{ji} \,|\, x\rangle$. For all $x,y \in \mathfrak{g}$,

$$\langle \nabla_x \xi_{ij} \,|\, y\rangle = \langle \mathrm{d}_x \xi_{ij}, y\rangle = \langle \xi_{ij}, y\rangle = \langle E_{ji} \,|\, y\rangle ,$$

so that $\nabla_x \xi_{ij} = \nabla \xi_{ij}$ is independent of x and is given by $\nabla \xi_{ij} = E_{ji}$. It follows that the Poisson bracket of two such linear functions on $\mathfrak{g}$ is given, at $x \in \mathfrak{g}$, by

$$\begin{aligned}\left\{\xi_{ij}, \xi_{k\ell}\right\}_{\mathfrak{g}}(x) &= \langle x \,|\, [E_{ji}, E_{\ell k}]\rangle = \langle x \,|\, \delta_{i,\ell}E_{jk} - \delta_{j,k}E_{\ell i}\rangle \\ &= (\delta_{i,\ell}\xi_{kj} - \delta_{j,k}\xi_{i\ell})(x) ,\end{aligned}$$

so that $\left\{\xi_{ij}, \xi_{k\ell}\right\}_{\mathfrak{g}} = \delta_{i,\ell}\xi_{kj} - \delta_{j,k}\xi_{i\ell}$. Let us show that for every $p \in \mathbb{N}^*$, the function $H_p : \mathfrak{g} \to \mathbb{F}$, defined by $H_p(x) := \frac{1}{p}\,\text{Trace}\, x^p$, is a Casimir function for $\{\cdot\,,\cdot\}_{\mathfrak{g}}$. For all $x,y \in \mathfrak{g}$,

$$\langle \nabla_x H_p \,|\, y\rangle = \frac{\mathrm{d}}{\mathrm{d}t}_{|t=0} H_p(x+ty) = \text{Trace}\left(x^{p-1}y\right) = \langle x^{p-1} \,|\, y\rangle .$$

It follows that $\nabla_x H_p = x^{p-1}$, which implies that $[\nabla_x H_p, x] = 0$, for every $x \in \mathfrak{g}$. In view of (7.15), this means that the functions H_p are Casimir functions of $\{\cdot\,,\cdot\}_{\mathfrak{g}}$.

7.3 Properties of the Lie–Poisson Structure

In this section we first study the symplectic foliation of a Lie–Poisson structure and show how it is related to the coadjoint action. We then explain how the Poisson cohomology of a Lie–Poisson structure is related to Lie algebra cohomology. In Section 7.3.3, we relate the modular class of a Lie–Poisson structure to the modular form of a Lie algebra.

7.3.1 The Symplectic Foliation: Coadjoint Orbits

We have seen in Section 1.3.3 that a Poisson manifold naturally decomposes into symplectic manifolds. We now show that in the case of a Lie–Poisson structure on $\mathfrak{g}^*$, these symplectic manifolds are the coadjoint orbits of $\mathbf{G}$ in $\mathfrak{g}^*$, where $\mathbf{G}$ is any connected Lie group whose Lie algebra is $\mathfrak{g}$; we also show that the Casimir functions are the Ad^*-invariant functions. See Section 5.1.3 for the definition and the basic properties of the coadjoint action and of Ad^*-invariant functions on $\mathfrak{g}^*$.

Proposition 7.7. *Let* $(\mathfrak{g},[\cdot\,,\cdot])$ *be a finite-dimensional Lie algebra and let* $\mathbf{G}$ *be an arbitrary connected Lie group whose Lie algebra is* $\mathfrak{g}$. *The symplectic leaves of the Lie–Poisson structure on* $\mathfrak{g}^*$ *are the coadjoint orbits of* $\mathbf{G}$ *and the Casimir functions are the* Ad^**-invariant functions.*

Proof. We first show that at each point $\xi \in \mathfrak{g}^*$ the vector space of all fundamental vector fields of the coadjoint action of $\mathbf{G}$ on $\mathfrak{g}^*$ at ξ coincides with $\mathrm{Ham}_\xi(\mathfrak{g}^*,\{\cdot\,,\cdot\})$, the vector space of all Hamiltonian vector fields of the Lie–Poisson structure on $\mathfrak{g}^*$, at ξ. To show it, rewrite (7.5) as

$$(\mathscr{X}_G)_\xi\,[F] = \left\langle \xi, -\,\mathrm{ad}_{\mathrm{d}_\xi G}\,\mathrm{d}_\xi F \right\rangle = \left\langle \mathrm{ad}^*_{\mathrm{d}_\xi G}\,\xi, \mathrm{d}_\xi F \right\rangle$$

to find that the Hamiltonian vector field $\mathscr{X}_G$ is given, at $\xi \in \mathfrak{g}^*$, by

$$(\mathscr{X}_G)_\xi = \mathrm{ad}^*_{\mathrm{d}_\xi G}\,\xi\ .$$

For every $x \in \mathfrak{g}$, the linear function x^*, defined by $x^*(\xi) := \langle \xi, x\rangle$ for all $\xi \in \mathfrak{g}^*$, satisfies $\mathrm{d}_\xi x^* = x^*$, for every $\xi \in \mathfrak{g}^*$. It follows that

$$\mathrm{Ham}_\xi\,(\mathfrak{g}^*) = \left\{ \mathrm{ad}^*_{\mathrm{d}_\xi G}\,\xi \mid G \in \mathscr{F}(\mathfrak{g}^*) \right\} = \left\{ \mathrm{ad}^*_x\,\xi \mid x \in \mathfrak{g} \right\}\ ,$$

as we needed to show. Since the coadjoint orbits of $\mathbf{G}$ are connected, this shows that the leaves of the symplectic foliation of the Lie–Poisson structure on $\mathfrak{g}^*$ are precisely the coadjoint orbits. It follows that the Ad^*-invariant functions are precisely the functions which are constant on every leaf of the symplectic foliation. According to Proposition 1.31, they are the Casimir functions of the Lie–Poisson structure on $\mathfrak{g}^*$. □

Suppose now, as in Section 7.2, that we have a non-degenerate symmetric bilinear form $\langle\cdot\,|\,\cdot\rangle$ on $\mathfrak{g}$, and let $\chi : \mathfrak{g} \to \mathfrak{g}^*$ be the resulting isomorphism (see (7.7)). Since χ realizes an isomorphism between the Lie–Poisson structure on $\mathfrak{g}^*$ and the Lie–Poisson structure on $\mathfrak{g}$ with respect to $\langle\cdot\,|\,\cdot\rangle$, Proposition 7.7 leads to the following result.

Proposition 7.8. *Let* $(\mathfrak{g},[\cdot\,,\cdot])$ *be a finite-dimensional quadratic Lie algebra, whose bilinear form is denoted by* $\langle\cdot\,|\,\cdot\rangle$. *Let* $\mathbf{G}$ *be an arbitrary connected Lie group whose Lie algebra is* $\mathfrak{g}$. *The symplectic leaves of the Lie–Poisson structure on* $\mathfrak{g}$ *with respect to* $\langle\cdot\,|\,\cdot\rangle$ *are the adjoint orbits of* $\mathbf{G}$ *and the Casimir functions are the* Ad*-invariant functions.*

The fact that the symplectic foliation of the dual of a Lie algebra is given by the orbits of the coadjoint action shows that, even in the case of a linear Poisson struc-

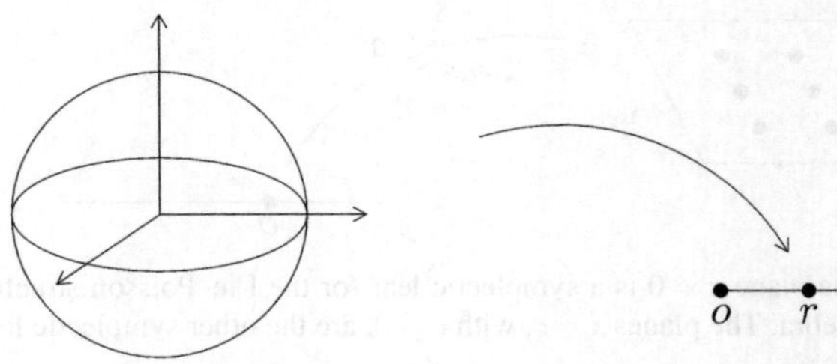

Fig. 7.1 For $r \geqslant 0$ the sphere $x^2+y^2+z^2=r^2$ is a symplectic leaf of the Lie–Poisson structure on $(\mathfrak{so}_3(\mathbb{R}))^*$.

ture, the symplectic foliation can be rather complicated. This is illustrated in the following examples of (real) three-dimensional Lie algebras.

Example 7.9. We start with the Lie–Poisson structure of $(\mathfrak{so}_3(\mathbb{R}))^*$, which is defined by the following brackets

$$[x,y]=z\,,\quad [y,z]=x\,,\quad [z,x]=y\,.$$

The Poisson matrix is given by

$$\begin{pmatrix} 0 & z & -y \\ -z & 0 & x \\ y & -x & 0 \end{pmatrix},$$

and $C:=x^2+y^2+z^2$ is a Casimir function. The rank of the Poisson structure is two at every point, different from the origin o, and at such a point the differential of the Casimir function C is non-zero. According to Proposition 1.32, the symplectic leaves in $\mathbb{R}^3\setminus\{o\}$ are the connected components of the fibers of C, i.e., the spheres $x^2+y^2+z^2=r^2$, with $r\neq 0$. Of course, $\{o\}$ is itself a symplectic leaf, so the symplectic leaves are the concentric spheres $x^2+y^2+z^2=r^2$, with $r\in\mathbb{R}_{\geqslant 0}$, where the origin appears as a sphere of radius $r=0$. See Fig. 7.1.

Example 7.10. We next consider the Lie–Poisson structure of $(\mathfrak{sl}_2(\mathbb{R}))^*$, where the standard $\mathfrak{sl}_2(\mathbb{R})$ brackets

$$[x,y]=z\,,\quad [y,z]=2y\,,\quad [z,x]=2x\,,$$

lead to the following Poisson matrix

$$\begin{pmatrix} 0 & z & -2x \\ -z & 0 & 2y \\ 2x & -2y & 0 \end{pmatrix}.$$

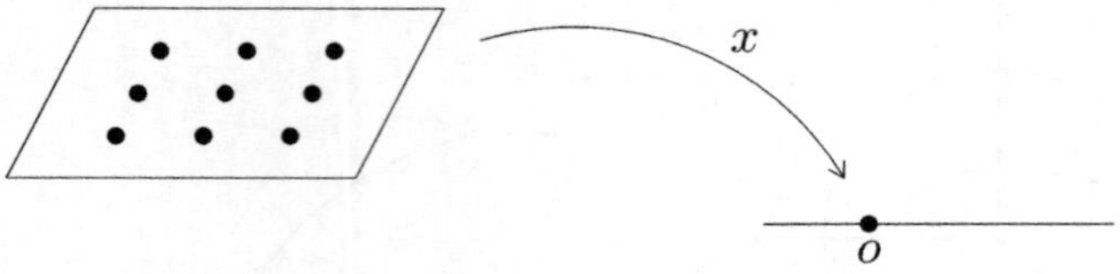

Fig. 7.2 Every point of the plane $x = 0$ is a symplectic leaf for the Lie–Poisson structure on the dual of the Heisenberg algebra. The planes $x = c$, with $c \neq 0$, are the other symplectic leaves.

Clearly, $C := 4xy + z^2$ is a Casimir function and the rank of the Poisson structure is two at every point, except at the origin; also, the differential of C is non-zero, except at the origin. As in the previous example, Proposition 1.32 implies that the symplectic leaves are the connected components of the fibers of C, namely,

(1) The one-sheeted hyperboloids $4xy + z^2 = c$, for $c > 0$;
(2) The two components of the two-sheeted hyperboloids $4xy + z^2 = c$, for $c < 0$;
(3) The two components of the cone $z^2 = -4xy$ minus the origin;
(4) The origin.

Notice that the previous example and the present example are geometrically very different, yet the underlying Lie algebras have the same complexification, namely $\mathfrak{sl}_2(\mathbb{C})$.

Example 7.11. Our third example is the Heisenberg algebra, with Lie brackets

$$[x,y] = 0\,, \quad [y,z] = x\,, \quad [z,x] = 0\,.$$

The Poisson matrix is now given by

$$\begin{pmatrix} 0 & 0 & 0 \\ 0 & 0 & x \\ 0 & -x & 0 \end{pmatrix}.$$

It is immediate that the Casimir functions are precisely the functions which are independent of y and of z. The Poisson structure has now rank zero on the plane $x = 0$, so that each point of this plane is a symplectic leaf; together, these points form the zero locus of the Casimir function x. The other symplectic leaves are two-dimensional, they are the planes $x = c$, with $c \in \mathbb{R}^*$. See Fig. 7.2.

Example 7.12. In the fourth example we show that, for a linear Poisson structure on $\mathbb{R}^3$, non-trivial smooth Casimir functions may even not exist locally (at some points). Take on $\mathbb{R}^3$, with respect to the system of coordinates (x,y,z), the following Poisson matrix,

$$\begin{pmatrix} 0 & 0 & x \\ 0 & 0 & y \\ -x & -y & 0 \end{pmatrix}. \tag{7.16}$$

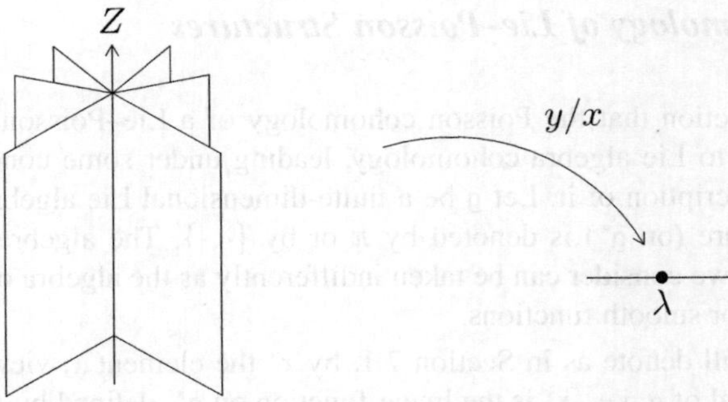

Fig. 7.3 For the open book foliation, which is the symplectic foliation of the Lie–Poisson structure, given by the Poisson matrix (7.16) the leaves are the points of the Z-axis and the half-planes, obtained by removing the Z-axis from the planes passing through it.

On $\mathbb{R}^3$, minus the plane $x = 0$, y/x is a Casimir function, while on $\mathbb{R}^3$, minus the plane $y = 0$, x/y is a Casimir function. The symplectic leaves of dimension 0 are the points on the Z-axis, while the symplectic leaves of dimension 2 are the half-planes, obtained by removing the Z-axis from all the planes which pass through the Z-axis, as follows again from Proposition 1.32. This symplectic foliation is called the *open book foliation*. Taking an arbitrary point on the Z-axis, it is clear that on no neighborhood of it is there a non-constant Casimir function. See Fig. 7.3.

Example 7.13. We consider, in the last example, for a fixed $\alpha \in \mathbb{C}$, the basic Lie brackets

$$[x,y] = 0\,, \quad [y,z] = y\,, \quad [z,x] = \alpha x\,,$$

which lead to the following Poisson matrix for the corresponding Lie–Poisson structure on $\mathbb{C}^3$,

$$\begin{pmatrix} 0 & 0 & -\alpha x \\ 0 & 0 & y \\ \alpha x & -y & 0 \end{pmatrix}$$

A holomorphic function F, defined on an open subset $U \subset \mathbb{C}^3$, is a Casimir function for this Poisson structure if and only if

$$\frac{\partial F}{\partial z} = \alpha x \frac{\partial F}{\partial x} - y \frac{\partial F}{\partial y} = 0\,,$$

on U. A solution of these equations is easily found, giving $F = cxy^{\alpha} + d$, where c and d are integration constants. If $\alpha \notin \mathbb{Q}$, then the level sets of F are not even algebraic varieties.

7.3.2 The Cohomology of Lie–Poisson Structures

We show in this section that the Poisson cohomology of a Lie–Poisson structure is naturally related to Lie algebra cohomology, leading under some conditions to a fairly explicit description of it. Let $\mathfrak{g}$ be a finite-dimensional Lie algebra, whose Lie–Poisson structure (on $\mathfrak{g}^*$) is denoted by π or by $\{\cdot,\cdot\}$. The algebra of functions $\mathscr{F}(\mathfrak{g}^*)$ which we consider can be taken indifferently as the algebra of polynomial, holomorphic or smooth functions.

For $x \in \mathfrak{g}$, we will denote as in Section 7.1, by x^* the element x, viewed as an element of the bidual of $\mathfrak{g}$, i.e., x^* is the linear function on $\mathfrak{g}^*$, defined by $\langle x^*, \xi\rangle := \langle \xi, x\rangle$ for all $\xi \in \mathfrak{g}^*$. Consider the map $\mathfrak{g} \to \mathfrak{X}^1(\mathfrak{g}^*)$, which associates to $x \in \mathfrak{g}$ the Hamiltonian vector field of $-x^*$ with respect to the Lie–Poisson structure. It is a Lie algebra homomorphism, since

$$\mathscr{X}_{[x,y]^*} = \mathscr{X}_{\{x^*,y^*\}} = -\left[\mathscr{X}_{x^*}, \mathscr{X}_{y^*}\right],$$

for all $x, y \in \mathfrak{g}$. Equivalently, the assignment $\mathfrak{g} \times \mathscr{F}(\mathfrak{g}^*) \to \mathscr{F}(\mathfrak{g}^*)$ given by $(x,F) \mapsto -\mathscr{X}_{x^*}[F] = \{x^*, F\}$ defines a representation of $\mathfrak{g}$ on $\mathscr{F}(\mathfrak{g}^*)$. We can therefore consider $H_L^\bullet(\mathfrak{g}, \mathscr{F}(\mathfrak{g}^*))$, the $\mathscr{F}(\mathfrak{g}^*)$-valued Lie algebra cohomology of $\mathfrak{g}$. Recall from Section 4.1.1 that for every $q \in \mathbb{N}$, the space of $\mathscr{F}(\mathfrak{g}^*)$-valued q-cochains of $\mathfrak{g}$ is given by

$$C^q(\mathfrak{g}; \mathscr{F}(\mathfrak{g}^*)) = \operatorname{Hom}(\wedge^q \mathfrak{g}, \mathscr{F}(\mathfrak{g}^*)),$$

while the differential is given, for all $c \in C^q(\mathfrak{g}; \mathscr{F}(\mathfrak{g}^*))$ and all $x_0, \dots, x_q \in \mathfrak{g}$, by

$$\begin{aligned} \delta_L^q(c)(x_0,\dots,x_q) &= \sum_{i=0}^q (-1)^i \left\{x_i^*, c(x_0,\dots,\widehat{x_i},\dots,x_q)\right\} \\ &+ \sum_{0\leqslant i<j\leqslant q} (-1)^{i+j} c\Big([x_i,x_j], x_0,\dots,\widehat{x_i},\dots,\widehat{x_j},\dots,x_q\Big). \end{aligned} \tag{7.17}$$

We show in the following proposition that, for every $q \in \mathbb{N}$, the cohomology space $H_L^q(\mathfrak{g}, \mathscr{F}(\mathfrak{g}^*))$ is canonically isomorphic to the q-th Poisson cohomology space of $(\mathfrak{g}^*, \pi)$.

Proposition 7.14. *Let $\mathfrak{g}$ be a finite-dimensional Lie algebra. The dual $\mathfrak{g}^*$ of $\mathfrak{g}$ is equipped with its algebra $\mathscr{F}(\mathfrak{g}^*)$ of polynomial, holomorphic or smooth functions. The Lie–Poisson bracket on $\mathscr{F}(\mathfrak{g}^*)$ is denoted by π. For every $q \in \mathbb{N}$ there is a canonical isomorphism*

$$H_L^q(\mathfrak{g}, \mathscr{F}(\mathfrak{g}^*)) \simeq H_\pi^q(\mathfrak{g}^*).$$

Proof. We show that the $\mathscr{F}(\mathfrak{g}^*)$-valued Lie algebra cohomology complex of $\mathfrak{g}$ is isomorphic to the Poisson cohomology complex of $(\mathfrak{g}^*, \pi)$. Let Q be a Poisson q-cochain of $(\mathscr{F}(\mathfrak{g}^*), \pi)$, i.e., $Q \in \mathfrak{X}^q(\mathfrak{g}^*)$. We associate to Q the element $\Psi_q(Q) \in C^q(\mathfrak{g}; \mathscr{F}(\mathfrak{g}^*)) = \operatorname{Hom}(\wedge^q \mathfrak{g}, \mathscr{F}(\mathfrak{g}^*))$, defined by

$$\begin{aligned} \Psi_q(Q): \quad \wedge^q \mathfrak{g} &\to \mathscr{F}(\mathfrak{g}^*) \\ x_1 \wedge \dots \wedge x_q &\mapsto Q[x_1^*, \dots, x_q^*]. \end{aligned} \tag{7.18}$$

It is easy to see that for every $q \in \mathbb{N}$ the resulting linear map $\Psi_q : \mathfrak{X}^q(\mathfrak{g}^*) \to \mathrm{Hom}(\wedge^q \mathfrak{g}, \mathscr{F}(\mathfrak{g}^*))$ is an isomorphism. Moreover, a direct comparison of (4.4) and (7.17), using $\left\{x_i^*, x_j^*\right\} = [x_i, x_j]^*$, shows that the following diagram is commutative.

$$\begin{array}{ccc} \mathfrak{X}^q(\mathfrak{g}^*) & \xrightarrow{\ \delta_\pi^q\ } & \mathfrak{X}^{q+1}(\mathfrak{g}^*) \\ \Big\downarrow \Psi_q & & \Big\downarrow \Psi_{q+1} \\ \mathrm{Hom}(\wedge^q \mathfrak{g}, \mathscr{F}(\mathfrak{g}^*)) & \xrightarrow{\ \delta_L^q\ } & \mathrm{Hom}(\wedge^{q+1} \mathfrak{g}, \mathscr{F}(\mathfrak{g}^*)) \end{array}$$

The isomorphisms $(\Psi_q)_{q \in \mathbb{N}}$ therefore induce isomorphisms in cohomology. □

For a representation (ρ, V) of $\mathfrak{g}$, recall that $V^{\mathfrak{g}}$ stands for the space of all $\mathfrak{g}$-invariant elements, i.e., elements $v \in V$ such that $\rho(x, v) = 0$ for every $x \in \mathfrak{g}$. Recall also that we denote by $H_L^\bullet(\mathfrak{g})$ the trivial Lie algebra cohomology (see Example 4.2). Since $V^{\mathfrak{g}}$ is a subrepresentation of V, on which $\mathfrak{g}$ acts trivially, we have linear maps

$$H_L^q(\mathfrak{g}) \otimes V^{\mathfrak{g}} \longrightarrow H_L^q(\mathfrak{g}, V^{\mathfrak{g}}) \longrightarrow H_L^q(\mathfrak{g}, V)\,, \tag{7.19}$$

where the first map is an isomorphism, but the second map is in general neither injective nor surjective. In the case of the above representation of $\mathfrak{g}$ on $\mathscr{F}(\mathfrak{g}^*)$, it leads for every $q \in \mathbb{N}$ to the linear map

$$\iota_q : H_L^q(\mathfrak{g}) \otimes \mathrm{Cas}(\mathfrak{g}^*) \to H_\pi^q(\mathfrak{g}^*)\,, \tag{7.20}$$

since the invariant elements of $\mathscr{F}(\mathfrak{g}^*)$ are the functions $F \in \mathscr{F}(\mathfrak{g}^*)$ such that $\{x^*, F\} = 0$ for every $x \in \mathfrak{g}$, i.e., the Casimir functions of $(\mathfrak{g}^*, \pi)$. Recall from Proposition 7.7 that these functions are the Ad^*-invariant functions. For a semi-simple Lie algebra, a non-trivial property on its Lie algebra cohomology (with coefficients in a finite-dimensional representation) implies that ι_q is an isomorphism, for every $q \in \mathbb{N}$, as we show in the following proposition.

Proposition 7.15. *Let $\mathfrak{g}$ be a semi-simple Lie algebra. We denote by $\mathscr{F}(\mathfrak{g}^*)$ the algebra of polynomial functions on $\mathfrak{g}^*$ and by $\mathrm{Cas}(\mathfrak{g}^*) \subset \mathscr{F}(\mathfrak{g}^*)$ the subalgebra of polynomial Casimir functions. For every $q \in \mathbb{N}$, the map ι_q, given by (7.20), is an isomorphism:*

$$\iota_q : H_L^q(\mathfrak{g}) \otimes \mathrm{Cas}(\mathfrak{g}^*) \simeq H_\pi^q(\mathfrak{g}^*)\,.$$

Proof. We denote by $\mathscr{F}_k(\mathfrak{g}^*)$ the vector space of homogeneous polynomials of degree k on $\mathfrak{g}^*$, where $k \in \mathbb{N}$. It is a subrepresentation of $\mathscr{F}(\mathfrak{g}^*)$, and

$$\mathscr{F}(\mathfrak{g}^*) = \bigoplus_{k \in \mathbb{N}} \mathscr{F}_k(\mathfrak{g}^*)\,,$$

which implies that the $\mathscr{F}(\mathfrak{g}^*)$-valued Lie algebra cohomology of $\mathfrak{g}$ decomposes as a direct sum,

$$H_L^q(\mathfrak{g}, \mathscr{F}(\mathfrak{g}^*)) = \bigoplus_{k \in \mathbb{N}} H_L^q(\mathfrak{g}, \mathscr{F}_k(\mathfrak{g}^*))\,,$$

for every $q \in \mathbb{N}$. Since the Lie–Poisson structure on $\mathfrak{g}^*$ is linear, a function $F \in \mathscr{F}(\mathfrak{g}^*)$ is a Casimir function if and only if for every $k \in \mathbb{N}$ its homogeneous component of degree k is a Casimir function. Setting $\mathrm{Cas}_k(\mathfrak{g}^*) := \mathrm{Cas}(\mathfrak{g}^*) \cap \mathscr{F}_k(\mathfrak{g}^*)$, it follows that

$$H_L^q(\mathfrak{g}) \otimes \mathrm{Cas}(\mathfrak{g}^*) = \bigoplus_{k \in \mathbb{N}} H_L^q(\mathfrak{g}) \otimes \mathrm{Cas}_k(\mathfrak{g}^*) ,$$

for every $q \in \mathbb{N}$. Moreover, for every $q, k \in \mathbb{N}$, the linear map ι_q restricts to yield a map

$$\iota_{q,k} : H_L^q(\mathfrak{g}) \otimes \mathrm{Cas}_k(\mathfrak{g}^*) \to H_L^q(\mathfrak{g}, \mathscr{F}_k(\mathfrak{g}^*)) .$$

Now, according to [95, Th. 11], for every finite-dimensional representation V of a semi-simple Lie algebra $\mathfrak{g}$, the natural linear map $H_L^q(\mathfrak{g}) \otimes V^{\mathfrak{g}} \to H_L^q(\mathfrak{g}, V)$ described in (7.19) is bijective for every $q \in \mathbb{N}$. In particular, for every $k \in \mathbb{N}$, $\iota_{q,k}$ is an isomorphism, which, in turn, implies that ι_q is an isomorphism, as was to be shown. □

7.3.3 The Modular Class of a Lie–Poisson Structure

The *modular form* of a finite-dimensional Lie algebra $\mathfrak{g}$ is the linear form $\mu_{\mathfrak{g}}$ on $\mathfrak{g}$, defined for $x \in \mathfrak{g}$ by

$$\mu_{\mathfrak{g}}(x) := \mathrm{Trace}(\mathrm{ad}_x) .$$

A Lie algebra is said to be *unimodular* when its modular form vanishes. For $x, y \in \mathfrak{g}$,

$$\mu_{\mathfrak{g}}([x,y]) = \mathrm{Trace}(\mathrm{ad}_{[x,y]}) = \mathrm{Trace}([\mathrm{ad}_x, \mathrm{ad}_y]) = 0 , \tag{7.21}$$

so that $\mu_{\mathfrak{g}}$ vanishes on $[\mathfrak{g}, \mathfrak{g}]$. Every Lie algebra which satisfies $[\mathfrak{g}, \mathfrak{g}] = \mathfrak{g}$ is therefore unimodular; in particular, if $\mathfrak{g}$ is semi-simple, then $\mathfrak{g}$ is unimodular. If $\mathfrak{g}$ is nilpotent, then $\mathfrak{g}$ is unimodular, since ad_x is nilpotent, hence traceless, for every $x \in \mathfrak{g}$. For a different reason, quadratic Lie algebras are unimodular as well. However, solvable Lie algebras are not unimodular in general, for instance, the two-dimensional Lie algebra defined, on a basis (e, f) of $\mathfrak{g}$ by $[e, f] := f$, is not unimodular, since $\mathrm{Trace}(\mathrm{ad}_e) = 1$.

In the following proposition, we relate the modular form $\mu_{\mathfrak{g}}$ of $\mathfrak{g}$ with the modular vector field Φ of the Poisson manifold $(\mathfrak{g}^*, \pi)$, where π stands for the canonical Lie–Poisson structure on $\mathfrak{g}^*$. The volume form which we choose on $\mathfrak{g}^*$ is any translation invariant volume form (see Remark 4.13). Φ is a constant vector field on $\mathfrak{g}^*$, hence corresponds in a canonical way to an element of $\mathfrak{g}^*$.

Proposition 7.16. *Let $\mathfrak{g}$ be a finite-dimensional Lie algebra. Interpreted as an element of $\mathfrak{g}^*$, the modular vector field of $(\mathfrak{g}^*, \pi)$ with respect to any translation invariant volume form is, up to a sign the modular form of $\mathfrak{g}$.*

Proof. Let $(e_1, \dots, e_d)$ be a basis of $\mathfrak{g}$ and let $(\xi_1, \dots, \xi_d)$ be the dual basis of $\mathfrak{g}^*$. We use the functions $x_1 = e_1^*, \dots, x_d = e_d^*$ as linear coordinates on $\mathfrak{g}^*$. According to (4.22), the divergence of the Lie–Poisson structure π on $\mathfrak{g}^*$ is given by

$$\mathrm{Div}(\pi) = \sum_{1\leqslant i,j\leqslant d} \frac{\partial}{\partial x_j}\{x_i,x_j\}\frac{\partial}{\partial x_i} = \sum_{1\leqslant i,j\leqslant d}\left\langle [e_i,e_j]^*,\xi_j\right\rangle\frac{\partial}{\partial x_i}\;.$$

Since $\Phi = -\mathrm{Div}(\pi)$ (see Proposition 4.17), it follows that

$$\Phi(x) = -\sum_{1\leqslant i,j\leqslant d}\left\langle [e_i,e_j]^*,\xi_j\right\rangle x_i = -\sum_{1\leqslant j\leqslant d}\left\langle \xi_j,[x,e_j]\right\rangle = -\mathrm{Trace}(\mathrm{ad}_x)\;,$$

for every $x = \sum_{i=1}^d x_i e_i \in \mathfrak{g}$. This shows that $\Phi = -\mu_{\mathfrak{g}}$. □

7.4 Affine Poisson Structures

We now turn to a combination of constant and linear Poisson structures, which we call affine Poisson structures.

Definition 7.17. A Poisson structure $\{\cdot,\cdot\}$ on a finite-dimensional vector space V is called an *affine Poisson structure* if, for every pair of affine functions F and G on V, their Poisson bracket $\{F,G\}$ is an affine function on V.

Equivalently, for every pair of linear functions F and G on V, their Poisson bracket $\{F,G\}$ is an affine function on V.

Proposition 7.18. *For every finite-dimensional vector space V, there is a one-to-one correspondence between affine Poisson structures on V and pairs of compatible Poisson structures (π_0,π_1) on V, where π_0 is a constant Poisson structure and π_1 is a linear Poisson structure.*

Proof. Suppose that π is a skew-symmetric biderivation of $\mathscr{F}(V)$ such that for every pair of linear functions F and G on V, one has that $\pi[F,G]$ is an affine function on V. For such functions, let $\pi_0[F,G]$ and $\pi_1[F,G]$ denote the constant, respectively linear parts of $\pi[F,G]$. We view π_0 and π_1 as bilinear maps on V^* and we extend them to biderivations of $\mathscr{F}(V)$, still denoted by π_0 and π_1, so that $\pi = \pi_0 + \pi_1$. According to Chapter 6, π_0 is a constant Poisson structure. In terms of the Schouten bracket $[\cdot,\cdot]_S$, one has that π is a Poisson structure if and only if $[\pi,\pi]_S = 0$, which is equivalent to the vanishing of $[\pi,\pi]_S[F,G,H]$ for all linear functions F,G and H on V. Since $[\pi_0,\pi_0]_S = 0$, because π_0 is a Poisson structure, we have that

$$[\pi,\pi]_S[F,G,H] = [\pi_1,\pi_1]_S[F,G,H] + 2\,[\pi_0,\pi_1]_S[F,G,H]\;.$$

In view of (3.37), the first term in this expression is a linear function on V, while the second term is a constant function on V. Thus, $[\pi,\pi]_S = 0$ if and only if $[\pi_1,\pi_1]_S = 0$ (i.e., π_1 is a linear Poisson structure) and $[\pi_0,\pi_1]_S = 0$ (i.e., π_0 and π_1 are compatible). It follows that the map, which sends an affine Poisson structure π to its constant and linear parts (π_0,π_1), yields the announced one-to-one correspondence. □

In view of Proposition 7.18, it is natural to look for a Lie algebraic interpretation of affine Poisson structures. Indeed, if π is an affine Poisson structure on V^*, then V^* is the dual of a Lie algebra V and π_0, restricted to V, is a skew-symmetric bilinear form on V.

Proposition 7.19. *For every finite-dimensional vector space V, there is a one-to-one correspondence between affine Poisson structures on V^* and pairs $([\cdot,\cdot],c)$ where $[\cdot,\cdot]$ is a Lie algebra structure on V and c is a 2-cocycle in the trivial Lie algebra cohomology of $(V,[\cdot,\cdot])$.*

Proof. According to Proposition 7.18, we need to show that there is a one-to-one correspondence between pairs (π_0,π_1) of compatible Poisson structures on V^*, where π_0 is constant and π_1 is linear, and pairs $([\cdot,\cdot],c)$, where $[\cdot,\cdot]$ is a Lie algebra structure on V and c is a 2-cocycle in the trivial Lie algebra cohomology of $(V,[\cdot,\cdot])$. Let (π_0,π_1) be a pair of Poisson structures on V^*, where π_0 is constant and π_1 is linear. Let $[\cdot,\cdot]$ be the Lie algebra structure on V, defined[2] by $[v,w]^* := \pi_1[v^*,w^*]$, for all $v,w\in V$, and let $c(v,w) := \pi_0[v^*,w^*]$, which defines a linear map $c : V\wedge V \to \mathbb{F}$. We show that π_0 and π_1 are compatible if and only if c is a 2-cocycle in the trivial Lie algebra cohomology of $\mathfrak{g} := V$; since the Poisson structures π_0 and π_1 can be reconstructed from the pair $([\cdot,\cdot],c)$, this will prove the proposition. We have for all $u,v,w\in V=\mathfrak{g}$,

$$\begin{aligned}[\pi_0,\pi_1]_S(u^*,v^*,w^*) &= \pi_0[\pi_1[u^*,v^*],w^*] + \circlearrowleft (u,v,w)\\ &= c([u,v],w) + \circlearrowleft (u,v,w)\\ &= -\delta_L^2(c)(u,v,w)\,,\end{aligned}$$

see Example 4.2, in particular formula (4.2) for the Lie algebra coboundary operator δ_L, in the case of a trivial representation. It follows that $[\pi_0,\pi_1]_S=0$ if and only if c is a Lie 2-cocycle, $\delta_L^2(c)=0$. □

In view of the above proposition, affine Poisson structures are also called *modified canonical Poisson structures* or *modified Lie–Poisson structures*.

If $\mathfrak{g}$ is semisimple, then $H_L^2(\mathfrak{g})$ is trivial (see Lemma 4.1), so that, if c is a Lie 2-cocycle, then c is a Lie 2-coboundary, $c=\delta_L^1(c')$, written out, $c(x_i,x_j)=c'([x_j,x_i])$ and we see that the affine Poisson structure, corresponding to $([\cdot,\cdot],c)$, is nothing but the Lie–Poisson structure associated to $[\cdot,\cdot]$, with x_i^* replaced by $x_i^*+c'(x_i)$, i.e., both Poisson structures are the same up to an *affine* change of variables (a translation).

Affine Poisson structures are, in general, also closely related to Lie–Poisson structures in a different way. Namely, given a Lie algebra $(\mathfrak{g},[\cdot,\cdot])$ and a Lie 2-cocycle c, as in Proposition 7.19, we consider on $\mathfrak{g}\times\mathbb{F}$ the Lie bracket

$$[(x,a),(y,b)]_c := ([x,y],c(x,y))\,. \tag{7.22}$$

For every $(x_i,a_i)\in\mathfrak{g}\times\mathbb{F}$, where $i=1,2,3$, one has that

[2] As in the case of Lie algebras, if $v\in V$, then v^* denotes its image under the canonical isomorphism $V\simeq(V^*)^*$.

$$[[(x_1,a_1),(x_2,a_2)]_c,(x_3,a_3)]_c = [([x_1,x_2],c(x_1,x_2)),(x_3,a_3)]_c$$
$$= ([[x_1,x_2],x_3],c([x_1,x_2],x_3)) ,$$

so that the Jacobi identity for $[\cdot,\cdot]$ and the cocycle condition for c imply the Jacobi identity for $[\cdot,\cdot]_c$, i.e., $(\mathfrak{g}\times\mathbb{F},[\cdot,\cdot]_c)$ is a Lie algebra. In the following proposition we show that the Lie–Poisson structure $\{\cdot,\cdot\}_c$ on $(\mathfrak{g}\times\mathbb{F})^*$, restricted to a hyperplane, is isomorphic to the affine Poisson structure $\{\cdot,\cdot\}$ on $\mathfrak{g}$, defined by the 2-cocycle c.

Proposition 7.20. *Let $(\mathfrak{g},[\cdot,\cdot])$ be a finite-dimensional Lie algebra and let c be a 2-cocycle in the trivial Lie algebra cohomology of $\mathfrak{g}$. Denote the Lie–Poisson structure on $(\mathfrak{g}\times\mathbb{F})^*$, associated to $(\mathfrak{g}\times\mathbb{F},[\cdot,\cdot]_c)$, by $\{\cdot,\cdot\}_c$, and denote by $\{\cdot,\cdot\}$ the affine Poisson structure on $\mathfrak{g}^*$, with constant term c.*

(1) *The linear function $1^* : (\mathfrak{g}\times\mathbb{F})^* \to \mathbb{F}$, defined by $\phi \mapsto \phi(0,1)$, is a Casimir function of $\{\cdot,\cdot\}_c$;*
(2) *The hyperplane $N := \{\phi \in (\mathfrak{g}\times\mathbb{F})^* \mid \phi(0,1) = 1\}$ is a Poisson submanifold of $(\mathfrak{g}\times\mathbb{F})^*$;*
(3) *The isomorphism $\Psi : N \to \mathfrak{g}^*$, defined for $\phi \in N$ by $\phi \mapsto \phi(\cdot,0)$, is a Poisson isomorphism between $(N,\{\cdot,\cdot\}_c)$ and $(\mathfrak{g}^*,\{\cdot,\cdot\})$.*

Proof. We first show that 1^* is a Casimir function of $\{\cdot,\cdot\}_c$. Notice that $1^* = (0,1)$, as an element of $\mathfrak{g}\times\mathbb{F}$. For $(\xi,a) \in \mathfrak{g}^*\times\mathbb{F}$, it follows that $\mathrm{d}_{(\xi,a)}1^* = 1^* = (0,1)$, which belongs to the center of $[\cdot,\cdot]_c$ (see (7.22)). It follows that for every function F on $(\mathfrak{g}\times\mathbb{F})^*$,

$$\{F,1^*\}_c(\xi,a) = \langle(\xi,a),[\mathrm{d}_{(\xi,a)}F,(0,1)]_c\rangle = 0 .$$

This shows *(1)*. Since the hyperplane N, as defined in *(2)*, is a level set of the Casimir function 1^*, it is a Poisson submanifold of $(\mathfrak{g}\times\mathbb{F})^*$, which yields *(2)*. It is clear that Ψ is an isomorphism between the affine space N and the vector space $\mathfrak{g}^*$. In order to prove that Ψ is a Poisson morphism, it is sufficient to prove that $\{x^*\circ\Psi,y^*\circ\Psi\}_c = \{x^*,y^*\}\circ\Psi$, for every $x,y\in\mathfrak{g}$, where, as above, stars stand for the corresponding linear functions on $\mathfrak{g}^*$. By a slight abuse of notation, we will also consider x^* as a linear function on $(\mathfrak{g}\times\mathbb{F})^*$, where we put $x^*(\phi) := \phi(x,0)$, for all $\phi\in(\mathfrak{g}\times\mathbb{F})^*$. With this abuse of notation, $x^*\circ\Psi = x^*$, and $\{x^*,y^*\}\circ\Psi = [x,y]^* + c(x,y)$. For $\{x^*\circ\Psi,y^*\circ\Psi\}_c$, which is now written as $\{x^*,y^*\}_c$, let $\phi\in N$ and compute

$$\{x^*,y^*\}_c(\phi) = \langle\phi,[\mathrm{d}_\phi x^*,\mathrm{d}_\phi y^*]_c\rangle = \langle\phi,[(x,0),(y,0)]_c\rangle$$
$$= \langle\phi,([x,y],c(x,y))\rangle = [x,y]^*(\phi) + c(x,y) ,$$

which is $\{x^*,y^*\}\circ\Psi$, evaluated at ϕ. This shows that Ψ is a Poisson morphism; since Ψ is an isomorphism, it is a Poisson isomorphism, which is the content of *(3)*.
□

As we have shown earlier in this section, every affine Poisson structure which is constructed by using a Lie 2-cocycle which is a coboundary, is isomorphic to the original Lie–Poisson structure, by an affine change of variables, which implies that both Poisson manifolds have the same symplectic foliation, up to a translation. We show in the following example that when the Lie 2-cocycle is not a coboundary, these symplectic foliations may be quite different.

Example 7.21. Consider on $\mathbb{R}^3$ the linear and affine Poisson structures π_1 and π, whose Poisson matrices are given by

$$\begin{pmatrix} 0 & 0 & -x \\ 0 & 0 & y \\ x & -y & 0 \end{pmatrix} \quad \text{and} \quad \begin{pmatrix} 0 & 1 & -x \\ -1 & 0 & y \\ x & -y & 0 \end{pmatrix} .$$

The underlying Lie algebra of both Poisson structures is defined by the brackets $[x,y]=0$, and $[y,z]=y$ and $[z,x]=x$. The corresponding Lie 2-cocycle is given by $c(x,y)=1$ and $c(x,z)=c(y,z)=0$. Notice that π is a regular Poisson structure, of rank two, while the rank of π_1 vanishes at the origin. Clearly, xy is a Casimir function of π_1, while $xy+z$ is a Casimir function of π. The symplectic leaves of π_1 are the points on the Z-axis, the two half-planes obtained by removing the Z-axis from the plane $x=0$, the two half-planes obtained by removing the Z-axis from the plane $y=0$, and the connected components of the hyperbolic cylinders $xy=c$, with $c\in\mathbb{R}^*$. For the affine Poisson structure π, the symplectic leaves are the parallel hyperboloids $xy+z=c$, with $c\in\mathbb{R}$.

7.5 The Linearization of Poisson Structures

According to Weinstein's splitting theorem (Theorem 1.25), every Poisson manifold (M,π) is, in the neighborhood of each of its points, isomorphic to the product of a symplectic manifold and a Poisson manifold whose Poisson structure vanishes at a point. According to Darboux's theorem (Theorem 1.26), the symplectic Poisson structure can be written in a canonical form in well-chosen coordinates. For Poisson structures which vanish at a point m, there is no canonical form; we will explain in this section that, under some conditions, such a Poisson structure is isomorphic in a neighborhood of m to a linear Poisson structure, which is canonically associated to the Poisson structure at m. This linear Poisson structure is introduced in the following proposition.

Proposition 7.22. *Let (M,π) be a real or complex Poisson manifold, and suppose that $m\in M$ is a point at which π vanishes. There exists a unique linear Poisson structure $\{\cdot,\cdot\}_1=\pi_1$ on T_mM, such that, for every neighborhood U of m in M and for all $F,G\in\mathscr{F}(U)$,*

$$\{\mathrm{d}_mF,\mathrm{d}_mG\}_1=\mathrm{d}_m\{F,G\}\ , \tag{7.23}$$

where the elements d_mF, d_mG and $\mathrm{d}_m\{F,G\}$ of T^*_mM are viewed as (linear) functions on T_mM.

Suppose that $\Psi : (M,\pi) \to (M',\pi')$ is a Poisson map, where (M',π') is also a Poisson manifold. Then π' vanishes at $\Psi(m)$ and $\mathrm{d}_m\Psi : (T_mM,\pi_1) \to (T_{\Psi(m)}M',\pi_1')$ is a Poisson map, where the linear Poisson structure π_1' on $T_{\Psi(m)}M'$ is defined as in (7.23).

We call the linear Poisson structure π_1 on T_mM the *linearized Poisson structure* of π at m.

Proof. Every linear function on T_mM is an element of T_m^*M, and is therefore of the form d_mF for some function F, defined in a neighborhood of m in M. Since a Poisson structure on a vector space (here T_mM) is uniquely determined by its value on linear functions, it follows that there exists at most one Poisson structure π_1 on T_mM, satisfying (7.23). In order to establish the existence of this Poisson structure, we use (7.23); namely for $\xi,\eta \in T_m^*M$, which we view as linear functions on T_mM, we define $\{\xi,\eta\}_1 := \mathrm{d}_m\{F,G\}$, where F and G are arbitrary local functions, whose differentials at m are ξ and η respectively. In order to show that $\mathrm{d}_m\{F,G\}$ does not depend on the choice of F and G, we use a system of local coordinates $(x_1,\dots,x_d)$, centered at m: writing

$$\{F,G\} := \sum_{i,j=1}^{d} x_{ij}\frac{\partial F}{\partial x_i}\frac{\partial G}{\partial x_j}, \tag{7.24}$$

where $x_{ij} := \{x_i,x_j\}$ for all $1 \leqslant i,j \leqslant d$, the assumption that π vanishes at m amounts to $x_{ij}(m) = 0$ for all i,j, so that the differential of both sides of (7.24) at m is given by

$$\mathrm{d}_m\{F,G\} = \sum_{i,j=1}^{d} \frac{\partial F}{\partial x_i}(m)\frac{\partial G}{\partial x_j}(m)\,\mathrm{d}_m x_{ij}\,.$$

This gives the required independence of the chosen functions F and G, since

$$\frac{\partial F}{\partial x_i}(m) = \left\langle \mathrm{d}_mF, \left(\frac{\partial}{\partial x_i}\right)_m \right\rangle = \left\langle \xi, \left(\frac{\partial}{\partial x_i}\right)_m \right\rangle,$$

and similarly for the partial derivatives of G at m. The Jacobi identity for the Poisson structure π implies the Jacobi identity for π_1 since it follows at once from (7.23) that

$$\{\xi,\{\eta,\zeta\}_1\}_1 = \mathrm{d}_m\{F,\{G,H\}\}\ ,$$

where F,G and H are arbitrary functions, defined in a neighborhood of m, whose differentials at m are ξ,η and ζ, respectively. This shows that there exists a (unique) Poisson structure π_1 on T_m^*M, satisfying (7.23) for all open subsets U of M which contain m and for all $F,G \in \mathscr{F}(U)$.

Suppose now that (M,π) and (M',π') are two Poisson manifolds and that $\Psi : M \to M'$ is a Poisson map. If $\pi_m = 0$, then $\pi'_{\Psi(m)} = 0$ and we can consider the linearized Poisson structures at T_mM and at $T_{\Psi(m)}M'$; we denote them by $\{\cdot,\cdot\}_1$ and $\{\cdot,\cdot\}_1'$ respectively. We show that $\mathrm{d}_m\Psi : (T_mM,\{\cdot,\cdot\}_1) \to (T_{\Psi(m)}M',\{\cdot,\cdot\}_1')$ is a Poisson map. To do this, it suffices to show that for all linear functions L_1,L_2

on $T_{\Psi(m)}M'$,

$$\{L_1,L_2\}_1' \circ \mathrm{d}_m\Psi = \{L_1 \circ \mathrm{d}_m\Psi, L_2 \circ \mathrm{d}_m\Psi\}_1 . \tag{7.25}$$

Given such functions L_1 and L_2, there exist functions F and G, defined on a neighborhood of $\Psi(m)$ in M', such that $\mathrm{d}_{\Psi(m)}F = L_1$ and $\mathrm{d}_{\Psi(m)}G = L_2$. Since Ψ is a Poisson map, $\{F,G\}' \circ \Psi = \{F \circ \Psi, G \circ \Psi\}$, which we differentiate at m, giving

$$\mathrm{d}_{\Psi(m)}\{F,G\}' \circ \mathrm{d}_m\Psi = \mathrm{d}_m\{F \circ \Psi, G \circ \Psi\} .$$

Written in terms of the linearized Poisson structures, this becomes

$$\begin{aligned}\left\{\mathrm{d}_{\Psi(m)}F, \mathrm{d}_{\Psi(m)}G\right\}_1' \circ \mathrm{d}_m\Psi &= \{\mathrm{d}_m(F \circ \Psi), \mathrm{d}_m(G \circ \Psi)\}_1 \\ &= \left\{\mathrm{d}_{\Psi(m)}F \circ \mathrm{d}_m\Psi, \mathrm{d}_{\Psi(m)}G \circ \mathrm{d}_m\Psi\right\}_1 ,\end{aligned}$$

which is (7.25), since $L_1 = \mathrm{d}_{\Psi(m)}F$ and $L_2 = \mathrm{d}_{\Psi(m)}G$. □

Definition 7.23. Let (M,π) be a Poisson manifold, suppose that m is a point of M at which π vanishes and denote the linearized Poisson structure of π at m by π_1. We say that π is *linearizable* at m, if there exists a Poisson diffeomorphism Ψ, between a neighborhood of m in (M,π) and of the origin o in (T_mM,π_1), with $\Psi(m) = o$.

Example 7.24. Let m be a point of a Poisson manifold (M,π). Suppose that π vanishes at m and that the linearized Poisson structure of π at m is the trivial Poisson structure. Then π is linearizable at m if and only if π is zero in a neighborhood of m.

Example 7.25. If π is a linear Poisson structure on a finite-dimensional vector space V (over $\mathbb{R}$ or $\mathbb{C}$), then π is linearizable at the origin o. In fact, the canonical isomorphism $V \simeq T_oV$ is a Poisson isomorphism, where V is equipped with π and where T_oV is equipped with the linearized Poisson structure of π at o.

Example 7.26. Let $\pi = \{\cdot,\cdot\}$ be a non-trivial Poisson structure on a finite-dimensional vector space V, such that for every pair of linear functions on V, their Poisson bracket is either zero or it is a homogeneous function of degree $k \geqslant 2$ (in the language of Section 8.1, π is a homogeneous Poisson structure of degree k). Then π vanishes at the origin $o \in V$ and the linearized Poisson structure $\{\cdot,\cdot\}_1$ is trivial, since for all linear functions F, G on V,

$$\{\mathrm{d}_oF, \mathrm{d}_oG\}_1 = \mathrm{d}_o\{F,G\} = 0 ,$$

because $\{F,G\}$ is either zero or is homogeneous of degree $k \geqslant 2$. Since π is non-trivial and is homogeneous, there is no neighborhood of o on which π is the trivial Poisson bracket. It follows from Example 7.24 that π is not linearizable at o.

Example 7.27. Let F be a smooth or holomorphic function, defined on a neighborhood U of the origin o of $\mathbb{F}^2$. We assume that F vanishes at o, but that the restriction of F to every neighborhood of o is different from zero. Consider the Poisson structure on U, defined in terms of the standard coordinates x,y of $\mathbb{F}^2$ by $\{x,y\} = F$. We claim that π is linearizable at o if and only if $\mathrm{d}_oF \neq 0$.

Let us prove the claim. First, if $\mathrm{d}_o F = 0$, then $\{\mathrm{d}_o x, \mathrm{d}_o y\}_1 = \mathrm{d}_o F = 0$, so that the linearized Poisson structure is trivial; in view of Example 7.24, π is in this case not linearizable at o. Assume now that $\mathrm{d}_o F \neq 0$ and consider the vector field

$$\mathscr{V} := \frac{\partial F}{\partial x}\frac{\partial}{\partial y} - \frac{\partial F}{\partial y}\frac{\partial}{\partial x},$$

which does not vanish at o. By the straightening theorem (Theorem B.7), there exists a function G, defined in a neighborhood of o, such that $\mathscr{V}[G] = 1$. Since

$$\frac{\partial F}{\partial x}\frac{\partial G}{\partial y} - \frac{\partial F}{\partial y}\frac{\partial G}{\partial x} = \mathscr{V}[G] = 1,$$

we have on the one hand that $\mathrm{d}_o F$ and $\mathrm{d}_o G$ are linearly independent, and on the other hand that $\{F, G\} = F$. Thus, (F, G) is a system of coordinates on a neighborhood of o and π is linear in terms of these coordinates. According to Example 7.25, π is linearizable at o.

Remark 7.28. For $i = 1, 2$, let (M_i, π_i) be a Poisson manifold, whose Poisson structure vanishes at $m_i \in M_i$. Let $\Psi : M_1 \to M_2$ be an isomorphism of Poisson manifolds with $\Psi(m_1) = m_2$. Then the tangent map $T_m\Psi : T_{m_1}M_1 \to T_{m_2}M_2$ is an isomorphism of Poisson manifolds, where both tangent spaces are equipped with the linearized Poisson structures of π_1 and π_2.

In particular, if two finite-dimensional vector spaces V_1 and V_2, equipped with a linear Poisson structure, are isomorphic as Poisson manifolds via a map Ψ which sends the origin o of V_1 to the origin of V_2, then the linear map $\mathrm{d}_o\Psi : V_1 \simeq T_oV_1 \to T_oV_2 \simeq V_2$ is a Poisson isomorphism. In other words, if there exists (in the neighborhood of the origin) a Poisson isomorphism between two linear Poisson structures, then a linear Poisson isomorphism between them also exists.

The condition that a Poisson structure π which vanishes at $m \in M$ is linearizable at m can be expressed as the existence of a neighborhood U of m in M and of a diffeomorphism Ψ between U and a neighborhood of the origin o in T_mM, with $\Psi(m) = o$, such that

$$\{\xi \circ \Psi, \eta \circ \Psi\} = \{\xi, \eta\}_1 \circ \Psi,$$

for all $\xi, \eta \in T_m^*M$. Denoting for $\ell \in \mathbb{N}$ by $\mathscr{I}_m^{(\ell+1)}(U)$ the ideal of $\mathscr{F}(U)$, consisting of all functions which vanish, together with all their partial derivatives up to order ℓ, at the point m, leads to the following weaker notion of linearizability.

Definition 7.29. Let (M, π) be a Poisson manifold, and suppose that $m \in M$ is a point at which π vanishes. Then π is said to be *linearizable* at m up to *order* $\ell \in \mathbb{N}^*$ if there exists a neighborhood U of m in M and a diffeomorphism Ψ between U and a neighborhood of the origin o in T_mM, with $\Psi(m) = o$, such that

$$\{\xi \circ \Psi, \eta \circ \Psi\} - \{\xi, \eta\}_1 \circ \Psi \in \mathscr{I}_m^{(\ell+1)}(U),$$

for all $\xi, \eta \in T_m^*M$.

Example 7.30. It is clear from the definition that every Poisson structure which is linearizable at a point is at that point linearizable up to order ℓ, for all $\ell \in \mathbb{N}^*$. However, the converse is not true, in general. To show this, we define a (smooth) Poisson structure π on $\mathbb{R}^2$ by $\{x,y\} := f(x)$, where $f(x) := e^{-x^{-2}}$ for $x \neq 0$ and $f(0) := 0$. The fact that f vanishes at 0, together with all its derivatives at 0, implies that $\{x,y\} \in \mathscr{I}_o^{(\ell+1)}(\mathbb{R}^2)$, for all $\ell \in \mathbb{N}$, where o denotes the origin of $\mathbb{R}^2$. It follows on the one hand that the linearized Poisson structure at o is trivial, and on the other hand that π is linearizable up to order ℓ, for all $\ell \in \mathbb{N}^*$. However, the Poisson structure π is not linearizable at o, since there is no neighborhood of o on which π is zero.

In order to clarify and compare the different notions of linearizability at a point m in a Poisson manifold (M,π), we consider a coordinate chart (U,x) centered at m, we define, for $1 \leqslant i,j,k \leqslant d$, the constants $c_{ij}^k := \frac{\partial}{\partial x_k}\{x_i,x_j\}(m)$ and we consider the following four statements.

$$\{x_i,x_j\} - \sum_{k=1}^{d} c_{ij}^k x_k \in \mathscr{I}_m^{(2)}(U)\,, \tag{7.26}$$

$$\{\tilde{x}_i,\tilde{x}_j\}_1 = \sum_{k=1}^{d} c_{ij}^k \tilde{x}_k\,, \tag{7.27}$$

$$\{y_i,y_j\} = \sum_{k=1}^{d} c_{ij}^k y_k\,, \tag{7.28}$$

$$\{y_i,y_j\} - \sum_{k=1}^{d} c_{ij}^k y_k \in \mathscr{I}_m^{(\ell+1)}(V)\,. \tag{7.29}$$

The inclusion (7.26) holds for all $1 \leqslant i,j \leqslant d$ if and only if π vanishes at m. In this case, the linearized Poisson structure π_1 at m is given by (7.27), where $\tilde{x}_1,\dots,\tilde{x}_d$ are the linear coordinates on T_mM, defined by $\tilde{x}_i := \mathrm{d}_m x_i$, for $i = 1,\dots,d$. Still assuming that π vanishes at m, there exists a coordinate chart (V,y) for M, centered at $m \in M$, such that (7.28) (respectively (7.29)) holds for all $1 \leqslant i,j \leqslant d$, if and only if π is linearizable at m (respectively π is linearizable at m up to order ℓ).

According to Proposition 7.3, there is a one-to-one correspondence between linear Poisson structures on T_mM and Lie algebra brackets on T_m^*M. It follows that there corresponds to the linearized Poisson structure at m (where m is a point at which π vanishes) a Lie algebra structure $[\cdot\,,\cdot]$ on T_m^*M. It satisfies, by construction,

$$[\mathrm{d}_m F, \mathrm{d}_m G] = \{\mathrm{d}_m F, \mathrm{d}_m G\}_1 = \mathrm{d}_m\{F,G\}\,,$$

for all functions F and G, defined on an arbitrary neighborhood of m in M. The structure of this Lie algebra is a key element in Conn's linearization results, which we formulate in the following theorem.

Theorem 7.31 (Conn's linearization theorem). *Let (M,π) be a Poisson manifold, and let $m \in M$ be a point at which π vanishes. Let $[\cdot,\cdot]$ denote the Lie bracket on T_m^*M, associated to the linearized Poisson structure at m.*

(1) In the holomorphic case, if $(T_m^*M,[\cdot,\cdot])$ is a semi-simple complex Lie algebra, then π is linearizable at m;

(2) In the real case, if $(T_m^*M,[\cdot,\cdot])$ is a compact semi-simple Lie algebra, then π is linearizable at m;

(3) In both cases, if $(T_m^*M,[\cdot,\cdot])$ is a semi-simple Lie algebra, then π is linearizable at m up to order ℓ, for all $\ell \in \mathbb{N}^*$.

Proof. We do not prove Conn's results *(1)* and *(2)*, as the proof is long and rather technical (see [45, 46]); instead, we prove the analogous (but easier to prove) result *(3)*. We do this by recursion on ℓ. Every Poisson structure which vanishes at m is linearizable up to order 1, so that the statement holds true for $\ell = 1$. Assume now that π is linearizable up to order ℓ. According to Definition 7.29, there exists a neighborhood U of m in M and a diffeomorphism Ψ_ℓ between U and a neighborhood of the origin o in T_mM, with $\Psi_\ell(m) = o$, such that

$$\{x \circ \Psi_\ell, y \circ \Psi_\ell\} - [x,y] \circ \Psi_\ell \in \mathscr{I}_m^{(\ell+1)}(U)\,, \tag{7.30}$$

for all $x,y \in T_m^*M$; we write here, and in the rest of the proof, the elements of T_m^*M with latin letters x,y,z, because we view $T_m^*M = \mathfrak{g}$ as a Lie algebra. For $x \in \mathfrak{g}$, we write F_x as a shorthand for the element $x \circ \Psi_\ell$ of $\mathscr{F}(U)$ and we denote $\mathscr{I} := \mathscr{I}_m^{(\ell+1)}(U)$ and $\mathscr{I}' := \mathscr{I}_m^{(\ell+2)}(U)$. Notice that $\mathscr{I}/\mathscr{I}'$ is finite-dimensional, because $\mathscr{I}/\mathscr{I}'$ is isomorphic to the space of homogeneous polynomials of degree $\ell+1$ in $\dim M$ variables. With these notations, (7.30) can be written as

$$\{F_x,F_y\} - F_{[x,y]} \in \mathscr{I}\,. \tag{7.31}$$

Since π vanishes at m, we have that $\{\mathscr{I},\mathscr{F}(U)\} \subset \mathscr{I}$ and $\{\mathscr{I}',\mathscr{F}(U)\} \subset \mathscr{I}'$. It follows that we can define a map

$$\begin{aligned} \chi : \mathfrak{g} \times (\mathscr{I}/\mathscr{I}') &\to \mathscr{I}/\mathscr{I}' \\ (x, p(F)) &\mapsto x \cdot p(F) := p(\{F_x, F\})\,, \end{aligned} \tag{7.32}$$

where $p : \mathscr{I} \to \mathscr{I}/\mathscr{I}'$ denotes the canonical projection. We claim that χ is a (finite-dimensional) representation of $\mathfrak{g}$. To prove this, we need to show that for all $x,y \in \mathfrak{g}$ and for every $F \in \mathscr{I}$,

$$[x,y] \cdot p(F) = x \cdot (y \cdot p(F)) - y \cdot (x \cdot p(F))\,. \tag{7.33}$$

On the one hand, (7.31) and the inclusion $\{\mathscr{I},\mathscr{I}\} \subset \mathscr{I}'$ imply that

$$[x,y] \cdot p(F) = p(\{F_{[x,y]},F\}) = p(\{\{F_x,F_y\},F\})\,,$$

while, on the other hand,

$$x \cdot (y \cdot p(F)) - y \cdot (x \cdot p(F)) = p(\{F_x, \{F_y, F\}\}) - p(\{F_y, \{F_x, F\}\}) ,$$

so that (7.33) is a consequence of the Jacobi identity for π. Since $\mathfrak{g}$ is semi-simple and $\mathscr{I}/\mathscr{I}'$ is finite-dimensional, the cohomology space $H^2_L(\mathfrak{g}; \mathscr{I}/\mathscr{I}')$ is trivial (for a proof, see [102, Th. III.13]; Lie algebra cohomology is recalled in Section 4.1.1), so every Lie 2-cocycle is a Lie 2-coboundary. The Lie 2-cocycle $\bar{c}$ which we consider is constructed from (7.31):

$$\begin{aligned} \bar{c} : \mathfrak{g} \wedge \mathfrak{g} &\to \mathscr{I}/\mathscr{I}' \\ (x,y) &\mapsto p(\{F_x, F_y\} - F_{[x,y]}) . \end{aligned} \tag{7.34}$$

We have that $\bar{c}$ is a Lie 2-cocycle, which means that for all $x, y, z \in \mathfrak{g}$,

$$x \cdot \bar{c}(y,z) + \bar{c}(x, [y,z]) + \circlearrowleft (x,y,z) = 0,$$

because, using (7.31),

$$x \cdot \bar{c}(y,z) + \bar{c}(x, [y,z]) = p(\{F_x, \{F_y, F_z\}\}) - p(F_{[x,[y,z]]}) .$$

Since $H^2_L(\mathfrak{g}; \mathscr{I}/\mathscr{I}')$ is trivial, there exists a linear map $\bar{b} : \mathfrak{g} \to \mathscr{I}/\mathscr{I}'$, such that $\bar{c}$ is the coboundary of $\bar{b}$,

$$\bar{c}(x,y) = x \cdot \bar{b}_y - y \cdot \bar{b}_x - \bar{b}_{[x,y]} \tag{7.35}$$

for all $x, y \in \mathfrak{g}$. For every $x \in \mathfrak{g}$, let $F'_x := F_x - b_x$, where the linear map $b : \mathfrak{g} \to \mathscr{I}$ is an arbitrary lift of $\bar{b}$. We claim that

$$\{F'_x, F'_y\} - F'_{[x,y]} \in \mathscr{I}' , \tag{7.36}$$

for all $x, y \in \mathfrak{g}$. Since

$$\{F'_x, F'_y\} = \{F_x, F_y\} - \{F_x, b_y\} - \{b_x, F_y\} + \{b_x, b_y\} ,$$

and since $\{b_x, b_y\}$ is an element of $\mathscr{I}^{(2\ell+1)}_m(U) \subset \mathscr{I}^{(\ell+2)}_m(U) = \mathscr{I}'$, as $\ell \geqslant 1$, (7.36) amounts to saying that

$$\{F_x, F_y\} - F_{[x,y]} + b_{[x,y]} - \{F_x, b_y\} - \{b_x, F_y\} , \tag{7.37}$$

which is an element of $\mathscr{I}$, belongs to the kernel of p. According to the definition of $\bar{c}$, the image of (7.37) under p is given by

$$\begin{aligned} p\left(\{F_x, F_y\} - F_{[x,y]}\right) + p(b_{[x,y]}) - x \cdot p(b_y) + y \cdot p(b_x) \\ = \bar{c}(x,y) + \bar{b}_{[x,y]} - x \cdot \bar{b}_y + y \cdot \bar{b}_x , \end{aligned}$$

which is zero in view of (7.35). This proves (7.36). Since $x \mapsto F'_x$ is a linear map, there exists a map $\Psi_{\ell+1}$ from U to a neighborhood of o in $T_m M$, such that $F'_x = x \circ \Psi_{\ell+1}$, for every $x \in \mathfrak{g}$. Since $\mathrm{d}_m \Psi_\ell = \mathrm{d}_m \Psi_{\ell+1}$, there exists a neighborhood $U' \subset U$

of m in M on which $\Psi_{\ell+1}$ is a diffeomorphism. In terms of $\Psi_{\ell+1}$, (7.36) is written as

$$\{x \circ \Psi_{\ell+1}, y \circ \Psi_{\ell+1}\} - [x,y] \circ \Psi_{\ell+1} \in \mathscr{I}_m^{(\ell+2)}(U') , \tag{7.38}$$

for all $x, y \in \mathfrak{g}$, which means that π is linearizable up to order $\ell+1$ at m. □

Remark 7.32. It is clear from the above proof that the open subsets on which the diffeomorphisms Ψ_ℓ are defined form a sequence which decreases with ℓ, and that the intersection of all these open subsets may be reduced to a point. Notice that, even if this intersection is not reduced to a point, the Poisson structure may not be linearizable at m (see Example 7.30).

7.6 Notes

The fact that Lie–Poisson structures are in one-to-one correspondence with Lie algebra structures is implicit in Lie's works, in which one also finds the relation to the coadjoint representation of the adjoint group of the Lie algebra. In the first half of the twentieth century, the theory of Lie algebras and Lie groups became a subject of its own, with focus on applications to theoretical physics and differential geometry. The cited observations of Lie were rediscovered and reinforced in the works of Kirillov [105, 106], Kostant [117] and Souriau [185], who clearly exhibit the symplectic structure on the coadjoint orbits which is inherited from the Lie–Poisson structure; this structure is nowadays known as the Kostant–Kirillov–Souriau symplectic structure. For general information about Lie algebras and their representation theory, we refer to Jacobson [102] or to the first chapter of Bourbaki [27].

Exactly as Lie algebra structures are in one-to-one correspondence with linear Poisson structures, Lie algebroids are in one-to-one correspondence with fiberwise linear Poisson structures [64], i.e., Poisson structures on a vector bundle which have the property that the bracket of any two fiberwise linear functions is a fiberwise linear function, while the bracket of a fiberwise linear function with a basic function is a basic function. See [53] for an introduction to Lie algebroids.

The linearization problem was first posed and studied by Weinstein [199]. The main results, which give conditions under which a Poisson structure is linearizable, were obtained by Conn [45, 46]; a geometrical proof of these results is given in [52]. See Dufour–Zung [63] for a detailed account on the linearization problem.

Chapter 8
Higher Degree Poisson Structures

Constant and linear Poisson structures, studied in the previous two chapters, are particular cases of a bigger class of Poisson structures, namely the (weight) homogeneous Poisson structures on a finite-dimensional vector space V: constant and linear Poisson structures are the homogeneous Poisson structures of degree zero and one, respectively. Like in the constant or linear case, (weight) homogeneous Poisson structures are easier to study than general Poisson structures. For example, homogeneity of a Poisson structure permits one to split the determination of its Poisson cohomology into its homogeneous parts. Also, weight homogeneous Poisson structures turn out to be useful for studying general Poisson structures. We will see an example of this in Section 9.1.3 of the next chapter, when classifying Poisson structures on $\mathbb{F}^2$, with a simple singularity.

Among distinguished classes of weight homogeneous Poisson structures, one finds – besides constant and linear Poisson structures – quadratic Poisson structures (i.e., homogeneous Poisson structures of degree two), rank 2 Poisson structures which arise from Nambu–Poisson structures and the transverse Poisson structures to adjoint orbits in a semi-simple Lie algebra.

Section 8.1 is devoted to the study of general properties of (weight) homogeneous Poisson structures, while we study in Section 8.2 a few special properties and characterizations of quadratic Poisson structures: their multiplicativity, their modular class and the decomposition of a quadratic Poisson structure, using a related unimodular Poisson structure. In Section 8.3 we define the notion of a Nambu–Poisson structure and we explain the construction which leads to a large class of (weight homogeneous) rank 2 Poisson structures. Finally, Section 8.4 deals with the transverse Poisson structure to a nilpotent orbit in a semi-simple Lie algebra.

Unless otherwise stated, $\mathbb{F}$ denotes an arbitrary field of characteristic zero.

C. Laurent-Gengoux et al., *Poisson Structures*,
Grundlehren der mathematischen Wissenschaften 347,
DOI 10.1007/978-3-642-31090-4_8, © Springer-Verlag Berlin Heidelberg 2013

8.1 Polynomial and (Weight) Homogeneous Poisson Structures

In this section, we define and study polynomial Poisson structures on a vector space, with special emphasis on homogeneous and weight homogeneous Poisson structures.

Throughout the section, we consider Poisson structures and, more generally, multivector fields on a finite-dimensional vector space V, which is equipped with its algebra of functions, denoted by $\mathscr{F}(V)$. If $\mathbb{F}$ is an arbitrary field, this algebra is the algebra of polynomial functions on V; however, if $\mathbb{F}=\mathbb{R}$ (respectively $\mathbb{F}=\mathbb{C}$), $\mathscr{F}(V)$ may stand either for the algebra of smooth (respectively holomorphic) or polynomial functions on V, depending on the context. It is clear that, in either case, $\mathscr{F}(V)$ contains all polynomial functions on V. For $p\in\mathbb{N}$, the corresponding space of p-vector fields on V (skew-symmetric p-derivations of $\mathscr{F}(V)$) is denoted by $\mathfrak{X}^p(V)$. The origin of V will be denoted by o.

8.1.1 Polynomial Poisson Structures

Let $\mathscr{V}$ be a vector field on a finite-dimensional vector space V. Then, $\mathscr{V}$ is said to be a *polynomial vector field* on V if, for every polynomial function F on V, $\mathscr{V}[F]$ is a polynomial function on V. More generally, a p-vector field P on V ($p\in\mathbb{N}^*$) is said to be a *polynomial p-vector field* on V if $P[F_1,\dots,F_p]$ is a polynomial function on V, for all polynomial functions $F_1,\dots,F_p$ on V. In the case of Poisson structures on V, this leads to the following definition.

Definition 8.1. A Poisson structure π on a finite-dimensional vector space V is said to be a *polynomial Poisson structure* if for all polynomial functions F and G on V, their Poisson bracket $\{F,G\}$ is a polynomial function on V.

In terms of arbitrary linear coordinates $x_1,\dots,x_d$ of V, every p-vector field P on V can be written as

$$P=\sum_{1\leqslant i_1<\cdots<i_p\leqslant d} P[x_{i_1},\dots,x_{i_p}]\frac{\partial}{\partial x_{i_1}}\wedge\cdots\wedge\frac{\partial}{\partial x_{i_p}}, \tag{8.1}$$

so that P is a polynomial p-vector field if and only if $P[x_{i_1},\dots,x_{i_p}]$ is a polynomial function, for all $1\leqslant i_1<\cdots<i_p\leqslant d$. In this case, the maximum of the degrees of all $P[x_{i_1},\dots,x_{i_p}]$ is called the *degree* of P. In particular, every polynomial Poisson structure on V is of the form

$$\pi=\sum_{1\leqslant i<j\leqslant d} x_{ij}\frac{\partial}{\partial x_i}\wedge\frac{\partial}{\partial x_j},$$

where, for every $1\leqslant i<j\leqslant d$, the function $x_{ij}:=\{x_i,x_j\}$ is a polynomial function on V, and the degree of π is the maximum of the degrees of the polynomials x_{ij}.

Example 8.2. Constant Poisson structures (see Chapter 6) are polynomial Poisson structures of degree zero, while linear and affine Poisson structures (see Chapter 7) are polynomial Poisson structures of degree one.

It is clear that when $\mathscr{F}(V)$ is the algebra of polynomial functions on V, then all Poisson structures on V are polynomial Poisson structures on V.

8.1.2 Homogeneous Poisson Structures

Let V be a finite-dimensional vector space. A function $F \in \mathscr{F}(V)$ is said to be *homogeneous* if there exists an integer $r \in \mathbb{N}$, such that for every $\lambda \in \mathbb{F}$ and for every $v \in V$, we have

$$F(\lambda v) = \lambda^r F(v) . \tag{8.2}$$

The integer r, which is unique if $F \neq 0$, is then called the *degree* of F, denoted $\deg(F)$. If F is expressed in terms of a system of *linear* coordinates $(x_1, \ldots, x_d)$ for V, which we write as $F = F(x_1, \ldots, x_d)$, then condition (8.2) is equivalent to the fact that

$$F(\lambda x_1, \ldots, \lambda x_d) = \lambda^r F(x_1, \ldots, x_d) , \tag{8.3}$$

for every $\lambda \in \mathbb{F}$. A homogeneous function of degree zero is constant, since the condition $F(\lambda v) = F(v)$ for all $\lambda \in \mathbb{F}$ and $v \in V$ implies that $F(v) = F(0)$ for every $v \in V$.

The following proposition states that smooth or holomorphic functions which are homogeneous, are polynomial functions.

Proposition 8.3. *Let V be a finite-dimensional real (respectively complex) vector space. If F is a smooth (respectively holomorphic) function on V and F is homogeneous of degree r, then F is a polynomial function on V, whose degree is r.*

Proof. Let $F \in \mathscr{F}(V)$ be a smooth or holomorphic function on V, satisfying (8.2) for every $\lambda \in \mathbb{F}$ and every $v \in V$. By differentiating, for a fixed λ, both sides of this equation in the direction of a vector $u \in V$, we obtain that, for every $u \in V$, the directional derivative of F along the vector u is a homogeneous function of degree $r-1$. By iterating this procedure, we obtain that the directional derivatives of F of order r are all homogeneous of degree 0, i.e., are constant functions. This implies that F is a polynomial function on V, of degree less than or equal to r. Then, because of equation (8.2), we can conclude that the degree of F is r. □

According to Proposition 8.3, the notions of "homogeneous function" and "homogeneous polynomial function" do not have to be distinguished and the notion of degree coincides with the usual notion of degree of a polynomial.

A (smooth or holomorphic) vector field $\mathscr{V}$ on a finite-dimensional vector space V is said to be a *homogeneous vector field* on V if, for every $F \in \mathscr{F}(V)$ which is homogeneous, the function $\mathscr{V}[F] \in \mathscr{F}(V)$ is also homogeneous. More generally, a (smooth or holomorphic) p-vector field $P \in \mathfrak{X}^p(V)$ is said to be *homogeneous* if, for all $F_1,\ldots,F_p \in \mathscr{F}(V)$ which are homogeneous, the function $P[F_1,\ldots,F_p]$ is homogeneous. Since every polynomial function is the sum of homogeneous functions, Proposition 8.3 implies that every homogeneous p-vector field on V is a polynomial p-vector field on V.

In the following proposition, we give a few characterizations of homogeneous multivector fields. For one of them, we use the *Euler vector field* on V, which is denoted by $\mathscr{E}$ and is defined as the unique vector field on V which acts as the identity on all linear functions on V; in terms of linear coordinates $x_1,\ldots,x_d$ on V, it is given by

$$\mathscr{E} = x_1 \frac{\partial}{\partial x_1} + \cdots + x_d \frac{\partial}{\partial x_d} . \tag{8.4}$$

Clearly, $\mathscr{E}$ is a homogeneous vector field on V. Notice that, by differentiating (8.3) with respect to λ, at $\lambda = 1$, we obtain that a function F on V is homogeneous of degree $r \in \mathbb{N}$ if and only if

$$\mathscr{E}[F] = \mathscr{L}_{\mathscr{E}} F = rF . \tag{8.5}$$

This formula is called the *Euler formula*.

Proposition 8.4. *Let P be a non-zero p-vector field on a finite-dimensional vector space V, where $p \in \mathbb{N}$. The following conditions on P are equivalent.*

(i) *P is homogeneous;*
(ii) *There exists an integer $r \in \mathbb{N}$, such that for all linear functions $F_1,\ldots,F_p$ on V, one has that $P[F_1,\ldots,F_p]$ is zero, or is a homogeneous function of degree r;*
(iii) *There exists an integer $r \in \mathbb{N}$, such that for all homogeneous functions $F_1,\ldots,F_p$ on V, one has that $P[F_1,\ldots,F_p]$ is zero, or is a homogeneous function of degree $\deg(F_1)+\cdots+\deg(F_p)+r-p$;*
(iv) *There exists an integer $r \in \mathbb{N}$, such that $\mathscr{L}_{\mathscr{E}} P = (r-p)P$.*

The integer r, which appears in (ii)–(iv) is the same and is called[1] *the* degree *of P, denoted* $\deg(P)$, *while $r-p$ is called the* weight *of P, denoted $\varpi(P)$.*

Proof. Let P be a non-zero p-vector field, $P \in \mathfrak{X}^p(V)$. We assume that $p \geqslant 1$, since the proposition is clear for $p = 0$. Condition *(ii)* implies obviously condition *(i)*, in view of (8.1). We only show that *(i)* implies *(ii)* for $p = 1$, since the generalization of the proof to arbitrary $p \geqslant 1$ will be clear. Thus, we assume that P is a homogeneous vector field, and we consider two linear functions F and G on V, satisfying $P[F] \neq 0$ and $P[G] \neq 0$. For a generic $t \in \mathbb{F}$ the polynomial $P[(1-t)F + tG]$ is non-zero and homogeneous, with a degree r which is independent of t; in particular, since $P[F]$ and $P[G]$ are different from zero, they have the same degree r, which proves *(ii)*. The equivalence between conditions *(ii)* and *(iii)* follows immediately from the fact that a p-vector field P is a derivation in each of its arguments. Finally, let $F_1,\ldots,F_p$ be homogeneous elements of $\mathscr{F}(V)$ and let $r \in \mathbb{N}$. Using (3.7) and (8.5), we have

[1] In order that the formulas which involve the degree or weight are valid also for the zero p-vector field, we adopt the convention that both the degree and weight of the zero p-vector field are $-\infty$.

$$\begin{aligned}
&(\mathscr{L}_{\mathscr{E}} P-(r-p)P)[F_1,\dots,F_p] \\
&\quad=\mathscr{E}\left[P[F_1,\dots,F_p]\right]-\sum_{\ell=1}^{p} P[F_1,\dots,\mathscr{E}[F_\ell],\dots,F_p]-(r-p)P[F_1,\dots,F_p] \\
&\quad=\mathscr{E}\left[P[F_1,\dots,F_p]\right]-\left(\sum_{\ell=1}^{p}\deg(F_\ell)-(r-p)\right)P[F_1,\dots,F_p]\,.
\end{aligned}$$

Using (8.5), we conclude that there exists $r\in\mathbb{N}$ such that $\mathscr{L}_{\mathscr{E}}P=(r-p)P$ if and only if there exists $r\in\mathbb{N}$ such that the following holds: for all homogeneous elements $F_1,\dots,F_p$ of $\mathscr{F}(V)$, we have that $P[F_1,\dots,F_p]$ is zero, or is homogeneous of degree $\sum_{\ell=1}^{p}\deg(F_\ell)-(r-p)$. This shows the equivalence *(iii)* $\Leftrightarrow$ *(iv)*. □

As a corollary, every polynomial p-vector field P on V can be written uniquely as a finite sum of homogeneous p-vector fields on V, whose degrees are pairwise different.

Notice that, according to *(iii)* of Proposition 8.4, if $F_1,\dots,F_p\in\mathscr{F}(V)$ and $P\in\mathfrak{X}^p(V)$ are homogeneous, then the polynomial $P[F_1,\dots,F_p]$ is either zero, or is homogeneous of degree

$$\deg\left(P[F_1,\dots,F_p]\right)=\deg(F_1)+\cdots+\deg(F_p)+\deg(P)-p\,, \tag{8.6}$$

which, in terms of weights, becomes

$$\varpi(P[F_1,\dots,F_p])=\varpi(F_1)+\cdots+\varpi(F_p)+\varpi(P)\,. \tag{8.7}$$

Since $\mathscr{L}_{\mathscr{E}}\mathscr{E}=[\mathscr{E},\mathscr{E}]=0$, the Euler vector field is weight homogeneous of degree one, i.e., of weight zero.

Specializing the definition of homogeneous bivector fields to Poisson structures on V, we obtain the following definition.

Definition 8.5. Let V be a finite-dimensional vector space. A Poisson structure π on V is said to be *homogeneous* if for all $F,G\in\mathscr{F}(V)$, which are homogeneous, their Poisson bracket $\{F,G\}\in\mathscr{F}(V)$ is also homogeneous.

According to Proposition 1.35, every Poisson structure π on V can be written in the following form:

$$\pi=\sum_{1\leqslant i<j\leqslant d} x_{ij}\frac{\partial}{\partial x_i}\wedge\frac{\partial}{\partial x_j}\,;$$

according to Proposition 8.4, it is homogeneous (of degree r) if and only if all the functions $x_{ij}:=\{x_i,x_j\}$ are homogeneous of the same degree r; in this case, $\mathscr{L}_{\mathscr{E}}\pi=(r-2)\pi$.

Example 8.6. Let π be a Poisson structure on a finite-dimensional vector space V. Then π is homogeneous of degree zero (equivalently, of weight -2) if and only if π is a constant Poisson structure. Also, π is homogeneous of degree one (equivalently, of weight -1) if and only if π is a linear Poisson structure.

Example 8.7. Consider the bivector field $\pi := \mathscr{V}_1 \wedge \mathscr{V}_2$, where $\mathscr{V}_1$ and $\mathscr{V}_2$ are two homogeneous vector fields on a finite-dimensional vector space V. Then π is homogeneous of weight $\varpi(\mathscr{V}_1) + \varpi(\mathscr{V}_2)$, because

$$\mathscr{L}_{\mathscr{E}}(\pi) = \mathscr{L}_{\mathscr{E}}(\mathscr{V}_1) \wedge \mathscr{V}_2 + \mathscr{V}_1 \wedge \mathscr{L}_{\mathscr{E}}(\mathscr{V}_2) = (\varpi(\mathscr{V}_1) + \varpi(\mathscr{V}_2))\, \mathscr{V}_1 \wedge \mathscr{V}_2 \,.$$

Since

$$[\pi,\pi]_S = [\mathscr{V}_1 \wedge \mathscr{V}_2, \mathscr{V}_1 \wedge \mathscr{V}_2]_S = -2\,\mathscr{V}_1 \wedge \mathscr{V}_2 \wedge [\mathscr{V}_1, \mathscr{V}_2] \,,$$

we have that π is a (homogeneous) Poisson structure on V if and only if the vector fields $\mathscr{V}_1, \mathscr{V}_2, [\mathscr{V}_1, \mathscr{V}_2]$ are linearly dependent at all points of V. This condition is automatically satisfied when $\mathscr{V}_1$ is the Euler vector field $\mathscr{E}$ and $\mathscr{V}_2 = \mathscr{V}$ is an arbitrary homogeneous vector field, since then $[\mathscr{E}, \mathscr{V}] = \mathscr{L}_{\mathscr{E}} \mathscr{V} = \varpi(\mathscr{V})\, \mathscr{V}$, which is a multiple of $\mathscr{V}$. The Poisson structure $\mathscr{E} \wedge \mathscr{V}$ is homogeneous of weight $\varpi(\mathscr{V})$, since $\varpi(\mathscr{E}) = 0$.

To finish this section, we give a proposition whose main content is that an isomorphism between two homogeneous Poisson structures on a vector space can always be realized by a linear isomorphism.

Proposition 8.8. *Let π_1 and π_2 be two homogeneous Poisson structures on a finite-dimensional vector space V. Then the following are equivalent:*

(i) *There exists a Poisson isomorphism $L : (V, \pi_1) \to (V, \pi_2)$ which is a linear map;*
(ii) *There exists a Poisson isomorphism $\Psi : (V, \pi_1) \to (V, \pi_2)$ with $\Psi(o) = o$;*
(iii) *There exists a Poisson isomorphism $\Psi : (U_1, \pi_1) \to (U_2, \pi_2)$ with $\Psi(o) = o$, where U_1, U_2 are neighborhoods of the origin o in V.*

Proof. The implications $(i) \Longrightarrow (ii) \Longrightarrow (iii)$ are clear. Assume now that (iii) holds, and let $\Psi : (U_1, \pi_1) \to (U_2, \pi_2)$ be a Poisson isomorphism with $\Psi(o) = o$. We are going to prove that the differential $L = \mathrm{d}_o \Psi$ of Ψ at the origin is a linear Poisson isomorphism between (V, π_1) and (V, π_2). This can be done by comparing the Taylor expansions at o of both sides of the following identity, which is valid for all functions $F, G \in \mathscr{F}(V_2)$

$$\{F, G\}_2 \circ \Psi = \{F \circ \Psi, G \circ \Psi\}_1$$

where $\{\cdot,\cdot\}_1 = \pi_1$ and $\{\cdot,\cdot\}_2 = \pi_2$ are the Poisson brackets corresponding to π_1 and π_2. Since π_1 and π_2 are homogeneous of degree k, the Taylor expansions at the origin o of both sides of this equality vanish for degrees $0, \dots, k-1$, while the components of degree k are equal to $\{\mathrm{d}_o F, \mathrm{d}_o G\}_2 \circ L$ and $\{\mathrm{d}_o F \circ L, \mathrm{d}_o G \circ L\}_1$ respectively. Since every linear function on V is the differential at the origin of a function in $\mathscr{F}(V)$, we obtain that for all linear functions $F', G' \in \mathscr{F}(V)$,

$$\{F', G'\}_2 \circ L = \{F' \circ L, G' \circ L\}_1$$

which, in turn, implies that L is a Poisson map. Moreover, since $L = \mathrm{d}_o \Psi$ is invertible, L is in fact a Poisson isomorphism. □

We will use the above proposition in Section 9.2.3, when we classify the set of all quadratic Poisson structures on $\mathbb{C}^3$.

8.1.3 Weight Homogeneous Poisson Structures

Many known examples of (non-homogeneous) Poisson structures on a vector space $\mathbb{F}^d$ become homogeneous Poisson structures upon assigning specific (positive integer) weights to the natural coordinates $x_1, \ldots, x_d$ on $\mathbb{F}^d$. This means that many properties and techniques of homogeneous Poisson structures carry over to this more general class of Poisson structures, which we will call weight homogeneous Poisson structures.

We first introduce the notion of weight homogeneity for functions and for multivector fields on $\mathbb{F}^d$. A d-tuple of integers $\varpi := (\varpi_1, \ldots, \varpi_d)$ will be called a *weight vector* (for $\mathbb{F}^d$); unless otherwise stated, we will always assume that all integers ϖ_i are positive. Generalizing (8.3), a function F on $\mathbb{F}^d$ is called *weight homogeneous* with respect to ϖ if there exists an integer $r \in \mathbb{N}$, such that

$$F(\lambda^{\varpi_1} x_1, \ldots, \lambda^{\varpi_d} x_d) = \lambda^r F(x_1, \ldots, x_d) , \tag{8.8}$$

for all $\lambda \in \mathbb{F}$. In this case, the integer r, which is unique if $F \neq 0$, is called the *weight of F* and is denoted by $\varpi(F)$. It is easy to see that, if $F, G \in \mathscr{F}(\mathbb{F}^d)$ are weight homogeneous with respect to ϖ, then their product FG is also weight homogeneous with respect to ϖ, of weight $\varpi(FG) = \varpi(F) + \varpi(G)$.

Remark 8.9. It is clear that a function F on $\mathbb{F}^d$ is homogeneous if and only if it is weight homogeneous with respect to $\varpi = (1, \ldots, 1)$, and in this case, the weight of F is equal to the degree of F.

Taking $F := x_i$ in (8.8), we see that each coordinate function x_i, with $1 \leqslant i \leqslant d$ is weight homogeneous of weight ϖ_i, so that we often refer to ϖ_i as being the weight of x_i.

Similarly to the case of homogeneous functions (see Proposition 8.3), we show in the following proposition that a smooth or holomorphic function which is weight homogeneous, is a polynomial function.

Proposition 8.10. *If F is a smooth function on $\mathbb{R}^d$ (respectively a holomorphic function on $\mathbb{C}^d$) and F is weight homogeneous with respect to some weight vector, then F is a polynomial function.*

Proof. Let $\varpi = (\varpi_1, \ldots, \varpi_d)$ be a weight vector for $\mathbb{F}^d$ ($\mathbb{F} = \mathbb{R}$ or $\mathbb{F} = \mathbb{C}$) and assume that $F \in \mathscr{F}(\mathbb{F}^d)$ satisfies (8.8), for all $\lambda \in \mathbb{F}$. Let us consider $s \in \mathbb{N}^*$, such that, for all $1 \leqslant j_1, \ldots, j_s \leqslant d$, we have that $\varpi_{j_1} + \cdots + \varpi_{j_s} > r$. By differentiating both sides of equation (8.3) with respect to $x_{i_1}, \ldots, x_{i_s}$, for all $1 \leqslant i_1, \ldots, i_s \leqslant d$, we obtain that all functions $\frac{\partial^s F}{\partial x_{i_1} \ldots \partial x_{i_s}}$ have negative weight, hence are zero on $\mathbb{F}^d$. This implies that F is necessarily a polynomial function on $\mathbb{F}^d$. □

A p-vector field P on $\mathbb{F}^d$ is said to be *weight homogeneous* (with respect to ϖ), if for all $F_1,\ldots,F_p \in \mathscr{F}(\mathbb{F}^d)$, which are weight homogeneous with respect to ϖ, the function $P[F_1,\ldots,F_p] \in \mathscr{F}(\mathbb{F}^d)$ is weight homogeneous with respect to ϖ. One shows, as in the case of homogeneous multivector fields, that a weight homogeneous multivector field on $\mathbb{F}^d$ is a *polynomial* multivector field.

Remark 8.11. It is clear that a multivector field on $\mathbb{F}^d$ is homogeneous if and only if it is weight homogeneous with respect to $\varpi = (1,\ldots,1)$.

We will give a few characterizations of the weight homogeneity of a multivector field, which generalize the characterizations of the homogeneity of a multivector field, given in Proposition 8.4. To do this, we introduce the *weighted Euler vector field* $\mathscr{E}_\varpi$ on $\mathbb{F}^d$, which is defined by

$$\mathscr{E}_\varpi := \varpi_1 x_1 \frac{\partial}{\partial x_1} + \cdots + \varpi_d x_d \frac{\partial}{\partial x_d} . \tag{8.9}$$

Notice that the weighted Euler vector field $\mathscr{E}_\varpi$ is weight homogeneous with respect to ϖ. It is clear from (8.8) that a function F on $\mathbb{F}^d$ is weight homogeneous of weight r if and only if

$$\mathscr{E}_\varpi[F] = \mathscr{L}_{\mathscr{E}_\varpi} F = rF . \tag{8.10}$$

This formula, which generalizes the classical Euler formula (8.5), is called the *weighted Euler formula*.

Proposition 8.12. *Let $\varpi = (\varpi_1,\ldots,\varpi_d)$ be a weight vector for $\mathbb{F}^d$ and let P be a non-zero p-vector field on $\mathbb{F}^d$, where $p \in \mathbb{N}$. The following conditions on P are equivalent.*

- *(i) P is weight homogeneous with respect to ϖ;*
- *(ii) There exists an integer $r \in \mathbb{Z}$, such that for all $1 \leqslant i_1 < \cdots < i_p \leqslant d$, the function $P[x_{i_1},\ldots,x_{i_p}]$ is zero, or is weight homogeneous of weight $r + \varpi_{i_1} + \cdots + \varpi_{i_p}$;*
- *(iii) There exists an integer $r \in \mathbb{Z}$, such that for all weight homogeneous elements $F_1,\ldots,F_p$ of $\mathscr{F}(\mathbb{F}^d)$, one has that $P[F_1,\ldots,F_p]$ is zero, or is weight homogeneous of weight $r + \varpi(F_1) + \cdots + \varpi(F_p)$;*
- *(iv) There exists an integer $r \in \mathbb{Z}$, such that $\mathscr{L}_{\mathscr{E}_\varpi} P = rP$.*

The (possibly negative) integer r, which appears in (ii)–(iv) is the same and is called[2] *the* weight *of P, denoted $\varpi(P)$.*

Proof. Since the proof is very similar to the proof of Proposition 8.4, we will not repeat it here; we only focus on the implication *(i)⇒(ii)*, which requires some extra explanation. Suppose that $\mathscr{V}$ is a vector field on $\mathbb{F}^d$ and that for some $1 \leqslant i < j \leqslant d$ the functions $\mathscr{V}[x_i]$ and $\mathscr{V}[x_j]$ are different from zero. We claim that if $\mathscr{V}$ is weight

[2] In order that the formulas which involve the weight are valid also for the zero p-vector field, we adopt, as for homogeneous multivector fields, the convention that both the degree and weight of the zero p-vector field are $-\infty$.

homogeneous, then the integers $\varpi(\mathscr{V}[x_i]) - \varpi_i$ and $\varpi(\mathscr{V}[x_j]) - \varpi_j$ are equal. In order to prove this property, we consider for $t \in \mathbb{F}$ the function $\mathscr{V}\left[(1-t)x_i^{\varpi_j} + tx_j^{\varpi_i}\right]$, which is weight homogeneous, because $(1-t)x_i^{\varpi_j} + tx_j^{\varpi_i}$ and $\mathscr{V}$ are weight homogeneous. As in the proof of Proposition 8.4, it follows that $\mathscr{V}\left[x_i^{\varpi_j}\right]$ and $\mathscr{V}\left[x_j^{\varpi_i}\right]$ have the same weight. Since $\mathscr{V}[x_i^{\varpi_j}] = \varpi_j x_i^{\varpi_j - 1}\mathscr{V}[x_i]$ and $\mathscr{V}[x_j^{\varpi_i}] = \varpi_i x_j^{\varpi_i - 1}\mathscr{V}[x_j]$, we have that

$$\varpi_i(\varpi_j - 1) + \varpi(\mathscr{V}[x_i]) = \varpi_j(\varpi_i - 1) + \varpi(\mathscr{V}[x_j]),$$

and hence $\varpi(\mathscr{V}[x_i]) - \varpi_i = \varpi(\mathscr{V}[x_j]) - \varpi_j$, which is what we wanted to show. □

It follows, as in the case of homogeneous multivector fields, that every polynomial p-vector field $P \in \mathfrak{X}^p(\mathbb{F}^d)$ can be written in a unique way as a finite sum of weight homogeneous p-vector fields, whose weights are pairwise different. Also, if $P \in \mathfrak{X}^p(\mathbb{F}^d)$ and $F_1, \dots, F_p \in \mathscr{F}(F^d)$ are weight homogeneous, then the polynomial $P[F_1, \dots, F_p]$ is weight homogeneous, of weight

$$\varpi(P[F_1, \dots, F_p]) = \varpi(P) + \varpi(F_1) + \cdots + \varpi(F_p). \tag{8.11}$$

Since $\mathscr{L}_{\mathscr{E}_\varpi}\mathscr{E}_\varpi = [\mathscr{E}_\varpi, \mathscr{E}_\varpi] = 0$, the weighted Euler vector field is weight homogeneous of weight zero.

Finally, we specialize the definition of weight homogeneity to the case of Poisson structures.

Definition 8.13. A Poisson structure π on $\mathbb{F}^d$ is said to be *weight homogeneous* if for all $F, G \in \mathscr{F}(\mathbb{F}^d)$ which are weight homogeneous with respect to ϖ, their Poisson bracket $\{F, G\}$ is weight homogeneous with respect to ϖ.

Writing a Poisson structure π on $\mathbb{F}^d$ in the form

$$\pi = \sum_{1 \leqslant i < j \leqslant d} x_{ij} \frac{\partial}{\partial x_i} \wedge \frac{\partial}{\partial x_j},$$

it follows that π is weight homogeneous (of weight $\varpi(\pi)$) with respect to ϖ if and only if each one of the functions $x_{ij} := \{x_i, x_j\}$ is a weight homogeneous polynomial, of weight $\varpi(\pi) + \varpi_i + \varpi_j$; this is also equivalent to the equality

$$\mathscr{L}_{\mathscr{E}_\varpi}[\pi] = \varpi(\pi)\,\pi.$$

8.2 Quadratic Poisson Structures

A bivector field on a finite-dimensional vector space V is said to be *quadratic*, if it is homogeneous of degree 2. A *quadratic Poisson structure* on V is a Poisson bivector field on V which is quadratic. In terms of an arbitrary system of linear coordinates $(x_1,\ldots,x_d)$ on V, every quadratic Poisson structure π takes the form

$$\pi = \sum_{1\leqslant i<j\leqslant d} x_{ij}\frac{\partial}{\partial x_i}\wedge\frac{\partial}{\partial x_j}, \tag{8.12}$$

where all $x_{ij} := \{x_i,x_j\}$, with $1\leqslant i<j\leqslant d$, are homogeneous polynomial functions of degree 2 on V. Notice that a Poisson structure π on V is quadratic if and only if the Poisson bracket of every pair of linear functions on V is a homogeneous polynomial function of degree 2 on V.

To start with, we give two important classes of quadratic Poisson structures.

Example 8.14. Let $A=(a_{ij})\in \mathrm{Mat}_d(\mathbb{F})$ be an arbitrary skew-symmetric $d\times d$ matrix. A quadratic bivector field $\pi=\{\cdot\,,\cdot\}$ is defined on $\mathbb{F}^d$ by setting

$$\pi := \sum_{1\leqslant i<j\leqslant d} a_{ij}x_ix_j\frac{\partial}{\partial x_i}\wedge\frac{\partial}{\partial x_j}. \tag{8.13}$$

A direct computation from (8.13) gives, for all $1\leqslant i<j<k\leqslant d$,

$$\{\{x_i,x_j\},x_k\}+\circlearrowleft(i,j,k) = a_{ij}(a_{ik}+a_{jk})x_ix_jx_k+\circlearrowleft(i,j,k)\,,$$

which is zero since A is skew-symmetric. This proves the Jacobi identity for every triple of coordinate functions, which shows, in view of Proposition 1.36, that the bivector field π is a Poisson structure on $\mathbb{F}^d$. A quadratic Poisson structure of the form (8.13) is called a *diagonal Poisson structure*.

Notice that the rank of this Poisson structure is equal to the rank of the matrix A. More precisely, for every $m\in\mathbb{F}^d$, the rank of π at m is equal to the rank of the matrix, obtained by removing from the matrix A the k-th line and the k-th column for all k for which $x_k(m)=0$.

Example 8.15. If $\mathscr{V}$ is a linear vector field on a finite-dimensional vector space V, then $\mathscr{E}\wedge\mathscr{V}$ is a quadratic bivector field, where $\mathscr{E}$ is the Euler vector field (8.4). Since

$$[\mathscr{E}\wedge\mathscr{V},\mathscr{E}\wedge\mathscr{V}]_S = -2\mathscr{E}\wedge\mathscr{V}\wedge[\mathscr{E},\mathscr{V}] = 0\,,$$

the bivector field $\mathscr{E}\wedge\mathscr{V}$ is a quadratic Poisson structure. It is a particular case of the class of homogeneous Poisson structures, constructed in Example 8.7.

In the following proposition, we give two characterizations of quadratic Poisson structures. We recall from Section 1.3.2 that, by definition, a vector field $\mathscr{V}$ on (V,π) is a Poisson vector field if the Lie derivative of π with respect to $\mathscr{V}$ is zero.

Proposition 8.16. *For a Poisson structure π on a finite-dimensional vector space V, the following conditions are equivalent:*

(i) *The Poisson structure π is quadratic;*
(ii) *The Euler vector field $\mathscr{E}$ is a Poisson vector field;*

(iii) *For every $a \in \mathbb{F}^*$, the scalar multiplication by a is a Poisson automorphism of V.*

Proof. The equivalence between *(i)* and *(ii)* is a particular case of the equivalence between *(i)* and *(iv)* in Proposition 8.4 with $p = 2$. Let $\pi = \{\cdot,\cdot\}$ be a Poisson structure on V. The multiplication by $a \in \mathbb{F}^*$ is a Poisson map if and only if

$$\{F,G\}(av) = \{aF, aG\}(v) = a^2\{F,G\}(v) ,$$

for all linear functions $F, G \in V^*$ and all $v \in V$, which is equivalent to saying that $\{F,G\}$ is a homogeneous polynomial function of degree 2, for all $F, G \in V^*$. By Proposition 8.4, the latter means that π is quadratic. This shows that *(i)* and *(iii)* are equivalent. □

The projective space $\mathbb{P}(V)$, which is the variety of 1-dimensional subspaces of V, is naturally identified with the quotient space $(V\backslash\{o\})/\mathbb{F}^*$ of the free action of $\mathbb{F}^*$ by scalar multiplication. By Propositions 5.33 and 8.16, if π is a quadratic Poisson structure on V, then $\mathbb{P}(V)$ admits a (unique) Poisson structure, such that the canonical projection map $V \setminus \{o\} \to \mathbb{P}(V)$ is a Poisson map.

8.2.1 Multiplicative Poisson Structures

Let V be a finite-dimensional vector space and let $\mu : V \times V \to V$ be a bilinear map. We refer in this section to μ as a *product* on V, even if we do not demand that μ is associative. We say that μ is a product with *unit*, if V admits an element e, such that $\mu(v,e) = v = \mu(e,v)$ for all $v \in V$. A Poisson structure π on V is said to be *multiplicative* with respect to μ if $\mu : V \times V \to V$ is a Poisson map, where $V \times V$ is endowed with the product Poisson structure.

We show in this section that Poisson structures, which are multiplicative with respect to a product with unit, are quadratic. We will do this by using the following elementary lemma.

Lemma 8.17. *The only polynomial p in one variable, satisfying*

$$p(ab) = a^2 p(b) + b^2 p(a) , \tag{8.14}$$

for all $a, b \in \mathbb{F}$, is the zero polynomial.

Proof. Let p be a polynomial which satisfies (8.14), for all $a, b \in \mathbb{F}$. The degree r of p is at most 2, because $(a,b) \mapsto p(ab)$ is a polynomial function (in two variables) of total degree $2r$, while the total degree of $a^2 p(b) + b^2 p(a)$ is at most $r + 2$. But (8.14) implies that $p(0) = p(1) = p(-1) = 0$, so that p is the zero polynomial. □

We now establish the link between multiplicative and quadratic Poisson structures.

Proposition 8.18. *Let V be a finite-dimensional vector space and let $\mu : V \times V \to V$ be a product with unit. Every polynomial Poisson structure on V which is multiplicative with respect to μ, is quadratic.*

Proof. Let π be a polynomial Poisson structure on V, which is multiplicative. In view of (2.12), multiplicativity of $\{\cdot,\cdot\} = \pi$ implies that, for all $F, G \in \mathscr{F}(V)$, and all $v, w \in V$:

$$\{F,G\}(\mu(v,w)) = \{F \circ R_w, G \circ R_w\}(v) + \{F \circ L_v, G \circ L_v\}(w), \tag{8.15}$$

where $R_w : V \to V$ and $L_v : V \to V$ are the endomorphisms of V, defined for $u \in V$ by $R_w(u) := \mu(u,w)$ and $L_v(u) := \mu(v,u)$ respectively. By bilinearity of μ, for every $a \in \mathbb{F}$ and $v \in V$, the equalities $\mu(v,ae) = av = \mu(ae,v)$ hold, as well as the dual equalities $F \circ R_{ae} = aF = F \circ L_{ae}$, valid for all linear functions F on V and all $a \in \mathbb{F}$. Let F, G be two linear functions on V. For all $a, b \in \mathbb{F}$, (8.15) specializes to

$$\{F,G\}(abe) = \{F,G\}(\mu(ae,be)) = a^2\{F,G\}(be) + b^2\{F,G\}(ae). \tag{8.16}$$

The map $p : \mathbb{F} \to \mathbb{F}$, defined for all $a \in \mathbb{F}$ by $p(a) := \{F,G\}(ae)$, is a polynomial function, since π is a polynomial Poisson structure, and (8.16) states that p satisfies condition (8.14) of Lemma 8.17. Therefore, $\{F,G\}(ae) = 0$ for all $a \in \mathbb{F}$. Specializing (8.15), we obtain therefore that

$$\{F,G\}(av) = \{F,G\}(\mu(v,ae)) = a^2\{F,G\}(v),$$

for all $a \in \mathbb{F}$ and $v \in V$. This shows that $\{F,G\}$ is a homogeneous polynomial of degree 2. Since this holds for all linear functions F and G on V, the Poisson structure π is quadratic. □

Proposition 8.18 takes in the smooth or holomorphic context the following form.

Proposition 8.19. *Let V be a finite-dimensional vector space over $\mathbb{R}$ or over $\mathbb{C}$ and let $\mu : V \times V \to V$ be a product with unit. Every smooth or holomorphic Poisson structure on V, multiplicative with respect to μ, is quadratic.*

Proof. The proof of the proposition goes along the same lines as the proof of Proposition 8.18; the only difference is that one needs to reprove Lemma 8.17 in that context, namely, to prove that there exists no non-zero smooth or holomorphic function p satisfying (8.14). Deriving condition (8.14) three times with respect to a, yields that if p satisfies (8.14) for all $a, b \in \mathbb{F}$, it also satisfies $bp'''(ab) = p'''(a)$, for all $a, b \in \mathbb{F}$. Plugging $b = 0$ in it, we obtain that $p'''(a) = 0$ for all $a \in \mathbb{F}$, and therefore that p is a polynomial. According to Lemma 8.17, p is the zero polynomial. □

Remark 8.20. It is clear that the only differentiability condition which we use on the function p in Proposition 8.19 is that p is of class C^3 in a neighborhood of o in V. Therefore, Proposition 8.19 can be easily generalized to multiplicative bivector fields of class C^k, with $k \geqslant 3$, where for $k \in \mathbb{N}^*$, we call *bivector field of class* C^k on $\mathbb{R}^d$ every bi-differential operator π of the form

$$\pi = \sum_{1 \leqslant i < j \leqslant d} x_{ij} \frac{\partial}{\partial x_i} \wedge \frac{\partial}{\partial x_j},$$

where all the functions x_{ij} are of class C^k on $\mathbb{R}^d$. Notice that, if π is of class C^k on $\mathbb{R}^d$ (with $k \in \mathbb{N}^*$), then for every pair of smooth functions (F,G) on V, the bracket $\{F,G\}$ is a function of class C^k, so one can still consider the Jacobi identity for smooth functions and define Poisson structures of class C^k, with $k \in \mathbb{N}^*$, by saying that a bivector field π on $\mathbb{R}^d$ is a Poisson structure of class C^k if π satisfies the Jacobi identity for all triples of *smooth* functions F,G,H on V.

For Poisson structures of class C^2, it can be shown that Proposition 8.19 still holds, but for Poisson structures of class C^1 this is not the case (in general), as is shown in the following example.

Example 8.21. On $\mathbb{R}^2$ we consider the product (with unit) μ which corresponds to the natural product on $\mathbb{R}[h]/h^2$, via the vector space isomorphism, defined by $(a_1,a_2) \mapsto a_1 + ha_2$. Explicitly, μ is given by

$$\mu((a_1,a_2),(b_1,b_2)) = (a_1b_1, a_1b_2 + a_2b_1)\,,$$

and the unit for μ is $(1,0)$. It follows that $x_1 \circ \mu = y_1z_1$ and $x_2 \circ \mu = y_1z_2 + z_1y_2$, where (x_1,x_2), respectively (y_1,y_2,z_1,z_2), are the natural systems of coordinates on $\mathbb{R}^2$, respectively on $\mathbb{R}^2 \times \mathbb{R}^2 \simeq \mathbb{R}^4$.

A Poisson structure π of class C^1 on $\mathbb{R}^2$ is defined by $\{x_1,x_2\} := x_1^2 \ln(|x_1|)$. We denote the product Poisson structure on $\mathbb{R}^4 = \mathbb{R}^2 \times \mathbb{R}^2$ by π'. In order to show that π is multiplicative with respect to μ, notice that $\{x_1 \circ \mu, x_2 \circ \mu\}' = \{x_1,x_2\} \circ \mu$, which follows from

$$\{x_1,x_2\} \circ \mu = (x_1^2 \ln(|x_1|)) \circ \mu = (y_1z_1)^2 \ln(|y_1z_1|)$$

and

$$\begin{aligned}\{x_1 \circ \mu, x_2 \circ \mu\}' &= \{y_1z_1, y_1z_2 + z_1y_2\}' \\ &= z_1^2\{y_1,y_2\} + y_1^2\{z_1,z_2\} \\ &= z_1^2y_1^2 \ln(|y_1|) + y_1^2z_1^2 \ln(|z_1|)\,.\end{aligned}$$

However, π is not quadratic.

Example 8.22. Let $V := \mathrm{Mat}_d(\mathbb{F})$ the vector space of square matrices of size d. For $1 \leqslant i,j \leqslant d$, we consider on V the linear function, defined for all $x = (x_{ij})_{i,j=1}^d \in V$ by $\xi_{ij}(x) := x_{ij}$. These functions provide a set of linear coordinates on V. For i and j as above, let $\varepsilon_{i,j} := 1$ if $i > j$ and $\varepsilon_{i,j} := -1$ if $i < j$ and $\varepsilon_{i,j} := 0$ if $i = j$. Then the brackets, defined for $1 \leqslant i,j,k,\ell \leqslant d$, by

$$\{\xi_{ij}, \xi_{k\ell}\} := (\varepsilon_{i,k} + \varepsilon_{j,\ell})\xi_{i\ell}\xi_{kj}$$

define a multiplicative quadratic Poisson structure on V. This can be checked by direct computation; see Example 11.14 below for a more conceptual proof.

8.2.2 The Modular Vector Field of a Quadratic Poisson Structure

On a finite-dimensional vector space V, which is equipped with a Poisson structure, there is a vector field, which is naturally associated to the Poisson structure: the modular vector field, with respect to an arbitrary translation invariant volume form λ on V. Recall from (4.24) that the modular vector field Φ_π of a Poisson structure π on V is, up to a sign, the divergence $\operatorname{Div}(\pi)$ of π, i.e., $\Phi_\pi = -\operatorname{Div}(\pi)$. As a first application, we have, in view of the explicit formula (4.22) for the divergence of a bivector field, that the modular vector field is homogeneous of degree $r-1$, if the Poisson structure is homogeneous of degree r; in the particular case of quadratic Poisson structures, this leads to the following proposition.

Proposition 8.23. *Let λ be an arbitrary translation invariant volume form on a finite-dimensional vector space V. The modular vector field of every quadratic Poisson structure on V, with respect to λ, is a linear vector field on V.*

In the study of quadratic Poisson structures, the modular vector field plays a particular rôle, which we now underline, and which will lead in Section 9.2.3 to a classification of all quadratic Poisson structures on $\mathbb{C}^3$.

Example 8.24. We compute the modular vector field Φ_π of the diagonal Poisson structure π on $\mathbb{F}^d$, introduced in Example 8.14 with respect to the translation invariant volume form $\lambda = \mathrm{d}x_1 \wedge \cdots \wedge \mathrm{d}x_d$. By using the formula $\Phi_\pi = -\operatorname{Div}(\pi)$ and the explicit expression of the divergence of a bivector field (4.22), we obtain

$$\Phi_\pi = -2\sum_{i=1}^{d}\left(\sum_{j=1}^{d} a_{ij}\right) x_i \frac{\partial}{\partial x_i}\ . \tag{8.17}$$

Example 8.25. Let V be a d-dimensional vector space and let $\mathscr{V}$ be a linear vector field on V. As we have seen in Example 8.15, $\pi := \mathscr{E} \wedge \mathscr{V}$ is a quadratic Poisson structure on V. In order to compute its divergence, notice that $\operatorname{Div}(\mathscr{E}) = d$, which follows from (4.21), and that $[\mathscr{E}, \mathscr{V}] = 0$, because the weight of a linear vector field is 0. Substituted in equation (4.23) for the divergence of a wedge product, one finds

$$\operatorname{Div}(\mathscr{E} \wedge \mathscr{V}) = \operatorname{Div}(\mathscr{V})\mathscr{E} - d\mathscr{V}\ , \tag{8.18}$$

so that the modular vector field Φ_π of the Poisson structure $\pi = \mathscr{E} \wedge \mathscr{V}$ is $\Phi_\pi = d\mathscr{V} - \operatorname{Div}(\mathscr{V})\mathscr{E}$.

We show in the following proposition that every quadratic Poisson structure π can be decomposed as a sum of two compatible quadratic Poisson structures: one of the terms involves only the modular vector field of π, while the other term is unimodular. This decomposition is very useful, as we will see in Section 9.2.3.

We first recall two more facts from Section 4.4.3, namely that, given a volume form λ on V, the star operator $\star$ is the family of isomorphisms

$$\star : \mathfrak{X}^q(V) \to \Omega^{d-q}(V),$$

defined for $Q \in \mathfrak{X}^q(V)$ by $\star Q = \iota_Q \lambda$ as in (4.19), and that Div is defined in terms of the de Rham differential d by

$$\mathrm{Div} = \star^{-1} \circ \mathrm{d} \circ \star .$$

Since $\Phi_\pi = -\mathrm{Div}(\pi)$, it follows that π is unimodular if and only if the differential $(d-2)$-form $\star\pi$ is closed.

Proposition 8.26. *Let V be a d-dimensional vector space and let λ be a translation invariant volume form on V. Every quadratic Poisson structure π on V decomposes as follows*

$$\pi = \star^{-1}\alpha + \frac{1}{d}\mathscr{E} \wedge \Phi_\pi ,$$

where Φ_π is the modular vector field of π with respect to λ, and where α is a closed differential $(d-2)$-form which satisfies $\mathscr{L}_{\Phi_\pi}\alpha = 0$. Also, $\star^{-1}\alpha$ has the following properties: it is a quadratic unimodular Poisson structure, compatible with π, and it satisfies $\mathscr{L}_{\Phi_\pi}(\star^{-1}\alpha) = 0$.

Proof. Since Φ_π is a linear vector field, $\mathscr{E} \wedge \Phi_\pi$ is a quadratic Poisson structure (see Example 8.15). This Poisson structure is compatible with π, since

$$[\mathscr{E} \wedge \Phi_\pi, \pi]_S = (\mathscr{L}_{\mathscr{E}}\pi) \wedge \Phi_\pi - \mathscr{E} \wedge (\mathscr{L}_{\Phi_\pi}\pi) = 0 ,$$

where we have used in the last step that both $\mathscr{E}$ and Φ_π are Poisson vector fields (with respect to π). The modular vector field $\Phi_{\pi'}$ of the Poisson structure $\pi' := \mathscr{E} \wedge \Phi_\pi$ is given by

$$\Phi_{\pi'} = -\mathrm{Div}(\mathscr{E} \wedge \Phi_\pi) = d\,\Phi_\pi - \mathrm{Div}(\Phi_\pi)\mathscr{E} = d\,\Phi_\pi , \tag{8.19}$$

where we have used (8.18), together with the fact that $\mathrm{Div}(\Phi_\pi) = 0$, which holds because Φ_π is itself the divergence of a bivector field. Since the quadratic Poisson structures π and $\pi' = \mathscr{E} \wedge \Phi_\pi$ are compatible, and since their modular vector fields are proportional (see (8.19)), there exists a linear combination of π and $\mathscr{E} \wedge \Phi_\pi$ which is a unimodular Poisson structure: $R := \pi - \frac{1}{d}\mathscr{E} \wedge \Phi_\pi$ is a Poisson structure on V which is quadratic, compatible with π and unimodular.

Let us denote by α the differential $(d-2)$-form on V, defined by $\alpha := \star R = \iota_R \lambda$. It is a closed form since R is unimodular. It remains to be checked that $\mathscr{L}_{\Phi_\pi} R = 0$ and that $\mathscr{L}_{\Phi_\pi}\alpha = 0$. First, we have that $\mathscr{L}_{\Phi_\pi}\pi = 0$ (since Φ_π is a Poisson vector field), and we also have that $\mathscr{L}_{\Phi_\pi}(\mathscr{E} \wedge \Phi_\pi) = 0$ (since the vector fields $\mathscr{E}$ and Φ_π commute). We therefore have $\mathscr{L}_{\Phi_\pi} R = 0$. It follows from this, upon using *(2)* of

Proposition 3.11, that

$$\mathscr{L}_{\Phi_\pi}\alpha = \mathscr{L}_{\Phi_\pi}(\iota_R\lambda) = \iota_R(\mathscr{L}_{\Phi_\pi}\lambda) + \iota_{[\Phi_\pi,R]_S}\lambda = 0\,,$$

where we used in the last step that $[\Phi_\pi, R]_S = \mathscr{L}_{\Phi_\pi}R = 0$ and that $\mathscr{L}_{\Phi_\pi}\lambda = \mathrm{Div}(\Phi_\pi)\,\lambda = 0$ (see (4.20)). This finishes the proof of the announced properties of α and of $R = \star^{-1}\alpha$. □

The quadratic Poisson structures whose modular vector field satisfies a genericity assumption, stated below, are diagonal, when expressed in a system of well-chosen coordinates. We devote the end of this section to explain and to prove this. First, we introduce the required notions. Let $\mathscr{V}$ be a linear vector field on a d-dimensional vector space V. Since $\mathscr{V}$ sends linear functions on V (elements of V^*) to linear functions on V, we can restrict $\mathscr{V}$ to V^*, which yields an endomorphism $\mathscr{V}^{(1)}$ of V^*,

$$\begin{aligned} \mathscr{V}^{(1)} : V^* &\to V^* \\ F &\mapsto \mathscr{V}[F]\,. \end{aligned} \tag{8.20}$$

Definition 8.27. Let $\mathscr{V}$ be a linear vector field on a finite-dimensional vector space V. We say that $\mathscr{V}$ is *diagonalizable* if the endomorphism $\mathscr{V}^{(1)}$ of V^* is diagonalizable. In this case, the eigenvalues, respectively eigenvectors of $\mathscr{V}^{(1)}$ are called *eigenvalues*, respectively *eigenvectors* of $\mathscr{V}$.

It is clear that, if $(x_1,\dots,x_d)$ is a basis of eigenvectors of a diagonalizable linear vector field $\mathscr{V}$, with corresponding eigenvalues $a_1,\dots,a_d \in \mathbb{F}$, then $(x_1,\dots,x_d)$ is a system of linear coordinates on V, and that $\mathscr{V}$ takes, in these coordinates, the following form:

$$\mathscr{V} = \sum_{i=1}^{d} a_i x_i \frac{\partial}{\partial x_i}\,. \tag{8.21}$$

It is also clear that, conversely, every vector field on V, which is of the form (8.21) is a diagonalizable linear vector field.

Comparing (8.17) and (8.21), we see that the modular vector field of every diagonal quadratic Poisson structure is diagonalizable. As we will see, the converse is almost true: up to a technical assumption on the eigenvalues of the modular vector field, a quadratic Poisson structure is diagonal (in some system of linear coordinates) if its modular vector field is diagonalizable. It motivates the following definition of a diagonalizable Poisson structure.

Definition 8.28. A quadratic Poisson structure π on a finite-dimensional vector space V is said to be *diagonalizable*, if there exist linear coordinates $x_1,\dots,x_d$ on V in terms of which π is diagonal, i.e.,

$$\pi = \sum_{1\leqslant i<j\leqslant d} a_{ij}x_ix_j \frac{\partial}{\partial x_i}\wedge\frac{\partial}{\partial x_j}\,, \tag{8.22}$$

where $(a_{ij}) \in \mathrm{Mat}_d(\mathbb{F})$ is a skew-symmetric matrix. The system of linear coordinates $(x_1,\dots,x_d)$ is then said to be *log-canonical* for π.

The above-mentioned converse, which yields a sufficient condition for the diagonalizability of a quadratic Poisson structure is given in the following proposition.

Proposition 8.29. *Let π be a quadratic Poisson structure on a d-dimensional vector space V. Suppose that*

(1) The modular vector field Φ_π of π is diagonalizable, with eigenvalues $a_1,\dots,a_d$, which belong to $\mathbb{F}$;
(2) The $\binom{d+1}{2}$ sums a_i+a_j with $1\leqslant i\leqslant j\leqslant d$ are all different.

Then π is a diagonalizable Poisson structure on V. Moreover, every system of linear coordinates on V, consisting of eigenvectors of Φ_π, is log-canonical for π.

Proof. Since the modular vector field Φ_π is linear, it restricts to a linear endomorphism $\Phi_\pi^{(2)}$ of the space $\mathscr{A}_2$ of all homogeneous polynomial functions of degree 2 on V,

$$\begin{aligned}\Phi_\pi^{(2)}:\mathscr{A}_2&\to\mathscr{A}_2\\ F&\mapsto\Phi_\pi[F]\,.\end{aligned}$$

Suppose that Φ_π is diagonalizable and let $x_1,\dots,x_d$ be independent eigenvectors of Φ_π, with eigenvalues $a_1,\dots,a_d$. We use $x_1,\dots,x_d$ as linear coordinates on V. The $\binom{d+1}{2}$ polynomial functions $\{x_ix_j\mid 1\leqslant i\leqslant j\leqslant d\}$ form a basis of $\mathscr{A}_2$, and each one of them is an eigenvector of $\Phi_\pi^{(2)}$, because Φ_π is a derivation: x_ix_j is an eigenvector of $\Phi_\pi^{(2)}$, corresponding to the eigenvalue a_i+a_j, for all $1\leqslant i\leqslant j\leqslant d$. If we assume that the $\binom{d+1}{2}$ sums $\left\{a_i+a_j\mid 1\leqslant i\leqslant j\leqslant d\right\}$ are all different (which is assumption *(2)*), then all eigenvalues of $\Phi_\pi^{(2)}$ are different, hence the eigenspace decomposition of $\mathscr{A}_2$ with respect to $\Phi_\pi^{(2)}$ is given by

$$\mathscr{A}_2=\bigoplus_{1\leqslant i\leqslant j\leqslant d}\mathbb{C}x_ix_j\,. \tag{8.23}$$

We show that this implies that π is diagonalizable. To do this, we use that the modular vector field Φ_π is a Poisson vector field: writing $\{\cdot\,,\cdot\}$ for π and specializing

$$\Phi_\pi\left[\{F,G\}\right]=\{\Phi_\pi[F],G\}+\{F,\Phi_\pi[G]\}\ ,$$

to $F=x_i$ and $G=x_j$, where $1\leqslant i,j\leqslant d$, we obtain that

$$\Phi_\pi[\{x_i,x_j\}]=\left\{\Phi_\pi[x_i],x_j\right\}+\left\{x_i,\Phi_\pi[x_j]\right\}=(a_i+a_j)\left\{x_i,x_j\right\}\,.$$

Therefore, $\left\{x_i,x_j\right\}$ is zero or it is an eigenvector of $\Phi_\pi^{(2)}$ with respect to the eigenvalue a_i+a_j. In view of the eigenspace decomposition (8.23), it follows, in either case, that $\left\{x_i,x_j\right\}$ is proportional to x_ix_j, i.e., there exists a unique constant $a_{ij}\in\mathbb{F}$, such that $\left\{x_i,x_j\right\}=a_{ij}x_ix_j$. By construction, $a_{ij}=-a_{ji}$, and π is of the diagonal form (8.22) in the system of coordinates $(x_1,\dots,x_d)$. □

8.3 Nambu–Poisson Structures

Let V be a d-dimensional vector space, equipped with linear coordinates $x_1,\ldots,x_d$ and denote the algebra of functions on V by $\mathscr{F}(V)$, as before. Let $\chi,\varphi_1,\cdots,\varphi_{d-2}$ be $d-1$ functions on V and define, for $F,G\in\mathscr{F}(V)$, a function on V by

$$\{F,G\} := \chi\left|\frac{\partial(F,G,\varphi_1,\ldots,\varphi_{d-2})}{\partial(x_1,\ldots,x_d)}\right| = \chi\begin{vmatrix} \frac{\partial F}{\partial x_1} & \frac{\partial G}{\partial x_1} & \frac{\partial \varphi_1}{\partial x_1} & \cdots & \frac{\partial \varphi_{d-2}}{\partial x_1} \\ \vdots & & \vdots & & \vdots \\ \frac{\partial F}{\partial x_d} & \frac{\partial G}{\partial x_d} & \frac{\partial \varphi_1}{\partial x_d} & \cdots & \frac{\partial \varphi_{d-2}}{\partial x_d} \end{vmatrix} . \tag{8.24}$$

It leads to a map $\mathscr{F}(V)\times\mathscr{F}(V)\to\mathscr{F}(V)$, which is a skew-symmetric biderivation of $\mathscr{F}(V)$ and which, up to a non-zero constant, does not depend on the chosen linear coordinates $x_1,\ldots,x_d$, since (8.24) can also be written as

$$\{F,G\} = \chi\frac{\mathrm{d}F\wedge\mathrm{d}G\wedge\mathrm{d}\varphi_1\wedge\cdots\wedge\mathrm{d}\varphi_{d-2}}{\mathrm{d}x_1\wedge\mathrm{d}x_2\wedge\cdots\wedge\mathrm{d}x_d} . \tag{8.25}$$

Moreover, it defines a Poisson structure on V, a fact which we will prove in this section. When the functions $\varphi_1,\ldots,\varphi_{d-2}$ are polynomial or weight homogeneous, then so is the corresponding Poisson structure. In order to prove that $\{\cdot,\cdot\}$ defines a Poisson structure on V, we briefly introduce the notion of a Nambu–Poisson structure, which generalizes the notion of a Poisson structure.

Definition 8.30. For $q\geqslant 2$, a q-vector field $\{\cdot,\ldots,\cdot\}$ on a finite-dimensional vector space V is called a *Nambu–Poisson structure* of order q on V if it satisfies the so-called *fundamental identity*:

$$\begin{aligned}\{F_1,\ldots,F_{q-1},\{G_1,\ldots,G_q\}\} = \\ \sum_{i=1}^{q}\{G_1,\ldots,G_{i-1},\{F_1,\ldots,F_{q-1},G_i\},G_{i+1},\ldots,G_q\} ,\end{aligned} \tag{8.26}$$

for all functions $F_1,\ldots,F_{q-1},G_1,\ldots,G_q\in\mathscr{F}(V)$.

Remark 8.31. The definition is easily generalized to arbitrary (real or complex) manifolds by demanding that the q-vector field satisfies the fundamental identity for local functions, which yields the notion of a *Nambu–Poisson manifold*. Replacing the algebra of functions $\mathscr{F}(V)$ by an arbitrary commutative associative algebra, leads to the abstract notion of a *Nambu–Poisson algebra*, which generalizes the notion of a Poisson algebra, as given in Definition 1.1.

Example 8.32. Specializing Definition 8.30 to the case of $q=2$, we find that a Nambu–Poisson structure of order 2 is a bivector field $\{\cdot,\cdot\}$ which satisfies

$$\{F,\{G_1,G_2\}\} = \{\{F,G_1\},G_2\}+\{G_1,\{F,G_2\}\} ,$$

for all $F, G_1, G_2 \in \mathscr{F}(V)$, which is precisely the Jacobi identity for $\{\cdot,\cdot\}$. Therefore, a Nambu–Poisson structure of order 2 is the same as a Poisson structure. In this sense, Nambu–Poisson structures are a generalization of Poisson structures.

Given a Nambu–Poisson structure $\{\cdot,\dots,\cdot\}$ of order $q \geqslant 2$ on V, one can associate to $q-1$ elements $F_1,\dots,F_{q-1}$ of $\mathscr{F}(V)$, a vector field $\mathscr{X}_{F_1,\dots,F_{q-1}}$ on V by defining, for $G \in \mathscr{F}(V)$:

$$\mathscr{X}_{F_1,\dots,F_{q-1}}[G] := \{G, F_1, \dots, F_{q-1}\}.$$

It is called the *Hamiltonian vector field* associated to $F_1,\dots,F_{q-1}$. The fundamental identity (8.26) says that all Hamiltonian vector fields $\mathscr{X}_{F_1,\dots,F_{q-1}}$ are derivations of $(\mathscr{F}(V), \{\cdot,\dots,\cdot\})$.

We show in the following proposition how Nambu–Poisson structures of order $q-1$ can be constructed from Nambu–Poisson structures of order q.

Proposition 8.33. *Let $\{\cdot,\dots,\cdot\}$ be a Nambu–Poisson structure of order $q \geqslant 3$ on a finite-dimensional vector space V. Given $H \in \mathscr{F}(V)$, define a $(q-1)$-vector field on V by*

$$\{\cdot,\dots,\cdot\}_H := \{\cdot,\dots,\cdot,H\}.$$

Then $\{\cdot,\dots,\cdot\}_H$ is a Nambu–Poisson structure of order $q-1$ on V.

Proof. The fundamental identity (8.26) for $\{\cdot,\dots,\cdot\}_H$ takes the form

$$\begin{aligned}\{F_1,\dots,F_{q-2},\{G_1,\dots,G_{q-1}\}_H\}_H = \\ \sum_{i=1}^{q-1}\{G_1,\dots,G_{i-1},\{F_1,\dots,F_{q-2},G_i\}_H,G_{i+1},\dots,G_{q-1}\}_H,\end{aligned} \tag{8.27}$$

for all $F_1,\dots,F_{q-2},G_1,\dots,G_{q-1} \in \mathscr{F}(V)$. To prove (8.27), substitute $F_{q-1} = H$ and $G_q = H$ in the fundamental identity (8.26), to find

$$\begin{aligned}\{F_1,\dots,F_{q-2},H,\{G_1,\dots,G_{q-1},H\}\} = \\ \sum_{i=1}^{q-1}\{G_1,\dots,G_{i-1},\{F_1,\dots,F_{q-2},H,G_i\},G_{i+1},\dots,G_{q-1},H\} \\ + \{G_1,\dots,G_{q-1},\{F_1,\dots,F_{q-2},H,H\}\},\end{aligned} \tag{8.28}$$

which is precisely (8.27), since the skew-symmetry of $\{\cdot,\dots,\cdot\}$ implies that the last term in (8.28) is equal to zero. □

We show in the following proposition that for d-vector fields on a d-dimensional vector space V, the fundamental identity is automatically satisfied.

Proposition 8.34. *Let V be a d-dimensional vector space, with $d \geqslant 2$. Every d-vector field on V is a Nambu–Poisson structure of order d on V.*

Proof. Let $\{\cdot,\dots,\cdot\}$ be a d-vector field on V. We need to prove that $\{\cdot,\dots,\cdot\}$ satisfies the fundamental identity (8.26). To do this, we consider the multilinear map $\mathscr{N} : \mathscr{F}(V)^{2d-1} \to \mathscr{F}(V)$, which is defined by

$$\begin{aligned}\mathcal{N}(F_1,\ldots,F_{d-1},G_1,\ldots,G_d)\\ &:= \{F_1,\ldots,F_{d-1},\{G_1,\ldots,G_d\}\}\\ &\quad - \sum_{i=1}^{d}\{G_1,\ldots,G_{i-1},\{F_1,\ldots,F_{d-1},G_i\},G_{i+1},\ldots,G_d\}\,,\end{aligned}$$

for $F_1,\ldots,F_{d-1},G_1,\ldots,G_d \in \mathcal{F}(V)$. Obviously, the vanishing of $\mathcal{N}$ is equivalent to the fundamental identity (8.26) for $\{\cdot,\ldots,\cdot\}$. Moreover, since for all $F_1,\ldots,F_{d-1} \in \mathcal{F}(V)$, the map

$$\begin{array}{rl}\mathcal{F}(V)^d & \to \mathcal{F}(V)\\ (G_1,\ldots,G_d) & \mapsto \mathcal{N}(F_1,\ldots,F_{d-1},G_1,\ldots,G_d)\end{array} \tag{8.29}$$

is a skew-symmetric d-derivation of $\mathcal{F}(V)$, as is easily checked directly from the definition of $\mathcal{N}$, the vanishing of $\mathcal{N}$ is equivalent to

$$\mathcal{N}(F_1,\ldots,F_{d-1},x_1,\ldots,x_d) = 0\,, \tag{8.30}$$

for all $F_1,\ldots,F_{d-1} \in \mathcal{F}(V)$, where $(x_1,\ldots,x_d)$ is an arbitrary system of linear coordinates on V. If we denote $\chi := \{x_1,\ldots,x_d\}$, then the d-vector field $\{\cdot,\ldots,\cdot\}$ can be written, according to (8.1), as

$$\{\cdot,\ldots,\cdot\} = \chi\frac{\partial}{\partial x_1}\wedge\cdots\wedge\frac{\partial}{\partial x_d}\,, \tag{8.31}$$

so that (8.30), which we need to prove, takes the form

$$\{F_1,\ldots,F_{d-1},\chi\} = \chi\sum_{i=1}^{d}\frac{\partial}{\partial x_i}\{F_1,\ldots,F_{d-1},x_i\}\,. \tag{8.32}$$

In order to prove (8.32) we compute its left-hand side from (8.31):

$$\begin{aligned}\{F_1,\ldots,F_{d-1},\chi\} &= \chi\begin{vmatrix}\frac{\partial F_1}{\partial x_1} & \ldots & \frac{\partial F_{d-1}}{\partial x_1} & \frac{\partial \chi}{\partial x_1}\\ \vdots & & \vdots & \vdots\\ \frac{\partial F_1}{\partial x_d} & \ldots & \frac{\partial F_{d-1}}{\partial x_d} & \frac{\partial \chi}{\partial x_d}\end{vmatrix}\\ &= \chi\,\frac{\mathrm{d}F_1\wedge\cdots\wedge\mathrm{d}F_{d-1}\wedge\mathrm{d}\chi}{\mathrm{d}x_1\wedge\cdots\wedge\mathrm{d}x_d}\\ &= (-1)^{d-1}\chi\,\frac{\mathrm{d}\left(\chi\,\mathrm{d}F_1\wedge\cdots\wedge\mathrm{d}F_{d-1}\right)}{\mathrm{d}x_1\wedge\cdots\wedge\mathrm{d}x_d}\\ &= (-1)^{d-1}\chi\sum_{i=1}^{d}\frac{\mathrm{d}\left(C_i\,\mathrm{d}x_1\wedge\cdots\wedge\widehat{\mathrm{d}x_i}\wedge\cdots\wedge\mathrm{d}x_d\right)}{\mathrm{d}x_1\wedge\cdots\wedge\mathrm{d}x_d}\end{aligned}$$

$$= \chi \sum_{i=1}^{d} (-1)^{d-i} \frac{\partial C_i}{\partial x_i},$$

where

$$C_i := \chi\,(\mathrm{d}F_1 \wedge \cdots \wedge \mathrm{d}F_{d-1}) \left(\frac{\partial}{\partial x_1}, \ldots, \widehat{\frac{\partial}{\partial x_i}}, \ldots, \frac{\partial}{\partial x_d} \right)$$

$$= \chi\,(\mathrm{d}F_1 \wedge \cdots \wedge \mathrm{d}F_{d-1} \wedge \mathrm{d}x_i) \left(\frac{\partial}{\partial x_1}, \ldots, \widehat{\frac{\partial}{\partial x_i}}, \ldots, \frac{\partial}{\partial x_d}, \frac{\partial}{\partial x_i} \right)$$

$$= (-1)^{d-i} \chi\,(\mathrm{d}F_1 \wedge \cdots \wedge \mathrm{d}F_{d-1} \wedge \mathrm{d}x_i) \left(\frac{\partial}{\partial x_1}, \ldots, \frac{\partial}{\partial x_d} \right)$$

$$= (-1)^{d-i} \{F_1, \ldots, F_{d-1}, x_i\}\,.$$

We thus have obtained (8.32), so that $\{\cdot, \ldots, \cdot\}$ satisfies the fundamental identity and is a Nambu–Poisson structure on V. □

Combining Propositions 8.33 and 8.34, we find the following proposition, which yields the Poisson structures of rank 2, which we introduced in (8.24).

Proposition 8.35. *Let V be a d-dimensional vector space and let $\chi, \varphi_1, \ldots, \varphi_{d-2} \in \mathscr{F}(V)$ be $d-1$ functions on V. A Poisson structure π is defined on V by*

$$\{F, G\} := \chi \left| \frac{\partial (F, G, \varphi_1, \ldots, \varphi_{d-2})}{\partial (x_1, \ldots, x_d)} \right|, \tag{8.33}$$

for all $F, G \in \mathscr{F}(V)$, where $(x_1, \ldots, x_d)$ is an arbitrary system of linear coordinates on V. The given functions $\varphi_1, \ldots, \varphi_{d-2}$ are Casimir functions for π and the rank of π is 2 at all points of V, except at the zeros of χ and at the points $m \in V$ where $\mathrm{d}_m\varphi_1, \ldots, \mathrm{d}_m\varphi_{d-2}$ are linearly dependent (in those points the rank is zero).

Proof. Repeated use of Proposition 8.33 on the d-vector field (Nambu–Poisson structure of order d) $\{\cdot, \ldots, \cdot\} := \chi \frac{\partial}{\partial x_1} \wedge \cdots \wedge \frac{\partial}{\partial x_d}$ on V, yields a Nambu–Poisson structure π of order 2, which is given by

$$\{F, G\} := \{F, G, \varphi_1, \ldots, \varphi_{d-2}\} = \chi \left| \frac{\partial (F, G, \varphi_1, \ldots, \varphi_{d-2})}{\partial (x_1, \ldots, x_d)} \right|. \tag{8.34}$$

Since π is a Nambu–Poisson structure of order 2, it is a Poisson structure (see Example 8.32). The skew-symmetry of $\{\cdot, \ldots, \cdot\}$ implies that $\{F, \varphi_i\} = 0$ for all $F \in \mathscr{F}(V)$ and $1 \leqslant i \leqslant d-2$, so that each one of $\varphi_1, \ldots, \varphi_{d-2}$ is a Casimir function of π. If $m \in V$ with $\chi(m) = 0$ or $\mathrm{d}_m\varphi_1 \wedge \cdots \wedge \mathrm{d}_m\varphi_{d-2} = 0$, then it is easy to see from (8.25) that $\{F, G\}(m) = 0$ for all $F, G \in \mathscr{F}(V)$, so that the Poisson structure vanishes at m. Suppose now that m is a point of V for which $\chi(m) \neq 0$ and $\mathrm{d}_m\varphi_1 \wedge \cdots \wedge \mathrm{d}_m\varphi_{d-2} \neq 0$. Then there exist i, j such that $\mathrm{d}_m x_i \wedge \mathrm{d}_m x_j \wedge \mathrm{d}_m\varphi_1 \wedge \cdots \wedge \mathrm{d}_m\varphi_{d-2} \neq 0$, so that $\{x_i, x_j\}(m) \neq 0$. This shows that $\mathrm{Rk}_m \pi \geqslant 2$; since we have $d-2$ independent Casimir functions at m, the rank is at most 2 at m, so it is precisely 2. □

It follows at once from Eq. (8.33) that, if the functions $\chi,\varphi_1,\dots,\varphi_{d-2}$ are polynomial functions on V, then the Poisson structure π is a polynomial Poisson structure on V. If, moreover, for some fixed weights $\varpi_1,\dots,\varpi_d$ of the linear coordinates $x_1,\dots,x_d$ on V, the functions $\chi,\varphi_1,\dots,\varphi_{d-2}$ are weight homogeneous, then it also follows from this formula that the Poisson structure π is a weight homogeneous Poisson structure on V, of weight

$$\varpi(\pi)=\varpi(\chi)+\sum_{i=1}^{d-2}\varpi(\varphi_i)-\sum_{i=1}^{d}\varpi_i\;.$$

Remark 8.36. In the case of a d-dimensional manifold M, every d-vector field on M defines a Nambu–Poisson structure of order d and hence, upon fixing $d-2$ functions $\varphi_1,\dots,\varphi_{d-2}$ on M, a Poisson structure of rank two on M (assuming that there is at least one point where the differentials of these functions are independent). In local coordinates $x_1,\dots,x_d$, a d-vector field $\{\cdot,\dots,\cdot\}$ takes the form

$$\{\cdot,\dots,\cdot\}=\chi\frac{\partial}{\partial x_1}\wedge\cdots\wedge\frac{\partial}{\partial x_d}\;,$$

and the formula for the Poisson structure is given by

$$\{F,G\}:=\chi\left|\frac{\partial\left(F,G,\varphi_1,\dots,\varphi_{d-2}\right)}{\partial\left(x_1,\dots,x_d\right)}\right|\;,$$

so it is in local coordinates formally given by the same expression as in (8.33). A coordinate-free description of the Poisson structure can be given when M is an orientable manifold, so that there exists a volume form λ on M. Then a Nambu–Poisson structure $\{\cdot,\dots,\cdot\}$ of order d is defined on M by letting

$$\{F_1,\dots,F_d\}:=\chi\frac{\mathrm{d}F_1\wedge\cdots\wedge\mathrm{d}F_d}{\lambda}\;,$$

for all (local) functions $F_1,\dots,F_d$ on M. In particular, if we fix $d-1$ functions $\chi,\varphi_1,\dots,\varphi_{d-2}$ on M and a volume form λ, then a Poisson structure π on M is defined by

$$\{F,G\}:=\chi\frac{\mathrm{d}F\wedge\mathrm{d}G\wedge\mathrm{d}\varphi_1\wedge\cdots\wedge\mathrm{d}\varphi_{d-2}}{\lambda}\;,$$

for all (local) functions F,G on M. The functions $\varphi_1,\dots,\varphi_{d-2}$ are Casimir functions of π and the rank of π is two at all points of M, except at the zeros of χ and at the points where the differentials of the latter Casimir functions are dependent. In this case, a simple proof of the fact that π is a Poisson structure can be given by extending, at points where the rank is two, the Casimir functions $\varphi_1,\dots,\varphi_{d-2}$ to a system of local coordinates.

8.4 Transverse Poisson Structures to Adjoint Orbits

In this section, we show that, in terms of a system of natural coordinates which we will construct, the transverse Poisson structure to a nilpotent orbit in a semi-simple Lie algebra is weight homogeneous (of weight -2).

8.4.1 S-triples in Semi-Simple Lie Algebras

In this section, we recall a few facts which we will use about complex semi-simple Lie algebra (see [99, 190] for details and proofs). Let $(\mathfrak{g},[\cdot\,,\cdot])$ be a semi-simple Lie algebra over $\mathbb{F}=\mathbb{C}$; we denote the algebra of polynomial functions or of holomorphic functions on $\mathfrak{g}$ by $\mathscr{F}(\mathfrak{g})$. The Killing form of $\mathfrak{g}$ is the symmetric bilinear form on $\mathfrak{g}$, defined for all $x,y\in\mathfrak{g}$ by

$$\langle x\,|\,y\rangle := \operatorname{Trace}(\operatorname{ad}_x\circ\operatorname{ad}_y)\,,$$

where we recall that $\operatorname{ad}_x:\mathfrak{g}\to\mathfrak{g}$ is the linear map, defined by $\operatorname{ad}_x z:=[x,z]$, for all $z\in\mathfrak{g}$. For a subspace $\mathfrak{n}$ of $\mathfrak{g}$, the orthogonal complement of $\mathfrak{n}$ in $\mathfrak{g}$ with respect to the Killing form is denoted by $\mathfrak{n}^\perp$. For $x\in\mathfrak{g}$, we denote $\mathfrak{g}(x):=\{y\in\mathfrak{g}\mid[x,y]=0\}$, the *centralizer* of x in $\mathfrak{g}$. An element $x\in\mathfrak{g}$ is said to be *nilpotent* if the endomorphism ad_x of $\mathfrak{g}$ is nilpotent, i.e., if there exists a $k\in\mathbb{N}$, such that $\operatorname{ad}_x^k=0$. According to the Jacobson–Morozov theorem, if $e\in\mathfrak{g}$ is nilpotent, then there exist $h,f\in\mathfrak{g}$ such that

$$[h,e]=2e\,,\qquad [h,f]=-2f\,,\qquad [e,f]=h\,. \tag{8.35}$$

Every non-zero triple (h,e,f), satisfying (8.35), is called an *S-triple* of $\mathfrak{g}$. In view of (8.35), the vectors h,e and f generate a Lie subalgebra of $\mathfrak{g}$, isomorphic to $\mathfrak{sl}_2(\mathbb{C})$, which we denote by $\mathfrak{s}$. It leads to two decompositions of $\mathfrak{g}$:

(1) A decomposition of $\mathfrak{g}$ into eigenspaces relative to ad_h, making $\mathfrak{g}$ into a graded Lie algebra. Since each eigenvalue of ad_h is an integer, we have

$$\mathfrak{g}=\bigoplus_{i\in\mathbb{Z}}\mathfrak{g}_i\,,$$

where $\mathfrak{g}_i$ is the eigenspace of ad_h which corresponds to the eigenvalue i. For example, $e\in\mathfrak{g}_2$ and $f\in\mathfrak{g}_{-2}$. For $x\in\mathfrak{g}_i$ and $y\in\mathfrak{g}_j$, we have an inclusion $\operatorname{ad}_x\circ\operatorname{ad}_y(\mathfrak{g}_k)\subset\mathfrak{g}_{i+j+k}$, for every k, so that

$$\langle\mathfrak{g}_i\,|\,\mathfrak{g}_j\rangle=0\quad\text{if}\quad i+j\neq 0\,. \tag{8.36}$$

(2) A decomposition of $\mathfrak{g}$ as an $\mathfrak{s}$-module:

$$\mathfrak{g}=\bigoplus_{j=1}^{s}V_{n_j}\,, \tag{8.37}$$

where each V_{n_j} is a simple $\mathfrak{s}$-module, with ad_h-eigenvalues (also called ad_h-weights) $n_j, n_j-2, n_j-4, \ldots, -n_j$, so that $\dim V_{n_j} = n_j + 1$. It follows that

$$\sum_{j=1}^{s} n_j = \dim \mathfrak{g} - s\,. \tag{8.38}$$

Moreover, $s = \dim \mathfrak{g}(e)$, since the centralizer $\mathfrak{g}(e)$ of e is generated by the vectors of highest ad_h-weight of each V_{n_j}.

8.4.2 Weights Associated to an S-Triple

We show in this section that every S-triple (h,e,f) of $\mathfrak{g}$ leads to a system of linear coordinates on $\mathfrak{g}$, centered at e, such that the Lie–Poisson structure on $\mathfrak{g}$ is weight homogeneous with respect to some weight vector ϖ, which has a natural Lie algebraic interpretation: up to a shift of 2, it consists of the weights of $\mathfrak{g}$ as an $\mathfrak{s}$-module. For every $t \in \mathbb{C}^*$, choose a complex number λ_t such that $\exp(-\lambda_t) = t$. Such a choice cannot be made in such a way that λ_t depends in a continuous way on t, but that will be irrelevant for what follows. We denote the adjoint group of $\mathfrak{g}$ by $\mathbf{G}$ and we consider the map

$$\begin{aligned} \lambda : \mathbb{C}^* &\to \mathbf{G} \\ t &\mapsto \exp(\lambda_t h)\,. \end{aligned}$$

For a fixed t, the map $\mathrm{Ad}_{\lambda(t)} : \mathfrak{g} \to \mathfrak{g}$ is well-defined (independent of the choice made for λ_t) and it acts on each of the eigenspaces $\mathfrak{g}_i$ of ad_h as a homothecy with ratio t^{-i},

$$\mathrm{Ad}_{\lambda(t)}\, x = t^{-i} x\,, \tag{8.39}$$

for all x in $\mathfrak{g}_i$. Indeed, for all $t \in \mathbb{C}^*$ and $x \in \mathfrak{g}_i$,

$$\begin{aligned} \mathrm{Ad}_{\lambda(t)}\, x &= \mathrm{Ad}_{\exp(\lambda_t h)}\, x = \exp(\mathrm{ad}_{\lambda_t h}) x = \exp(\lambda_t\, \mathrm{ad}_h) x \\ &= \sum_{k=0}^{\infty} (i\lambda_t)^k x = \exp(i\lambda_t) x = t^{-i} x\,; \end{aligned}$$

also, replacing in this computation λ_t by $\lambda_t + 2\pi n\sqrt{-1}$, where $n \in \mathbb{Z}$, clearly leads to the same result, which shows that the above choice of λ_t is irrelevant. It follows that $t \mapsto \mathrm{Ad}_{\lambda(t)}$ is a (linear) action of $\mathbb{C}^*$ on $\mathfrak{g}$. Since $e \in \mathfrak{g}_2$, the action ρ of $\mathbb{C}^*$ on $\mathfrak{g}$, defined for $t \in \mathbb{C}^*$ and for $y \in \mathfrak{g}$ by $\rho_t \cdot y := t^2\, \mathrm{Ad}_{\lambda(t)}\, y$, fixes e; the action ρ is known as *Slodowy's action*. Let us denote for $x \in \mathfrak{g}$ by F_x the affine function on $\mathfrak{g}$, defined by

$$F_x(z) := \langle z - e \,|\, x \rangle\ ,$$

for all $z \in \mathfrak{g}$. We consider a basis $(x_1, \ldots, x_d)$ of $\mathfrak{g}$, where each x_k belongs to some eigenspace $\mathfrak{g}_{i_k}$, and we denote the function F_{x_k} which corresponds to x_k simply by F_k. For $k = 1, \ldots, d$, the functions F_k define linear coordinates on $\mathfrak{g}$, centered at e. For

$x \in \mathfrak{g}_i$, (8.39) and Ad-invariance of the Killing form imply that

$$\begin{aligned}(\rho_t^* F_x)(z) &= \langle \rho_{t^{-1}} \cdot z - e \,|\, x \rangle = t^{-2} \left\langle \mathrm{Ad}_{\lambda(t^{-1})}(z-e) \,|\, x \right\rangle \\ &= t^{-2} \langle z - e \,|\, \mathrm{Ad}_{\lambda(t)} x \rangle = t^{-2} \langle z - e \,|\, t^{-i} x \rangle = t^{-i-2} F_x(z) \,.\end{aligned}$$

We claim that, if we define for $x \in \mathfrak{g}_i$ the weight $\varpi(F_x)$ of F_x to be equal to $i+2$, then the Lie–Poisson structure $\{\cdot,\cdot\}_\mathfrak{g}$, which is an affine Poisson structure in terms of the coordinates $F_1, \dots, F_d$, is weight homogeneous of degree -2 with respect to the weight vector $(\varpi(F_1), \dots, \varpi(F_d)) = (i_1+2, \dots, i_d+2)$. To prove this, first recall from Section 7.2 that the Lie–Poisson structure $\{\cdot,\cdot\}_\mathfrak{g}$ on $\mathfrak{g}$ is given, for $F, G \in \mathscr{F}(\mathfrak{g})$, by

$$\{F, G\}_\mathfrak{g}(z) = \langle z \,|\, [\nabla_z F, \nabla_z G] \rangle \,, \tag{8.40}$$

where we recall from (7.11) that $\nabla_z F$, the gradient of F at z (with respect to $\langle \cdot \,|\, \cdot \rangle$) is defined, for $F \in \mathscr{F}(\mathfrak{g})$ and $z \in \mathfrak{g}$ by

$$\langle \nabla_z F \,|\, u \rangle = \langle \mathrm{d}_z F, u \rangle \,,$$

for all $u \in \mathfrak{g}$. Since $\nabla_z F_x = x$, for all $x, z \in \mathfrak{g}$, we have, for all x, y, $z \in \mathfrak{g}$,

$$\{F_x, F_y\}_\mathfrak{g}(z) = \langle z \,|\, [x,y] \rangle = F_{[x,y]}(z) + \langle e \,|\, [x,y] \rangle \,. \tag{8.41}$$

If $x \in \mathfrak{g}_i$ and $y \in \mathfrak{g}_j$, with $i + j \neq -2$, then (8.36) implies that $\langle e \,|\, [x,y] \rangle = 0$, so that

$$\begin{aligned}\varpi(\{F_x, F_y\}_\mathfrak{g}) - \varpi(F_x) - \varpi(F_y) &= \varpi(F_{[x,y]}) - \varpi(F_x) - \varpi(F_y) \\ &= i + j + 2 - (i+2) - (j+2) = -2 \,.\end{aligned}$$

This result extends to the case $i + j = -2$, since then $\varpi(F_{[x,y]}) = i + j + 2 = 0$, which is the weight of the constant function $\langle e \,|\, [x,y] \rangle$, and then $\varpi(\{F_x, F_y\}_g) - \varpi(F_x) - \varpi(F_y) = -(i+2) - (j+2) = -2$. This shows, according to (8.11), that the Lie–Poisson structure on $\mathfrak{g}$ is weight homogeneous of degree -2 with respect to the weight vector $(i_1+2, \dots, i_d+2)$. Notice that this weight vector contains, in general, both positive and negative integers.

8.4.3 Transverse Poisson Structures to Nilpotent Orbits

In this section we consider the transverse Poisson structure to a *nilpotent orbit* $\mathscr{O}$ in a semi-simple Lie algebra $\mathfrak{g}$. Such an orbit is by definition the orbit $\mathbf{G} \cdot e$ of a nilpotent element $e \in \mathfrak{g}$ under the adjoint action of $\mathbf{G}$ (the adjoint group of $\mathfrak{g}$) on $\mathfrak{g}$; in this case, all elements of $\mathscr{O}$ are nilpotent. Since $\mathscr{O}$ is a symplectic leaf of the Lie–Poisson structure on $\mathfrak{g}$ (see Section 7.3.1), the transverse Poisson structure at $e \in \mathscr{O}$ is independent of the chosen point on the orbit; we speak of the *transverse structure to the orbit*. Recall from Theorem 5.28 that the transverse Poisson structure at e is the reduced Poisson structure on an arbitrary Poisson–Dirac manifold which is transversal to the symplectic leaf $\mathscr{O}$ at e. We show in the following proposition that, in terms of coordinates which we will construct, the transverse Poisson structure to a nilpotent orbit is polynomial and weight homogeneous of weight -2, just like the Lie–Poisson structure on $\mathfrak{g}$. Recall from Section 8.4.2 that the weight homogeneity of the Lie–Poisson structure on $\mathfrak{g}$ was determined by the choice of an S-triple (h,e,f) of $\mathfrak{g}$ which contains e, that this S-triple generates a Lie subalgebra $\mathfrak{s}$ of $\mathfrak{g}$, isomorphic to $\mathfrak{sl}_2(\mathbb{C})$ and that up to a shift of 2, the weight vector consists of the weights of $\mathfrak{g}$, as an $\mathfrak{s}$-module.

Proposition 8.37. *Let $\mathscr{O} = \mathbf{G}\cdot e$ be a nilpotent orbit in a semi-simple Lie algebra $\mathfrak{g}$ and let (h,e,f) be an S-triple of $\mathfrak{g}$ which contains e. If $\mathfrak{n}$ is a complementary subspace to $\mathfrak{g}(e)$ in $\mathfrak{g}$, which is* ad_h*-invariant,* $[h,\mathfrak{n}]\subset\mathfrak{n}$*, then*

(1) $N := e+\mathfrak{n}^\perp$ *is transverse to $\mathscr{O}$ at e;*
(2) *N is a Poisson–Dirac submanifold of $\mathfrak{g}$;*
(3) *The transverse Poisson structure on N is a polynomial Poisson structure which is weight homogeneous of weight -2, with respect to the weight vector $(n_1+2,\dots,n_s+2)$.*

The integers $n_1,\dots,n_s$ are the highest weights of $\mathfrak{g}$ as an $\mathfrak{s}$-module, where $\mathfrak{s}$ denotes the Lie subalgebra of $\mathfrak{g}$, generated by h,e and f.

Proof. First, let us point out that an ad_h-invariant complementary subspace $\mathfrak{n}$ to $\mathfrak{g}(e)$ exists, because the centralizer $\mathfrak{g}(e)$ is generated by the highest weight vectors of each V_{n_j} in the decomposition (8.37) of $\mathfrak{g}$. Also, it is clear that

$$T_e\,\mathfrak{g} = T_e\mathscr{O}\oplus T_eN\;,$$

since under the natural isomorphism $T_e\mathfrak{g}\simeq\mathfrak{g}$, the subspace $T_e\mathscr{O}=T_e(\mathbf{G}\cdot e)$ corresponds to $\mathrm{ad}_{\mathfrak{g}}\,e=\mathfrak{g}(e)^\perp$ and T_eN corresponds to $\mathfrak{n}^\perp$. In particular, N is transverse to $\mathscr{O}$ at e, which is *(1)*.

In order to show *(2)* and *(3)*, we use a natural basis of $\mathfrak{g}=\mathfrak{g}(e)\oplus\mathfrak{n}$. As we already pointed out, the centralizer $\mathfrak{g}(e)$ of e in $\mathfrak{g}$ admits a basis $(x_1,\dots,x_s)$, consisting of the highest weight vectors of $V_{n_1},\dots,V_{n_s}$ in the decomposition (8.37) of $\mathfrak{g}$; recall also that the $\mathrm{ad}_{\mathfrak{h}}$-weight of x_j is n_j. Since $\mathfrak{n}$, which has dimension $2r:=\dim\mathscr{O}$, is ad_h-invariant, we can also choose a basis $(x_{s+1},\dots,x_{s+2r})$ for $\mathfrak{n}$, where each vector $x_{s+\ell}$ is an eigenvector of ad_h; we denote the ad_h-weight of $x_{s+\ell}$ by ν_ℓ, where $1\leqslant\ell\leqslant 2r$. The basis $(x_1,\dots,x_{s+2r})$ of $\mathfrak{g}$ leads to linear coordinates $F_1,\dots,F_{s+2r}$ on $\mathfrak{g}$, centered at e, defined by $F_k(z):=\langle z-e\,|\,x_k\rangle$, for $k=1,\dots,s+2r$ and $z\in\mathfrak{g}$. Since $\nabla_zF_k=x_k$ for $k=1,\dots,s+2r$, it follows from (8.40) that the Poisson matrix of $\{\cdot,\cdot\}_{\mathfrak{g}}$ at $z\in\mathfrak{g}$ is given by

$$\left(\{F_i,F_j\}_{\mathfrak{g}}(z)\right)_{1\leqslant i,j\leqslant s+2r}=\begin{pmatrix}A(z) & B(z)\\ -B(z)^\top & D(z)\end{pmatrix},$$

where

$$
\begin{aligned}
A_{ij}(z) &= \langle z \,|\, [x_i, x_j] \rangle & & 1 \leqslant i, j \leqslant s\,;\\
B_{im}(z) &= \langle z \,|\, [x_i, x_{s+m}] \rangle & & 1 \leqslant i \leqslant s, \quad 1 \leqslant m \leqslant 2r\,;\\
D_{\ell m}(z) &= \langle z \,|\, [x_{s+\ell}, x_{s+m}] \rangle & & 1 \leqslant \ell, m \leqslant 2r\,.
\end{aligned}
$$

At e, both matrices A and B vanish, so the rank of $D(e)$ is the rank of $\{\cdot,\cdot\}_{\mathfrak{g}}$ at e, which equals $\dim \mathscr{O} = 2r$. It follows that $D(e)$ is invertible and hence that $D(z)$ is invertible for z in a neighborhood $V \subset N$ of e in $\mathfrak{g}$. According to Proposition 5.27, V is a Poisson–Dirac submanifold of $\mathfrak{g}$ and the Poisson matrix X_N of the reduced Poisson structure $\{\cdot,\cdot\}_N$ is given, for $n \in N$, by

$$X_N(n) = A(n) + B(n)D(n)^{-1}B(n)^\top\,. \tag{8.42}$$

The functions $F_1, \dots, F_s$, restricted to N, will be denoted by $q_1, \dots, q_s$; they yield linear coordinates (centered at e) for N and all other functions F_k vanish on N, so that N can also be defined as the submanifold defined by $F_{s+1} = \cdots = F_{s+2r} = 0$. Since the weight of F_i is $n_i + 2$, we define the weight of q_i to be $\varpi(q_i) := n_i + 2$, for $i = 1, \dots, s$. Notice that these weights are all positive.

We prove that the determinant of D is a non-zero constant function on N, which shows that the transverse Poisson structure $\{\cdot,\cdot\}_N$ is polynomial, in view of (8.42). Since the restriction of the polynomial function $\det(D)$ on $\mathfrak{g}$ to N is obtained by setting $F_{s+1} = \cdots = F_{s+2r} = 0$ and since the remaining coordinates $q_1, \dots, q_s$ have positive weight, it suffices to show that the restriction of $\det(D)$ to N is a weight homogeneous function of weight 0. A typical term of $\det(D)$ is, up to a sign, of the form $D_{1,\alpha_1} \dots D_{2r,\alpha_{2r}}$, where $\{\alpha_1, \dots, \alpha_{2r}\} = \{1, \dots, 2r\}$, so that its weight is given by

$$\sum_{\ell=1}^{2r} \varpi(D_{\ell,\alpha_\ell}) = \sum_{\ell=1}^{2r} (\nu_\ell + \nu_{\alpha_\ell} + 2) = 2\sum_{\ell=1}^{2r} (\nu_\ell + 1)\,.$$

Since $\sum_{\ell=1}^{2r} \nu_\ell + \sum_{i=1}^{s} n_i$ is the sum of all ad_h-weights, it is zero, which yields in combination with (8.38)

$$\sum_{\ell=1}^{2r} (\nu_\ell + 1) = 2r - \sum_{i=1}^{s} n_i = 2r - (\dim \mathfrak{g} - s) = 0\,.$$

This shows that every term of $\det(D)$, restricted to N, is of weight zero, hence $\det(D)$ is constant; since $\det(D(e)) \neq 0$, this constant is different from zero and the transverse Poisson structure is polynomial.

To finish, we show that the Poisson structure $\{\cdot,\cdot\}_N$ is weight homogeneous of weight -2 with respect to $\varpi = (n_1 + 2, \dots, n_s + 2)$. According to (8.42) we need to show that for all $1 \leqslant i, j \leqslant s$ the functions A_{ij} and $(BD^{-1}B^\top)_{ij}$ are weight homogeneous of degree $\varpi(q_i) + \varpi(q_j) - 2 = n_i + n_j + 2$. For A_{ij} this is clear, since A is part of the Poisson matrix of the Lie–Poisson structure on $\mathfrak{g}$, which we have seen to be weight homogeneous of weight -2. Similarly, we have that $\varpi(B_{im}) = n_i + \nu_m + 2$. Since

$$\varpi(B_{im} D_{m\ell}^{-1} B_{j\ell}) = n_i + n_j + \nu_m + \nu_\ell + 4 + \varpi(D_{m\ell}^{-1})\,,$$

it means that we need to show that

$$\varpi(D^{-1}_{m\ell}) = -\nu_m - \nu_\ell - 2\,. \tag{8.43}$$

A typical term of the element $D^{-1}_{m\ell}$ of the inverse matrix D^{-1} is of the form $\frac{D'_{\alpha\beta}}{\det(D)}$, where $D'_{\alpha\beta} = D_{\alpha_1\beta_1}\dots D_{\alpha_{2r-1}\beta_{2r-1}}$, with

$$\{\alpha_1,\dots,\alpha_{2r-1}\} = \{1,\dots,2r\}\setminus\{\ell\} \text{ and } \{\beta_1,\dots,\beta_{2r-1}\} = \{1,\dots,2r\}\setminus\{m\}\,.$$

Since $\det(D)$ has weight zero, we need to show that $D'_{\alpha\beta}$ has weight $-\nu_m - \nu_\ell - 2$, which is done in precisely the same way as the above proof that $\det(D)$ has weight zero. This gives us (8.43). □

8.5 Notes

For higher order Poisson structures, starting with quadratic Poisson structures, there is no general theory and there is no immediate interpretation, as there is in the case of constant Poisson structures (in terms of bivectors) and in the case of linear Poisson structures (in terms of Lie brackets). For a few subclasses of quadratic Poisson structures there is, however, a quite general theory and an interpretation. Namely, multiplicative quadratic Poisson structures are closely related to Poisson–Lie groups: the datum of a skew-symmetric r-matrix on an associative algebra leads to multiplicative quadratic Poisson structure, which makes the set of invertible elements into a Poisson–Lie group; see Section 11.1.6 for details. It also leads to a cubic Poisson structure, whose virtue seems at this point still very mysterious. Poisson structures coming from r-matrices or R-matrices are the subject of Chapter 10; see Section 10.3.

See Dufour–Zung [63] for an extensive account on Nambu–Poisson structures. Transverse Poisson structures to general adjoint orbits in a semi-simple Lie algebra, with special emphasis to transverse Poisson structures to subregular orbits, are studied in Damianou–Sabourin–Vanhaecke [56].

Chapter 9
Poisson Structures in Dimensions Two and Three

Low-dimensional Poisson manifolds are often used as toy models, to obtain a better understanding of the theory of Poisson manifolds, as well as to illustrate their unexpected complexity. In the two-dimensional case, for example, describing all Poisson structures on the affine space $\mathbb{F}^2$ is deceivingly simple, because the Jacobi identity is satisfied for all bivector fields; however, their local classification is non-trivial, and has up to now only been accomplished under quite strong regularity assumptions on the singular locus of the Poisson structure, which can be identified with the zero locus of a local function on the manifold. As a result, the study of Poisson structures in two dimensions already takes us to the non-trivial singularity theory of functions of two variables!

Dimension three is the smallest dimension in which the Jacobi identity for a bivector field is not always satisfied. In this dimension, the Jacobi identity can be stated as an integrability condition of a distribution, which eventually leads to the symplectic foliation, or as the integrability condition of a differential one-form, dual to the Poisson structure with respect to a volume form (assuming that the manifold is orientable). Despite the extra complexity which comes from the extra dimension, Poisson structures in dimension three have one key property in common with Poisson structures in dimension two: their rank is equal to two (except when the Poisson structure is trivial). Many results about three-dimensional Poisson manifolds are essentially true because the Poisson structure is of rank two. A key example of such a result is the characterization of a regular Poisson structure of rank two in terms of its symplectic foliation.

Poisson structures in dimensions two and three are presented in different sections (Sections 9.1 and 9.2). In both cases, we pay special attention to the global/local and geometrical/algebraic aspects of these Poisson structures and we prove at the end of each section a (partial) classification result.

Unless otherwise stated, $\mathbb{F}$ is an arbitrary field of characteristic zero.

C. Laurent-Gengoux et al., *Poisson Structures*,
Grundlehren der mathematischen Wissenschaften 347,
DOI 10.1007/978-3-642-31090-4_9, © Springer-Verlag Berlin Heidelberg 2013

9.1 Poisson Structures in Dimension Two

In this section, we study Poisson structures on manifolds or affine varieties of dimension two. Since a Poisson structure is a bivector field or a biderivation, the smallest dimension which one can consider for studying non-trivial Poisson structures is dimension two. In this case, the Jacobi identity is always satisfied, so that every bivector field or skew-symmetric biderivation is a Poisson structure. We present Poisson structures in dimension two first from the global point of view, then from the local point of view and finally we discuss their formal classification.

9.1.1 Global Point of View

First of all, let M be a manifold (real or complex, depending on whether $\mathbb{F}=\mathbb{R}$ or $\mathbb{F}=\mathbb{C}$) of dimension two and let $\mathscr{F}(M)$ be the algebra of smooth or holomorphic functions on M. According to Proposition 3.5, every bivector field π on M is a Poisson structure because $[\pi,\pi]_S$ is a trivector field on M, hence is zero. We conclude that the space of all Poisson structures on M coincides with the vector space $\mathfrak{X}^2(M)$ of all bivector fields on M. In particular, two Poisson structures on M are always compatible (see Section 3.3.2). For every point m of M, the rank of π at m is zero or two. Discarding the case of the trivial Poisson structure, it follows that the singular locus of (M,π) coincides with the set of points of M where π vanishes.

Next, let us consider a complex affine surface $M\subset\mathbb{C}^d$ of dimension two, equipped with its algebra $\mathscr{F}(M)=\mathbb{C}[x_1,\dots,x_d]/\mathscr{I}$ of regular functions. We show that every skew-symmetric biderivation π of $\mathscr{F}(M)$ is a Poisson structure of M. To do this, we prove that every skew-symmetric triderivation of $\mathscr{F}(M)$ is zero; since $[\pi,\pi]_S$ is a skew-symmetric triderivation of $\mathscr{F}(M)$, it follows that π is a Poisson structure on M. Let us denote by M^{sm} the set of all smooth points of M. Let P be a skew-symmetric triderivation on M and suppose that P is different from zero at some point $m\in M$. Since P can be written as

$$P=\sum_{1\leqslant i<j<k\leqslant d} P_{ijk}\frac{\partial}{\partial x_i}\wedge\frac{\partial}{\partial x_j}\wedge\frac{\partial}{\partial x_k},$$

where $P_{ijk}:=P[x_i,x_j,x_k]$, this means that $P_{ijk}(m)\neq 0$, for some i,j,k. It follows that P is different from zero in a neighborhood of m in M. Since M^{sm} is dense in M, this means that P is non-zero in the neighborhood of some smooth point of M. However, as we have seen in Section 2.3.2, the set of smooth points of M carries a natural structure of a complex manifold, and every skew-symmetric triderivation of $\mathscr{F}(M)$ corresponds on it to a trivector field. Since M^{sm} is a two-dimensional manifold, every trivector field on M^{sm} is zero, hence P is zero on M^{sm}, and we arrive at a contradiction. This shows that P vanishes at every point of M, hence that every skew-symmetric biderivation π of $\mathscr{F}(M)$ is a Poisson structure on M. As in the case of two-dimensional Poisson manifolds, the rank of π can only take the

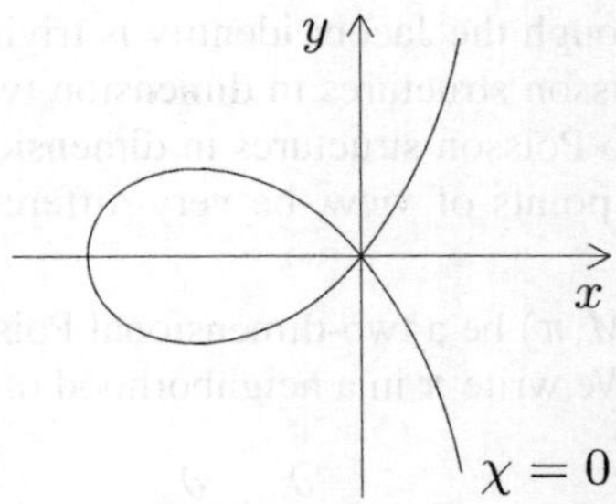

Fig. 9.1 The singular locus of the Poisson structure $\pi = \chi \frac{\partial}{\partial x} \wedge \frac{\partial}{\partial y}$ is the zero locus of χ, i.e., the curve defined by $\chi = 0$.

values zero and two at the points of M, and the singular locus of (M, π) coincides with the set of points where π vanishes (assuming that π is non-trivial).

Finally, consider $\mathbb{F}^2$, with standard coordinates (x, y), equipped with its algebra of functions $\mathscr{F}(\mathbb{F}^2)$; recall that $\mathscr{F}(\mathbb{F}^2)$ denotes the algebra of polynomial functions on $\mathbb{F}^2$ if $\mathbb{F}$ is an arbitrary field, but can also be the algebra of smooth or holomorphic functions on $\mathbb{F}^2$ if $\mathbb{F} = \mathbb{R}$ or $\mathbb{F} = \mathbb{C}$. Every skew-symmetric biderivation on $\mathscr{F}(\mathbb{F}^2)$ or bivector field on $\mathbb{F}^2$ (i.e., every Poisson structure on $\mathbb{F}^2$) is of the form

$$\pi = \chi \frac{\partial}{\partial x} \wedge \frac{\partial}{\partial y}, \quad \text{where } \chi \in \mathscr{F}(\mathbb{F}^2), \tag{9.1}$$

and by the above, every such π is a Poisson structure on $\mathbb{F}^2$. Hence, there is a one-to-one correspondence between the following three objects: the algebra $\mathscr{F}(\mathbb{F}^2)$ of all functions on $\mathbb{F}^2$, the vector space $\mathfrak{X}^2(\mathbb{F}^2)$ of all bivector fields on $\mathbb{F}^2$ and the set of all Poisson structures on $\mathbb{F}^2$. The singular locus of π coincides with the zero locus of χ, which is the curve $\{\chi = 0\} \subset \mathbb{F}^2$. See Fig. 9.1.

9.1.2 Local Point of View

Let M be a (real or complex) manifold of dimension two and let $\mathscr{F}(M)$ denote its algebra of smooth or holomorphic functions. Let $(U, (x, y))$ be a coordinate chart of M. According to (1.23), every bivector field on U can be written as

$$\pi = \chi \frac{\partial}{\partial x} \wedge \frac{\partial}{\partial y}, \quad \text{where } \chi = \pi[x, y] \in \mathscr{F}(U). \tag{9.2}$$

We have seen in Section 9.1.1 that every bivector field on U is a Poisson structure. The rank of π is two at all points $m \in U$ where $\chi(m) \neq 0$ and is zero at all other points of U.

Let us show that, although the Jacobi identity is trivially satisfied in dimension two, and although all Poisson structures in dimension two are locally given by the simple formula (9.2), two Poisson structures in dimension two may, both from the algebraic and geometric points of view, be very different in the neighborhood of their singularities.

To illustrate this, let (M,π) be a two-dimensional Poisson manifold and let m be an arbitrary point of M. We write π in a neighborhood of m in M as

$$\pi = \chi \frac{\partial}{\partial x} \wedge \frac{\partial}{\partial y},$$

where x,y are local coordinates, vanishing at m, and where $\chi := \pi[x,y]$. We distinguish the following three cases:

(1) If $\chi(m) \neq 0$, which means that the rank of π at m is two, then there exists, according to Weinstein's splitting theorem (Theorem 1.25), a coordinate neighborhood U of m in M, with coordinates x,y, centered at m, such that, on U,

$$\pi = \frac{\partial}{\partial x} \wedge \frac{\partial}{\partial y}.$$

(2) If $\chi(m) = 0$ but $\mathrm{d}_m\chi \neq 0$, Example 7.27 tells us that π is linearizable at m, i.e., there exists a coordinate chart for M, still denoted by $(U,(x,y))$ and centered at m, such that

$$\pi = x \frac{\partial}{\partial x} \wedge \frac{\partial}{\partial y}.$$

(3) If $\chi(m) = 0$ and $\mathrm{d}_m\chi = 0$, then the zero locus of the function χ has a singularity at m. Let us consider the case of quadratic Poisson structures on $\mathbb{C}^2$. Two examples are given by

$$\pi_1 := (x^2 - y^2) \frac{\partial}{\partial x} \wedge \frac{\partial}{\partial y} \quad \text{and} \quad \pi_2 := x^2 \frac{\partial}{\partial x} \wedge \frac{\partial}{\partial y}.$$

Since the singular locus of π_1 consists of the pair of lines $y = \pm x$, while the singular locus of π_2 consists of the (double) line $x^2 = 0$, it is clear that these Poisson structures are (even locally, in the neighborhood of a point of their singular locus) not isomorphic. This simple example illustrates the fact that there are at least as many non-isomorphic Poisson structures on the plane, as there are non-isomorphic algebraic curves in the plane. In fact there are more. In order to see this, consider the following two Poisson structures:

$$\pi_1 := (x^2 - y^2) \frac{\partial}{\partial x} \wedge \frac{\partial}{\partial y}, \quad \text{and} \quad \pi_3 := 2(x^2 - y^2) \frac{\partial}{\partial x} \wedge \frac{\partial}{\partial y}.$$

Since they are proportional, they have the same zero locus, but they are not isomorphic as Poisson structures, a fact which can easily be checked by using Proposition 8.8.

As we will see in the next section, there is a partial local classification of Poisson structures on $\mathbb{C}^2$ (or $\mathbb{R}^2$), under the strong hypothesis that the singular locus of the Poisson structure admits a simple singularity (at the origin).

9.1.3 Classification in Dimension Two

In this section, we consider the classification problem for Poisson structures in dimension two. Since the Jacobi identity is trivially satisfied in dimension two, the classification amounts to the simpler problem of classifying bivector fields in dimension two, yet only partial results are known. The classification theorem which we will present below has the following restrictions:

(1) The Poisson structures and the coordinate transformations which are considered are *formal*;
(2) The Poisson structures which are considered, are assumed to have a singular locus which admits a simple singularity at the origin;
(3) The ground field is $\mathbb{C}$.

As is made explicit in the notes at the end of the chapter, some of these restrictions can be weakened.

We first give an overview of the results and the ideas which will be presented in this section; full proofs will be given in Subsections A, B, C and D below. We begin by giving the definition of a formal Poisson structure on $\mathbb{C}^2$.

Definition 9.1. A *formal Poisson structure* on $\mathbb{C}^2$ is a Poisson bracket on the algebra of all formal power series in two variables $\mathbb{C}[[x,y]]$.

As we show in Subsection 9.1.3.1, every formal Poisson structure on $\mathbb{C}^2$ is of the form $\alpha\frac{\partial}{\partial x}\wedge\frac{\partial}{\partial y}$, with $\alpha\in\mathbb{C}[[x,y]]$, and every such biderivation is a formal Poisson structure on $\mathbb{C}^2$.

By definition, a *formal coordinate transformation* of $\mathbb{C}^2$ is an algebra isomorphism $\Phi:\mathbb{C}[[x,y]]\to\mathbb{C}[[x,y]]$ such that the formal power series $\Phi(x)$ and $\Phi(y)$ have no constant term. Two formal power series $\alpha,\beta\in\mathbb{C}[[x,y]]$ are said to be *equivalent* if there exists a formal coordinate transformation Φ of $\mathbb{C}^2$, such that $\Phi(\alpha)=\beta$. The following formal version of the inverse function theorem holds: if Φ is an algebra homomorphism such that the formal power series $\Phi(x)$ and $\Phi(y)$ have no constant term and whose Jacobian

$$\operatorname{Jac}\Phi := \det\begin{pmatrix}\dfrac{\partial\Phi(x)}{\partial x} & \dfrac{\partial\Phi(x)}{\partial y}\\[2mm] \dfrac{\partial\Phi(y)}{\partial x} & \dfrac{\partial\Phi(y)}{\partial y}\end{pmatrix}$$

is invertible, i.e., its constant term is different from zero, then Φ is a formal coordinate transformation.

Let Φ be a formal coordinate transformation of $\mathbb{C}^2$. Given a formal Poisson structure π on $\mathbb{C}^2$, there exists a unique formal Poisson structure on $\mathbb{C}^2$, denoted $\Phi_*\pi$, such that $\Phi : \mathbb{C}[[x,y]] \to \mathbb{C}[[x,y]]$

$$\Phi : (\mathbb{C}[[x,y]], \Phi_*\pi) \to (\mathbb{C}[[x,y]], \pi)$$

is a morphism (hence isomorphism) of Poisson algebras. Explicitly, $\Phi_*\pi$ is given by

$$(\Phi_*\pi)\,[\beta,\gamma] := \Phi^{-1}\left(\pi\,[\Phi(\beta),\Phi(\gamma)]\right)\,,$$

for all $\beta,\gamma \in \mathbb{C}[[x,y]]$. Since

$$\Phi_*\pi = (\Phi_*\pi)[x,y]\frac{\partial}{\partial x}\wedge\frac{\partial}{\partial y},$$

it follows that, if we write π in the form $\pi = \alpha\frac{\partial}{\partial x}\wedge\frac{\partial}{\partial y}$, then $\Phi_*\pi$ takes the form

$$\Phi_*\pi = \Phi^{-1}\left(\alpha\,\mathrm{Jac}(\Phi)\right)\frac{\partial}{\partial x}\wedge\frac{\partial}{\partial y}\,. \tag{9.3}$$

It leads to the notion of equivalence of formal Poisson structures (on $\mathbb{C}^2$).

Definition 9.2. Two formal Poisson structures π_1 and π_2 on $\mathbb{C}^2$ are said to be *formally isomorphic* if there exists a formal coordinate transformation of $\mathbb{C}^2$, such that $\pi_2 = \Phi_*\pi_1$.

We consider formal Poisson structures on $\mathbb{C}^2$, with a simple singularity at the origin, i.e., formal Poisson structures of the form $\alpha\frac{\partial}{\partial x}\wedge\frac{\partial}{\partial y}$, where $\alpha \in \mathbb{C}[[x,y]]$ is a formal power series, without constant term, which admits a simple singularity at the origin (see Subsection 9.1.3.1 below for a precise definition of this notion). According to a theorem by Arnold, such a formal power series α is equivalent to a weight homogeneous polynomial $\chi \in \mathbb{C}[x,y]$ (see Theorem 9.7 below). Combined with (9.3), this implies that every formal Poisson structure on $\mathbb{C}^2$ is formally isomorphic to a Poisson structure of the form

$$\chi\beta\frac{\partial}{\partial x}\wedge\frac{\partial}{\partial y}\,, \tag{9.4}$$

where $\chi \in \mathbb{C}[x,y]$ is a weight homogeneous polynomial and $\beta \in \mathbb{C}[[x,y]]$ is a formal power series whose constant term is different from zero. The next step in the classification is the following proposition, which states that the formal power series β in (9.4) can be replaced, modulo a non-zero additive constant, by a weight homogeneous polynomial.

Proposition 9.3. *Let $\pi = \chi\beta\frac{\partial}{\partial x}\wedge\frac{\partial}{\partial y}$ be a formal Poisson structure on $\mathbb{C}^2$, where $\chi \in \mathbb{C}[x,y]$ is a non-constant polynomial, which is weight homogeneous with respect to the weight vector $\varpi = (\varpi_1,\varpi_2)$, and $\beta \in \mathbb{C}[[x,y]]$ is a formal power series, with constant term* 1. *Then π is formally isomorphic to a Poisson structure of the form*

$$\chi(1+\psi)\,\frac{\partial}{\partial x}\wedge\frac{\partial}{\partial y}\,,$$

where $\psi \in \mathbb{C}[x,y]$ is a weight homogeneous polynomial of weight $\varpi(\chi)-\varpi_1-\varpi_2$.

See Subsection 9.1.3.3 below for a proof of this proposition. It follows that a Poisson structure which admits a simple singularity at the origin is formally isomorphic, up to a non-zero multiplicative constant, to a Poisson structure of the form

$$\chi(1+\psi)\,\frac{\partial}{\partial x}\wedge\frac{\partial}{\partial y}\,,$$

where χ is one of the weight homogeneous polynomials which appear in Arnold's classification theorem of formal power series with a simple singularity (Theorem 9.7) and $\psi \in \mathbb{C}[x,y]$ is an arbitrary weight homogeneous polynomial of weight $\varpi(\chi)-\varpi_1-\varpi_2$. For each χ in Arnold's list, one easily determines all possibilities for ψ, which leads to the following classification theorem of formal Poisson structures on $\mathbb{C}^2$, which admit a simple singularity at the origin.

Theorem 9.4. *Let $\pi = \alpha\frac{\partial}{\partial x}\wedge\frac{\partial}{\partial y}$ be a formal Poisson structure on $\mathbb{C}^2$, where $\alpha \in \mathbb{C}[[x,y]]$ is a formal power series, which has a simple singularity at the origin. Then, up to a non-zero multiplicative constant, π is formally isomorphic to a Poisson structure of the form $\eta\frac{\partial}{\partial x}\wedge\frac{\partial}{\partial y}$, where $\eta \in \mathbb{C}[x,y]$ is one the polynomials which appear in the third column of the following table.*

Type	k	η
A_{2k}	$k \geqslant 1$	x^2+y^{2k+1}
A_{2k-1}	$k \geqslant 1$	$(x^2+y^{2k})\,(1+\lambda y^{k-1})$
D_{2k}	$k \geqslant 2$	$(x^2y+y^{2k-1})\,(1+\lambda x+\mu y^{k-1})$
D_{2k+1}	$k \geqslant 2$	$(x^2y+y^{2k})\,(1+\lambda x)$
E_6		x^3+y^4
E_7		$(x^3+xy^3)\,(1+\lambda y^2)$
E_8		x^3+y^5

Remark 9.5. The Poisson structures which appear in the classification theorem are not claimed to be all non-isomorphic. Special care has to be taken with respect to the phrase "up to a non-zero multiplicative constant", which appears in its statement. In general, a formal Poisson structure π on $\mathbb{C}^2$ is formally isomorphic to every Poisson structure of the form $\lambda\pi$, with $\lambda \in \mathbb{C}^*$, but there are exceptions. We already gave in Section 9.1.2 an example of two proportional quadratic Poisson structures which are non-isomorphic. In order to make a more general statement, let χ be a weight homogeneous polynomial of $\mathbb{C}[x,y]$, with respect to the weight vector (ϖ_1,ϖ_2) and

of weight $\varpi(\chi)$. Let us consider the Poisson structure on $\mathbb{C}^2$ corresponding to χ, namely $\pi := \chi \frac{\partial}{\partial x} \wedge \frac{\partial}{\partial y}$. Define, for arbitrary $a \in \mathbb{C}^*$, the formal coordinate transformation of $\mathbb{C}^2$, which amounts to rescaling x and y as follows:

$$\Phi_a(x) := a^{\varpi_1} x, \quad \Phi_a(y) := a^{\varpi_2} y .$$

According to (8.8) and (9.3), we have

$$(\Phi_a)_* \pi = a^{\varpi_1 + \varpi_2 - \varpi(\chi)} \chi \frac{\partial}{\partial x} \wedge \frac{\partial}{\partial y} .$$

This formula permits us to conclude that, if $\varpi(\chi) \neq \varpi_1 + \varpi_2$, then every Poisson structure of the form $\lambda\pi$, with $\lambda \in \mathbb{C}^*$ is formally isomorphic to π. However, if $\varpi(\chi) = \varpi_1 + \varpi_2$, then each of the formal coordinate transformations Φ_a acts trivially on π; in fact, as can be checked by using Proposition 8.8, the homogeneous Poisson structures π and $\lambda\pi$ are, for $\lambda \neq \pm 1$, non-isomorphic.

9.1.3.1 A – Formal Power Series and Simple Singularities

In this subsection, we recall some basic notions about formal power series (in two variables) and we introduce some concepts which are needed to state the theorem of Arnold about the classification of formal power series with a simple singularity.

A formal power series $\alpha \in \mathbb{C}[[x,y]]$ can be written in a unique way as $\alpha = \sum_{i,j \in \mathbb{N}} \alpha_{ij} x^i y^j$, so the datum of a family $(\alpha_{ij})_{i,j \in \mathbb{N}}$ in $\mathbb{C}$ is equivalent to the datum of a formal power series in $\mathbb{C}[[x,y]]$. This elementary fact is important when dealing with the question whether certain operations yield well-defined power series: a formal power series is well-defined if for every $i, j \in \mathbb{N}$ the coefficient a_{ij} is well-defined. For example, the product of two power series is well-defined, because each term of the product is given by a finite sum. Similarly, if $\alpha \in \mathbb{C}[[x,y]]$ is a formal power series without constant term, then

$$\exp(\alpha) := \sum_{i \in \mathbb{N}} \frac{\alpha^i}{i!}$$

is a well-defined power series. A formal power series $\alpha \in \mathbb{C}[[x,y]]$ can be evaluated at $(x,y) = (0,0)$, which yields the constant term of the series, denoted $\alpha(0,0)$, but can in general not be evaluated at other values of x, y. With the standard addition and product of power series, it is clear that $\mathbb{C}[[x,y]]$ forms an associative algebra, whose invertible elements are the series α for which $\alpha(0,0) \neq 0$.

We define the *order* of a formal power series $\alpha \in \mathbb{C}[[x,y]]$ as the minimum of the total degrees of each of its terms, denoted $\mathrm{ord}(\alpha)$. When we are in a weight homogeneous context, we will also consider the order of a power series α with respect to a weight vector $\varpi = (\varpi_1, \varpi_2)$, where ϖ_1 and ϖ_2 are the weights of the variables x and y (see Section 8.1.3). Then $\operatorname{ord}_{\varpi}(\alpha)$, the *order* of α with respect to ϖ, is by definition the minimum of the weights of each of the terms of α.

As we said previously, a formal coordinate transformation of $\mathbb{C}^2$ is an algebra isomorphism $\Phi : \mathbb{C}[[x,y]] \to \mathbb{C}[[x,y]]$ so that the constant terms of $\Phi(x)$ and of $\Phi(y)$ are zero. Also, recall that two formal power series $\alpha, \beta \in \mathbb{C}[[x,y]]$ are said to be equivalent if there exists a formal coordinate transformation Φ of $\mathbb{C}^2$ such that $\Phi(\alpha) = \beta$. Notice that if α and β are two equivalent formal series of $\mathbb{C}[[x,y]]$, they have the same constant term.

Definition 9.6. A formal power series $\alpha \in \mathbb{C}[[x,y]]$ satisfying $\alpha(0,0) = 0$ is said to have a *simple singularity* at the origin, if a small enough neighborhood of α (for the Whitney topology, see [17]) intersects only a finite number of orbits associated to the equivalence of formal power series.

We are now able to state Arnold's classification in the case of formal power series in $\mathbb{C}[[x,y]]$, with a simple singularity. The names of the singularity types are borrowed from the classification theory of simple Lie algebras; see [147] for an explanation.

Theorem 9.7 (Arnold's theorem). *Let $\alpha \in \mathbb{C}[[x,y]]$ be a formal power series without constant term, which has a simple singularity at $(0,0)$. Then α is equivalent to one of the weight homogeneous polynomials χ which appear in the third column of the following table.*

Type	k	χ
A_k	$k \geqslant 1$	$x^2 + y^{k+1}$
D_k	$k \geqslant 4$	$x^2y + y^{k-1}$
E_6		$x^3 + y^4$
E_7		$x^3 + xy^3$
E_8		$x^3 + y^5$

9.1.3.2 B – Formal Poisson Structures on $\mathbb{C}^2$

We have defined a formal Poisson structure on $\mathbb{C}^2$ as a Poisson bracket on the algebra $\mathbb{C}[[x,y]]$ of formal power series in two variables. Since a Poisson bracket is in particular a skew-symmetric biderivation, we first study the algebra of skew-symmetric multi-derivations of $\mathbb{C}[[x,y]]$. As we will see, the results are very similar to the case of the algebra of polynomial or holomorphic functions on $\mathbb{C}^2$.

The formal partial derivatives $\frac{\partial}{\partial x}$ and $\frac{\partial}{\partial y}$ clearly yield derivations of $\mathbb{C}[[x,y]]$. Also, their wedge, $\frac{\partial}{\partial x} \wedge \frac{\partial}{\partial y}$ yields a skew-symmetric biderivation of this algebra. More general examples are constructed by taking $\alpha\frac{\partial}{\partial x} + \beta\frac{\partial}{\partial y}$, respectively $\alpha\frac{\partial}{\partial x} \wedge \frac{\partial}{\partial y}$, where $\alpha, \beta \in \mathbb{C}[[x,y]]$ are arbitrary. We show in the following proposition that there are no other skew-symmetric k-derivations of $\mathbb{C}[[x,y]]$ ($k \geqslant 1$).

Proposition 9.8. *The spaces of skew-symmetric multi-derivations of* $\mathbb{C}[[x,y]]$ *are given by*

$$\begin{aligned}
\mathfrak{X}^1\left(\mathbb{C}[[x,y]]\right) &= \left\{\alpha\frac{\partial}{\partial x} + \beta\frac{\partial}{\partial y} \mid \alpha, \beta \in \mathbb{C}[[x,y]]\right\}, \\
\mathfrak{X}^2\left(\mathbb{C}[[x,y]]\right) &= \left\{\alpha\frac{\partial}{\partial x} \wedge \frac{\partial}{\partial y} \mid \alpha \in \mathbb{C}[[x,y]]\right\}, \\
\mathfrak{X}^k\left(\mathbb{C}[[x,y]]\right) &= \{0\}, \text{ for all } k \geqslant 3.
\end{aligned}$$

Proof. Let us prove this proposition for the case of the derivations of $\mathbb{C}[[x,y]]$, the case of the skew-symmetric biderivations and skew-symmetric k-derivations of $\mathbb{C}[[x,y]]$ (for $k \geqslant 3$) being analogous. To do that, we first point out that every derivation $\mathscr{W}$ of $\mathbb{C}[[x,y]]$ satisfies the following property: if $\gamma \in \mathbb{C}[[x,y]]$, then $\mathrm{ord}(\mathscr{W}[\gamma])$ is either $-\infty$ (i.e. $\mathscr{W}[\gamma] = 0$) or $\mathrm{ord}(\mathscr{W}[\gamma]) \geqslant \mathrm{ord}(\gamma) - 1$. This can be done easily by recursion on $\mathrm{ord}(\gamma)$, first in the particular case where $\gamma \in \mathbb{C}[[x]]$ or $\gamma \in \mathbb{C}[[y]]$, and secondly for an arbitrary $\gamma \in \mathbb{C}[[x,y]]$, using the fact that every $\gamma \in \mathbb{C}[[x,y]]$, with $\mathrm{ord}(\gamma) \geqslant 1$, can be written in at least one of the following two forms:

- $\gamma = x\gamma_1 + \gamma_2$, with $\gamma_1 \in \mathbb{C}[[x,y]]$, $\mathrm{ord}(\gamma_1) = \mathrm{ord}(\gamma) - 1$ and $\gamma_2 \in \mathbb{C}[[y]]$;
- $\gamma = y\gamma_1 + \gamma_2$, with $\gamma_1 \in \mathbb{C}[[x,y]]$, $\mathrm{ord}(\gamma_1) = \mathrm{ord}(\gamma) - 1$ and $\gamma_2 \in \mathbb{C}[[x]]$.

Now, let $\mathscr{V} \in \mathfrak{X}^1(\mathbb{C}[[x,y]])$ and consider the formal power series $\alpha := \mathscr{V}[x]$ and $\beta := \mathscr{V}[y]$. We prove that $\mathscr{V} = \alpha\frac{\partial}{\partial x} + \beta\frac{\partial}{\partial y}$, by showing that $\mathscr{W} := \mathscr{V} - \alpha\frac{\partial}{\partial x} - \beta\frac{\partial}{\partial y}$ is zero. First, $\mathscr{W}$ is a derivation of $\mathbb{C}[[x,y]]$ and $\mathscr{W}[x] = \mathscr{W}[y] = 0$, hence $\mathscr{W}$ vanishes on every polynomial in $\mathbb{C}[x,y]$. Let now $\gamma \in \mathbb{C}[[x,y]]$ be an arbitrary formal power series. Suppose that $\mathscr{W}[\gamma] \neq 0$ and let $s := \mathrm{ord}(\mathscr{W}[\gamma])$. We decompose the formal power series γ as follows: $\gamma = \gamma_s + \hat{\gamma}_s$, where $\gamma_s \in \mathbb{C}[x,y]$ is a polynomial of total degree at most $s+1$ and $\hat{\gamma}_s \in \mathbb{C}[[x,y]]$ is a formal power series of order at least $s+2$. Since $\mathscr{W}$ vanishes on all polynomials, $\mathscr{W}[\gamma] = \mathscr{W}[\gamma_s] + \mathscr{W}[\hat{\gamma}_s] = \mathscr{W}[\hat{\gamma}_s]$, so that

$$\mathrm{ord}(\mathscr{W}[\gamma]) = \mathrm{ord}(\mathscr{W}[\hat{\gamma}_s]) \geqslant \mathrm{ord}(\hat{\gamma}_s) - 1 \geqslant s + 1 = \mathrm{ord}(\mathscr{W}[\gamma]) + 1,$$

which is impossible. Therefore, $\mathscr{W}[\gamma] = 0$ for all formal power series γ and the equality $\mathscr{V} = \alpha\frac{\partial}{\partial x} + \beta\frac{\partial}{\partial y}$ follows. $\square$

It follows from Proposition 9.8 that every formal Poisson structure π on $\mathbb{C}^2$ is of the form:

$$\pi = \alpha\frac{\partial}{\partial x} \wedge \frac{\partial}{\partial y}, \tag{9.5}$$

for some formal power series $\alpha \in \mathbb{C}[[x,y]]$. Moreover, $\mathfrak{X}^3(\mathbb{C}[[x,y]]) = \{0\}$, so that the Schouten bracket of every skew-symmetric biderivation of $\mathbb{C}[[x,y]]$ with itself is zero, which implies that every such biderivation defines a formal Poisson structure on $\mathbb{C}^2$. Thus, there is a one-to-one correspondence between the following three objects: the algebra $\mathbb{C}[[x,y]]$ of all formal power series in two variables, the vector space $\mathfrak{X}^2(\mathbb{C}[[x,y]])$ of all skew-symmetric biderivations of $\mathbb{C}[[x,y]]$ and the set of all formal Poisson structures on $\mathbb{C}^2$. Notice that the series α in (9.5) is given by $\alpha = \pi[x,y]$, so that a formal Poisson structure π on $\mathbb{C}^2$ is completely determined by its value on the variables x and y. Given the above one-to-one correspondence, it is natural to transport the property that a function has a simple singularity, to the case of a formal Poisson structure on $\mathbb{C}^2$.

Definition 9.9. Let $\pi = \alpha \frac{\partial}{\partial x} \wedge \frac{\partial}{\partial y}$ be a formal Poisson structure of $\mathbb{C}^2$, satisfying $\alpha(0,0) = 0$. If α admits a simple singularity at the origin, then π is said to admit a *simple singularity* at the origin.

To finish this subsection, we introduce a certain class of formal coordinate transformations of $\mathbb{C}^2$, which are constructed from derivations of $\mathbb{C}[[x,y]]$. Let $\mathscr{V}$ be a derivation of $\mathbb{C}[[x,y]]$ and assume that for every $\beta \in \mathbb{C}[[x,y]]$ the series $\mathscr{V}[\beta]$ is either zero or of order larger than the order of β, with respect to a weight vector ϖ. For every $k \geqslant 1$ and for every $\beta \in \mathbb{C}[[x,y]]$, the formal power series $\mathscr{V}^k[\beta]$ is either zero or of order at least $k + \mathrm{ord}_\varpi(\beta)$. Therefore, for every $i, j \in \mathbb{N}$, the coefficient of $x^i y^j$ in the infinite sum

$$\exp(\mathscr{V})(\beta) := \sum_{k \in \mathbb{N}} \frac{1}{k!} \mathscr{V}^k[\beta]$$

is determined by a finite number of terms, and we have a well-defined map

$$\exp(\mathscr{V}) : \mathbb{C}[[x,y]] \to \mathbb{C}[[x,y]] .$$

Clearly, the formal power series $\exp(\mathscr{V})(x)$ and $\exp(\mathscr{V})(y)$ vanish at the origin. We show that $\exp(\mathscr{V})$ is an algebra homomorphism; i.e., that $\exp(\mathscr{V})(\beta\gamma) = \exp(\mathscr{V})(\beta)\exp(\mathscr{V})(\gamma)$, for all $\beta, \gamma \in \mathbb{C}[[x,y]]$. Since $\mathscr{V}$ is a derivation of $\mathbb{C}[[x,y]]$, we have for all $\beta, \gamma \in \mathbb{C}[[x,y]]$,

$$\mathscr{V}^i(\beta\gamma) = \sum_{k=0}^{i} \binom{i}{k} \mathscr{V}^{i-k}(\beta)\mathscr{V}^k(\gamma) ,$$

so that

$$\begin{aligned}
\exp(\mathscr{V})(\beta\gamma) &= \sum_{i \in \mathbb{N}} \frac{1}{i!} \mathscr{V}^i(\beta\gamma) = \sum_{i \in \mathbb{N}} \sum_{k=0}^{i} \frac{1}{i!} \binom{i}{k} \mathscr{V}^{i-k}(\beta)\mathscr{V}^k(\gamma) \\
&= \sum_{i \in \mathbb{N}} \sum_{k=0}^{i} \frac{1}{(i-k)!\,k!} \mathscr{V}^{i-k}(\beta)\mathscr{V}^k(\gamma) \\
&= \sum_{j,k \in \mathbb{N}} \frac{1}{j!\,k!} \mathscr{V}^j(\beta)\mathscr{V}^k(\gamma) = \exp(\mathscr{V})(\beta)\exp(\mathscr{V})(\gamma) .
\end{aligned}$$

The equality which is proven in the above computation is an equality of formal power series (for a given β and γ); since at every step, each term of the series is determined by a finite number of terms, the above manipulations of infinite sums are valid. Since the Jacobian of $\exp(\mathscr{V})$ equals 1 at $(0,0)$, $\exp(\mathscr{V})$ defines a formal coordinate transformation of $\mathbb{C}^2$.

An important class of derivations $\mathscr{V}$ which satisfy the above property, consists of the derivations $\alpha\mathscr{E}_\varpi$, where α is a formal power series without constant term and ϖ is a weight vector. In this case, we have for every non-constant formal power series β that $\mathrm{ord}_\varpi(\alpha\mathscr{E}_\varpi[\beta]) = \mathrm{ord}_\varpi(\alpha) + \mathrm{ord}_\varpi(\beta)$, so that $\exp(\alpha\mathscr{E}_\varpi)$ is a formal coordinate transformation of $\mathbb{C}^2$. Fixing a formal power series α without constant term, we will refer to the family $(\exp(t\alpha\mathscr{E}_\varpi))_{t\in\mathbb{C}}$ as the *formal flow* of the derivation $\alpha\mathscr{E}_\varpi$. For fixed t, the formal coordinate transformation $\exp(t\alpha\mathscr{E}_\varpi)$ of $\mathbb{C}^2$ is called the *formal flow of $\alpha\mathscr{E}_\varpi$ at t*.

9.1.3.3 C – Proof of Proposition 9.3

The aim of this subsection is to prove Proposition 9.3. Throughout the subsection, $\varpi = (\varpi_1, \varpi_2)$ is a fixed weight vector; weight homogeneity of a polynomial and the order of a formal power series are to be understood as being with respect to ϖ. As before, $\mathscr{E}_\varpi$ denotes the weighted Euler derivation and we denote $|\varpi| := \varpi_1 + \varpi_2$.

We start with three technical lemmas.

Lemma 9.10. *Let $\pi = \chi\beta\,\frac{\partial}{\partial x}\wedge\frac{\partial}{\partial y}$ be a formal Poisson structure on $\mathbb{C}^2$, where χ is a (non-zero) weight homogeneous polynomial and β is a formal power series, satisfying $\beta(0,0) = 1$. Let $\alpha \in \mathbb{C}[[x,y]]$ be a formal power series without constant term and let $(\Phi_t)_{t\in\mathbb{C}}$ denote the formal flow of $\alpha\,\mathscr{E}_\varpi$. For every $t \in \mathbb{C}$, the polynomial $\chi \in \mathbb{C}[x,y]$ divides the skew-symmetric biderivation $(\Phi_t)_*\pi$.*

Proof. According to (9.3), $(\Phi_t)_*\pi = \gamma\frac{\partial}{\partial x}\wedge\frac{\partial}{\partial y}$, where $\gamma \in \mathbb{C}[[x,y]]$ is given by

$$\gamma = \Phi_t^{-1}(\chi\beta\,\mathrm{Jac}\,\Phi_t) = \Phi_t^{-1}(\beta\,\mathrm{Jac}\,\Phi_t)\;\Phi_t^{-1}(\chi)\,. \tag{9.6}$$

It therefore suffices to show that χ divides $\Phi_t^{-1}(\chi)$ for every $t \in \mathbb{C}$. To do this, we will show that there exists a one-parameter family $(\lambda_t)_{t\in\mathbb{C}}$ of formal power series in $\mathbb{C}[[x,y]]$, such that

$$\Phi_t^{-1}(\chi) = \lambda_t\chi\,,$$

and such that moreover each coefficient of the power series λ_t depends polynomially on $t \in \mathbb{C}$. Suppose that $(\lambda_t)_{t\in\mathbb{C}}$ is a one-parameter family of formal power series in $\mathbb{C}[[x,y]]$, which satisfies $\lambda_0 = 1$ and $\lambda_t(0,0) \neq 0$, for every $t \in \mathbb{C}$. Then we have the following equivalences:

$$\Phi_t^{-1}(\chi) = \lambda_t\,\chi \iff \mu_t\,\chi = \Phi_t(\chi)\,,$$

where $\mu_t := \dfrac{1}{\Phi_t(\lambda_t)}$ is a formal power series in $\mathbb{C}[[x,y]]$ (depending on t), since $\Phi_t(\lambda_t)(0,0) = \lambda_t(0,0) \neq 0$ for all t. Our problem is therefore reduced to constructing a one-parameter family $(\mu_t)_{t\in\mathbb{C}}$ of formal power series in $\mathbb{C}[[x,y]]$, such that $\Phi_t(\chi) = \mu_t\chi$, such that each coefficient of the power series μ_t depends polynomially on t, and such that $\mu_0 = 1$ and $\mu_t(0,0) \neq 0$ for all $t \in \mathbb{C}$. Under the latter hypothesis on μ_t, we have the following equivalences:[1]

$$\begin{aligned}
\mu_t\chi = \Phi_t(\chi) &\Longleftrightarrow \left(\frac{1}{\mu_t}\Phi_t(\chi)\right)^{\dot{}} = 0\\
&\Longleftrightarrow -\frac{\dot{\mu}_t}{\mu_t^2}\Phi_t(\chi) + \frac{1}{\mu_t}\Phi_t\left(\alpha\mathscr{E}_\varpi[\chi]\right) = 0\\
&\Longleftrightarrow \alpha\mathscr{E}_\varpi[\chi] = \Phi_t^{-1}\left(\frac{\dot{\mu}_t}{\mu_t}\right)\chi\\
&\Longleftrightarrow \varpi(\chi)\alpha = \Phi_t^{-1}\left(\frac{\dot{\mu}_t}{\mu_t}\right),
\end{aligned}$$

where we have used in the second equivalence that $(\Phi_t)_{t\in\mathbb{C}}$ is the formal flow of $\alpha\mathscr{E}_\varpi$, and the weighted Euler formula (8.10) in the last equivalence. Our problem is thereby reduced to solving the linear differential equation

$$\dot{\mu}_t = \varpi(\chi)\Phi_t(\alpha)\mu_t\;, \tag{9.7}$$

with initial condition $\mu_0 = 1$. As in calculus,

$$\mu_t := \exp\left(\varpi(\chi)\int_0^t \Phi_\tau(\alpha)\mathrm{d}\tau\right)$$

is the unique solution. It is for each t a well-defined formal power series (since the formal power series $\int_0^t \Phi_\tau(\alpha)\mathrm{d}\tau$ does not have a constant term) whose coefficients depend polynomially on t, and it satisfies $\mu_t(0,0) = 1$, for all $t \in \mathbb{C}$. This proves the announced existence of the family $(\mu_t)_{t\in\mathbb{C}}$, and hence of the family $(\lambda_t)_{t\in\mathbb{C}}$, which proves that χ divides $(\Phi_t)_*\pi$, for all t. □

Lemma 9.11. *Let $\pi = \alpha\frac{\partial}{\partial x}\wedge\frac{\partial}{\partial y}$ be a formal Poisson structure on $\mathbb{C}^2$ and let us write α as*

$$\alpha = \chi_{s_0+|\varpi|} + \cdots + \chi_{s_r+|\varpi|} + \hat{\alpha}\;,$$

where $s_0 < s_1 < \cdots < s_r \in \mathbb{Z}$ and where $\chi_{s_0+|\varpi|}, \ldots, \chi_{s_r+|\varpi|}$ are the first $r+1$ non-zero weight homogeneous terms of the formal power series α, of weight $s_0+|\varpi|, \ldots, s_r+|\varpi|$ and $\hat{\alpha}$ is a formal power series in $\mathbb{C}[[x,y]]$ of order at least

[1] For a family $(F_t)_{t\in\mathbb{C}}$ of power series in $\mathbb{C}[[x,y]]$, whose coefficients are polynomial functions of t, we denote by $\dot{F}_t$, respectively by $\int_0^t F_\tau\mathrm{d}\tau$, the formal power series obtained by differentiating, respectively integrating, each coefficient with respect to t.

$s_r+|\varpi|+1$. *Assume that there exists a weight homogeneous derivation* $\mathscr{V}=\mathscr{V}_1\frac{\partial}{\partial x}+\mathscr{V}_2\frac{\partial}{\partial y}$ *of* $\mathbb{C}[[x,y]]$, *such that*

$$\mathscr{L}_{\mathscr{V}}\left(\chi_{s_0+|\varpi|}\frac{\partial}{\partial x}\wedge\frac{\partial}{\partial y}\right)=\chi_{s_r+|\varpi|}\frac{\partial}{\partial x}\wedge\frac{\partial}{\partial y}\,. \tag{9.8}$$

Let Φ *be a formal coordinate transformation of* $\mathbb{C}^2$, *defined by* $\Phi:=\exp(\mathscr{V})$. *Then* $\Phi_*\pi=\beta\frac{\partial}{\partial x}\wedge\frac{\partial}{\partial y}$, *where* β *is of the form*

$$\beta=\chi_{s_0+|\varpi|}+\cdots+\chi_{s_{r-1}+|\varpi|}+\hat{\beta}\,, \tag{9.9}$$

with $\hat{\beta}\in\mathbb{C}[[x,y]]$ *and of order at least* $s_r+|\varpi|+1$. *Stated briefly,* Φ *eliminates the term* $\chi_{s_r+|\varpi|}$ *without modifying the terms of lower weight.*

Proof. Let $\pi=\alpha\frac{\partial}{\partial x}\wedge\frac{\partial}{\partial y}$ and suppose that there exists a weight homogeneous derivation $\mathscr{V}$, satisfying (9.8), where $\chi_{s_0+|\varpi|}$ and $\chi_{s_r+|\varpi|}$ are the parts of α of weight $s_0+|\varpi|$, respectively of weight $s_r+|\varpi|$. Then $\mathscr{V}_1$ and $\mathscr{V}_2$ are weight homogeneous polynomials of weight $\varpi(\mathscr{V}_1)=s_r-s_0+\varpi_1$ and $\varpi(\mathscr{V}_2)=s_r-s_0+\varpi_2$. It implies that $\Phi:=\exp(\mathscr{V})$ is a well-defined formal coordinate transformation of $\mathbb{C}^2$; explicitly, $\Phi(x)$ and $\Phi(y)$ are of the form

$$\begin{cases}\Phi(x)=x+\mathscr{V}_1+\hat{\mathscr{V}}_1\,,\\ \Phi(y)=y+\mathscr{V}_2+\hat{\mathscr{V}}_2\,,\end{cases} \tag{9.10}$$

where $\hat{\mathscr{V}}_i$ is a formal power series in $\mathbb{C}[[x,y]]$ with $\operatorname{ord}_\varpi(\hat{\mathscr{V}}_i)>s_r-s_0+\varpi_i$ for $i=1,2$, so that

$$\varpi\left(\frac{\partial\mathscr{V}_1}{\partial x}\right)=\varpi\left(\frac{\partial\mathscr{V}_2}{\partial y}\right)=s_r-s_0\,,$$

$$\operatorname{ord}_\varpi\left(\frac{\partial\hat{\mathscr{V}}_1}{\partial x}\right)>s_r-s_0,\quad \operatorname{ord}_\varpi\left(\frac{\partial\hat{\mathscr{V}}_2}{\partial y}\right)>s_r-s_0\,,$$

and

$$\operatorname{Jac}\Phi=\begin{vmatrix}1+\dfrac{\partial\mathscr{V}_1}{\partial x}+\dfrac{\partial\hat{\mathscr{V}}_1}{\partial x} & \dfrac{\partial\mathscr{V}_1}{\partial y}+\dfrac{\partial\hat{\mathscr{V}}_1}{\partial y}\\[2ex] \dfrac{\partial\mathscr{V}_2}{\partial x}+\dfrac{\partial\hat{\mathscr{V}}_2}{\partial x} & 1+\dfrac{\partial\mathscr{V}_2}{\partial y}+\dfrac{\partial\hat{\mathscr{V}}_2}{\partial y}\end{vmatrix}.$$

It follows that the terms of $\alpha\operatorname{Jac}\Phi$, whose weights are at most $s_r+|\varpi|$, are precisely the following:

$$\chi_{s_0+|\varpi|},\cdots,\chi_{s_{r-1}+|\varpi|},\chi_{s_r+|\varpi|}+\frac{\partial\mathscr{V}_1}{\partial x}\chi_{s_0+|\varpi|}+\frac{\partial\mathscr{V}_2}{\partial y}\chi_{s_0+|\varpi|}\,.$$

Let β be the formal power series, defined by $\Phi_* \pi = \beta \frac{\partial}{\partial x} \wedge \frac{\partial}{\partial y}$. We denote its term of weight i by ψ_i. According to (9.3), $\Phi(\beta) = \alpha \operatorname{Jac} \Phi$, with Φ of the form (9.10). The terms of $\Phi(\beta)$, and hence of β, of weight i, smaller than $s_r + |\varpi|$, coincide with the corresponding terms of $\alpha \operatorname{Jac} \Phi$. Specifically, all are zero except for

$$\chi_{s_0+|\varpi|} = \psi_{s_0+|\varpi|}, \dots, \chi_{s_{r-1}+|\varpi|} = \psi_{s_{r-1}+|\varpi|}. \tag{9.11}$$

It follows that the term of weight $s_r + |\varpi|$ of $\Phi(\beta)$ is given by

$$\psi_{s_r+|\varpi|} + \mathscr{V}\left[\psi_{s_0+|\varpi|}\right] = \psi_{s_r+|\varpi|} + \mathscr{V}\left[\chi_{s_0+|\varpi|}\right].$$

Equating these terms with the terms of weight $s_r + |\varpi|$ in $\alpha \operatorname{Jac} \Phi$, we obtain

$$\chi_{s_r+|\varpi|} + \left(\frac{\partial \mathscr{V}_1}{\partial x} + \frac{\partial \mathscr{V}_2}{\partial y}\right) \chi_{s_0+|\varpi|} = \psi_{s_r+|\varpi|} + \mathscr{V}\left[\chi_{s_0+|\varpi|}\right]. \tag{9.12}$$

Recall that $\mathscr{V}$ satisfies (9.8). Applied to x and y, this condition means that

$$\mathscr{V}\left[\chi_{s_0+|\varpi|}\right] - \left(\frac{\partial \mathscr{V}_1}{\partial x} + \frac{\partial \mathscr{V}_2}{\partial y}\right) \chi_{s_0+|\varpi|} = \chi_{s_r+|\varpi|},$$

which leads, according to equation (9.12), to $\psi_{s_r+|\varpi|} = 0$. Combined with (9.11), this shows that β is indeed of the desired form (9.9). □

The following lemma is an analog of Euler's weighted formula for bivector fields on $\mathbb{C}^2$.

Lemma 9.12. *Let $\psi \in \mathbb{C}[x,y]$ be a polynomial and let $\pi = \chi \frac{\partial}{\partial x} \wedge \frac{\partial}{\partial y}$ be a Poisson structure on $\mathbb{C}^2$ If ψ and π are weight homogeneous with respect to some weight vector $\varpi = (\varpi_1, \varpi_2)$, then*

$$\mathscr{L}_{\psi \mathscr{E}_\varpi} \pi = (\varpi(\chi) - |\varpi| - \varpi(\psi))\, \psi \pi. \tag{9.13}$$

Proof. Under the stated assumptions on ψ and π, we need to show that

$$\left(\mathscr{L}_{\psi \mathscr{E}_\varpi} \pi\right)[x,y] = (\varpi(\chi) - |\varpi| - \varpi(\psi))\, \psi \chi.$$

Using several times the weighted Euler formula (8.10), we compute

$$\begin{aligned}
\left(\mathscr{L}_{\psi \mathscr{E}_\varpi} \pi\right)[x,y] &= \psi \mathscr{E}_\varpi[\chi] - \pi[\psi \mathscr{E}_\varpi[x], y] - \pi[x, \psi \mathscr{E}_\varpi(y)] \\
&= \psi \varpi(\chi) \chi - \varpi_1 \pi[\psi x, y] - \varpi_2 \pi[x, \psi y] \\
&= (\varpi(\chi) - \varpi_1 - \varpi_2) \psi \chi - \left(\varpi_1 x \frac{\partial \psi}{\partial x} + \varpi_2 y \frac{\partial \psi}{\partial y}\right) \chi \\
&= (\varpi(\chi) - |\varpi| - \varpi(\psi))\, \psi \chi,
\end{aligned}$$

which yields the desired result. □

Using the above three lemmas, we prove Proposition 9.3.

Proof (of Proposition 9.3). Let $\pi = \chi\beta \frac{\partial}{\partial x}\wedge\frac{\partial}{\partial y}$ be a formal Poisson structure on $\mathbb{C}^2$, where $\chi \in \mathbb{C}[x,y]$ and $\beta \in \mathbb{C}[[x,y]]$ are as stated. Since β verifies $\beta(0,0) = 1$, we can write $\beta = 1+\psi_s+\beta_s$, where $\psi_s \in \mathbb{C}[x,y]$ is a weight homogeneous polynomial of weight s and $\beta_s \in \mathbb{C}[[x,y]]$ is a formal power series of order at least $s+1$. Let us first show that s can be assumed to be at least equal to $\varpi(\chi)-|\varpi|$. Assume that $s < \varpi(\chi)-|\varpi|$ and consider the Lie derivative

$$\mathscr{L}_{\psi_s\mathscr{E}_\varpi}\left(\chi\frac{\partial}{\partial x}\wedge\frac{\partial}{\partial y}\right) = (\varpi(\chi)-|\varpi|-s)\,\chi\psi_s\frac{\partial}{\partial x}\wedge\frac{\partial}{\partial y},$$

which follows easily from Lemma 9.12. Letting $\mathscr{V} := \psi_s\mathscr{E}_\varpi/(\varpi(\chi)-|\varpi|-s)$, we have that (9.8) is satisfied, with $r:=1$ and $\chi_{s_0+|\varpi|} := \chi$ and $\chi_{s_r+|\varpi|} := \chi\psi_s$. We can therefore apply Lemma 9.11, which shows that π is formally isomorphic to $(\chi+\hat\alpha)\frac{\partial}{\partial x}\wedge\frac{\partial}{\partial y}$, where $\hat\alpha$ is a formal power series, of order at least $\varpi(\chi)+s+1$. The formal coordinate transformation (9.10) is in this case of the form $\exp\mathscr{V}$, with $\mathscr{V}$ a polynomial times the weighted Euler vector field. Therefore, Lemma 9.10 applies, showing that $(\chi+\hat\alpha)\frac{\partial}{\partial x}\wedge\frac{\partial}{\partial y}$ is a multiple of χ. In conclusion, π is formally isomorphic to $\chi(1+\beta_s)\frac{\partial}{\partial x}\wedge\frac{\partial}{\partial y}$, where β_s is a formal power series of order at least $s+1$. A repeated use of this procedure shows that π is formally isomorphic to $\chi(1+\beta)\frac{\partial}{\partial x}\wedge\frac{\partial}{\partial y}$, where β is a formal power series of order at least $\varpi(\chi)-|\varpi|$.

Let us now show that, using the same technique, the formal power series β in $\chi(1+\beta)\frac{\partial}{\partial x}\wedge\frac{\partial}{\partial y}$ can be replaced by its terms of weight $\varpi(\chi)-|\varpi|$. To do this, let us write $\beta = \psi+\varphi+\hat\beta$, where ψ is a weight homogeneous polynomial of weight $\varpi(\chi)-|\varpi|$, where φ is a weight homogeneous polynomial of weight at least $\varpi(\chi)-|\varpi|+1$, and where $\hat\beta$ is a formal power series, with $\operatorname{ord}_\varpi\hat\beta > \varpi(\varphi)$. We first show how to remove the term φ from β, without changing ψ. Setting $r:=2$ and $\chi_{s_0+|\varpi|} := \chi$ and $\chi_{s_2+|\varpi|} := \chi\varphi$ it is verified, as above, that the vector field $\mathscr{V} := \varphi\mathscr{E}_\varpi/(\varpi(\chi)-|\varpi|-\varpi(\varphi))$ satisfies (9.8), so Lemma 9.11 applies again, showing as above that π is formally isomorphic to $\chi(1+\psi+\hat\alpha)\frac{\partial}{\partial x}\wedge\frac{\partial}{\partial y}$, where ψ has weight $\varpi(\chi)-|\varpi|$ and the order of $\hat\alpha$ is at least $\varpi(\chi)-|\varpi|$.

Repeating this procedure, we can make the order of $\hat\alpha$ arbitrarily large. To make $\hat\alpha$ disappear, the procedure needs to be repeated an infinite number of times, i.e., we need to consider the algebra homomorphism $\Phi : \mathbb{C}[[x,y]] \to \mathbb{C}[[x,y]]$ which is obtained as the composition of an infinite number of the above formal coordinate transformations. However, at each step, the coordinate transformation is of the form (9.10), with $\mathscr{V}_1$ and $\mathscr{V}_2$ of higher order than the $\mathscr{V}_1$ and $\mathscr{V}_2$ which were used in the previous step, each term in $\Phi(x)$ and in $\Phi(y)$ will only depend on a finite number of the formal coordinate transformations. Therefore, π is formally isomorphic to $\chi(1+\psi)\frac{\partial}{\partial x}\wedge\frac{\partial}{\partial y}$. □

9.1.3.4 D – Proof of the Classification Theorem

We are finally ready to give the proof of the classification theorem of formal Poisson structures on $\mathbb{C}^2$, which admit a simple singularity.

Proof (of Theorem 9.4). Let $\pi = \alpha \frac{\partial}{\partial x} \wedge \frac{\partial}{\partial y}$ be a formal Poisson structure on $\mathbb{C}^2$, where $\alpha \in \mathbb{C}[[x,y]]$ is a formal power series, which has a simple singularity at the origin. According to Arnold's theorem (Theorem 9.7) there exists a formal coordinate transformation Φ, such that $\Phi^{-1}(\alpha) = \chi$, where χ is a weight homogeneous polynomial, with respect to some weight vector $\varpi = (\varpi_1, \varpi_2)$. According to (9.3), $\Phi_*\pi = \chi\beta \frac{\partial}{\partial x} \wedge \frac{\partial}{\partial y}$, where $\beta \in \mathbb{C}[[x,y]]$ is a formal power series, with $\beta(0,0) \neq 0$. Using Proposition 9.3, we can conclude that, up to a constant, π is isomorphic to $\chi(1+\psi)\frac{\partial}{\partial x} \wedge \frac{\partial}{\partial y}$, where χ is one of the weight homogeneous polynomials which appear in Theorem 9.7 and where ψ is a weight homogeneous polynomial of weight $\varpi(\chi) - |\varpi|$. It suffices now to consider χ in the list of weight homogeneous polynomials of Theorem 9.7 and construct the most general weight homogeneous polynomial $\psi \in \mathbb{C}[x,y]$ of weight $\varpi(\chi) - |\varpi|$, which amounts to finding all monomials $x^i y^j$, with $i, j \in \mathbb{N}$ satisfying

$$\varpi_1 i + \varpi_2 j = \varpi(\chi) - \varpi_1 - \varpi_2 . \tag{9.14}$$

The result is displayed in Table 9.1. □

Table 9.1 The different types of formal simple singularities which appear in Theorem 9.7, with the corresponding weight vector $\varpi = (\varpi_1, \varpi_2)$, the integers ϖ_1 and ϖ_2 being the weights of the variables x and y. For each type, equation (9.14) and the monomials corresponding to the solutions of this equation are given.

Type	k	χ	(ϖ_1, ϖ_2)	Equation (9.14)	$x^i y^j$
A_{2k}	$k \geqslant 1$	$x^2 + y^{2k+1}$	$(2k+1, 2)$	$(2k+1)i + 2j = 2k-1$	–
A_{2k-1}	$k \geqslant 1$	$x^2 + y^{2k}$	$(k, 1)$	$ki + j = k-1$	y^{k-1}
D_{2k}	$k \geqslant 2$	$x^2 y + y^{2k-1}$	$(k-1, 1)$	$(k-1)i + j = k-1$	x, y^{k-1}
D_{2k+1}	$k \geqslant 2$	$x^2 y + y^{2k}$	$(2k-1, 2)$	$(2k-1)i + 2j = 2k-1$	x
E_6		$x^3 + y^4$	$(4, 3)$	$4i + 3j = 5$	–
E_7		$x^3 + xy^3$	$(3, 2)$	$3i + 2j = 4$	y^2
E_8		$x^3 + y^5$	$(5, 3)$	$5i + 3j = 7$	–

9.2 Poisson Structures in Dimension Three

We now consider Poisson structures on manifolds of dimension three. In the two-dimensional case, we have seen that every bivector field is a Poisson structure, because the Jacobi identity is trivially satisfied. This is not the case in dimension three. However, like in the two-dimensional case, the rank of a bivector field, in particular of a Poisson structure, is at most two (hence zero or two), at every point. Section 9.2.1 is devoted to Poisson structures on $\mathbb{F}^3$, while Section 9.2.2 enlarges the discussion to real or complex manifolds of dimension three. Quadratic Poisson structures on $\mathbb{C}^3$ will be classified in Section 9.2.3. In Section 9.2.4, we will consider singular Poisson surfaces in $\mathbb{C}^3$.

9.2.1 Poisson Structures on $\mathbb{F}^3$

In this section, we study Poisson structures on the affine space $\mathbb{F}^3$, which is equipped with its algebra of functions, denoted by $\mathscr{F}(\mathbb{F}^3)$. If $\mathbb{F}$ is an arbitrary field, this algebra is the algebra of polynomial functions on $\mathbb{F}^3$; however, if $\mathbb{F}=\mathbb{R}$ (respectively $\mathbb{F}=\mathbb{C}$), then $\mathscr{F}(\mathbb{F}^3)$ may stand either for the algebra of smooth (respectively holomorphic) or polynomial functions on $\mathbb{F}^3$, depending on the context. The standard coordinates on $\mathbb{F}^3$ are denoted by (x,y,z). Every bivector field π on $\mathbb{F}^3$ decomposes as

$$\pi = P_1\frac{\partial}{\partial y}\wedge\frac{\partial}{\partial z}+P_2\frac{\partial}{\partial z}\wedge\frac{\partial}{\partial x}+P_3\frac{\partial}{\partial x}\wedge\frac{\partial}{\partial y}\,, \tag{9.15}$$

where $P_1:=\{y,z\}$, $P_2:=\{z,x\}$ and $P_3:=\{x,y\}$. It follows from item *(iii)* in Proposition 1.16 that this bivector field satisfies the Jacobi identity, and is therefore a Poisson structure on $\mathbb{F}^3$, if and only if

$$\{x,\{y,z\}\}+\{y,\{z,x\}\}+\{z,\{x,y\}\}=0\,,$$

which is equivalent to

$$P_1\left(\frac{\partial P_3}{\partial y}-\frac{\partial P_2}{\partial z}\right)+P_2\left(\frac{\partial P_1}{\partial z}-\frac{\partial P_3}{\partial x}\right)+P_3\left(\frac{\partial P_2}{\partial x}-\frac{\partial P_1}{\partial y}\right)=0\,. \tag{9.16}$$

Condition (9.16) can be compactly restated in vector notation. To do this, define a vector field $\overrightarrow{P}$ on $\mathbb{F}^3$ by $\overrightarrow{P}:=(P_1,P_2,P_3)$, and denote by $\langle\cdot\,|\,\cdot\rangle$ the standard inner product on $\mathbb{F}^3$ and by curl the curl operator. Then (9.16) can be rewritten as

$$\langle\overrightarrow{P}\mid\operatorname{curl}\overrightarrow{P}\rangle=0\,. \tag{9.17}$$

Condition (9.17) provides an elementary way to produce examples and to establish results about Poisson structures on $\mathbb{F}^3$ by using well-known properties of vector analysis. For instance, recall the classical identity

$$\operatorname{curl}\left(\chi\overrightarrow{P}\right) = \chi \operatorname{curl}\overrightarrow{P} + \operatorname{grad}\chi \times \overrightarrow{P}\,, \tag{9.18}$$

which is valid for every function $\chi \in \mathscr{F}(\mathbb{F}^3)$ and for every vector field $\overrightarrow{P}$ on $\mathbb{F}^3$; in this formula, $\times$ stands for the vector product in $\mathbb{F}^3$ and grad is the gradient operator. We use (9.18) to prove the following proposition.

Proposition 9.13. *Let π be a Poisson structure on $\mathbb{F}^3$. For every function $\chi \in \mathscr{F}(\mathbb{F}^3)$, the bivector field $\chi\pi$ is a Poisson structure on $\mathbb{F}^3$.*

Proof. We denote, as above, by $\overrightarrow{P}$ the vector field on $\mathbb{F}^3$, associated to π, so that $\chi\overrightarrow{P}$ is the vector field associated to $\chi\pi$. One computes, with the help of (9.18),

$$\langle \chi\overrightarrow{P} \mid \operatorname{curl}(\chi\overrightarrow{P})\rangle = \chi^2\langle \overrightarrow{P} \mid \operatorname{curl}\overrightarrow{P}\rangle + \chi\langle \overrightarrow{P} \mid \operatorname{grad}\chi \times \overrightarrow{P}\rangle\,.$$

By (9.17), the assumption that π is a Poisson structure gives the vanishing of the first term on the right-hand side of this equation, while the vanishing of the second term follows from the identity $\langle A \mid B \times A\rangle = 0$, valid for all $A, B \in \mathbb{F}^3$. This shows that condition (9.17) is satisfied for $\chi\overrightarrow{P}$, so that the bivector field $\chi\pi$ is a Poisson structure on $\mathbb{F}^3$. □

Another classical identity which involves the curl operator is

$$\operatorname{curl}\operatorname{grad}\varphi = 0\,,$$

valid for every function $\varphi \in \mathscr{F}(\mathbb{F}^3)$. It implies, in view of (9.17), that when the vector field $\overrightarrow{P}$, associated to a bivector field π, is the gradient of a function, $\overrightarrow{P} = \operatorname{grad}\varphi$, then π is a Poisson structure. Substituted in (9.15), this means that

$$\pi_\varphi := \frac{\partial\varphi}{\partial x}\frac{\partial}{\partial y} \wedge \frac{\partial}{\partial z} + \frac{\partial\varphi}{\partial y}\frac{\partial}{\partial z} \wedge \frac{\partial}{\partial x} + \frac{\partial\varphi}{\partial z}\frac{\partial}{\partial x} \wedge \frac{\partial}{\partial y} \tag{9.19}$$

is a Poisson structure on $\mathbb{F}^3$, for every function $\varphi \in \mathscr{F}(\mathbb{F}^3)$. Let $\lambda = \mathrm{d}x \wedge \mathrm{d}y \wedge \mathrm{d}z$ denote the standard volume form on $\mathbb{F}^3$. Then it is clear from (9.19) that π_φ and $\mathrm{d}\varphi$ are related by

$$\imath_{\pi_\varphi}\lambda = \mathrm{d}\varphi\,, \tag{9.20}$$

which yields an alternative definition for π_φ. Also, this Poisson bracket comes from a standard Nambu–Poisson structure of order 3 on $\mathbb{F}^3$, upon using the function φ, as described in Proposition 8.35. Therefore, the Poisson bracket $\{F,G\}_\varphi$ of two arbitrary functions $F, G \in \mathscr{F}(\mathbb{F}^3)$ can also be written as

$$\{F,G\}_\varphi = \frac{\mathrm{d}F \wedge \mathrm{d}G \wedge \mathrm{d}\varphi}{\mathrm{d}x \wedge \mathrm{d}y \wedge \mathrm{d}z}\,.$$

In particular, φ is a Casimir function for π_φ. Proposition 9.13 implies the following statement.

Proposition 9.14. *Let $\varphi \in \mathscr{F}(\mathbb{F}^3)$ be an arbitrary function on $\mathbb{F}^3$ and consider the bivector field π_φ on $\mathbb{F}^3$, defined by (9.19). For every function $\chi \in \mathscr{F}(\mathbb{F}^3)$, the bivector field $\chi\pi_\varphi$ is a Poisson structure on $\mathbb{F}^3$ and it admits φ as a Casimir function.*

This proposition motivates the next theorem. By a slight abuse of terminology, we call *Casimir function of a bivector field* $\pi = \{\cdot,\cdot\}$ (not assumed to be a Poisson structure) every function $\varphi \in \mathscr{F}(\mathbb{F}^3)$ such that $\{F,\varphi\} = 0$ for all $F \in \mathscr{F}(\mathbb{F}^3)$.

Proposition 9.15. *Let π be a bivector field on $\mathbb{F}^3$ and suppose that π admits a Casimir function φ.*

(1) If the differential of φ is different from zero in a dense subset of $\mathbb{F}^3$, then π is a Poisson structure on $\mathbb{F}^3$;

(2) If the bivector field π is a polynomial Poisson structure, and if the Casimir function φ is a polynomial function for which the partial derivatives $\frac{\partial\varphi}{\partial x}$, $\frac{\partial\varphi}{\partial y}$ and $\frac{\partial\varphi}{\partial z}$ are relatively prime, then there exists a polynomial function $\chi \in \mathscr{F}(\mathbb{F}^3)$, such that $\pi = \chi\pi_\varphi$, where π_φ is the Poisson structure defined in (9.19).

Proof. The bivector field π is a Poisson structure on $\mathbb{F}^3$ if and only if the Schouten bracket $[\pi,\pi]_S$ is zero. Since

$$\frac{1}{2}[\pi,\pi]_S[F,G,H] = \{\{F,G\},H\} + \{\{G,H\},F\} + \{\{H,F\},G\},$$

for all functions $F,G,H \in \mathscr{F}(\mathbb{F}^3)$, we have that, if φ is a Casimir function of π, then $[\pi,\pi]_S[F,G,\varphi] = 0$, for all $F,G \in \mathscr{F}(\mathbb{F}^3)$. Let us show that this implies that, if φ is a Casimir function of π and $\mathrm{d}\varphi$ is non-zero on a dense subset of $\mathbb{F}^3$, then $[\pi,\pi]_S = 0$, so that π is a Poisson structure on $\mathbb{F}^3$. Since $[\pi,\pi]_S$ is a trivector field on $\mathbb{F}^3$, it can be written as

$$[\pi,\pi]_S = \psi\frac{\partial}{\partial x} \wedge \frac{\partial}{\partial y} \wedge \frac{\partial}{\partial z}, \tag{9.21}$$

where $\psi \in \mathscr{F}(\mathbb{F}^3)$. Plugging $F := x$ and $G := y$ in $[\pi,\pi]_S[F,G,\varphi] = 0$ we obtain, in view of (9.21), $\psi\frac{\partial\varphi}{\partial z} = 0$. Similarly, one obtains $\psi\frac{\partial\varphi}{\partial x} = \psi\frac{\partial\varphi}{\partial y} = 0$, by plugging in the two other pairs of variables. Since, by assumption, $\mathrm{d}\varphi$ is non-zero on a dense subset of $\mathbb{F}^3$, it follows that $\psi = 0$ on $\mathbb{F}^3$, hence that the trivector field $[\pi,\pi]_S$ is zero on $\mathbb{F}^3$. This proves *(1)*.

We next prove *(2)*. We assume that π is a polynomial Poisson structure on $\mathbb{F}^3$ and we define polynomial functions $P_1,P_2,P_3 \in \mathscr{F}(\mathbb{F}^3)$ by

$$\pi = P_1\frac{\partial}{\partial y} \wedge \frac{\partial}{\partial z} + P_2\frac{\partial}{\partial z} \wedge \frac{\partial}{\partial x} + P_3\frac{\partial}{\partial x} \wedge \frac{\partial}{\partial y}.$$

Suppose that φ is a polynomial Casimir function of π. We apply the equation $\{F,\varphi\} = 0$, which is valid for all $F \in \mathscr{F}(\mathbb{F}^3)$, successively to $F := x$, $F := y$ and $F := z$. It gives the following relations between the functions P_1, P_2 and P_3:

$$P_2\frac{\partial\varphi}{\partial z} = P_3\frac{\partial\varphi}{\partial y}, \quad P_3\frac{\partial\varphi}{\partial x} = P_1\frac{\partial\varphi}{\partial z}, \quad P_1\frac{\partial\varphi}{\partial y} = P_2\frac{\partial\varphi}{\partial x}. \tag{9.22}$$

Let us assume that the partial derivatives $\frac{\partial\varphi}{\partial x}, \frac{\partial\varphi}{\partial y}, \frac{\partial\varphi}{\partial z}$ are relatively prime, and let Q denote the greatest common divisor of $\frac{\partial\varphi}{\partial x}$ and $\frac{\partial\varphi}{\partial y}$. Then it follows from the third equation in (9.22) that there exists $\chi' \in \mathscr{F}(\mathbb{F}^3)$, such that

$$QP_1 = \chi'\frac{\partial\varphi}{\partial x} \quad \text{and} \quad QP_2 = \chi'\frac{\partial\varphi}{\partial y} \,. \tag{9.23}$$

In view of the first (or the second) equality in (9.22), this implies that

$$QP_3 = \chi'\frac{\partial\varphi}{\partial z} \,. \tag{9.24}$$

Since Q and $\frac{\partial\varphi}{\partial z}$ are relatively prime, Q divides χ'. Define χ by $Q\chi = \chi'$. We obtain in view of (9.23) and (9.24):

$$P_1 = \chi\frac{\partial\varphi}{\partial x}\,, \quad P_2 = \chi\frac{\partial\varphi}{\partial y} \quad \text{and} \quad P_3 = \chi\frac{\partial\varphi}{\partial z}\,.$$

This proves *(2)*. □

Remark 9.16. When φ is a non-constant weight homogeneous polynomial which is square-free, then the partial derivatives $\frac{\partial\varphi}{\partial x}, \frac{\partial\varphi}{\partial y}, \frac{\partial\varphi}{\partial z}$ are relatively prime, so that the assumptions about φ in item *(2)* of Proposition 9.15 are satisfied. Let us prove this claim. If φ is weight homogeneous of weight $\varpi(\varphi) \neq 0$, it satisfies the weighted Euler formula (8.10),

$$\varpi(\varphi)\varphi = \varpi_1 x\frac{\partial\varphi}{\partial x} + \varpi_2 y\frac{\partial\varphi}{\partial y} + \varpi_3 z\frac{\partial\varphi}{\partial z}\,, \tag{9.25}$$

where ϖ_1, ϖ_2 and ϖ_3 are the weights of the variables x, y and z. Suppose that $Q \in \mathscr{F}(\mathbb{F}^3)$ is a non-constant irreducible polynomial function which divides the partial derivatives $\frac{\partial\varphi}{\partial x}, \frac{\partial\varphi}{\partial y}$ and $\frac{\partial\varphi}{\partial z}$. It follows from (9.25) that Q divides φ. Since Q is non-constant, at least one of the partial derivatives $\frac{\partial Q}{\partial x}, \frac{\partial Q}{\partial y}$ and $\frac{\partial Q}{\partial z}$, say $\frac{\partial Q}{\partial x}$, is different from zero. Since the irreducible polynomial Q divides φ and $\frac{\partial\varphi}{\partial x}$, its square Q^2 also divides φ, so that φ is not square-free. Therefore, if φ is non-constant, weight homogeneous and square-free, then $\frac{\partial\varphi}{\partial x}, \frac{\partial\varphi}{\partial y}$ and $\frac{\partial\varphi}{\partial z}$ are relatively prime.

We finish this section with an example of Poisson structure on $\mathbb{F}^3$ which is not of the form $\chi\pi_\varphi$, and which in fact does not admit Casimir functions defined on $\mathbb{F}^3$, except constant functions. Consider the bivector field on $\mathbb{F}^3$, defined by

$$\pi := \left(x\frac{\partial}{\partial x} + y\frac{\partial}{\partial y}\right) \wedge \frac{\partial}{\partial z}\,.$$

Note that it appeared already in Example 7.12. The associated vector field is given by $\overrightarrow{P} : (y, -x, 0)$, whose curl is $(0, 0, -2)$, so that $\langle \overrightarrow{P} \mid \operatorname{curl} \overrightarrow{P} \rangle = 0$, which means, in view of (9.17), that π is a Poisson structure on $\mathbb{F}^3$. A function $\varphi \in \mathscr{F}(M)$ is a Casimir function for π if and only if

$$\frac{\partial \varphi}{\partial z} = 0 \quad \text{and} \quad x\frac{\partial \varphi}{\partial x} + y\frac{\partial \varphi}{\partial y} = 0\,.$$

The first equation means that φ does not depend on the variable z, while the second one implies, in view of the Euler formula (8.5), that φ is homogeneous of degree 0 in the two remaining variables. According to Proposition 8.3, every homogeneous function of degree zero is constant, so every Casimir function of π is constant. It follows from Proposition 9.14 that π is not of the form $\chi\pi_\varphi$, with $\chi, \varphi \in \mathscr{F}(\mathbb{F}^3)$.

9.2.2 Poisson Manifolds of Dimension Three

We now turn to Poisson structures on three-dimensional manifolds. Since the rank of a Poisson structure is an even number, smaller than or equal to the dimension of the manifold, the rank of every Poisson structure on a three-dimensional manifold is equal to two (except when the trivial Poisson structure is considered, in which case the rank is zero). Many properties of such Poisson structures hold true because their rank is two, rather than because the dimension of the manifold is three. We will therefore, in Subsection 9.2.2.1, first study the properties of rank two Poisson structures on manifolds of arbitrary dimension.

9.2.2.1 A – Poisson Structures of Rank Two

Let π be a bivector field on a manifold M. For every $m \in M$, we consider the associated linear map $\pi_m^\sharp : T_m^*M \to T_mM$. As in the case of Poisson structures, the rank of $\pi_m^\sharp$, which is always even, is called the *rank* of π at m, and π is said to be a bivector field of *rank two* if the rank of π is at most two at all points of M and is two at least one point of M. We denote by $M_{(2)}$ the open subset of M of points at which the rank of π is two. If π is a bivector field of rank two on M, then the subset $\{\pi_m^\sharp(T_m^*M) \mid m \in M_{(2)}\}$ of the tangent bundle is a distribution of rank two on $M_{(2)}$. We show in the following proposition that the Jacobi identity for π is equivalent to the involutivity of this distribution.

Proposition 9.17. *Let $\pi = \{\cdot\,,\cdot\}$ be a bivector field of rank two on a manifold M, and let $M_{(2)} \subset M$ denote the open subset of points at which the rank of π is two.*

(1) Let U be an open subset of M and let F and G be functions on U. Then, on U, we have

$$\{F,G\}\,\pi = \mathscr{X}_F \wedge \mathscr{X}_G\,; \tag{9.26}$$

(2) The bivector field π is a Poisson structure on M if and only if the distribution $\{\pi_m^\sharp(T_m^*M) \mid m \in M_{(2)}\}$ on $M_{(2)}$ is involutive.

Proof. Suppose that π is a bivector field of rank two on M. Let F and G be arbitrary functions, defined on a non-empty open subset U of M. Let m be an arbitrary point of U. We show that

$$\{F,G\}(m)\,\pi_m = (\mathscr{X}_F)_m \wedge (\mathscr{X}_G)_m \,. \tag{9.27}$$

Suppose first that the tangent vectors $(\mathscr{X}_F)_m$ and $(\mathscr{X}_G)_m$ are proportional, say $(\mathscr{X}_G)_m = a(\mathscr{X}_F)_m$ for some $a \in \mathbb{F}$. Then the right-hand side of (9.27) vanishes at m, while

$$\{F,G\}(m) = (\mathscr{X}_G)_m[F] = a(\mathscr{X}_F)_m[F] = a\{F,F\}(m) = 0\,,$$

so the left-hand side of (9.27) also vanishes at m. Thus, equation (9.27) is satisfied.

Assume now that the tangent vectors $(\mathscr{X}_F)_m$ and $(\mathscr{X}_G)_m$ are not proportional. Since the rank of π at m is two, these tangent vectors span $\pi_m^\sharp(T_m^*M)$, so that for every function H, defined on a neighborhood of $m \in M$, there exist $a,b \in \mathbb{F}$ such that

$$(\mathscr{X}_H)_m = a(\mathscr{X}_F)_m + b(\mathscr{X}_G)_m \,. \tag{9.28}$$

Applying (9.28) to F yields

$$(\mathscr{X}_H)_m[F] = a(\mathscr{X}_F)_m[F] + b(\mathscr{X}_G)_m[F] = b\{F,G\}(m)\,,$$

and we obtain that $\{F,H\}(m) = b\{F,G\}(m)$. Similarly, applying (9.28) to G yields $\{G,H\}(m) = -a\{F,G\}(m)$. Substituted in (9.28), we obtain

$$\{F,G\}(m)(\mathscr{X}_H)_m = \{F,H\}(m)(\mathscr{X}_G)_m - \{G,H\}(m)(\mathscr{X}_F)_m \,.$$

Since H is arbitrary, this proves (9.27) also when $(\mathscr{X}_F)_m$ and $(\mathscr{X}_G)_m$ are not proportional, thereby proving *(1)*.

We now prove *(2)*. If π is a Poisson structure of rank 2 on M, then $\{\pi_m^\sharp(T_m^*M) \mid m \in M_{(2)}\}$, the distribution on $M_{(2)}$, defined by all Hamiltonian vector fields, is involutive because $[\mathscr{X}_F, \mathscr{X}_G] = -\mathscr{X}_{\{F,G\}}$ for all functions F and G. Conversely, let π be a bivector field of rank two on M and assume that the distribution $\{\pi_m^\sharp(T_m^*M) \mid m \in M_{(2)}\}$ is involutive. Let m be a point of M. If π_m is zero, then the Schouten bracket $[\pi,\pi]_S$, evaluated at m, vanishes as well. If π_m is different from zero, so that $m \in M_{(2)}$, then there exist functions F,G, defined in a neighborhood U of m, such that $\{F,G\} \neq 0$ on U. According to *(1)*, $\pi = \mathscr{V} \wedge \mathscr{X}_G$ on U, where $\mathscr{V} := \frac{\mathscr{X}_F}{\{F,G\}}$. A direct computation, with the help of the graded Leibniz identity (3.42), gives

$$\frac{1}{2}[\pi,\pi]_S = \mathscr{V} \wedge [\mathscr{V}, \mathscr{X}_G] \wedge \mathscr{X}_G \,.$$

Since $\mathscr{X}_G$ and $\mathscr{V}$ are tangent to the distribution $\{\pi_m^\sharp(T_m^*M) \mid m \in M_{(2)}\}$, assumed to be involutive, the bracket $[\mathscr{V}, \mathscr{X}_G]$ is also tangent to the latter distribution. Hence, $([\pi,\pi]_S)_m \in \wedge^3 \pi_m^\sharp(T_m^*M)$, for every $m \in M_{(2)}$. But the latter space is zero, since

$\pi_m^\sharp(T_m^*M)$ is a two-dimensional vector space. In conclusion, the Schouten bracket of π with itself vanishes at every point in M, so that π is a Poisson structure. □

As a corollary, we obtain the following analog of Proposition 9.13.

Proposition 9.18. *Let π be a Poisson structure of rank two on a manifold M. For every function $\varphi \in \mathscr{F}(M)$, the bivector field $\varphi\pi$ is a Poisson structure.*

Proof. A proof can be given by picking local coordinates and using Proposition 9.13. We give a more geometrical proof. Consider the open subsets $M_{(2)}$ and $M'_{(2)}$ of M consisting of all points where the rank of π, respectively of $\varphi\pi$, is two. It is clear that $M'_{(2)} \subset M_{(2)}$ and that for every point $m \in M'_{(2)}$, the vector spaces

$$(\varphi\pi)_m^\sharp(T_m^*M) \quad \text{and} \quad \pi_m^\sharp(T_m^*M)$$

coincide. The distribution on $M'_{(2)}$ defined by the latter vector spaces is involutive, in view of item *(2)* in Proposition 9.17. Hence, the distribution on $M'_{(2)}$ defined by the former is also involutive, which in view of the same proposition, implies that $\varphi\pi$ is a Poisson structure on M. □

If in Proposition 9.18 the function φ is nowhere vanishing, the vector spaces $(\varphi\pi)_m^\sharp(T_m^*M)$ and $\pi_m^\sharp(T_m^*M)$ coincide for all $m \in M$, so that both Poisson structures have the same symplectic foliation. For regular Poisson structures of rank 2, the following inverse statement holds.

Proposition 9.19. *Let π_1 and π_2 be two regular Poisson structures of rank 2 on a manifold M. If π_1 and π_2 have the same symplectic leaves, then there exists a (unique) nowhere vanishing function $\chi \in \mathscr{F}(M)$, such that $\pi_1 = \chi\pi_2$.*

Proof. Suppose that π_1 and π_2 are Poisson structures which have the same symplectic leaves. For $m \in M$, denote by $\mathscr{S}_m$ their common symplectic leaf through m. If both Poisson structures are regular of rank 2, then on the one hand, $T_m\mathscr{S}_m$ is a two-dimensional vector space, so that $\wedge^2 T_m\mathscr{S}_m$ is one-dimensional, while on the other hand both $(\pi_1)_m$ and $(\pi_2)_m$ are non-zero elements of $\wedge^2 T_m\mathscr{S}_m$, so there exists a unique $\chi(m) \in \mathbb{F}^*$, such that $\pi_1(m) = \chi(m)\pi_2(m)$. Thus, under the above assumptions, there exists a nowhere vanishing function χ on M, such that $\pi_1 = \chi\pi_2$; since χ is given locally as the quotient of two non-vanishing functions, χ belongs to $\mathscr{F}(M)$. □

9.2.2.2 B – Poisson Manifolds of Dimension Three and One-Forms

Let M be a real orientable manifold of dimension three and let λ be a volume form on M. Recall from (4.19) that there is a natural $\mathscr{F}(M)$-linear isomorphism $\star : \mathfrak{X}^k(M) \to \Omega^{3-k}(M)$, defined for $Q \in \mathfrak{X}^k(M)$ by

$$\star Q := \imath_Q \lambda \; .$$

In particular, to a bivector field π on M corresponds a differential one-form $\alpha := \star\pi = \imath_\pi\lambda$ on M. In this subsection we wish to express the properties of π, for example the condition that π be a Poisson structure on M, in terms of the differential one-form α. Of course, this differential one-form depends on the chosen volume form: replacing λ by $F\lambda$, with F a nowhere vanishing function, amounts to replacing α by $F\alpha$.

Proposition 9.20. *Let M be a real orientable manifold of dimension three. Suppose that π is a bivector field on M.*

(1) Let λ be an arbitrary volume form on M, and let $\alpha := \imath_\pi\lambda$. Then π is a Poisson structure on M if and only if

$$\alpha \wedge \mathrm{d}\alpha = 0 . \tag{9.29}$$

In this case, for every $m \in M$ at which the rank of π is two, the kernel of α_m is the tangent space of the symplectic leaf of π through m;

(2) The bivector field π is a unimodular Poisson structure if and only if there exists a volume form λ on M, such that the differential one-form $\alpha := \imath_\pi\lambda$ is closed.

Proof. Recall from Cartan's formula (3.38) that if P and Q are two multivector fields on M, then

$$[[\imath_P, \mathrm{d}], \imath_Q] = \imath_{[P,Q]_S} .$$

Specializing to $P := \pi$ and $Q := \pi$, and applied to the volume form λ, it yields

$$2\imath_\pi \mathrm{d}\imath_\pi\lambda = \imath_{[\pi,\pi]_S}\lambda . \tag{9.30}$$

Multiplied by the volume form λ, and expressed in terms of α, (9.30) becomes

$$2(\imath_\pi \mathrm{d}\alpha)\lambda = (\imath_{[\pi,\pi]_S}\lambda)\lambda . \tag{9.31}$$

Now, for every bivector field P and for every differential 2-form β, the fact that λ is a top form implies that[2]

$$(\imath_P\beta)\lambda = \beta \wedge \imath_P\lambda . \tag{9.32}$$

In particular, $(\imath_\pi \mathrm{d}\alpha)\lambda = \mathrm{d}\alpha \wedge \imath_\pi\lambda = \mathrm{d}\alpha \wedge \alpha$, so that (9.31) gives eventually:

$$2\mathrm{d}\alpha \wedge \alpha = \left(\imath_{[\pi,\pi]_S}\lambda\right)\lambda .$$

As a consequence, $\mathrm{d}\alpha \wedge \alpha = 0$ if and only if $[\pi,\pi]_S = 0$, i.e., if and only if π is a Poisson structure on M. This proves the first part of item *(1)*.

Let us prove the second part of *(1)*, i.e., that if the rank of π at m is two, then the spaces $\mathrm{Ker}(\alpha_m)$ and $T_m\mathscr{S}_m$ coincide, where $\mathscr{S}_m$ is the symplectic leaf of π, passing through m. Since both vector spaces are two-dimensional, it suffices to show

[2] In the actual case where the dimension of M is 3, a simple proof of (9.32) can be given by taking local coordinates in which $\lambda = \mathrm{d}x \wedge \mathrm{d}y \wedge \mathrm{d}z$.

the inclusion $T_m\mathscr{S}_m \subset \operatorname{Ker}\alpha_m$. The tangent space at m to $\mathscr{S}_m$ is spanned by the Hamiltonian vector fields $\mathscr{X}_F$ at m, where F is an arbitrarily function, defined in a neighborhood of m. Now,

$$\alpha(\mathscr{X}_F) = (\imath_\pi\lambda)(\mathscr{X}_F) = \lambda(\pi \wedge \mathscr{X}_F),$$

and $\pi \wedge \mathscr{X}_F = 0$ because the rank of π is 2 (see (9.26)), so that $(\mathscr{X}_F)_m \in \operatorname{Ker}\alpha_m$ for every $m \in M$ and for every function F, defined in a neighborhood of m in M. This completes the proof of *(1)*.

We now prove *(2)*. Recall from (4.24) that the modular vector field Φ of π with respect to a given volume form λ, can be expressed in terms of the divergence of π as $\Phi = -\operatorname{Div}(\pi)$. According to the definition of Div (see (4.18)) and the definition $\alpha := \imath_\pi\lambda$, we have that

$$\star\Phi = -\star\operatorname{Div}(\pi) = -\mathrm{d}\star\pi = -\mathrm{d}\imath_\pi\lambda = -\mathrm{d}\alpha,$$

so that $\Phi = 0$ if and only if α is closed. Since a Poisson structure is unimodular if and only if there exists a volume form λ with respect to which the modular vector field Φ is zero, this means that π is unimodular if and only if there exists a volume form λ such that $\alpha := \imath_\pi\lambda$ is closed. This shows *(2)*. □

9.2.3 Quadratic Poisson Structures on $\mathbb{F}^3$

In this section, we present a classification of quadratic Poisson structures on the affine space $\mathbb{C}^3$. Recall from Proposition 8.8 that two quadratic Poisson structures on $\mathbb{F}^3$ are isomorphic if and only if there exists a *linear* isomorphism of $\mathbb{F}^3$, which sends one of the quadratic Poisson structures to the other one. As before, we denote the standard coordinates on $\mathbb{F}^3$ by x,y,z, and we denote by λ the standard volume form

$$\lambda := \mathrm{d}x \wedge \mathrm{d}y \wedge \mathrm{d}z.$$

Recall from Section 8.2.2 that the modular vector field, with respect to λ, of a quadratic Poisson structure is a linear vector field (Proposition 8.23), and recall from Proposition 4.17 that its divergence with respect to λ is zero.

Specializing and adapting Proposition 8.26 to the three-dimensional case, yields the next proposition. By definition, a *cubic polynomial* function on $\mathbb{F}^d$ is a *homogeneous* polynomial function of degree three on $\mathbb{F}^d$.

Proposition 9.21. *There is a one-to-one correspondence between quadratic Poisson structures on* $\mathbb{F}^3$ *and pairs* (Φ,φ)*, with* Φ *a linear vector field on* $\mathbb{F}^3$*, and* φ *a cubic polynomial function on* $\mathbb{F}^3$*, satisfying*

$$\operatorname{Div}(\Phi) = 0 \quad \text{and} \quad \Phi[\varphi] = 0. \tag{9.33}$$

This correspondence assigns to such a pair (Φ,φ) *the Poisson structure, defined by*

$$\pi_{\Phi,\varphi} := \pi_\varphi - \frac{1}{3}\mathscr{E} \wedge \Phi\,, \tag{9.34}$$

where $\mathscr{E}$ is the Euler vector field, and π_φ is the Poisson structure defined in (9.19).

Proof. To start with, let us show that, given a linear vector field Φ on $\mathbb{F}^3$ and a cubic function φ on $\mathbb{F}^3$, which satisfy (9.33), the bivector field $\pi_{\Phi,\varphi}$ defined by (9.34) is a Poisson structure. First, both $\mathscr{E} \wedge \Phi$ and π_φ are Poisson structures according to Example 8.15 and Proposition 9.14 respectively. As a consequence,

$$[\pi_{\Phi,\varphi}, \pi_{\Phi,\varphi}]_S = -\frac{2}{3}[\pi_\varphi, \mathscr{E} \wedge \Phi]_S = -\frac{2}{3}(\mathscr{L}_{\mathscr{E}}\pi_\varphi) \wedge \Phi + \frac{2}{3}(\mathscr{L}_\Phi \pi_\varphi) \wedge \mathscr{E}\,.$$

It suffices to prove that both of the remaining terms vanish. For the first term, by Proposition 8.4, the Lie derivative with respect to the Euler vector field of every quadratic Poisson structure is zero, so that $\mathscr{L}_{\mathscr{E}}\pi_\varphi = 0$. The vanishing of the second term follows at once by applying the following identity, valid for every vector field $\mathscr{V}$ on $\mathbb{F}^3$,

$$\mathscr{L}_{\mathscr{V}}\pi_\varphi = \mathrm{Div}(\mathscr{V})\,\pi_\varphi + \pi_{\mathscr{V}[\varphi]} \tag{9.35}$$

to $\mathscr{V} := \Phi$, upon using (9.33). We prove the identity (9.35) by showing that

$$\imath_{\mathscr{L}_{\mathscr{V}}\pi_\varphi}\lambda = \mathrm{Div}(\mathscr{V})\,\imath_{\pi_\varphi}\lambda + \imath_{\pi_{\mathscr{V}[\varphi]}}\lambda\,. \tag{9.36}$$

First, by item *(2)* in Proposition 3.11 and by (4.20), the left-hand side of (9.36) is given by

$$\imath_{\mathscr{L}_{\mathscr{V}}\pi_\varphi}\lambda = \imath_{[\mathscr{V},\pi_\varphi]_S}\lambda = \mathscr{L}_{\mathscr{V}}\imath_{\pi_\varphi}\lambda + \imath_{\pi_\varphi}\mathscr{L}_{\mathscr{V}}\lambda = \mathscr{L}_{\mathscr{V}}\imath_{\pi_\varphi}\lambda + \mathrm{Div}(\mathscr{V})\,\imath_{\pi_\varphi}\lambda\,. \tag{9.37}$$

Therefore, it remains to be shown that $\mathscr{L}_{\mathscr{V}}\imath_{\pi_\varphi}\lambda = \imath_{\pi_{\mathscr{V}[\varphi]}}\lambda$, which means, in view of (9.20), that $\mathscr{L}_{\mathscr{V}}\mathrm{d}\varphi = \mathrm{d}(\mathscr{V}[\varphi])$; but this is clear since $\mathscr{L}_{\mathscr{V}}$ and d commute. This proves (9.35), and hence that the bivector field $\pi_{\Phi,\varphi}$ defined by (9.34) is a Poisson structure.

Next, we show that the assignment $(\Phi,\varphi) \mapsto \pi_{\Phi,\varphi}$, defined by (9.34), is surjective. Let π be a quadratic Poisson bracket, and let Φ_π denote its modular vector field. Proposition 8.26 implies that there exists a closed differential one-form α such that:

$$\pi = \star^{-1}\alpha - \frac{1}{3}\mathscr{E} \wedge \Phi_\pi\,.$$

By the algebraic Poincaré lemma, there exists a function φ such that $\mathrm{d}\varphi = \alpha$; since α is a quadratic differential one-form, φ can be chosen (in a unique way) to be a cubic function. In view of (9.20), we have $\star^{-1}\alpha = \star^{-1}\mathrm{d}\varphi = \pi_\varphi$. Hence, π is given as in (9.34). We still have to check that the pair (Φ_π, φ) satisfies (9.33). According to Proposition 4.17, the divergence of the modular vector field Φ_π is zero. According to Proposition 8.26, we have that $\mathscr{L}_{\Phi_\pi}\alpha = 0$, so that $\mathrm{d}(\Phi_\pi[\varphi]) = 0$. Since $\Phi_\pi[\varphi]$ is a cubic function, this implies $\Phi_\pi[\varphi] = 0$.

We are left with the task of showing that the assignment, defined by (9.34) is injective. Since the divergence of π_φ vanishes and according to Example 8.25,

$$\mathrm{Div}(\pi_{\Phi,\varphi}) = \mathrm{Div}\left(-\frac{1}{3}\mathscr{E}\wedge\Phi\right) = \Phi - \frac{1}{3}\mathrm{Div}(\Phi)\mathscr{E} = \Phi\ ,$$

so that we can recover Φ as the modular vector field of $\pi_{\Phi,\varphi}$. Since the cubic polynomial φ can also be uniquely recovered from π_φ, both Φ and φ can be uniquely recovered from $\pi_{\Phi,\varphi}$, which proves injectivity, hence bijectivity of the assignment, defined by (9.34). This completes the proof. □

Proposition 9.21 leads to an explicit description of the family of all quadratic Poisson structures on $\mathbb{F}^3$, as given in the following theorem.

Theorem 9.22. *Let π be a quadratic Poisson structure on $\mathbb{C}^3$. There exist linear coordinates on $\mathbb{C}^3$ in which π takes the form*

$$\pi = \pi_\varphi - \frac{1}{3}\mathscr{E}\wedge\Phi\ ,$$

where (Φ,φ) corresponds to one of the types I–VII *in Table 9.2, where π_φ is the Poisson structure*

$$\pi_\varphi := \frac{\partial\varphi}{\partial x}\frac{\partial}{\partial y}\wedge\frac{\partial}{\partial z} + \frac{\partial\varphi}{\partial y}\frac{\partial}{\partial z}\wedge\frac{\partial}{\partial x} + \frac{\partial\varphi}{\partial z}\frac{\partial}{\partial x}\wedge\frac{\partial}{\partial y}\ ,$$

and where $\mathscr{E}$ is the Euler vector field on $\mathbb{C}^3$.

In order to prove this theorem, we first establish, in the following lemma, the list of all divergence-free linear vector fields on $\mathbb{C}^3$. Recall that in this section the divergence is always taken with respect to the standard volume form $\lambda = \mathrm{d}x\wedge\mathrm{d}y\wedge\mathrm{d}z$ on $\mathbb{C}^3$.

Lemma 9.23. *Let Φ be a divergence-free linear vector field on $\mathbb{C}^3$. There exist linear coordinates x,y,z on $\mathbb{C}^3$ such that Φ takes one of the forms, given in the second column of Table 9.3.*

Proof. Recall from Section 8.2 that linear vector fields on a finite-dimensional vector space V are in one-to-one correspondence with endomorphisms of the dual vector space V^*: more precisely, to a linear vector field Φ on V corresponds, as in (8.20), the endomorphism $\mathscr{V}^{(1)}$ of V^* defined by

$$\begin{aligned}\Phi^{(1)}:V^* &\to V^*\\ F &\mapsto \Phi[F]\ .\end{aligned} \tag{9.38}$$

If we write Φ in terms of linear coordinates $x_1,\dots,x_d$ on V as

$$\Phi = \sum_{i,j=1}^{d} a_{ij}x_i\frac{\partial}{\partial x_j}\ ,$$

Table 9.2 Up to linear isomorphism, every quadratic Poisson structure on $\mathbb{C}^3$ is of the form $\pi_{\Phi,\varphi}$, where Φ and φ are given in the second and third columns.

Type	Φ	φ	Parameters
I	$a_1x\frac{\partial}{\partial x}+a_2y\frac{\partial}{\partial y}+a_3z\frac{\partial}{\partial z}$	$bxyz$	$a_1,a_2,a_3 \in \mathbb{C}^*$ all a_i different $a_1+a_2+a_3=0$; $b\in\mathbb{C}$
II	$ax\frac{\partial}{\partial x}+ay\frac{\partial}{\partial y}-2az\frac{\partial}{\partial z}$	$zG(x,y)$	$a\in\mathbb{C}^*$ G quadratic function
III	$ax\frac{\partial}{\partial x}-ay\frac{\partial}{\partial y}$	$bxyz+cz^3$	$a\in\mathbb{C}^*$; $b,c\in\mathbb{C}$
IV	$ax\frac{\partial}{\partial x}+(ay+x)\frac{\partial}{\partial y}-2az\frac{\partial}{\partial z}$	bx^2z	$a\in\mathbb{C}^*$; $b\in\mathbb{C}$
V	$x\frac{\partial}{\partial y}+y\frac{\partial}{\partial z}$	$bx^3+cx(y^2-2xz)$	$b,c\in\mathbb{C}$
VI	$x\frac{\partial}{\partial z}$	$G(x,y)$	G cubic function
VII	0	$G(x,y,z)$	G cubic function

then $A := (a_{ij}) \in \mathrm{Mat}_d(\mathbb{C})$ is the matrix of $\Phi^{(1)}$ in terms of the basis $(x_1,\dots,x_d)$ for V^*. Using the classical formula (4.21) for the divergence of a vector field, we find

$$\mathrm{Div}(\Phi) = \sum_{i,j=1}^{d} \frac{\partial}{\partial x_j}(a_{ij}x_i) = \sum_{i=1}^{d} a_{ii}\,,$$

so that the divergence of Φ is a constant function, whose value is the trace of the endomorphism $\Phi^{(1)}$. In particular, a linear vector field Φ on $\mathbb{C}^3$ is divergence-free if and only if the corresponding endomorphism $\Phi^{(1)}$ of $\mathbb{C}^3$ is traceless. There are seven types of such endomorphisms:

- If all eigenvalues of $\Phi^{(1)}$ are different, then there are two possibilities, according to whether one or zero of these eigenvalues are zero, leading to types I and III in Table 9.3;

Table 9.3 There are, up to isomorphism, seven types of traceless endomorphisms of $\mathbb{C}^3$. For each type, its matrix takes in terms of a well-chosen basis the form indicated in the second column of the table. The range of the parameters which appear in the matrices is indicated in the third column.

Type	Matrix	Parameters
I	$\mathrm{diag}(a_1,a_2,a_3)$	$a_1,a_2,a_3 \in \mathbb{C}^*$ all a_i different $a_1+a_2+a_3=0$
II	$\mathrm{diag}(a,a,-2a)$	$a \in \mathbb{C}^*$
III	$\mathrm{diag}(a,-a,0)$	$a \in \mathbb{C}^*$
IV	$\begin{pmatrix} a & 1 & 0 \\ 0 & a & 0 \\ 0 & 0 & -2a \end{pmatrix}$	$a \in \mathbb{C}^*$
V	$\begin{pmatrix} 0 & 1 & 0 \\ 0 & 0 & 1 \\ 0 & 0 & 0 \end{pmatrix}$	–
VI	$\begin{pmatrix} 0 & 0 & 1 \\ 0 & 0 & 0 \\ 0 & 0 & 0 \end{pmatrix}$	–
VII	0	–

- If $\Phi^{(1)}$ has precisely two different eigenvalues, then they are both different from zero, leading to types II or IV, according to whether $\Phi^{(1)}$ is diagonalizable or not;
- If all eigenvalues of $\Phi^{(1)}$ are equal, then they are all zero, so that $\Phi^{(1)}$ is a nilpotent endomorphism. This leads to three types, the types V–VII.

The conditions on the parameters in Table 9.3 are chosen such that every traceless endomorphism of $\mathbb{C}^3$ appears exactly once in the table. □

It is clear that the matrices which are given in the second column of Table 9.3 correspond to the linear vector fields on $\mathbb{C}^3$, listed in the second column of Table 9.2.

Combining Proposition 9.21 and Lemma 9.23 leads to the announced classification of the set of quadratic Poisson structures on $\mathbb{C}^3$.

Proof (of Theorem 9.22). Let π be a quadratic Poisson structure on $\mathbb{C}^3$. According to Proposition 9.21, there exists a pair (Φ,φ), with Φ a divergence-free linear vector field on $\mathbb{C}^3$, and φ a cubic polynomial function on $\mathbb{C}^3$, satisfying $\Phi[\varphi]=0$, such that π is of the form (9.34). According to Lemma 9.23, there exist linear coordinates x,y,z on $\mathbb{C}^3$ in which Φ takes the form corresponding to one of the types I–VII in Table 9.3. It therefore suffices to determine for each of these types the vector space of all cubic functions $\varphi\in\mathscr{F}(\mathbb{C}^3)$, satisfying $\Phi[\varphi]=0$. It is easy to see, by direct computation, that for each line of the table, the given function φ satisfies indeed $\Phi[\varphi]=0$. Let us show that if Φ is of one of the types I–V, there are no other such φ (for Φ of the other types VI and VII this is obvious).

For the types I–III, Φ is the weighted Euler operator, with (not necessarily integral) weight vector (a_1,a_2,a_3) with $a_1+a_2+a_3=0$ for type I, with weight vector $(a,a,-2a)$ for type II and weight vector $(a,-a,0)$ for type III. Therefore, for each of these types, the vector space of all polynomials annihilated by Φ is spanned by all monomials F satisfying $\Phi[F]=0$. Taking a cubic monomial $F=x^{\alpha_1}y^{\alpha_2}z^{\alpha_3}$, this leads for type I to the equation $\alpha_1a_1+\alpha_2a_2+\alpha_3a_3=0$, which has $(1,1,1)$ as its unique solution, under the constraint $\alpha_1+\alpha_2+\alpha_3=3$ (because $\deg(F)=3$), with all $a_i\in\mathbb{C}^*$ being different. This leads to the monomial $F=xyz$. Similarly, the equation $\Phi[F]=0$ leads for type II to the equation $a(\alpha_1+\alpha_2)-2a\alpha_3=0$, under the constraint $\alpha_1+\alpha_2+\alpha_3=3$, which has as solution all triples $(\alpha_1,\alpha_2,1)$, with $\alpha_1+\alpha_2=2$, corresponding to all monomials of the form $zG(x,y)$, where G is a quadratic polynomial. Similarly, for type III, one looks for all triples $(\alpha_1,\alpha_2,\alpha_3)$, satisfying $\alpha_1-\alpha_2=0$, which has $(1,1,1)$ and $(0,0,3)$ as its only solutions under the constraint $\alpha_1+\alpha_2+\alpha_3=3$; it corresponds to the monomials xyz and z^3.

For Φ of type IV, one computes easily the images under Φ of the 10 monomials $x^{\alpha_1}y^{\alpha_2}z^{\alpha_3}$, with $\alpha_1+\alpha_2+\alpha_3=3$, by using the fact that Φ is the sum of the weighted Euler operator (with weights $(a,a,-2a)$) and the operator $x\frac{\partial}{\partial y}$. One sees that, besides for $\Phi(x^2z)$ (which is zero), these images are linearly independent, proving that the kernel of Φ (restricted to the space of cubic polynomials) is spanned by x^2z. Similarly, for Φ of type V, one easily checks that the images of the eight cubic monomials $x^{\alpha_1}y^{\alpha_2}z^{\alpha_3}$, with $(\alpha_1,\alpha_2,\alpha_3)\neq(3,0,0)$ and $(\alpha_1,\alpha_2,\alpha_3)\neq(2,0,1)$ are linearly independent. The kernel of Φ is therefore two-dimensional, and is spanned by x^3 and $x(y^2-2xz)$. □

9.2.4 Poisson Surfaces in $\mathbb{C}^3$ and Du Val Singularities

We have seen two different ways of constructing (singular) Poisson surfaces.

(1) Let $\mathbf{G}$ be a finite subgroup of $\mathbf{SL}_2(\mathbb{C})$. The standard action of $\mathbf{G}$ on $\mathbb{C}^2$ induces on the quotient surface $\mathbb{C}^2/\mathbf{G}$ a Poisson structure $\{\cdot\,,\cdot\}_{\mathbb{C}^2/\mathbf{G}}$, making it into a (singular) Poisson surface (see Section 6.3.4).

(2) Let φ be an irreducible polynomial in three variables. The Poisson structure $\pi_\varphi=\{\cdot\,,\cdot\}_\varphi$ on $\mathbb{C}^3$, introduced in (9.19), given by

$$\pi_\varphi := \frac{\partial\varphi}{\partial x}\frac{\partial}{\partial y}\wedge\frac{\partial}{\partial z}+\frac{\partial\varphi}{\partial y}\frac{\partial}{\partial z}\wedge\frac{\partial}{\partial x}+\frac{\partial\varphi}{\partial z}\frac{\partial}{\partial x}\wedge\frac{\partial}{\partial y}, \qquad (9.39)$$

admits φ as a Casimir, hence the singular affine surface $\Sigma_\varphi \subset \mathbb{C}^3$, given by $\varphi = 0$, is a (singular) Poisson surface.

The purpose of the present section is to show how these two constructions are related. Namely, we will show that every Poisson surface $\mathbb{C}^2/\mathbf{G}$ of the first type is also of the second type Σ_φ, for a polynomial φ which will be given explicitly (for each $\mathbf{G}$).

Before doing this, let us recall a few facts about the classification of finite subgroups of $\mathbf{SL}_2(\mathbb{C})$ (see [183] for proofs and details). Consider the list of subgroups A_k, for $k \geqslant 1$, D_k for $k \geqslant 4$, or E_6, E_7, E_8, whose generators are given in Table 9.4.

Table 9.4 Every finite subgroup of $\mathbf{SL}_2(\mathbb{C})$ is conjugated to precisely one of the subgroups, given in terms of generators in the second column of the table, where $\varepsilon := e^{\frac{2\pi\sqrt{-1}}{8}}$ and $\eta := e^{\frac{2\pi\sqrt{-1}}{5}}$.

Type of $\mathbf{G}$	Generators of $\mathbf{G} \subset \mathbf{SL}_2(\mathbb{C})$
$A_k \quad k \geqslant 1$	$\begin{pmatrix} e^{\frac{2\sqrt{-1}\pi}{k+1}} & 0 \\ 0 & e^{\frac{-2\sqrt{-1}\pi}{k+1}} \end{pmatrix}$
$D_k \quad k \geqslant 4$	$\begin{pmatrix} e^{\frac{\sqrt{-1}\pi}{k}} & 0 \\ 0 & e^{\frac{-\sqrt{-1}\pi}{k}} \end{pmatrix}$ and $\begin{pmatrix} 0 & 1 \\ -1 & 0 \end{pmatrix}$
E_6	generators of D_4 and $\frac{1}{\sqrt{2}}\begin{pmatrix} \varepsilon^7 & \varepsilon^7 \\ \varepsilon^5 & \varepsilon \end{pmatrix}$
E_7	generators of E_6 and $\begin{pmatrix} \varepsilon & 0 \\ 0 & \varepsilon^7 \end{pmatrix}$
E_8	$\begin{pmatrix} -\eta^3 & 0 \\ 0 & -\eta^2 \end{pmatrix}$ and $\frac{1}{\eta^2-\eta^3}\begin{pmatrix} \eta+\eta^4 & 1 \\ 1 & -\eta-\eta^4 \end{pmatrix}$

Up to conjugation, every subgroup $\mathbf{G}$ of $\mathbf{SL}_2(\mathbb{C})$ is conjugated to one of these subgroups $A_1, A_2, \ldots, E_8$. We then say that $\mathbf{G}$ is of *type* $A_1, A_2, \ldots, E_8$. The quotient surface $\mathbb{C}^2/\mathbf{G}$ is singular at the image in $\mathbb{C}^2/\mathbf{G}$ of the origin o of $\mathbb{C}^2$, and at no other point; a singularity obtained in this way is called a *Du Val singularity* or *Kleinian singularity*, depending on the reference. It is a classical result [183, 188] that the affine variety $\mathbb{C}^2/\mathbf{G}$ is, as an affine variety, isomorphic to Σ_φ, where φ is the polynomial, depending on the type of $\mathbf{G}$, which appears in the second column of Table 9.5.

Table 9.5 For each of the finite subgroups $\mathbf{G}$ of $\mathbf{SL}_2(\mathbb{C})$, the quotient surface $\mathbb{C}^2/\mathbf{G}$ is the zero locus of a weight homogeneous polynomial, given in the second column of the table. The third column indicates the weight vector with respect to which each of these polynomials is weight homogeneous.

Type of $\mathbf{G}$	φ	Weights
$A_k \quad k \geqslant 1$	$x^2+y^2-z^{k+1}$	$(k+1,k+1,2)$
$D_k \quad k \geqslant 4$	$x^2+y^2z+z^{k-1}$	$(k-1,k-2,2)$
E_6	$x^2+y^3+z^4$	$(6,4,3)$
E_7	$x^2+y^3+yz^3$	$(9,6,4)$
E_8	$x^2+y^3+z^5$	$(15,10,6)$

Note that this means that, for each group $\mathbf{G}$ which appears in the table, the algebra of $\mathbf{G}$-invariant polynomials is an algebra with three weight homogeneous generators P_1,P_2,P_3, and that every polynomial $R \in \mathbb{C}[X,Y,Z]$, satisfying $R(P_1,P_2,P_3)=0$, is a multiple of φ. The isomorphism $\mathbb{C}^2/\mathbf{G} \simeq \Sigma_\varphi$ is induced by the map $\mathbb{C}^2 \to \mathbb{C}^3$, given by (P_1,P_2,P_3), as expressed in the following commutative diagram:

$$\begin{array}{ccc} \mathbb{C}^2 & \xrightarrow{(P_1,P_2,P_3)} & \mathbb{C}^3 \\ {\scriptstyle p}\downarrow & & \uparrow{\scriptstyle \imath} \\ \mathbb{C}^2/\mathbf{G} & \xrightarrow[\simeq]{} & \Sigma_\varphi \end{array} \tag{9.40}$$

Recall from Section 5.1.2 that $\mathscr{F}(\mathbb{C}^2)^{\mathbf{G}} \simeq \mathscr{F}(\mathbb{C}^2/\mathbf{G})$ (since $\mathbf{G}$ is a finite group). On the other hand, since Σ_φ is the affine surface of $\mathbb{C}^3$, defined by the polynomial φ, its algebra of functions $\mathscr{F}(\Sigma_\varphi)$ is given by $\mathbb{C}[x,y,z]/\langle\varphi\rangle$. It follows that we have an algebra isomorphism

$$\mathscr{F}(\mathbb{C}^2)^{\mathbf{G}} \simeq \mathbb{C}[x,y,z]/\langle\varphi\rangle\,. \tag{9.41}$$

We show in the following proposition that the isomorphism $\mathbb{C}^2/\mathbf{G} \simeq \Sigma_\varphi$ is actually an isomorphism of Poisson varieties.

Proposition 9.24. *Let $\mathbf{G}$ be a finite subgroup of $\mathbf{SL}_2(\mathbb{C})$ and let $\{\cdot\,,\cdot\}_{\mathbb{C}^2/\mathbf{G}}$ denote the Poisson structure on $\mathbb{C}^2/\mathbf{G}$, induced by the action of $\mathbf{G}$ on $\mathbb{C}^2$. The Poisson sur-*

face $(\mathbb{C}^2/\mathbf{G},\{\cdot,\cdot\}_{\mathbb{C}^2/\mathbf{G}})$ *is isomorphic, as a Poisson variety, to the Poisson surface* $(\Sigma_\varphi,\{\cdot,\cdot\}_\varphi)$*, where* φ *is the polynomial, associated to* $\mathbf{G}$*, as indicated in Table 9.5.*

Proof. Consider the isomorphism $\Xi:\mathbb{C}^2/\mathbf{G}\to\Sigma_\varphi$, given in (9.40). We show that Ξ is a Poisson isomorphism, with respect to the stated Poisson structures (up to a constant). According to Corollary 6.20, $\mathbb{C}^2/\mathbf{G}$ has a unique singular point, the point $p(o)$, where o is the origin of $\mathbb{C}^2$. Since the origin of $\mathbb{C}^3$ is clearly a singular point of Σ_φ, it follows that it is the image of $p(o)$ under Ξ and that it is the only singular point of Σ_φ (the latter fact can also easily be verified by direct computation). According to the latter corollary, the Poisson structure $\{\cdot,\cdot\}_{\mathbb{C}^2/\mathbf{G}}$ on $\mathbb{C}^2/\mathbf{G}$ has rank two at every point of $\mathbb{C}^2/\mathbf{G}$, except for the singular point $p(o)$, where its rank is zero. The Poisson matrix of $\{\cdot,\cdot\}_\varphi$ is given, in terms of the generators $\bar{x},\bar{y}$ and $\bar{z}$ of $\mathscr{F}(\Sigma_\varphi)$ as

$$\begin{pmatrix} 0 & \dfrac{\partial\varphi}{\partial z} & -\dfrac{\partial\varphi}{\partial y} \\ -\dfrac{\partial\varphi}{\partial z} & 0 & \dfrac{\partial\varphi}{\partial x} \\ \dfrac{\partial\varphi}{\partial y} & -\dfrac{\partial\varphi}{\partial x} & 0 \end{pmatrix}.$$

The rank of this matrix is two, except at the point(s) of Σ_φ for which $\frac{\partial\varphi}{\partial x}=\frac{\partial\varphi}{\partial y}=\frac{\partial\varphi}{\partial z}=0$, i.e., at the singular points of Σ_φ. Since the origin of Σ_φ is its only singular point, the rank of $\{\cdot,\cdot\}_\varphi$ is two at all points, except at the singular point of Σ_φ (where the rank is zero).

Using the isomorphism $\Xi:\mathbb{C}^2/\mathbf{G}\to\Sigma_\varphi$, we can transport the Poisson structure $\{\cdot,\cdot\}_{\mathbb{C}^2/\mathbf{G}}$ on $\mathbb{C}^2/\mathbf{G}$ to a Poisson structure on Σ_φ, which we denote by $\{\cdot,\cdot\}'_\varphi$. On the smooth surface $\Sigma_\varphi\setminus\{p(o)\}$ both Poisson structures $\{\cdot,\cdot\}_\varphi$ and $\{\cdot,\cdot\}'_\varphi$ are of rank 2, hence according to Proposition 9.19 there exists a nowhere vanishing holomorphic function χ on $\Sigma_\varphi\setminus\{p(o)\}$ such that $\{\cdot,\cdot\}_\varphi=\chi\{\cdot,\cdot\}'_\varphi$. For arbitrary $F,G\in\mathscr{F}(\Sigma_\varphi)$ such that $\{F,G\}'_\varphi\neq 0$, we have that

$$\chi=\frac{\{F,G\}_\varphi}{\{F,G\}'_\varphi},$$

so that χ is actually a rational function on Σ_φ. Since its divisor of zeros and its divisor of poles are both discrete (they contain at most the point $p(o)$), χ is constant. This implies that $(\Sigma_\varphi,\{\cdot,\cdot\}_\varphi)$ and $(\mathbb{C}^2/\mathbf{G},\mu\{\cdot,\cdot\}_{\mathbb{C}^2/\mathbf{G}})$ are isomorphic Poisson varieties for some $\mu\in\mathbb{C}^*$. In view of Remark 6.19, this constant μ can be chosen equal to 1. □

9.3 Notes

The classification in dimension two given in this chapter follows Monnier [150, 152], a variant of Arnold's proof [16]. In these references, the proof is more general than the proof given here, since the case when $\mathbb{R}$ is taken as the ground field is also considered, and with some extra work it is shown that the classification is actually non-formal, i.e., the formal coordinate transformations can be realized by smooth (or analytic) *local* coordinate transformations, when the given Poisson structure is smooth (or analytic). The question of having a global rather than a local classification in the context of varieties or manifolds remains a real challenge; for an important contribution in this direction, see [173], in which a classification of topologically stable Poisson structures on real surfaces is given.

The classification of quadratic Poisson structures in dimension three, detailed in Section 9.2.3 above, has been obtained independently by Dufour–Haraki [60] and Liu–Xu [129]. For the relation between Poisson structures and differential forms, in the presence of a volume form, see [88].

Chapter 10
R-Brackets and *r*-Brackets

The theory of integrable systems gave birth to interesting constructions, which yield new Poisson structures on a given (finite-dimensional) Lie algebra $\mathfrak{g}$, upon using some extra structure on the Lie algebra, roughly speaking a *matrix*. There are two distinct formalisms for doing this, depending on whether this matrix is viewed as a linear map $R : \mathfrak{g} \to \mathfrak{g}$, in which case R is called an R-matrix, or as an element $r \in \mathfrak{g} \otimes \mathfrak{g}$, in which case r is called an r-matrix. In their basic form, the new Poisson structures are Lie–Poisson structures, so it will be assumed that the reader is familiar with the basic theory of Lie–Poisson structures, as developed in Chapter 7.

For a given linear map $R : \mathfrak{g} \to \mathfrak{g}$, which satisfies a condition which will be explained in detail, a new Lie bracket on $\mathfrak{g}$ is defined by letting $[x,y]_R := \frac{1}{2}([Rx,y] + [x,Ry])$, for all $x,y \in \mathfrak{g}$. The new Lie bracket on $\mathfrak{g}$ leads to a Poisson structure on $\mathfrak{g}^*$, which is referred to as an R-bracket on $\mathfrak{g}^*$. Upon identifying $\mathfrak{g}$ with $\mathfrak{g}^*$, which is usually done by using a non-degenerate symmetric bilinear form, the R-bracket on $\mathfrak{g}^*$ becomes a Lie–Poisson structure on $\mathfrak{g}$.

A tensor $r \in \mathfrak{g} \otimes \mathfrak{g}$ can be viewed as a zero-chain on $\mathfrak{g}$, with values in $\mathfrak{g} \otimes \mathfrak{g}$, so its coboundary $\delta_L^0(r) : \mathfrak{g} \to \mathfrak{g} \otimes \mathfrak{g}$ admits as transpose a linear map $\gamma_r : \mathfrak{g}^* \otimes \mathfrak{g}^* \to \mathfrak{g}^*$. When r satisfies a condition, which will also be explained in detail, γ_r defines a Lie algebra structure on $\mathfrak{g}^*$, hence leads directly to a Lie–Poisson bracket on $\mathfrak{g}$, i.e., no identification of $\mathfrak{g}$ with $\mathfrak{g}^*$ is needed to define this Lie–Poisson structure on $\mathfrak{g}$. This Lie–Poisson bracket is called an r-bracket.

Section 10.1 is devoted to the construction of the R-bracket on $\mathfrak{g}$ and on $\mathfrak{g}^*$, with special emphasis on the main example, which comes from Lie algebra splittings. In Section 10.2, we discuss the case of r-brackets and we show how R-brackets and r-brackets are related. The construction of higher degree brackets, from an R-matrix or r-matrix, is given in Section 10.3. For the connection with Poisson–Lie groups, we refer to Chapter 11.

In this chapter, $\mathbb{F}$ is an arbitrary field of characteristic zero, $\mathfrak{g}$ is a finite-dimensional Lie algebra over $\mathbb{F}$ and $\mathscr{F}(\mathfrak{g})$ denotes the algebra of polynomial functions on $\mathfrak{g}$, but it may also be taken as the algebra of smooth, respectively holomorphic, functions on $\mathfrak{g}$, when $\mathbb{F} = \mathbb{R}$, respectively when $\mathbb{F} = \mathbb{C}$.

C. Laurent-Gengoux et al., *Poisson Structures*,
Grundlehren der mathematischen Wissenschaften 347,
DOI 10.1007/978-3-642-31090-4_10, © Springer-Verlag Berlin Heidelberg 2013

10.1 Linear R-Brackets

As we have seen in Chapter 7, a Lie algebra structure on a finite-dimensional vector space leads to a Poisson structure on its dual vector space. We consider in this section Lie algebras which are equipped with a second Lie algebra structure, obtained from the original one by using an R-matrix, as explained below. In this case, the dual vector space comes equipped with a second Lie–Poisson structure, whose properties are in general quite different from the properties of the original Lie–Poisson structure.

10.1.1 R-Matrices and the Yang–Baxter Equation

We first give the general definition of an R-matrix and discuss its relation with the Yang–Baxter equation.

Definition 10.1. Let $(\mathfrak{g},[\cdot\,,\cdot])$ be a Lie algebra and let $R:\mathfrak{g}\to\mathfrak{g}$ be a linear map. If the skew-symmetric bilinear map $[\cdot\,,\cdot]_R:\mathfrak{g}\times\mathfrak{g}\to\mathfrak{g}$, defined for $x,y\in\mathfrak{g}$ by

$$[x,y]_R := \frac{1}{2}\left([Rx,y]+[x,Ry]\right)\,, \tag{10.1}$$

defines a Lie bracket on $\mathfrak{g}$, then R is called an *R-matrix* for $\mathfrak{g}$. Then $\mathfrak{g}$, equipped with the two Lie algebra structures $[\cdot\,,\cdot]$ and $[\cdot\,,\cdot]_R$, is called a *double Lie algebra.*

The standard terminology R-matrix and the notation Rx instead of $R(x)$ are used because one often thinks of R (and also x) as being a matrix.

We show in the following proposition that the conditions which express that a linear map $R:\mathfrak{g}\to\mathfrak{g}$ is an R-matrix for $\mathfrak{g}$, can be rewritten in a symmetric form, using a bilinear map on $\mathfrak{g}$, which depends quadratically on R.

Proposition 10.2. *Let $(\mathfrak{g},[\cdot,\cdot])$ be a Lie algebra. A linear map $R:\mathfrak{g}\to\mathfrak{g}$ is an R-matrix for $\mathfrak{g}$ if and only if*

$$\forall x,y,z\in\mathfrak{g}:\quad [B_R(x,y),z]+\circlearrowleft(x,y,z)=0\,, \tag{10.2}$$

where $B_R\in\operatorname{Hom}(\mathfrak{g}\wedge\mathfrak{g},\mathfrak{g})$ is defined, for $x,y\in\mathfrak{g}$, by

$$B_R(x,y) := [Rx,Ry]-R\left([Rx,y]+[x,Ry]\right)\,. \tag{10.3}$$

Proof. For $x,y,z\in\mathfrak{g}$, we have

$$\begin{aligned}
&4\,[[x,y]_R\,,z]_R+\circlearrowleft(x,y,z)\\
&=2\,[[Rx,y]+[x,Ry]\,,z]_R+\circlearrowleft(x,y,z)\\
&=[R\left([Rx,y]+[x,Ry]\right),z]+[[Rx,y]+[x,Ry]\,,Rz]+\circlearrowleft(x,y,z)\\
&=[R\left([Rx,y]+[x,Ry]\right),z]+[[Ry,z]\,,Rx]+[[z,Rx]\,,Ry]+\circlearrowleft(x,y,z)
\end{aligned}$$

$$= [R([Rx,y]+[x,Ry]) - [Rx,Ry],z] + \circlearrowleft (x,y,z) \,,$$

where we have used the Jacobi identity for $[\cdot\,,\cdot]$ in the last step. This shows that the Jacobi identity for $[\cdot\,,\cdot]_R$ is equivalent to (10.2). □

It follows from the proposition that, if there exists a constant $c \in \mathbb{F}$ such that $B_R(x,y) = -c\,[x,y]$ for all $x,y \in \mathfrak{g}$, then R is an R-matrix for $\mathfrak{g}$. Given $c \in \mathbb{F}$, the equation

$$\forall x,y \in \mathfrak{g}: \quad B_R(x,y) = -c\,[x,y] \,, \tag{10.4}$$

viewed as an equation in R, is called the *modified Yang–Baxter equation*. When $\mathfrak{g}$ is a complex (respectively real) Lie algebra and $c \neq 0$, then a rescaling of R in (10.4) yields $c = 1$ (respectively $c = \pm 1$). When $c = 0$, one speaks of the *Yang–Baxter equation*. Sometimes one adds the adjective *classical*, to distinguish these equations from the quantum Yang–Baxter equation. We warn the reader that the terminology *Yang–Baxter equation*, with or without adjectives, is used for many equations which are closely related to, but not equivalent to, the above Yang–Baxter equation (see [108]).

Given a Lie algebra $(\mathfrak{g},[\cdot\,,\cdot])$, a linear map $T:\mathfrak{g} \to \mathfrak{g}$ is called an *intertwining operator* of $\mathfrak{g}$ if $T[x,y] = [Tx,y] = [x,Ty]$, for all $x,y \in \mathfrak{g}$, where we have written Tx as a shorthand for $T(x)$. An intertwining operator of $\mathfrak{g}$ can be used to construct a new R-matrix for $\mathfrak{g}$, starting from a given R-matrix for $\mathfrak{g}$, as given in the following proposition.

Proposition 10.3. *Let $(\mathfrak{g},[\cdot\,,\cdot])$ be a Lie algebra and let R be an R-matrix for $\mathfrak{g}$. If $T:\mathfrak{g} \to \mathfrak{g}$ is an intertwining operator of $\mathfrak{g}$, then the linear map $R \circ T$ is an R-matrix for $\mathfrak{g}$.*

Proof. Suppose that R is an R-matrix for $\mathfrak{g}$ and that T is an intertwining operator of $\mathfrak{g}$. According to Proposition 10.2, we need to show that

$$[B_R(Tx,Ty),z] + \circlearrowleft (x,y,z) = 0 \,, \tag{10.5}$$

for all $x,y,z \in \mathfrak{g}$, where we have used that $B_{R\circ T}(x,y) = B_R(Tx,Ty)$, which is valid because T is an intertwining operator of $\mathfrak{g}$. If T is invertible, then (10.5) is equivalent to

$$T\,[B_R(Tx,Ty),z] + \circlearrowleft (x,y,z) = 0 \,, \tag{10.6}$$

which holds for all $x,y,z \in \mathfrak{g}$, because T is an intertwining operator of $\mathfrak{g}$ and in view of (10.2). If T is not invertible, pick an arbitrary $\alpha \in \mathbb{F}$ and replace in (10.6) T by $T + \alpha \mathbb{1}_{\mathfrak{g}}$. The identity

$$(T+\alpha\mathbb{1}_{\mathfrak{g}})\,[B_R(Tx+\alpha x, Ty+\alpha y),z] + \circlearrowleft (x,y,z) = 0 \tag{10.7}$$

holds because $T + \alpha\mathbb{1}_g$ is also an intertwining operator. Since (10.7) is a cubic polynomial in α and since it holds for all $\alpha \in \mathbb{F}$, all of its coefficients as a polynomial in α are equal to zero. This gives the following equations:

$$\begin{aligned}
&[B_R(x,y),z] + \circlearrowleft(x,y,z) = 0\,,\\
&[B_R(Tx,y) + B_R(x,Ty),z] + T\,[B_R(x,y),z] + \circlearrowleft(x,y,z) = 0\,,\\
&[B_R(Tx,Ty),z] + T([B_R(Tx,y) + B_R(x,Ty),z]) + \circlearrowleft(x,y,z) = 0\,,\\
&T\,[B_R(Tx,Ty),z] + \circlearrowleft(x,y,z) = 0\,.
\end{aligned}$$

In view of the first line, the second line becomes

$$[B_R(Tx,y) + B_R(x,Ty),z] + \circlearrowleft(x,y,z) = 0\,;$$

next, substituting this in the third line yields (10.5). □

If T is an intertwining operator of $\mathfrak{g}$ and R is a solution of the Yang–Baxter equation, then the R-matrix $R \circ T$ is also a solution of the Yang–Baxter equation, since $B_{R\circ T}(x,y) = B_R(Tx,Ty)$ for all $x,y \in \mathfrak{g}$. However, the corresponding property for solutions of the modified Yang–Baxter equation does not hold in general.

10.1.2 R-Matrices and Lie Algebra Splittings

The main, and historically first, example of an R-matrix comes from a Lie algebra splitting. If $(\mathfrak{g}, [\cdot\,,\cdot])$ is a Lie algebra and $\mathfrak{g}$ is written as a vector space direct sum $\mathfrak{g} = \mathfrak{g}_+ \oplus \mathfrak{g}_-$ of two Lie subalgebras $\mathfrak{g}_+$ and $\mathfrak{g}_-$ of $\mathfrak{g}$, then $\mathfrak{g}_+ \oplus \mathfrak{g}_-$ is called a *Lie algebra splitting*. It leads to two projections $P_+ : \mathfrak{g} \to \mathfrak{g}_+$ and $P_- : \mathfrak{g} \to \mathfrak{g}_-$. For $x \in \mathfrak{g}$, we abbreviate $P_+(x)$ to x_+ and $P_-(x)$ to x_-. A Lie algebra splitting leads to an R-matrix for $\mathfrak{g}$, as described by the following proposition.

Proposition 10.4. *Let $(\mathfrak{g}, [\cdot\,,\cdot])$ be a Lie algebra and let $\mathfrak{g} = \mathfrak{g}_+ \oplus \mathfrak{g}_-$ be a Lie algebra splitting, with corresponding projections $P_\pm : \mathfrak{g} \to \mathfrak{g}_\pm$. The involutive endomorphism $R := P_+ - P_-$ of $\mathfrak{g}$ is a solution of the modified Yang–Baxter equation (10.4), with* $c = 1$, *namely*

$$[Rx,Ry] - R([Rx,y] + [x,Ry]) = -[x,y]\,, \tag{10.8}$$

for all $x,y \in \mathfrak{g}$. In particular, R is an R-matrix for $\mathfrak{g}$.

Proof. Let $\mathfrak{g} = \mathfrak{g}_+ \oplus \mathfrak{g}_-$ be a Lie algebra splitting and let $R := P_+ - P_-$ be the difference of the corresponding projection maps. For $x,y \in \mathfrak{g}$, we have

$$[Rx,Ry] + [x,y] = 2([x_+,y_+] + [x_-,y_-])\,,$$

while

$$R([Rx,y] + [x,Ry]) = R(2\,[x_+,y_+] - 2\,[x_-,y_-]) = 2([x_+,y_+] + [x_-,y_-])\,,$$

where we have used in the last step that $\mathfrak{g}_+$ and $\mathfrak{g}_-$ are Lie subalgebras of $\mathfrak{g}$. This proves that R satisfies (10.8). □

Notice that, when $R = P_+ - P_-$ comes from a Lie algebra splitting, then an alternative formula for the R-bracket $[\cdot,\cdot]_R$ is given, for $x, y \in \mathfrak{g}$, by

$$[x,y]_R = [x_+, y_+] - [x_-, y_-] \ . \tag{10.9}$$

This elementary fact has important consequences, as we will see in the next section. Notice that (10.9) can be used to give an elementary direct proof of the Jacobi identity for $[\cdot,\cdot]_R$, without using Propositions 10.2 and 10.4.

Remark 10.5. More generally, suppose that $\mathfrak{g} = \mathfrak{g}_+ \oplus \mathfrak{g}_0 \oplus \mathfrak{g}_-$ is a vector space decomposition of a Lie algebra $\mathfrak{g}$, where $\mathfrak{g}_+$ and $\mathfrak{g}_-$ are Lie subalgebras, $\mathfrak{g}_0$ is abelian and is contained in the *normalizer* of $\mathfrak{g}_+$ and of $\mathfrak{g}_-$, i.e., $[\mathfrak{g}_0, \mathfrak{g}_\pm] \subset \mathfrak{g}_\pm$. Let $R \in \operatorname{End}(\mathfrak{g})$ be defined for $x \in \mathfrak{g}$ by $Rx := x_+ - x_-$, where x_+, x_0 and x_- stand for the projections of x on $\mathfrak{g}_+, \mathfrak{g}_0$ and $\mathfrak{g}_-$ respectively. Then one finds, as in the proof of Proposition 10.4, that

$$\begin{aligned}[Rx, Ry] + [x,y] &= 2\,[x_+, y_+] + 2\,[x_-, y_-] + [x_0, y] + [x, y_0] \\ &= R([Rx, y] + [x, Ry]) \ ,\end{aligned}$$

for all $x, y \in \mathfrak{g}$. It follows that, as in the case of a Lie algebra splitting, R is a solution of the modified Yang–Baxter equation (10.4), with $c = 1$.

10.1.3 Linear R-Brackets on $\mathfrak{g}^*$ and on $\mathfrak{g}$

If we assume that the Lie algebra $(\mathfrak{g}, [\cdot,\cdot])$ is finite-dimensional, every R-matrix for $\mathfrak{g}$ leads to a Lie–Poisson structure on $\mathfrak{g}^*$, called an *R-bracket*; the trivial case $R = \mathbb{1}_{\mathfrak{g}}$ is usually excluded, because it corresponds to the canonical Lie–Poisson structure on $\mathfrak{g}^*$. According to (7.5), it is given for all $F, G \in \mathscr{F}(\mathfrak{g}^*)$ at $\xi \in \mathfrak{g}^*$ by

$$\begin{aligned}\{F,G\}_R(\xi) &:= \left\langle \xi, \left[\mathrm{d}_\xi F, \mathrm{d}_\xi G\right]_R \right\rangle \\ &= \frac{1}{2}\left\langle \xi, \left[R\left(\mathrm{d}_\xi F\right), \mathrm{d}_\xi G\right] + \left[\mathrm{d}_\xi F, R\left(\mathrm{d}_\xi G\right)\right]\right\rangle \ ,\end{aligned}$$

where we recall that $\mathrm{d}_\xi F$ is viewed in this formula as an element of $\mathfrak{g}$, under the canonical isomorphism between $\mathfrak{g}$ and its bidual. Also, $\langle\cdot,\cdot\rangle$ stands for the canonical pairing between $\mathfrak{g}^*$ and $\mathfrak{g}$. When $R = P_+ - P_-$ comes from a Lie algebra splitting $\mathfrak{g} = \mathfrak{g}_+ \oplus \mathfrak{g}_-$, this formula can, according to (10.9), also be written as

$$\{F,G\}_R(\xi) = \left\langle \xi, \left[\left(\mathrm{d}_\xi F\right)_+, \left(\mathrm{d}_\xi G\right)_+\right]\right\rangle - \left\langle \xi, \left[\left(\mathrm{d}_\xi F\right)_-, \left(\mathrm{d}_\xi G\right)_-\right]\right\rangle \ . \tag{10.10}$$

Upon using, as in Section 7.2, a non-degenerate symmetric bilinear form $\langle\cdot\,|\,\cdot\rangle$ on $\mathfrak{g}$, we can transfer the R-bracket to $\mathfrak{g}$, which leads according to (7.10) for a general R-matrix R to

$$\{F,G\}_{R,\mathfrak{g}}(x) = \frac{1}{2}\langle x \,|\, [R(\nabla_x F), \nabla_x G] + [\nabla_x F, R(\nabla_x G)]\rangle ,$$

for $F,G \in \mathscr{F}(\mathfrak{g})$ and $x \in \mathfrak{g}$. When R comes from a Lie algebra splitting $\mathfrak{g} = \mathfrak{g}_+ \oplus \mathfrak{g}_-$, this bracket becomes

$$\{F,G\}_{R,\mathfrak{g}}(x) = \langle x \,|\, [(\nabla_x F)_+, (\nabla_x G)_+]\rangle - \langle x \,|\, [(\nabla_x F)_-, (\nabla_x G)_-]\rangle , \qquad (10.11)$$

for $F,G \in \mathscr{F}(\mathfrak{g})$ and $x \in \mathfrak{g}$.

An important property of the latter R-bracket is that it restricts to two natural subspaces of $\mathfrak{g}$, namely the orthogonal complement $\mathfrak{g}_+^\perp$ of $\mathfrak{g}_+$, respectively $\mathfrak{g}_-^\perp$ of $\mathfrak{g}_-$, with respect to $\langle \cdot \,|\, \cdot \rangle$ (see (7.13)). It is given in the following proposition.

Proposition 10.6. *Let* $(\mathfrak{g}, [\cdot\,,\cdot])$ *be a finite-dimensional Lie algebra, equipped with a non-degenerate symmetric bilinear form* $\langle \cdot \,|\, \cdot \rangle$, *and let R be the R-matrix for* $\mathfrak{g}$, *associated with a Lie algebra splitting* $\mathfrak{g} = \mathfrak{g}_+ \oplus \mathfrak{g}_-$. *Then the orthogonal complements* $\mathfrak{g}_+^\perp$ *and* $\mathfrak{g}_-^\perp$ *are Poisson submanifolds of* $(\mathfrak{g}, \{\cdot\,,\cdot\}_{R,\mathfrak{g}})$. *An explicit formula for the restricted Poisson bracket on* $\mathfrak{g}_-^\perp$ *is given, for* $F,G \in \mathscr{F}(\mathfrak{g}_-^\perp)$ *at* $x \in \mathfrak{g}_-^\perp$, *by*

$$\{F,G\}_-(x) = \left\langle x \,\middle|\, \left[(\nabla_x \tilde{F})_+, (\nabla_x \tilde{G})_+\right]\right\rangle , \qquad (10.12)$$

where $\tilde{F}$ *and* $\tilde{G}$ *are arbitrary extensions of F and G to* $\mathfrak{g}$. *An analogous formula holds for the restricted Poisson bracket on* $\mathfrak{g}_+^\perp$.

Proof. It is clear from formula (10.9) for the R-bracket which comes from a Lie algebra splitting that $[\mathfrak{g}_+, \mathfrak{g}]_R \subset \mathfrak{g}_+$ and that $[\mathfrak{g}_-, \mathfrak{g}]_R \subset \mathfrak{g}_-$. Therefore, $\mathfrak{g}_+$ and $\mathfrak{g}_-$ are both Lie ideals of $(\mathfrak{g}, [\cdot\,,\cdot]_R)$ and we may conclude, according to Proposition 7.5, that $\mathfrak{g}_+^\perp$ and $\mathfrak{g}_-^\perp$ are both Poisson submanifolds of $(\mathfrak{g}, \{\cdot\,,\cdot\}_{R,\mathfrak{g}})$. Equation (10.12) follows at once upon combining Eqs. (7.14) and (10.11). □

We finish this section by giving two simple, but important, examples of R-brackets.

Example 10.7. Consider the Lie algebra splitting of $\mathfrak{g} := \mathfrak{gl}_d(\mathbb{F})$, given by

$$\mathfrak{g} = \Delta_d^{\geqslant} \oplus \Delta_d^{<} = \mathfrak{g}_+ \oplus \mathfrak{g}_- ,$$

where $\mathfrak{g}_+ := \Delta_d^{\geqslant}$ (respectively $\mathfrak{g}_- := \Delta_d^{<}$) denotes the Lie algebra of upper triangular matrices (respectively of strictly lower triangular matrices). Letting $\langle x \,|\, y \rangle := \operatorname{Trace}(xy)$, we have that $\mathfrak{g}_-^\perp = \left(\Delta_d^{<}\right)^\perp = \Delta_d^{\leqslant}$, the vector space of all lower triangular matrices in $\mathfrak{g}$. If F is a linear function on $\mathfrak{g}$, then $\nabla_x F = \nabla F$ is independent of $x \in \mathfrak{g}$ and it follows from (7.12) that it is given by $\langle \nabla F \,|\, y \rangle = F(y)$, for all $y \in \mathfrak{g}$. Denoting by x_{ij} the linear function which picks the element (i,j) of a matrix, it follows that $\nabla x_{ij} = E_{ji}$, the matrix whose only non-zero entry is a 1 at position (j,i). According to (10.12), the Poisson bracket on $\Delta_d^{\leqslant}$ is therefore given by

$$\{x_{ij}, x_{k\ell}\}_-(x) = \langle x \,|\, [E_{ji}, E_{\ell k}]\rangle ,$$

where $j \leqslant i$ and $\ell \leqslant k$.

This example is easily generalized to arbitrary semi-simple Lie algebras: upon choosing a Cartan subalgebra $\mathfrak{h}$, such a Lie algebra $\mathfrak{g}$ can be decomposed as $\mathfrak{g} = \mathfrak{n}_- \oplus \mathfrak{h} \oplus \mathfrak{n}_+$, where each one of the four subspaces $\mathfrak{n}_\pm$ and $\mathfrak{b}_\pm := \mathfrak{n}_\pm \oplus \mathfrak{h}$ is a Lie subalgebra of $\mathfrak{g}$. Thus, $\mathfrak{g} = \mathfrak{b}_+ \oplus \mathfrak{n}_-$ is a Lie algebra splitting. With respect to the Killing form, $\mathfrak{n}_-^\perp = \mathfrak{b}_-$ and we have on the latter vector space a linear Poisson structure.

Example 10.8. Another natural Lie algebra splitting of $\mathfrak{gl}_d(\mathbb{F})$ is given by $\mathfrak{gl}_d(\mathbb{F}) = \Delta_d^{\geqslant} \oplus \mathfrak{g}_-$, where $\mathfrak{g}_-$ stands for the Lie subalgebra of skew-symmetric matrices. Using again $\langle x \,|\, y \rangle := \operatorname{Trace}(xy)$, we have that $\mathfrak{g}_-^\perp$ is the vector space of all symmetric $d \times d$ matrices, which inherits a Poisson structure from the Lie algebra splitting.

10.2 Linear r-Brackets

In the previous section, we have seen that a linear map $R : \mathfrak{g} \to \mathfrak{g}$ leads, under certain conditions, to a new Lie–Poisson structure on $\mathfrak{g}^*$, and hence on $\mathfrak{g}$, when $\mathfrak{g}^*$ and $\mathfrak{g}$ have been identified. In the present section, we describe a similar construction, starting from an element $r \in \mathfrak{g} \otimes \mathfrak{g}$, which gives, under certain conditions, directly a Lie–Poisson structure on $\mathfrak{g}$.

10.2.1 Coboundary Lie Bialgebras and r-Matrices

Let r be an element of $\mathfrak{g} \otimes \mathfrak{g}$, where $(\mathfrak{g}, [\cdot\,,\cdot])$ is a finite-dimensional Lie algebra. In this section, we explain how there is naturally associated with r a linear map $\mathfrak{g} \to \mathfrak{g} \otimes \mathfrak{g}$, whose transpose defines, under some conditions, a Lie algebra structure $[\cdot\,,\cdot]_r : \mathfrak{g}^* \otimes \mathfrak{g}^* \to \mathfrak{g}^*$. To do this, we need a natural representation of $\mathfrak{g}$ on the tensor algebra $T^\bullet\mathfrak{g}$ of $\mathfrak{g}$ and we need to make explicit the Lie algebra cohomology of $\mathfrak{g}$, with values in this representation; see Section A.1 in Appendix A and Section 4.1.1 for the basic definitions of tensor algebra, respectively of Lie algebra cohomology.

A natural representation of a Lie algebra $\mathfrak{g}$ on its tensor algebra $T^\bullet\mathfrak{g}$ is obtained by extending, for every $x \in \mathfrak{g}$, the linear map $\operatorname{ad}_x : \mathfrak{g} \to \mathfrak{g}$ to a (graded) derivation (of degree zero) of $T^\bullet\mathfrak{g}$. Explicitly, for a monomial $y_1 \otimes y_2 \otimes \cdots \otimes y_p \in T^\bullet\mathfrak{g}$,

$$\operatorname{ad}_x(y_1 \otimes y_2 \otimes \cdots \otimes y_p) := \sum_{i=1}^{p} y_1 \otimes \cdots \otimes [x, y_i] \otimes \cdots \otimes y_p\,. \tag{10.13}$$

The Jacobi identity for $[\cdot\,,\cdot]$ implies that the linear map $\operatorname{ad} : \mathfrak{g} \times T^\bullet\mathfrak{g} \to T^\bullet\mathfrak{g}$, defined by (10.13), is a representation of $\mathfrak{g}$ on $T^\bullet\mathfrak{g}$. Combined with the natural projection map $T^\bullet\mathfrak{g} \to \wedge^\bullet\mathfrak{g}$, we obtain the adjoint representation of $\mathfrak{g}$ on $\wedge^\bullet\mathfrak{g}$ which, as we have seen in Section 5.1.3, can be written in terms of the algebraic Schouten bracket: $\operatorname{ad}_x Y = [\![x, Y]\!]$ for all $x \in \mathfrak{g}$ and all $Y \in \wedge^\bullet\mathfrak{g}$. Elements Y of $T^\bullet\mathfrak{g}$ or of $\wedge^\bullet\mathfrak{g}$ for which

$\mathrm{ad}_x Y = 0$, for all $x \in \mathfrak{g}$, are called ad-*invariant*; for example, elements of degree one are ad-invariant if and only if they belong to the center of $\mathfrak{g}$.

It is clear from (10.13) that ad_x can be restricted to $\mathfrak{g} \otimes \mathfrak{g}$, which yields a representation of $\mathfrak{g}$ on $\mathfrak{g} \otimes \mathfrak{g}$. It allows us to consider, for $p \in \mathbb{N}$, the vector space $C^p(\mathfrak{g}) := \mathrm{Hom}(\wedge^p \mathfrak{g}, \mathfrak{g} \otimes \mathfrak{g})$, of skew-symmetric p-linear maps on $\mathfrak{g}$, with values in the representation space $\mathfrak{g} \otimes \mathfrak{g}$. According to Section 4.1.1, $C^\bullet(\mathfrak{g})$ carries a coboundary operator $\delta_L : C^\bullet(\mathfrak{g}) \to C^{\bullet+1}(\mathfrak{g})$. Applied to $r \in \mathfrak{g} \otimes \mathfrak{g} \simeq C^0(\mathfrak{g})$, we obtain a linear map $\gamma_r := \delta_L^0(r) : \mathfrak{g} \to \mathfrak{g} \otimes \mathfrak{g}$, which is given by

$$\gamma_r(x) = \mathrm{ad}_x r ,$$

for all $x \in \mathfrak{g}$. The transpose to γ_r is a linear map $\gamma_r^\top : \mathfrak{g}^* \otimes \mathfrak{g}^* \simeq (\mathfrak{g} \otimes \mathfrak{g})^* \to \mathfrak{g}^*$, which we view as a bilinear map $\mathfrak{g}^* \times \mathfrak{g}^* \to \mathfrak{g}^*$, still denoted by $\gamma_r^\top$. Explicitly,

$$\left\langle \gamma_r^\top(\xi, \eta), x \right\rangle = (\xi \otimes \eta)(\mathrm{ad}_x r) \tag{10.14}$$

for all $x \in \mathfrak{g}$ and $\xi, \eta \in \mathfrak{g}^*$. We will be interested in the case in which the bilinear map $\gamma_r^\top$ defines a Lie algebra structure on $\mathfrak{g}^*$.

Definition 10.9. Let $(\mathfrak{g}, [\cdot\,,\cdot])$ be a finite-dimensional Lie algebra. A bivector $r \in \mathfrak{g} \otimes \mathfrak{g}$ is called an *r-matrix* for $\mathfrak{g}$ if the transpose $\gamma_r^\top$ of $\gamma_r = \delta_L^0(r) : \mathfrak{g} \to \mathfrak{g} \otimes \mathfrak{g}$ is a Lie bracket on $\mathfrak{g}^*$. In this case, r leads to a Lie bracket on $\mathfrak{g}^*$, denoted by $[\cdot\,,\cdot]_r$, and defined by[1]

$$\langle [\xi, \eta]_r, x \rangle := \langle \xi \wedge \eta, \mathrm{ad}_x r \rangle \tag{10.15}$$

for all $\xi, \eta \in \mathfrak{g}^*$ and $x \in \mathfrak{g}$. This bracket is called the Lie bracket, *associated* with r, and $(\mathfrak{g}, [\cdot\,,\cdot], [\cdot\,,\cdot]_r)$ is called a *coboundary Lie bialgebra.*

As the terminology suggests, coboundary Lie bialgebras are a particular case of Lie bialgebras, but we will only consider this more general class when we discuss Poisson–Lie groups, see Chapter 11.

We will give necessary and sufficient conditions on $r \in \mathfrak{g} \otimes \mathfrak{g}$ so that $(\mathfrak{g}, [\cdot\,,\cdot], [\cdot\,,\cdot]_r)$ is a coboundary Lie bialgebra. Our proof will be based on the following proposition.

Proposition 10.10. *Let $\phi : V \to V \wedge V$ be a linear map, where V is a finite-dimensional vector space. Denote by $\tilde{\phi} : \wedge^\bullet V \to \wedge^{\bullet+1} V$ the extension of ϕ to a graded derivation of $\wedge^\bullet V$ of degree 1 and let $\phi^\top : V^* \wedge V^* \to V^*$ denote the transpose map to ϕ. Then $\phi^\top$ satisfies the Jacobi identity (so that $\phi^\top$ defines a Lie algebra structure on V^*) if and only if $\tilde{\phi} \circ \tilde{\phi} = 0$.*

Proof. Recall from Section A.5 in Appendix A, in particular Example A.3, that ϕ extends uniquely to a derivation $\tilde{\phi}$ of $\wedge^\bullet V$ of degree 1, where the derivation property means that

[1] We use here and below the natural pairing $\langle \cdot\,,\cdot \rangle$ between $\wedge^p \mathfrak{g}^*$ and $\wedge^p \mathfrak{g}$, introduced in (5.10). When $p = 2$, it is given, for all $\xi_1, \xi_2 \in \mathfrak{g}^*$ and for all $x_1, x_2 \in \mathfrak{g}$, by $\langle \xi_1 \wedge \xi_2, x_1 \wedge x_2 \rangle = \xi_1(x_1)\xi_2(x_2) - \xi_1(x_2)\xi_2(x_1)$.

$$\tilde{\phi}(X \wedge Y) = \tilde{\phi}(X) \wedge Y + (-1)^p X \wedge \tilde{\phi}(Y)\,, \tag{10.16}$$

for $X \in \wedge^p V$ and $Y \in \wedge^\bullet V$. It follows from (10.16) that $\tilde{\phi} \circ \tilde{\phi} = 0$ if and only if $\tilde{\phi}(\phi(v)) = 0$ for all $v \in V$. Now, for all $\xi, \eta, \zeta \in V^*$,

$$\begin{aligned}\langle \tilde{\phi}(\phi(v)), \xi \wedge \eta \wedge \zeta \rangle &\overset{(*)}{=} \left\langle \phi(v), \phi^\top(\xi \wedge \eta) \wedge \zeta \right\rangle + \circlearrowleft(\xi, \eta, \zeta) \\ &= \left\langle v, \phi^\top\left(\phi^\top(\xi \wedge \eta) \wedge \zeta\right) + \circlearrowleft(\xi, \eta, \zeta)\right\rangle,\end{aligned}$$

so that $\tilde{\phi} \circ \tilde{\phi} = 0$ if and only if $\phi^\top$ satisfies the Jacobi identity. Step $(*)$ requires a proof. Let $X \in \wedge^2 V$, which we suppose to be of the form $X = v \wedge w$, with $v, w \in V$ (in general, an element of $\wedge^2 V$ is a linear combination of such monomials). In view of the graded derivation property (10.16) we have, for all $\xi, \eta, \zeta \in V^*$, that

$$\begin{aligned}&\langle \tilde{\phi}(X), \xi \wedge \eta \wedge \zeta \rangle \\ &\quad = \langle \phi(v) \wedge w - \phi(w) \wedge v, \xi \wedge \eta \wedge \zeta \rangle \\ &\quad = \langle \phi(v), \xi \wedge \eta \rangle \langle w, \zeta \rangle - \langle \phi(w), \xi \wedge \eta \rangle \langle v, \zeta \rangle + \circlearrowleft(\xi, \eta, \zeta) \\ &\quad = \left\langle v, \phi^\top(\xi \wedge \eta)\right\rangle \langle w, \zeta \rangle - \left\langle w, \phi^\top(\xi \wedge \eta)\right\rangle \langle v, \zeta \rangle + \circlearrowleft(\xi, \eta, \zeta) \\ &\quad = \left\langle v \wedge w, \phi^\top(\xi \wedge \eta) \wedge \zeta \right\rangle + \circlearrowleft(\xi, \eta, \zeta) \\ &\quad = \left\langle X, \phi^\top(\xi \wedge \eta) \wedge \zeta \right\rangle + \circlearrowleft(\xi, \eta, \zeta).\end{aligned}$$

This proves $(*)$. □

We are now ready to give necessary and sufficient conditions on r, implying that $(\mathfrak{g}, [\cdot\,,\cdot], [\cdot\,,\cdot]_r)$ is a coboundary Lie bialgebra. They will be expressed in terms of the algebraic Schouten bracket $[\![\cdot\,,\cdot]\!]$ on $\wedge^\bullet \mathfrak{g}$ (see Section 3.3.3). We denote for $s \in \mathfrak{g} \otimes \mathfrak{g}$ by s^+ (respectively s^-) its symmetric (respectively skew-symmetric) part. For example, if $x, y \in \mathfrak{g}$, then $(x \otimes y)^- = x \wedge y$, viewed as an element of $\wedge^2 \mathfrak{g}$.

Proposition 10.11. *Let $(\mathfrak{g}, [\cdot\,,\cdot])$ be a finite-dimensional Lie algebra and let r be an element of $\mathfrak{g} \otimes \mathfrak{g}$. Then r is an r-matrix for $\mathfrak{g}$ if and only if r satisfies the following two conditions:*

(1) r^+ is ad*-invariant;*
(2) $[\![r^-, r^-]\!]$ is ad*-invariant.*

In this case, the Lie bracket associated with r is given by

$$\langle [\xi, \eta]_r, x \rangle = \langle \xi \wedge \eta, \mathrm{ad}_x r^- \rangle\,, \tag{10.17}$$

for all $\xi, \eta \in \mathfrak{g}^*$ and $x \in \mathfrak{g}$.

Proof. Let $\xi, \eta \in \mathfrak{g}^*$ and let $x \in \mathfrak{g}$. Then

$$\left\langle \gamma_r^\top(\xi, \eta) + \gamma_r^\top(\eta, \xi), x \right\rangle = (\xi \otimes \eta)(\mathrm{ad}_x r) + (\eta \otimes \xi)(\mathrm{ad}_x r) = 2(\xi \otimes \eta)(\mathrm{ad}_x r^+)\,, \tag{10.18}$$

where we have used that $(\mathrm{ad}_x r)^\pm = \mathrm{ad}_x r^\pm$; to verify the latter, write $r^\pm$ as a sum of elements of the form $y \otimes z \pm z \otimes y$. It follows from (10.18) that $\gamma_r^\top$ is skew-symmetric if and only if r^+ is ad-invariant.

Suppose now that r^+ is ad-invariant. Then $\gamma_r(x) = \mathrm{ad}_x r = \mathrm{ad}_x r^- = \gamma_{r^-}(x)$ for all $x \in \mathfrak{g}$, so that γ_r depends only on $a := r^-$. Thus we are given a skew-symmetric element $a \in \mathfrak{g} \wedge \mathfrak{g}$ and we wish to prove that the Jacobi identity for γ_a is equivalent to the ad-invariance of $[\![a,a]\!]$. In view of Proposition 10.10, it suffices to show that the ad-invariance of $[\![a,a]\!]$ is equivalent to $\tilde{\gamma}_a \circ \tilde{\gamma}_a = 0$ (on $\wedge^\bullet \mathfrak{g}$; equivalently, on $\mathfrak{g}$). Combining (10.13) and (3.46), we find that

$$\gamma_a(x) = \mathrm{ad}_x a = [\![x,a]\!] = -[\![a,x]\!] ,$$

for all $x \in \mathfrak{g}$. Since $\tilde{\gamma}_a$ and $-[\![a,\cdot]\!]$ are both graded derivations of degree 1 of $\wedge^\bullet \mathfrak{g}$, it follows that $\tilde{\gamma}_a(X) = -[\![a,X]\!]$, for all $X \in \wedge^\bullet \mathfrak{g}$. Therefore, $\tilde{\gamma}_a \circ \tilde{\gamma}_a = 0$ if and only if $[\![a,[\![a,x]\!]]\!] = 0$, for all $x \in \mathfrak{g}$. In view of the graded Jacobi identity for $[\![\cdot\,,\cdot]\!]$,

$$[\![x,[\![a,a]\!]]\!] + [\![a,[\![a,x]\!]]\!] - [\![a,[\![x,a]\!]]\!] = 0 ,$$

so that $[\![a,[\![a,x]\!]]\!] = 0$, for all $x \in \mathfrak{g}$, is equivalent to $[\![x,[\![a,a]\!]]\!] = 0$, for all $x \in \mathfrak{g}$, which is precisely the ad-invariance of $[\![a,a]\!]$. □

The simplest way to satisfy both conditions in Proposition 10.11 is to demand that r is skew-symmetric and satisfies $[\![r,r]\!] = 0$.

10.2.2 Linear r-Brackets on $\mathfrak{g}$

Suppose that $r \in \mathfrak{g} \otimes \mathfrak{g}$ is an r-matrix for a finite-dimensional Lie algebra $(\mathfrak{g}, [\cdot\,,\cdot])$. Recall from Definition 10.9 and Proposition 10.11 that this means that the bilinear map, defined for $x \in \mathfrak{g}$ and $\xi, \eta \in \mathfrak{g}^*$ by

$$\langle [\xi,\eta]_r, x\rangle := \langle \xi \wedge \eta, \mathrm{ad}_x r\rangle = \left\langle \xi \wedge \eta, \mathrm{ad}_x r^-\right\rangle , \tag{10.19}$$

defines a Lie algebra structure on $\mathfrak{g}^*$. Then $\mathfrak{g}$ inherits a linear Poisson structure $\{\cdot\,,\cdot\}_r$ from $[\cdot\,,\cdot]_r$, since $\mathfrak{g} \simeq (\mathfrak{g}^*)^*$ is then the dual of a Lie algebra. According to (7.4), the Poisson bracket of two functions $F, G \in \mathscr{F}(\mathfrak{g})$ is given by

$$\{F,G\}_r(x) = \langle [\mathrm{d}_x F, \mathrm{d}_x G]_r, x\rangle = \left\langle \mathrm{d}_x F \wedge \mathrm{d}_x G, \mathrm{ad}_x r^-\right\rangle , \tag{10.20}$$

for all $x \in \mathfrak{g}$. This leads to the following proposition.

Proposition 10.12. *Let $(\mathfrak{g}, [\cdot\,,\cdot])$ be a finite-dimensional Lie algebra and let r be an r-matrix for $\mathfrak{g}$. Let $(e_1,\dots,e_d)$ be a basis for $\mathfrak{g}$ and let $(\xi_1,\dots,\xi_d)$ be the dual basis. The Poisson matrix of $\{\cdot\,,\cdot\}_r$ at $x \in \mathfrak{g}$, with respect to the coordinates $\xi_1,\dots,\xi_d$, is the matrix of $\mathrm{ad}_x r^-$ in the basis $(e_1,\dots,e_d)$.*

Proof. By definition, the matrix of $\mathrm{ad}_x r^- \in \mathfrak{g}\wedge\mathfrak{g}$ in the basis $(e_1,\dots,e_d)$ is the skew-symmetric matrix $(a_{ij})_{1\leqslant i,j\leqslant d}$, defined by

$$\mathrm{ad}_x r^- = \sum_{1\leqslant i<j\leqslant d} a_{ij}\, e_i\wedge e_j\,.$$

Combined with (10.20), this means that the coefficients of the matrix can be computed by

$$a_{ij} = \left\langle \xi_i\wedge\xi_j, \mathrm{ad}_x r^-\right\rangle = \left\{\xi_i,\xi_j\right\}_r(x)\,,$$

as was to be shown. □

A compact formula for the collection of Poisson brackets $\left\{\xi_i,\xi_j\right\}_r$ can be given when $\mathfrak{g}$ is the Lie algebra of an associative algebra (or is a subalgebra of such a Lie algebra[2]). To show this, let us assume that we have an associative product on $\mathfrak{g}$ and that $[x,y]=xy-yx$ for all $x,y\in\mathfrak{g}$. Then

$$\mathrm{ad}_x r^- = [\![x,r^-]\!] = \left[\mathbb{1}\otimes x + x\otimes\mathbb{1}, r^-\right]\,,$$

where the latter bracket stands for the commutator in $\mathfrak{g}\otimes\mathfrak{g}$. Then

$$\left\{\Xi \overset{\otimes}{,} \Xi\right\}_r(x) = \left[\mathbb{1}\otimes x + x\otimes\mathbb{1}, r^-\right]\,, \tag{10.21}$$

where the left-hand side is the bivector, defined by

$$\left\{\Xi \overset{\otimes}{,} \Xi\right\}_r(x) := \sum_{1\leqslant i<j\leqslant d} \left\{\xi_i,\xi_j\right\}_r(x)\, e_i\wedge e_j\,.$$

In this formula, $(e_1,\dots,e_d)$ is a basis of $\mathfrak{g}$ and $(\xi_1,\dots,\xi_d)$ is its dual basis. Equation (10.21) is known as the *first Russian formula*.

The first Russian formula is very useful for explicit computations, in particular when $\mathfrak{g}$ is realized as a Lie subalgebra of $\mathfrak{gl}_N(\mathbb{F})\simeq \mathrm{Mat}_N(\mathbb{F})$. Then this formula can be thought of as an equality of matrices of size N^2. Indeed, there is a standard algebra isomorphism between $\mathrm{Mat}_N(\mathbb{F})\otimes\mathrm{Mat}_N(\mathbb{F})$ and $\mathrm{Mat}_{N^2}(\mathbb{F})$; to describe it, we use the classical elementary matrices E_{ij}, whose only non-zero entry is a 1 at position (i,j), but we use the convention that numbering starts from zero, in order to simplify the formula. Then the N^2 elements E_{ij} with $0\leqslant i,j<N$ form a basis of $\mathrm{Mat}_N(\mathbb{F})$ and the N^4 elements $E_{ij}\otimes E_{k\ell}$, with $0\leqslant i,j,k,\ell<N$ are a basis of $\mathrm{Mat}_N(\mathbb{F})\otimes\mathrm{Mat}_N(\mathbb{F})$. The isomorphism $\mathrm{Mat}_N(\mathbb{F})\otimes\mathrm{Mat}_N(\mathbb{F})\to\mathrm{Mat}_{N^2}(\mathbb{F})$ is given by mapping $E_{ij}\otimes E_{k\ell}$ to $E_{iN+k,jN+\ell}$. It is an isomorphism of associative algebras, since

$$(E_{ij}\otimes E_{k\ell})(E_{i'j'}\otimes E_{k'\ell'}) = \delta_{j,i'}\delta_{\ell,k'}\, E_{ij'}\otimes E_{k\ell'}\,,$$

while

$$E_{iN+k,jN+\ell}E_{i'N+k',j'N+\ell'} = \delta_{j,i'}\delta_{\ell,k'}\, E_{iN+k,j'N+\ell'}\,.$$

[2] Since $\mathfrak{g}$ is finite-dimensional, $\mathfrak{g}$ is isomorphic to a subalgebra of $\mathfrak{gl}_N(\mathbb{F})$, for some N, which is the Lie algebra of the associative algebra $\mathrm{Mat}_n(\mathbb{F})$ of square matrices of size N.

10.2.3 From r-Matrices to R-Matrices

In this section, we show how R-matrices and r-matrices on a finite-dimensional quadratic Lie algebra $\mathfrak{g}$ are related. Denoting the bilinear form of $\mathfrak{g}$ by $\langle\cdot\,|\,\cdot\rangle$, we have an isomorphism $\chi : \mathfrak{g} \to \mathfrak{g}^*$, defined by $\langle \chi(x), y\rangle := \langle x\,|\,y\rangle$, for all $x, y \in \mathfrak{g}$. This leads also to an isomorphism

$$\chi \otimes \mathbb{1}_{\mathfrak{g}} : \mathfrak{g}\otimes\mathfrak{g} \to \mathfrak{g}^*\otimes\mathfrak{g} \simeq \operatorname{End}(\mathfrak{g}) ,$$

which allows us to associate with every element of $\mathfrak{g}\otimes\mathfrak{g}$ a linear map $\mathfrak{g}\to\mathfrak{g}$. We recall that, for a given linear map $R : \mathfrak{g}\to\mathfrak{g}$, we denote its adjoint with respect to $\langle\cdot\,|\,\cdot\rangle$ by R^*. In formulas, this means that

$$\langle Rx\,|\,y\rangle = \langle x\,|\,R^*y\rangle ,$$

for all $x, y \in \mathfrak{g}$. If $R^* = R$ (respectively $R^* = -R$), then R is said to be *symmetric* (respectively *skew-symmetric*). Obviously, R can be decomposed uniquely as $R = R_+ + R_-$, where R_+ is symmetric and R_- is skew-symmetric.

Proposition 10.13. *Let* $(\mathfrak{g}, [\cdot\,,\cdot])$ *be a finite-dimensional quadratic Lie algebra and let* $\chi : \mathfrak{g}\to\mathfrak{g}^*$ *be the isomorphism which is induced by the bilinear form on* $\mathfrak{g}$. *Let* $r \in \mathfrak{g}\otimes\mathfrak{g}$ *and* $R \in \mathfrak{g}^*\otimes\mathfrak{g} \simeq \operatorname{End}(\mathfrak{g})$ *be two elements related to each other by*

$$R = 4(\chi\otimes\mathbb{1}_{\mathfrak{g}})(r) \tag{10.22}$$

(or, equivalently, $r = \frac{1}{4}(\chi^{-1}\otimes\mathbb{1}_{\mathfrak{g}})(R)$*). Then,* R *is a skew-symmetric R-matrix for* $\mathfrak{g}$ *if and only if* r *is a skew-symmetric r-matrix for* $\mathfrak{g}$. *In this case,* $\chi : (\mathfrak{g}, [\cdot\,,\cdot]_R) \to (\mathfrak{g}^*, [\cdot\,,\cdot]_r)$ *is a Lie algebra isomorphism, where* $[\cdot\,,\cdot]_R$ *is the Lie bracket on* $\mathfrak{g}$*, associated with* R*, and* $[\cdot\,,\cdot]_r$ *the Lie bracket on* $\mathfrak{g}^*$*, associated with* r.

Proof. We decompose $r = \sum_\alpha s_\alpha \otimes t_\alpha$, which we write in this proof as $s_\alpha\otimes t_\alpha$ in order to simplify the formulas: in the sequel of the proof, a sum over α is implicit in all formulas which contain α. Using this notation, it follows from (10.22) that $Rx = 4\,\langle s_\alpha\,|\,x\rangle\, t_\alpha$ for $x \in \mathfrak{g}$. It implies that $R^*x = 4\,\langle t_\alpha\,|\,x\rangle\, s_\alpha$ and hence that r is skew-symmetric if and only if $R^* = -R$, i.e., R is also skew-symmetric. Let us suppose that R (and hence r) is skew-symmetric. We show that

$$\chi([x,y]_R) = [\chi(x), \chi(y)]_r , \tag{10.23}$$

for all $x, y \in \mathfrak{g}$, which is, in view of the definition (10.1) of $[\cdot\,,\cdot]_R$, equivalent to showing that, for all $x, y, z \in \mathfrak{g}$,

$$\frac{1}{2}\langle [Rx, y] + [x, Ry]\,|\,z\rangle = \langle [\chi(x), \chi(y)]_r, z\rangle . \tag{10.24}$$

With the above formulas for R, the left-hand side in (10.24) can be written as

$$2\langle\langle s_\alpha\,|\,x\rangle\,[t_\alpha,y]+\langle t_\alpha\,|\,y\rangle\,[s_\alpha,x]\,|\,z\rangle=2(\langle s_\alpha\,|\,x\rangle\,\langle[t_\alpha,y]\,|\,z\rangle+\langle t_\alpha\,|\,y\rangle\,\langle[s_\alpha,x]\,|\,z\rangle)\,, \tag{10.25}$$

while the right-hand side in (10.24) is given, in view of (10.19), by

$$\begin{aligned}\langle[\chi(x),\chi(y)]_r,z\rangle &= \left\langle\chi(x)\wedge\chi(y),\mathrm{ad}_z r^-\right\rangle\\ &= \langle\chi(x)\wedge\chi(y),[z,s_\alpha]\otimes t_\alpha+s_\alpha\otimes[z,t_\alpha]\rangle\\ &= \langle x|\,[z,s_\alpha]\rangle\langle y\,|\,t_\alpha\rangle+\langle x\,|\,s_\alpha\rangle\,\langle y|\,[z,t_\alpha]\rangle\\ &\quad-\langle y|\,[z,s_\alpha]\rangle\,\langle x\,|\,t_\alpha\rangle-\langle y\,|\,s_\alpha\rangle\,\langle x|\,[z,t_\alpha]\rangle\,,\end{aligned}$$

which is the same as (10.25), because $\langle\cdot\,|\,\cdot\rangle$ is ad-invariant and r is skew-symmetric. This proves (10.23), which yields at the same time that $[\cdot,\cdot]_R$ is a Lie bracket if and only if $[\cdot,\cdot]_r$ is a Lie bracket, and that χ is then an isomorphism of Lie algebras. This completes the proof, since saying that $[\cdot,\cdot]_R$ (respectively $[\cdot,\cdot]_r$) is a Lie bracket is tantamount to saying that R is an R-matrix (respectively r is an r-matrix). □

We leave it as an exercise for the reader to check that, under the assumptions of Proposition 10.13, the operator B_R, defined in (10.3), is expressible in terms of the bracket $[\![r^-,r^-]\!]$, defined in (3.46), as

$$\langle z\,|\,B_R(x,y)\rangle=\left\langle\chi(x)\wedge\chi(y)\wedge\chi(z),[\![r^-,r^-]\!]\right\rangle\,,$$

for all $x,y,z\in\mathfrak{g}$. It implies that $[\![r^-,r^-]\!]=0$ if and only if $B_R(x,y)=0$ for all $x,y\in\mathfrak{g}$, i.e., if and only if R is a solution of the Yang–Baxter equation.

Example 10.14. Let $(\mathfrak{g},[\cdot,\cdot])$ be a finite-dimensional Lie algebra, equipped with a non-degenerate symmetric bilinear form $\langle\cdot\,|\,\cdot\rangle$. Suppose that $\mathfrak{g}=\mathfrak{g}_+\oplus\mathfrak{g}_-$ is a Lie algebra splitting, where each of the factors is *isotropic*: $\langle\mathfrak{g}_+\,|\,\mathfrak{g}_+\rangle=\langle\mathfrak{g}_-\,|\,\mathfrak{g}_-\rangle=\{0\}$. Let R be the linear endomorphism of $\mathfrak{g}$, given by $R:=P_+-P_-$, where P_+ and P_- are the projections onto $\mathfrak{g}_+$ and $\mathfrak{g}_-$ respectively. According to Proposition 10.4, R is an R-matrix. For $x\in\mathfrak{g}$, we denote $x_+:=P_+(x)$ and $x_-:=P_-(x)$. Then the Lie bracket associated with R is given, for all $x,y\in\mathfrak{g}$, by

$$[x,y]_R=[x_+,y_+]-[x_-,y_-]\,.$$

Since $\mathfrak{g}_+$ and $\mathfrak{g}_-$ are isotropic, R is skew-symmetric. Let $(\varepsilon_1,\dots,\varepsilon_d)$ be a basis of $\mathfrak{g}_+$. There exists a (unique) basis $(e_1,\dots,e_d)$ of $\mathfrak{g}_-$ which is dual to it with respect to $\langle\cdot\,|\,\cdot\rangle$, i.e., $\langle e_i\,|\,\varepsilon_j\rangle=\delta_{i,j}$ for $i,j=1,\dots,d$; indeed, both $\mathfrak{g}_+$ and $\mathfrak{g}_-$ are isotropic and $\langle\cdot\,|\,\cdot\rangle$ is non-degenerate. The endomorphism R can be described, for all $x\in\mathfrak{g}$, by

$$R(x)=\sum_{i=1}^d\langle e_i\,|\,x\rangle\,\varepsilon_i-\sum_{i=1}^d\langle\varepsilon_i\,|\,x\rangle\,e_i\,,$$

so that $R=\sum_{i=1}^d\chi(e_i)\otimes\varepsilon_i-\sum_{i=1}^d\chi(\varepsilon_i)\otimes e_i$, as an element of $\mathfrak{g}^*\otimes\mathfrak{g}$. It follows that the r-matrix $r:=\frac{1}{4}(\chi^{-1}\otimes\mathbb{1}_{\mathfrak{g}})(R)$, associated with R, is given by

$$r=\frac{1}{2}\sum_{i=1}^{d}e_i\wedge\varepsilon_i\,. \tag{10.26}$$

Example 10.15. The above example can be easily generalized to the case of a decomposition $\mathfrak{g}=\mathfrak{g}_+\oplus\mathfrak{g}_0\oplus\mathfrak{g}_-$, as in Remark 10.5. It is assumed, as in the cited remark, that $\mathfrak{g}_+$ and $\mathfrak{g}_-$ are Lie subalgebras of $\mathfrak{g}$ and that $\mathfrak{g}_0$ is abelian and is contained in the normalizer of $\mathfrak{g}_+$ and of $\mathfrak{g}_-$. Moreover, it is assumed that $\mathfrak{g}$ comes equipped with a non-degenerate symmetric bilinear form $\langle\cdot\,|\,\cdot\rangle$ with respect to which both $\mathfrak{g}_+$ and $\mathfrak{g}_-$ are isotropic and orthogonal to $\mathfrak{g}_0$. Recall that the endomorphism R of $\mathfrak{g}$, defined for $x\in\mathfrak{g}$ by $Rx:=x_+-x_-$, where x_+,x_0 and x_- stand for the projections of x on $\mathfrak{g}_+,\mathfrak{g}_0$ and $\mathfrak{g}_-$ respectively, is a solution of the modified Yang–Baxter equation (10.4), with $c=1$. In view of the above assumptions,

$$\langle Rx\,|\,y\rangle=\langle x_+-x_-\,|\,y\rangle=\langle x_+\,|\,y_-\rangle-\langle x_-\,|\,y_+\rangle=\langle x\,|\,y_--y_+\rangle=-\langle x\,|\,Ry\rangle\,,$$

for all $x,y\in\mathfrak{g}$, so that R is skew-symmetric. Let $(\varepsilon_1,\dots,\varepsilon_d)$ be a basis of $\mathfrak{g}_+$. There exists in view of the above assumptions a (unique) basis $(e_1,\dots,e_d)$ of $\mathfrak{g}_-$ which is dual to it with respect to $\langle\cdot\,|\,\cdot\rangle$; a basis of $\mathfrak{g}_0$ is not needed in the formulas which follow. As in the previous example, R and the corresponding r are explicitly given by

$$R=\sum_{i=1}^{d}\chi(e_i)\otimes\varepsilon_i-\sum_{i=1}^{d}\chi(\varepsilon_i)\otimes e_i\,,$$
$$r=\frac{1}{2}\sum_{i=1}^{d}e_i\wedge\varepsilon_i\,.$$

Example 10.16. Starting from a particular case of Example 10.15, we construct an R-matrix for $\mathfrak{su}_d$ and we compute the corresponding r-matrix. We consider $\mathfrak{sl}_d(\mathbb{C})$ decomposed as $\mathfrak{sl}_d(\mathbb{C})=\mathfrak{g}_+\oplus\mathfrak{g}_0\oplus\mathfrak{g}_-$, where $\mathfrak{g}_+$ (respectively $\mathfrak{g}_-$) consists of the strictly upper (respectively lower) triangular matrices in $\mathfrak{sl}_d(\mathbb{C})$ and $\mathfrak{g}_0$ consists of the diagonal matrices in $\mathfrak{sl}_d(\mathbb{C})$. Viewed as a real vector space, $\mathfrak{sl}_d(\mathbb{C})$ contains

$$\mathfrak{su}_d:=\left\{x\in\mathfrak{sl}_d(\mathbb{C})\mid\bar{x}=-x^\top\right\}$$

as a Lie subalgebra. Consider on $\mathfrak{sl}_d(\mathbb{C})$ the R-matrix associated to the above decomposition, given in Example 10.15 and let $R':=2\sqrt{-1}R$, which is an R-matrix for $\mathfrak{sl}_d(\mathbb{C})$, because R' is a multiple of R. Notice that R' leaves $\mathfrak{su}_d$ invariant, since $\bar{x}+x^\top=0$ implies that $\bar{x}_++x_-^\top=0$, which in turns yields $\overline{\sqrt{-1}Rx}+\sqrt{-1}(Rx)^\top=0$. It follows that R' can be restricted to $\mathfrak{su}_d$ and is an R-matrix for $\mathfrak{su}_d$. We consider the symmetric bilinear form on $\mathfrak{su}_d$, given by $(x,y)\mapsto\langle x\,|\,y\rangle:=\Re(\mathrm{Trace}(xy))$. It is non-degenerate, hence induces an isomorphism χ between $\mathfrak{su}_d$ and its dual, and R' is skew-symmetric with respect to $\langle\cdot\,|\,\cdot\rangle$. Define, for all $1\leqslant i<j\leqslant d$, the skew-symmetric matrices $A_{ij}:=\frac{1}{\sqrt{2}}(E_{ij}-E_{ji})$, the symmetric matrices $S_{ij}:=\frac{\sqrt{-1}}{\sqrt{2}}(E_{ij}+E_{ji})$ and the diagonal matrices $H_i:=\frac{\sqrt{-1}}{\sqrt{2}}(E_{ii}-E_{i+1,i+1})$. These matrices form a basis of $\mathfrak{su}_d$, with the set of all S_{ij} and $A_{k\ell}$ forming an orthogonal set,

with $\langle S_{ij} \,|\, S_{ij} \rangle = \langle A_{ij} \,|\, A_{ij} \rangle = -1$. The restriction of R' to $\mathfrak{su}_d$ vanishes on diagonal matrices, and satisfies

$$R'(A_{ij}) = 2S_{ij} \text{ and } R'(S_{ij}) = -2A_{ij} \;,$$

so that it is given by

$$R' = 2 \sum_{1\leqslant i<j\leqslant d} \chi(S_{ij}) \otimes A_{ij} - 2 \sum_{1\leqslant i<j\leqslant d} \chi(A_{ij}) \otimes S_{ij} \;.$$

As a consequence, the corresponding r-matrix r' is given by

$$r' = \sum_{1\leqslant i<j\leqslant d} S_{ij} \wedge A_{ij} \;.$$

10.3 *R*-Brackets and *r*-Brackets of Higher Degree

In this section we show that, under some assumptions, an R-matrix leads not only to a linear Poisson structure (see Section 10.1), but also to a quadratic and a cubic Poisson structure. For each of these Poisson structures, the finite-dimensional Lie algebra $\mathfrak{g}$, which underlies the construction, is assumed to be the Lie algebra of an associative algebra: we suppose that we have an associative product on $\mathfrak{g}$ and the Lie bracket on $\mathfrak{g}$ is defined, for all $x,y \in \mathfrak{g}$, by $[x,y] := xy - yx$. It is moreover assumed that $\mathfrak{g}$ comes equipped with a non-degenerate symmetric bilinear map $\langle \cdot \,|\, \cdot \rangle$, which satisfies, for all $x,y,z \in \mathfrak{g}$,

$$\langle xy \,|\, z \rangle = \langle x \,|\, yz \rangle \;. \tag{10.27}$$

Notice that this property implies that $\langle \cdot \,|\, \cdot \rangle$ is ad-invariant.

Example 10.17. The key example which the reader should keep in mind is that of $\mathfrak{g} := \mathfrak{gl}_d(\mathbb{F})$, which is the Lie algebra of $\mathrm{Mat}_d(\mathbb{F})$, the associative algebra of $d \times d$ matrices with coefficients in $\mathbb{F}$. The symmetric bilinear form $\langle \cdot \,|\, \cdot \rangle$, which is defined by $\langle x \,|\, y \rangle := \mathrm{Trace}(xy)$, for all $x,y \in \mathfrak{g}$, satisfies (10.27).

We start with the construction of the quadratic R-bracket.

Proposition 10.18. *Let $(\mathfrak{g}, [\cdot\,,\cdot])$ be a finite-dimensional Lie algebra, associated with an associative algebra, equipped with a non-degenerate symmetric bilinear map $\langle \cdot \,|\, \cdot \rangle$, which satisfies $\langle xy \,|\, z \rangle = \langle x \,|\, yz \rangle$, for all $x,y,z \in \mathfrak{g}$. Let $R : \mathfrak{g} \to \mathfrak{g}$ be an arbitrary linear map. If R and R_- are solutions of the modified Yang–Baxter equation (10.4), with the same constant, then the bivector field on $\mathfrak{g}$, defined for $F,G \in \mathscr{F}(\mathfrak{g})$ and $x \in \mathfrak{g}$ by*

$$\begin{aligned}\{F,G\}_R^Q(x) := {} & \tfrac{1}{2} \langle [x, \nabla_x F] \,|\, R(x \nabla_x G + \nabla_x G x) \rangle \\ & - \tfrac{1}{2} \langle [x, \nabla_x G] \,|\, R(x \nabla_x F + \nabla_x F x) \rangle \;,\end{aligned} \tag{10.28}$$

is a quadratic Poisson structure on $\mathfrak{g}$. *If* H *is an* Ad-*invariant function on* $\mathfrak{g}$, *then its Hamiltonian vector field with respect to this Poisson structure is given by*

$$\dot{x} = [R(x\nabla_x H), x] \ . \tag{10.29}$$

Proof. It is clear that $\{\cdot,\cdot\}_R^Q$, as defined by (10.28), is a quadratic skew-symmetric biderivation of $\mathscr{F}(\mathfrak{g})$, i.e., it defines a bivector field on $\mathfrak{g}$. We need to show that $\{\cdot,\cdot\}_R^Q$ satisfies the Jacobi identity, which amounts to verifying that, if F_1, F_2 and F_3 are linear functions on $\mathfrak{g}$, then for all $x \in \mathfrak{g}$,

$$\left\{\{F_1,F_2\}_R^Q, F_3\right\}_R^Q (x) + \circlearrowleft (1,2,3) = 0 \ . \tag{10.30}$$

In order to organize this verification, it is useful to write $R = R_+ + R_-$, where R_+ is symmetric and R_- is skew-symmetric. Since (10.28) depends linearly on R, we have that $\{\cdot,\cdot\}_R^Q = \{\cdot,\cdot\}_+ + \{\cdot,\cdot\}_-$, where $\{\cdot,\cdot\}_+$ and $\{\cdot,\cdot\}_-$ are the bivector fields, defined as in (10.28), but using $R_\pm$ instead of R. Since $R_+^* = R_+$ and $R_-^* = -R_-$, these two bivector fields can be written as

$$\begin{aligned} \{F,G\}_+(x) &= \langle R_+(x\nabla_x F) \mid \nabla_x G x\rangle - \langle R_+(\nabla_x F x) \mid x\nabla_x G\rangle \ , \\ \{F,G\}_-(x) &= \langle R_-(\nabla_x F x) \mid \nabla_x G x\rangle - \langle R_-(x\nabla_x F) \mid x\nabla_x G\rangle \ , \end{aligned} \tag{10.31}$$

for all $F, G \in \mathscr{F}(\mathfrak{g})$. For the linear functions F_i $(i = 1,2,3)$, their gradient $\nabla_x F_i \in \mathfrak{g}$ is independent of x; we therefore simply denote $\nabla_x F_i$ by f_i. When computing (10.30), we will need $\nabla_x \{F_1,F_2\}_+$ and $\nabla_x \{F_1,F_2\}_-$. Since $\{F_1,F_2\}_+$ and $\{F_1,F_2\}_-$ depend quadratically on x, these gradients are easily computed from (10.31), giving

$$\begin{aligned} \nabla_x \{F_1,F_2\}_+ &= R_+(xf_1)f_2 + f_1R_+(f_2x) - f_2R_+(f_1x) - R_+(xf_2)f_1 \ , \\ \nabla_x \{F_1,F_2\}_- &= R_-(f_1x)f_2 - R_-(f_2x)f_1 + f_1R_-(xf_2) - f_2R_-(xf_1) \ . \end{aligned} \tag{10.32}$$

Using in two subsequent steps that R_- is skew-symmetric and then that R_- satisfies the modified Yang–Baxter equation, we compute

$$\begin{aligned} &\{\{F_1,F_2\}_-, F_3\}_-(x) + \circlearrowleft (1,2,3) \\ &\quad = \langle [R_-(xf_1), R_-(xf_2)] \mid xf_3\rangle - \langle [R_-(f_1x), R_-(f_2x)] \mid f_3x\rangle + \circlearrowleft (1,2,3) \\ &\quad = \left\langle B_{R_-}(xf_1, xf_2) \mid xf_3\right\rangle - \left\langle B_{R_-}(f_1x, f_2x) \mid f_3x\right\rangle \\ &\quad = -c\,\langle [xf_1, xf_2] \mid xf_3\rangle + c\,\langle [f_1x, f_2x] \mid f_3x\rangle \\ &\quad = 0 \ , \end{aligned}$$

which shows that $\{\cdot,\cdot\}_-$ is a (quadratic) Poisson structure. In particular, when we write (10.30) in terms of R_+ and R_-, only three terms survive: the Jacobiator of $\{\cdot,\cdot\}_+$,

$$\begin{aligned}&\left\{\{F_1,F_2\}_+,F_3\right\}_+ + \circlearrowleft(1,2,3)\\ &\quad= \langle [R_+(f_1x),R_+(f_2x)]\,|\,xf_3\rangle - \langle [R_+(xf_1),R_+(xf_2)]\,|\,f_3x\rangle + \circlearrowleft(1,2,3)\,,\end{aligned} \tag{10.33}$$

and the following two other terms:

$$\begin{aligned}&\left\{\{F_1,F_2\}_+,F_3\right\}_- + \left\{\{F_1,F_2\}_-,F_3\right\}_+ + \circlearrowleft(1,2,3)\\ &\quad= 2\left\langle R_+[xf_1,xf_2]_{R_-}\,|\,f_3x\right\rangle - 2\left\langle R_+[f_1x,f_2x]_{R_-}\,|\,xf_3\right\rangle + \circlearrowleft(1,2,3)\,,\end{aligned} \tag{10.34}$$

where the Lie bracket $[\cdot\,,\cdot]_{R_-}$ is defined as in (10.1), using R_- instead of R: for $x,y\in\mathfrak{g}$,

$$[x,y]_{R_-} := \frac{1}{2}([R_-x,y]+[x,R_-y])\,.$$

Clearly, (10.33) and (10.34) sum up to zero (which yields (10.30)), if

$$\frac{1}{2}R_+ : (\mathfrak{g},[\cdot\,,\cdot]_{R_-}) \to (\mathfrak{g},[\cdot\,,\cdot])$$

is a homomorphism of Lie algebras. Written out, this means that it suffices to prove that

$$[R_+x,R_+y] - R_+[R_-x,y] - R_+[x,R_-y] = 0\,, \tag{10.35}$$

for all $x,y\in\mathfrak{g}$. We define a linear map $B_R^*:\mathfrak{g}\otimes\mathfrak{g}\to\mathfrak{g}$ by requiring that for all $x,y,z\in\mathfrak{g}$,

$$\langle B_R(x,y)\,|\,z\rangle = \langle x\,|\,B_R^*(y,z)\rangle\,. \tag{10.36}$$

Explicitly, B_R^* is given, for $x,y\in\mathfrak{g}$, by

$$B_R^*(x,y) = R^*[Rx,y] - R^*[x,R^*y] - [Rx,R^*y]\,.$$

Notice that

$$\begin{aligned}&B_R^*(y,x) - B_R^*(x,y)\\ &\quad= 2([R_+x,R_+y] - [R_-x,R_-y] - R^*[R_-x,y] - R^*[x,R_-y])\,,\end{aligned}$$

so that

$$\begin{aligned}&B_{R_-}(x,y) + \frac{1}{2}B_R^*(y,x) - \frac{1}{2}B_R^*(x,y)\\ &\quad= [R_+x,R_+y] - R_+[R_-x,y] - R_+[x,R_-y]\,,\end{aligned} \tag{10.37}$$

which is the left-hand side of (10.35). Since R_- is a solution of the modified Yang–Baxter equation (with constant c), we have that $B_{R_-}(x,y) = -c\,[x,y]$, for all $x,y\in\mathfrak{g}$. But R is also a solution of the modified Yang–Baxter equation (with the same constant c) which, combined with (10.36) yields that $B_R^*(x,y) = -c\,[x,y]$, for all $x,y\in\mathfrak{g}$. Substituted in (10.37), this yields (10.35). Thus, (10.33) and (10.34) sum up to zero, which proves (10.30), hence $\{\cdot\,,\cdot\}_R^Q$ is a (quadratic) Poisson structure.

Finally, let H be an Ad-invariant function on $\mathfrak{g}$. We show that its Hamiltonian vector field is given by the Lax equation (10.29). To do this, recall from Section 5.1.4 that the Ad-invariance of H implies that $[\nabla_x H, x] = 0$ for all $x \in \mathfrak{g}$, hence that $\frac{1}{2}(x\nabla_x H + \nabla_x H x) = x\nabla_x H$. Therefore, if $F \in \mathscr{F}(\mathfrak{g})$ and $x \in \mathfrak{g}$, then

$$\begin{aligned}
\langle \mathrm{d}_x F, (\mathscr{X}_H)_x \rangle = \{F,H\}_R^Q(x) &= \frac{1}{2}\langle [x, \nabla_x F] \,|\, R(x\nabla_x H + \nabla_x H x) \rangle \\
&= \langle [x, \nabla_x F] \,|\, R(x\nabla_x H) \rangle = \langle \nabla_x F \,|\, [R(x\nabla_x H), x] \rangle \\
&= \langle \mathrm{d}_x F, [R(x\nabla_x H), x] \rangle ,
\end{aligned}$$

which proves (10.29). □

If, in addition to the hypotheses of Proposition 10.18, R is assumed to be skew-symmetric, then the formula (10.28) for the quadratic bracket takes the simpler form

$$\{F,G\}_R^Q(x) = \langle R(\nabla_x F x) \,|\, \nabla_x G x \rangle - \langle R(x\nabla_x F) \,|\, x\nabla_x G \rangle , \qquad (10.38)$$

for all $F, G \in \mathscr{F}(\mathfrak{g})$. In the particular case where R comes from an r-matrix $r \in \mathfrak{g} \otimes \mathfrak{g}$, as explained in Section 10.2.3, this formula takes a particularly simple form. In order to establish it, we decompose $r^- = \sum_\alpha s_\alpha \otimes t_\alpha$ and use, as in the proof of Proposition 10.13, the convention that a sum over α is implicit in all formulas which contain α. Recall that with this notation,

$$\langle Rx \,|\, y \rangle = 4 \langle s_\alpha \,|\, x \rangle \langle t_\alpha \,|\, y \rangle , \qquad (10.39)$$

for all $x, y \in \mathfrak{g}$. For $F, G \in \mathscr{F}(\mathfrak{g})$, the bracket (10.38) can be rewritten, using (10.39), as

$$\begin{aligned}
\{F,G\}_R^Q(x) &= 4 \langle s_\alpha \,|\, \nabla_x F x \rangle \langle t_\alpha \,|\, \nabla_x G x \rangle - 4 \langle s_\alpha \,|\, x\nabla_x F \rangle \langle t_\alpha \,|\, x\nabla_x G \rangle \\
&= 4 \langle \nabla_x F \,|\, x s_\alpha \rangle \langle \nabla_x G \,|\, x t_\alpha \rangle - 4 \langle \nabla_x F \,|\, s_\alpha x \rangle \langle \nabla_x G \,|\, t_\alpha x \rangle \\
&= 4 \langle \mathrm{d}_x F, x s_\alpha \rangle \langle \mathrm{d}_x G, x t_\alpha \rangle - 4 \langle \mathrm{d}_x F, s_\alpha x \rangle \langle \mathrm{d}_x G, t_\alpha x \rangle \\
&= 2 \langle \mathrm{d}_x F \wedge \mathrm{d}_x G, x s_\alpha \otimes x t_\alpha - s_\alpha x \otimes t_\alpha x \rangle \\
&= 2 \langle \mathrm{d}_x F \wedge \mathrm{d}_x G, [x \otimes x, s_\alpha \otimes t_\alpha] \rangle \\
&= 2 \big\langle \mathrm{d}_x F \wedge \mathrm{d}_x G, \big[x \otimes x, r^-\big] \big\rangle .
\end{aligned}$$

We choose a basis $(e_1, \dots, e_d)$ of $\mathfrak{g}$, with dual basis $(\xi_1, \dots, \xi_d)$ of $\mathfrak{g}^*$. Then $\left\{\xi_i, \xi_j\right\}_R^Q(x) = 2\left\langle \xi_i \wedge \xi_j, [x \otimes x, r^-] \right\rangle$, so that $\left\{\xi_i, \xi_j\right\}_R^Q(x)$ is twice the coefficient of $e_i \otimes e_j$ in $[x \otimes x, r^-]$. It follows that, if we define

$$\left\{\Xi \overset{\otimes}{,} \Xi\right\}_r^Q(x) := \sum_{1 \leqslant i < j \leqslant d} \left\{\xi_i, \xi_j\right\}_R^Q(x)\, e_i \wedge e_j ,$$

then we find the following compact formula

$$\left\{\Xi \overset{\otimes}{,} \Xi\right\}_r^Q(x) = 2\left[x \otimes x, r^-\right] , \qquad (10.40)$$

which is known as the *second Russian formula*. As in the case of the first Russian formula, this formula can be thought of in terms of matrices in $\operatorname{Mat}_{N^2}(\mathbb{F})$, upon using the standard algebra isomorphism between $\operatorname{Mat}_N(\mathbb{F}) \otimes \operatorname{Mat}_N(\mathbb{F})$ and $\operatorname{Mat}_{N^2}(\mathbb{F})$.

For the cubic R-bracket, the context is the same as in the case of the quadratic R-bracket, but the assumptions are weaker.

Proposition 10.19. *Let $(\mathfrak{g},[\cdot\,,\cdot])$ be a finite-dimensional Lie algebra, associated with an associative algebra, equipped with a non-degenerate symmetric bilinear map $\langle\cdot\,|\,\cdot\rangle$, which satisfies $\langle xy\,|\,z\rangle = \langle x\,|\,yz\rangle$, for all $x,y,z\in\mathfrak{g}$. Suppose that $R:\mathfrak{g}\to\mathfrak{g}$ is a linear map, which is a solution of the modified Yang–Baxter equation (10.4). The bivector field on $\mathfrak{g}$, defined for $F,G\in\mathscr{F}(\mathfrak{g})$ and $x\in\mathfrak{g}$ by*

$$\{F,G\}^C_R(x) := \langle [x,\nabla_x F]\,|\,R(x\nabla_x G x)\rangle - \langle [x,\nabla_x G]\,|\,R(x\nabla_x F x)\rangle\;, \tag{10.41}$$

is a cubic Poisson structure on $\mathfrak{g}$. If H is an Ad*-invariant function on $\mathfrak{g}$, then its Hamiltonian vector field with respect to this Poisson structure is given by*

$$\dot{x} = [R(x\nabla_x H x),x]\;.$$

Proof. The proposition can be proved in the same way as in the case of the quadratic R-bracket (Proposition 10.18). Actually, the proof of the Jacobi identity for $\{\cdot\,,\cdot\}^C_R$ requires less computation, as one does not need to split up R into its symmetric and skew-symmetric parts and as the formula for the cubic bracket (10.41) has only half as many terms as the formula for the quadratic bracket (10.28). For linear functions $F_1,F_2,F_3\in\mathscr{F}(\mathfrak{g})$, whose gradients (at any point) are denoted by f_1,f_2,f_3, first compute from (10.41) that at $x\in\mathfrak{g}$ one has

$$\nabla_x\{F_1,F_2\}^C_R = R^*[x,f_1]xf_2 + f_2xR^*[x,f_1] + [f_1,R(xf_2x)] - (f_1\leftrightarrow f_2)\;,$$

which yields, with some extra work,

$$\begin{aligned}
&\left\{\{F_1,F_2\}^C_R,F_3\right\}^C_R + \circlearrowleft(1,2,3)\\
&= \left\langle\left[x,\nabla_x\{F_1,F_2\}^C_R\right]\,|\,R(xf_3x)\right\rangle - \left\langle [x,f_3]\,\middle|\,R\left(x\nabla_x\{F_1,F_2\}^C_R x\right)\right\rangle\\
&\qquad + \circlearrowleft(1,2,3)\\
&= \langle [x,f_1]\,|\,B_R(xf_2x,xf_3x)\rangle + \circlearrowleft(1,2,3)\\
&= -c\langle [x,f_1]\,|\,[xf_2x,xf_3x]\rangle + \circlearrowleft(1,2,3)\\
&= 0\,.
\end{aligned}$$

This completes the proof. □

To finish this section, we show how the linear, quadratic and cubic R-brackets are related. The Lie algebra $\mathfrak{g}$ is, as above, the Lie algebra of an associative algebra, with unit e. On $\mathfrak{g}$, we can therefore consider the vector fields $\mathscr{V}_0,\mathscr{V}_1$ and $\mathscr{V}_2$, given at all $x\in\mathfrak{g}$ by

$$(\mathscr{V}_0)_x := e\,,\qquad (\mathscr{V}_1)_x := x\,,\qquad (\mathscr{V}_2)_x := x^2\,.$$

For a linear function F on $\mathfrak{g}$, this means that

$$(\mathscr{V}_0)_x[F] = F(e)\,,\qquad (\mathscr{V}_1)_x[F] = F(x)\,,\qquad (\mathscr{V}_2)_x[F] = F(x^2)\,.$$

for every $x \in \mathfrak{g}$. In particular, $\mathscr{V}_1$ is the Euler vector field, which was defined in Section 8.1.2. For $i = 0,1,2$, we write $\mathscr{L}_i$ for the Lie derivative with respect to $\mathscr{V}_i$.

Proposition 10.20. *Let $(\mathfrak{g}, [\cdot\,,\cdot])$ be the Lie algebra of an associative algebra, with unit e, which is equipped with a non-degenerate symmetric bilinear map $\langle\cdot\,|\,\cdot\rangle$, which satisfies $\langle xy\,|\,z\rangle = \langle x\,|\,yz\rangle$, for all $x,y,z \in \mathfrak{g}$. Let R be a solution of the modified Yang–Baxter equation, with constant $c \in \mathbb{F}$, and suppose that its skew-symmetric part R_- is also a solution of the modified Yang–Baxter equation, with the same constant c. Then the linear, quadratic and cubic Poisson structures on $\mathfrak{g}$, associated with R, namely $\{\cdot\,,\cdot\}_{R,\mathfrak{g}}$, $\{\cdot\,,\cdot\}^Q_R$ and $\{\cdot\,,\cdot\}^C_R$, are related by Lie derivatives, as indicated in the following commutative diagram:*

$$\begin{array}{ccccccc}
\{\cdot\,,\cdot\}_{R,\mathfrak{g}} & \xrightarrow{-\mathscr{L}_2} & \{\cdot\,,\cdot\}^Q_R & \xrightarrow{-\mathscr{L}_2} & \{\cdot\,,\cdot\}^C_R & \xrightarrow{\mathscr{L}_2} & 0 \\
\Big\downarrow{\scriptstyle -\mathscr{L}_1} & & & & \Big\downarrow{\scriptstyle \mathscr{L}_1} & & \\
\{\cdot\,,\cdot\}_{R,\mathfrak{g}} & \xleftarrow{\frac12\mathscr{L}_0} & \{\cdot\,,\cdot\}^Q_R & \xleftarrow{\frac12\mathscr{L}_0} & \{\cdot\,,\cdot\}^C_R & &
\end{array}\qquad 0 \xleftarrow{\mathscr{L}_0} \{\cdot\,,\cdot\}_{R,\mathfrak{g}} \tag{10.42}$$

Moreover, for a fixed such R, the three Poisson structures $\{\cdot\,,\cdot\}_{R,\mathfrak{g}}$, $\{\cdot\,,\cdot\}^Q_R$ and $\{\cdot\,,\cdot\}^C_R$ are compatible.

Proof. Since $\mathscr{V}_1$ is the Euler vector field, the Lie derivatives of the linear, quadratic and cubic structures with respect to $\mathscr{V}_1$ give a multiple of these Poisson structures, up to a constant, which is respectively -1, 0 and 1 (see Proposition 8.4). This yields the two vertical arrows in the diagram. For each of the Lie derivatives $\mathscr{L}_0$ and $\mathscr{L}_2$, the computation is very similar, so we will only do it for one of them, namely we show that $-\mathscr{L}_2\{\cdot\,,\cdot\}_{R,\mathfrak{g}} = \{\cdot\,,\cdot\}^Q_R$. Let F_1, F_2 be linear functions on $\mathfrak{g}$ and let $f_1 := \nabla F_1$ and $f_2 := \nabla F_2$, as before. According to the formula (3.7) for the Lie derivative of a biderivation ($p = 2$), we need to show that

$$\{F_1,F_2\}^Q_R = -\mathscr{V}_2\left[\{F_1,F_2\}_{R,\mathfrak{g}}\right] + \{\mathscr{V}_2[F_1],F_2\}_{R,\mathfrak{g}} + \{F_1,\mathscr{V}_2[F_2]\}_{R,\mathfrak{g}}\,. \tag{10.43}$$

To do this, first notice that for a linear function F on $\mathfrak{g}$, one has $\langle \nabla_x F\,|\,y\rangle = F(y)$, independently of $x \in \mathfrak{g}$ (so we write f for $\nabla_x F$) and $\mathscr{V}_2[F](x) = F(x^2) = \langle f\,|\,x^2\rangle$. It follows that, for all $x,y \in \mathfrak{g}$,

$$\langle \nabla_x(\mathscr{V}_2[F])\,|\,y\rangle = \frac{\mathrm{d}}{\mathrm{d}t}_{|t=0} F((x+ty)^2) = F(xy) + F(yx) = \langle xf + fx\,|\,y\rangle\,.$$

Therefore, for all linear functions F on $\mathfrak{g}$, we have

$$\nabla_x(\mathscr{V}_2[F]) = xf + fx\,.$$

One computes similarly that $\nabla_x(\mathscr{V}_1[F]) = f$ and that $\nabla_x(\mathscr{V}_0[F]) = 0$, but these formulas are only needed for the verification of the other cases. Since $\{F_1, F_2\}_{R,\mathfrak{g}}$ is a linear function on $\mathfrak{g}$, the right-hand side of (10.43), evaluated at x, is given by

$$\begin{aligned}
&-\langle x^2 \mid [f_1, f_2]_R\rangle + \langle x \mid [\nabla_x(\mathscr{V}_2[F_1]), f_2]_R\rangle + \langle x \mid [f_1, \nabla_x(\mathscr{V}_2[F_2])]_R\rangle \\
&= -\langle x^2 \mid [f_1, f_2]_R\rangle + \langle x \mid [xf_1 + f_1x, f_2]_R\rangle + \langle x \mid [f_1, xf_2 + f_2x]_R\rangle \\
&= \frac{1}{2}\langle [x, f_1] \mid R(xf_2 + f_2x)\rangle - \frac{1}{2}\langle [x, f_2] \mid R(xf_1 + f_1x)\rangle\,,
\end{aligned}$$

which is precisely $\{F_1, F_2\}_R^Q$ (see (10.28)).

The compatibility of the three Poisson structures follows easily from the relations in the diagram (10.42) and the properties of the Schouten bracket. Recall from Section 3.3 that, in terms of the Schouten bracket, two Poisson structures π and π' are compatible if $[\pi, \pi']_S = 0$, while the Lie derivative with respect to a vector field $\mathscr{V}$ is expressed as $\mathscr{L}_{\mathscr{V}}\pi = [\mathscr{V}, \pi]_S$. Clearly, $[\mathscr{L}_{\mathscr{V}}\pi, \pi]_S = [[\mathscr{V}, \pi]_S, \pi]_S = 0$, since $[\cdot, \pi]_S$ is a coboundary operator (since π is a Poisson structure, see Section 4.1.2). Thus, on the one hand, $\{\cdot,\cdot\}_{R,\mathfrak{g}}$ and $\{\cdot,\cdot\}_R^Q$ are compatible, on the other hand, $\{\cdot,\cdot\}_R^Q$ and $\{\cdot,\cdot\}_R^C$ are compatible. In order to see that[3] $\{\cdot,\cdot\}_{R,\mathfrak{g}}$ and $\{\cdot,\cdot\}_R^C$ are also compatible, let us denote the three Poisson structures by π_R, π_R^Q and π_R^C and compute, using the graded Jacobi identity for the Schouten bracket (Proposition 3.7),

$$[\pi_R, \pi_R^C]_S = -\left[\pi_R, \left[\mathscr{V}_2, \pi_R^Q\right]_S\right]_S = -\left[\mathscr{V}_2, \left[\pi_R^Q, \pi_R\right]_S\right]_S - \left[\pi_R^Q, [\pi_R, \mathscr{V}_2]_S\right]_S = 0\,,$$

where we used in the last step that π_R^Q and π_R are compatible, and that $[\mathscr{V}_2, \pi_R]_S = -\pi_R^Q$ (which is a Poisson structure!). □

10.4 Notes

R-matrices and r-matrices have their origins in the theory of integrable systems, but also play a fundamental rôle in the theory of quantum groups. They first appeared in the contexts of the Adler–Kostant–Symes theorem, exhibiting the Lie algebraic structures which underlie the integrability of both the Toda lattices and the Korteweg–de Vries equation, see Adler [2].

The term "quantum group" is a generic name which is used for several types of non-commutative algebras, usually Hopf algebras, i.e., algebras (rather than groups!) endowed with an algebra structure on the dual. Quantum groups are in

[3] In general, the notion "is compatible with" is not transitive.

general constructed from solutions of the quantum Yang–Baxter equation, just as Poisson–Lie groups are constructed from solutions of the modified classical Yang–Baxter equation. These constructions are related, since a solution of the Yang–Baxter equation admits an r-matrix as its first order term, while Poisson–Lie groups appear as limits of quantum groups. For the theory of quantum groups, we refer to the classical books of Kassel [103] or Majid [142].

The notion of an r-matrix has been generalized to the general context of Lie algebroids. The resulting object is called a (classical) dynamical r-matrix; it depends on parameters, turning some of the conditions which it has to satisfy into differential equations. See Etingof–Varchenko [69, 70] and Liu–Xu [130].

Chapter 11
Poisson–Lie Groups

A Poisson–Lie group $(\mathbf{G}, \pi)$ is a Lie group $\mathbf{G}$ which is equipped with a Poisson structure π. The Poisson structure is demanded to be compatible with the group structure, in the sense that one requires the product map $\mathbf{G} \times \mathbf{G} \to \mathbf{G}$ to be a morphism of Poisson manifolds, where $\mathbf{G} \times \mathbf{G}$ is equipped with the product Poisson structure.

To a Lie group $\mathbf{G}$ there corresponds a Lie algebra $\mathfrak{g}$, which is the tangent space to $\mathbf{G}$ at e, where the Lie bracket $[\cdot\,,\cdot]_{\mathfrak{g}}$ on $\mathfrak{g}$ is inherited from the Lie bracket on the space of (left-invariant) vector fields on $\mathbf{G}$. If, in addition, $\mathbf{G}$ comes equipped with a Poisson structure, the Lie algebra $\mathfrak{g}$ inherits the structure of a Lie coalgebra, which amounts to a Lie algebra structure $[\cdot\,,\cdot]_{\mathfrak{g}^*}$ on the dual vector space $\mathfrak{g}^*$. In the case of a Poisson–Lie group $(\mathbf{G}, \pi)$, the fact that the multiplication in $\mathbf{G}$ and the Poisson structure on $\mathbf{G}$ are compatible, implies a compatibility between the two Lie brackets $[\cdot\,,\cdot]_{\mathfrak{g}}$ and $[\cdot\,,\cdot]_{\mathfrak{g}^*}$: the transpose to $[\cdot\,,\cdot]_{\mathfrak{g}^*}$ is a derivation of the Lie bracket $[\cdot\,,\cdot]_{\mathfrak{g}}$. Formalizing this property leads to the notion of a Lie bialgebra.

Thus, to every Poisson–Lie group, there corresponds a Lie bialgebra. Moreover, to every homomorphism of Poisson–Lie groups there naturally corresponds a homomorphism of Lie bialgebras, hence the correspondence between Poisson–Lie groups and Lie bialgebras is functorial. Most importantly, the converse also holds. As we know from Lie's third theorem, every finite-dimensional Lie algebra is the Lie algebra of some Lie group, which can be chosen to be connected and simply connected. This Lie group can be made into a Poisson–Lie group if the Lie algebra is a Lie bialgebra.

Poisson–Lie groups are introduced in Section 11.1, while their infinitesimal analogs, Lie bialgebras, are introduced in Section 11.2. The above correspondence between Lie bialgebras and Poisson–Lie groups is discussed in Section 11.3. Using dressing actions, we study the symplectic leaves of Poisson–Lie groups in Section 11.4.

All Lie groups in this chapter are real ($\mathbb{F} = \mathbb{R}$) or complex ($\mathbb{F} = \mathbb{C}$).

C. Laurent-Gengoux et al., *Poisson Structures*,
Grundlehren der mathematischen Wissenschaften 347,
DOI 10.1007/978-3-642-31090-4_11,

11.1 Multiplicative Poisson Structures and Poisson–Lie Groups

A Poisson structure on a Lie group $\mathbf{G}$ makes $\mathbf{G}$ into a Poisson manifold. For reasons which will be given later in this chapter, one demands the following compatibility relation between the Poisson structure on $\mathbf{G}$ and the group structure of $\mathbf{G}$.

Definition 11.1. A Poisson structure π on a (real or complex) Lie group $\mathbf{G}$ is said to be *multiplicative* if the product map $\mu : \mathbf{G}\times\mathbf{G}\to\mathbf{G}$ is a Poisson map, where $\mathbf{G}\times\mathbf{G}$ is endowed with the product Poisson structure. The pair $(\mathbf{G},\pi)$ is then called a *Poisson–Lie group*.

A map $\Phi : \mathbf{G}\to\mathbf{H}$ between two Poisson–Lie groups $(\mathbf{G},\pi)$ and $(\mathbf{H},\pi')$ is called a *Poisson–Lie group homomorphism* if it is both a Poisson map and a group homomorphism.

11.1.1 The Condition of Multiplicativity

We give in the following proposition two useful characterizations of the multiplicativity of a Poisson structure on a Lie group. Recall that for $g\in\mathbf{G}$ the maps of left and right translation L_g and R_g in $\mathbf{G}$ are defined by $L_g(h) := gh$ and $R_g(h) := hg$, for all $h\in\mathbf{G}$.

Proposition 11.2. *Let $\mathbf{G}$ be a Lie group. For a Poisson structure π on $\mathbf{G}$, the following three conditions are equivalent:*

(i) *π is multiplicative;*

(ii) *For all $g,h\in\mathbf{G}$:*

$$\pi_{gh} = \wedge^2(T_gR_h)\,\pi_g + \wedge^2(T_hL_g)\,\pi_h\,; \tag{11.1}$$

(iii) *The map $\Psi : \mathbf{G}\to\wedge^2\mathfrak{g}$, which is defined for all $g\in\mathbf{G}$ by*

$$\Psi(g) := \wedge^2(T_gR_{g^{-1}})\,\pi_g\,, \tag{11.2}$$

is a cocycle of $\mathbf{G}$, with respect to the adjoint representation of $\mathbf{G}$ on $\wedge^2\mathfrak{g}$, i.e.,

$$\Psi(gh) = \Psi(g) + \mathrm{Ad}_g\,\Psi(h)\,, \tag{11.3}$$

for all $g,h\in\mathbf{G}$.

Proof. We first prove that *(i)* and *(ii)* are equivalent, which amounts to proving that $\mu : \mathbf{G}\times\mathbf{G}\to\mathbf{G}$ is a Poisson map if and only if (11.1) holds for all $g,h\in\mathbf{G}$. Let us denote the product Poisson structure on $\mathbf{G}\times\mathbf{G}$ by Π; according to (2.11), the value of Π at $(g,h)\in\mathbf{G}\times\mathbf{G}$ is given by the bivector

$$\Pi_{(g,h)} = \wedge^2(T_g\imath_h)\,\pi_g + \wedge^2(T_h\imath'_g)\,\pi_h\,, \tag{11.4}$$

where ι_h and ι'_g are the inclusions of $\mathbf{G}$ into $\mathbf{G}\times\mathbf{G}$ given by $\iota_h : g \mapsto (g,h)$ and $\iota'_g : h \mapsto (g,h)$ respectively. According to (1.32), μ is a Poisson map if and only if

$$\pi_{gh} = \wedge^2(T_{(g,h)}\mu)\,\Pi_{(g,h)}\,, \tag{11.5}$$

for all $g,h \in \mathbf{G}$. According to (11.4), and using the identities $\mu \circ \iota_h = R_h$ and $\mu \circ \iota'_g = L_g$, the right-hand side of (11.5) is given by

$$\begin{aligned}\wedge^2(T_{(g,h)}\mu)\,\Pi_{(g,h)} &= \wedge^2(T_{(g,h)}\mu)\,(\wedge^2(T_g\iota_h)\,\pi_g) + \wedge^2(T_{(g,h)}\mu)\,(\wedge^2(T_h\iota'_g)\,\pi_h)\\ &= \wedge^2(T_g(\mu\circ\iota_h))\,\pi_g + \wedge^2(T_h(\mu\circ\iota'_g))\,\pi_h\\ &= \wedge^2(T_gR_h)\,\pi_g + \wedge^2(T_hL_g)\,\pi_h\,,\end{aligned}$$

which is precisely the right-hand side of (11.1). This proves that *(i)* and *(ii)* are equivalent.

In order to show that *(ii)* and *(iii)* are equivalent, we fix $g,h \in \mathbf{G}$ and we apply the isomorphism $\wedge^2(T_{gh}R_{(gh)^{-1}})$ to both sides of (11.1), giving

$$\begin{aligned}&\wedge^2(T_{gh}R_{(gh)^{-1}})\,\pi_{gh}\\ &\quad = \wedge^2(T_{gh}R_{(gh)^{-1}})\,\wedge^2(T_gR_h)\,\pi_g + \wedge^2(T_{gh}R_{(gh)^{-1}})\,\wedge^2(T_hL_g)\,\pi_h\,,\end{aligned}$$

and we show that the latter equation is, term by term, equation (11.3). For the first two terms, this follows from the definition of Ψ and from the identity $R_{(gh)^{-1}} \circ R_h = R_{g^{-1}}$, namely

$$\begin{aligned}\wedge^2(T_{gh}R_{(gh)^{-1}})\,\pi_{gh} &= \Psi(gh)\,,\\ \wedge^2(T_{gh}R_{(gh)^{-1}})(\wedge^2(T_gR_h)\,\pi_g) &= \wedge^2(T_gR_{g^{-1}})\,\pi_g = \Psi(g)\,.\end{aligned}$$

For the third and final term, recall that the map C_g is conjugation by g, so that $C_g = R_{g^{-1}} \circ L_g$, leading to the identity $R_{(gh)^{-1}} \circ L_g = C_g \circ R_{h^{-1}}$, which in turn yields

$$\begin{aligned}\wedge^2(T_{gh}R_{(gh)^{-1}})(\wedge^2(T_hL_g)\,\pi_h) &= \wedge^2(T_h(R_{(gh)^{-1}}\circ L_g))\,\pi_h\\ &= \wedge^2(T_h(C_g\circ R_{h^{-1}}))\,\pi_h = \wedge^2(T_eC_g)(\wedge^2(T_hR_{h^{-1}})\,\pi_h)\\ &= \mathrm{Ad}_g\,\Psi(h)\,,\end{aligned}$$

where we have used in the last step that $\wedge(T_eC_g)(x) = \mathrm{Ad}_g\,x$ for all $x \in \mathfrak{g}$ and $g \in \mathbf{G}$. This proves that condition (11.1) is equivalent to condition (11.3). □

In item *(iii)* of Proposition 11.2, we used the term *cocycle*, which is borrowed from the terminology used when introducing the concept of cohomology for Lie groups. We will not develop Lie group cohomology in this book, we rather restrict ourselves to giving the general definition of a cocycle of a group (strictly speaking a 1-cocycle) and proving the one result on group cocycles which we will use.

Definition 11.3. Let $\rho : \mathbf{G} \times V \to V$ be a representation of a Lie group $\mathbf{G}$ on a finite-dimensional vector space V. A *cocycle* of $\mathbf{G}$ in V is a smooth (or holomorphic) function $\Psi : \mathbf{G} \to V$, satisfying

$$\Psi(gh) = \Psi(g) + \rho(g, \Psi(h)) , \tag{11.6}$$

for all $g, h \in \mathbf{G}$.

Proposition 11.4. *Let $\mathbf{G}$ be a connected Lie group, let $\rho : \mathbf{G} \times V \to V$ be a representation of $\mathbf{G}$ on a finite-dimensional vector space V and let $W \subset V$ be a $\mathbf{G}$-invariant subspace. Suppose that $\Psi : \mathbf{G} \to V$ is a cocycle of $\mathbf{G}$ in V.*

(1) If $T_e\Psi(x) = 0$ for all $x \in \mathfrak{g}$, then $\Psi = 0$;
(2) If $T_e\Psi(x) \in W$ for all $x \in \mathfrak{g}$, then $\Psi(g) \in W$ for all $g \in \mathbf{G}$.

Proof. The tangent map of the cocycle relation (11.6) with respect to the variable h at the point $h = e$ is given by

$$T_g\Psi \circ T_eL_g = \rho_g \circ T_e\Psi ,$$

so that the vanishing of $T_e\Psi$ implies the vanishing of $T_g\Psi \circ T_eL_g$, and therefore of $T_g\Psi$, for all $g \in \mathbf{G}$. Since $\mathbf{G}$ is assumed to be connected, this implies that Ψ is a constant function. Since $\Psi(e) = 0$, as follows from (11.6), Ψ vanishes at each point, which shows *(1)*. If W is a $\mathbf{G}$-invariant subspace, then we have an induced representation $\mathbf{G} \times V/W \to V/W$ and Ψ induces a cocycle of $\mathbf{G}$ in V/W. In view of *(1)*, the latter cocycle is zero as soon as its tangent map at e is zero, i.e., as soon as $T_e\Psi(x) \in W$ for all $x \in \mathfrak{g}$, which proves *(2)*. □

See [181] for more information on *group cohomology*.

11.1.2 Basic Properties of Poisson–Lie Groups

We start with an interesting feature of Poisson–Lie groups. Let $(\mathbf{G}, \pi)$ be a Poisson–Lie group with Lie algebra $\mathfrak{g}$. The Poisson structure π vanishes at the unit e of $\mathbf{G}$, hence leads to a linear Poisson structure on $\mathfrak{g}$, i.e., a Lie bracket on $\mathfrak{g}^*$. This structure will play a mayor rôle in the sequel of this chapter.

Proposition 11.5. *Let $(\mathbf{G}, \pi)$ be a Poisson–Lie group.*

(1) π vanishes at e;
(2) The linearized Poisson structure of π at e is the linear Poisson structure π_1 on $\mathfrak{g} = T_e\mathbf{G}$, whose value at $x \in \mathfrak{g}$ is given by $(\pi_1)_x = T_e\Psi(x)$, where $\Psi : \mathbf{G} \to \wedge^2\mathfrak{g}$ is the cocycle, defined as in (11.2);
(3) The inverse map $\mathrm{inv} : \mathbf{G} \to \mathbf{G}$ is an anti-Poisson map.

Proof. Substituting e for g and h in (11.1) amounts to $\pi_e = \pi_e + \pi_e$, hence $\pi_e = 0$, which yields *(1)*. According to Proposition 7.22, $\mathfrak{g} = T_e\mathbf{G}$ inherits from $\pi = \{\cdot,\cdot\}$

a linear Poisson structure $\pi_1 = \{\cdot,\cdot\}_1$. For functions F, G on $\mathfrak{g}$, defined in a neighborhood of the origin $o \in \mathfrak{g}$, the linearized Poisson bracket $\{F,G\}_1$ and the original Poisson bracket $\{F,G\}$ are, according to Proposition 7.22, related for every $x \in \mathfrak{g}$ by[1]

$$\langle \mathrm{d}_e \{F,G\}, x\rangle = \{\mathrm{d}_e F, \mathrm{d}_e G\}_1 (x) = \langle \mathrm{d}_e F \wedge \mathrm{d}_e G, (\pi_1)_x\rangle \,, \tag{11.7}$$

where $\mathrm{d}_e F$, $\mathrm{d}_e G$ and $\mathrm{d}_e \{F,G\}$ are viewed as (linear) functions on $\mathfrak{g} = T_e\mathbf{G}$, and $(\pi_1)_x$ is identified with an element of $\wedge^2\mathfrak{g}$. Alternatively, in terms of the map $\Psi : \mathbf{G} \to \wedge^2\mathfrak{g}$ defined in (11.2), the Poisson bracket $\{F,G\}$ at $g \in \mathbf{G}$ can be written as

$$\{F,G\}(g) = \left\langle \mathrm{d}_g F \wedge \mathrm{d}_g G, \pi_g \right\rangle = \left\langle \mathrm{d}_g F \wedge \mathrm{d}_g G, \wedge^2 (T_e R_g)\Psi(g) \right\rangle \,,$$

which yields, upon putting $g := \exp(tx)$ and taking the derivative with respect to t at $t = 0$,

$$\langle \mathrm{d}_e \{F,G\}, x\rangle = \langle \mathrm{d}_e F \wedge \mathrm{d}_e G, T_e\Psi(x)\rangle \,, \tag{11.8}$$

where we have used that $\Psi(e) = 0$. Comparing (11.7) with (11.8) leads to the desired equality

$$(\pi_1)_x = T_e\Psi(x) \,, \tag{11.9}$$

proving *(2)*. To say that inv is an *anti-Poisson map* means that inv : $(\mathbf{G},\pi) \to (\mathbf{G},-\pi)$ is a Poisson map: for every $g \in \mathbf{G}$, $\wedge^2(T_g \,\mathrm{inv})\,\pi_g = -\pi_{g^{-1}}$. To prove that the latter holds, let $g \in \mathbf{G}$, substitute g^{-1} for h in (11.1), and use $\pi_e = 0$, to find that

$$0 = \wedge^2(T_g R_{g^{-1}})\,\pi_g + \wedge^2(T_{g^{-1}} L_g)\,\pi_{g^{-1}} \,.$$

Combined with $T_g \,\mathrm{inv} = -T_e L_{g^{-1}} \circ T_g R_{g^{-1}}$ (see (5.2)), this leads to

$$\begin{aligned}
\wedge^2(T_g \,\mathrm{inv})\,\pi_g &= \wedge^2(T_e L_{g^{-1}})(\wedge^2(T_g R_{g^{-1}})\,\pi_g) \\
&= -\wedge^2(T_e L_{g^{-1}})(\wedge^2(T_{g^{-1}} L_g)\,\pi_{g^{-1}}) \\
&= -\wedge^2(T_{g^{-1}}(L_{g^{-1}} \circ L_g))\,\pi_{g^{-1}} \\
&= -\pi_{g^{-1}}
\end{aligned}$$

which proves that inv is an anti-Poisson map. □

Comparing Definitions 5.32 and 11.1, we see that a Poisson structure π on a Lie group $\mathbf{G}$ is multiplicative if and only if left translation $L\colon \mathbf{G}\times\mathbf{G} \to \mathbf{G}$ is a Poisson action. We show in the following corollary of Proposition 11.2 that this condition does not imply, in general, that the map L_g (left translation by a given element $g \in \mathbf{G}$) is, for every $g \in \mathbf{G}$, a Poisson map.

Proposition 11.6. *Let $(\mathbf{G},\pi)$ be a Poisson–Lie group. For every $g \in \mathbf{G}$, the following conditions are equivalent:*

(i) $\pi_g = 0$;
(ii) *The map $L_g : \mathbf{G} \to \mathbf{G}$, left translation by g, is a Poisson map;*

[1] We use here and below the natural pairing $\langle\cdot,\cdot\rangle$ between $\wedge^p\mathfrak{g}^*$ and $\wedge^p\mathfrak{g}$, introduced in (5.10).

(iii) *The map $R_g : \mathbf{G} \to \mathbf{G}$, right translation by g, is a Poisson map.*

Proof. For $g \in \mathbf{G}$, left translation by g is a Poisson map if and only if, for all $h \in \mathbf{G}$,

$$\pi_{gh} = \wedge^2(T_h L_g)\,\pi_h\,. \tag{11.10}$$

In view of (11.1), which says that π is multiplicative, (11.10) is tantamount to $\wedge^2(T_g R_h)\,\pi_g = 0$, which is equivalent to $\pi_g = 0$, because the linear map $T_g R_h$ is an isomorphism. This proves the equivalence of *(i)* and *(ii)*. The proof of the equivalence of *(i)* and *(iii)* is similar. □

11.1.3 Poisson–Lie Subgroups

We consider in this section the notion of a Poisson–Lie subgroup of a Poisson–Lie group $(\mathbf{G}, \pi)$. Recall that a Lie subgroup of $\mathbf{G}$ need not be a closed subset of $\mathbf{G}$ and that there is associated to π a map $\Psi : \mathbf{G} \to \wedge^2\mathfrak{g}$, defined for $g \in \mathbf{G}$ by $\Psi(g) := \wedge^2(T_g R_{g^{-1}})\,\pi_g$.

Proposition 11.7. *Let $(\mathbf{G}, \pi)$ be a Poisson–Lie group. For a Lie subgroup $\mathbf{H}$ of $\mathbf{G}$, with Lie algebra $\mathfrak{h}$, the following conditions are equivalent:*

(i) $\mathbf{H}$ *is a Poisson submanifold of* $(\mathbf{G}, \pi)$*;*

(ii) *For every $h \in \mathbf{H}$, the bivector $\Psi(h)$ belongs to $\wedge^2\mathfrak{h}$.*

Under these conditions, the pair $(\mathbf{H}, \pi')$ is a Poisson–Lie group, where π' denotes the restriction of π to $\mathbf{H}$. One then says that $\mathbf{H}$ is a Poisson–Lie subgroup *of $\mathbf{G}$. When $\mathbf{H}$ is connected, the conditions (i) and (ii) are also equivalent to the following condition:*

(iii) *For every $x \in \mathfrak{h}$, the bivector $T_e\Psi(x)$ belongs to $\wedge^2\mathfrak{h}$.*

Proof. In view of Proposition 2.12, $\mathbf{H}$ is a Poisson submanifold of $\mathbf{G}$ if and only if $\pi_h \in \wedge^2 T_h\mathbf{H}$ for every $h \in \mathbf{H}$. Since $\wedge^2 T_h\mathbf{H} = \wedge^2(T_e R_h)\,(\wedge^2\mathfrak{h})$, the equivalence of *(i)* and *(ii)* follows.

Suppose that $\mathbf{H}$ is a Poisson submanifold of $\mathbf{G}$ and consider the product map on $\mathbf{H}$. It is the restriction of the product map on $\mathbf{G}$ to $\mathbf{H}$, hence it is a Poisson map. It follows that $\mathbf{H}$ itself is also a Poisson–Lie group.

Suppose now that $\mathbf{H}$ is connected. It is clear that *(ii)* implies *(iii)*. Let us show the inverse implication. Since $\Psi : \mathbf{G} \to \wedge^2\mathfrak{g}$ is a cocycle for the adjoint representation of $\mathbf{G}$ in $\wedge^2\mathfrak{g}$, its restriction to $\mathbf{H}$ is a cocycle for the adjoint representation of $\mathbf{H}$ in $\wedge^2\mathfrak{g}$, which admits $\wedge^2\mathfrak{h} \subset \wedge^2\mathfrak{g}$ as an $\mathbf{H}$-invariant subspace. By assumption, $T_e\Psi(x) \in \wedge^2\mathfrak{h}$ for all $x \in \mathfrak{h}$. According to Proposition 11.4, this implies that $\Psi(h) \in \wedge^2\mathfrak{h}$ for all $h \in \mathbf{H}$, which is condition *(ii)*. □

We show in the following proposition that the datum of a Poisson–Lie group and of a Poisson–Lie subgroup leads, under a topological assumption, to a Poisson structure on the (left) quotient space.

Proposition 11.8. *Let $(\mathbf{G},\pi)$ be a Poisson–Lie group and let $(\mathbf{H},\pi')$ be a Poisson–Lie subgroup of* **G**. *Suppose that* **H** *is a closed subgroup of* **G**, *so that the left coset space $\mathbf{H}\backslash\mathbf{G}$ is a smooth manifold. There exists a unique Poisson structure on $\mathbf{H}\backslash\mathbf{G}$, such that the canonical projection $p:\mathbf{G}\to\mathbf{H}\backslash\mathbf{G}$ is a Poisson map. Moreover, the right action of* **G** *on $\mathbf{H}\backslash\mathbf{G}$ is a (right) Poisson action.*

Proof. Consider the left action of **H** on **G** which is the restriction of the product map on **G**. Since **H** is a Poisson submanifold of **G**, the product $\mathbf{H}\times\mathbf{G}$ is a Poisson submanifold of $\mathbf{G}\times\mathbf{G}$, hence the left action of **H** on **G** is a Poisson action. Since the quotient space of this action is precisely $\mathbf{H}\backslash\mathbf{G}$, according to Proposition 5.33, the quotient space $\mathbf{H}\backslash\mathbf{G}$ inherits a unique Poisson structure from $(\mathbf{G},\pi)$ such that the canonical projection $p:\mathbf{G}\to\mathbf{H}\backslash\mathbf{G}$ is a Poisson map. Since the product map $\mu:\mathbf{G}\times\mathbf{G}\to\mathbf{G}$ is a Poisson map, the induced map $(\mathbf{H}\backslash\mathbf{G})\times\mathbf{G}\to(\mathbf{H}\backslash\mathbf{G})$ is also a Poisson map, hence the (right) action of **G** on $\mathbf{H}\backslash\mathbf{G}$ is a Poisson action. □

11.1.4 Linear Multiplicative Poisson Structures

Let V be a finite-dimensional vector space (over $\mathbb{R}$ or $\mathbb{C}$). Addition of vectors in V defines a group structure on V which makes V into an abelian Lie group. Using Proposition 11.2, we characterize multiplicative Poisson structures on V as linear Poisson structures on V.

Proposition 11.9. *Let V be a finite-dimensional vector space. A Poisson structure on the Lie group $(V,+)$ is multiplicative if and only if it is a linear Poisson structure.*

Proof. Recall that a Poisson structure $\pi=\{\cdot,\cdot\}$ on V is linear if and only if the Poisson bracket of every pair of linear functions on V is a linear function on V (see Section 7.1). By continuity, a real smooth (or complex holomorphic) function H on V, which satisfies
$$H(v+w)=H(v)+H(w)\;,$$
for all $v,w\in V$, is a linear function on V. Therefore, π is a linear Poisson structure on V, if and only if
$$\{F,G\}(v+w)=\{F,G\}(v)+\{F,G\}(w)\;, \tag{11.11}$$
for all $v,w\in V$ and for all linear functions $F,G\in V^*$. We have on the one hand that
$$\{F,G\}(v+w)=\langle \mathrm{d}_{v+w}F\wedge\mathrm{d}_{v+w}G,\pi_{v+w}\rangle \tag{11.12}$$
and on the other hand, since F and $F\circ R_w$ differ by a constant (namely $F(w)$), that
$$\begin{aligned}\{F,G\}(v)&=\langle \mathrm{d}_vF\wedge\mathrm{d}_vG,\pi_v\rangle\\&=\langle \mathrm{d}_v(F\circ R_w)\wedge\mathrm{d}_v(G\circ R_w),\pi_v\rangle\\&=\left\langle \mathrm{d}_{v+w}F\wedge\mathrm{d}_{v+w}G,\wedge^2(T_vR_w)\,\pi_v\right\rangle\;.\end{aligned} \tag{11.13}$$

Substituting (11.12) and (11.13) in (11.11) we conclude that π is a linear Poisson structure on V if and only if

$$\pi_{v+w} = \wedge^2(T_v R_w)\,\pi_v + \wedge^2(T_w L_v)\,\pi_w\,,$$

for all $v,w \in V$. Since the latter condition is precisely condition *(ii)* in Proposition 11.2, this shows that π is linear if and only if π is multiplicative. □

Since the dual of a finite-dimensional Lie algebra admits a canonical linear Poisson structure (the Lie–Poisson structure), it follows from the above proposition that the dual of a finite-dimensional Lie algebra is in a canonical way a Poisson–Lie group.

11.1.5 Coboundary Poisson–Lie Groups

An important class of Poisson–Lie groups are those constructed out of r-matrices. Recall from Proposition 10.11 that if $\mathfrak{g}$ is a Lie algebra and r is an r-matrix for $\mathfrak{g}$, then its skew-symmetric part $a := r^- \in \wedge^2\mathfrak{g}$ is such that $[\![a,a]\!]$ is ad-invariant. We show in the following proposition that this condition admits a natural connection with Poisson–Lie groups. Recall from Section 5.1 that if $\mathbf{G}$ is a Lie group with Lie algebra $\mathfrak{g}$, we denote for $x \in \mathfrak{g}$ by $\overrightarrow{x}$ (respectively $\overleftarrow{x}$) the right-invariant (respectively left-invariant) vector field on $\mathbf{G}$, whose value at e is x, and similarly for elements $X \in \wedge^\bullet\mathfrak{g}$.

Proposition 11.10. *Let* $\mathbf{G}$ *be a Lie group with Lie algebra* $(\mathfrak{g},[\cdot\,,\cdot])$ *and let* $a \in \wedge^2\mathfrak{g}$*. On* $\mathbf{G}$ *we consider the bivector field* π*, defined by* $\pi := \overleftarrow{a} - \overrightarrow{a}$*. The pair* $(\mathbf{G},\pi)$ *is a Poisson–Lie group if and only if* $[\![a,a]\!]$ *is an* Ad*-invariant element. In particular, if* $\mathbf{G}$ *is connected, then* $(\mathbf{G},\pi)$ *is a Poisson–Lie group if and only if* a *is an* r*-matrix.*

Proof. It follows immediately from Proposition 5.2 that the Schouten bracket of $\pi := \overleftarrow{a} - \overrightarrow{a}$ with itself is given by

$$\begin{aligned}
[\pi,\pi]_S &= [\overleftarrow{a} - \overrightarrow{a}, \overleftarrow{a} - \overrightarrow{a}]_S \\
&= [\overleftarrow{a}, \overleftarrow{a}]_S - [\overleftarrow{a}, \overrightarrow{a}]_S - [\overrightarrow{a}, \overleftarrow{a}]_S + [\overrightarrow{a}, \overrightarrow{a}]_S \\
&= [\overleftarrow{a}, \overleftarrow{a}]_S + [\overrightarrow{a}, \overrightarrow{a}]_S \\
&= \overleftarrow{[\![a,a]\!]} - \overrightarrow{[\![a,a]\!]}\,.
\end{aligned}$$

In view of item *(3)* in Proposition 5.3, we can conclude that π, as defined above, is a Poisson structure on $\mathbf{G}$ if and only if $[\![a,a]\!]$ is Ad-invariant. It remains to be shown that π is moreover multiplicative. The map $\Psi : \mathbf{G} \to \wedge^2\mathfrak{g}$ associated with π as in item *(iii)* in Proposition 11.2 (i.e. defined, for all $g \in \mathbf{G}$, by $\Psi(g) := \wedge^2(T_g R_{g^{-1}})\,\pi_g$) is given by

$$\Psi(g) = \mathrm{Ad}_g\, a - a\,, \tag{11.14}$$

as one computes using the following description of the bivector field $\overleftarrow{a} - \overrightarrow{a}$ at a point $g \in \mathbf{G}$:

$$\pi_g = \wedge^2(T_eL_g)a - \wedge^2(T_eR_g)a\,.$$

For every $g, h \in \mathbf{G}$, we can therefore check that

$$\Psi(gh) = \mathrm{Ad}_{gh}\, a - a = \mathrm{Ad}_{gh}\, a - \mathrm{Ad}_g\, a + \mathrm{Ad}_g\, a - a = \Psi(g) + \mathrm{Ad}_g\, \Psi(h)\,, \quad (11.15)$$

which, in view of the equivalence of items *(i)* and *(iii)* in Proposition 11.2, implies that π is multiplicative. Therefore, $(\mathbf{G}, \pi)$ is a Poisson–Lie group if and only if $[\![a,a]\!]$ is Ad-invariant. If $\mathbf{G}$ is connected, then this is, according to Proposition 5.3, equivalent to $[\![a,a]\!]$ being ad-invariant, i.e., to a being an r-matrix. □

The previous proposition leads to the following definition.

Definition 11.11. A Poisson–Lie group $(\mathbf{G}, \pi)$ is said to be a *coboundary Poisson–Lie group* if π is of the form $\pi = \overleftarrow{a} - \overrightarrow{a}$ for some element $a \in \wedge^2\mathfrak{g}$.

For a given coboundary Poisson–Lie group $(\mathbf{G}, \pi)$ there may exist two different elements $a_1, a_2 \in \wedge^2\mathfrak{g}$ such that $\pi = \overleftarrow{a_1} - \overrightarrow{a_1}$ and $\pi = \overleftarrow{a_2} - \overrightarrow{a_2}$. However, in such a case, we have

$$\overrightarrow{a_1 - a_2} = \overleftarrow{a_1 - a_2}$$

so that a_1 and a_2 differ by an Ad-invariant element in $\wedge^2\mathfrak{g}$. Therefore, for a given coboundary Poisson–Lie group, the element in $\wedge^2\mathfrak{g}$ which appears in Definition 11.11 is defined only up to addition of an Ad-invariant element in $\wedge^2\mathfrak{g}$.

Remark 11.12. The expression (11.14) of Ψ in terms of a, namely that $\Psi(g) = \mathrm{Ad}_g\, a - a$ for all $g \in \mathbf{G}$, has the following cohomological interpretation: it says that Ψ is a 1-coboundary (in fact, of a) for the chain complex which underlies the group cohomology of $\mathbf{G}$ with coefficients in $\wedge^2\mathfrak{g}$. It explains the terminology *coboundary Poisson–Lie group*, introduced above. Also, since every 1-coboundary is a cocycle, the cohomological interpretation of Ψ yields an alternative proof of (11.15), which shows that Ψ is a cocycle.

11.1.6 Multiplicative Poisson Structures on Vector Spaces

In this section, we explain how the notion of a multiplicative Poisson structure on a Lie group $\mathbf{G}$ is related to the notion of a multiplicative Poisson structure on a finite-dimensional vector space V, equipped with a product μ, a notion which was introduced in Section 8.2.1. For the present discussion, we will assume that the product μ is associative and admits a unit e; we will therefore refer to (V, μ) as an associative algebra with unit. An element $v \in V$ is said to be *invertible* if it admits a right inverse, i.e., there exists $w \in V$ such that $\mu(v, w) = e$.

Proposition 11.13. *Let (V, μ) be a finite-dimensional associative algebra with unit, and let V_{inv} denote the set of all invertible elements in V.*

(1) *V_{inv} is a (non-empty) Zariski open subset of V and the restriction of μ to V_{inv} defines the structure of a Lie group on V_{inv}, whose Lie algebra is isomorphic to V, with Lie bracket given by $[x,y] = \mu(x,y) - \mu(y,x)$ for all $x,y \in V$;*

(2) *If π is a multiplicative Poisson structure on (V,μ), in the sense of Section 8.2.1, then the restriction π' of π to V_{inv} is a multiplicative Poisson structure on the Lie group V_{inv}, making (V_{inv},π') into a Poisson–Lie group;*

(3) *Let $a \in \wedge^2 V$ be a skew-symmetric r-matrix for V. The quadratic bivector field π on V, whose value at $x \in V$ is given by*

$$\pi_x := 2\,[x \otimes x, a] \tag{11.16}$$

is a multiplicative Poisson structure on V. The Poisson–Lie group (V_{inv},π'), where π' denotes the restriction of π to V_{inv}, is the coboundary Poisson–Lie group $(V_{\text{inv}}, 2(\overleftarrow{a} - \overrightarrow{a}))$.

Proof. Since μ is assumed to be associative, the invertible elements are precisely the elements $v \in V$ for which the linear map $L_v : V \to V$, defined by $L_v(w) := \mu(v,w)$, is an isomorphism, i.e.,

$$V_{\text{inv}} = \{v \in V \mid \det L_v \neq 0\} \; .$$

Since μ is a bilinear map, $v \mapsto L_v$ is a linear map, and hence $v \mapsto \det L_v$ is a polynomial function on V (it is homogeneous of degree $\dim V$). It follows that V_{inv} is a Zariski open subset of V; it is non-empty since it contains the unit e of μ. Defining $R_v : V \to V$ by $R_v(w) := \mu(w,v)$, we have that R_v is an isomorphism if and only if L_v is an isomorphism. Indeed, if L_v is an isomorphism, so that there exists $w \in V$ with $\mu(v,w) = e$, then we have, in view of the associativity of μ,

$$\mu(\mu(w,v),w) = \mu(w,\mu(v,w)) = w \; ,$$

so that $\operatorname{Ker} R_v = \{0\}$, i.e., R_v is an isomorphism (recall that V is finite-dimensional). Therefore, the invertible elements of (V,μ) form a group, with e as unit. As an open subset of V, the group (V_{inv},μ) inherits a manifold structure from V; since μ is a bilinear map, it restricts to a smooth map μ' on V_{inv}, hence (V_{inv},μ') is a Lie group. For $x,y \in V \simeq T_eV$, we need to show that their Lie bracket $[x,y]$ is given by $\mu(x,y) - \mu(y,x)$. Under the canonical isomorphism $V \simeq T_eV$ we have, by definition, $[x,y] = [\overleftarrow{x}, \overleftarrow{y}]_e$ and $(\overleftarrow{\mu(x,y) - \mu(y,x)})_e = \mu(x,y) - \mu(y,x)$. It therefore suffices to show that

$$[\overleftarrow{x}, \overleftarrow{y}] = \overleftarrow{\mu(x,y) - \mu(y,x)} \; . \tag{11.17}$$

Let $F : V \to \mathbb{F}$ be a linear function. For $v \in V$ we have that

$$\overleftarrow{x}[F](v) = x_e[F \circ L_v] = \lim_{t \to 0} \frac{F(v + t\mu(v,x)) - F(v)}{t} = F(\mu(v,x)),$$

so that $\overleftarrow{x}[F] = F \circ R_x$. It follows that

$$
\begin{aligned}
[\overleftarrow{x}, \overleftarrow{y}][F] &= \overleftarrow{x}[\overleftarrow{y}[F]] - \overleftarrow{y}[\overleftarrow{x}[F]] \\
&= F \circ (R_y \circ R_x - R_x \circ R_y) \\
&= F \circ R_{\mu(x,y)-\mu(y,x)} \\
&= \overleftarrow{(\mu(x,y) - \mu(y,x))}[F],
\end{aligned}
$$

for all linear functions F on V, leading to (11.17). This shows *(1)*.

Let π be a multiplicative Poisson structure on (V,μ). Recall from Section 8.2.1 that this means that the product map μ is a Poisson map. Since V_{inv} is an open subset of V, the restriction π' of π to V_{inv} is a Poisson structure and the restriction μ' of μ to V_{inv}, i.e., the product map of the Lie group (V_{inv},μ'), is a Poisson map. This shows that (V_{inv},π') is a Poisson–Lie group, which is the content of *(2)*.

Let a be a skew-symmetric r-matrix for V, which is the Lie algebra of the connected Lie group (V_{inv},μ). According to Proposition 11.10, $(V_{\text{inv}}, \overleftarrow{a} - \overrightarrow{a})$ is a coboundary Poisson–Lie group. We show that the quadratic bivector field[2] π on V, defined by (11.16), takes at every point of V_{inv} the same value as the bivector field $2(\overleftarrow{a} - \overrightarrow{a})$. To do this, we identify all tangent spaces to V with V, so that the derivative of left translation by $x \in V_{\text{inv}}$ is the linear map $T_eL_x : V \to V$, given for $z \in V$ by $T_eL_x(z) = xz$, since L_x is a linear map. Thus, the left-invariant vector field $\overleftarrow{z}$ of $z \in V$ is given at $x \in V_{\text{inv}}$, by $\overleftarrow{z}_x = xz$. Similarly $\overrightarrow{z}_x = zx$. It follows that, for a bivector $z_1 \wedge z_2 \in \wedge^2 V$,

$$
\left(\overleftarrow{z_1 \wedge z_2}\right)_x = xz_1 \wedge xz_2\,, \qquad \left(\overrightarrow{z_1 \wedge z_2}\right)_x = z_1x \wedge z_2x\,,
$$

for all $x \in V_{\text{inv}}$, in particular

$$
\overleftarrow{a}_x - \overrightarrow{a}_x = (x \otimes x)a - a(x \otimes x) = [x \otimes x, a]\,. \tag{11.18}
$$

This shows our claim. Since π and $2(\overleftarrow{a} - \overrightarrow{a})$ have the same value at every point of V_{inv}, the restriction of π to V_{inv} is multiplicative, hence π itself is multiplicative, showing *(3)*. □

Example 11.14. We show that the bivector field on $V = \text{Mat}_d(\mathbb{F})$, which we discussed in Example 8.22 is associated with an r-matrix, which proves that it is a multiplicative Poisson structure. Let V be equipped with the non-degenerate ad-invariant bilinear form $\langle x \,|\, y\rangle = \text{Trace}(xy)$. We denote the elementary matrices of V by E_{ij}, and we introduce linear coordinates ξ_{ij} on V by defining, for a matrix $x \in \text{Mat}_d(\mathbb{F})$, $\xi_{ij}(x) := \langle E_{ji} \,|\, x\rangle$, so that $\xi_{ij}(x) = x_{ij}$, when x is written as $(x_{ij})_{1 \leqslant i,j \leqslant d}$. An arbitrary matrix can be decomposed as the sum of a strictly upper triangular, a diagonal, and a strictly lower triangular matrix: this decomposition satisfies the requirements of Example 10.15 so that

$$
r := \frac{1}{2} \sum_{1 \leqslant a < b \leqslant d} E_{ab} \wedge E_{ba} \tag{11.19}
$$

---

[2] We know from Section 10.3, in particular from the second Russian formula (10.40), that π is a Poisson structure, but this fact is not used in the proof.

is an r-matrix on the Lie algebra $\mathrm{Mat}_d(\mathbb{F})$. With the help of this r-matrix, a Poisson structure π on $\mathrm{Mat}_d(\mathbb{F})$ can be constructed as in Proposition 11.13. Let us express the coefficients of the Poisson matrix of π with respect to the linear coordinates ξ_{ij}. For all $i, j, k, \ell = 1, \ldots, d$, and all $x \in \mathrm{Mat}_d(\mathbb{F})$, we find using the explicit formula (11.19) for r and (11.18) for the corresponding Poisson structure,

$$\begin{aligned}\{\xi_{ij}, \xi_{k\ell}\}_r(x) &= 2\langle \xi_{ij} \wedge \xi_{kl}, [x \otimes x, r]\rangle \\ &= \sum_{a<b} \langle \xi_{ij} \wedge \xi_{k\ell}, [x \otimes x, E_{ab} \wedge E_{ba}]\rangle \\ &= \sum_{a<b} \langle \xi_{ij} \wedge \xi_{k\ell}, xE_{ab} \wedge xE_{ba} - E_{ab}x \wedge E_{ba}x\rangle \\ &= \sum_{a<b} (\delta_{j,b}\delta_{\ell,a} - \delta_{j,a}\delta_{\ell,b} - \delta_{i,a}\delta_{k,b} + \delta_{i,b}\delta_{k,a})\, \xi_{i\ell}(x)\xi_{kj}(x) \\ &= (\varepsilon_{j,\ell} + \varepsilon_{i,k})\, \xi_{i\ell}(x)\xi_{kj}(x),\end{aligned}$$

where $\varepsilon_{i,j}$ is 1 if $i > j$, is -1 if $i < j$ and is 0 if $i = j$. It is the quadratic Poisson structure which we met in Example 8.22.

11.2 Lie Bialgebras

In this section, we consider Lie algebra structures on a finite-dimensional vector space $\mathfrak{g}$, on its dual vector space $\mathfrak{g}^*$ and on their product $\mathfrak{d} := \mathfrak{g} \times \mathfrak{g}^*$. We will often view $\mathfrak{g}$ and $\mathfrak{g}^*$ as subspaces of $\mathfrak{d}$, via the natural identifications $\mathfrak{g} \simeq \mathfrak{g} \times \{0\}$ and $\mathfrak{g}^* \simeq \{0\} \times \mathfrak{g}^*$. By a slight abuse of notation, $(x, 0)$ and $(0, \xi)$ are often abbreviated to x and ξ. We will make a notational distinction between the three Lie brackets which will be considered: we will write $[\cdot,\cdot]_{\mathfrak{g}}$, $[\cdot,\cdot]_{\mathfrak{g}^*}$, and $[\cdot,\cdot]_{\mathfrak{d}}$ for the Lie bracket on $\mathfrak{g}$, on $\mathfrak{g}^*$ and on $\mathfrak{d}$ respectively. We will however use the same notation ad^* for the coadjoint actions of $\mathfrak{g}$ on $\mathfrak{g}^*$ and of $\mathfrak{g}^*$ on $\mathfrak{g}$. In view of the definition of the coadjoint action (see Section 5.1.3),

$$\left\langle \xi, [x,y]_{\mathfrak{g}} \right\rangle = -\langle \mathrm{ad}^*_x \xi, y\rangle \quad \text{and} \quad \left\langle [\xi,\eta]_{\mathfrak{g}^*}, x \right\rangle = -\left\langle \eta, \mathrm{ad}^*_\xi x \right\rangle, \tag{11.20}$$

for all $x, y \in \mathfrak{g}$ and for all $\xi, \eta \in \mathfrak{g}^*$.

We will find it often convenient to view the Lie algebra structure $[\cdot,\cdot]_{\mathfrak{g}^*}$ on $\mathfrak{g}^*$ as a structure on $\mathfrak{g}$. This is done by taking the transpose of $[\cdot,\cdot]_{\mathfrak{g}^*} : \mathfrak{g}^* \wedge \mathfrak{g}^* \to \mathfrak{g}^*$, which can be considered as a linear map $\delta : \mathfrak{g} \to \mathfrak{g} \wedge \mathfrak{g}$. Precisely, δ is defined by

$$\langle \xi \wedge \eta, \delta(x)\rangle = \left\langle [\xi,\eta]_{\mathfrak{g}^*}, x \right\rangle, \tag{11.21}$$

for all $x \in \mathfrak{g}$, and for all $\xi, \eta \in \mathfrak{g}^*$. We will use the natural bilinear form $\langle \cdot \,|\, \cdot \rangle_{\mathfrak{d}}$ on $\mathfrak{d}$, defined for all $x, y \in \mathfrak{g}$ and for all $\xi, \eta \in \mathfrak{g}^*$ by

$$\langle (x,\xi) \mid (y,\eta) \rangle_{\mathfrak{d}} := \langle \xi, y \rangle + \langle \eta, x \rangle \ . \tag{11.22}$$

It is clear that $\langle \cdot \mid \cdot \rangle_{\mathfrak{d}}$ is symmetric and non-degenerate.

11.2.1 Lie Bialgebras

The following proposition will justify the definition of a Lie bialgebra.

Proposition 11.15. *Let $\mathfrak{g}$ be a finite-dimensional vector space. Suppose that $[\cdot,\cdot]_{\mathfrak{g}}$ and $[\cdot,\cdot]_{\mathfrak{g}^*}$ are Lie brackets on $\mathfrak{g}$ and $\mathfrak{g}^*$ respectively. On $\mathfrak{d} := \mathfrak{g} \times \mathfrak{g}^*$, consider the non-degenerate symmetric bilinear form $\langle \cdot \mid \cdot \rangle_{\mathfrak{d}}$, given by (11.22). Then the following conditions are equivalent:*

(i) There exists a Lie bracket on $\mathfrak{d}$, with respect to which $\langle \cdot \mid \cdot \rangle_{\mathfrak{d}}$ is ad*-invariant, such that the natural inclusions $\mathfrak{g} \hookrightarrow \mathfrak{d}$ and $\mathfrak{g}^* \hookrightarrow \mathfrak{d}$ are Lie algebra homomorphisms;*

(ii) The skew-symmetric bilinear map $[\cdot,\cdot]_{\mathfrak{d}} : \mathfrak{d} \times \mathfrak{d} \to \mathfrak{d}$ given, for all $x,y \in \mathfrak{g}$ and for all $\xi,\eta \in \mathfrak{g}^*$, by

$$[(x,\xi),(y,\eta)]_{\mathfrak{d}} := \left([x,y]_{\mathfrak{g}} + \mathrm{ad}^*_{\xi} y - \mathrm{ad}^*_{\eta} x, [\xi,\eta]_{\mathfrak{g}^*} + \mathrm{ad}^*_x \eta - \mathrm{ad}^*_y \xi\right), \tag{11.23}$$

satisfies the Jacobi identity;

(iii) The transpose $\delta : \mathfrak{g} \to \wedge^2 \mathfrak{g}$ of the Lie bracket $[\cdot,\cdot]_{\mathfrak{g}^*} : \wedge^2 \mathfrak{g}^* \to \mathfrak{g}^*$, defined by (11.21), satisfies for $x,y \in \mathfrak{g}$ the derivation property

$$\delta\left([x,y]_{\mathfrak{g}}\right) = [\![\delta(x), y]\!] + [\![x, \delta(y)]\!] \ . \tag{11.24}$$

If any of these three equivalent conditions is satisfied, the Lie bracket on $\mathfrak{d}$, which satisfies the conditions demanded in (i), is unique and is given by (11.23); moreover, it makes $(\mathfrak{d}, \langle \cdot \mid \cdot \rangle_{\mathfrak{d}})$ into a quadratic Lie algebra.

The proof of this proposition will make use of the following lemma.

Lemma 11.16. *Let $\mathfrak{g}$ be a finite-dimensional vector space, equipped with Lie brackets $[\cdot,\cdot]_{\mathfrak{g}}$ and $[\cdot,\cdot]_{\mathfrak{g}^*}$ on $\mathfrak{g}$, respectively on $\mathfrak{g}^*$. The skew-symmetric bilinear map $[\cdot,\cdot]_{\mathfrak{d}} : \mathfrak{d} \times \mathfrak{d} \to \mathfrak{d}$, defined in (11.23), has the following properties:*

(1) For all $d_1, d_2, d_3 \in \mathfrak{d}$:

$$\langle d_1 \mid [d_2, d_3]_{\mathfrak{d}} \rangle_{\mathfrak{d}} = \langle [d_1, d_2]_{\mathfrak{d}} \mid d_3 \rangle_{\mathfrak{d}} \ ; \tag{11.25}$$

(2) $[\cdot,\cdot]_{\mathfrak{d}}$ satisfies the Jacobi identity if and only if for all $x,y \in \mathfrak{g} \hookrightarrow \mathfrak{d}$, and for all $\xi,\eta \in \mathfrak{g}^* \hookrightarrow \mathfrak{d}$:

$$\langle [x,y]_{\mathfrak{d}} \mid [\xi,\eta]_{\mathfrak{d}} \rangle_{\mathfrak{d}} + \langle [y,\xi]_{\mathfrak{d}} \mid [x,\eta]_{\mathfrak{d}} \rangle_{\mathfrak{d}} + \langle [\xi,x]_{\mathfrak{d}} \mid [y,\eta]_{\mathfrak{d}} \rangle_{\mathfrak{d}} = 0 \ . \tag{11.26}$$

Proof. Let $d_i = (x_i, \xi_i)$, for $i = 1,2,3$. In view of the definition of $[\cdot,\cdot]_{\mathfrak{d}}$, of $\langle \cdot \mid \cdot \rangle_{\mathfrak{d}}$ and of the coadjoint actions of $\mathfrak{g}$ on $\mathfrak{g}^*$ and of $\mathfrak{g}^*$ on $\mathfrak{g}$,

$$\langle d_1 \,|\, [d_2,d_3]_\mathfrak{d} \rangle_\mathfrak{d} = \left\langle \xi_1, [x_2,x_3]_\mathfrak{g} \right\rangle + \left\langle [\xi_1,\xi_2]_{\mathfrak{g}^*}, x_3 \right\rangle + \circlearrowleft (1,2,3)\,.$$

Since the latter expression is invariant with respect to a cyclic permutation of the indices $1,2,3$, it follows that

$$\langle d_1 \,|\, [d_2,d_3]_\mathfrak{d} \rangle_\mathfrak{d} = \langle d_3 \,|\, [d_1,d_2]_\mathfrak{d} \rangle_\mathfrak{d} = \langle [d_1,d_2]_\mathfrak{d} \,|\, d_3 \rangle_\mathfrak{d}\,,$$

which proves *(1)*. Let φ denote the 4-linear map from $\mathfrak{d}$ to itself, defined for all $d_1,\dots,d_4 \in \mathfrak{d}$ by

$$\varphi(d_1,d_2,d_3,d_4) := \langle \mathscr{J}_\mathfrak{d}(d_1,d_2,d_3) \,|\, d_4 \rangle_\mathfrak{d}\,,$$

where $\mathscr{J}_\mathfrak{d}$ is the Jacobiator of $[\cdot\,,\cdot]_\mathfrak{d}$, i.e., $\mathscr{J}_\mathfrak{d}$ is given, for all $d_1,d_2,d_3 \in \mathfrak{d}$, by

$$\mathscr{J}_\mathfrak{d}(d_1,d_2,d_3) := [[d_1,d_2]_\mathfrak{d}, d_3]_\mathfrak{d} + [[d_2,d_3]_\mathfrak{d}, d_1]_\mathfrak{d} + [[d_3,d_1]_\mathfrak{d}, d_2]_\mathfrak{d}\,. \tag{11.27}$$

Since $\mathscr{J}_\mathfrak{d} = 0$ if and only if $[\cdot\,,\cdot]_\mathfrak{d}$ satisfies the Jacobi identity, and since $\langle\cdot\,|\,\cdot\rangle_\mathfrak{d}$ is non-degenerate, $\varphi = 0$ if and only if $[\cdot\,,\cdot]_\mathfrak{d}$ satisfies the Jacobi identity.

In order to prove *(2)*, we therefore only need to show that (11.26) holds for all $x,y \in \mathfrak{g}$ and all $\xi,\eta \in \mathfrak{g}^*$, if and only if $\varphi = 0$. Since $\mathscr{J}_\mathfrak{d}(d_1,d_2,d_3) = 0$ when d_1,d_2,d_3 all belong to $\mathfrak{g}$ or all belong to $\mathfrak{g}^*$, we have that $\varphi(d_1,d_2,d_3,d_4) = 0$ when d_1,d_2,d_3 all belong to $\mathfrak{g}$ or all belong to $\mathfrak{g}^*$, whatever the value of $d_4 \in \mathfrak{d}$. The skew-symmetry of $\mathscr{J}_\mathfrak{d}$, combined with (11.25), implies that φ is skew-symmetric. Therefore, $\varphi(d_1,d_2,d_3,d_4) = 0$ as soon as at least three of the d_i all belong to $\mathfrak{g}$, or all belong to $\mathfrak{g}^*$. It follows, since φ is 4-linear, that $\varphi = 0$ if and only if $\varphi(d_1,d_2,d_3,d_4) = 0$, whenever two of the d_i belong to $\mathfrak{g}$ and the two other belong to $\mathfrak{g}^*$; but that is, in view of (11.25), precisely the condition that (11.26) holds for all $x,y \in \mathfrak{g}$ and all $\xi,\eta \in \mathfrak{g}^*$. This establishes the proof of *(2)*. □

We are now ready to give the proof of Proposition 11.15.

Proof. We first prove that *(i)* and *(ii)* are equivalent. Suppose first that *(ii)* is satisfied. Then the Lie bracket, defined by the explicit formula (11.23) has in view of *(1)* in Lemma 11.16 the properties announced in *(i)*. Suppose next that there exists a Lie bracket $[\cdot\,,\cdot]$ on $\mathfrak{d}$, with the properties announced in item *(i)*. We show that $[\cdot\,,\cdot] = [\cdot\,,\cdot]_\mathfrak{d}$, i.e., $[\cdot\,,\cdot]$ is given by the explicit formula (11.23). In view of *(i)*, $[\cdot\,,\cdot]$ and $[\cdot\,,\cdot]_\mathfrak{d}$ agree when both arguments are taken either from $\mathfrak{g}$ or from $\mathfrak{g}^*$, so we only need to show that

$$[x,\eta] = (-\operatorname{ad}^*_\eta x, \operatorname{ad}^*_x \eta)\,, \tag{11.28}$$

for all $x \in \mathfrak{g}$ and all $\eta \in \mathfrak{g}^*$. For $(z,\gamma) \in \mathfrak{g} \times \mathfrak{g}^*$, we have in view of (11.25), combined with the properties of $[\cdot\,,\cdot]$, assumed in *(i)*, that

$$\begin{aligned}
\langle [x,\eta] \,|\, (z,\gamma) \rangle_\mathfrak{d} &= \langle [(x,0),(0,\eta)] \,|\, (z,0) + (0,\gamma) \rangle_\mathfrak{d} \\
&= \left\langle \left([z,x]_\mathfrak{g}, 0 \right) \,|\, (0,\eta) \right\rangle_\mathfrak{d} + \left\langle \left(0, [\eta,\gamma]_{\mathfrak{g}^*} \right) \,|\, (x,0) \right\rangle_\mathfrak{d} \\
&= \left\langle \eta, [z,x]_\mathfrak{g} \right\rangle + \left\langle [\eta,\gamma]_{\mathfrak{g}^*}, x \right\rangle
\end{aligned}$$

$$= \left\langle (-\mathrm{ad}^*_\eta x, \mathrm{ad}^*_x \eta) | (z,\gamma) \right\rangle_{\mathfrak{d}} .$$

Since $\langle \cdot | \cdot \rangle_{\mathfrak{d}}$ is non-degenerate, this shows (11.28), hence that if there exists a Lie bracket on $\mathfrak{d}$, satisfying *(i)*, then it is given by (11.23). In particular, it shows that *(i)* and *(ii)* are equivalent.

We now prove the equivalence of items *(ii)* and *(iii)*. In view of item *(2)* of Lemma 11.16, we need to show that (11.26) holds for all $x, y \in \mathfrak{g}$, and for all $\xi, \eta \in \mathfrak{g}^*$, if and only if (11.24) holds for all $x, y \in \mathfrak{g}$. We claim that, to do this, it suffices to show that the following formulas hold, for all $x, y \in \mathfrak{g}$, and for all $\xi, \eta \in \mathfrak{g}^*$:

$$\langle \xi \wedge \eta, [\![\delta(x), y]\!] \rangle = \langle \mathrm{ad}^*_y \xi, \mathrm{ad}^*_\eta x \rangle - \left\langle \mathrm{ad}^*_y \eta, \mathrm{ad}^*_\xi x \right\rangle . \qquad (11.29)$$

Indeed, using (11.21), (11.29) and the definitions of $[\cdot,\cdot]_{\mathfrak{d}}$ and $\langle \cdot | \cdot \rangle_{\mathfrak{d}}$, we find that

$$\begin{aligned}
&\left\langle \xi \wedge \eta, \delta([x,y]_{\mathfrak{g}}) - [\![\delta(x), y]\!] - [\![x, \delta(y)]\!] \right\rangle \\
&= \left\langle [\xi,\eta]_{\mathfrak{g}^*}, [x,y]_{\mathfrak{g}} \right\rangle - \langle \mathrm{ad}^*_y \xi, \mathrm{ad}^*_\eta x \rangle + \left\langle \mathrm{ad}^*_y \eta, \mathrm{ad}^*_\xi x \right\rangle \\
&\qquad + \langle \mathrm{ad}^*_x \xi, \mathrm{ad}^*_\eta y \rangle - \left\langle \mathrm{ad}^*_x \eta, \mathrm{ad}^*_\xi y \right\rangle \\
&= \langle [x,y]_{\mathfrak{d}} \mid [\xi,\eta]_{\mathfrak{d}} \rangle_{\mathfrak{d}} + \langle [y,\xi]_{\mathfrak{d}} \mid [x,\eta]_{\mathfrak{d}} \rangle_{\mathfrak{d}} + \langle [\xi,x]_{\mathfrak{d}} \mid [y,\eta]_{\mathfrak{d}} \rangle_{\mathfrak{d}} ,
\end{aligned}$$

for all $x, y \in \mathfrak{g}$, and for all $\xi, \eta \in \mathfrak{g}^*$. In order to show (11.29), use (5.13) and the derivation property of the coadjoint action of $\mathfrak{g}$ on $\wedge^2 \mathfrak{g}^*$:

$$\begin{aligned}
\langle \xi \wedge \eta, [\![\delta(x), y]\!] \rangle &= -\langle \xi \wedge \eta, \mathrm{ad}_y \delta(x) \rangle = \langle \mathrm{ad}^*_y(\xi \wedge \eta), \delta(x) \rangle \\
&= \langle \mathrm{ad}^*_y \xi \wedge \eta + \xi \wedge \mathrm{ad}^*_y \eta, \delta(x) \rangle \\
&= \left\langle \left[\mathrm{ad}^*_y \xi, \eta\right]_{\mathfrak{g}^*} + \left[\xi, \mathrm{ad}^*_y \eta\right]_{\mathfrak{g}^*}, x \right\rangle \\
&= \langle \mathrm{ad}^*_y \xi, \mathrm{ad}^*_\eta x \rangle - \left\langle \mathrm{ad}^*_y \eta, \mathrm{ad}^*_\xi x \right\rangle .
\end{aligned}$$

This proves (11.29) and hence also the equivalence of *(ii)* and *(iii)*. □

Proposition 11.15 motivates the following definition.

Definition 11.17. Let $\mathfrak{g}$ be a finite-dimensional vector space, let $[\cdot,\cdot]_{\mathfrak{g}}$ be a Lie bracket on $\mathfrak{g}$ and let $[\cdot,\cdot]_{\mathfrak{g}^*}$ be a Lie bracket on $\mathfrak{g}^*$. The triple $(\mathfrak{g}, [\cdot,\cdot]_{\mathfrak{g}}, [\cdot,\cdot]_{\mathfrak{g}^*})$ is said to be a *Lie bialgebra* if it satisfies any one of the equivalent conditions given in Proposition 11.15. In this case, the Lie algebra $(\mathfrak{d}, [\cdot,\cdot]_{\mathfrak{d}})$, where $[\cdot,\cdot]_{\mathfrak{d}}$ is the Lie bracket on $\mathfrak{d} := \mathfrak{g} \times \mathfrak{g}^*$, given in (11.23), is called the *double* of the Lie bialgebra $(\mathfrak{g}, [\cdot,\cdot]_{\mathfrak{g}}, [\cdot,\cdot]_{\mathfrak{g}^*})$.

Suppose that $(\mathfrak{g}, [\cdot,\cdot]_{\mathfrak{g}}, [\cdot,\cdot]_{\mathfrak{g}^*})$ and $(\mathfrak{h}, [\cdot,\cdot]_{\mathfrak{h}}, [\cdot,\cdot]_{\mathfrak{h}^*})$ are two (finite-dimensional) Lie bialgebras. A linear map $\phi : \mathfrak{g} \to \mathfrak{h}$ is called a *homomorphism of Lie bialgebras* if both $\phi : (\mathfrak{g}, [\cdot,\cdot]_{\mathfrak{g}}) \to (\mathfrak{h}, [\cdot,\cdot]_{\mathfrak{h}})$ and its transpose $\phi^\top : (\mathfrak{h}^*, [\cdot,\cdot]_{\mathfrak{h}^*}) \to (\mathfrak{g}^*, [\cdot,\cdot]_{\mathfrak{g}^*})$ are Lie algebra homomorphisms.

Example 11.18. Let $(\mathfrak{g},[\cdot,\cdot]_{\mathfrak{g}})$ be a finite-dimensional Lie algebra. When $\mathfrak{g}^*$ is equipped with the trivial Lie bracket ($[\cdot,\cdot]_{\mathfrak{g}^*} := 0$), then the triple $(\mathfrak{g},[\cdot,\cdot]_{\mathfrak{g}},[\cdot,\cdot]_{\mathfrak{g}^*})$ is a Lie biagebra, since item *(iii)* in Proposition 11.15 is trivially satisfied when $\delta = 0$. Dually, let V be any finite-dimensional vector space, whose dual vector space V^* is equipped with a Lie bracket $[\cdot,\cdot]_{V^*}$. Making V in a trivial way into a Lie algebra by putting $[\cdot,\cdot]_V := 0$, the triple $(V,[\cdot,\cdot]_V,[\cdot,\cdot]_{V^*})$ is a Lie bialgebra.

Example 11.19. Let $\mathfrak{g}$ be a vector space of dimension two and let $[\cdot,\cdot]_{\mathfrak{g}}$ and $[\cdot,\cdot]_{\mathfrak{g}^*}$ be Lie brackets on $\mathfrak{g}$ and on $\mathfrak{g}^*$ respectively. The triple $(\mathfrak{g},[\cdot,\cdot]_{\mathfrak{g}},[\cdot,\cdot]_{\mathfrak{g}^*})$ is a Lie bialgebra. In view of Example 11.18, this is clear when at least one of the Lie brackets $[\cdot,\cdot]_{\mathfrak{g}}$ or $[\cdot,\cdot]_{\mathfrak{g}^*}$ is the trivial bracket. Suppose therefore that none of these brackets is trivial and denote by V, respectively W the image of the maps $[\cdot,\cdot]_{\mathfrak{g}} : \wedge^2\mathfrak{g} \to \mathfrak{g}$ and $[\cdot,\cdot]_{\mathfrak{g}^*} : \wedge^2\mathfrak{g}^* \to \mathfrak{g}^*$. Both V and W are one-dimensional because $\mathfrak{g}$ is two-dimensional. If W is the annihilator of V, i.e., $\langle W,V\rangle = 0$, there exists a basis (x,y) of $\mathfrak{g}$ in which

$$[x,y]_{\mathfrak{g}} = x \quad \text{and} \quad [x^*,y^*]_{\mathfrak{g}^*} = y^* ,$$

where (x^*,y^*) is the basis of $\mathfrak{g}^*$ which is dual to (x,y). Then $\delta(x) = 0$ and $\delta(y) = x\wedge y$ and one sees that condition *(iii)* in Proposition 11.15 is satisfied. Otherwise, there exists a basis (x,y) of $\mathfrak{g}$ in which

$$[x,y]_{\mathfrak{g}} = x \quad \text{and} \quad [x^*,y^*]_{\mathfrak{g}^*} = \lambda x^* , \tag{11.30}$$

for some $\lambda \in \mathbb{F}^*$. In this case $\delta(x) = \lambda x \wedge y$ and $\delta(y) = 0$, so that we arrive at the same conclusion.

Remark 11.20. For every $\lambda,\mu \in \mathbb{F}$ and every Lie bialgebra $(\mathfrak{g},[\cdot,\cdot]_{\mathfrak{g}},[\cdot,\cdot]_{\mathfrak{g}^*})$, the triple $(\mathfrak{g},\lambda[\cdot,\cdot]_{\mathfrak{g}},\mu[\cdot,\cdot]_{\mathfrak{g}^*})$ is a Lie bialgebra. It is clear that it is isomorphic to the original one if $\lambda = \mu \in \mathbb{F}^*$, under the isomorphism $\lambda\mathbb{1}_{\mathfrak{g}} : \mathfrak{g} \to \mathfrak{g}$. However, in general, it is not the case when $\lambda \neq \mu$, even if both λ and μ are different from 0. For instance, for two different values of the parameter λ, the Lie bialgebra structures described by (11.30) are not isomorphic.

11.2.2 Lie Sub-bialgebras

The notion of a homomorphism of Lie bialgebras, given in Definition 11.17, leads naturally to the notion of a Lie sub-bialgebra.

Definition 11.21. Let $(\mathfrak{g},[\cdot,\cdot]_{\mathfrak{g}},[\cdot,\cdot]_{\mathfrak{g}^*})$ be a finite-dimensional Lie bialgebra. A subspace $\mathfrak{h}$ of $\mathfrak{g}$ is called a *Lie sub-bialgebra* if $\mathfrak{h}$ admits the structure of a Lie bialgebra, such that the inclusion $\mathfrak{h} \hookrightarrow \mathfrak{g}$ is a homomorphism of Lie bialgebras.

It is clear that the Lie bialgebra structure on a Lie subalgebra is completely determined by the Lie bialgebra structure on the ambient Lie bialgebra. The fact that the transpose of the inclusion map $\mathfrak{h} \hookrightarrow \mathfrak{g}$ needs to be a Lie algebra homomorphism, admits a natural reformulation, which leads to the following proposition.

Proposition 11.22. *Let $(\mathfrak{g}, [\cdot,\cdot]_{\mathfrak{g}}, [\cdot,\cdot]_{\mathfrak{g}^*})$ be a finite-dimensional Lie bialgebra. A subspace $\mathfrak{h}$ of $\mathfrak{g}$ is a Lie sub-bialgebra of $(\mathfrak{g}, [\cdot,\cdot]_{\mathfrak{g}}, [\cdot,\cdot]_{\mathfrak{g}^*})$ if and only if the following two conditions are satisfied:*

(1) The vector space $\mathfrak{h}$ is a Lie subalgebra of $(\mathfrak{g}, [\cdot,\cdot]_{\mathfrak{g}})$;
(2) The annihilator $\mathfrak{h}^\perp$ of $\mathfrak{h}$ is a Lie ideal of $(\mathfrak{g}^*, [\cdot,\cdot]_{\mathfrak{g}^*})$.

Proof. For $\mathfrak{h}$ a subspace of $\mathfrak{g}$, let $\iota : \mathfrak{h} \hookrightarrow \mathfrak{g}$ denote the inclusion map, with transpose $\iota^\top : \mathfrak{g}^* \to \mathfrak{h}^*$. If a Lie sub-bialgebra structure on $\mathfrak{h}$ exists, making the inclusion map into a homomorphism of Lie bialgebras, then the Lie bracket on $\mathfrak{h}$ is the restriction of the Lie bracket on $\mathfrak{g}$ (and condition *(1)* is satisfied); also, $\iota^\top : \mathfrak{g}^* \to \mathfrak{h}^*$ is a Lie algebra homomorphism, so that its kernel $\mathfrak{h}^\perp$ is a Lie ideal of $\mathfrak{g}^*$ (and condition *(2)* is satisfied).

Conversely, if the two conditions are satisfied, the Lie brackets on $\mathfrak{g}$ and on $\mathfrak{g}^*$ induce Lie brackets on $\mathfrak{h}$ (by restriction) and on $\mathfrak{h}^*$ (by taking the quotient with respect to $\mathfrak{h}^\perp$), hence equip $\mathfrak{h}$ and $\mathfrak{h}^*$ with two Lie brackets $[\cdot,\cdot]_{\mathfrak{h}}$ and $[\cdot,\cdot]_{\mathfrak{h}^*}$. Since the transpose δ' of $[\cdot,\cdot]_{\mathfrak{h}^*}$ is the restriction of δ to $\mathfrak{h}$, item *(iii)* in Proposition 11.15 implies that these brackets define a Lie bialgebra structure on $\mathfrak{h}$. By construction, ι is a homomorphism of Lie bialgebras. □

11.2.3 Duality for Lie Bialgebras

We now come to the notion of duality for Lie bialgebras.

Proposition 11.23. *Let $(\mathfrak{g}, [\cdot,\cdot]_{\mathfrak{g}}, [\cdot,\cdot]_{\mathfrak{g}^*})$ be a Lie bialgebra, where $\mathfrak{g}$ is a finite-dimensional vector space. The triple $(\mathfrak{g}^*, [\cdot,\cdot]_{\mathfrak{g}^*}, [\cdot,\cdot]_{\mathfrak{g}})$ is a Lie bialgebra, called the* dual *of the Lie bialgebra $(\mathfrak{g}, [\cdot,\cdot]_{\mathfrak{g}}, [\cdot,\cdot]_{\mathfrak{g}^*})$. The natural isomorphism $S : \mathfrak{g} \times \mathfrak{g}^* \to \mathfrak{g}^* \times \mathfrak{g}$, given for all $(x,\xi) \in \mathfrak{g} \times \mathfrak{g}^*$ by $S(x,\xi) := (\xi, x)$ is a Lie algebra isomorphism of the doubles of these two Lie bialgebras.*

Proof. In item *(i)* in Proposition 11.15, the Lie algebras $(\mathfrak{g}, [\cdot,\cdot]_{\mathfrak{g}})$ and $(\mathfrak{g}^*, [\cdot,\cdot]_{\mathfrak{g}^*})$ play a symmetric rôle, hence the first part of the proposition is clear. The isomorphism $S : \mathfrak{g} \times \mathfrak{g}^* \to \mathfrak{g}^* \times \mathfrak{g}$ which sends (x,ξ) to (ξ, x) is then clearly a Lie algebra isomorphism between the doubles of the Lie bialgebras $(\mathfrak{g}, [\cdot,\cdot]_{\mathfrak{g}}, [\cdot,\cdot]_{\mathfrak{g}^*})$ and $(\mathfrak{g}^*, [\cdot,\cdot]_{\mathfrak{g}^*}, [\cdot,\cdot]_{\mathfrak{g}})$, which is the second part of the proposition. Notice that S arises as the composition of the isomorphism $\mathfrak{g} \times \mathfrak{g}^* \simeq (\mathfrak{g} \times \mathfrak{g}^*)^*$, induced by $\langle\cdot|\cdot\rangle_{\eth}$, with the natural isomorphism $(\mathfrak{g} \times \mathfrak{g}^*)^* \simeq \mathfrak{g}^* \times \mathfrak{g}$. □

It is clear that the dual of the dual of a (finite-dimensional) Lie bialgebra is just the Lie bialgebra itself. It is also obvious from Definition 11.17 that the transpose of a homomorphism of Lie bialgebras is a Lie bialgebra homomorphism between their dual Lie bialgebras. We leave it as an exercise to the reader to show that, if $\mathfrak{h}$ is a Lie sub-bialgebra of $\mathfrak{g}$, then there exists a (unique) Lie bialgebra structure on the quotient space $\mathfrak{g}^*/\mathfrak{h}^\perp$ such that the projection map $\mathfrak{g}^* \to \mathfrak{g}^*/\mathfrak{h}^\perp$ is a Lie bialgebra homomorphism.

11.2.4 Coboundary Lie Bialgebras and r-Matrices

In Section 10.2.1, the notion of a coboundary Lie bialgebra was introduced. As the terminology suggests, coboundary Lie bialgebras are Lie bialgebras, as we will see in the next proposition. Recall that, given $(\mathfrak{g}, [\cdot\,,\cdot]_{\mathfrak{g}})$ a Lie algebra, an element $r \in \mathfrak{g} \otimes \mathfrak{g}$ is a called an r-matrix when the transpose of the map $\delta : x \mapsto \mathrm{ad}_x r$, which is a linear map $\mathfrak{g}^* \otimes \mathfrak{g}^* \to \mathfrak{g}^*$, defines a Lie bracket on $\mathfrak{g}^*$. Recall also that the Lie bracket on $\mathfrak{g}^*$ which is associated to an r-matrix r is denoted by $[\cdot\,,\cdot]_r$ and is explicitly given by (10.15). The corresponding triple $(\mathfrak{g}, [\cdot\,,\cdot]_{\mathfrak{g}}, [\cdot\,,\cdot]_r)$ was called a coboundary Lie bialgebra in Definition 10.9.

Proposition 11.24. *Let $(\mathfrak{g}, [\cdot\,,\cdot]_{\mathfrak{g}})$ be a Lie algebra, and let $r \in \mathfrak{g} \otimes \mathfrak{g}$ be an r-matrix of $\mathfrak{g}$. Then the coboundary Lie bialgebra $(\mathfrak{g}, [\cdot\,,\cdot]_{\mathfrak{g}}, [\cdot\,,\cdot]_r)$ is a Lie bialgebra.*

Proof. Consider the coboundary Lie bialgebra $(\mathfrak{g}, [\cdot\,,\cdot]_{\mathfrak{g}}, [\cdot\,,\cdot]_r)$, where r is an r-matrix of $\mathfrak{g}$. Since both $[\cdot\,,\cdot]_{\mathfrak{g}}$ and $[\cdot\,,\cdot]_r$ are Lie brackets, we only need to verify that condition *(iii)* in Proposition 11.15 holds, where we recall from (11.21) that $\delta : \mathfrak{g} \to \mathfrak{g} \wedge \mathfrak{g}$ is the linear map, defined by

$$\langle \xi \wedge \eta, \delta(x) \rangle = \langle [\xi, \eta]_r, x \rangle \ ,$$

for all $x \in \mathfrak{g}$ and all $\xi, \eta \in \mathfrak{g}^*$. According to (10.17),

$$\langle [\xi, \eta]_r, x \rangle = \left\langle \xi \wedge \eta, \mathrm{ad}_x r^- \right\rangle \ ,$$

so that $\delta(x) = \mathrm{ad}_x r^-$, where r^- denotes the skew-symmetric part of r. Using the fact that ad is a Lie algebra representation of $\mathfrak{g}$ (on $\mathfrak{g} \wedge \mathfrak{g}$),

$$\delta([x,y]_{\mathfrak{g}}) = \mathrm{ad}_{[x,y]_{\mathfrak{g}}} r^- = \mathrm{ad}_x \mathrm{ad}_y r^- - \mathrm{ad}_y \mathrm{ad}_x r^- = \mathrm{ad}_x \delta(y) - \mathrm{ad}_y \delta(x)$$

for all $x, y \in \mathfrak{g}$. This proves that condition *(iii)* in Proposition 11.15 holds, hence that $(\mathfrak{g}, [\cdot\,,\cdot]_{\mathfrak{g}}, [\cdot\,,\cdot]_r)$ is a Lie bialgebra. □

Not every Lie bialgebra is a coboundary Lie bialgebra. When the Lie algebra $\mathfrak{g}$ is abelian, recall from Example 11.18 that $(\mathfrak{g}^*, [\cdot\,,\cdot]_{\mathfrak{g}} = 0, [\cdot\,,\cdot]_{\mathfrak{g}^*})$ is a Lie bialgebra for every Lie bracket $[\cdot\,,\cdot]_{\mathfrak{g}^*}$ on $\mathfrak{g}^*$. Such a Lie bialgebra cannot be a coboundary Lie bialgebra, unless $[\cdot\,,\cdot]_{\mathfrak{g}^*}$ is trivial.

We show in the following proposition that the double of a Lie bialgebra is a coboundary Lie bialgebra. Thus we can embed every Lie bialgebra in a coboundary Lie bialgebra, a fact which will be useful when we prove, in the next section, that every Lie bialgebra is the Lie bialgebra of a Poisson–Lie group.

Proposition 11.25. *Let $(\mathfrak{g}, [\cdot\,,\cdot]_{\mathfrak{g}}, [\cdot\,,\cdot]_{\mathfrak{g}^*})$ be a finite-dimensional Lie bialgebra. On its double $(\mathfrak{d}, [\cdot\,,\cdot]_{\mathfrak{d}})$, there exists a skew-symmetric r-matrix $a \in \wedge^2 \mathfrak{d}$, with corresponding Lie bracket $[\cdot\,,\cdot]_a$ on $\mathfrak{d}^*$, such that $(\mathfrak{g}, [\cdot\,,\cdot]_{\mathfrak{g}}, [\cdot\,,\cdot]_{\mathfrak{g}^*})$ is a Lie sub-bialgebra of $(\mathfrak{d}, [\cdot\,,\cdot]_{\mathfrak{d}}, [\cdot\,,\cdot]_a)$. In terms of a pair of dual bases $(e_1, \dots, e_d)$ and $(\varepsilon_1, \dots, \varepsilon_d)$ for $\mathfrak{g}$ and for $\mathfrak{g}^*$ respectively, a is given by $a = \frac{1}{2} \sum_{i=1}^d e_i \wedge \varepsilon_i$.*

Proof. There is a natural Lie algebra splitting of $\mathfrak{d}$, given by $\mathfrak{d} = \mathfrak{g}^* \oplus \mathfrak{g}$, where $\mathfrak{g}$ and $\mathfrak{g}^*$ have been identified with Lie subalgebras of $\mathfrak{d}$ via $\mathfrak{g} \simeq \mathfrak{g} \times \{0\}$ and $\mathfrak{g}^* \simeq \{0\} \times \mathfrak{g}^*$. The corresponding R-matrix of $\mathfrak{d}$ is given by $R(x,\xi) = \xi - x$, for $(x,\xi) \in \mathfrak{d}$. It is skew-symmetric because both $\mathfrak{g}$ and $\mathfrak{g}^*$ are isotropic subspaces of $\mathfrak{d}$ with respect to the bilinear form $\langle \cdot \mid \cdot \rangle_{\mathfrak{d}}$ (see (11.22)). Using the isomorphism between $\mathfrak{d}$ and its dual $\mathfrak{d}^*$, induced by $\langle \cdot \mid \cdot \rangle_{\mathfrak{d}}$, the R-matrix becomes an element $a \in \mathfrak{g} \otimes \mathfrak{g}$, which is according to Proposition 10.13 a skew-symmetric r-matrix, whose associated Lie bracket $[\cdot,\cdot]_a$ makes $(\mathfrak{d}, [\cdot,\cdot]_{\mathfrak{d}}, [\cdot,\cdot]_a)$ into a (coboundary) Lie bialgebra. As we have seen in Example 10.14, a is given in terms of dual bases $(e_1,\dots,e_d)$ and $(\varepsilon_1,\dots,\varepsilon_d)$ for $\mathfrak{g}$ and for $\mathfrak{g}^*$ respectively, by $a = \frac{1}{2}\sum_{i=1}^d e_i \wedge \varepsilon_i$. We show that $(\mathfrak{d}, [\cdot,\cdot]_{\mathfrak{d}}, [\cdot,\cdot]_a)$ contains $(\mathfrak{g}, [\cdot,\cdot]_{\mathfrak{g}}, [\cdot,\cdot]_{\mathfrak{g}^*})$ as a Lie sub-bialgebra. The natural inclusion $\imath : \mathfrak{g} \simeq \mathfrak{g} \times \{0\} \to \mathfrak{d}$ is, by definition of the bracket on $\mathfrak{d}$, a Lie algebra homomorphism $(\mathfrak{g}, [\cdot,\cdot]_{\mathfrak{g}}) \to (\mathfrak{d}, [\cdot,\cdot]_{\mathfrak{d}})$. Under the natural identification between $\mathfrak{d}^*$ and $\mathfrak{g}^* \times \mathfrak{g}$, the transpose $\imath^\top$ to $\imath$ is given by $\imath^\top(\xi,x) = \xi$, for $(\xi,x) \in \mathfrak{g}^* \times \mathfrak{g}$. Therefore, showing that $\imath^\top$ is also a Lie algebra homomorphism, amounts to showing that

$$\langle [(\xi,y),(\eta,z)]_a, (x,0) \rangle = \left\langle [\xi,\eta]_{\mathfrak{g}^*}, x \right\rangle,$$

for all $(\xi,y),(\eta,z) \in \mathfrak{g}^* \times \mathfrak{g}$ and for all $x \in \mathfrak{g}$. This is done by explicitly computing the left-hand side of the latter equation, namely

$$\begin{aligned}
&\langle [(\xi,y),(\eta,z)]_a, (x,0) \rangle \\
&= \frac{1}{2} \left\langle (\xi,y) \wedge (\eta,z), \operatorname{ad}_{(x,0)} \sum_{i=1}^d e_i \wedge \varepsilon_i \right\rangle \\
&= \frac{1}{2} \sum_{i=1}^d \langle (\xi,y) \wedge (\eta,z), [(x,0),(e_i,0)]_{\mathfrak{d}} \wedge (0,\varepsilon_i) + (e_i,0) \wedge [(x,0),(0,\varepsilon_i)]_{\mathfrak{d}} \rangle \\
&= \frac{1}{2} \sum_{i=1}^d \left\langle (\xi,y) \wedge (\eta,z), ([x,e_i]_{\mathfrak{g}},0) \wedge (0,\varepsilon_i) + (e_i,0) \wedge (-\operatorname{ad}^*_{\varepsilon_i} x, \operatorname{ad}^*_x \varepsilon_i) \right\rangle \\
&= \frac{1}{2} \sum_{i=1}^d \left(\left\langle \xi, [x,e_i]_{\mathfrak{g}} \right\rangle \langle \varepsilon_i, z \rangle - \left\langle \eta, [x,e_i]_{\mathfrak{g}} \right\rangle \langle \varepsilon_i, y \rangle \right) \\
&\quad + \frac{1}{2} \sum_{i=1}^d \left(\langle \xi, e_i \rangle \left(\langle \operatorname{ad}^*_x \varepsilon_i, z \rangle - \langle \eta, \operatorname{ad}^*_{\varepsilon_i} x \rangle \right) - \langle \eta, e_i \rangle \left(\langle \operatorname{ad}^*_x \varepsilon_i, y \rangle - \langle \xi, \operatorname{ad}^*_{\varepsilon_i} x \rangle \right) \right) \\
&= -\frac{1}{2} \sum_{i=1}^d \left(\langle \xi, e_i \rangle \langle \eta, \operatorname{ad}^*_{\varepsilon_i} x \rangle - \langle \eta, e_i \rangle \langle \xi, \operatorname{ad}^*_{\varepsilon_i} x \rangle \right) \\
&= \left\langle [\xi,\eta]_{\mathfrak{g}^*}, x \right\rangle .
\end{aligned}$$

It follows that $\imath^\top$ is a Lie algebra homomorphism, so that $\imath$ is a Lie bialgebra homomorphism. $\square$

Proposition 11.26. *Let* $(\mathfrak{g}, [\cdot,\cdot]_{\mathfrak{g}}, [\cdot,\cdot]_{\mathfrak{g}^*})$ *be a Lie bialgebra. If* $(\mathfrak{g}, [\cdot,\cdot]_{\mathfrak{g}})$ *is semi-simple, then* $(\mathfrak{g}, [\cdot,\cdot]_{\mathfrak{g}}, [\cdot,\cdot]_{\mathfrak{g}^*})$ *is a coboundary Lie bialgebra.*

Proof. Condition *(iii)* in Proposition 11.15 means that the map $\delta : \mathfrak{g} \to \wedge^2\mathfrak{g}$ is a 1-cocycle in the complex computing the cohomology of the Lie algebra $\mathfrak{g}$ with values on $\wedge^2\mathfrak{g}$, which is acted upon by adjoint action. According to Whitehead's lemma (Lemma 4.1), δ is a coboundary, which, in the present case means that $\delta(x) = \mathrm{ad}_x a$ for some $a \in \wedge^2\mathfrak{g}$. Since δ is the transpose of a Lie algebra bracket, a is a (skew-symmetric) r-matrix, hence $(\mathfrak{g}, [\cdot,\cdot]_\mathfrak{g}, [\cdot,\cdot]_{\mathfrak{g}^*})$ is a coboundary Lie bialgebra. □

11.2.5 Manin Triples

Making abstraction of the structure of the double $\mathfrak{d} := \mathfrak{g} \times \mathfrak{g}^*$ of a Lie bialgebra $(\mathfrak{g}, [\cdot,\cdot]_\mathfrak{g}, [\cdot,\cdot]_{\mathfrak{g}^*})$ leads to the notion of a Manin triple, defined as follows.

Definition 11.27. Let $(E, [\cdot\,,\cdot])$ be a finite-dimensional quadratic Lie algebra, whose bilinear form is denoted by $\langle\cdot\,|\,\cdot\rangle$. Let V and W be vector subspaces of E. The triple[3] $((E, [\cdot\,,\cdot], \langle\cdot\,|\,\cdot\rangle), V, W)$ is said to be a *Manin triple* if

(1) $E = V \oplus W$;
(2) V and W are Lie subalgebras of E;
(3) V and W are isotropic with respect to $\langle\cdot\,|\,\cdot\rangle$.

We show in the following proposition that there is a natural one-to-one correspondence between Manin triples and Lie bialgebras.

Proposition 11.28.

(1) *Let $(\mathfrak{g}, [\cdot,\cdot]_\mathfrak{g}, [\cdot,\cdot]_{\mathfrak{g}^*})$ be a finite-dimensional Lie bialgebra. Define $\mathfrak{d} := \mathfrak{g} \times \mathfrak{g}^*$, the double Lie algebra, with Lie bracket $[\cdot\,,\cdot]_\mathfrak{d}$ and with bilinear form $\langle\cdot\,|\,\cdot\rangle_\mathfrak{d}$ given by (11.23) and (11.22) respectively. The triple $((\mathfrak{d}, [\cdot\,,\cdot]_\mathfrak{d}, \langle\cdot\,|\,\cdot\rangle_\mathfrak{d}), \mathfrak{g}, \mathfrak{g}^*)$ is a Manin triple.*
(2) *Conversely, let $((E, [\cdot\,,\cdot]_E, \langle\cdot\,|\,\cdot\rangle_E), V, W)$ be a Manin triple, where E is assumed to be finite-dimensional. Denote by $[\cdot\,,\cdot]_V$ and $[\cdot\,,\cdot]_W$ the restriction of $[\cdot\,,\cdot]_E$ to V and W respectively, and by $\chi_*([\cdot\,,\cdot]_W)$ the Lie algebra bracket on V^* obtained by transporting $[\cdot\,,\cdot]_W$ under the isomorphism $\chi : V^* \simeq W$, induced by $\langle\cdot\,|\,\cdot\rangle_E$. Then $(V, [\cdot\,,\cdot]_V, \chi_*([\cdot\,,\cdot]_W))$ is a Lie bialgebra.*

Proof. Let $(\mathfrak{g}, [\cdot,\cdot]_\mathfrak{g}, [\cdot,\cdot]_{\mathfrak{g}^*})$ be a finite-dimensional Lie bialgebra, with double $\mathfrak{d} := \mathfrak{g} \times \mathfrak{g}^*$. Identify, as before, $\mathfrak{g}$ and $\mathfrak{g}^*$ as Lie subalgebras of $\mathfrak{d}$, via the natural identifications $\mathfrak{g} \simeq \mathfrak{g} \times \{0\}$ and $\mathfrak{g}^* \simeq \{0\} \times \mathfrak{g}^*$, so that $\mathfrak{d} = \mathfrak{g} \oplus \mathfrak{g}^*$. Clearly, both $\mathfrak{g}$ and $\mathfrak{g}^*$ are isotropic with respect to $\langle\cdot\,|\,\cdot\rangle_\mathfrak{d}$, so that $((\mathfrak{d}, [\cdot\,,\cdot]_\mathfrak{d}, \langle\cdot\,|\,\cdot\rangle_\mathfrak{d}), \mathfrak{g}, \mathfrak{g}^*)$ is a Manin triple. For the converse, let $((E, [\cdot\,,\cdot]_E, \langle\cdot\,|\,\cdot\rangle_E), V, W)$ be a Manin triple, with E finite-dimensional. Then $\langle\cdot\,|\,\cdot\rangle_E$ induces an isomorphism between V^* and W, so we obtain a Lie algebra structure on V and on V^*. In view of *(i)* in Proposition 11.15 and because $(E, \langle\cdot\,|\,\cdot\rangle_E)$ is quadratic, $(V, [\cdot\,,\cdot]_V, \chi_*([\cdot\,,\cdot]_W))$ is a Lie bialgebra. It is easily verified that both constructions are inverse to each other. □

[3] The triple is usually simply written as (E, V, W).

Notice that, inverting the rôles of V and W in a Manin triple, corresponds at the bialgebra level to taking the dual.

Example 11.29. We construct a natural Manin triple which corresponds to Example 10.15. Thus, we assume that $(\mathfrak{g}, [\cdot\,,\cdot])$ is a finite-dimensional Lie algebra and that $\mathfrak{g} = \mathfrak{g}_+ \oplus \mathfrak{g}_0 \oplus \mathfrak{g}_-$ is a Lie algebra decomposition, where $\mathfrak{g}_+$ and $\mathfrak{g}_-$ are Lie subalgebras of $\mathfrak{g}$ and that $\mathfrak{g}_0$ is abelian and is contained in the normalizer of $\mathfrak{g}_+$ and in the normalizer of $\mathfrak{g}_-$. Moreover, it is assumed that $\mathfrak{g}$ comes equipped with a non-degenerate symmetric bilinear form $\langle\cdot\,|\,\cdot\rangle$ with respect to which both $\mathfrak{g}_+$ and $\mathfrak{g}_-$ are isotropic and orthogonal to $\mathfrak{g}_0$. Recall that the endomorphism R of $\mathfrak{g}$, defined for $x \in \mathfrak{g}$ by $Rx := x_+ - x_-$, where x_+, x_0 and x_- stand for the projections of x on $\mathfrak{g}_+, \mathfrak{g}_0$ and $\mathfrak{g}_-$ respectively, is a skew-symmetric solution to the modified Yang–Baxter equation (10.4), with $c = 1$, hence leads to a skew-symmetric r-matrix for $\mathfrak{g}$, making $(\mathfrak{g}, [\cdot\,,\cdot], [\cdot\,,\cdot]_r)$ into a Lie bialgebra. Let $E := \mathfrak{g} \times \mathfrak{g}$, equipped with the product Lie bracket, denoted by $[\cdot\,,\cdot]'$, and with the ad-invariant symmetric non-degenerate bilinear form $\langle\cdot\,|\,\cdot\rangle'$, which is defined, for all $x_1, x_2, y_2, y_2 \in \mathfrak{g}$, by

$$\langle (x_1, y_1) \,|\, (x_2, y_2) \rangle' := \langle x_1 \,|\, x_2 \rangle - \langle y_1 \,|\, y_2 \rangle \ . \tag{11.31}$$

Let Δ be the subspace of E, defined by $\Delta := \{(x,x) \mid x \in \mathfrak{g}\}$ and let W denote the subspace of E, defined by

$$W := \{(x+z, y-z) \mid x \in \mathfrak{g}_+,\ y \in \mathfrak{g}_- \text{ and } z \in \mathfrak{g}_0\} \ .$$

It is clear that $E = \Delta \oplus W$. Moreover, the assumed properties on the factors of $\mathfrak{g}$ imply at once that Δ and W are both subalgebras of E and are both isotropic with respect to $\langle\cdot\,|\,\cdot\rangle'$. It follows that (E, Δ, W) is a Manin triple. In order to show that this Manin triple corresponds to the coboundary Lie bialgebra $(\mathfrak{g}, [\cdot\,,\cdot], [\cdot\,,\cdot]_r)$ we need to show that the restriction of $[\cdot\,,\cdot]'$ to Δ, respectively to W, yields the Lie brackets $[\cdot\,,\cdot]$ and $[\cdot\,,\cdot]_r$ upon identifying Δ and $\mathfrak{g}$, respectively W and $\mathfrak{g}^*$; the former identification is done by the diagonal map, which maps $x \in \mathfrak{g}$ to $(x,x) \in \mathfrak{g} \times \mathfrak{g}$. It is clear that $[\cdot\,,\cdot]'$, restricted to Δ, coincides under this identification with $[\cdot\,,\cdot]$. Consider the map $\psi : W \to \mathfrak{g}$, defined for $(x+z, y-z) \in W$ by $\psi(x+z, y-z) := x - y + 2z$. It is the transpose of the isomorphism $\mathfrak{g} \to W^*$, induced by $\langle\cdot\,|\,\cdot\rangle'$, composed with the isomorphism $\mathfrak{g}^* \to \mathfrak{g}$, induced by $\langle\cdot\,|\,\cdot\rangle$. We claim that, under ψ, the Lie algebra structure on W corresponds to the R-bracket $[\cdot\,,\cdot]_R$ on $\mathfrak{g}$, showing our claim, since the r-bracket on $\mathfrak{g}^*$ corresponds to the R-bracket on $\mathfrak{g}$ under the isomorphism induced by $\langle\cdot\,|\,\cdot\rangle$. This follows from the following computation, where we use that both $\mathfrak{g}_+$ and $\mathfrak{g}_-$ are Lie subalgebras of $\mathfrak{g}$, that $\mathfrak{g}_0$ is contained in the normalizer of $\mathfrak{g}_+$ and in the normalizer of $\mathfrak{g}_-$, and that $\mathfrak{g}_0$ is abelian: for all $(x+z, y-z) \in W$ and $(x'+z', y'-z') \in W$:

$$\begin{aligned}
&\left[\psi(x+z, y-z), \psi(x'+z', y'-z')\right]_R \\
&= \left[x - y + 2z, x' - y' + 2z'\right]_R \\
&= \frac{1}{2}\left(\left[x+y, x' - y' + 2z'\right] + \left[x - y + 2z, x' + y'\right]\right)
\end{aligned}$$

$$\begin{aligned}
&= [x,x'] - [y,y'] + [x,z'] + [z,x'] + [y,z'] + [z,y'] \\
&= [x+z, x'+z'] - [y-z, y'-z'] \\
&= \psi([x+z,x'+z'], [y-z,y'-z']) \\
&= \psi([(x+z,y-z),(x'+z',y'-z')]') \,.
\end{aligned}$$

Notice that, considering the particular case of $\mathfrak{g}_0 = \{0\}$, the constructed Manin triple corresponds to the case of a Lie algebra splitting (see Example 10.14).

Example 11.30. We also construct the Manin triple which corresponds to Example 10.16. To do this, we consider the complex Lie algebra $\mathfrak{sl}_d(\mathbb{C})$ as a quadratic Lie algebra over $\mathbb{R}$, equipped with the bilinear form, defined for $x,y \in \mathfrak{sl}_d(\mathbb{C})$ by $\langle x \,|\, y \rangle_{\mathfrak{I}} := \mathfrak{I}(\mathrm{Trace}(xy))$. We consider two Lie subalgebras: $\mathfrak{t}_+$, the space of upper triangular matrices admitting only real coefficients on the diagonal, and $\mathfrak{su}_d$. We have that $\mathfrak{sl}_d(\mathbb{C}) = \mathfrak{su}_d \oplus \mathfrak{t}_+$. For all $x,y \in \mathfrak{su}_d$,

$$\overline{\mathrm{Trace}(xy)} = \mathrm{Trace}(\bar{x}\bar{y}) = \mathrm{Trace}\left(x^\top y^\top\right) = \mathrm{Trace}(yx) = \mathrm{Trace}(xy) \,,$$

so that $\langle x \,|\, y \rangle_{\mathfrak{I}} = \mathfrak{I}(\mathrm{Trace}(xy)) = 0$. This shows that $\mathfrak{su}_d$ is isotropic with respect to $\langle \cdot \,|\, \cdot \rangle_{\mathfrak{I}}$. For every $x,y \in \mathfrak{t}_+$, it is clear that $xy \in \mathfrak{t}_+$, hence $\mathrm{Trace}(xy)$ has vanishing imaginary part and $\mathfrak{t}_+$ is also isotropic with respect to $\langle \cdot \,|\, \cdot \rangle_{\mathfrak{I}}$. It follows that $(\mathfrak{sl}_d(\mathbb{C}), \mathfrak{su}_d, \mathfrak{t}_+)$ is a Manin triple. We show that it is the Manin triple corresponding to the Lie bialgebra discussed in Example 10.16. To do this, recall from the latter example that we consider on $\mathfrak{su}_d$ the non-degenerate symmetric bilinear form $\langle \cdot \,|\, \cdot \rangle$, defined for $x,y \in \mathfrak{su}_d$ by $\langle x \,|\, y \rangle := \mathfrak{R}(\mathrm{Trace}(xy))$. Let $\psi : \mathfrak{t}_+ \to \mathfrak{su}_d$ be the composition of the isomorphisms $\mathfrak{t}_+ \simeq \mathfrak{su}_d^* \simeq \mathfrak{su}_d$, where the first one is induced by $\langle \cdot \,|\, \cdot \rangle_{\mathfrak{I}}$ and the second one by $\langle \cdot \,|\, \cdot \rangle$. Explicitly, $\psi(x) = -\frac{\sqrt{-1}}{2}(x + \bar{x}^\top)$, for $x \in \mathfrak{t}_+$, as is computed from $\langle x \,|\, y \rangle_{\mathfrak{I}} = \langle \psi(x) \,|\, y \rangle$, for all $y \in \mathfrak{su}_d$. For $x,y \in \mathfrak{t}_+$, one checks by direct computation that $\psi([x,y]) = [\psi(x), \psi(y)]_{R'}$, which confirms that the Lie bialgebra, associated to the above Manin triple, is the one from Example 10.16.

We have seen in Proposition 11.25 that the double of a Lie bialgebra is also a Lie bialgebra. We construct in the following proposition the Manin triple of this double Lie bialgebra in terms of the Manin triple of the original bialgebra.

Proposition 11.31. *Let $((E, [\cdot,\cdot], \langle \cdot \,|\, \cdot \rangle), V, W)$ be a Manin triple, where E is finite-dimensional. Let $E \times E$ be equipped with the product Lie bracket, which we denote by $[\cdot,\cdot]'$, and with the non-degenerate* ad*-invariant symmetric bilinear form $\langle \cdot \,|\, \cdot \rangle'$, given for $(x_1,y_1),(x_2,y_2) \in E \times E$ by*

$$\langle (x_1,y_1) \,|\, (x_2,y_2) \rangle' := \langle x_1 \,|\, x_2 \rangle - \langle y_1 \,|\, y_2 \rangle \,. \tag{11.32}$$

Let

$$\begin{aligned}
V' &:= \{(u,u) \in E \times E \mid u \in E\} \,, \\
W' &:= W \times V \,.
\end{aligned} \tag{11.33}$$

Then $((E\times E,[\cdot\,,\cdot]',\langle\cdot\,|\,\cdot\rangle'),V',W')$ *is a Manin triple and the Lie bialgebra associated to it is the double of the Lie bialgebra associated to* (E,V,W).

Proof. Let $((E,[\cdot\,,\cdot],\langle\cdot\,|\,\cdot\rangle),V,W)$ be a Manin triple and let $(\mathfrak{g},[\cdot\,,\cdot]_{\mathfrak{g}},[\cdot\,,\cdot]_{\mathfrak{g}^*})$ be its associated Lie bialgebra. According to Proposition 11.28, we may identify $(E,[\cdot\,,\cdot])$ with $(\mathfrak{d}=\mathfrak{g}\times\mathfrak{g}^*,[\cdot\,,\cdot]_{\mathfrak{d}})$; under this identification, $(V,[\cdot\,,\cdot]_{|V})$ and $(W,[\cdot\,,\cdot]_{|W})$ are identified with $(\mathfrak{g},[\cdot\,,\cdot]_{\mathfrak{g}})$ and $(\mathfrak{g}^*,[\cdot\,,\cdot]_{\mathfrak{g}^*})$ respectively; the inner product $\langle\cdot\,|\,\cdot\rangle$ on E corresponds to the natural inner product (11.22) on $\mathfrak{d}$. According to Proposition 11.25, $\mathfrak{d}$ is a coboundary Lie bialgebra, where the Lie bracket on $\mathfrak{d}^*$ is the r-bracket a which is associated to the Lie algebra splitting $\mathfrak{d}=\mathfrak{g}^*\oplus\mathfrak{g}$, i.e., the Lie algebra splitting $E=W\oplus V$. According to Example 11.29, in the particular case of $\mathfrak{g}_0=\{0\}$, the Manin triple of such a coboundary Lie algebra is given by $((E\times E,[\cdot\,,\cdot]',\langle\cdot\,|\,\cdot\rangle'),V',W')$, where $[\cdot\,,\cdot]'$ is the product Lie bracket on $E\times E$ and where $\langle\cdot\,|\,\cdot\rangle'$ and V', W' are given by (11.32) and (11.33). □

11.2.6 Lie Bialgebras and Poisson Structures

We give in the following proposition a different description of the notion of a Lie bialgebra, by interpreting one of the Lie algebra structures as a linear Poisson structure: according to Proposition 7.3, to a Lie bracket $[\cdot\,,\cdot]_{\mathfrak{g}^*}$ on $\mathfrak{g}^*$ corresponds a linear Poisson structure π on $\mathfrak{g}$ (and vice versa).

Proposition 11.32. *Let* $(\mathfrak{g},[\cdot\,,\cdot]_{\mathfrak{g}})$ *be a finite-dimensional Lie algebra and let* π *be a linear Poisson structure on* $\mathfrak{g}$. *Then the following conditions are equivalent:*

(i) *Denoting by* $[\cdot\,,\cdot]_{\mathfrak{g}^*}$ *the Lie bracket on* $\mathfrak{g}^*$, *corresponding to* π, *the triple* $(\mathfrak{g},[\cdot\,,\cdot]_{\mathfrak{g}},[\cdot\,,\cdot]_{\mathfrak{g}^*})$ *is a Lie bialgebra;*

(ii) *For every* $x,y\in\mathfrak{g}$, *the bivector* $\pi_x\in\wedge^2T_x\mathfrak{g}$, *considered as an element of* $\wedge^2\mathfrak{g}$, *satisfies*
$$\pi_{[x,y]_{\mathfrak{g}}}=\mathrm{ad}_x\,\pi_y-\mathrm{ad}_y\,\pi_x\,. \tag{11.34}$$

Proof. Let $[\cdot,\cdot]_{\mathfrak{g}^*}$ be the Lie bracket on $\mathfrak{g}^*$ which corresponds to the linear Poisson structure $\pi=\{\cdot\,,\cdot\}$ on $\mathfrak{g}$. For $F,G\in\mathscr{F}(\mathfrak{g})$, their Poisson bracket at $x\in\mathfrak{g}$ is, according to (7.4) and (11.21), given by
$$\{F,G\}(x)=\left\langle[\mathrm{d}_xF,\mathrm{d}_xG]_{\mathfrak{g}^*},x\right\rangle=\langle\mathrm{d}_xF\wedge\mathrm{d}_xG,\delta(x)\rangle\;,$$
where we recall that in this formula d_xF and d_xG are viewed as elements of $\mathfrak{g}^*$. It follows that $\delta(x)=\pi_x$ for every $x\in\mathfrak{g}$, where the latter bivector is considered as an element of $\wedge^2\mathfrak{g}$. The equivalence between *(i)* and *(ii)* then follows from item *(iii)* in Proposition 11.15 and Definition 11.17. □

If $(\mathfrak{g},[\cdot\,,\cdot]_{\mathfrak{g}},[\cdot\,,\cdot]_{\mathfrak{g}^*})$ and $(\mathfrak{h},[\cdot\,,\cdot]_{\mathfrak{h}},[\cdot\,,\cdot]_{\mathfrak{h}^*})$ are two finite-dimensional Lie bialgebras, with corresponding Poisson structures $\{\cdot\,,\cdot\}_{\mathfrak{g}}$ and $\{\cdot\,,\cdot\}_{\mathfrak{h}}$ on $\mathfrak{g}$ and $\mathfrak{h}$ respectively, then a linear map $\psi:\mathfrak{g}\to\mathfrak{h}$ is a homomorphism of Lie bialgebras if and only if

$\psi : (\mathfrak{g}, [\cdot,\cdot]_\mathfrak{g}) \to (\mathfrak{h}, [\cdot,\cdot]_\mathfrak{h})$ is a homomorphism of Lie algebras and $\psi : (\mathfrak{g}, \{\cdot,\cdot\}_\mathfrak{g}) \to (\mathfrak{h}, \{\cdot,\cdot\}_\mathfrak{h})$ is a Poisson map. Indeed, the latter condition is equivalent to the fact that $\psi^\top : (\mathfrak{h}^*, [\cdot,\cdot]_{\mathfrak{h}^*}) \to (\mathfrak{g}^*, [\cdot,\cdot]_{\mathfrak{g}^*})$ is a Lie algebra homomorphism. It follows that Lie sub-bialgebras can be characterized as follows:

Proposition 11.33. *Let $(\mathfrak{g}, [\cdot,\cdot]_\mathfrak{g}, [\cdot,\cdot]_{\mathfrak{g}^*})$ be a Lie bialgebra, and denote by π the linear Poisson structure on $\mathfrak{g}$ corresponding to $[\cdot,\cdot]_{\mathfrak{g}^*}$. A subspace $\mathfrak{h} \subset \mathfrak{g}$ is a Lie sub-bialgebra if and only if the following two conditions are satisfied:*

(1) $\mathfrak{h}$ is a Lie subalgebra of the Lie algebra $(\mathfrak{g}, [\cdot,\cdot]_\mathfrak{g})$;
(2) $\mathfrak{h}$ is a Poisson submanifold of the Poisson manifold $(\mathfrak{g}, \pi)$.

11.3 Poisson–Lie Groups and Lie Bialgebras

In this section, we relate the objects which appeared in the two previous sections, namely we establish a natural correspondence between Poisson–Lie groups and finite-dimensional Lie bialgebras. We first show that the tangent space to the unit of a Poisson–Lie group inherits the structure of a Lie bialgebra (Section 11.3.1). We then show that every finite-dimensional Lie bialgebra is the Lie bialgebra of a Poisson–Lie group. This is first done in Section 11.3.2 for coboundary Lie bialgebras and then in general in Section 11.3.3, using the fact that every finite-dimensional Lie bialgebra can be embedded in a coboundary Lie bialgebra (its double, see Proposition 11.25).

11.3.1 The Lie Bialgebra of a Poisson–Lie Group

Let $(\mathbf{G}, \pi)$ be a Poisson–Lie group, whose unit is denoted by e. First, since $\mathbf{G}$ is a Lie group, the tangent space $\mathfrak{g} := T_e\mathbf{G}$ is equipped with a Lie algebra bracket $[\cdot,\cdot]_\mathfrak{g}$ (see Theorem 5.1). Second, according to item *(1)* in Proposition 11.5, the Poisson structure π vanishes at e, leading to a linear Poisson structure π_1 on $T_e\mathbf{G} = \mathfrak{g}$ (the linearized Poisson structure, see Section 7.5), hence a Lie bracket $[\cdot,\cdot]_{\mathfrak{g}^*}$ on $\mathfrak{g}^*$ (see Proposition 7.3), which we call the *linearized bracket* of $(\mathbf{G}, \pi)$. We show in the following proposition that the Lie structures $[\cdot,\cdot]_\mathfrak{g}$ and $[\cdot,\cdot]_{\mathfrak{g}^*}$ make $\mathfrak{g}$ into a Lie bialgebra, so that we can naturally associate to every Poisson–Lie group a Lie bialgebra.

Proposition 11.34. *Let $(\mathbf{G}, \pi)$ be a Poisson–Lie group, with Lie algebra $(\mathfrak{g}, [\cdot,\cdot]_\mathfrak{g})$, and with linearized bracket $[\cdot,\cdot]_{\mathfrak{g}^*}$.*

(1) The triple $(\mathfrak{g}, [\cdot,\cdot]_\mathfrak{g}, [\cdot,\cdot]_{\mathfrak{g}^*})$ is a Lie bialgebra;
(2) Let Φ be a Poisson–Lie group homomorphism from $(\mathbf{G}, \pi)$ to a Poisson–Lie group $(\mathbf{H}, \pi')$, with corresponding Lie bialgebra $(\mathfrak{h}, [\cdot,\cdot]_\mathfrak{h}, [\cdot,\cdot]_{\mathfrak{h}^*})$. The tangent map $T_e\Phi$ of Φ at the unit e of $\mathbf{G}$ is a Lie bialgebra homomorphism from $(\mathfrak{g}, [\cdot,\cdot]_\mathfrak{g}, [\cdot,\cdot]_{\mathfrak{g}^*})$ to $(\mathfrak{h}, [\cdot,\cdot]_\mathfrak{h}, [\cdot,\cdot]_{\mathfrak{h}^*})$.

Proof. Let $(\mathbf{G},\pi)$ be a Poisson–Lie group with Lie algebra $(\mathfrak{g},[\cdot\,,\cdot]_{\mathfrak{g}})$ and let Ψ denote the cocycle $\mathbf{G}\to\wedge^2\mathfrak{g}$, defined by (11.2). We show that the cocycle condition (11.3) implies condition *(ii)* in Proposition 11.32, which guarantees that the triple $(\mathfrak{g},[\cdot\,,\cdot]_{\mathfrak{g}},[\cdot\,,\cdot]_{\mathfrak{g}^*})$ is a Lie bialgebra, where $[\cdot\,,\cdot]_{\mathfrak{g}^*}$ denotes the linearized bracket of $(\mathbf{G},\pi)$. Consider the identity

$$\Psi(ghg^{-1})=\Psi(g)+\mathrm{Ad}_g\Psi(h)-\mathrm{Ad}_{ghg^{-1}}\Psi(g)\,, \tag{11.35}$$

which is valid for all $g,h\in\mathbf{G}$, as follows from the cocycle condition (11.3), upon using that $\Psi(e)=0$. Differentiating (11.35) leads to an identity for the linearized Poisson bracket π_1, since

$$(\pi_1)_x=T_e\Psi(x) \tag{11.36}$$

for $x\in\mathfrak{g}$ (see *(2)* in Proposition 11.5). In order to obtain this identity, replace h by $\exp(ty)$ in (11.35) and take the derivative at $t=0$. In view of (11.36), this leads to

$$(\pi_1)_{\mathrm{Ad}_g y}=\mathrm{Ad}_g(\pi_1)_y-\mathrm{ad}_{\mathrm{Ad}_g y}\Psi(g)\,, \tag{11.37}$$

for all $y\in\mathfrak{g}$ and $g\in\mathbf{G}$. We replace now in (11.37) g by $\exp(tx)$ and we take the derivative at $t=0$; we claim that we obtain

$$(\pi_1)_{[x,y]_{\mathfrak{g}}}=\mathrm{ad}_x(\pi_1)_y-\mathrm{ad}_y(\pi_1)_x\,, \tag{11.38}$$

which in view of Proposition 11.32, shows that the triple $(\mathfrak{g},[\cdot\,,\cdot]_{\mathfrak{g}}),[\cdot\,,\cdot]_{\mathfrak{g}^*})$ is a Lie bialgebra.

Let us show in detail how (11.38) is obtained from (11.37). Since π_1 is a linear Poisson structure, the map $\mathfrak{g}\to\wedge^2\mathfrak{g}$, defined by $x\mapsto(\pi_1)_x$ is a linear map. It follows that, for every function $z:\mathbb{F}\to\mathfrak{g}$,

$$\left.\frac{\mathrm{d}}{\mathrm{d}t}\right|_{t=0}(\pi_1)_{z(t)}=(\pi_1)_{\frac{\mathrm{d}}{\mathrm{d}t}|_{t=0}z(t)}\,.$$

Applied to the left-hand side of (11.37), with $g=\exp(tx)$, this yields the left-hand side of (11.38), in view of the definition of ad. Also,

$$\left.\frac{\mathrm{d}}{\mathrm{d}t}\right|_{t=0}\mathrm{Ad}_{\exp(tx)}(\pi_1)_y=\mathrm{ad}_x(\pi_1)_y\,,$$

which yields the first term in the right-hand side of (11.37). Finally, since $\Psi(e)=0$,

$$\begin{aligned}\left.\frac{\mathrm{d}}{\mathrm{d}t}\right|_{t=0}\mathrm{ad}_{\mathrm{Ad}_{\exp(tx)}y}\Psi(\exp(tx))&=\mathrm{ad}_y\left.\frac{\mathrm{d}}{\mathrm{d}t}\right|_{t=0}\Psi(\exp(tx))\\&=\mathrm{ad}_y\,T_e\Psi(x)=\mathrm{ad}_y\,(\pi_1)_x\,,\end{aligned}$$

which is the second term in the right-hand side of (11.37). This completes the proof of (11.38), and hence of item *(1)*.

We turn to the proof of *(2)*. Since $\Phi : \mathbf{G} \to \mathbf{H}$ is a Lie group homomorphism, the tangent map $T_e\Phi : \mathfrak{g} \to \mathfrak{h}$ is a Lie algebra homomorphism (Theorem 5.1). Similarly, since $\Phi : \mathbf{G} \to \mathbf{H}$ is a Poisson map, with π vanishing at e and π' vanishing at $\Phi(e)$, the linear map $T_e\Phi : \mathfrak{g} \to \mathfrak{h}$ is a Poisson map (Proposition 7.22). According to Section 11.2.6, this shows that ϕ is a Lie bialgebra homomorphism. □

The second item of the previous proposition implies that the Lie bialgebra of a Poisson–Lie subgroup of $(\mathbf{G}, \pi)$ is a Lie sub-bialgebra of the bialgebra of $(\mathbf{G}, \pi)$. This is stated, together with its converse, in the following proposition.

Proposition 11.35. *Let $(\mathbf{G}, \pi)$ be a Poisson–Lie group, whose Lie bialgebra is denoted by $(\mathfrak{g}, [\cdot\,,\cdot]_{\mathfrak{g}}, [\cdot\,,\cdot]_{\mathfrak{g}^*})$. For a connected Lie subgroup $\mathbf{H}$ of $\mathbf{G}$, with Lie algebra $\mathfrak{h}$, the following conditions are equivalent:*

(i) *$\mathbf{H}$ is a Poisson–Lie subgroup of $(\mathbf{G}, \pi)$;*
(ii) *$\mathfrak{h}$ is a Lie sub-bialgebra of $(\mathfrak{g}, [\cdot\,,\cdot]_{\mathfrak{g}}, [\cdot\,,\cdot]_{\mathfrak{g}^*})$.*

Assuming that these equivalent conditions hold, denote the Lie brackets of the Lie bialgebra $\mathfrak{h}$ by $[\cdot,\cdot]_{\mathfrak{h}}$ and $[\cdot,\cdot]_{\mathfrak{h}^*}$, and denote the Poisson structure on $\mathbf{H}$, which makes $\mathbf{H}$ into a Poisson submanifold of $(\mathbf{G}, \pi)$, by π'. Then $(\mathfrak{h}, [\cdot\,,\cdot]_{\mathfrak{h}}, [\cdot\,,\cdot]_{\mathfrak{h}^*})$ is the Lie bialgebra of $(\mathbf{H}, \pi')$.

Proof. The implication *(i)* $\Rightarrow$ *(ii)* follows at once from Proposition 11.34, since the inclusion map of a Poisson–Lie subgroup is a Poisson–Lie group homomorphism. Conversely, assume that $\mathfrak{h}$ is a Lie sub-bialgebra of the Lie bialgebra $(\mathfrak{g}, [\cdot\,,\cdot]_{\mathfrak{g}}, [\cdot\,,\cdot]_{\mathfrak{g}^*})$ of $(\mathbf{G}, \pi)$. Then the transpose $\delta : \mathfrak{g} \to \wedge^2\mathfrak{g}$ of $[\cdot,\cdot]_{\mathfrak{g}^*}$ maps $\mathfrak{h}$ to $\wedge^2\mathfrak{h}$. According to item *(2)* in Proposition 11.5 and the proof of Proposition 11.32, $\delta(x) = T_e\Psi(x)$ for all $x \in \mathfrak{g}$, where Ψ is defined as usual by (11.2). It follows that $T_e\Psi(x) \in \wedge^2\mathfrak{h}$ for all $x \in \mathfrak{h}$. According to Proposition 11.7, this shows that the connected Lie subgroup $\mathbf{H}$ of $\mathbf{G}$ is a Poisson–Lie subgroup. Thus, condition *(ii)* implies condition *(i)*. □

11.3.2 Coboundary Poisson–Lie Groups and Lie Bialgebras

We show in this section that the Lie bialgebra of a coboundary Poisson–Lie group is a coboundary Lie bialgebra, which justifies why the same adjective "coboundary" is used for both structures. We also show the following converse: every finite-dimensional coboundary Lie bialgebra is the Lie bialgebra of a coboundary Poisson–Lie group.

We first show that there corresponds to every coboundary Poisson–Lie group a coboundary Lie bialgebra. Let $(\mathbf{G}, \pi)$ be a coboundary Poisson–Lie group. By definition, there exists $a \in \wedge^2\mathfrak{g}$ such that $\pi = \overleftarrow{a} - \overrightarrow{a}$, where $(\mathfrak{g}, [\cdot\,,\cdot])$ is the Lie algebra of $\mathbf{G}$. According to Proposition 11.10, $[\![a,a]\!]$ is Ad-invariant, hence is ad-invariant. Proposition 10.11 implies that a is a skew-symmetric r-matrix, i.e., the associated bracket $[\cdot\,,\cdot]_a$ is a Lie bracket on $\mathfrak{g}^*$, making $(\mathfrak{g}, [\cdot\,,\cdot], [\cdot\,,\cdot]_a)$ into a coboundary Lie bialgebra (see Definition 10.9). The next proposition shows that this coboundary Lie bialgebra is the Lie bialgebra of the Poisson–Lie group $(\mathbf{G},\pi)$.

Proposition 11.36. *Let $(\mathbf{G},\pi)$ be a coboundary Poisson–Lie group, with $\pi = \overleftarrow{a} - \overrightarrow{a}$ for some r-matrix $a \in \wedge^2\mathfrak{g}$, where $(\mathfrak{g},[\cdot\,,\cdot]_\mathfrak{g})$ denotes the Lie algebra of $\mathbf{G}$. The Lie bialgebra of $(\mathbf{G},\pi)$ is the coboundary Lie bialgebra $(\mathfrak{g},[\cdot\,,\cdot]_\mathfrak{g},[\cdot\,,\cdot]_a)$.*

Proof. According to (11.14), the map $\Psi : \mathbf{G} \to \wedge^2\mathfrak{g}$, defined by $g \mapsto \wedge^2(T_g R_{g^{-1}})\,\pi_g$, is given by $\Psi(g) = \mathrm{Ad}_g\, a - a$. The tangent map of Ψ at e is therefore given by $T_e\Psi(x) = \mathrm{ad}_x\, a$ for all $x \in \mathfrak{g}$. According to item *(2)* in Proposition 11.5, the linearized Poisson structure π_1 of π at the unit e of G is given by $(\pi_1)_x = T_e\Psi(x)$, for all $x \in \mathfrak{g}$, so that the linearized Poisson structure π_1 is given at $x \in \mathfrak{g}$ by $(\pi_1)_x = \mathrm{ad}_x\, a$, whose transpose is precisely the Lie bracket $[\cdot\,,\cdot]_a$. This shows that $(\mathfrak{g},[\cdot\,,\cdot]_\mathfrak{g},[\cdot\,,\cdot]_a)$ is the Lie bialgebra of $(\mathbf{G},\pi)$. □

Corollary 11.37. *Let $(\mathfrak{g},[\cdot\,,\cdot]_\mathfrak{g},[\cdot\,,\cdot]_a)$ be a finite-dimensional coboundary Lie bialgebra. If $\mathbf{G}$ is a connected Lie group, with Lie algebra $(\mathfrak{g},[\cdot\,,\cdot]_\mathfrak{g})$, then $\mathbf{G}$ admits the structure of a coboundary Poisson–Lie group, whose Lie bialgebra is $(\mathfrak{g},[\cdot\,,\cdot]_\mathfrak{g},[\cdot\,,\cdot]_a)$. In particular, every finite-dimensional coboundary Lie bialgebra is the Lie bialgebra of a coboundary Poisson–Lie group.*

Proof. We only need to prove the first part of the corollary, since the last statement follows from it at once upon using the fact that every Lie algebra is the Lie algebra of a connected Lie group (see Theorem 5.1). Let $(\mathfrak{g},[\cdot\,,\cdot]_\mathfrak{g},[\cdot\,,\cdot]_a)$ be a finite-dimensional coboundary Lie bialgebra, with $a \in \wedge^2\mathfrak{g}$ so that $[\![a,a]\!]$ is ad-invariant. Let $\mathbf{G}$ be a connected Lie group whose Lie algebra is $\mathfrak{g}$. Since $\mathbf{G}$ is assumed to be connected, $[\![a,a]\!]$ is Ad-invariant. Setting $\pi := \overleftarrow{a} - \overrightarrow{a}$, Propositions 11.10 and 11.36 imply that $(\mathbf{G},\pi)$ is a Poisson–Lie group, whose Lie bialgebra is $(\mathfrak{g},[\cdot\,,\cdot]_\mathfrak{g},[\cdot\,,\cdot]_a)$, as was to be shown. □

Corollary 11.38. *Let $(\mathfrak{g},[\cdot\,,\cdot]_\mathfrak{g},[\cdot\,,\cdot]_{\mathfrak{g}^*})$ be a finite-dimensional Lie bialgebra and let $\mathbf{D}$ be a connected Lie group whose Lie algebra is the double $(\mathfrak{d},[\cdot\,,\cdot]_\mathfrak{d})$ of $(\mathfrak{g},[\cdot\,,\cdot]_\mathfrak{g},[\cdot\,,\cdot]_{\mathfrak{g}^*})$. Then $\mathbf{D}$ admits a unique Poisson structure $\pi_\mathbf{D}$, making $(\mathbf{D},\pi_\mathbf{D})$ into a coboundary Poisson–Lie group, whose Lie bialgebra is $(\mathfrak{d},[\cdot\,,\cdot]_\mathfrak{d},[\cdot\,,\cdot]_a)$. In terms of a pair of dual bases $(e_1,\dots,e_d)$ and $(\varepsilon_1,\dots,\varepsilon_d)$ for $\mathfrak{g}$ and for $\mathfrak{g}^*$ respectively, the skew-symmetric r-matrix a of $\mathfrak{d}$ is given by $a = \frac{1}{2}\sum_{i=1}^d e_i \wedge \varepsilon_i$.*

Proof. According to Proposition 11.25, the double $(\mathfrak{d},[\cdot\,,\cdot]_\mathfrak{d})$ is a coboundary Lie bialgebra; as before, we denote the corresponding Lie structure on $\mathfrak{d}^*$ by $[\cdot\,,\cdot]_a$. If $\mathbf{D}$ is a connected Lie group with Lie algebra $(\mathfrak{d},[\cdot\,,\cdot]_\mathfrak{d})$, then $\mathbf{D}$ admits, according to Corollary 11.37, the structure of a coboundary Poisson–Lie group, whose Lie bialgebra is $(\mathfrak{d},[\cdot\,,\cdot]_\mathfrak{d},[\cdot\,,\cdot]_a)$. As we will see in Theorem 11.39, the Poisson structure on $\mathbf{D}$ which integrates the given Lie bialgebra structure is unique. □

11.3.3 The Integration of Lie Bialgebras

We show in this section that every finite-dimensional Lie bialgebra can be integrated. We mean by this that there exists for every finite-dimensional Lie bialgebra $(\mathfrak{g}, [\cdot\,,\cdot]_{\mathfrak{g}}, [\cdot\,,\cdot]_{\mathfrak{g}^*})$ a (not necessarily unique) Poisson–Lie group $(\mathbf{G}, \pi)$ whose Lie bialgebra is isomorphic to $(\mathfrak{g}, [\cdot\,,\cdot]_{\mathfrak{g}}, [\cdot\,,\cdot]_{\mathfrak{g}^*})$. For the particular case of a coboundary Lie bialgebras, this was proved in Corollary 11.37.

Theorem 11.39. *Let* $(\mathfrak{g}, [\cdot\,,\cdot]_{\mathfrak{g}}, [\cdot\,,\cdot]_{\mathfrak{g}^*})$ *be a finite-dimensional Lie bialgebra and let* $\mathbf{G}$ *be a connected Lie group with Lie algebra* $(\mathfrak{g}, [\cdot\,,\cdot]_{\mathfrak{g}})$.

(1) There exists at most one Poisson structure π *on* $\mathbf{G}$ *such that* $(\mathbf{G}, \pi)$ *is a Poisson–Lie group, whose Lie bialgebra is* $(\mathfrak{g}, [\cdot\,,\cdot]_{\mathfrak{g}}, [\cdot\,,\cdot]_{\mathfrak{g}^*})$;

(2) If $\mathbf{G}$ *is simply connected, then there exists on* $\mathbf{G}$ *a unique Poisson structure* π, *such that* $(\mathbf{G}, \pi)$ *is a Poisson–Lie group, whose Lie bialgebra is* $(\mathfrak{g}, [\cdot\,,\cdot]_{\mathfrak{g}}, [\cdot\,,\cdot]_{\mathfrak{g}^*})$;

(3) In particular, every finite-dimensional Lie bialgebra is the Lie bialgebra of a Poisson–Lie group.

Proof. Let $\pi^{(1)}$ and $\pi^{(2)}$ be two multiplicative Poisson structures on the connected Lie group $\mathbf{G}$, leading to the same Lie bialgebra, i.e., their linearized Poisson structures at e coincide. In view of *(2)* in Proposition 11.5, $T_e\Psi^{(1)}(x) = T_e\Psi^{(2)}(x)$, for all $x \in \mathfrak{g}$, where $\Psi^{(i)} : \mathbf{G} \to \wedge^2\mathfrak{g}$ is defined by $\Psi^{(i)}(g) := \wedge^2(T_g R_{g^{-1}})\pi_g^{(i)}$, for $i = 1, 2$. It follows that $\Psi^{(1)} - \Psi^{(2)}$, which is a cocycle (since both $\Psi^{(1)}$ and $\Psi^{(2)}$ are cocycles), has zero tangent map at e. By Proposition 11.4, this implies that $\Psi^{(1)} - \Psi^{(2)} = 0$, and hence that $\pi^{(1)} = \pi^{(2)}$. This shows item *(1)*.

Let $(\mathfrak{g}, [\cdot\,,\cdot]_{\mathfrak{g}}, [\cdot\,,\cdot]_{\mathfrak{g}^*})$ be a finite-dimensional Lie bialgebra. According to Proposition 11.25, $(\mathfrak{g}, [\cdot\,,\cdot]_{\mathfrak{g}}, [\cdot\,,\cdot]_{\mathfrak{g}^*})$ is a Lie sub-bialgebra of a coboundary Lie bialgebra, which we denote by $(\mathfrak{d}, [\cdot\,,\cdot]_{\mathfrak{d}}, [\cdot\,,\cdot]_a)$, because the underlying Lie algebra is the double of $(\mathfrak{g}, [\cdot\,,\cdot]_{\mathfrak{g}}, [\cdot\,,\cdot]_{\mathfrak{g}^*})$. In view of Proposition 11.37, $(\mathfrak{d}, [\cdot\,,\cdot]_{\mathfrak{d}}, [\cdot\,,\cdot]_a)$ can be integrated into a Poisson–Lie group $(\mathbf{D}, \pi_{\mathbf{D}})$, which is the only fact about $\mathfrak{d}$ which we will use in this proof. We denote by $\hat{\mathbf{G}}$ the unique connected Lie subgroup of $\mathbf{D}$, whose Lie algebra is $\mathfrak{g}$. According to Proposition 11.35, the subgroup $\hat{\mathbf{G}}$ inherits a multiplicative Poisson structure $\hat{\pi}$ from $\pi_{\mathbf{D}}$, whose corresponding Lie bialgebra is $(\mathfrak{g}, [\cdot\,,\cdot]_{\mathfrak{g}}, [\cdot\,,\cdot]_{\mathfrak{g}^*})$. This shows *(3)*.

We show now that $(\hat{\mathbf{G}}, \hat{\pi})$ can be replaced by the connected and simply connected Lie group $\mathbf{G}$ whose Lie algebra is $\mathfrak{g}$, which is the content of *(2)*. In view of Lie's theorem (see item *(4)* of Theorem 5.1), there exists a Lie group homomorphism $\Phi : \mathbf{G} \to \hat{\mathbf{G}}$, whose induced Lie algebra homomorphism is the identity map. We claim that there is a unique Poisson structure π on $\mathbf{G}$ which turns $\Phi : \mathbf{G} \to \hat{\mathbf{G}}$ into a Poisson map, and we show that π is multiplicative. Then $(\mathbf{G}, \pi)$ is a Poisson–Lie group and Φ is a homomorphism of Poisson–Lie groups, so that the induced Lie bialgebra homomorphism, which is the identity map, is a homomorphism of Lie bialgebras, which means that the Lie bialgebra of $\mathbf{G}$ is $(\mathfrak{g}, [\cdot\,,\cdot]_{\mathfrak{g}}, [\cdot\,,\cdot]_{\mathfrak{g}^*})$.

Let us prove the claim. Differentiating the identity $\Phi = R_{\Phi(g)}\Phi R_{g^{-1}}$ at $g \in \mathbf{G}$ shows that the tangent maps of Φ at $g \in \mathbf{G}$ and at e are related by $T_g\Phi = T_{\hat{e}}R_{\Phi(g)} \circ$

$T_e\Phi \circ T_g R_{g^{-1}}$ (in this formula, e and $\hat{e}$ are the units of the Lie groups $\mathbf{G}$ and $\hat{\mathbf{G}}$ respectively). Since the three linear maps which appear in the right-hand side of this formula are invertible, so is $T_g\Phi$, hence Φ is a local diffeomorphism. We can therefore define π for all $g \in \mathbf{G}$ by setting

$$\pi_g := \wedge^2 (T_g\Phi)^{-1} \hat{\pi}_{\Phi(g)} . \tag{11.39}$$

According to Proposition 1.19, if there exists a Poisson structure on $\mathbf{G}$ which makes Φ into a Poisson map, then it must be given by (11.39), which gives the desired uniqueness of π. Also, the vertical maps of the following commutative diagram are local diffeomorphisms (here μ and $\hat{\mu}$ stand for the group products of $\mathbf{G}$ and $\hat{\mathbf{G}}$).

$$\begin{array}{ccc} \mathbf{G}\times\mathbf{G} & \xrightarrow{\ \mu\ } & \mathbf{G} \\ {\scriptstyle \Phi\times\Phi}\big\downarrow & & \big\downarrow{\scriptstyle \Phi} \\ \hat{\mathbf{G}}\times\hat{\mathbf{G}} & \xrightarrow[\ \hat{\mu}\ ]{} & \hat{\mathbf{G}} \end{array}$$

Combined with the fact that μ is a Poisson map, this shows that $\hat{\mu}$ is Poisson map. As a consequence, π is multiplicative, as remained to be shown. □

Having established the fact that every Lie bialgebra can be integrated into a Poisson–Lie group, we now show that every Lie bialgebra homomorphism can be integrated into a morphism of Poisson–Lie groups.

Theorem 11.40. *Let $(\mathfrak{g}, [\cdot,\cdot]_{\mathfrak{g}}, [\cdot,\cdot]_{\mathfrak{g}^*})$ and $(\mathfrak{h}, [\cdot,\cdot]_{\mathfrak{h}}, [\cdot,\cdot]_{\mathfrak{h}^*})$ be two finite-dimensional Lie bialgebras and let $(\mathbf{G},\pi)$ and $(\mathbf{H},\pi')$ be Poisson–Lie groups which integrate them, where $\mathbf{G}$ is chosen connected and simply connected. For every Lie bialgebra homomorphism $\phi : \mathfrak{g} \to \mathfrak{h}$, there exists a unique Poisson–Lie group homomorphism $\Phi : \mathbf{G} \to \mathbf{H}$ whose induced Lie bialgebra homomorphism is ϕ, i.e., such that $T_e\Phi = \phi$, where e denotes the unit element of $\mathbf{G}$.*

Proof. Suppose that $\phi : \mathfrak{g} \to \mathfrak{h}$ is a homomorphism of Lie bialgebras. In view of Lie's theorem (item *(4)* of Theorem 5.1), there exists a unique Lie group homomorphism $\Phi : \mathbf{G} \to \mathbf{H}$ such that $T_e\Phi = \phi$. It remains to be shown that Φ is a Poisson map. To do this, we consider the map $\Xi : \mathbf{G} \to \wedge^2\mathfrak{h}$, defined for all $g \in \mathbf{G}$ by

$$\Xi(g) := \wedge^2\phi\,(\Psi(g)) - \Psi'(\Phi(g)) ,$$

where $\Psi : \mathbf{G} \to \wedge^2\mathfrak{g}$ and $\Psi' : \mathbf{H} \to \wedge^2\mathfrak{h}$ denote the group cocycles, associated to $(\mathbf{G},\pi)$ and to $(\mathbf{H},\pi')$ respectively (see item *(iii)* in Proposition 11.2). The map Ξ is a cocycle of $\mathbf{G}$ with respect to the adjoint representation of $\mathbf{G}$ on $\wedge^2\mathfrak{h}$, i.e.,

$$\Xi(gh) = \Xi(g) + \mathrm{Ad}_{\Phi(g)}\,\Xi(h) ,$$

for all $g,h \in \mathbf{G}$. We first show that $\Xi = 0$. In view of item *(1)* of Proposition 11.4, it suffices to show that $T_e\Xi(x) = 0$ for all $x \in \mathfrak{g}$. To prove the latter, recall from item *(2)* in Proposition 11.5 that $T_e\Psi(x) = (\pi_1)_x$, for all $x \in \mathfrak{g}$, where π_1 denotes the linearized Poisson structure of π at e, and similarly for Ψ' and π'_1. It allows us to compute

$$\begin{aligned} T_e\Xi(x) &= \wedge^2\phi\,(T_e\Psi(x)) - T_{e'}\Psi'(T_e\Phi(x)) \\ &= \wedge^2\phi\,((\pi_1)_x) - (\pi'_1)_{\phi(x)} = 0\,, \end{aligned}$$

where the last equality, which states that ϕ is a Poisson map, is tantamount to the fact that the transpose map $\phi^\top : (\mathfrak{h}^*, [\cdot\,,\cdot]_{\mathfrak{h}^*}) \to (\mathfrak{g}^*, [\cdot\,,\cdot]_{\mathfrak{g}^*})$ is a Lie algebra homomorphism (see Section 11.2.6). It follows that $\Xi = 0$, leading to the equality

$$\wedge^2\phi\,(\Psi(g)) = \Psi'(\Phi(g))\,,$$

valid for all $g \in \mathbf{G}$, which can be written, using $T_e\Phi = \phi$ and using the definition (11.2) of the cocycle Ψ, as

$$\wedge^2\,(T_g(\Phi \circ R_{g^{-1}}))\pi_g = \wedge^2\left(T_{\Phi(g)}R_{\Phi(g)^{-1}}\right)\pi'_{\Phi(g)}\,. \tag{11.40}$$

Now Φ is a group homomorphism, so that

$$\Phi \circ R_{g^{-1}} = R_{\Phi(g^{-1})} \circ \Phi\,,$$

which, substituted in (11.40), leads to $\wedge^2(T_g\Phi)\pi_g = \pi'_{\Phi(g)}$, since in a Lie group, right translation by an element of the group is a diffeomorphism. In view of Proposition 1.19, the latter equality implies that Φ is a Poisson map, as was to be shown. □

The notion of duality for Lie bialgebra leads to a notion of duality for connected, simply connected Poisson–Lie groups.

Definition 11.41. Two Poisson–Lie groups $(\mathbf{G},\pi)$ and $(\mathbf{G}^*,\pi^*)$ are said to be *dual* to each other if their Lie bialgebras are dual to each other.

Let $(\mathbf{G},\pi)$ be a Poisson–Lie group and let $(\mathfrak{g}, [\cdot\,,\cdot]_{\mathfrak{g}}, [\cdot\,,\cdot]_{\mathfrak{g}^*})$ denote its Lie bialgebra, whose dual bialgebra is $(\mathfrak{g}^*, [\cdot\,,\cdot]_{\mathfrak{g}^*}, [\cdot\,,\cdot]_{\mathfrak{g}})$. Every Poisson–Lie group $(\mathbf{G}^*,\pi^*)$ which integrates the latter Lie bialgebra is dual to $(\mathbf{G},\pi)$. A canonical dual is picked by demanding that $\mathbf{G}^*$ be connected and simply connected. Indeed, there is up to isomorphism a unique connected and simply connected Lie group $\mathbf{G}^*$ whose Lie algebra is $(\mathfrak{g}^*, [\cdot\,,\cdot]_{\mathfrak{g}^*})$ (item *(3)* of Theorem 5.1) and on $\mathbf{G}^*$ there exists a unique Poisson structure π^* such that $(\mathbf{G}^*,\pi^*)$ integrates $(\mathfrak{g}, [\cdot\,,\cdot]_{\mathfrak{g}}, [\cdot\,,\cdot]_{\mathfrak{g}^*})$. This Poisson–Lie group is called *the* dual of $(\mathbf{G},\pi)$. For $(\mathbf{G},\pi)$ a connected and simply connected Poisson–Lie group, taking twice the dual gives back the Poisson–Lie group $(\mathbf{G},\pi)$.

Example 11.42. Let $\mathbf{G}$ be a connected, simply connected Lie group, with Lie algebra $(\mathfrak{g}, [\cdot\,,\cdot]_{\mathfrak{g}})$. Consider $\mathfrak{g}^*$ as a Lie group, where the group operation is addition. In view of Corollary 11.9, $\mathfrak{g}^*$ becomes a Poisson–Lie group, when it is equipped with the linear Poisson structure π on $\mathfrak{g}^*$ corresponding to the Lie bracket $[\cdot\,,\cdot]_{\mathfrak{g}}$. Its Lie bialgebra is $(\mathfrak{g}^*, 0, [\cdot,\cdot]_{\mathfrak{g}})$, with dual Lie bialgebra $(\mathfrak{g}, [\cdot,\cdot]_{\mathfrak{g}}, 0)$. It follows that the (connected and simply connected) dual to the Poisson–Lie group $(\mathfrak{g}^*, \pi)$ is the Lie group $\mathbf{G}$, equipped with the trivial Poisson structure.

11.4 Dressing Actions and Symplectic Leaves

We show in this section that the symplectic leaves of a Poisson–Lie group can be described by using dressing actions, a notion which we first define in the general context of Lie groups.

Let $\mathbf{G}$ be a connected Lie group and let $\mathbf{G}_1$ and $\mathbf{G}_2$ be two Lie subgroups of $\mathbf{G}$. The triple $(\mathbf{G}, \mathbf{G}_1, \mathbf{G}_2)$ is said to be *complete* if the map $\mathbf{G}_1 \times \mathbf{G}_2 \to \mathbf{G}$ given by $(g_1, g_2) \mapsto g_1 g_2$ is a diffeomorphism. Since $(g_1 g_2)^{-1} = g_2^{-1} g_1^{-1}$, for all $g_1, g_2 \in \mathbf{G}$, the triple $(\mathbf{G}, \mathbf{G}_1, \mathbf{G}_2)$ is complete if and only if the triple $(\mathbf{G}, \mathbf{G}_2, \mathbf{G}_1)$ is complete. It follows that, when $(\mathbf{G}, \mathbf{G}_1, \mathbf{G}_2)$ is complete, there exists for every $(g_1, g_2) \in \mathbf{G}_1 \times \mathbf{G}_2$ a unique pair $(g_1', g_2') \in \mathbf{G}_1 \times \mathbf{G}_2$, such that $g_1 g_2 = g_2' g_1'$. It is customary to denote the elements g_1' and g_2' by $g_1^{g_2}$ and $g_2^{g_1}$ respectively.

The associativity of the group structure on $\mathbf{G}$ implies that the assignments $(g,h) \mapsto h^g$ and $(g,h) \mapsto g^{h^{-1}}$ are group actions. Indeed, the relation $(g_1 h_1) g_2 = g_1 (h_1 g_2)$ implies, for $g_1, h_1 \in \mathbf{G}_1$ and $g_2 \in \mathbf{G}_2$, that

$$g_2^{g_1 h_1} (g_1 h_1)^{g_2} = g_1 g_2^{h_1} h_1^{g_2} = (g_2^{h_1})^{g_1} g_1^{(g_2^{h_1})} h_1^{g_2} ,$$

so that, by uniqueness of the decomposition, we can derive the relation

$$g_2^{g_1 h_1} = (g_2^{h_1})^{g_1} , \tag{11.41}$$

valid for all $g_1, h_1 \in \mathbf{G}_1$ and $g_2 \in \mathbf{G}_2$. This shows that the assignment $(g_1, g_2) \mapsto g_2^{g_1}$ defines a group action of $\mathbf{G}_1$ on (the manifold) $\mathbf{G}_2$. Similarly, one derives from the relation $g_1(g_2 h_2) = (g_1 g_2) h_2$, for $g_1 \in \mathbf{G}_1$ and $g_2, h_2 \in \mathbf{G}_2$, the relation

$$g_1^{g_2 h_2} = (g_1^{g_2})^{h_2} , \tag{11.42}$$

which says that the assignment $(g_2, g_1) \mapsto g_1^{g_2^{-1}}$ defines a group action of $\mathbf{G}_2$ on (the manifold) $\mathbf{G}_1$. These two group actions are respectively called the *dressing action* of $\mathbf{G}_1$ on $\mathbf{G}_2$ and the *dressing action* of $\mathbf{G}_2$ on $\mathbf{G}_1$.

Example 11.43. According to a standard theorem of linear algebra (essentially the Gram–Schmidt orthogonalization process), every complex invertible matrix of determinant 1 can be decomposed in a unique way as the product of an element in $\mathbf{SU}_d$ and an element in $\mathbf{T}_+$, the group of upper triangular matrices of determinant 1, admitting only positive entries on the diagonal. It follows that $(\mathbf{SL}_d(\mathbb{C}), \mathbf{SU}_d, \mathbf{T}_+)$ is complete.

In the case of Poisson–Lie groups, the notion of dressing action leads to the following definition.

Definition 11.44. Let $(\mathbf{G},\pi)$ be a connected Poisson–Lie group with Lie bialgebra $(\mathfrak{g},[\cdot,\cdot]_{\mathfrak{g}},[\cdot,\cdot]_{\mathfrak{g}^*})$. A Lie group $\mathbf{D}$ is called a *dressing group* of $(\mathbf{G},\pi)$ if the following three conditions are satisfied.

(1) The Lie algebra of $\mathbf{D}$ is the double $(\mathfrak{d},[\cdot,\cdot]_{\mathfrak{d}})$ of $(\mathfrak{g},[\cdot,\cdot]_{\mathfrak{g}},[\cdot,\cdot]_{\mathfrak{g}^*})$;
(2) $\mathbf{G}$ is the connected Lie subgroup of $\mathbf{D}$ corresponding to the Lie subalgebra $\mathfrak{g}\subset\mathfrak{d}$;
(3) $(\mathbf{D},\mathbf{G},\mathbf{G}^*)$ is complete, where $\mathbf{G}^*$ stands for the connected Lie subgroup of $\mathbf{D}$ corresponding to the Lie subalgebra $\mathfrak{g}^*\subset\mathfrak{d}$.

The Lie subgroup $\mathbf{G}^*$ of $\mathbf{D}$ is then called the *dual* of $(\mathbf{G},\pi)$ with respect to the dressing group $\mathbf{D}$. The Poisson structure on $\mathbf{G}^*$ as a Poisson submanifold of the Poisson–Lie group $(\mathbf{D},\pi_{\mathbf{D}})$ is denoted by π^*.

Example 11.45. A Lie bialgebra structure on $\mathfrak{g}:=\mathfrak{su}_d$ was given in Example 10.16, We showed in Example 11.30 that its double $\mathfrak{d}=\mathfrak{g}\oplus\mathfrak{g}^*$ is isomorphic to the Lie algebra $\mathfrak{sl}_d(\mathbb{C})$ of traceless complex matrices, identification under which the Lie algebras $\mathfrak{g}$ and $\mathfrak{g}^*$ become the Lie algebras $\mathfrak{su}_d$ and $\mathfrak{t}_+$, where $\mathfrak{t}_+$ is the Lie algebra of $\mathbf{T}_+$, i.e., the vector space of upper triangular matrices of trace zero with real coefficients on the diagonal. Consider the Poisson bracket π on $\mathbf{SU}_d$ which makes $(\mathbf{SU}_d,\pi)$ into a Poisson–Lie group, integrating the latter bialgebra structure on $\mathfrak{su}_d$. Since $(\mathbf{SL}_d(\mathbb{C}),\mathbf{SU}_d,\mathbf{T}_+)$ is complete, as we have seen in Example 11.43, conditions (1)–(3) in the above definition are verified, so that $\mathbf{SL}_d(\mathbb{C})$ is a dressing group of $(\mathbf{SU}_d,\pi)$.

Proposition 11.46. *Let $(\mathbf{G},\pi)$ be a connected Poisson–Lie group and suppose that $\mathbf{D}$ is a dressing group of $(\mathbf{G},\pi)$, and denote by $(\mathbf{G}^*,\pi^*)$ the dual of $(\mathbf{G},\pi)$ with respect to $\mathbf{D}$.*

(1) The diffeomorphism $\mathbf{G}\times\mathbf{G}^*\to\mathbf{G}^*\times\mathbf{G}$, defined by $(g,h)\mapsto(h^g,g^h)$ is a Poisson map;
(2) The dressing action of $\mathbf{G}$ on $\mathbf{G}^*$ is a Poisson action;
(3) The dressing action of $\mathbf{G}^*$ on $\mathbf{G}$ is also a Poisson action, provided that we replace the Poisson structure π^* by its opposite $-\pi^*$;
(4) For $\alpha\in\mathfrak{g}^*$, the fundamental vector field $\underline{\alpha}$ of the dressing action of $\mathbf{G}^*$ on $\mathbf{G}$ is given, at $g\in\mathbf{G}$, by

$$\underline{\alpha}_g=-T_eR_g\left(P_{\mathfrak{g}}\left(\mathrm{Ad}_g\,\alpha\right)\right)\,, \tag{11.43}$$

where $P_{\mathfrak{g}}:\mathfrak{d}\to\mathfrak{g}$ is the projection from $\mathfrak{d}=\mathfrak{g}\oplus\mathfrak{g}^*$ onto $\mathfrak{g}$, and where Ad *stands for the adjoint action of $\mathbf{D}$ on its Lie algebra $\mathfrak{d}$.*

Proof. The inclusion map $\mathbf{G}\times\mathbf{G}^*\hookrightarrow\mathbf{D}\times\mathbf{D}$ is a Poisson map because $\mathbf{G}$ and $\mathbf{G}^*$ are Poisson submanifolds of $\mathbf{D}$. Since the product map $\mathbf{D}\times\mathbf{D}\to\mathbf{D}$ is a Poisson map (because $\mathbf{D}$ is a Poisson–Lie group), the diffeomorphism $\mathbf{G}\times\mathbf{G}^*\to\mathbf{D}$, given by

$(g,h) \mapsto gh$, is a Poisson diffeomorphism. Since the same is true for the diffeomorphism $\mathbf{G}^* \times \mathbf{G} \to \mathbf{D}$, given by $(h,g) \mapsto hg$, the map $\mathbf{G} \times \mathbf{G}^* \to \mathbf{G}^* \times \mathbf{G}$, defined by $(g,h) \mapsto (h^g, g^h)$ is a Poisson diffeomorphism. This proves *(1)*. As a corollary, the map $\mathbf{G} \times \mathbf{G}^* \to \mathbf{G}^*$, defined by $(g,h) \mapsto h^g$, is a Poisson map, meaning that the dressing action of $\mathbf{G}$ on $\mathbf{G}^*$ is a Poisson map, proving *(2)*. Since the inverse map $h \mapsto h^{-1}$ is a Poisson map from $(\mathbf{G}^*, -\pi^*)$ to $(\mathbf{G}^*, \pi^*)$ (see Proposition 11.5, item *(3)*), the assignment $(g,h) \mapsto g^{h^{-1}}$ is also a Poisson map, provided that $\mathbf{G} \times \mathbf{G}^*$ is endowed with the product Poisson structure $(\pi, -\pi^*)$, which means precisely that the dressing action of $\mathbf{G}^*$ on $\mathbf{G}$ is Poisson, provided that we replace the Poisson structure π^* of $\mathbf{G}^*$ by its opposite. This proves *(3)*.

We now prove *(4)*. For all $g \in \mathbf{G}$ and $t \in \mathbb{F}$ we have, by definition of the dressing action

$$g\exp(-t\alpha) = h(t) g^{\exp(-t\alpha)} = h(t) g^{\exp(t\alpha)^{-1}}, \tag{11.44}$$

where $h(t) = \exp(-t\alpha)^g$. Note that $h(0) = e$, and that $h(t)$ is a path in $\mathbf{G}^*$. Multiplying (11.44) by g^{-1} and taking the derivative with respect to t at $t=0$, we obtain

$$\begin{aligned} -\operatorname{Ad}_g \alpha &= \left.\frac{\mathrm{d}}{\mathrm{d}t}\right|_{t=0} g^{\exp(t\alpha)^{-1}} g^{-1} + \left.\frac{\mathrm{d}}{\mathrm{d}t}\right|_{t=0} h(t) \\ &= T_g R_{g^{-1}} \underline{\alpha}_g + \left.\frac{\mathrm{d}}{\mathrm{d}t}\right|_{t=0} h(t), \end{aligned} \tag{11.45}$$

where the last equality follows from the definition of the fundamental vector field of a Lie group action. Since $h(t)$ is a path in $\mathbf{G}^*$, (11.45) is the decomposition of $-\operatorname{Ad}_g \alpha$ in $\mathfrak{g} \oplus \mathfrak{g}^*$, so that $-P_{\mathfrak{g}}(\operatorname{Ad}_g \alpha) = T_g R_{g^{-1}} \underline{\alpha}_g$, which amounts to (11.43). □

We can now describe the symplectic leaves of a Poisson–Lie group which admits a dressing group, in terms of the dressing action.

Theorem 11.47. *Let $(\mathbf{G}, \pi)$ be a connected Poisson–Lie group. Assume that $\mathbf{D}$ is a dressing group of $(\mathbf{G}, \pi)$, and let $\mathbf{G}^* \subset \mathbf{D}$ be the dual of $\mathbf{G}$ with respect to $\mathbf{D}$. Then the symplectic leaves of $(\mathbf{G}, \pi)$ are the orbits of the dressing action of $\mathbf{G}^*$ on $\mathbf{G}$.*

Proof. Recall from Corollary 11.38 that $\mathbf{D}$ admits a Poisson structure $\pi_{\mathbf{D}}$, making $(\mathbf{D}, \pi_{\mathbf{D}})$ into a coboundary Poisson–Lie group. This Poisson structure is associated to the skew-symmetric r-matrix $a := \frac{1}{2}\sum_{i=1}^d e_i \wedge \varepsilon_i \in \wedge^2 \mathfrak{d}$, where $(e_1, \dots, e_d)$ and $(\varepsilon_1, \dots, \varepsilon_d)$ are dual bases of $\mathfrak{g}$ and $\mathfrak{g}^*$. In view of (11.14), one has for every $h \in \mathbf{D}$,

$$\wedge^2 (T_h R_{h^{-1}}) (\pi_{\mathbf{D}})_h = \operatorname{Ad}_h a - a.$$

Since $(\mathbf{G}, \pi)$ is a Poisson–Lie subgroup of $(\mathbf{D}, \pi_{\mathbf{D}})$, it follows for every $g \in \mathbf{G}$ that

$$\wedge^2 (T_g R_{g^{-1}}) \pi_g = \frac{1}{2} \sum_{i=1}^d \operatorname{Ad}_g e_i \wedge \operatorname{Ad}_g \varepsilon_i - \frac{1}{2} \sum_{i=1}^d e_i \wedge \varepsilon_i.$$

The left-hand side of this equation belongs to $\wedge^2 \mathfrak{g}$, hence is fixed by the linear map $\wedge^2 P_{\mathfrak{g}}$ (with $P_{\mathfrak{g}}$ the projection of $\mathfrak{d} = \mathfrak{g} \oplus \mathfrak{g}^*$ onto $\mathfrak{g}$). It follows that

$$\wedge^2 (T_g R_{g^{-1}})\, \pi_g = \frac{1}{2} \sum_{i=1}^{d} \operatorname{Ad}_g e_i \wedge P_{\mathfrak{g}}(\operatorname{Ad}_g \varepsilon_i)\,, \tag{11.46}$$

which, in turn, gives the following expression of the Poisson bivector π, evaluated at the point g,

$$\pi_g = \frac{1}{2} \sum_{i=1}^{d} (\overleftarrow{e_i})_g \wedge \underline{\varepsilon_i}_g\,, \tag{11.47}$$

where we used item *(3)* of Proposition 11.46 as well as the identity $T_e R_g(\operatorname{Ad}_g x) = T_e L_g(x) = \overleftarrow{x}_g$ for every $x \in \mathfrak{g}$.

We use the explicit expression (11.47) of π_g to show that the symplectic leaf of π through g coincides with the orbit through g of the dressing action of $\mathbf{G}^*$ on $\mathbf{G}$. Since $\mathbf{G}^*$ is connected and since $(\varepsilon_1, \dots, \varepsilon_d)$ is a basis of $\mathfrak{g}^*$, it amounts to showing that the image of the linear map $\pi_g^\sharp : T_g^*\mathbf{G} \to T_g\mathbf{G}$, defined by π_g, is the vector subspace of $T_g\mathbf{G}$ generated by $\underline{\varepsilon_1}_g, \dots, \underline{\varepsilon_d}_g$. To show this, it suffices according to (11.47) to show that the matrix $(a_{ij})_{i,j=1,\dots,d}$ expressing the elements $\underline{\varepsilon_i}_g$, $i = 1, \dots, d$, in terms of the basis of tangent vectors $(\overleftarrow{e_1})_g, \dots, (\overleftarrow{e_d})_g$ is skew-symmetric, since these elements are then the columns of the Poisson matrix of π at g, expressed in terms of the latter basis. Notice that, in view of (11.46), this matrix is also the matrix which expresses the elements $P_{\mathfrak{g}}(\operatorname{Ad}_g \varepsilon_i)$, $i = 1, \dots, g$, in terms of the vectors $\operatorname{Ad}_g e_1, \dots, \operatorname{Ad}_g e_d$,

$$P_{\mathfrak{g}}(\operatorname{Ad}_g \varepsilon_i) = \sum_{j=1}^{d} a_{ij} \operatorname{Ad}_g e_j\,,$$

so that

$$a_{ij} = \left\langle \operatorname{Ad}_g^* \varepsilon_j, P_{\mathfrak{g}}(\operatorname{Ad}_g \varepsilon_i) \right\rangle = \left\langle P_{\mathfrak{g}^*}(\operatorname{Ad}_g \varepsilon_j), P_{\mathfrak{g}}(\operatorname{Ad}_g \varepsilon_i) \right\rangle\,,$$

where $P_{\mathfrak{g}^*}$ is the projection of $\mathfrak{d} = \mathfrak{g} \oplus \mathfrak{g}^*$ onto $\mathfrak{g}^*$; we also used the fact that $\operatorname{ad}_x^* \varepsilon_j = P_{\mathfrak{g}^*}([x, (y, \varepsilon_j)]_{\mathfrak{d}})$ for all $x, y \in \mathfrak{g}$, which follows from the definition (11.23) of $[\cdot,\cdot]_{\mathfrak{d}}$. Hence, to show that the matrix $(a_{ij})_{i,j=1,\dots,d}$ is skew-symmetric, we need to show that

$$\left\langle P_{\mathfrak{g}^*}(\operatorname{Ad}_g \varepsilon_i), P_{\mathfrak{g}}(\operatorname{Ad}_g \varepsilon_j) \right\rangle + \left\langle P_{\mathfrak{g}^*}(\operatorname{Ad}_g \varepsilon_j), P_{\mathfrak{g}}(\operatorname{Ad}_g \varepsilon_i) \right\rangle = 0 \tag{11.48}$$

for $1 \leqslant i, j \leqslant d$. According to the definition (11.22) of $\langle \cdot \,|\, \cdot \rangle_{\mathfrak{d}}$, the left-hand side of this equation can be written as $\left\langle \operatorname{Ad}_g \varepsilon_i \,|\, \operatorname{Ad}_g \varepsilon_j \right\rangle_{\mathfrak{d}}$. Now

$$\left\langle \operatorname{Ad}_g \varepsilon_i \,|\, \operatorname{Ad}_g \varepsilon_j \right\rangle_{\mathfrak{d}} = \left\langle \varepsilon_i \,|\, \varepsilon_j \right\rangle_{\mathfrak{d}} = 0\,,$$

in view of the Ad-invariance of $\langle \cdot \,|\, \cdot \rangle_{\mathfrak{d}}$. This proves (11.48), hence yields that the matrix $(a_{ij})_{i,j=1,\dots,d}$ is skew-symmetric, which was to be shown. □

11.5 Notes

Poisson structures can be seen as the semi-classical limit of quantum structures, i.e., as being what remains of quantization when $\hbar^2$ (but not $\hbar$) is considered to be small enough to be ignored (see Chapter 13). This is especially true for Poisson–Lie groups which first appeared as semi-classical limits of quantum groups, in the pioneering work [58] of Drinfel'd. A second motivation for the introduction of Poisson–Lie groups was the study of integrable systems associated to infinite-dimensional Lie algebras, such as the Korteweg–de Vries equation, where Poisson–Lie groups provide the hidden symmetry, see Semenov–Tian–Shansky [180].

Poisson–Lie groups have proved to be interesting on their own. First, they naturally appear when studying the singularities of several highly singular spaces associated to semi-simple Lie groups, like Bruhat and Schubert cells, see Lu–Weinstein [132, 136]. Also, a Poisson–Lie group is the first instance of a more general object, namely a Poisson–Lie groupoid, see Mackenzie–Xu [140], which appears naturally when studying the dynamical Yang–Baxter equation. Exactly like Poisson–Lie groups and Lie bialgebras are in one-to-one correspondence, Poisson–Lie groupoids are in one-to-one correspondence with Lie bialgebroids, see Mackenzie–Xu [140].

One of the first systematic studies of Poisson–Lie groups is the PhD thesis of Lu [132], which remains an excellent introduction to the subject, especially to those who want to learn more about the relation with Lie groupoids and enlarged definitions of moment maps. It also deals with a slightly more general object: affine Poisson structures on Lie groups. Another useful reference is the review article by Kosmann–Schwarzbach [110], which insists in particular on the link with Lie bialgebras.

Part III
Applications

Chapter 12
Liouville Integrable Systems

In this chapter we present the main application of Poisson structures: the theory of integrable Hamiltonian systems.

Poisson's theorem, which states that the Poisson bracket of two constants of motion (of a Hamiltonian system) is a constant of motion, is from the modern point of view a direct consequence of the formalism, but was in Poisson's time considered as a fundamental step towards the explicit integration of Hamilton's equations, as for doing this by the known methods, a large number of constants of motion was required.

The fundamental rôle of Poisson structures in integrability was emphasized by Liouville. As he showed, having half of the dimension of phase space of independent constants of motion is sufficient for integrating the equations of motion, under the assumption that these constants of motion are in involution (Poisson commute). Precisely, integration means here integration by quadratures. A geometrical counterpart to this result states that, under some topological assumptions, the motion of such a Hamiltonian system evolves on tori (the so-called Liouville tori), whose dimension is half the dimension of phase space, and the motion on them is quasi-periodic. A more elaborate version of this theorem is the action-angle theorem, which yields a canonical model for Liouville integrable systems, in the neighborhood of a Liouville torus.

Liouville's theorems and the action-angle theorems will be presented here in the context of general Poisson manifolds. These manifolds generalize the phase spaces, considered in the classical results, namely the cotangent bundle of the configuration space, equipped with its canonical symplectic structure. We give the basic definitions and properties of functions in involution and of the map, associated to them, in Section 12.1. Several constructions of functions in involution are given in Section 12.2: Poisson's theorem, the Hamiltonian form of Noether's theorem, bi-Hamiltonian vector fields, Thimm's method, Lax equations and the Adler–Kostant–Symes theorem. The action-angle theorem for Poisson manifolds is stated and proved in Section 12.3.

C. Laurent-Gengoux et al., *Poisson Structures*,
Grundlehren der mathematischen Wissenschaften 347,
DOI 10.1007/978-3-642-31090-4_12, © Springer-Verlag Berlin Heidelberg 2013

12.1 Functions in Involution

In this section we give the basic definitions and the (geometrical) properties of functions in involution. Throughout the section, all manifolds considered are real ($\mathbb{F}=\mathbb{R}$) or complex ($\mathbb{F}=\mathbb{C}$). For such a manifold M, we denote its algebra of smooth or holomorphic functions by $\mathscr{F}(M)$.

Definition 12.1. Let (M,π) be a Poisson manifold and let F_1 and F_2 be two elements of $\mathscr{F}(M)$. We say that F_1 and F_2 are *in involution* if their Poisson bracket is zero, $\{F_1,F_2\}=0$. More generally, let $\mathbf{F}=(F_1,\dots,F_s)$ be an s-tuple of elements of $\mathscr{F}(M)$. We say that $\mathbf{F}$ is *involutive* if for all $1\leqslant i,j\leqslant s$, the functions F_i and F_j are in involution.

Classically, a function $F\in\mathscr{F}(M)$ which is constant on all the integral curves of a Hamiltonian vector field $\mathscr{X}_H$, where H is a given function on a Poisson manifold (M,π), is called a *constant of motion* (for H). Since $\mathscr{X}_H[F]=\{F,H\}$, we have that F is a constant of motion for H if and only if F and H are in involution. Therefore, the notion of involutive s-tuples generalizes the notion of constant of motion.

An s-tuple $\mathbf{F}=(F_1,\dots,F_s)$ of elements of $\mathscr{F}(M)$ defines a map $M\to\mathbb{F}^s$, which will be denoted by the same letter $\mathbf{F}$. We say that $\mathbf{F}$ is *independent* if the open subset $\mathscr{U}_\mathbf{F}$ of M, defined by

$$\mathscr{U}_\mathbf{F}:=\{m\in M\mid \mathrm{d}_mF_1\wedge\cdots\wedge\mathrm{d}_mF_s\neq 0\}\ , \tag{12.1}$$

is a dense subset of M. Besides the open subset $\mathscr{U}_\mathbf{F}$, defined by (12.1), we will in this chapter also make extensive use of the open subset

$$M_{(r)}:=\{m\in M\mid \mathrm{Rk}_m\,\pi=2r\}\ ,$$

where $2r$ stands for the rank of π. Notice that, if $\mathbf{F}$ is independent, then the open subset $\mathscr{U}_\mathbf{F}\cap M_{(r)}$ of M is non-empty.

When dealing with involutive functions, the reader should have the geometrical picture in mind, described by the following proposition (see also Fig. 12.1).

Proposition 12.2. *Let (M,π) be a Poisson manifold and assume that $\mathbf{F}=(F_1,\dots,F_s)$ is an involutive and independent s-tuple of elements of $\mathscr{F}(M)$. The Hamiltonian vector fields $\mathscr{X}_{F_i}$, $1\leqslant i\leqslant s$, commute and they are tangent to the smooth fibers of the map $\mathbf{F}:M\to\mathbb{F}^s$.*

Proof. The Hamiltonian vector fields, which are associated to functions in involution, commute, because $[\mathscr{X}_F,\mathscr{X}_G]=-\mathscr{X}_{\{F,G\}}$ for all $F,G\in\mathscr{F}(M)$. Assume that $\mathbf{F}=(F_1,\dots,F_s)$ is involutive and let $1\leqslant i\leqslant s$; since $\mathscr{X}_{F_i}[F_j]=\{F_j,F_i\}=0$ for all $1\leqslant j\leqslant s$, the vector field $\mathscr{X}_{F_i}$ is tangent to the smooth level sets of $F_1,\dots,F_s$, that is, to the smooth fibers of $\mathbf{F}$. □

The number of independent functions on a manifold M is bounded by the dimension of M. In the case of a Poisson manifold, the upper bound for the number of independent functions *in involution* is much lower, as given by the following proposition.

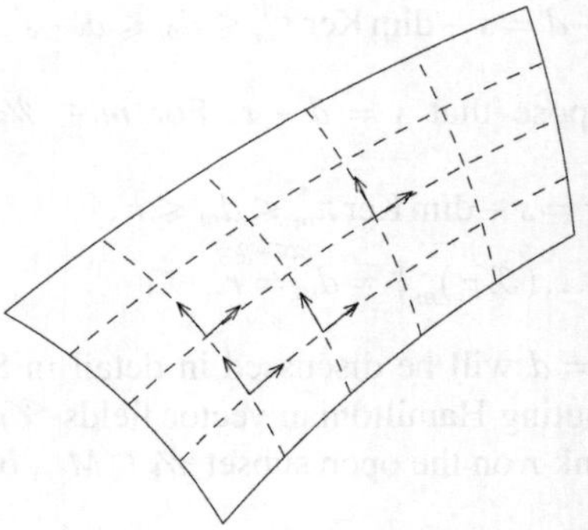

Fig. 12.1 The Hamiltonian vector fields of the components of $\mathbf{F} = (F_1, \ldots, F_s)$ are tangent to the smooth fibers $\mathbf{F}^{-1}(c)$ of $\mathbf{F}$ and they commute.

Proposition 12.3. *Let (M, π) be a Poisson manifold of dimension d and rank $2r$ and suppose that $\mathbf{F} = (F_1, \ldots, F_s)$ is an independent s-tuple of elements of $\mathscr{F}(M)$.*
(1) If $F_1, \ldots, F_s$ are Casimirs, then $s \leqslant d - 2r$;
(2) If $\mathbf{F}$ is involutive, then $s \leqslant d - r$;
(3) If $\mathbf{F}$ is involutive, with $s = d - r$, then

$$\dim \operatorname{span}\left\{ (\mathscr{X}_{F_1})_m, \ldots, (\mathscr{X}_{F_s})_m \right\} = r$$

for every $m \in \mathscr{U}_{\mathbf{F}} \cap M_{(r)}$.

Proof. For $m \in M$, let $\pi_m^\sharp$ denote the linear map $T_m^*M \to T_mM$, which corresponds to the Poisson structure. Explicitly, for $H \in \mathscr{F}(M)$,

$$\pi_m^\sharp(\mathrm{d}_m H) = -(\mathscr{X}_H)_m \,.$$

Recall that the rank of $\pi_m^\sharp$ is $\mathrm{Rk}_m \pi$. For every $H \in \mathrm{Cas}(M)$ the cotangent vector $\mathrm{d}_m H$ belongs to $\mathrm{Ker}\, \pi_m^\sharp$, whose dimension is $d - \mathrm{Rk}_m \pi$. Let $\mathbf{F} = (F_1, \ldots, F_s)$ be independent and let m be an element of the non-empty (open) subset $\mathscr{U}_{\mathbf{F}} \cap M_{(r)}$ of M. Suppose first that each element of $\mathbf{F}$ is a Casimir. Since $\mathrm{d}_m F_1, \ldots, \mathrm{d}_m F_s$ are independent elements of $\mathrm{Ker}\, \pi_m^\sharp$, we have that

$$s \leqslant \dim \mathrm{Ker}\, \pi_m^\sharp = d - 2r$$

and *(1)* follows. Next, suppose that $\mathbf{F}$ is involutive and consider $\mathbf{F}_m$, the fiber of $\mathbf{F}$, passing through m. Since $m \in \mathscr{U}_{\mathbf{F}} \cap M_{(r)}$, the restriction of $\mathbf{F}_m$ to a neighborhood U of m is a submanifold of dimension $d - s$ of U, passing through m. This dimension is an upper bound for the dimension d_m of $\operatorname{span}\left\{ (\mathscr{X}_{F_1})_m, \ldots, (\mathscr{X}_{F_s})_m \right\}$, because these s vectors are tangent to that fiber at m. Moreover, $d_m \geqslant s - \dim \mathrm{Ker}\, \pi_m^\sharp = s + 2r - d$, because the differential one-forms $\mathrm{d}F_1, \ldots, \mathrm{d}F_s$ are independent at m. Combining the two inequalities for d_m, we get

$$s+2r-d=s-\dim\operatorname{Ker}\pi_m^\sharp \leqslant d_m \leqslant d-s\,, \tag{12.2}$$

leading to *(2)*. Third, suppose that $s=d-r$. For $m\in\mathscr{U}_\mathbf{F}\cap M_{(r)}$ we deduce from (12.2) that

$$r=s-\dim\operatorname{Ker}\pi_m^\sharp \leqslant d_m \leqslant r\,,$$

so that $\dim\operatorname{span}\left\{(\mathscr{X}_{F_1})_m,\dots,(\mathscr{X}_{F_s})_m\right\}=d_m=r$. □

The extreme case $r+s=d$ will be discussed in detail in Section 12.3. Notice that, in this case, the commuting Hamiltonian vector fields $\mathscr{X}_{F_1},\dots,\mathscr{X}_{F_s}$ define an integrable distribution of rank r on the open subset $\mathscr{U}_\mathbf{F}\cap M_{(r)}$ of M.

12.2 Constructions of Functions in Involution

In this section we will give a list of constructions of functions in involution. The first two of them (Poisson's theorem and Noether's theorem (in its Hamiltonian form)) are very classical, the other ones were discovered quite recently. All manifolds which are considered in this section are real or complex; accordingly, $\mathbb{F}$ stands for $\mathbb{R}$ or for $\mathbb{C}$.

12.2.1 Poisson's Theorem

Poisson's theorem yields new constants of motion (of a given Hamiltonian H), starting from two given constants of motion.

Theorem 12.4 (Poisson's theorem). *Let (M,π) be a Poisson manifold and let F,G and H be elements of $\mathscr{F}(M)$. If F and G are constants of motion for H, then their Poisson bracket $\{F,G\}$ is also a constant of motion for H.*

Proof. The proof is an easy consequence of the Jacobi identity, valid for all $F,G,H\in\mathscr{F}(M)$:

$$\{F,\{G,H\}\}+\{G,\{H,F\}\}+\{H,\{F,G\}\}=0\,. \tag{12.3}$$

Namely, if F and G are constants of motion for H, then the first two terms in (12.3) vanish. Then the remaining term also vanishes, that is, $\{F,G\}$ is a constant of motion for H. □

In practice, Poisson's theorem often does not yield a function which is independent of the given constants of motion; as we will see, it happens in many important cases that the Poisson bracket of two constants of motion is actually equal to zero.

12.2.2 Noether's Theorem

A second classical theorem which leads to constants of motion is Noether's theorem, which we give in a Hamiltonian form (the classical version is in a Lagrangian form). Let $\mathbf{G}$ be a Lie group, whose Lie algebra is denoted by $\mathfrak{g}$. Suppose that $\mathbf{G}$ acts on a Poisson manifold (M,π) and suppose that the action is Hamiltonian (Definition 5.38). Recall that this means in particular that one can construct a co-momentum map, which is a linear map $\tilde{\mu}:\mathfrak{g}\to\mathscr{F}(M)$, having the property that for every $x\in\mathfrak{g}$ the function $-\tilde{\mu}_x\in\mathscr{F}(M)$ is a Hamiltonian for the fundamental vector field $\underline{x}$, i.e.,

$$\underline{x}=-\mathscr{X}_{\tilde{\mu}_x}\ .$$

Noether's theorem, in its Hamiltonian form, then states that the co-momentum map yields constants of motion for every $\mathbf{G}$-invariant Hamiltonian.

Theorem 12.5 (Noether's theorem). *Let* $\mathbf{G}$ *be a group which acts on a Poisson manifold* (M,π) *and assume that the action is Hamiltonian, with co-momentum map* $\tilde{\mu}$. *If* H *is a* $\mathbf{G}$*-invariant function on* M, *then for every* $x\in\mathfrak{g}$ *the function* $\tilde{\mu}_x$ *is a constant of motion for* $\mathscr{X}_H$.

Proof. Let $x\in\mathfrak{g}$ and $m\in M$ be arbitrary. We denote for every $g\in\mathbf{G}$ by $\chi_g:M\to M$ the diffeomorphism which is induced by the action and we write $\{\cdot,\cdot\}$ for π. By $\mathbf{G}$-invariance of H, that is, $H\circ\chi_g=H$ for every $g\in\mathbf{G}$, we have that

$$\begin{aligned}\{H,\tilde{\mu}_x\}(m)&=\left(\mathscr{X}_{\tilde{\mu}_x}\right)_m[H]=-\underline{x}_m[H]\\&=-\frac{\mathrm{d}}{\mathrm{d}t}_{|t=0}H\left(\chi_{\exp(-tx)}(m)\right)=-\frac{\mathrm{d}}{\mathrm{d}t}_{|t=0}H(m)=0\,.\end{aligned}$$

Since $m\in M$ is arbitrary, this shows that the function $\tilde{\mu}_x$ is a constant of motion for H. □

12.2.3 Bi-Hamiltonian Vector Fields

A *bi-Hamiltonian manifold* (M,π_1,π_2) is a manifold M, equipped with two compatible Poisson structures $\pi_i=\{\cdot,\cdot\}_i$, $i=1,2$. Let $\mathscr{V}$ be a vector field on a bi-Hamiltonian manifold (M,π_1,π_2) and suppose that there exist functions F and G on M for which

$$\mathscr{V}=\{\cdot,F\}_1=\{\cdot,G\}_2\ . \tag{12.4}$$

Then $\mathscr{V}$ is called a *bi-Hamiltonian vector field*. Equation (12.4) implies that F and G are in involution with respect to both Poisson structures since $\{F,G\}_2=\{F,F\}_1=0$ and $\{G,F\}_1=\{G,G\}_2=0$. They are in involution with respect to every linear combination of these Poisson structures, which is itself a Poisson structure, because π_1 and π_2 are compatible.

More generally, suppose that we have a *bi-Hamiltonian hierarchy* on a bi-Hamiltonian manifold (M, π_1, π_2), i.e., a sequence of functions $\mathbf{F} = (F_i)_{i\in\mathbb{Z}}$ such that

$$\{\cdot, F_{i+1}\}_1 = \{\cdot, F_i\}_2 , \tag{12.5}$$

for every $i \in \mathbb{Z}$. In this case one has, for all $i < j \in \mathbb{Z}$,

$$\begin{aligned}\{F_i, F_j\}_1 &= \{F_i, F_{j-1}\}_2 \\ &= \{F_{i+1}, F_{j-1}\}_1 \\ &= \quad \cdots \\ &= \{F_j, F_i\}_1 ,\end{aligned}$$

so that $\{F_i, F_j\}_1 = 0$ by skew-symmetry. It follows that $\mathbf{F}$ is involutive with respect to π_1. Also, $\mathbf{F}$ is involutive with respect to π_2, because $\{F_i, F_j\}_2 = \{F_i, F_{j+1}\}_1$.

Such bi-Hamiltonian hierarchies may be constructed, under some assumptions, as follows. Given two compatible Poisson structures π_1 and π_2, one first considers, for arbitrary $\lambda \in \mathbb{F}$, the Poisson structure $\pi_\lambda := \pi_1 - \lambda\pi_2$. We set, as before, $\{\cdot,\cdot\}_\lambda := \pi_\lambda$. The one-dimensional family of Poisson structures $(\pi_\lambda)_{\lambda\in\mathbb{F}}$ is called a *Poisson pencil*. Suppose that there exists a Laurent polynomial in λ,

$$F_\lambda = \sum_{i=-k}^{\ell} \lambda^i F_i ,$$

such that $\{\cdot, F_\lambda\}_\lambda = 0$ for all $\lambda \in \mathbb{F}$. Such a Laurent polynomial is called a *Casimir for the Poisson pencil*. Setting $F_i := 0$ for $i > \ell$ and for $i < -k$, we can write

$$\{\cdot, F_\lambda\}_\lambda = \sum_{i\in\mathbb{Z}} \lambda^{i+1} (\{\cdot, F_{i+1}\}_1 - \{\cdot, F_i\}_2) .$$

It follows that, since F_λ is a Casimir for the Poisson pencil, the sequence of functions $(F_i)_{i\in\mathbb{Z}}$ satisfies (12.5) for all $i \in \mathbb{Z}$, hence is a bi-Hamiltonian hierarchy. By the above, $(F_i)_{i\in\mathbb{Z}}$ is involutive.

12.2.4 Poisson Maps and Thimm's Method

Thimm's method [191] for constructing large families of functions in involution is based on a few elementary observations, cleverly combined. The first one is that if $\Psi : (M_1, \{\cdot,\cdot\}_1) \to (M_2, \{\cdot,\cdot\}_2)$ is a Poisson map (between two Poisson manifolds), then the pull-back of every involutive s-tuple on M_2 yields an involutive s-tuple on M_1. This is simply because if F and G are functions on M_2 and $\{F, G\}_2 = 0$, then

$$\{F\circ\Psi, G\circ\Psi\}_1 = \{F, G\}_2 \circ \Psi = 0 .$$

The second one is that Lie algebra homomorphisms between finite-dimensional Lie algebras induce Poisson maps: let $\phi : \mathfrak{g}_1 \to \mathfrak{g}_2$ be a Lie algebra homomorphism, where $(\mathfrak{g}_1, [\cdot,\cdot]_1)$ and $(\mathfrak{g}_2, [\cdot,\cdot]_2)$ are finite-dimensional Lie algebras and consider the transpose map $\phi^\top : \mathfrak{g}_2^* \to \mathfrak{g}_1^*$, which assigns to $\xi_2 \in \mathfrak{g}_2^*$ the linear form $\phi^\top(\xi_2) : \mathfrak{g}_1 \to \mathbb{F}$, defined for all $x \in \mathfrak{g}_1$ by

$$\left\langle \phi^\top(\xi_2), x \right\rangle := \langle \xi_2, \phi(x) \rangle ,$$

where we recall that $\langle \cdot,\cdot \rangle$ stands for the canonical pairing between a vector space and its dual. Then $\phi^\top$ is a Poisson map, when both $\mathfrak{g}_1^*$ and $\mathfrak{g}_2^*$ are equipped with their Lie–Poisson structure (denoted $\{\cdot,\cdot\}_i$, for $i = 1,2$). To check this, it is sufficient to check that if $x, y \in \mathfrak{g}_1 \simeq (\mathfrak{g}_1^*)^*$, then

$$\left\{ x \circ \phi^\top, y \circ \phi^\top \right\}_2 = \{x, y\}_1 \circ \phi^\top . \tag{12.6}$$

Recall that under the canonical isomorphism $\mathfrak{g}_1 \simeq (\mathfrak{g}_1^*)^*$, the Lie–Poisson bracket $[x,y]_1$ corresponds to $\{x,y\}_1$, and similarly for $\mathfrak{g}_2$. Also, under the isomorphism $(\mathfrak{g}_2^*)^* \simeq \mathfrak{g}_2$, the linear map $x \circ \phi^\top : \mathfrak{g}_2^* \to \mathbb{F}$ corresponds to $\phi(x)$, for all $x \in \mathfrak{g}_1$. Therefore, (12.6) is equivalent to saying that ϕ is a Lie algebra homomorphism.

Combining both observations, it is clear that Lie algebra inclusions lead to functions in involution (on the dual of the larger Lie algebra). It has the following non-trivial consequence: if $\mathfrak{g}_0$ is a Lie subalgebra of $\mathfrak{g}$, then the Ad^*-invariant functions on $\mathfrak{g}_0^*$ provide involutive functions on $\mathfrak{g}^*$; notice that while the Hamiltonian vector fields of these functions are trivial on $\mathfrak{g}_0^*$, they are non-trivial, in general, on $\mathfrak{g}^*$.

Applied to a sequence of subalgebras, a family of functions in involution can be constructed on the dual of a Lie algebra (equipped with its Lie–Poisson structure). One usually refers to this technique as *Thimm's method*. Explicitly, let $\mathfrak{g}$ be a finite-dimensional Lie algebra and suppose that

$$\mathfrak{g}_1 \subset \mathfrak{g}_2 \subset \cdots \subset \mathfrak{g}_k \subset \mathfrak{g}_{k+1} = \mathfrak{g}$$

where $\mathfrak{g}_1, \ldots, \mathfrak{g}_k$ are Lie subalgebras of $\mathfrak{g}$. By the above, the functions on $\mathfrak{g}^*$, obtained by pulling back to $\mathfrak{g}^*$ all Ad^*-invariant functions on each of the spaces $\mathfrak{g}_1, \ldots, \mathfrak{g}_{k+1}$, are in involution. Notice that to this involutive family, the pull-back to $\mathfrak{g}^*$ of an arbitrary involutive set on $\mathfrak{g}_1^*$ can be added.

Thimm [191] also analyzes the case of two Lie subalgebras $\mathfrak{g}_1, \mathfrak{g}_2$ of the same Lie algebra $\mathfrak{g}$ (the above case corresponds to $\mathfrak{g}_1 \subset \mathfrak{g}_2$).

Proposition 12.6. *Let $\mathfrak{g}$ be a finite-dimensional Lie algebra and let $\mathfrak{g}_1$ and $\mathfrak{g}_2$ be two Lie subalgebras of $\mathfrak{g}$, with inclusion maps ι_i, whose transpose maps $\mathfrak{g}^* \to \mathfrak{g}_i^*$ are denoted by $\iota_i^\top$ $(i = 1,2)$. Suppose that*

$$[\mathfrak{g}_1, \mathfrak{g}_2] \subset \mathfrak{g}_2 . \tag{12.7}$$

Then for every function F_1 on $\mathfrak{g}_1^*$ and for every Ad^**-invariant function F_2 on $\mathfrak{g}_2^*$, the functions $F_1 \circ \iota_1^\top$ and $F_2 \circ \iota_2^\top$ on $\mathfrak{g}^*$ are in involution.*

Proof. Let us denote the Lie–Poisson brackets on $\mathfrak{g}$, respectively on $\mathfrak{g}_i$, by $\{\cdot,\cdot\}$, respectively $\{\cdot,\cdot\}_i$ $(i = 1,2)$. For all $\xi \in \mathfrak{g}^*$ we have that

$$\{F_1 \circ \iota_1^\top, F_2 \circ \iota_2^\top\}(\xi) = \langle \xi, [\mathrm{d}_\xi(F_1 \circ \iota_1^\top), \mathrm{d}_\xi(F_2 \circ \iota_2^\top)] \rangle
= \left\langle \xi, \left[\mathrm{d}_{\iota_1^\top(\xi)} F_1, \mathrm{d}_{\iota_2^\top(\xi)} F_2\right] \right\rangle .$$

In view of (12.7), the latter Lie bracket belongs to $\mathfrak{g}_2$, so we can replace ξ by $\iota_2^\top(\xi)$, which is the restriction of ξ to $\mathfrak{g}_2$. It follows that

$$\left\{F_1 \circ \iota_1^\top, F_2 \circ \iota_2^\top\right\}(\xi) = \left\langle \iota_2^\top(\xi), \left[\mathrm{d}_{\iota_1^\top(\xi)} F_1, \mathrm{d}_{\iota_2^\top(\xi)} F_2\right] \right\rangle = 0 ,$$

where we used Ad-invariance of F_2 in the last step (see (5.15)). □

12.2.5 Lax Equations

Let $M \subseteq \mathfrak{gl}_d(\mathbb{F})$ be an affine subspace of the Lie algebra of $(d \times d)$-matrices with entries in $\mathbb{F}$ and let $\mathscr{V}$ be a vector field on M. Suppose that there exists a matrix-valued function $\phi : M \to \mathfrak{gl}_d(\mathbb{F})$ such that

$$\mathscr{V}_x = [x, \phi(x)] \tag{12.8}$$

for all $x \in M$. Then $\mathscr{V}$ is said to be written in *Lax form* and (12.8) is called a *Lax equation*. As we have seen in Section 7.2, if $\mathfrak{g}$ is a quadratic Lie algebra, identified with its dual using its bilinear form, and equipped with its Lie–Poisson structure, then every Hamiltonian vector field on $\mathfrak{g}$ is of this form. We claim that all coefficients of the characteristic polynomial $\det(\lambda \mathbb{1}_d - x)$, which we view as functions (of x) on M, are constants of motion of $\mathscr{V}$. To see this, let $x(t)$ be an integral curve of $\mathscr{V}$, defined for t in a neighborhood of 0, and denote by $y(t)$ the value of ϕ at $x(t) \in M$. Take an arbitrary $i > 0$ and use $\mathrm{Trace}(AB) = \mathrm{Trace}(BA)$ to compute

$$\frac{\mathrm{d}}{\mathrm{d}t} \mathrm{Trace}\, x^i(t) = i\, \mathrm{Trace}\left(x^{i-1}(t) \frac{\mathrm{d}x}{\mathrm{d}t}(t)\right)
= i\, \mathrm{Trace}\left(x^i(t) y(t) - x^{i-1}(t) y(t) x(t)\right) = 0 .$$

Thus, the functions $H_i : x \mapsto \mathrm{Trace}\, x^i$ are constants of motion of $\mathscr{V}$. The same is true for each coefficient of the characteristic polynomial of x, viewed as a function on M, because each such coefficient is a polynomial in the functions H_i. In particular, if $\mathscr{V} = \mathscr{X}_H$ is a Hamiltonian vector field, with Hamiltonian H, then H is in involution with each H_i, and with each coefficient of the characteristic polynomial of x.

12.2.6 The Adler–Kostant–Symes Theorem

The Adler–Kostant–Symes theorem is a fundamental theorem in the modern theory of integrable systems, which allows one to construct, from Lie algebra splittings, on the one hand large families of independent functions in involution and on the other hand explicit solutions to Hamilton's equations, with either one of these functions as Hamiltonian. We will focus on the case in which the Lie algebra is quadratic, because in the known examples where the Adler–Kostant–Symes theorem is applied, the Lie algebra is usually quadratic.

Let $\mathfrak{g}$ be a finite-dimensional Lie algebra. We recall from Section 10.1 that a linear map $R : \mathfrak{g} \to \mathfrak{g}$ is called an R-matrix of $\mathfrak{g}$ if the skew-symmetric bilinear map $[\cdot\,,\cdot]_R : \mathfrak{g} \times \mathfrak{g} \to \mathfrak{g}$, defined for $x, y \in \mathfrak{g}$ by

$$[x,y]_R := \frac{1}{2}\left([Rx,y] + [x,Ry]\right) ,$$

defines a Lie bracket on $\mathfrak{g}$. If $\mathfrak{g}$ is quadratic, i.e., $\mathfrak{g}$ is equipped with an ad-invariant non-degenerate symmetric bilinear form $\langle\cdot\,|\,\cdot\rangle$, then the original Lie bracket $[\cdot\,,\cdot]$ and the R-bracket $[\cdot\,,\cdot]_R$ lead to two Lie–Poisson structures $\{\cdot\,,\cdot\}$ and $\{\cdot\,,\cdot\}_R$ on $\mathfrak{g}$. In order to describe these, recall that to a function $F \in \mathscr{F}(\mathfrak{g})$ and an element $x \in \mathfrak{g}$, an element $\nabla_x F$ of $\mathfrak{g}$ is associated by setting $\langle \nabla_x F \,|\, \cdot\rangle = \mathrm{d}_x F$, that is,

$$\langle \nabla_x F \,|\, y\rangle = \langle \mathrm{d}_x F, y\rangle ,$$

for all $y \in \mathfrak{g}$. Using this notation, the two Lie–Poisson structures on $\mathfrak{g}$ take the form

$$\begin{aligned} \{F,G\}(x) &= \langle x \,|\, [\nabla_x F, \nabla_x G]\rangle , \\ \{F,G\}_R(x) &= \langle x \,|\, [\nabla_x F, \nabla_x G]_R\rangle , \end{aligned} \tag{12.9}$$

for $F, G \in \mathscr{F}(\mathfrak{g})$ at $x \in \mathfrak{g}$ (see Section 7.2). The involutivity theorem yields functions in involution, coming from the R-bracket; moreover, their Hamiltonian vector fields can be written in a Lax form.

Theorem 12.7 (Involutivity theorem). *Let $(\mathfrak{g}, [\cdot\,,\cdot])$ be a finite-dimensional quadratic Lie algebra, whose bilinear form is denoted by $\langle\cdot\,|\,\cdot\rangle$. Let R be an R-matrix for $\mathfrak{g}$ and denote the Lie–Poisson bracket on $\mathfrak{g}$, which corresponds to the R-bracket, by $\{\cdot\,,\cdot\}_R$. If F and H are* Ad*-invariant functions on $\mathfrak{g}$, then*

(1) $\{F,H\}_R = 0$;
(2) The Hamiltonian vector field of H with respect to $\{\cdot\,,\cdot\}_R$ is given, at $x \in \mathfrak{g}$, by

$$(\mathscr{X}_H)_x = -\frac{1}{2}[x, R(\nabla_x H)] . \tag{12.10}$$

Proof. Recall that the Ad-invariance of H means that $[\nabla_x H, x] = 0$ for all $x \in \mathfrak{g}$ (see Section 5.1.4). For functions F and H on $\mathfrak{g}$, their R-bracket is given, at $x \in \mathfrak{g}$, by

$$\begin{aligned}\{F,H\}_R(x) &= \frac{1}{2}\langle x \mid [R(\nabla_x F), \nabla_x H]\rangle + \frac{1}{2}\langle x \mid [\nabla_x F, R(\nabla_x H)]\rangle \\ &= \frac{1}{2}\langle R(\nabla_x F) \mid [\nabla_x H, x]\rangle + \frac{1}{2}\langle R(\nabla_x H) \mid [x, \nabla_x F]\rangle \ ,\end{aligned}$$

which is zero when F and H are both Ad-invariant; if only H is assumed Ad-invariant, only the first term vanishes, leading to

$$\{F,H\}_R(x) = \frac{1}{2}\langle \nabla_x F \mid [R(\nabla_x H), x]\rangle = \frac{1}{2}\langle \mathrm{d}_x F, [R(\nabla_x H), x]\rangle \ ,$$

which proves (12.10). □

We will now refine the involutivity theorem to obtain the (historically older) Adler–Kostant–Symes theorem, in which the R-matrix comes from a Lie algebra splitting. Recall from Section 10.1.2 that if $(\mathfrak{g}, [\cdot\,,\cdot])$ is a finite-dimensional Lie algebra and $\mathfrak{g}$ is written as a vector space direct sum $\mathfrak{g} = \mathfrak{g}_+ \oplus \mathfrak{g}_-$ of two Lie subalgebras $\mathfrak{g}_+$ and $\mathfrak{g}_-$ of $\mathfrak{g}$, then $\mathfrak{g}_+ \oplus \mathfrak{g}_-$ is called a *Lie algebra splitting*, and that such a splitting leads to a natural R-matrix. Denoting the projections of $x \in \mathfrak{g}$ on $\mathfrak{g}_+$ and $\mathfrak{g}_-$ by x_+, respectively x_-, the endomorphism $R : \mathfrak{g} \to \mathfrak{g}$ is defined by

$$Rx := x_+ - x_- \ ,$$

for all $x \in \mathfrak{g}$. As in the case of the involutivity theorem, we assume that $\mathfrak{g}$ is a quadratic Lie algebra, with bilinear form $\langle \cdot \mid \cdot \rangle$. The Lie–Poisson bracket $\{\cdot\,,\cdot\}_R$ on $\mathfrak{g}$ can then also be written as

$$\{F,G\}_R(x) = \langle x \mid [(\nabla_x F)_+, (\nabla_x G)_+]\rangle - \langle x \mid [(\nabla_x F)_-, (\nabla_x G)_-]\rangle \ .$$

Non-degeneracy of $\langle \cdot \mid \cdot \rangle$ implies that $\mathfrak{g}$ admits another (vector space) direct sum decomposition, namely as $\mathfrak{g} = \mathfrak{g}_+^\perp \oplus \mathfrak{g}_-^\perp$ where $\mathfrak{g}_+^\perp$ (respectively $\mathfrak{g}_-^\perp$) is the orthogonal complement (with respect to $\langle \cdot \mid \cdot \rangle$) of $\mathfrak{g}_+$ (respectively $\mathfrak{g}_-$).

The Adler–Kostant–Symes theorem (AKS theorem), which we formulate and prove next, yields functions in involution on certain translates of $\mathfrak{g}_-^\perp$ (items *(1)–(3)* below) and the explicit integration of their Hamiltonian vector fields (item *(4)*).

Theorem 12.8 (AKS theorem). *Suppose that $\mathfrak{g} = \mathfrak{g}_+ \oplus \mathfrak{g}_-$ is a Lie algebra splitting, with associated R-matrix R, and that $\langle \cdot \mid \cdot \rangle$ is an* Ad*-invariant non-degenerate symmetric bilinear form on $\mathfrak{g}$. Let F and H be* Ad*-invariant functions on $\mathfrak{g}$ and suppose that $\varepsilon \in \mathfrak{g}$ satisfies*

$$[\varepsilon, \mathfrak{g}_+] \subset \mathfrak{g}_+^\perp, \qquad [\varepsilon, \mathfrak{g}_-] \subset \mathfrak{g}_-^\perp \ . \tag{12.11}$$

The restrictions of F and H to $\varepsilon + \mathfrak{g}_-^\perp$ are denoted by F_ε and H_ε.

(1) *The affine subspace $\varepsilon + \mathfrak{g}_-^\perp$ of $\mathfrak{g}$ is a Poisson submanifold of $(\mathfrak{g}, \{\cdot\,,\cdot\}_R)$; the restricted Poisson structure on $\varepsilon + \mathfrak{g}_-^\perp$ will be denoted by $\{\cdot\,,\cdot\}_\varepsilon$;*

(2) $\{F_\varepsilon, H_\varepsilon\}_\varepsilon = 0$;

(3) The Hamiltonian vector field of H_ε with respect to $\{\cdot,\cdot\}_\varepsilon$ is given by

$$\mathscr{X}_{H_\varepsilon}(y) = -\frac{1}{2}[y, R(\nabla_y H)] = \pm[y, (\nabla_y H)_\mp],$$

where $y \in \varepsilon + \mathfrak{g}_-^\perp$;

(4) Let $\mathbf{G}$ be a Lie group, whose Lie algebra is $\mathfrak{g}$, and let $\mathbf{G}_+$ and $\mathbf{G}_-$ denote the Lie subgroups of $\mathbf{G}$, corresponding to $\mathfrak{g}_+$, respectively $\mathfrak{g}_-$. For $x_0 \in \mathfrak{g}$ and for $|t|$ small, let $g_+(t)$ and $g_-(t)$ denote the smooth curves in $\mathbf{G}_+$ respectively $\mathbf{G}_-$ which solve the factorization problem[1]

$$\exp(-t\nabla_{x_0} H) = g_+(t)^{-1} g_-(t), \qquad g_\pm(0) = e. \tag{12.12}$$

Then the integral curve of $\mathscr{X}_H := \{\cdot, H\}_R$, which starts at $x_0 \in \mathfrak{g}$ (or in particular of $\mathscr{X}_{H_\varepsilon}$, starting at $x_0 \in \varepsilon + \mathfrak{g}_-^\perp$), is given, for $|t|$ small, by

$$x(t) = \mathrm{Ad}_{g_+(t)} x_0 = \mathrm{Ad}_{g_-(t)} x_0. \tag{12.13}$$

Proof. We first show that $\varepsilon + \mathfrak{g}_-^\perp$ is a Poisson submanifold of $(\mathfrak{g}, \{\cdot,\cdot\}_R)$. Let $x \in \mathfrak{g}_-^\perp$ and let H be a function, defined in a neighborhood of $\varepsilon + x$ in $\mathfrak{g}$. We need to show that $(\mathscr{X}_H)_{\varepsilon+x} \in T_{\varepsilon+x}\mathfrak{g}_-^\perp \simeq \mathfrak{g}_-^\perp$. Let F be an arbitrary function, defined on a neighborhood of $\varepsilon + x$ in $\mathfrak{g}$, which is constant on $\varepsilon + \mathfrak{g}_-^\perp$. Then $\nabla_{\varepsilon+x} F \in \mathfrak{g}_-$ since

$$\langle \nabla_{\varepsilon+x} F \,|\, y\rangle = \frac{\mathrm{d}}{\mathrm{d}t}_{|t=0} F(\varepsilon + x + ty) = 0,$$

for all $y \in \mathfrak{g}_-^\perp$. It follows from ad-invariance of $\langle\cdot\,|\,\cdot\rangle$ and from the assumptions (12.11) on ε that

$$\begin{aligned}(\mathscr{X}_H)_{\varepsilon+x}[F] &= \{F, H\}_R(\varepsilon + x)\\ &= \langle \varepsilon + x \,|\, [(\nabla_{\varepsilon+x}F)_+, (\nabla_{\varepsilon+x}H)_+] - [(\nabla_{\varepsilon+x}F)_-, (\nabla_{\varepsilon+x}H)_-]\rangle\\ &= \langle x \,|\, [(\nabla_{\varepsilon+x}F)_+, (\nabla_{\varepsilon+x}H)_+]\rangle - \langle x \,|\, [(\nabla_{\varepsilon+x}F)_-, (\nabla_{\varepsilon+x}H)_-]\rangle.\end{aligned}$$

Both of the above terms are equal to zero: the first one because $\nabla_{\varepsilon+x} F \in \mathfrak{g}_-$ and the second one because $x \in \mathfrak{g}_-^\perp$, combined with the fact that $\mathfrak{g}_-$ is a Lie subalgebra (of $\mathfrak{g}$). This proves that all Hamiltonian vector fields $\mathscr{X}_H$ are tangent to $\varepsilon + \mathfrak{g}_-^\perp$, hence that $\varepsilon + \mathfrak{g}_-^\perp$ is a Poisson submanifold of $(\mathfrak{g}, \{\cdot,\cdot\}_R)$, which is the content of *(1)*. Since $(\varepsilon + \mathfrak{g}_-^\perp, \{\cdot,\cdot\}_\varepsilon)$ is a Poisson submanifold of $(\mathfrak{g}, \{\cdot,\cdot\}_R)$, the involutivity of two Ad-invariant functions F and H on $\mathfrak{g}$ (Theorem 12.7) implies the involutivity of F_ε and H_ε on $\varepsilon + \mathfrak{g}_-^\perp$, which proves *(2)*; similarly, the first equality in *(3)* follows from *(1)* and from the involutivity theorem (Theorem 12.7), while the second equality is a direct consequence of the obvious formulas

$$R(\nabla_x H) = (\nabla_x H)_+ - (\nabla_x H)_- = 2(\nabla_x H)_+ - \nabla_x H = \nabla_x H - 2(\nabla_x H)_-.$$

---

[1] For g in a neighborhood of the identity $e \in \mathbf{G}$, the factorization $g = g_+ g_-$, with $g_\pm \in \mathbf{G}_\pm$ exists and is unique.

We now turn to *(4)*, the integral curves of $\mathscr{X}_H$. We first show that $\mathrm{Ad}_{g_+(t)}x_0 = \mathrm{Ad}_{g_-(t)}x_0$ (the second equality in (12.13)). Since Ad is a group homomorphism, the factorization (12.12) implies that

$$\mathrm{Ad}_{\exp(-t\nabla_{x_0}H)}x_0 = \mathrm{Ad}_{g_+(t)^{-1}}\mathrm{Ad}_{g_-(t)}x_0\ ,$$

for all $x_0 \in \mathfrak{g}$. We have for every $x \in \mathfrak{g}$ that

$$\begin{aligned}\left\langle \mathrm{Ad}_{\exp(-t\nabla_{x_0}H)}x_0 \,|\, x\right\rangle &= \left\langle x_0 \,|\, \mathrm{Ad}_{\exp(t\nabla_{x_0}H)}x\right\rangle \\ &= \left\langle x_0 \,|\, \exp\left(t\,\mathrm{ad}_{\nabla_{x_0}H}\right)x\right\rangle \\ &= \langle x_0 \,|\, x\rangle + t\,\langle x_0 \,|\, [\nabla_{x_0}H, \star]\rangle \\ &= \langle x_0 \,|\, x\rangle + t\,\langle [x_0, \nabla_{x_0}H] \,|\, \star\rangle \\ &= \langle x_0 \,|\, x\rangle\ ,\end{aligned}$$

where the value of $\star$ (which depends on t) is irrelevant, in view of the Ad-invariance of H and $\langle\cdot\,|\,\cdot\rangle$. This shows that $\mathrm{Ad}_{\exp(-t\nabla_{x_0}H)}x_0 = x_0$, and hence that $\mathrm{Ad}_{g_+(t)}x_0 = \mathrm{Ad}_{g_-(t)}x_0$.

We now show that $x(t) := \mathrm{Ad}_{g_+(t)}x_0$ is (for small $|t|$) a solution to (12.10), which amounts to proving that

$$\frac{\mathrm{d}}{\mathrm{d}t}\mathrm{Ad}_{g_+(t)}x_0 = -\left[x(t), \left(\nabla_{x(t)}H\right)_+\right]\ . \tag{12.14}$$

In order to simplify the proof (mainly the notation), we will assume that **G** is a linear group, so that Ad is just given by conjugation and we will write $\dot{g}_\pm(t)$ for $\frac{dg_\pm}{dt}(t)$. Then the left-hand side of (12.14) is given by $\left[\dot{g}_+(t)g_+(t)^{-1}, x(t)\right]$, so it suffices to show that

$$\dot{g}_+(t)g_+(t)^{-1} = (\nabla_{x(t)}H)_+\ .$$

Differentiating the identity $g_+(t)\exp(-t\nabla_{x_0}H) = g_-(t)$ with respect to t and multiplying both sides of the result by $g_-(t)^{-1}$, we get

$$\dot{g}_+(t)g_+(t)^{-1} - g_-(t)\left(\nabla_{x_0}H\right)g_-(t)^{-1} = \dot{g}_-(t)g_-(t)^{-1}\ .$$

Taking the $+$ part of both sides of this equation, we find

$$\dot{g}_+(t)g_+(t)^{-1} = \left(g_-(t)\left(\nabla_{x_0}H\right)g_-(t)^{-1}\right)_+ = \left(\nabla_{x(t)}H\right)_+\ ,$$

where we have used Ad-invariance of H and of $\langle\cdot\,|\,\cdot\rangle$ in the last step. □

12.3 The Liouville Theorem and the Action-Angle Theorem

In this section, we formulate and prove two basic theorems in the theory of integrable systems: the Liouville theorem and the action-angle theorem. Classically, these theorems are considered for Liouville integrable systems on symplectic manifolds; we consider these theorems here in the more natural context of Poisson manifolds. We follow [3] in Section 12.3.1 and [120] in Sections 12.3.2 to 12.3.4. Throughout the section, all manifolds are real and $\mathscr{F}(M)$ stands for the algebra of smooth functions on the (real) manifold M.

12.3.1 Liouville Integrable Systems and Liouville's Theorem

We first give the definition of a Liouville integrable system on a real Poisson manifold.

Definition 12.9. Let (M,π) be a real Poisson manifold of rank $2r$ and of dimension d. Let $\mathbf{F}=(F_1,\dots,F_s)$ be an s-tuple of smooth functions on M and suppose that

(1) $\mathbf{F}$ is independent;
(2) $\mathbf{F}$ is in involution;
(3) $r+s=d$.

Then $(M,\pi,\mathbf{F})$ is called a *Liouville integrable system* of *dimension d* and *rank* $2r$. Viewed as a map, $\mathbf{F}:M\to\mathbb{R}^s$ is called the *momentum map* of $(M,\pi,\mathbf{F})$.

Suppose that $(M,\pi,\mathbf{F})$ is a Liouville integrable system, where (M,π) is a Poisson manifold of rank $2r$. We recall that $M_{(r)}$ denotes the open subset of M where the rank of π is equal to $2r$, and $\mathscr{U}_\mathbf{F}$ denotes the dense open subset of M, which consists of all points of M where the differentials of the elements of $\mathbf{F}$ are linearly independent. We consider on the non-empty open subset $\mathscr{U}_\mathbf{F}\cap M_{(r)}$ of M two (regular) foliations:

(1) The Hamiltonian vector fields $\mathscr{X}_{F_1},\dots,\mathscr{X}_{F_s}$ define, according to Propositions 12.2 and 12.3, on the non-empty open subset $\mathscr{U}_\mathbf{F}\cap M_{(r)}$ of M an integrable distribution $\mathscr{D}$ of rank r. The integral manifolds of $\mathscr{D}$ are the leaves of a foliation, which we denote by $\mathscr{F}$; the leaf of $\mathscr{F}$, passing through $m\in\mathscr{U}_\mathbf{F}\cap M_{(r)}$, is denoted by $\mathscr{F}_m$, and is called the *invariant manifold* of $\mathbf{F}$, through m;
(2) Since, by definition, the restriction $\overline{\mathbf{F}}$ of $\mathbf{F}$ to $\mathscr{U}_\mathbf{F}\cap M_{(r)}$ is a submersion, its fibers define a foliation on $\mathscr{U}_\mathbf{F}\cap M_{(r)}$, which is denoted by $\overline{\mathscr{F}}$. Thus, the leaves of $\overline{\mathscr{F}}$ are the connected components of the fibers of the map

$$\overline{\mathbf{F}}=(F_1,\dots,F_s):\mathscr{U}_\mathbf{F}\cap M_{(r)}\to\mathbb{R}^s\ .$$

We show in the following proposition that both foliations $\mathscr{F}$ and $\overline{\mathscr{F}}$ coincide, leading to a description of the invariant manifolds of $\mathbf{F}$.

Proposition 12.10. *Let $(M,\pi,\mathbf{F})$ be a Liouville integrable system and let $\mathscr{F}$ and $\overline{\mathscr{F}}$ denote the foliations, defined above.*

(1) The foliations $\mathscr{F}$ and $\overline{\mathscr{F}}$ coincide;
(2) For $m \in \mathscr{U}_{\mathbf{F}} \cap M_{(r)}$, the invariant manifold $\mathscr{F}_m$ of $\mathbf{F}$ is the connected component of the fiber of $\mathbf{F}$ which contains m. In particular, $\mathscr{F}_m$ is an (embedded) submanifold of M.

Proof. We prove item *(1)*; item *(2)* is an immediate consequence of it. Since all leaves of $\overline{\mathscr{F}}$ and of $\mathscr{F}$ are r-dimensional (and are, by definition, connected), it suffices to show for every $m \in \mathscr{U}_{\mathbf{F}} \cap M_{(r)}$ the following inclusion of tangent spaces: $T_m\mathscr{F} \subset T_m\overline{\mathscr{F}}$. According to Proposition 12.2, the fact that $F_1,\ldots,F_s$ are pairwise in involution implies that each of the Hamiltonian vector fields $\mathscr{X}_{F_1},\ldots,\mathscr{X}_{F_s}$, at m, belong to $T_m\overline{\mathscr{F}}$. Since $T_m\mathscr{F}$ is spanned by these vector fields, the required inclusion follows. □

We now come to the Liouville theorem for Liouville integrable systems on Poisson manifolds.

Theorem 12.11 (Liouville's theorem). *Let $(M,\pi,\mathbf{F})$ be a Liouville integrable system of dimension d and rank $2r$. For $m \in \mathscr{U}_{\mathbf{F}} \cap M_{(r)}$, let $\mathscr{F}_m$ denote the invariant manifold of $\mathbf{F} = (F_1,\ldots,F_s)$ which passes through m.*

(1) If $\mathscr{F}_m$ is compact, then there exists a diffeomorphism from $\mathscr{F}_m$ to the torus $\mathbb{T}^r = (\mathbb{R}/\mathbb{Z})^r$, under which the vector fields $\mathscr{X}_{F_1},\ldots,\mathscr{X}_{F_s}$ are mapped to constant (i.e., translation-invariant) vector fields;
(2) If $\mathscr{F}_m$ is not compact, but the flow of each of the vector fields $\mathscr{X}_{F_i}$ $(i = 1,\ldots,s)$ is complete on $\mathscr{F}_m$, then there exists a diffeomorphism from $\mathscr{F}_m$ to a cylinder $\mathbb{R}^{r-q} \times \mathbb{T}^q$ $(0 \leqslant q < r)$, under which the vector fields $\mathscr{X}_{F_i}$ are mapped to constant vector fields.

Proof. We suppose that the flow $\Phi^{(i)}$ of each of the integrable vector fields $\mathscr{X}_{F_i}$ is complete on $\mathscr{F}_m$, i.e., is defined for all $t \in \mathbb{R}$. We may order the functions F_i such that the r vector fields $\mathscr{X}_{F_1},\ldots,\mathscr{X}_{F_r}$ are independent at m. These vector fields are then independent at every point of $\mathscr{F}_m$. Indeed, since the vector fields $\mathscr{X}_{F_i}$ pairwise commute,

$$\mathscr{L}_{\mathscr{X}_{F_j}}(\mathscr{X}_{F_1} \wedge \cdots \wedge \mathscr{X}_{F_r}) = \sum_{i=1}^{r} \mathscr{X}_{F_1} \wedge \cdots \wedge \left[\mathscr{X}_{F_j}, \mathscr{X}_{F_i}\right] \wedge \cdots \wedge \mathscr{X}_{F_r} = 0\,,$$

for $j = 1,\ldots,s$. It means that $\mathscr{X}_{F_1} \wedge \cdots \wedge \mathscr{X}_{F_r}$ is conserved by the flow of each one of the vector fields $\mathscr{X}_{F_1},\ldots,\mathscr{X}_{F_s}$. In particular, since this r-vector field is non-vanishing at m, it is non-vanishing on the entire invariant manifold $\mathscr{F}_m$. As a consequence, $\mathscr{F}_m$ is a leaf of the distribution, defined by the first r vector fields $\mathscr{X}_{F_1},\ldots,\mathscr{X}_{F_r}$, in a neighborhood of $\mathscr{F}_m$.

The completeness and commutativity of the vector fields $\mathscr{X}_{F_1},\ldots,\mathscr{X}_{F_r}$ on $\mathscr{F}_m$ imply that we can define an action $\mathbb{R}^r \times \mathscr{F}_m \to \mathscr{F}_m$ by

$$((t_1,\ldots,t_r),m') \mapsto \Phi^{(1)}_{t_1} \circ \Phi^{(2)}_{t_2} \circ \cdots \circ \Phi^{(r)}_{t_r}(m') .$$

Since $\mathscr{F}_m$ is the integral manifold through m of the distribution defined by the first r integrable vector fields, the action is transitive on $\mathscr{F}_m$ and $\mathscr{F}_m$ becomes a homogeneous space. The action is also locally free, because the vector fields $\mathscr{X}_{F_i}$ are independent at every point of $\mathscr{F}_m$. Therefore the stabilizer is a discrete subgroup H_m of $\mathbb{R}^r$ and $\mathscr{F}_m$ is diffeomorphic to $\mathbb{R}^r/H_m$. If $\mathscr{F}_m$ is compact, then H_m must be a lattice, so $\mathbb{R}^r/H_m$ is a torus, smoothly embedded into M. Otherwise H_m is a discrete subgroup whose rank q is at most $r-1$ and $\mathbb{R}^r/H_m$ is isomorphic to $\mathbb{R}^{r-q}\times\mathbb{T}^q$. By construction, the vector fields $\mathscr{X}_{F_i}$ are mapped to translation-invariant vector fields in both cases. □

For what follows, we will be uniquely interested in the case in which $\mathscr{F}_m$ is compact. Then $\mathscr{F}_m$ is a torus; we call such a torus a *standard Liouville torus*.

12.3.2 Foliation by Standard Liouville Tori

We show in this section that, in some neighborhood of a standard Liouville torus, the invariant manifolds of a Liouville integrable system $(M,\pi,\mathbf{F})$ of dimension d and rank $2r$ form a trivial torus fibration over an open ball B^{d-r}.

Proposition 12.12. *Suppose that $\mathscr{F}_m$ is a standard Liouville torus of a Liouville integrable system $(M,\pi,\mathbf{F})$ of dimension d and rank $2r$. There exists an open subset $U \subset \mathscr{U}_{\mathbf{F}} \cap M_{(r)}$, containing $\mathscr{F}_m$, and there exists a diffeomorphism $\phi : U \to \mathbb{T}^r \times B^{d-r}$, which takes the foliation $\mathscr{F}$ to the foliation defined by the fibers of the canonical projection $p_B : \mathbb{T}^r \times B^{d-r} \to B^{d-r}$, leading to the following commutative diagram.*

$$\begin{array}{ccccc}
\overline{\mathscr{F}}_m = \mathscr{F}_m & \hookrightarrow & U & \xrightarrow[\simeq]{\ \phi\ } & \mathbb{T}^r \times B^{d-r} \\
 & & \downarrow {\scriptstyle \mathbf{F}_{|U}} & \swarrow {\scriptstyle p_B} & \\
 & & B^{d-r} & &
\end{array}$$

Moreover, U and ϕ can be chosen such that the last $d-2r$ components of ϕ are Casimirs of π, restricted to U.

Proof. Suppose that $\mathscr{F}_m$ is a standard Liouville torus. Setting $s := d-r$, as before, we first show that there exists a neighborhood U of $\mathscr{F}_m$ and a diffeomorphism $\phi : U \to \mathscr{F}_m \times B^s$, which sends the foliation $\mathscr{F}$ (which, according to Proposition 12.10, coincides with $\mathscr{F}$), restricted to U, to the foliation defined by p_B on $\mathscr{F}_m \times B^s$. The proof of this statement depends only on the fact that $\mathscr{F}_m$ is a compact component of a fiber of a submersion (namely $\overline{\mathbf{F}}$; see Proposition 12.10). Notice that since $\overline{\mathbf{F}}$ is a submersion, every point $m' \in \overline{\mathscr{F}}_m = \mathscr{F}_m$ has a neighborhood $U_{m'}$ in M, which is diffeomorphic to the product of a neighborhood $V_{m'}$ of m' in $\mathscr{F}_m$ times an open ball $B^s_{m'}$, centered at $\overline{\mathbf{F}}(m') = \overline{\mathbf{F}}(m)$ in $\mathbb{R}^s$; such a diffeomorphism $\phi_{m'}$, as provided by the implicit function theorem, is a lifting of $\overline{\mathbf{F}}$, that is, it leads to the following commutative diagram:

$$\begin{array}{ccc} U_{m'} & \overset{\phi_{m'}}{\dashrightarrow} & V_{m'} \times B^s_{m'} \\ & \underset{\overline{\mathbf{F}}}{\searrow} & \downarrow p_B \\ & & B^s_{m'} \end{array}$$

Since $\mathscr{F}_m$ is compact, it is covered by finitely many of the sets $V_{m'}$, say $V_{m_1}, \dots, V_{m_\ell}$. Thus, if every pair of the diffeomorphisms $\phi_{m_1}, \dots, \phi_{m_\ell}$ agrees on the intersection of their domain of definition (whenever non-empty), we can define a global diffeomorphism on a neighborhood U of $\mathscr{F}_m$, whose image is the intersection of the concentric balls $B^s_{m_1}, \dots, B^s_{m_\ell}$. In order to ensure that these diffeomorphisms agree, we need to chose them in a more specific way. This is done by choosing an arbitrary Riemannian metric on M. Using the exponential map, defined by the metric, we can identify a neighborhood of the zero section in the normal bundle of $\mathscr{F}_m$, with a neighborhood of $\mathscr{F}_m$ in M; in particular, for every $m' \in \mathscr{F}_m$ there exist neighborhoods $U_{m'}$ of m' in M and $V_{m'}$ of m' in $\mathscr{F}_m$, with smooth maps $\psi_{m'} : U_{m'} \to V_{m'}$, which have the important virtue that they agree on the intersection of their domains. Upon shrinking the open subsets $U_{m'}$, if necessary, the maps $\phi_{m'} := \psi_{m'} \times (F_1, \dots, F_s)$ are a choice of diffeomorphisms, defined on a neighborhood U of $\mathscr{F}_m$, with the required properties. This leads to the diffeomorphism $\phi : U \to \mathscr{F}_m \times B^s$, which we view as a diffeomorphism $\phi : U \to \mathbb{T}^r \times B^s$, using the fact that $\mathscr{F}_m$ is diffeomorphic with $\mathbb{T}^r$, as stated in the Liouville theorem.

It remains to be shown that ϕ can be chosen such that its last $d - 2r = s - r$ components are Casimirs. To do this, we consider on U the integrable distribution $\mathscr{D}'$ defined by all Hamiltonian vector fields on U; it has rank $2r$ and its leaves are the symplectic leaves of (U, π). It is clear that $\mathscr{D} \subset \mathscr{D}'$, where we recall that $\mathscr{D}$ is the distribution defined by the Hamiltonian vector fields $\mathscr{X}_{F_1}, \dots, \mathscr{X}_{F_s}$. Consider the submersive map

$$p_B \circ \phi : U \to \mathbb{T}^r \times B^s \to B^s ,$$

whose fibers are by assumption the leaves of $\mathscr{F}$, that is, the integral manifolds of $\mathscr{D}$ (restricted to U), so that the kernel of the tangent map $T(p_B \circ \phi)$ is precisely $\mathscr{D}$. The image of $\mathscr{D}'$ by $T(p_B \circ \phi)$ is therefore a (smooth) distribution $\mathscr{D}''$ of rank r on B^s, which is integrable, since $\mathscr{D}'$ is integrable. The foliation defined by the integral manifolds of $\mathscr{D}''$ is, in the neighborhood of the point $p_B(\phi(m))$, defined by $s - r = d - 2r$ independent functions $z'_1, \dots, z'_{d-2r}$. Pulling them back to M, we get functions $z_1, \dots, z_{d-2r}$ on a neighborhood $V \subset U$ of $\mathscr{F}_m$ which are Casimir functions, because they are constant on the leaves of $\mathscr{D}'$, which are the symplectic leaves of (U, π).

Since they have independent differentials on U, they can be chosen as $d-2r$ of the last $d-r$ components of ϕ, upon shrinking B^s into a smaller ball, if needed. □

12.3.3 Period Normalization

In this section, we study Liouville integrable systems of the form $(\mathbb{T}^r\times B^s,\pi,\mathbf{F})$, where $\mathbf{F}=p_B:\mathbb{T}^r\times B^s\to B^s$ is the projection on the second factor and π is a regular Poisson structure of rank $2r$ on $\mathbb{T}^r\times B^s$. Given such a Liouville integrable system, there is no reason why the Hamiltonian vector fields $\mathscr{X}_{F_i}$, which are tangent to the fibers of $\mathbf{F}$ (which are tori), would be *constant* on these fibers, and even if they were constant on these fibers, they might vary from one fiber to another in the sense that their flow may not come from an action of the torus $\mathbb{T}^r$ on $\mathbb{T}^r\times B^s$. We show in the following proposition that vector fields which satisfy the latter property can be constructed by taking well-chosen linear combinations of the Hamiltonian vector fields $\mathscr{X}_{F_i}$, with as coefficients $\mathbf{F}$-*basic functions*, i.e., functions of the form $\lambda\circ\mathbf{F}$, where $\lambda\in\mathscr{F}(B^s)$; equivalently, smooth functions on $\mathbb{T}^r\times B^s$ which are constant on the fibers of $\mathbf{F}$.

Proposition 12.13. *Let $(\mathbb{T}^r\times B^s,\pi,\mathbf{F})$ be a Liouville integrable system, where π has constant rank $2r$ and $\mathbf{F}=(F_1,\dots,F_s)$ is the projection on the second component. Suppose that among the s commuting vector fields $\mathscr{X}_{F_1},\dots,\mathscr{X}_{F_s}$, the first r are independent at every point of $\mathbb{T}^r\times B^s$. There exists a ball $B_0^s\subset B^s$, concentric with B^s, and there exist $\mathbf{F}$-basic functions $\lambda_i^j\in\mathscr{F}(B_0^s)$, such that the vector fields $\mathscr{V}_1,\dots,\mathscr{V}_r$, defined by*

$$\mathscr{V}_i:=\sum_{j=1}^r\lambda_i^j\,\mathscr{X}_{F_j}\,,$$

are the fundamental vector fields of a torus action of $\mathbb{T}^r$ on $\mathbb{T}^r\times B_0^s$.

Proof. The fibers of $\mathbf{F}=(F_1,\dots,F_s)$ are compact, so for $i=1,\dots,r$, the flow $\Phi^{(i)}_{t_i}$ of the Hamiltonian vector field $\mathscr{X}_{F_i}$ is complete and we can define a map,

$$\begin{aligned}\Phi:\mathbb{R}^r\times(\mathbb{T}^r\times B^s)&\to\mathbb{T}^r\times B^s\\ ((t_1,\dots,t_r),m)&\mapsto\Phi^{(1)}_{t_1}\circ\dots\circ\Phi^{(r)}_{t_r}(m)\,.\end{aligned}$$

Since the vector fields $\mathscr{X}_{F_i}$ are pairwise commuting, the flows $\Phi^{(i)}_{t_i}$ pairwise commute and Φ is an action of $\mathbb{R}^r$ on $\mathbb{T}^r\times B^s$. Since the vector fields $\mathscr{X}_{F_1},\dots,\mathscr{X}_{F_r}$ are independent at all points of $\mathbb{T}^r\times B^s$, the fibers of $\mathbf{F}$, which are r-dimensional tori, are the orbits of the action. For $c\in B^s$, let Λ_c denote the lattice of $\mathbb{R}^r$, which is the isotropy group of every point in $\mathbf{F}^{-1}(c)$; it is the period lattice of the action Φ, restricted to $\mathbf{F}^{-1}(c)$. In order to simplify the notation, we assume that B^s is centered at o.

Let m_o be an arbitrary point of $\mathbf{F}^{-1}(o)$ and choose a basis $(\lambda_1(o),\dots,\lambda_r(o))$ of the lattice Λ_o. For a fixed i, with $1\leqslant i\leqslant r$, for m in a neighborhood of m_o in $\mathbb{T}^r\times B^s$

and for L_i in a neighborhood of $\lambda_i(o)$ in $\mathbb{R}^r$, consider the equation $\Phi(L_i,m)=m$. Since $\mathbf{F}(\Phi(L_i,m))=\mathbf{F}(m)$ for all L_i and m, it is meaningful to write $\Phi(L_i,m)-m$ and solving the equation $\Phi(L_i,m)=m$ locally for L_i amounts to applying the implicit function theorem to the map

$$\mathbb{R}^r \times (\mathbb{T}^r \times B^s) \xrightarrow{\Phi(L_i,m)-m} \mathbb{T}^r \times B^s \longrightarrow \mathbb{T}^r .$$

Since the action is locally free, the Jacobian condition is satisfied and we get by solving for L_i around $\lambda_i(o)$ a smooth $\mathbb{R}^r$-valued function $\lambda_i(m)$, defined for m in a neighborhood W_i of m_o. Doing this for $i=1,\ldots,r$ and setting $W:=\cap_{i=1}^r W_i$, we have that W is a neighborhood of m_o, and on W we have functions $\lambda_1(m),\ldots,\lambda_r(m)$, with the property that $\Phi(\lambda_i(m),m)=m$ for all $m\in W$ and for all $1\leqslant i\leqslant r$. Thus, $\lambda_1(m),\ldots,\lambda_r(m)$ belong to the lattice $\Lambda_{\mathbf{F}(m)}$ for all $m\in W$ and they form a basis when $m=m_o$; by continuity, they form a basis of $\Lambda_{\mathbf{F}(m)}$ for all $m\in W$.

The functions λ_i can be extended to a neighborhood of the torus $\mathbf{F}^{-1}(o)$. In fact, the functions λ_i are $\mathbf{F}$-basic, hence extend uniquely to $\mathbf{F}$-basic functions on $\mathbf{F}^{-1}(\mathbf{F}(W))$. We will use in the sequel the same notation λ_i for these extensions and we write $\mathbf{F}^{-1}(\mathbf{F}(W))$ simply as W. Using these functions, we define the following smooth map:

$$\begin{array}{rccc} \tilde{\Phi}: & \mathbb{R}^r\times W & \to & W \\ & ((t_1,\ldots,t_r),m) & \mapsto & \Phi\left(\sum_{i=1}^r t_i\lambda_i(m),m\right). \end{array} \tag{12.15}$$

Since the functions λ_i are $\mathbf{F}$-basic, the fact that Φ is an action implies that $\tilde{\Phi}$ is an action. This action has the extra feature that the stabilizer of every point in W is $\mathbb{Z}^r$. Thus, $\tilde{\Phi}$ induces an action of $\mathbb{T}^r$ on W, which we still denote by $\tilde{\Phi}$. By shrinking W, if necessary, we may assume that W is of the form $\mathbf{F}^{-1}(B_0^s)$, where B_0^s is an open ball, concentric with B^s, and contained in it. Thus we have a torus action

$$\begin{array}{rccc} \tilde{\Phi}: & \mathbb{T}^r\times W & \to & W \\ & ((t_1,\ldots,t_r),m) & \mapsto & \Phi\left(\sum_{i=1}^r t_i\lambda_i(m),m\right). \end{array}$$

Looking at the action on a single point $m\in W$, it is clear that the fundamental vector fields of $\tilde{\Phi}$ are expressible as linear combinations of the Hamiltonian vector fields $\mathscr{X}_{F_i}$, with $\mathbf{F}$-basic functions as coefficients. □

12.3.4 The Existence of Action-Angle Coordinates

We formulate and prove in this section the action-angle theorem. The following lemma plays a key rôle in the proof.

Lemma 12.14. *Let M be a manifold, equipped with a vector field $\mathscr{V}$ and a bivector field P. Suppose that $\mathscr{V}$ is complete and that each one of its integral curves has period 1. If $\mathscr{L}_{\mathscr{V}}^2 P = 0$, then $\mathscr{L}_{\mathscr{V}} P = 0$.*

Proof. Let Q be the bivector field on M, defined by $Q := \mathscr{L}_{\mathscr{V}} P$. We pick an arbitrary point m and we show that $Q_m = 0$. Denoting the flow of $\mathscr{V}$ by Φ_t, we have for all t that

$$\frac{\mathrm{d}}{\mathrm{d}t}\left((\Phi_t)_* P_{\Phi_{-t}(m)}\right) = (\Phi_t)_*(\mathscr{L}_{\mathscr{V}} P)_{\Phi_{-t}(m)} = (\Phi_t)_* Q_{\Phi_{-t}(m)} = Q_m , \tag{12.16}$$

where we used in the last step that the bivector field Q satisfies $\mathscr{L}_{\mathscr{V}} Q = 0$. By integrating (12.16),

$$(\Phi_t)_* P_{\Phi_{-t}(m)} = P_m + t Q_m .$$

Evaluated at $t = 1$ this yields $Q_m = 0$, since $\Phi_1 = \Phi_{-1} = \mathbb{1}_M$, as $\mathscr{V}$ has period 1.
□

Theorem 12.15 (Action-angle theorem). *Let $(M, \pi, \mathbf{F})$ be a Liouville integrable system, where (M, π) is a Poisson manifold of dimension d and rank $2r$. Suppose that $\mathscr{F}_m$ is a standard Liouville torus, where $m \in \mathscr{U}_{\mathbf{F}} \cap M_{(r)}$. Then there exist $\mathbb{R}$-valued smooth functions $(p_1, \dots, p_{d-r})$ and $\mathbb{R}/\mathbb{Z}$-valued smooth functions $(\theta_1, \dots, \theta_r)$, defined in a neighborhood U of $\mathscr{F}_m$, such that*

(1) The map, defined by $(\theta_1, \dots, \theta_r, p_1, \dots, p_{d-r})$, is a diffeomorphism between U and $\mathbb{T}^r \times B^{d-r}$;
(2) The Poisson structure π can be written in terms of these coordinates as

$$\pi = \sum_{i=1}^{r} \frac{\partial}{\partial \theta_i} \wedge \frac{\partial}{\partial p_i} ,$$

in particular the functions $p_{r+1}, \dots, p_{d-r}$ are Casimirs of π (restricted to U);
(3) The leaves of the surjective submersion $\mathbf{F} = (F_1, \dots, F_{d-r})$ are given by the projection onto the second component $\mathbb{T}^r \times B^{d-r}$, in particular, the functions $F_1, \dots, F_{d-r}$ depend on the functions $p_1, \dots, p_{d-r}$ only.

The functions $\theta_1, \dots, \theta_r$ are called angle coordinates, *the functions $p_1, \dots, p_r$ are called* action coordinates *and the remaining functions $p_{r+1}, \dots, p_{d-r}$ are called* transverse coordinates.

Proof. We denote $s := d - r$, as before. The hypotheses imply, according to Proposition 12.12 that we can assume that $(M, \pi, \mathbf{F})$ is of the form $(\mathbb{T}^r \times B^s, \pi, p_B)$, where B^s is a ball, centered at the origin of $\mathbb{R}^s$. We do this, but we still write $\mathbf{F}$ for p_B, denoting its components as $(f_1, \dots, f_r, z_1, \dots, z_{s-r})$, since according to the latter proposition, the last components of $\mathbf{F}$ can be assumed to be Casimir functions. According to Proposition 12.13 there exist $\mathbf{F}$-basic functions $\lambda_i^j \in \mathscr{F}(\mathbb{T}^r \times B^s)$, such that the r vector fields

$$\mathscr{V}_i := \sum_{j=1}^{r} \lambda_i^j \mathscr{X}_{f_j} , \tag{12.17}$$

where $i = 1, \dots, r$, are the fundamental vector fields of an action of $\mathbb{T}^r$ on $\mathbb{T}^r \times B^s$.

We first prove that these vector fields are Poisson vector fields, i.e., that $\mathscr{L}_{\mathscr{V}_i}\pi = [\mathscr{V}_i, \pi]_S = 0$ for $i = 1, \dots, r$. Since each $\mathscr{V}_i$ has period 1 and is complete, it suffices to show, according to Lemma 12.14, that $\mathscr{L}^2_{\mathscr{V}_i}\pi = [\mathscr{V}_i, [\mathscr{V}_i, \pi]] = 0$, for $i = 1 \dots, r$. Since each of the coefficients λ_i^j which appear in (12.17) is an **F**-basic function, it suffices to show that if F, F', G and G' are **F**-basic functions on $\mathbb{T}^r \times B^s$, then $[F\mathscr{X}_G, [F'\mathscr{X}_{G'}, \pi]_S]_S = 0$. In fact, Hamiltonian vector fields leave π invariant, **F**-basic functions are in involution and their Hamiltonian vector fields commute, so that the graded Leibniz identity for the Schouten bracket yields

$$
\begin{aligned}
[F\mathscr{X}_G, [F'\mathscr{X}_{G'}, \pi]_S]_S &= [F\mathscr{X}_G, \mathscr{X}_{F'} \wedge \mathscr{X}_{G'}]_S \\
&= F[\mathscr{X}_G, \mathscr{X}_{F'} \wedge \mathscr{X}_{G'}]_S + \mathscr{X}_G[F, \mathscr{X}_{F'} \wedge \mathscr{X}_{G'}]_S = 0\,,
\end{aligned}
$$

as was to be shown.

We proceed to construct the action coordinates: we show that the Poisson vector fields $\mathscr{V}_i$ are actually Hamiltonian vector fields (upon shrinking B^r, if necessary), where the Hamiltonians can be taken as **F**-basic functions; these Hamiltonians will yield the action coordinates. According to Theorem 1.28, we can construct on a neighborhood U' of m in $\mathbb{T}^r \times B^s$ functions $g_1, \dots, g_r$ such that

$$
(f_1, \dots, f_r, g_1, \dots, g_r, z_1, \dots, z_{s-r})
$$

is a system of coordinates on U', in which the Poisson structure takes the form

$$
\pi = \sum_{i=1}^{r} \frac{\partial}{\partial f_i} \wedge \frac{\partial}{\partial g_i}\,.
$$

In what follows, we make no notational distinction between the **F**-basic functions $f_1, \dots, f_r, z_1, \dots, z_{s-r}, \lambda_i^j$, considered as functions on $\mathbf{F}(U') \subset B^s$, and the functions $f_1, \dots, f_r, z_1, \dots, z_{s-r}, \lambda_i^j$ themselves, defined on U'. With this notation, the vector fields $\mathscr{X}_{\lambda_i^j}$ and $\mathscr{V}_i$ can be written as

$$
\mathscr{X}_{\lambda_i^j} = \sum_{k=1}^{r} \frac{\partial \lambda_i^j}{\partial f_k} \mathscr{X}_{f_k}, \quad \text{and} \quad \mathscr{V}_i = \sum_{j=1}^{r} \lambda_i^j \mathscr{X}_{f_j} = -\sum_{j=1}^{r} \lambda_i^j \frac{\partial}{\partial g_j}\,.
$$

Expressing that $\mathscr{V}_i$ is a Poisson vector field, $[\mathscr{V}_i, \pi]_S = 0$, then takes the explicit form

$$
-\sum_{j,k=1}^{r} \left[\lambda_i^j \frac{\partial}{\partial g_j}, \frac{\partial}{\partial f_k} \wedge \frac{\partial}{\partial g_k} \right]_S = \sum_{j,k=1}^{r} \frac{\partial \lambda_i^j}{\partial f_k} \frac{\partial}{\partial g_j} \wedge \frac{\partial}{\partial g_k} = 0\,,
$$

which is rewritten in terms of the Hamiltonian vector fields $\mathscr{X}_{f_j}$ as

$$
\sum_{1 \leqslant j < k \leqslant r} \left(\frac{\partial \lambda_i^j}{\partial f_k} - \frac{\partial \lambda_i^k}{\partial f_j} \right) \mathscr{X}_{f_j} \wedge \mathscr{X}_{f_k} = 0\,. \tag{12.18}
$$

Since the vector fields $\mathscr{X}_{f_1}, \dots, \mathscr{X}_{f_r}$ are linearly independent at all points of U', all coefficients of (12.18) vanish and we get, for every $i, j, k \in \{1, \dots, r\}$:

$$\frac{\partial \lambda_i^j}{\partial f_k} = \frac{\partial \lambda_i^k}{\partial f_j} . \tag{12.19}$$

As in the proof of the classical Poincaré lemma, the smooth functions $p_1, \dots, p_r$ on $\mathbf{F}(U')$, defined by

$$p_i = p_i(f_1, \dots, f_s) := \sum_{j=1}^{r} \int_{t=0}^{1} \lambda_i^j(tf_1, \dots, tf_r, z_1, \dots, z_{s-r}) f_j \, \mathrm{d}t \tag{12.20}$$

satisfy

$$\lambda_i^j = \frac{\partial p_i}{\partial f_j} , \tag{12.21}$$

for all $1 \leqslant i, j \leqslant r$. As above, we view these functions as functions on U'. They are Hamiltonians of the vector fields $\mathscr{V}_i$, because

$$\mathscr{X}_{p_i} = \sum_{j=1}^{r} \frac{\partial p_i}{\partial f_j} \mathscr{X}_{f_j} = \sum_{j=1}^{r} \lambda_i^j \, \mathscr{X}_{f_j} = \mathscr{V}_i . \tag{12.22}$$

This shows that each one of the vector fields $\mathscr{V}_i$ is a Hamiltonian vector field on U'. Since the functions p_i and z_j are $\mathbf{F}$-basic, each admits a unique extension to an $\mathbf{F}$-basic function on $\mathbf{F}(\mathbf{F}^{-1}(U'))$, a neighborhood of $\mathscr{F}_m$ which we call U' in the sequel. On U', we still have $\mathscr{V}_i = \mathscr{X}_{p_i}$.

We proceed to the construction of the angle coordinates. In view of Theorem 1.28, there exist on a neighborhood $U'' \subset U'$ of m in $\mathbb{T}^r \times B^s$, smooth functions $\theta_1, \dots, \theta_r$ such that

$$\pi = \sum_{j=1}^{r} \frac{\partial}{\partial \theta_j} \wedge \frac{\partial}{\partial p_j} . \tag{12.23}$$

On U'', $\mathscr{X}_{p_j} = \frac{\partial}{\partial \theta_j}$, for $j = 1, \dots, r$; since each of these vector fields has period 1, it is natural to view the functions $\theta_1, \dots, \theta_r$ as $\mathbb{R}/\mathbb{Z}$-valued functions, which we will do without changing the notation. Notice that the functions $\theta_1, \dots, \theta_r$ are independent and pairwise in involution on U'', as a trivial consequence of (12.23). In particular, $\theta_1, \dots, \theta_r, p_1, \dots, p_s$ define local coordinates on U''. In these coordinates, the action of $\mathbb{T}^r$ is given by

$$(t_1, \dots, t_r) \cdot (\theta_1, \dots, \theta_r, p_1, \dots, p_s) = (\theta_1 + t_1, \dots, \theta_r + t_r, p_1, \dots, p_s) , \tag{12.24}$$

so that the functions θ_i uniquely extend to smooth $\mathbb{R}/\mathbb{Z}$-valued functions satisfying (12.24), on $U := \mathbf{F}^{-1}(\mathbf{F}(U''))$, which is an open neighborhood of $\mathscr{F}_m$ in M; the extended functions are still denoted by θ_i. It is clear that $\{\theta_i, p_j\} = \delta_{i,j}$ on U, for all $i, j = 1, \dots, r$. Combined with the Jacobi identity, this leads to

$$\mathscr{X}_{p_k}[\{\theta_i,\theta_j\}] = \{\{\theta_i,\theta_j\},p_k\} = \{\theta_i,\delta_{j,k}\} - \{\theta_j,\delta_{i,k}\} = 0\,,$$

which shows that the Poisson brackets $\{\theta_i,\theta_j\}$ are invariant under the $\mathbb{T}^r$-action; but the latter vanish on U'', hence these brackets vanish at every point of U. We may conclude that on U, the functions $(\theta_1,\dots,\theta_r,p_1,\dots,p_s)$ have independent differentials, so they define a diffeomorphism to $\mathbb{T}^r \times B^s$ where B^s is a (small) ball with center 0, and that the Poisson structure takes in terms of these coordinates the canonical form (12.23), as required. □

In terms of action-angle coordinates, the Hamiltonian vector fields $\mathscr{X}_{f_i}$ take a particularly simple form. In fact, since the functions F_i depend on the coordinates p_i only, the equations of motion, associated to the Hamiltonian F_i are given by

$$\begin{aligned}
\dot p_j &= \{p_j,F_i\} = -\frac{\partial F_i}{\partial \theta_j} = 0\,, && j=1,\dots,r\,,\\
\dot\theta_j &= \{\theta_j,F_i\} = \frac{\partial F_i}{\partial p_j} = c_j\,, && j=1,\dots,r\,,\\
\dot z_k &= 0\,, && k=1,\dots,s-2r\,,
\end{aligned}$$

where c_j depends on the p and z coordinates only. Integrating the first and third equations, we find that all p_j and z_k are constant (which is not surprising since the F_j are constants of motion and the z_k are Casimirs), so that the c_j are constant, which yields upon using the second equation that the angle coordinates evolve linearly, $\theta_j(t) = c_j t$, for $j = 1,\dots,r$. Thus, action-angle coordinates not only exhibit an integrable system locally as a Hamiltonian system with linear flow(s) on a family of tori, they are also the coordinates in which the equations of motion take their simplest possible form.

The action-angle theorem admits a natural generalization to the case of noncommutative integrable systems (see [120]). The classical examples of Liouville integrable systems can be found in [204]. For a modern treatment of Liouville integrability, including recently discovered examples, we refer to [3, 164].

12.4 Notes

The term "integrable system" appears in a large variety of mathematical and physical contexts, referring often more to phenomena than to a precise general definition. The notion of a Liouville integrable system, detailed in this chapter, is a notable exception. Liouville's observation that "Liouville integrable systems" can be solved by quadratures has been a main impulse to the development of Poisson geometry, for a long time restricted to symplectic geometry (the geometry of the phase space of a mechanical system). For an excellent account of what was classically known about integrable systems, the reader should consult Whittaker [204]. The books by Abraham–Marsden [1], Arnold [15] and Libermann–Marle [125] are all devoted partly to integrability in classical mechanics, in the setting of symplectic manifolds, although Lie groups and Lie algebras are present throughout these books, just as Poisson brackets are. See also the book by Kosmann–Schwarzbach [114], which focuses on the Noether theorems, with special emphasis on their historical developments, as well as on their influence in mathematics and physics.

Shortly before 1970, the remarkable discovery was made that the Korteweg–de Vries equation can be solved by the inverse spectral method, has an infinite number of constant of motion, has a (in fact many) Hamiltonian structure(s) and can be described as an isospectral flow of a Sturm–Liouville operator (as a Lax equation). It has very much revived the interest in integrable systems, in particular in Liouville integrable systems, on which the newly discovered infinite-dimensional tools (inverse spectral method, Lax equations, recursion operators), were successfully applied, thereby shedding a new light on some classical results, leading to the solution of some classical problems and to the discovery of some new integrable systems. See the books by Perelomov [164] and Fomenko [78], which together cover a large spectrum of examples and phenomena. From the Poisson point of view, the important observation by Adler that the Poisson structure of the Korteweg–de Vries equation comes from a Lie algebra splitting led to the notion of an R-bracket, which in turn was at the basis of a better understanding of the symmetries of quantum systems, culminating in the discovery of quantum groups, see Faddeev–Takhtajan [72] and Chari–Pressley [40].

The presence of algebraic geometry in integrable systems, already present in the classical works of Euler, Jacobi, Painlevé and Kowalevskaya, was reaffirmed by Its–Matveev, who showed that the theta functions of hyperelliptic curves lead to solutions of the Korteweg–de Vries equation, and by Adler–van Moerbeke, who showed upon building upon Kowalevskaya's intuition that basically all classical examples of integrable systems are algebraically integrable, see Adler–van Moerbeke–Vanhaecke [3, 196]. The underlying Lie and Poisson geometry of these systems is always very rich, but often only partly understood.

Chapter 13
Deformation Quantization

In this chapter we present a second application of Poisson structures: the theory of deformations of commutative associative algebras, also called deformation quantization. Geometric quantization and deformation quantization are two approaches to the process of forming a quantum mechanical system from a given classical system. Both are based on Dirac's observation that the natural quantization rules for Euclidean-invariant systems yield for the simplest variables (position and momentum) differential operators which obey commutation relations analogous to the Poisson bracket for the corresponding classical variables. We will only discuss deformation quantization here, which consists of considering for a given Poisson manifold $(M,\{\cdot,\cdot\})$ all associative products $\star$ on the vector space $C^\infty(M)[[\nu]]$, which are of the form:

$$F \star G = FG + \frac{1}{2}\{F,G\}\,\nu + C(F,G)\nu^2 + \cdots$$

Kontsevich's formality theorem implies that for every Poisson manifold, such a deformation quantization exists.

The problem of describing for a given commutative associative algebra $(\mathscr{A},\cdot)$ all associative structures on $\mathscr{A}[[\nu]]$, which deform the product "$\cdot$" is also interesting from the mathematical point of view. Since even in this more general context, any such deformation makes a Poisson bracket appear on $(\mathscr{A},\cdot)$, the question is most naturally asked when this Poisson bracket is fixed in advance, i.e., when $(\mathscr{A},\cdot,\{\cdot,\cdot\})$ is a Poisson algebra. As we will see, the problem of describing all Poisson brackets on $(\mathscr{A}[[\nu]],\cdot)$ which deform the Poisson bracket $\{\cdot,\cdot\}$ is related to the deformation quantization problem.

Formal deformations of associative products and of Poisson structures are introduced in Sections 13.1 and 13.2. Section 13.3 is devoted to the basics on differential graded Lie algebras, while the more advanced theory of L_∞-morphisms of differential graded Lie algebras is discussed in Section 13.4. Combined, these sections provide the material which allows the reader to understand and appreciate Kontsevich's formality theorem, which is stated, together with some of its consequences, in the final Section 13.5.

Unless stated otherwise, $\mathbb{F}$ is an arbitrary field of characteristic zero.

C. Laurent-Gengoux et al., *Poisson Structures*,
Grundlehren der mathematischen Wissenschaften 347,
DOI 10.1007/978-3-642-31090-4_13, © Springer-Verlag Berlin Heidelberg 2013

13.1 Deformations of Associative Products

In this section, we introduce the notion of a formal deformation and of a k-th order deformation of a commutative associative algebra and we show how these deformations are related to Hochschild cohomology. We pick an indeterminate ν and we denote by $\mathbb{F}^\nu$ the ring $\mathbb{F}[[\nu]]$ of formal power series in ν with coefficients in $\mathbb{F}$.

13.1.1 Formal and k-th Order Deformations

Let $\mathscr{A}$ be a commutative associative algebra over $\mathbb{F}$ with unit. The product on $\mathscr{A}$ will be denoted in its functional form by μ, or by a simple juxtaposition of the factors, so we write FG for $\mu(F,G)$ for all $F,G \in \mathscr{A}$. Associated to $\mathscr{A}$, we consider the $\mathbb{F}^\nu$-algebra $\mathscr{A}[[\nu]]$ of formal power series in ν, with coefficients in $\mathscr{A}$; we also denote this algebra by $\mathscr{A}^\nu$. Thus, an element of $\mathscr{A}^\nu$ is a formal power series $\sum_{i\in\mathbb{N}} F_i \nu^i$, where all $F_i \in \mathscr{A}$. The product which $\mathscr{A}^\nu$ inherits from $\mathscr{A}$ by "extension of scalars", as in Section 2.4.1, is given by

$$\left(\sum_{i\in\mathbb{N}} F_i \nu^i\right) \cdot \left(\sum_{j\in\mathbb{N}} G_j \nu^j\right) := \sum_{i,j\in\mathbb{N}} F_i G_j \, \nu^{i+j}, \tag{13.1}$$

which is well-defined, since for every $k \in \mathbb{N}$, the coefficient of ν^k in the right-hand side of (13.1) is a finite sum. This product is associative, $\mathbb{F}^\nu$-bilinear and reduces to the original product μ on $\mathscr{A}$ when setting $\nu = 0$. It is therefore a formal deformation of μ in the sense of the following definition.

Definition 13.1. Let $(\mathscr{A},\mu)$ be a commutative associative algebra. An $\mathbb{F}^\nu$-bilinear map $\mu_\star : \mathscr{A}^\nu \times \mathscr{A}^\nu \to \mathscr{A}^\nu$ is called a *formal deformation* of μ if $\mu_\star$ is associative and $\mu_\star = \mu \pmod{\nu}$, i.e., if $\mu_\star(F,G) - \mu(F,G) \in \nu\mathscr{A}^\nu$ for all $F,G \in \mathscr{A}$.

The question which we will address later in this chapter is the construction of *all* formal deformations of a commutative associative product μ. The above deformation (13.1), which is commutative, is a particular example, in fact a rather trivial one. We stress that the formal deformations which we consider, are not required to be commutative.

The notion of equivalence of deformations of a given commutative associative product μ is given in the following definition.

Definition 13.2. Let $(\mathscr{A},\mu)$ be a commutative associative algebra and let $\mu_\star$ and $\mu'_\star$ be two formal deformations of μ. Then $\mu_\star$ and $\mu'_\star$ are said to be *equivalent* if there exists an $\mathbb{F}^\nu$-linear map $\Phi : \mathscr{A}^\nu \to \mathscr{A}^\nu$, such that

(1) $\Phi(F) = F \pmod{\nu}$, for all $F \in \mathscr{A}$;
(2) $\Phi(\mu_\star(F,G)) = \mu'_\star(\Phi(F),\Phi(G))$, for all $F,G \in \mathscr{A}$.

The map Φ is then called an *equivalence* of formal deformations.

Condition (1) in the above definition implies that the map Φ is invertible, so the notion of equivalence of formal deformations defines an equivalence relation on the set of all formal deformations of a product μ. Also, if $\mu_\star$ is a formal deformation of μ and $\Phi : \mathscr{A}^\nu \to \mathscr{A}^\nu$ is an $\mathbb{F}^\nu$-linear map such that $\Phi(F) = F \pmod{\nu}$, for every $F \in \mathscr{A}$, then the map $\mu'_\star : \mathscr{A}^\nu \times \mathscr{A}^\nu \to \mathscr{A}^\nu$, defined for $F, G \in \mathscr{A}^\nu$ by

$$\mu'_\star(F,G) := \Phi\left(\mu_\star(\Phi^{-1}(F), \Phi^{-1}(G))\right),$$

is a formal deformation of μ, equivalent to $\mu_\star$.

Remark 13.3. A formal deformation $\mu_\star$ of μ can be seen as an element of $\mathrm{Hom}(\mathscr{A} \times \mathscr{A}, \mathscr{A})[[\nu]]$, because there is for every $k \in \mathbb{N}^*$ a natural isomorphism of $\mathbb{F}^\nu$-modules

$$\mathrm{Hom}_{\mathbb{F}}(\mathscr{A}^k, \mathscr{A})[[\nu]] \simeq \mathrm{Hom}_{\mathbb{F}^\nu}\left((\mathscr{A}^\nu)^k, \mathscr{A}^\nu\right).$$

We show this in the case of $k = 1$, the case of arbitrary k being only notationally more complicated. Suppose that $\Phi : \mathscr{A}^\nu \to \mathscr{A}^\nu$ is an $\mathbb{F}^\nu$-linear map. For $F \in \mathscr{A}$, we can write $\Phi(F)$ in a unique way as

$$\Phi(F) = \Phi_0(F) + \Phi_1(F)\nu + \cdots + \Phi_k(F)\nu^k + \cdots, \tag{13.2}$$

where every $\Phi_i(F)$ belongs to $\mathscr{A}$. It follows that we can associate to every element Φ of $\mathrm{Hom}_{\mathbb{F}^\nu}(\mathscr{A}^\nu, \mathscr{A}^\nu)$ a sequence of linear maps $(\Phi_i)_{i \in \mathbb{N}}$ from $\mathscr{A}$ to $\mathscr{A}$, that is, an element of $\mathrm{Hom}_{\mathbb{F}}(\mathscr{A}, \mathscr{A})[[\nu]]$. Conversely, given such a sequence of $\mathbb{F}$-linear maps $(\Phi_i)_{i \in \mathbb{N}}$, the map Φ, defined by (13.2), extends in a unique way to an $\mathbb{F}^\nu$-linear map. By a slight abuse of notation, we keep the notation Φ_i for the $\mathbb{F}^\nu$-linear extensions of the maps Φ_i to $\mathscr{A}^\nu$, which has the effect that (13.2) remains valid when F belongs to $\mathscr{A}^\nu$.

Formal deformations are often constructed term by term, which is done rigorously by considering k-th order deformations. For $k \in \mathbb{N}$, we consider the ring $\mathbb{F}^\nu_k := \mathbb{F}^\nu / \langle \nu^{k+1} \rangle$. Elements of this ring can be represented by polynomials in ν of degree at most k and the product in this ring is the ordinary multiplication of polynomials, followed by stripping off the terms whose degree is higher than k. Starting from a commutative associative algebra $\mathscr{A}$, we consider similarly

$$\mathscr{A}^\nu_k := \mathscr{A}^\nu / \langle \nu^{k+1} \rangle,$$

which we can view as an $\mathbb{F}^\nu$-algebra, or as an $\mathbb{F}^\nu_k$-algebra. We have that $\mathscr{A}^\nu_0 \simeq \mathscr{A}$, in a natural way; these algebras will be identified in the sequel. Notice also that in Remark 13.3, one can replace $\mathbb{F}^\nu$ by $\mathbb{F}^\nu_k$ and $\mathscr{A}^\nu$ by $\mathscr{A}^\nu_k$.

Definition 13.4. Let $(\mathscr{A}, \mu)$ be a commutative associative algebra. For $k \in \mathbb{N}$, an $\mathbb{F}^\nu_k$-bilinear map $\mu_{(k)} : \mathscr{A}^\nu_k \times \mathscr{A}^\nu_k \to \mathscr{A}^\nu_k$,

$$\mu_{(k)} = \mu_0 + \mu_1 \nu + \cdots + \mu_k \nu^k, \tag{13.3}$$

where each μ_i is an $\mathbb{F}$-bilinear map from $\mathscr{A}$ to itself, is called a *k-th order deformation* of μ if $\mu_0 = \mu$ and $\mu_{(k)}$ is associative.

Under the natural quotient maps $\mathscr{A}^\nu \to \mathscr{A}_k^\nu$ and $\mathscr{A}_{k+1}^\nu \to \mathscr{A}_k^\nu$, formal deformations (respectively $(k+1)$-th order deformations) are mapped to k-th order deformations. By a slight abuse of language, we say that an $\mathbb{F}^\nu$-bilinear map $\mu_\star : \mathscr{A}^\nu \times \mathscr{A}^\nu \to \mathscr{A}^\nu$ such that $\mu_\star = \mu \pmod{\nu}$ defines a k-th order deformation of μ if the induced $\mathbb{F}_k^\nu$-bilinear map $\mu_{(k)} : \mathscr{A}_k^\nu \times \mathscr{A}_k^\nu \to \mathscr{A}_k^\nu$ is a k-th order deformation of μ. Clearly, this is equivalent to saying that

$$\mu_\star(\mu_\star(F,G),H) = \mu_\star(F,\mu_\star(G,H)) \pmod{\nu^{k+1}}$$

for all $F,G,H \in \mathscr{A}$.

The notion of equivalence of k-th order deformations of μ is obtained by replacing $\mathbb{F}^\nu$ by $\mathbb{F}_k^\nu$ and $\mathscr{A}^\nu$ by $\mathscr{A}_k^\nu$ in Definition 13.2, which deals with the notion of equivalence of formal deformations. Again, an $\mathbb{F}^\nu$-linear map $\mathscr{A}^\nu \to \mathscr{A}^\nu$ which induces an equivalence of k-th order deformations will be said to define an equivalence of k-th order deformations.

Let $(\mathscr{A},\mu)$ be a commutative associative algebra and suppose that $\mu_{(1)} = \mu_0 + \mu_1\nu$ defines a first order deformation of $\mu = \mu_0$. For $a,b,c \in \mathscr{A}$, the coefficient in ν in the associativity equation $\mu_{(1)}(\mu_{(1)}(a,b),c) = \mu_{(1)}(a,\mu_{(1)}(b,c)) \pmod{\nu^2}$ is given by

$$\mu_1(ab,c) + \mu_1(a,b)c = a\mu_1(b,c) + \mu_1(a,bc)\;.$$

It follows that

$$\mu_1^-(ab,c) + \mu_1^-(a,b)c - a\mu_1^-(b,c) - \mu_1^-(a,bc) = 0\;,$$

where μ_1^- is the skew-symmetric part of μ_1, i.e.,

$$\mu_1^-(a,b) := \frac{1}{2}(\mu_1(a,b) - \mu_1(b,a))\;, \tag{13.4}$$

for all $a,b \in \mathscr{A}$. Performing a cyclic permutation of a,b,c in this equation leads to the following three equations.

$$\begin{aligned}
\mu_1^-(ab,c) + \mu_1^-(a,b)c - \mu_1^-(b,c)a - \mu_1^-(a,bc) = 0\;,\\
\mu_1^-(bc,a) + \mu_1^-(b,c)a - \mu_1^-(c,a)b - \mu_1^-(b,ac) = 0\;,\\
\mu_1^-(ac,b) + \mu_1^-(c,a)b - \mu_1^-(a,b)c - \mu_1^-(c,ab) = 0\;,
\end{aligned}$$

whose alternating sum simplifies to

$$\mu_1^-(ab,c) = \mu_1^-(a,c)b + \mu_1^-(b,c)a\;.$$

It follows that if $\mu_{(1)}$ is a first order deformation of $\mu = \mu_0$, then μ_1^- is a skew-symmetric biderivation of $(\mathscr{A},\mu)$. Suppose now that $\mu_{(2)} = \mu_0 + \mu_1\nu + \mu_2\nu^2$ is a second order deformation of $\mu = \mu_0$. The associativity of $\mu_{(2)}$ (up to ν^2) implies that the commutator

$$\begin{aligned} [a,b]_\star &:= \mu_\star(a,b) - \mu_\star(b,a) \\ &= 2\nu\mu_1^-(a,b) \pmod{\nu^2} . \end{aligned}$$

satisfies the Jacobi identity up to order ν^2,

$$[[a,b]_\star, c]_\star + \circlearrowleft(a,b,c) = 0 \pmod{\nu^3} ,$$

for all a,b and c in $\mathscr{A}$. It follows that

$$\mu_1^-(\mu_1^-(a,b),c) + \circlearrowleft(a,b,c) = 0 ,$$

for all a,b and c in $\mathscr{A}$, i.e., μ_1^- satisfies the Jacobi identity. In conclusion, if $\mu_{(2)} = \mu_0 + \mu_1\nu + \mu_2\nu^2$ is a second order deformation of an associative algebra $(\mathscr{A},\mu)$, then $(\mathscr{A},\mu,\mu_1^-)$ is a Poisson algebra!

13.1.2 Hochschild Cohomology

Deformations of associative products are governed by Hochschild cohomology. In the present section, we introduce this cohomology and we show in the next section how it is related to deformation theory. Throughout this section, we fix a commutative ring R.

Let V be a module over R. For $k \in \mathbb{N}^*$, we denote by $\mathrm{HC}^k(V) := \mathrm{Hom}(V^k, V)$ the space of all k-linear maps $V^k \to V$, while $\mathrm{HC}^0(V) := V$. We will define a bracket $[\cdot,\cdot]_G$, called the *Gerstenhaber bracket*, on

$$\mathrm{HC}^\bullet(V) := \bigoplus_{k\in\mathbb{N}} \mathrm{HC}^k(V) ,$$

which is a graded Lie bracket, just like the Schouten bracket, after a shift of degree (see Section 3.3.2). Thus, we denote $\bar{k} := k-1$, and we define $\overline{\mathrm{HC}}^{k-1}(V) := \mathrm{HC}^k(V)$ for $k \in \mathbb{N}$ so that $\overline{\mathrm{HC}}^{\bar{k}}(V) = \mathrm{HC}^k(V)$. Elements of the latter space have degree k and shifted degree $\bar{k}$. With respect to the shifted degree, the Gerstenhaber bracket is a graded Lie bracket, i.e., for all $k,\ell \in \mathbb{N}^*$,

$$[\cdot,\cdot]_G : \overline{\mathrm{HC}}^{\bar{k}}(V) \times \overline{\mathrm{HC}}^{\bar{\ell}}(V) \to \overline{\mathrm{HC}}^{\bar{k}+\bar{\ell}}(V)$$

and $[\cdot,\cdot]_G$ is graded skew-symmetric and satisfies the graded Jacobi identity (see Table A.2 in Appendix A).

Explicitly, the bracket $[\cdot,\cdot]_G$ is defined, for $\phi \in \overline{\mathrm{HC}}^{\bar{k}}(V)$ and $\psi \in \overline{\mathrm{HC}}^{\bar{\ell}}(V)$, by the following formula,

$$[\phi,\psi]_G(x_1,\ldots,x_{\bar{k}+\bar{\ell}+1}) \tag{13.5}$$
$$:= \sum_{i=1}^{k}(-1)^{\bar{i}\,\bar{\ell}}\phi\left(x_1,\ldots,x_{i-1},\psi\left(x_i,\ldots,x_{i+\bar{\ell}}\right),x_{i+\bar{\ell}+1},\ldots,x_{k+\bar{\ell}}\right)$$
$$-(-1)^{\bar{k}\bar{\ell}}\sum_{i=1}^{\ell}(-1)^{\bar{i}\,\bar{k}}\psi\left(x_1,\ldots,x_{i-1},\phi\left(x_i,\ldots,x_{i+\bar{k}}\right),x_{i+\bar{k}+1},\ldots,x_{\ell+\bar{k}}\right),$$

where $x_1,\ldots,x_{\bar{k}+\bar{\ell}+1}\in V$. In particular, if $\mu\in \mathrm{HC}^2(V)$, then

$$[\mu,\mu]_G(x_1,x_2,x_3) = 2\mu(\mu(x_1,x_2),x_3) - 2\mu(x_1,\mu(x_2,x_3))\,,$$

for all $x_1,x_2,x_3\in V$, so that μ defines an associative product on V if and only if $[\mu,\mu]_G = 0$.

Suppose now that $(\mathscr{A},\mu)$ is an associative (not necessarily commutative) algebra with unit, over R. Combined with the graded Jacobi identity, the associativity condition $[\mu,\mu]_G = 0$ implies that

$$\delta_\mu^k := -[\cdot,\mu]_G : \mathrm{HC}^k(\mathscr{A}) \to \mathrm{HC}^{k+1}(\mathscr{A}) \tag{13.6}$$

is a coboundary operator, i.e., $\delta_\mu^{k+1}\circ\delta_\mu^k = 0$ for every $k\in\mathbb{N}$. This operator is called the *Hochschild coboundary operator* and the associated complex

$$\cdots \longrightarrow \mathrm{HC}^{k-1}(\mathscr{A}) \xrightarrow{\delta_\mu^{k-1}} \mathrm{HC}^k(\mathscr{A}) \xrightarrow{\delta_\mu^{k}} \mathrm{HC}^{k+1}(\mathscr{A}) \xrightarrow{\delta_\mu^{k+1}} \cdots$$

is called the *Hochschild complex* of $\mathscr{A}$. The cohomology of this complex,

$$\mathrm{HH}_\mu^\bullet(\mathscr{A}) := \bigoplus_{k\in\mathbb{N}} \mathrm{HH}_\mu^k(\mathscr{A})$$

is called the *Hochschild cohomology* of $\mathscr{A}$, where

$$\mathrm{HH}_\mu^k(\mathscr{A}) := \mathrm{Ker}\,\delta_\mu^k / \mathrm{Im}\,\delta_\mu^{k-1}\,,$$

for $k\in\mathbb{N}^*$, while $\mathrm{HH}_\mu^0(\mathscr{A}) := \mathrm{Ker}\,\delta_\mu^0$. The elements of $\mathrm{Ker}\,\delta_\mu^k$ are called *Hochschild k-cocycles* and the elements of $\mathrm{Im}\,\delta_\mu^{k-1}$ are called *Hochschild k-coboundaries.*

An explicit formula for the Hochschild coboundary operator δ_μ is obtained by substituting μ for ψ in (13.5); writing FG for $\mu(F,G)$, it takes, for $\phi\in\mathrm{HC}^k(\mathscr{A})$, and $F_0,\ldots,F_k\in\mathscr{A}$, the simple form

$$\delta_\mu^k(\phi)(F_0,\ldots,F_k) := F_0\phi(F_1,\ldots,F_k) + (-1)^{k+1}\phi(F_0,\ldots,F_{k-1})F_k$$
$$+\sum_{i=1}^{k}(-1)^i\phi(F_0,\ldots,F_{i-1}F_i,\ldots,F_k)\,. \tag{13.7}$$

Since

$$\delta_\mu^0(F)(G) = GF - FG\,,$$

for $F,G \in \mathscr{A}$, one has that $\mathrm{HH}^0_\mu(\mathscr{A}) = \mathrm{Ker}\,\delta^0_\mu$ is the center of the associative algebra $(\mathscr{A},\mu)$. Since moreover

$$\delta^1_\mu(\phi)(F,G) = F\phi(G) - \phi(FG) + \phi(F)G\,, \tag{13.8}$$

for $F,G \in \mathscr{A}$ and $\phi \in \mathrm{HC}^1(\mathscr{A})$, the kernel of δ^1_μ consists of linear maps $\phi : \mathscr{A} \to \mathscr{A}$ which satisfy

$$\phi(FG) = F\phi(G) + \phi(F)G\,,$$

for all $F,G \in \mathscr{A}$, which is the non-commutative analog of the derivation property. It follows that the first Hochschild cohomology space associated to $(\mathscr{A},\mu)$ consists of the quotient of the space of all the derivations of $(\mathscr{A},\mu)$, by the space of all *inner derivations* of $(\mathscr{A},\mu)$ (i.e., derivations of $(\mathscr{A},\mu)$ of the form $[F,\cdot]$, for $F \in \mathscr{A}$, where $[F,G] := FG - GF$, for $G \in \mathscr{A}$). In particular, if $(\mathscr{A},\mu)$ is commutative, then $\mathrm{HH}^1_\mu(\mathscr{A})$ is the R-module $\mathfrak{X}^1(\mathscr{A})$ of all derivations of $\mathscr{A}$.

13.1.3 Deformations and Cohomology

We now explain how the Hochschild cohomology of a commutative associative algebra $(\mathscr{A},\mu)$ is related to the deformations of its product μ. As we already pointed out, a formal deformation (respectively, an ℓ-th order deformation) defines for every $k \in \mathbb{N}$ (respectively for every $k \leqslant \ell$) a k-th order deformation. Conversely, let $\mu_1,\dots,\mu_k \in \mathrm{HC}^2(\mathscr{A})$ and suppose that $\mu_{(k)} := \mu_0 + \mu_1\nu + \cdots + \mu_k\nu^k$ defines a k-th order deformation of $\mu = \mu_0$, so that $\left[\mu_{(k)},\mu_{(k)}\right]_G = 0 \pmod{\nu^{k+1}}$, where[1] $[\cdot,\cdot]_G$ denotes the $\mathbb{F}^\nu$-linear extension of $[\cdot,\cdot]_G$ to $\mathscr{A}^\nu$. Let μ_{k+1} be a bilinear map $\mathscr{A}\times\mathscr{A}\to\mathscr{A}$. The bilinear map $\mu_{(k+1)}$, defined by:

$$\mu_{(k+1)} := \mu_{(k)} + \mu_{k+1}\,\nu^{k+1}$$

defines a $(k+1)$-th order deformation of μ, if and only if $\left[\mu_{(k+1)},\mu_{(k+1)}\right]_G = 0 \pmod{\nu^{k+2}}$. Since $\left[\mu_{(k)},\mu_{(k)}\right]_G = 0$ (in $\mathscr{A}^\nu_k$), all that remains in the latter equation is

$$\sum_{\substack{i+j=k+1\\ i,j\geqslant 0}} [\mu_i,\mu_j]_G = 0\,,$$

which we can write in terms of the Hochschild coboundary operator (13.6), associated to $\mu = \mu_0$, as

$$\delta^2_\mu(\mu_{k+1}) = -[\mu_{k+1},\mu_0]_S = \frac{1}{2}\sum_{\substack{i+j=k+1\\ i,j\geqslant 1}} [\mu_i,\mu_j]_G\,. \tag{13.9}$$

[1] In other words, $[\cdot,\cdot]_G$ denotes the Gerstenhaber bracket on the $\mathbb{F}^\nu$-module $\mathrm{HC}^\bullet(\mathscr{A}^\nu)$.

This means that $\mu_{(k)}$ can be extended to a deformation of μ of order $k+1$, if and only if the right-hand side of (13.9) is a 3-coboundary for the Hochschild coboundary operator δ_μ^2. Notice that it is a 3-cocycle, since

$$\begin{aligned}
\delta_\mu^3\left(\sum_{\substack{i+j=k+1\\ i,j\geqslant 1}}[\mu_i,\mu_j]_G\right) &= \sum_{\substack{i+j=k+1\\ i,j\geqslant 1}}[\mu,[\mu_i,\mu_j]_G]_G\\
&= -\sum_{\substack{i+j=k+1\\ i,j\geqslant 1}}\left([\mu_i,[\mu_j,\mu]_G]_G+[\mu_j,[\mu,\mu_i]_G]_G\right)\\
&= 2\sum_{\substack{i+j=k+1\\ i,j\geqslant 1}}\left[\mu_i,\delta_\mu^2(\mu_j)\right]_G\\
&= \sum_{\substack{i+j+\ell=k+1\\ i,j,\ell\geqslant 1}}[\mu_i,[\mu_j,\mu_\ell]_G]_G\\
&= 0\,,
\end{aligned}$$

where we have used the graded Jacobi identity for $[\cdot,\cdot]_G$ to obtain the second and the last lines. It follows that the obstruction to extending a deformation of μ of some order to the next order lies in the third Hochschild cohomology space $\mathrm{HH}^3_\mu(\mathscr{A})$. This statement is the content of the following proposition.

Proposition 13.5. *Let $(\mathscr{A},\mu)$ be a commutative associative algebra and suppose that $\mu_{(k)}=\sum_{i=0}^k\mu_i\nu^i$ defines a k-th order deformation of μ. Then $\mu_{(k)}$ can be extended to a $(k+1)$-th order deformation of μ if and only if the Hochschild* 3*-cocycle $\sum_{i=1}^k[\mu_i,\mu_{k+1-i}]_G$ is a Hochschild* 3*-coboundary.*

It is clear from (13.9) that the term $\mu_{k+1}\nu^{k+1}$ which makes a k-th order deformation of μ into a $(k+1)$-th order deformation of μ is not unique (if it exists): one can always add a Hochschild 2-cocycle to μ_{k+1} and this is the only freedom in the choice of μ_{k+1}. If this 2-cocycle is a coboundary, then the two extended $(k+1)$-th order deformations are essentially the same, in the sense that they are equivalent (see Definition 13.2). More generally, we have the following proposition.

Proposition 13.6. *Let $(\mathscr{A},\mu)$ be a commutative associative algebra and suppose that $\mu_{(k)}=\sum_{i=0}^k\mu_i\nu^i$ defines a k-th order deformation of μ, where $k\in\mathbb{N}^*$. Let $\ell\leqslant k$, where $\ell\in\mathbb{N}^*$. Let $\phi:\mathscr{A}\to\mathscr{A}$ be an $\mathbb{F}$-linear map, which we extend to an $\mathbb{F}^\nu$-linear map $\mathscr{A}^\nu\to\mathscr{A}^\nu$, still denoted by ϕ, and let*

$$\mu'_{(\ell)}=\mu_0+\mu_1\nu+\mu_2\nu^2+\cdots+\mu_{\ell-1}\nu^{\ell-1}+(\mu_\ell+\delta_\mu^1(\phi))\,\nu^\ell\,. \qquad (13.10)$$

Then, $\mu'_{(\ell)}$ extends to a k-th order deformation $\mu'_{(k)}$ of μ, which is equivalent to $\mu_{(k)}$.

Proof. Consider the $\mathbb{F}^\nu$-linear map $\Phi:\mathscr{A}^\nu\to\mathscr{A}^\nu$, defined by

$$\begin{aligned}\Phi : \mathscr{A}^\nu &\to \mathscr{A}^\nu \\ F &\mapsto F - \phi(F)\,\nu^\ell\end{aligned}$$

whose inverse is given by $\Phi^{-1}(F) = \sum_{i\in\mathbb{N}} \phi^i(F)\,\nu^{i\ell}$. The $\mathbb{F}^\nu$-bilinear map $\mu'_{(k)}$, defined for all $F, G \in \mathscr{A}$ by

$$\mu'_{(k)}(F,G) := \Phi(\mu_{(k)}(\Phi^{-1}(F), \Phi^{-1}(G)))$$

is a k-th order deformation of μ, which is equivalent to $\mu_{(k)}$. Up to order ℓ, it is given by

$$\begin{aligned}\mu'_{(\ell)}(F,G) &= \Phi(\mu_{(\ell)}(F+\phi(F)\,\nu^\ell, G+\phi(G)\,\nu^\ell)) \\ &= \Phi(\mu_{(\ell)}(F,G) + \phi(F)G\,\nu^\ell + F\phi(G)\,\nu^\ell) \\ &= \mu_{(\ell)}(F,G) + (\phi(F)G - \phi(FG) + F\phi(G))\,\nu^\ell \\ &= \mu_{(\ell)}(F,G) + \delta^1_\mu(\phi)(F,G)\nu^\ell\end{aligned}$$

where we have used (13.8) in the last step. It follows that $\mu'_{(\ell)} = \mu_{(\ell)} + \delta^1_\mu(\phi)\,\nu^\ell$, as in (13.10). □

It follows from the proposition that in order to construct all possible deformations of μ, up to equivalence, one only has to consider, at every step k, as many possibilities as there are elements in $\mathrm{HH}^2_\mu(\mathscr{A})$.

13.1.4 The Maurer–Cartan Equation

Let $\mu_\star = \sum_{i=0}^\infty \mu_i \nu^i$ be an element of $\mathrm{HC}^2(\mathscr{A}^\nu)$, where $\mu_0 = \mu$. As above, $\mu_\star$ is associative if and only if $[\mu_\star, \mu_\star]_G = 0$, which is in turn equivalent to demanding that $\mu_\star - \mu \in \nu\,\mathrm{HC}^2(\mathscr{A}^\nu)$ is a solution of the following equation

$$\delta^2_\mu(x) - \frac{1}{2}[x,x]_G = 0\,, \tag{13.11}$$

where, by a slight abuse of notation, δ^2_μ stands for the $\mathbb{F}^\nu$-linear extension of δ^2_μ to $\mathrm{HC}^2(\mathscr{A}^\nu) \simeq \mathrm{HC}^2(\mathscr{A})[[\nu]]$. For future use, we stress that we consider only solutions x of this equation which are formal series without constant term. Precisely, $x \in \nu\,\mathrm{HC}^2(\mathscr{A}^\nu)$. In terms of the terminology of Section 13.3, equation (13.11) is called the *Maurer–Cartan equation* associated to the differential graded Lie algebra $(\overline{\mathrm{HC}^\bullet}(\mathscr{A}), [\cdot,\cdot]_G, \delta_\mu)$.

13.1.5 Star Products

As a particular case, we consider the algebra $\mathscr{A} = C^\infty(M)$ of smooth functions on a real manifold M, where the associative product μ is the usual product of functions. In this case, one usually considers a particular class of formal deformations of μ, where all Hochschild cochains are bi-differential operators, a notion which we define first.

A *bi-differential operator* on M is a map $\Phi : C^\infty(M) \times C^\infty(M) \to C^\infty(M)$, compatible with restrictions to open subsets and such that, on every coordinate chart $(U,(x_1,\dots,x_d))$ and for all smooth functions F,G on U, one can write:

$$\Phi(F,G) = \sum_{r,s\in\mathbb{N}^d} \phi_{r,s}\left(\frac{\partial^r F}{\partial x^r}\right)\left(\frac{\partial^s G}{\partial x^s}\right), \tag{13.12}$$

where for $r,s \in \mathbb{N}^d$ the coefficients $\phi_{r,s}$ are smooth functions on U, which vanish except for a finite number of (r,s). When all coefficients $\phi_{r,s}$ are constant, we speak of a *constant coefficient* bi-differential operator. For each multi-index $r = (r_1,\dots,r_d) \in \mathbb{N}^d$, we used in (13.12) the notation

$$\frac{\partial^r F}{\partial x^r} := \frac{\partial^{r_1+\cdots+r_d} F}{\partial x_1^{r_1}\cdots\partial x_d^{r_d}}.$$

The generalization of the definition of a bi-differential operator to a *multi-differential operator* is clear. Notice that μ is a bi-differential operator on M.

Definition 13.7. Let M be a real manifold.

(1) A *star product* on $C^\infty(M)$ is a formal deformation $\mu_\star = \sum_{i\in\mathbb{N}} \mu_i \nu^i$ of the product μ of $C^\infty(M)$, where each μ_i is a bi-differential operator on M;
(2) Two star products $\mu_\star$ and $\mu'_\star$ on $C^\infty(M)$ are said to be *equivalent star products* if they are equivalent as formal deformations, via an equivalence of the form $\Phi = \mathbb{1}_{C^\infty(M)} + \phi_1\nu + \cdots + \phi_k\nu^k + \cdots$, where each one of the $\mathbb{F}$-linear maps $\phi_i : C^\infty(M) \to C^\infty(M)$ is a differential operator;
(3) If π is a Poisson structure on M, then (M,π) is said to admit a *deformation quantization* if there exists a star product $\mu_\star = \sum_{i\in\mathbb{N}} \mu_i\nu^i$ on $C^\infty(M)$, such that $\mu_1 = \frac{\pi}{2}$. Then $\mu_\star$ is said to be a *star product* for (M,π).

We also use the infix notation "$\star$" for a star product. Thus, we often write $F \star G$ instead of $\mu_\star(F,G)$, for $F,G \in C^\infty(M)$.

Several authors demand that the constant power series "1" is a unit for the star product; this condition is not very essential, since it can be shown that every star product as defined above is equivalent to a star product which has "1" as its unit (see [39, Cor. 3.4.5]).

The description of star products in terms of cohomology is the same as for deformations, except that one considers as cochains only the *differential Hochschild cochains* of $C^\infty(M)$, i.e., the Hochschild cochains which are given by multi-differential operators. The space of all differential Hochschild cochains of $C^\infty(M)$ is denoted by $\mathrm{HC}^\bullet_{\mathrm{diff}}(M) := \bigoplus_{k\in\mathbb{N}} \mathrm{HC}^k_{\mathrm{diff}}(M)$. Notice that the Gerstenhaber bracket can be restricted to such cochains because the Gerstenhaber bracket of two multi-differential operators is a multi-differential operator.

13.1.6 A Star Product for Constant Poisson Structures

In this section we consider the case of a constant Poisson structure π on a finite-dimensional vector space V. We describe the Moyal–Weyl product, which is an explicit star product for (V,π).

In order to describe the Moyal–Weyl product, we first show how a constant Poisson structure leads naturally to an endomorphism of $\mathscr{F}(V)\otimes\mathscr{F}(V)$, where $\mathscr{F}(V) := C^\infty(V)$. Given two endomorphisms $\mathscr{V}_1$ and $\mathscr{V}_2$ of $\mathscr{F}(V)$, an endomorphism $\mathscr{V}_1\otimes\mathscr{V}_2$ of $\mathscr{F}(V)\otimes\mathscr{F}(V)$ is defined by letting

$$(\mathscr{V}_1\otimes\mathscr{V}_2)(F\otimes G) := \mathscr{V}_1(F)\otimes\mathscr{V}_2(G)\,,$$

for all $F,G\in\mathscr{F}(V)$. Given endomorphisms $\mathscr{V}_1,\ldots,\mathscr{V}_k,\mathscr{V}_1',\ldots,\mathscr{V}_k'$ of $\mathscr{F}(V)$ leads by linear extension to an endomorphism $\sum_{i=1}^k \mathscr{V}_i\otimes\mathscr{V}_i'$ of $\mathscr{F}(V)\otimes\mathscr{F}(V)$. In particular, given a constant Poisson structure on V, which we write as $\pi = \sum_{i=1}^k \mathscr{V}_i\wedge\mathscr{V}_i'$, with $\mathscr{V}_1,\ldots,\mathscr{V}_k,\mathscr{V}_1',\ldots,\mathscr{V}_k'$ constant vector fields on V, the previous construction yields an endomorphism of $\mathscr{F}(V)\otimes\mathscr{F}(V)$, which we also denote π. We can then write

$$\pi[F,G] = \mu(\pi(F\otimes G))\,, \tag{13.13}$$

where μ denotes the linear map $\mu:\mathscr{F}(V)\otimes\mathscr{F}(V)\to\mathscr{F}(V)$, which simply sends $F\otimes G$ to the product FG, for functions $F,G\in\mathscr{F}(V)$. The interest of writing the Poisson structure, and the other differential operators which will come up in the construction, as in the right-hand side of (13.13) is that although π cannot be reapplied to $\pi[F,G]$, it can be reapplied to $\pi(F\otimes G)$. It allows us to define, for $F,G\in\mathscr{F}(V)$ (or, by ν-linearity, for $F,G\in\mathscr{F}(V)^\nu$) the following formal power series:

$$e^{\nu\pi/2}(F\otimes G) := F\otimes G + \sum_{i\in\mathbb{N}^*}\frac{\nu^i}{2^i\,i!}\pi^i(F\otimes G)\,. \tag{13.14}$$

Proposition 13.8. *Let π be a constant Poisson structure on a finite-dimensional vector space V and let μ denote the linear map $\mu:\mathscr{F}(V)^\nu\otimes\mathscr{F}(V)^\nu\to\mathscr{F}(V)^\nu$, defined by $\mu(F\otimes G) := FG$, for all $F,G\in\mathscr{F}(V)^\nu$. The bilinear map $\mathscr{F}(V)^\nu\times\mathscr{F}(V)^\nu\to\mathscr{F}(V)^\nu$, defined by*

$$F\star G := \mu\left(e^{\nu\pi/2}(F\otimes G)\right), \tag{13.15}$$

is a star product for (V,π).

Proof. We first prove that $\star$ is associative: for all $F,G,H \in \mathscr{F}(V)$, we show that $F\star(G\star H) = (F\star G)\star H$. This can be done by direct computation, upon using the classical trick of representing a bi-differential operator as a composition of two differential operators: if Φ is a bi-differential operator, which we consider here with constant coefficients, then there exists a differential operator $\hat{\Phi}_{xy}$, such that

$$\Phi(F,G) = \hat{\Phi}_{xy}(F(x)G(y))_{|y=x} \;.$$

Explicitly, when Φ is of the form (13.12), with all $\phi_{r,s}$ constant, then $\hat{\Phi}_{xy}$ is given by

$$\hat{\Phi}_{xy} = \sum_{r,s\in\mathbb{N}^d} \phi_{r,s} \frac{\partial^{r+s}}{\partial x^r \partial y^s} \;.$$

With this notation, we can write (13.15) as

$$F\star G = e^{\nu\hat{\pi}_{xy}/2} F(x)G(y)_{|y=x} \;.$$

The associativity of $\star$ then follows from the following computation:

$$\begin{aligned}(F\star G)\star H &= e^{\nu\hat{\pi}_{xz}/2}(F\star G)(x)H(z)_{|z=x}\\ &= e^{\nu\hat{\pi}_{xz}/2+\nu\hat{\pi}_{yz}/2}\, e^{\nu\hat{\pi}_{xy}/2} F(x)G(y)H(z)_{|z=y=x}\\ &= e^{\nu(\hat{\pi}_{xz}+\hat{\pi}_{yz}+\hat{\pi}_{xy})/2} F(x)G(y)H(z)_{|z=y=x}\\ &= F\star(G\star H)\,.\end{aligned}$$

The computation demands some extra justification. We have used in the third equality that if ϕ_1 and ϕ_2 are commuting endomorphisms, then $e^{\phi_1+\phi_2} = e^{\phi_1}e^{\phi_2}$; this property follows at once from the definition given in (13.14). We used in the second equality that if $\mathscr{V} = \mathscr{V}_x$ is a derivation of $\mathscr{F}(V)$, then

$$\mathscr{V}\left(F(x,y)_{|y=x}\right) = \mathscr{V}_x F(x,y)_{|y=x} + \mathscr{V}_y F(x,y)_{|y=x} = (\mathscr{V}_x + \mathscr{V}_y)F(x,y)_{|y=x} \;,$$

which follows from the chain rule. This proves that $\star$ is associative, hence is a formal deformation of μ; since for $F,G \in \mathscr{F}(V)$ the coefficient of ν in $F\star G$ is $\frac{1}{2}\pi[F,G]$, it follows that $\star$ is a star product for (V,π). $\square$

When the rank $2r$ of the constant Poisson structure π is equal to the dimension of V, there exists a system of linear coordinates $(q_1,\dots,q_r,p_1,\dots,p_r)$ on V, in which the Poisson bracket of functions $F,G \in \mathscr{F}(V)$ takes the canonical form

$$\{F,G\} = \sum_{i=1}^{r}\left(\frac{\partial F}{\partial q_i}\frac{\partial G}{\partial p_i} - \frac{\partial F}{\partial p_i}\frac{\partial G}{\partial q_i}\right) \;;$$

see Proposition 6.5 for the existence of such coordinates. For such a choice of Poisson structure and such coordinates, the explicit formula (13.14) is referred to as the *Moyal–Weyl product*, or simply the *Moyal product*.

13.1.7 A Star Product for Symplectic Manifolds

In this section we explain how the Moyal–Weyl product can be used to prove that every symplectic manifold (M,ω) admits a star product; more precisely, the algebra of smooth functions on (M,ω), equipped with its canonical Poisson structure, admits a star product. The construction presented here is due to Fedosov; we refer to his original paper [73] for proofs and details.

We first need to elaborate on the star product in the case of a symplectic vector space (the Moyal–Weyl product, explained in Section 13.1.6 above). Let (V,ω) be a finite-dimensional symplectic real vector space and let $\pi = \{\cdot,\cdot\}$ denote the (constant) Poisson structure on V which corresponds to ω (see Section 6.3.1). We denote by $\hat{\mathscr{F}}(V)$ the space of all formal power series in the variables that constitute a basis for V^* and we denote by $\hat{\mathscr{F}}^{\geqslant j}(V)$ the subspace of $\hat{\mathscr{F}}(V)$, consisting of those formal power series whose valuation is at least $j \in \mathbb{N}$. We also consider, for $k \in \mathbb{N}$,

$$E^{\geqslant k}(V) := \sum_{\substack{i,j\in\mathbb{N}\\ 2i+j\geqslant k}} \nu^i \hat{\mathscr{F}}^{\geqslant j}(V)\,.$$

It is easy to see from the explicit formula (13.15) that the Moyal–Weyl product on $C^\infty(V)^\nu$ extends to $\hat{\mathscr{F}}(V)^\nu := \hat{\mathscr{F}}(V)[[\nu]]$; moreover, for all $k,\ell \in \mathbb{N}$, we have that

$$E^{\geqslant k}(V) \star E^{\geqslant \ell}(V) \subset E^{\geqslant k+\ell}(V)\,, \tag{13.16}$$

which follows immediately from the fact that $\{\hat{\mathscr{F}}^{\geqslant k}(V), \hat{\mathscr{F}}^{\geqslant \ell}(V)\} \subset \hat{\mathscr{F}}^{\geqslant k+\ell-2}(V)$, for all $k,\ell \in \mathbb{N}^*$. To V we associate the associative algebra $A(V)$, defined by

$$A(V) := E^{\geqslant 0}(V) \otimes \wedge^\bullet V^*\,,$$

equipped with the associative product $\circledast$ given by:

$$(F\otimes\xi) \circledast (G\otimes\eta) := (F\star G)\otimes(\xi\wedge\eta)\,,$$

for all $F,G \in E^{\geqslant 0}(V)$ and $\xi,\eta \in \wedge^\bullet V^*$. Note that there is a decreasing sequence of vector spaces

$$A^{\geqslant k}(V) := E^{\geqslant k}(V) \otimes \wedge^\bullet V^*$$

with $A^{\geqslant 0}(V) = A(V)$. Note also that $A(V)/A^{\geqslant k}(V)$ is a finite-dimensional vector space, for each $k \in \mathbb{N}$, and that there is on $A(V)$ a natural linear form $P_0 : A(V) \to \mathbb{R}$ defined by mapping $F\otimes\xi$ to 0 if $\xi \in \oplus_{i\geqslant 1} \wedge^i V^*$ and by mapping $F\otimes 1$ to $F(0)$ for all $F \in E^{\geqslant 0}(V)$, where 1 stands for the unit element of the algebra $\wedge^\bullet V^*$, and where $F(0)$ stands for the constant term of the formal power series F.

We now get back to the case of a symplectic manifold (M,ω). The canonical Poisson structure associated to ω is denoted by π and the algebra of smooth functions on M is denoted by $\mathscr{F}(M)$. Each tangent space T_mM to M is a symplectic vector space, associated to the Poisson structure π_m, so that we can consider as above a star product $\star_m$ on $\hat{\mathscr{F}}(T_mM)^\nu$ and an associative product $\circledast_m$ on $A(T_mM)$.

We say that a section of the canonical projection $\bigcup_{m\in M} A(T_mM) \to M$ mapping each element to its base point is smooth if its projection on the vector bundle

$$\bigcup_{m\in M} \frac{A(T_mM)}{A^{\geqslant k}(T_mM)} \to M$$

is smooth, for every $k \in \mathbb{N}$. For $k \in \mathbb{N}$, the space of all smooth sections which take values in $A^{\geqslant k}(T_mM)$, for every $m \in M$, is denoted by $\mathscr{A}^{\geqslant k}(M)$; we use the shorthand $\mathscr{A}(M)$ for $\mathscr{A}^{\geqslant 0}(M)$. The product of two sections $a \in \mathscr{A}^{\geqslant k}(M)$ and $b \in \mathscr{A}^{\geqslant \ell}(M)$ is the section of $\mathscr{A}^{\geqslant k+\ell}(M)$ whose value at $m \in M$ is $a(m) \circledast_m b(m)$. Combined, the linear maps $P_0 : A(T_mM) \to \mathbb{R}$ defined above lead to a linear map $\mathscr{A}(M) \to \mathscr{F}(M)^\nu$, which we also denote by P_0.

Theorem 13.9. *Let (M,ω) be a symplectic manifold and let $(\mathscr{A}(M),\circledast)$ be the algebra constructed as above. There exists a derivation $D : \mathscr{A}(M) \to \mathscr{A}(M)$ with the following property: for every $F \in \mathscr{F}(M)^\nu$, there exists a unique $a_F \in \mathscr{A}(M)$ such that $D(a_F) = 0$ and $P_0(a_F) = F$.*

Without going into details, we mention briefly how D is constructed, thereby sketching a proof of the theorem; see [73] for details. D is an operator of the following form:

$$D := \mathrm{d}^\nabla + [\hat{\omega} + \gamma, \cdot\,] \ . \tag{13.17}$$

The first term, d^∇, depends on the choice of a *symplectic connection*, i.e. a torsion-free connection ∇ on TM which satisfies

$$\iota_{\mathscr{Z}}\, \mathrm{d}\omega(\mathscr{V},\mathscr{W}) = \omega(\nabla_{\mathscr{Z}}\mathscr{V},\mathscr{W}) + \omega(\mathscr{V},\nabla_{\mathscr{Z}}\mathscr{W}) \, ,$$

for arbitrary vector fields $\mathscr{V},\mathscr{W},\mathscr{Z}$ on M. Such a symplectic connection induces a connection on the bundle $\bigcup_{m\in M} \hat{\mathscr{F}}(T_mM) \to M$, which is strictly speaking not a vector bundle (because the vector spaces $\hat{\mathscr{F}}(T_mM)$ are infinite-dimensional), yet it leads to a covariant derivative, which is $\mathrm{d}^\nabla : \mathscr{A}(M) \to \mathscr{A}(M)$. The bracket $[\cdot,\cdot]$ in (13.17) is the Lie bracket on $\mathscr{A}(M)$ induced from the bracket $[\cdot,\cdot]_m$ on the spaces $A(T_mM)$, with $m \in M$, where $[\cdot,\cdot]_m$ is defined by

$$[a\otimes\xi, b\otimes\eta]_m := (a\star b - b\star a)\otimes\xi\wedge\eta \, ,$$

for all $a,b \in \hat{\mathscr{F}}(T_mM)$ and $\xi,\eta \in \wedge^\bullet T_m^*M$. Third, $\hat{\omega}$ is the symplectic structure ω, viewed as an element of $\mathscr{A}(M)$. Finally, $\gamma \in \mathscr{A}^{\geqslant 3}(M)$ is a section of $\bigcup_{m\in M} \hat{\mathscr{F}}(T_mM)^\nu \otimes T_m^*M$, which is chosen such that $D^2 = 0$. Using the latter property, one proves the existence and uniqueness of a_F, as stated in Theorem 13.9. It is clear that D is a derivation, because both the adjoint action and the covariant derivatives are derivations.

Corollary 13.10. *Let (M,ω) be a symplectic manifold, let $(\mathscr{A}(M),\circledast)$ be the algebra constructed above and let D be a derivation of $\mathscr{A}(M)$ as in Theorem 13.9. Denote by $\star$ the bilinear map on $\mathscr{F}(M)^\nu$ defined for all $F,G \in \mathscr{F}(M)^\nu$ by*

$$F \star G := P_0(a_F \circledast a_G), \tag{13.18}$$

where a_F is the unique element in the kernel of D, for which $P_0(a_F) = F$, and similarly for a_G. Then $\star$ is a star product for (M,π), where π is the canonical Poisson structure on M, associated to ω.

Proof. We first prove that the product, defined by (13.18), is associative. Let $\mathscr{B}(M) \subset \mathscr{A}(M)$ be the kernel of D; it is a subalgebra of $\mathscr{A}(M)$ since D is a derivation. By Theorem 13.9, the assignment $F \mapsto a_F$ is a bijective linear map from $\mathscr{F}(M)^\nu$ to $\mathscr{B}(M)$ with inverse map the restriction of P_0 to $\mathscr{B}(M)$. It follows that the product $\star$ on $\mathscr{F}(M)^\nu$ is obtained from the associative product $\circledast$ on $\mathscr{B}(M)$, via a bijection, hence is associative. We prove that $\star$ is a star product for (M,π). Let F and G be smooth functions on M (in particular, they are independent of ν). A closer look at the construction of a_F and a_G in [73] implies that $a_F = (F + \mathrm{d}_m F + \mathrm{Hess}_m F) \otimes 1 + b_F$ and $a_G = (G + \mathrm{d}_m G + \mathrm{Hess}_m G) \otimes 1 + b_G$ for some $b_F, b_G \in \mathscr{A}^{\geqslant 3}(M)$, where $\mathrm{d}_m F$ and $\mathrm{d}_m G$ are viewed as linear functions on $T_m M$, and $\mathrm{Hess}_m F$ and $\mathrm{Hess}_m G$ are viewed as quadratic functions on $T_m M$. An explicit computation then yields $P_0(a_F \circledast a_G) = FG + \frac{\nu}{2}\{F,G\} \pmod{\nu^2}$, which completes the proof. $\square$

13.2 Deformations of Poisson Structures

The deformation theory of commutative associative algebras, described in Section 13.1, can quite easily be transcribed for Lie algebras. For Poisson algebras, a few extra adaptions are necessary, since Poisson algebras have not one but two algebra structures; in this section, we will deform only one of them, the Poisson bracket, but we need to keep in mind that the deformed Poisson bracket should be compatible with the associative product. As we will see, the deformation theory of Poisson algebras is, despite this constraint, formally very similar to the deformation theory of commutative associative algebras.

13.2.1 Formal and k-th Order Deformations

Let $(\mathscr{A},\cdot,\pi)$ be a Poisson algebra. On $\mathscr{A}^\nu$ we consider, as in (13.1), the $\mathbb{F}^\nu$-linear extension of the product "$\cdot$", which makes $\mathscr{A}^\nu$ into a commutative associative algebra (over the ring $\mathbb{F}^\nu$).

Definition 13.11. Let $(\mathscr{A},\cdot,\pi)$ be a Poisson algebra. An $\mathbb{F}^\nu$-linear skew-symmetric biderivation $\pi_\star$ of $(\mathscr{A}^\nu,\cdot)$ is called a *formal deformation* of π if $\pi_\star$ satisfies the Jacobi identity and $\pi_\star = \pi \pmod{\nu}$.

When $\pi_\star$ is a formal deformation of a Poisson structure π, the triple $(\mathscr{A}^\nu,\cdot,\pi_\star)$ is a Poisson algebra over the ring $\mathbb{F}^\nu$. Strictly speaking, our Definition 1.1 of a Poisson algebra is valid only for algebras over a field, but it is clear that one can define Poisson algebras over an arbitrary commutative ring (with unit) R, by replacing in Definition 1.1 "$\mathbb{F}$-vector space $\mathscr{A}$" by "R-module $\mathscr{A}$". It is also clear that the Jacobi identity for $\pi_\star$ is satisfied if and only if $[\pi_\star, \pi_\star]_S = 0$, where the Schouten bracket $[\cdot,\cdot]_S$ is still given by (3.36), but for $\mathbb{F}^\nu$-multi-linear maps, rather than $\mathbb{F}$-multi-linear maps; this is the main fact which we will use about this more general context.

Definition 13.12. Let $(\mathscr{A},\cdot,\pi)$ be a Poisson algebra and let $\pi_\star$ and $\pi'_\star$ be two formal deformations of π. Then $\pi_\star$ and $\pi'_\star$ are said to be *equivalent deformations* if there exists an $\mathbb{F}^\nu$-linear map $\Phi : \mathscr{A}^\nu \to \mathscr{A}^\nu$, satisfying for all $F, G \in \mathscr{A}$ the following conditions:

(1) $\Phi(F) = F \pmod{\nu}$;
(2) $\Phi(FG) = \Phi(F)\Phi(G)$;
(3) $\Phi(\pi_\star[F,G]) = \pi'_\star[\Phi(F), \Phi(G)]$.

Condition (1) implies that Φ is invertible; then (2) means that Φ is an automorphism of the associative algebra $(\mathscr{A},\cdot)$, hence of $(\mathscr{A}^\nu,\cdot)$ and (3) means that Φ is a Lie algebra isomorphism $(\mathscr{A}^\nu, \pi_\star) \to (\mathscr{A}^\nu, \pi'_\star)$.

It is clear that, if $\pi_\star$ is a formal deformation of π and $\Phi : \mathscr{A}^\nu \to \mathscr{A}^\nu$ is an $\mathbb{F}^\nu$-linear map satisfying items (1) and (2) of Definition 13.12, then the map $\pi'_\star : \mathscr{A}^\nu \times \mathscr{A}^\nu \to \mathscr{A}^\nu$, defined for $F, G \in \mathscr{A}^\nu$ by

$$\pi'_\star[F,G] := \Phi\left(\pi_\star[\Phi^{-1}(F), \Phi^{-1}(G)]\right) , \tag{13.19}$$

is a formal deformation of π, equivalent to $\pi_\star$.

As in the associative case, we define the notion of a k-th order deformation of a Poisson structure.

Definition 13.13. Let $(\mathscr{A},\cdot,\pi)$ be a Poisson algebra. For $k \in \mathbb{N}^*$, an $\mathbb{F}_k^\nu$-bilinear skew-symmetric biderivation $\pi_{(k)}$ of $\mathscr{A}_k^\nu$,

$$\pi_{(k)} = \pi_0 + \pi_1\nu + \cdots + \pi_k\nu^k , \tag{13.20}$$

is called a *k-th order deformation* of π, if $\pi_0 = \pi$ and if $\pi_{(k)}$ satisfies the Jacobi identity.

One defines the notion of equivalence of k-th order deformations by replacing $\mathbb{F}^\nu$ by $\mathbb{F}_k^\nu$ and $\mathscr{A}^\nu$ by $\mathscr{A}_k^\nu$ in Definition 13.12.

13.2.2 Deformations and Cohomology

As in the associative case, it is clear that a formal deformation can be truncated to a k-th order deformation, for every $k \in \mathbb{N}$. Similarly, an ℓ-th deformation can be truncated to a k-th order deformation, for every $k \leqslant \ell$. This suggests to approach also in this case the construction and the study of all (formal) deformations of a Poisson bracket, by analyzing the extensibility of a k-th order deformation to a $(k+1)$-th order deformation.

Therefore, let $(\mathscr{A},\cdot,\pi)$ be a Poisson algebra and suppose that $\pi_{(k)} = \pi_0 + \pi_1 \nu + \cdots + \pi_k \nu^k$ defines a k-th order deformation of π, where $\pi_1,\ldots,\pi_k$ are elements of $\mathfrak{X}^2(\mathscr{A})$ and $\pi_0 = \pi$. Given a skew-symmetric bilinear map $\pi_{k+1} : \mathscr{A}\times\mathscr{A}\to\mathscr{A}$, the skew-symmetric $\mathbb{F}^\nu_{k+1}$-bilinear map $\pi_{(k+1)}$, defined by $\pi_{(k+1)} := \pi_{(k)} + \pi_{k+1}\,\nu^{k+1}$, is a $(k+1)$-th order deformation of π if and only if $\left[\pi_{(k+1)},\pi_{(k+1)}\right]_S = 0 \pmod{\nu^{k+2}}$, i.e.,

$$\delta^2_\pi(\pi_{k+1}) = -[\pi_{k+1},\pi_0]_S = \frac{1}{2}\sum_{\substack{i+j=k+1\\ i,j\geqslant 1}} [\pi_i,\pi_j]_S\ , \tag{13.21}$$

where we have used the Poisson coboundary operator (4.5), associated to the Poisson bracket $\pi = \pi_0$. As in the associative case (see Section 13.1.3), one checks by using the graded Jacobi identity for $[\cdot,\cdot]_S$, that the right-hand side in (13.21) is a Poisson 3-cocycle, so that the obstruction for extending a deformation of the Poisson bracket π, of some order to the next order lies in the third Poisson cohomology space $H^3_\pi(\mathscr{A})$. This shows the following proposition.

Proposition 13.14. *Let $(\mathscr{A},\cdot,\pi)$ be a Poisson algebra and suppose that $\pi_{(k)} = \sum_{i=0}^k \pi_i \nu^i$ is a k-th order deformation of π. Then $\pi_{(k)}$ can be extended to a $(k+1)$-th order deformation of π if and only if the Poisson 3-cocycle $\sum_{i=1}^k [\pi_i,\pi_{k+1-i}]_S$ is a Poisson 3-coboundary.*

It is clear from (13.21) that the term $\pi_{k+1}\nu^{k+1}$ which makes a k-th order deformation of π into a $(k+1)$-th order deformation of π is not unique (if it exists): one can always add an arbitrary Poisson 2-cocycle (for π) to π_{k+1}. If this cocycle is a coboundary, then the two extended $(k+1)$-th order deformations are essentially the same, in the sense that they are equivalent (see Definition 13.12). A stronger statement is given in the following proposition, which is the analog of Proposition 13.6.

Proposition 13.15. *Let $(\mathscr{A},\cdot,\pi)$ be a Poisson algebra and suppose that $\pi_{(k)} = \sum_{i=0}^k \pi_i \nu^i$ is a k-th order deformation of π, where $k\in\mathbb{N}^*$ and let $\ell\leqslant k$, where $\ell\in\mathbb{N}^*$. Let $\varphi\in\mathfrak{X}^1(\mathscr{A})$, which we extend to an $\mathbb{F}^\nu$-linear map $\mathscr{A}^\nu\to\mathscr{A}^\nu$, still denoted by φ, and let*

$$\pi'_{(\ell)} = \pi_0 + \pi_1\nu + \pi_2\nu^2 + \cdots + \pi_{\ell-1}\nu^{\ell-1} + (\pi_\ell + \delta^1_\pi(\varphi))\,\nu^\ell\ . \tag{13.22}$$

Then, $\pi'_{(\ell)}$ extends to a k-th order deformation $\pi'_{(k)}$ of π, which is equivalent to $\pi_{(k)}$.

Proof. Let us consider the map

$$\begin{aligned} \Phi : \mathscr{A}^\nu &\to \mathscr{A}^\nu \\ F &\mapsto e^{-\nu^\ell\varphi}(F) = \sum_{k\in\mathbb{N}} \frac{(-1)^k}{k!}\nu^{k\ell}\varphi^k(F)\ . \end{aligned} \tag{13.23}$$

It is clear that Φ is $\mathbb{F}^\nu$-linear and that $\Phi(F) = F \pmod{\nu}$, for all $F\in\mathscr{A}$. We show that $\Phi(FG) = \Phi(F)\Phi(G)$ for all $F,G\in\mathscr{A}$. To do this, we use that for all $k\in\mathbb{N}$,

$$\varphi^k(FG) = \sum_{\substack{r+s=k \\ r,s\geqslant 0}} \binom{k}{s} \varphi^r(F)\,\varphi^s(G)\,, \tag{13.24}$$

which is a consequence of the fact that φ is a derivation of $\mathscr{A}$. Then,

$$\begin{aligned}
\Phi(FG) = e^{-\nu^\ell \varphi}(FG) &= \sum_{k\in\mathbb{N}} \frac{(-1)^k}{k!} \nu^{\ell k} \varphi^k(FG) \\
&= \sum_{k\in\mathbb{N}} \sum_{\substack{r+s=k \\ r,s\geqslant 0}} \frac{(-1)^k}{k!} \binom{k}{s} \nu^{\ell k} \varphi^r(F)\,\varphi^s(G) \\
&= \sum_{r\in\mathbb{N}} \sum_{s\in\mathbb{N}} \frac{(-1)^{r+s}}{s!r!} \nu^{\ell(r+s)} \varphi^r(F)\,\varphi^s(G) \\
&= \left(\sum_{r\in\mathbb{N}} \frac{(-1)^r}{r!} \nu^{\ell r} \varphi^r(F)\right) \left(\sum_{s\in\mathbb{N}} \frac{(-1)^s}{s!} \nu^{\ell s} \varphi^s(G)\right) \\
&= \Phi(F)\,\Phi(G)\,.
\end{aligned}$$

As in (13.19), we have that $\pi'_{(k)}$, which is defined for all $F,G \in \mathscr{A}^\nu$ by

$$\pi'_{(k)}[F,G] := \Phi\left(\pi_{(k)}\left[\Phi^{-1}(F), \Phi^{-1}(G)\right]\right),$$

is a k-th order deformation of π which is equivalent to $\pi_{(k)}$. Using the formula for the inverse of Φ, namely

$$\Phi^{-1}(F) = e^{\nu^\ell \varphi}(F) = \sum_{k\in\mathbb{N}} \frac{\nu^{\ell k}}{k!} \varphi^k(F)\,,$$

we can make the following computation in the algebra $\mathscr{A}_\ell^\nu$ (i.e., all equalities are modulo $\nu^{\ell+1}$):

$$\begin{aligned}
\pi'_{(\ell)}(F,G) &= \Phi\left(\pi_{(\ell)}\left[F + \varphi[F]\nu^\ell, G + \varphi[G]\nu^\ell\right]\right) \\
&= \Phi\left(\pi_{(\ell)}[F,G] + (\pi[\varphi[F],G] + \pi[F,\varphi[G]])\nu^\ell\right) \\
&= \pi_{(\ell)}[F,G] + (\pi[\varphi[F],G] + \pi[F,\varphi[G]] - \varphi[\pi[F,G]])\nu^\ell \\
&= \pi_{(\ell)}[F,G] + \delta^1_\pi(\varphi)[F,G]\,\nu^\ell\,.
\end{aligned}$$

It shows that $\pi'_{(\ell)} = \pi_{(\ell)} + \delta^1_\pi(\varphi)\,\nu^\ell$, which is the content of (13.22). □

It follows from the proposition that in order to construct all possible deformations of π, up to equivalence, one only has to consider, at every step k, as many possibilities as there are elements in $H^2_\pi(\mathscr{A})$. Notice that the associative analog of the above proposition, which was given in Proposition 13.6, is easier to prove; the reason for this is that in the Poisson case the map Φ which defines the equivalence needs to be an automorphism of the associative algebra $(\mathscr{A}_k^\nu, \cdot)$.

13.2.3 The Maurer–Cartan Equation

Let $(\mathscr{A}, \cdot, \pi)$ be a Poisson algebra and let $\pi_\star = \sum_{i=0}^{\infty} \pi_i \nu^i$ be an element of $\mathfrak{X}^2(\mathscr{A}^\nu)$, where $\pi_0 = \pi$. As above, $\pi_\star$ is a Poisson structure if and only if $[\pi_\star, \pi_\star]_S = 0$. Denoting by δ_π^2 the $\mathbb{F}^\nu$-linear extension of δ_π^2 to $\mathfrak{X}^2(\mathscr{A}^\nu) \simeq \mathfrak{X}^2(\mathscr{A})[[\nu]]$, it follows that $\pi_\star$ is a Poisson structure if and only if $\pi_\star - \pi$ is a solution of the following equation

$$\delta_\pi^2(x) - \frac{1}{2}\,[x,x]_S = 0\,. \tag{13.25}$$

In the terminology of Section 13.3, this equation is called the *Maurer–Cartan equation* associated to the differential graded Lie algebra $(\mathfrak{X}^\bullet(\mathscr{A}), [\cdot\,,\cdot]_S, \delta_\pi)$.

13.3 Differential Graded Lie Algebras

Differential graded Lie algebras are graded Lie algebras which are equipped with a derivation of degree one whose square is zero. A prime example, which we have already encountered several times in this book, is the algebra of skew-symmetric multi-derivations of a Poisson algebra (or multivector fields on a Poisson manifold), where the graded Lie bracket is the Schouten bracket and the differential is (up to a sign) the Poisson coboundary operator. Since in the study of a differential graded Lie algebra its symmetric algebra plays an important rôle we first give the basic definitions and properties of the symmetric algebra of a graded vector space.

13.3.1 The Symmetric Algebra of a Graded Vector Space

Let $V = \bigoplus_{i\in\mathbb{Z}} V_i$ be a graded vector space. We denote the degree of a homogeneous element $x \in V$ by $d^0(x)$. The *symmetric algebra* of V is the associative algebra $S^\bullet V$, which is obtained by dividing the tensor (graded associative) algebra $T^\bullet V = \bigoplus_{\ell\in\mathbb{N}} V^{\otimes \ell}$ of V by the two-sided ideal generated by all elements of the form $x\otimes y - (-1)^{d^0(x)d^0(y)} y\otimes x$, with x and y homogeneous. The induced product is denoted by juxtaposition of elements. With this notation,

$$xy = (-1)^{d^0(x)d^0(y)} yx, \tag{13.26}$$

for all homogeneous elements x, y of V. Note that if x is a homogeneous element of V of odd degree, then it follows from (13.26) that $x^2 = -x^2$, so that $x^2 = 0$. We call an element of $S^\bullet V$ of the form $x_1 \dots x_\ell$ with all x_i homogeneous, a *homogeneous monomial* of $S^\bullet V$. Let $i_1, \dots, i_\ell$ be integers for which there exists a non-zero homogeneous monomial $X = x_1 \dots x_\ell$ where $x_1 \in V_{i_1}, \dots, x_\ell \in V_{i_\ell}$. Then we can define the sign $\operatorname{sgn}(\sigma; i_1, \dots, i_\ell) \in \{-1, 1\}$, associated to $(i_1, \dots, i_\ell)$ and $\sigma \in S_\ell$, by the equality

$$x_{\sigma(1)} \dots x_{\sigma(\ell)} = \operatorname{sgn}(\sigma; i_1, \dots, i_\ell)\, x_1 \dots x_\ell\,, \tag{13.27}$$

since it only depends on the ℓ-tuple $(i_1, \dots, i_\ell)$ and not on the choice of $x_1, \dots, x_\ell$. With X as above, we often write $\operatorname{sgn}(\sigma; X)$ as a shorthand for $\operatorname{sgn}(\sigma; i_1, \dots, i_\ell)$. For indices $i_1, \dots, i_\ell$ such that $V_{i_1} V_{i_2} \cdots V_{i_\ell} = 0$ we define $\operatorname{sgn}(\sigma; i_1, \dots, i_\ell) := 1$ for all $\sigma \in S_\ell$.

In what follows, the graded vector space of which we consider the symmetric algebra will always be a graded Lie algebra, but in which we shift the grading by one. Thus, if $\mathfrak{g} = \bigoplus_{\ell \in \mathbb{N}} \mathfrak{g}_\ell$ is a graded Lie algebra, we consider on $\mathfrak{g}$ two gradings: for $x \in \mathfrak{g}_\ell$ we define $|x| := \ell$ and $d^0(x) := \ell - 1 = |x| - 1$. Then $\mathfrak{g}$ is a graded Lie algebra with respect to the grading $|\cdot|$, but when the grading d^0 is used, it is only a graded vector space; when we use the latter grading, we will write $\bar{\mathfrak{g}}$ instead of $\mathfrak{g}$. The symmetric algebra which plays an important rôle in what follows is the graded commutative algebra $S^\bullet \bar{\mathfrak{g}}$, where the grading on $S^\bullet \bar{\mathfrak{g}}$ is given, for a homogeneous monomial $x_1 \dots x_\ell$, by

$$d^0(x_1 \dots x_\ell) := \sum_{i=1}^{\ell} |x_i| - \ell = \sum_{i=1}^{\ell} d^0(x_i)\,. \tag{13.28}$$

Of course, $S^\bullet \bar{\mathfrak{g}}$ inherits also a grading from the tensor algebra, which assigns to a monomial $x_1 \dots x_\ell$ the degree ℓ; we will not use this grading, except that we denote, for $\ell \in \mathbb{N}$, by $S^\ell \bar{\mathfrak{g}}$ the span of all monomials of the form $x_1 \dots x_\ell$, where $x_1, \dots, x_\ell \in \mathfrak{g}$.

13.3.2 Differential Graded Lie Algebras

In this section, we introduce the notion of a differential graded Lie algebra and of a morphism of differential graded Lie algebras. The basic examples, given below, are the Poisson complex, associated to a Poisson algebra, and the Hochschild complex, associated to an associative algebra.

Definition 13.16. A *differential graded Lie algebra* is a triple $(\mathfrak{g}, [\cdot\,,\cdot], D)$, where $(\mathfrak{g}, [\cdot\,,\cdot])$ is a graded Lie algebra and $D : \mathfrak{g} \to \mathfrak{g}$ is a *differential* of $\mathfrak{g}$, i.e., D is a graded linear map of degree one from $\mathfrak{g}$ to $\mathfrak{g}$, D satisfies $D \circ D = 0$ and D is a graded derivation of degree one of $(\mathfrak{g}, [\cdot\,,\cdot])$.

Expressed in terms of formulas, the fact that $(\mathfrak{g} = \bigoplus_{k \in \mathbb{Z}} \mathfrak{g}_k, [\cdot\,,\cdot])$ is a graded Lie algebra means that $[\cdot\,,\cdot]$ satisfies, for all homogeneous elements x, y and z of $\mathfrak{g}$ and for all $i, j \in \mathbb{Z}$, the following properties:

(1) $[\mathfrak{g}_i, \mathfrak{g}_j] \subset \mathfrak{g}_{i+j}$;

(2) $[x,y] = -(-1)^{|x||y|}[y,x]$;
(3) $(-1)^{|x||z|}[[x,y],z] + \circlearrowleft(x,y,z) = 0$;

while the fact that D is a differential on $\mathfrak{g}$ means that D satisfies, for all homogeneous elements x and y of $\mathfrak{g}$ and for all $i \in \mathbb{Z}$, the following properties:

(4) $D(\mathfrak{g}_i) \subset \mathfrak{g}_{i+1}$;
(5) $D([x,y]) = [D(x),y] + (-1)^{|x|}[x,D(y)]$;
(6) $D \circ D = 0$.

Let $(\mathfrak{g},[\cdot\,,\cdot])$ be a graded Lie algebra and let z be an element of degree one, $z \in \mathfrak{g}_1$. The graded Jacobi identity for $[\cdot\,,\cdot]$, i.e., the above identity (3), implies that the degree one map D_z, defined by

$$D_z(y) := [z,y]$$

for all $y \in \mathfrak{g}$, is a graded derivation of $\mathfrak{g}$ of degree one. Moreover, if z satisfies $[z,z]=0$, the graded Jacobi identity implies that $D_z \circ D_z = 0$, hence $(\mathfrak{g},[\cdot\,,\cdot],D_z)$ is a differential graded Lie algebra. We then call the triple $(\mathfrak{g},[\cdot\,,\cdot],z)$ a *pointed differential graded Lie algebra*. The differential of a pointed differential graded Lie algebra $(\mathfrak{g},[\cdot\,,\cdot],z)$ will, without further mention, always be denoted by D_z.

Example 13.17. Let $(\mathscr{A},\cdot,\pi)$ be a Poisson algebra and consider the graded Lie algebra $(\overline{\mathfrak{X}}^\bullet(\mathscr{A}),[\cdot\,,\cdot]_S)$, where we recall that $\overline{\mathfrak{X}}^\bullet(\mathscr{A})$ denotes the graded vector space of skew-symmetric multi-derivations of $\mathscr{A}$, with shifted degree, and $[\cdot\,,\cdot]_S$ denotes the Schouten bracket on $\overline{\mathfrak{X}}^\bullet(\mathscr{A})$ (see Chapter 3). Since the shifted degree of π is one and since $[\pi,\pi]_S = 0$ because π is a Poisson bracket, $(\overline{\mathfrak{X}}^\bullet(\mathscr{A}),[\cdot\,,\cdot]_S,\pi)$ is a pointed differential graded Lie algebra.

Example 13.18. The previous example can also be considered in the smooth category. Namely, let (M,π) be a real Poisson manifold and recall that we denote by $(\overline{\mathfrak{X}}^\bullet(M),[\cdot\,,\cdot]_S)$ the graded Lie algebra of multivector fields on M, equipped with the Schouten bracket $[\cdot\,,\cdot]_S$. Then $(\overline{\mathfrak{X}}^\bullet(M),[\cdot\,,\cdot]_S,\pi)$ is a pointed differential graded Lie algebra. Notice that this example can be considered for any real manifold M by taking the trivial Poisson structure on M, i.e., for every real manifold M, the triple $(\overline{\mathfrak{X}}^\bullet(M),[\cdot\,,\cdot]_S,0)$ is a pointed differential graded Lie algebra.

Example 13.19. Let $(\mathscr{A},\mu)$ be a commutative associative algebra and consider the graded Lie algebra $(\overline{\mathrm{HC}}^\bullet(\mathscr{A}),[\cdot\,,\cdot]_G)$, where we recall from Section 13.1.2 that $\overline{\mathrm{HC}}^k(\mathscr{A}) = \mathrm{Hom}(\mathscr{A}^{k+1},\mathscr{A})$ is the space of all $(k+1)$-linear maps from $\mathscr{A}$ to itself and the graded Lie bracket $[\cdot\,,\cdot]_G$ is the Gerstenhaber bracket (defined in (13.5)). Since μ is associative, $[\mu,\mu]_G = 0$ and hence $(\overline{\mathrm{HC}}^\bullet(\mathscr{A}),[\cdot\,,\cdot]_G,\mu)$ is a pointed differential graded Lie algebra.

Example 13.20. The previous example can also be considered when $(\mathscr{A},\mu)$ is the algebra of smooth functions on a real manifold M, equipped with the usual product of functions μ. Consider the graded Lie algebra $(\overline{\mathrm{HC}}^\bullet_{\mathrm{diff}}(M),[\cdot\,,\cdot]_G)$, which is the subalgebra of $(\overline{\mathrm{HC}}^\bullet(C^\infty(M)),[\cdot\,,\cdot]_G)$ which consists of the differential Hochschild cochains (see Section 13.1.5). Then $(\overline{\mathrm{HC}}^\bullet_{\mathrm{diff}}(M),[\cdot\,,\cdot]_G,\mu)$ is a pointed differential graded Lie algebra.

Example 13.21. Every differential graded Lie algebra $(\mathfrak{g},[\cdot\,,\cdot],D)$ leads to a cohomology space, which is itself a pointed differential graded Lie algebra. To see this, consider $D^k : \mathfrak{g}_k \to \mathfrak{g}_{k+1}$, the restriction of the differential D to $\mathfrak{g}_k$. Since $D \circ D = 0$, we have a complex

$$\cdots \longrightarrow \mathfrak{g}_{k-1} \xrightarrow{D^{k-1}} \mathfrak{g}_k \xrightarrow{D^k} \mathfrak{g}_{k+1} \xrightarrow{D^{k+1}} \cdots$$

The cohomology of this complex is denoted by $H_D^\bullet(\mathfrak{g}) = \bigoplus_{k\in\mathbb{Z}} H_D^k(\mathfrak{g})$, where, for all $k \in \mathbb{Z}$,

$$H_D^k(\mathfrak{g}) := \operatorname{Ker} D^k / \operatorname{Im} D^{k-1} .$$

D is a graded derivation of degree one of $(\mathfrak{g},[\cdot\,,\cdot])$, hence $[\cdot\,,\cdot]$ induces a Lie bracket on $H_D^\bullet(\mathfrak{g})$, also denoted by $[\cdot\,,\cdot]$ and the triple $(H_D^\bullet(\mathfrak{g}),[\cdot\,,\cdot],0)$ is a pointed differential graded Lie algebra.

We now show that the differential and the bracket of a differential graded Lie algebra $(\mathfrak{g},[\cdot\,,\cdot],D)$ can be encoded in a single differential Λ on $S^\bullet\bar{\mathfrak{g}}$. Let $\mathfrak{g}$ be a graded vector space, let $[\cdot\,,\cdot] : \mathfrak{g}\times\mathfrak{g} \to \mathfrak{g}$ be a graded skew-symmetric bilinear map of degree zero and let $D : \mathfrak{g} \to \mathfrak{g}$ be a graded linear map of degree one. Two linear maps from $S^\bullet\bar{\mathfrak{g}}$ to itself are defined by setting for every homogeneous monomial $X = x_1 \ldots x_\ell \in S^\ell\bar{\mathfrak{g}}$,

$$\begin{aligned} \Lambda_D(X) &:= \sum_{i=1}^{\ell} \operatorname{sgn}(\sigma_i;X) D(x_i)\hat{X}_i , \\ \Lambda_{[\cdot,\cdot]}(X) &:= \sum_{1\leqslant i<j\leqslant \ell} \operatorname{sgn}(\sigma_{ij};X)(-1)^{d^0(x_i)}\,[x_i,x_j]\,\hat{X}_{ij} , \end{aligned} \tag{13.29}$$

where $\hat{X}_i$ and $\hat{X}_{ij}$ are obtained by removing from X the element x_i, respectively the elements x_i and x_j. For example, $\hat{X}_i = x_1 \ldots x_{i-1}x_{i+1} \ldots x_\ell$. The permutation $\sigma_i \in S_\ell$ is the permutation which sends 1 to i and subtracts one from the integers $2,\ldots,i$ and fixes the other integers, i.e., it is the cycle $(i,i-1,\ldots,1)$; similarly, σ_{ij} is the permutation which sends 1 to i, sends 2 to j, removes two from the integers $3,\ldots,i+1$, removes one from the integers $i+2,\ldots,j$ and fixes the other integers. We leave it as an exercise to the reader to check that the maps Λ_D and $\Lambda_{[\cdot,\cdot]}$ are well-defined. We define Λ to be the sum of the two linear maps in (13.29),

$$\Lambda := \Lambda_D + \Lambda_{[\cdot,\cdot]} . \tag{13.30}$$

Taking a homogeneous monomial $X \in S^\bullet\bar{\mathfrak{g}}$, it is easy to see that $d^0(\Lambda_D(X)) = d^0(\Lambda_{[\cdot,\cdot]}(X)) = d^0(X)+1$, so that Λ has degree 1. Notice however that $\Lambda_D : S^\ell\bar{\mathfrak{g}} \to S^\ell\bar{\mathfrak{g}}$, while $\Lambda_{[\cdot,\cdot]} : S^\ell\bar{\mathfrak{g}} \to S^{\ell-1}\bar{\mathfrak{g}}$.

Proposition 13.22. *Let $(\mathfrak{g},[\cdot\,,\cdot],D)$ be a triple, where $\mathfrak{g} = \bigoplus_{k\in\mathbb{Z}}\mathfrak{g}_k$ is a graded vector space, $D : \mathfrak{g} \to \mathfrak{g}$ a graded linear map of degree one and $[\cdot\,,\cdot] : \wedge^2\mathfrak{g} \to \mathfrak{g}$ a graded skew-symmetric bilinear map of degree zero. Let Λ_D, $\Lambda_{[\cdot,\cdot]}$ and Λ denote the linear maps defined in (13.29) and (13.30). Then the following statements are equivalent.*

(i) $(\mathfrak{g},[\cdot\,,\cdot],D)$ *is a differential graded Lie algebra;*

(ii) $\Lambda_D^2 = \Lambda_{[\cdot,\cdot]}^2 = 0$ *and* $\Lambda_D \circ \Lambda_{[\cdot,\cdot]} + \Lambda_{[\cdot,\cdot]} \circ \Lambda_D = 0$;

(iii) $\Lambda^2 = 0$.

For a given differential graded Lie algebra $(\mathfrak{g}, [\cdot,\cdot], D)$, *we call the linear map* $\Lambda : S^\bullet\bar{\mathfrak{g}} \to S^\bullet\bar{\mathfrak{g}}$ *the* total differential *of* $(\mathfrak{g}, [\cdot,\cdot], D)$.

Proof. As we already pointed out, for $X \in S^\ell\bar{\mathfrak{g}}$, we have that $\Lambda_D(X) \in S^\ell\bar{\mathfrak{g}}$, while $\Lambda_{[\cdot,\cdot]}(X) \in S^{\ell-1}\bar{\mathfrak{g}}$. Therefore, $\Lambda_D^2(X) \in S^\ell\bar{\mathfrak{g}}$, while $\Lambda_{[\cdot,\cdot]}^2(X) \in S^{\ell-2}\bar{\mathfrak{g}}$ and $(\Lambda_D \circ \Lambda_{[\cdot,\cdot]} + \Lambda_{[\cdot,\cdot]} \circ \Lambda_D)(X) \in S^{\ell-1}\bar{\mathfrak{g}}$. Since

$$\Lambda^2 = \Lambda_D^2 + \Lambda_D \circ \Lambda_{[\cdot,\cdot]} + \Lambda_{[\cdot,\cdot]} \circ \Lambda_D + \Lambda_{[\cdot,\cdot]}^2 ,$$

the equivalence of *(ii)* and *(iii)* follows. Let $X = x_1 \ldots x_\ell$ be a homogeneous monomial of $S^\ell\bar{\mathfrak{g}}$. Then

$$\Lambda_D^2(X) = \sum_{i=1}^{\ell} \operatorname{sgn}(\sigma_i; X) D^2(x_i)\hat{X}_i ,$$

so that $\Lambda_D^2 = 0$ if and only if $D^2 = 0$. Similarly,

$$\Lambda_{[\cdot,\cdot]}^2(X) = \sum_{1 \leqslant i<j<k \leqslant \ell} \operatorname{sgn}(\sigma_{ijk}; X)(-1)^{|x_i||x_k|+|x_j|} \alpha_{ijk}(X)\hat{X}_{ijk} ,$$

where σ_{ijk} and $\hat{X}_{ijk}$ are defined similarly to σ_{ij} and $\hat{X}_{ij}$, defined earlier in this section, and

$$\alpha_{ijk}(X) := (-1)^{|x_i||x_k|} [[x_i, x_j], x_k] + \circlearrowleft (i,j,k) .$$

It follows that $\Lambda_{[\cdot,\cdot]}^2 = 0$ if and only if $[\cdot,\cdot]$ satisfies the graded Jacobi identity. Finally,

$$(\Lambda_D \circ \Lambda_{[\cdot,\cdot]} + \Lambda_{[\cdot,\cdot]} \circ \Lambda_D)(X) = -\sum_{1 \leqslant i<j \leqslant \ell} \operatorname{sgn}(\sigma_{ij}; X)(-1)^{|x_i|} \beta_{ij}(X)\hat{X}_{ij} ,$$

where

$$\beta_{ij}(X) := D([x_i, x_j]) - [D(x_i), x_j] - (-1)^{|x_i|} [x_i, D(x_j)] ;$$

as a consequence, $\Lambda_D \circ \Lambda_{[\cdot,\cdot]} + \Lambda_{[\cdot,\cdot]} \circ \Lambda_D = 0$ if and only if D is a derivation of $[\cdot,\cdot]$. The last three equivalences, combined, prove the equivalence of *(i)* and *(ii)*. □

We finish this section by giving the definition of a morphism of differential graded Lie algebras.

Definition 13.23. Let $(\mathfrak{g}, [\cdot,\cdot]_\mathfrak{g}, D_\mathfrak{g})$ and $(\mathfrak{h}, [\cdot,\cdot]_\mathfrak{h}, D_\mathfrak{h})$ be two differential graded Lie algebras. A graded linear map $\phi : \mathfrak{g} \to \mathfrak{h}$ of degree zero is said to be a *morphism* of differential graded Lie algebras, if ϕ is a homomorphism of graded Lie algebras and $\phi \circ D_\mathfrak{g} = D_\mathfrak{h} \circ \phi$.

Let us denote by $\Lambda_\mathfrak{g}$ and $\Lambda_\mathfrak{h}$ the total differentials of $(\mathfrak{g}, [\cdot,\cdot]_\mathfrak{g}, D_\mathfrak{g})$, respectively of $(\mathfrak{h}, [\cdot,\cdot]_\mathfrak{h}, D_\mathfrak{h})$. It follows easily from the definitions that a graded linear map of degree zero $\phi : \mathfrak{g} \to \mathfrak{h}$ is a morphism of differential graded Lie algebras if and only if $S^\bullet\phi \circ \Lambda_\mathfrak{g} = \Lambda_\mathfrak{h} \circ S^\bullet\phi$, where $S^\bullet\phi : S^\bullet\bar{\mathfrak{g}} \to S^\bullet\bar{\mathfrak{h}}$ is defined by

$$S^\bullet\phi\,(x_1\ldots x_\ell) := \phi(x_1)\ldots\phi(x_\ell) \tag{13.31}$$

for all $x_1,\ldots,x_\ell \in \mathfrak{g}$.

13.3.3 The Maurer–Cartan Equation

In this section, we define the Maurer–Cartan equation associated to a general differential graded Lie algebra. We have already met this equation in the case of the pointed differential graded Lie algebras $(\overline{\mathrm{HC}}^\bullet(\mathscr{A}),[\cdot\,,\cdot]_G,\mu)$ and $(\overline{\mathfrak{X}}^\bullet(\mathscr{A}),[\cdot\,,\cdot]_S,\pi)$ (see (13.11) and (13.25)); as we explained, these equations govern deformations of associative products and of Poisson brackets. We will come back to these particular cases in Section 13.3.6 below.

As in the case of deformations of associative products and of Poisson brackets, we consider formal power series with coefficients in a (differential graded Lie) algebra. Thus, given a differential graded Lie algebra $(\mathfrak{g},[\cdot\,,\cdot],D)$, we consider $\mathfrak{g}^\nu := \mathfrak{g}[[\nu]]$, the $\mathbb{F}^\nu$-module of formal power series with coefficients in $\mathfrak{g}$. Every element x of $\mathfrak{g}^\nu$ is of the form $x=\sum_{k=0}^\infty x_k\nu^k$; even without writing explicitly x in this form, we will often write x_k (or $(x)_k$ in the case of ambiguity) for the coefficient of ν^k in x, a notation which comes in particularly handy when picking a particular coefficient of the series of a complex expression.

We keep the same notations $[\cdot\,,\cdot]$ and D for the $\mathbb{F}^\nu$-linear extension of $[\cdot\,,\cdot]$ and of D to $\mathfrak{g}^\nu$. Similarly, when Λ denotes the total differential on $S^\bullet\bar{\mathfrak{g}}$, then Λ will still denote its extension to $(S^\bullet\bar{\mathfrak{g}})^\nu := (S^\bullet\bar{\mathfrak{g}})[[\nu]]$. Consider the subspace $\nu\mathfrak{g}_1^\nu := \nu\mathfrak{g}_1[[\nu]]$ of $\mathfrak{g}^\nu$, which consists of all formal power series in ν without constant term and with coefficients in $\mathfrak{g}_1$, the degree one component of $\mathfrak{g}$. On $\nu\mathfrak{g}_1^\nu$ we consider the equation

$$D(x)+\frac{1}{2}\,[x,x]=0\,, \tag{13.32}$$

which is called the *Maurer–Cartan equation* associated to $(\mathfrak{g},[\cdot\,,\cdot],D)$ (or to $\mathfrak{g}$). We denote by $\mathrm{MC}(\mathfrak{g})$ the set of all elements x of $\nu\mathfrak{g}_1^\nu$ which are solutions of the Maurer–Cartan equation associated to $\mathfrak{g}$. We will give in Proposition 13.26 below two characterizations of the solutions of the Maurer–Cartan equation. To do this, we define, for $x\in\nu\mathfrak{g}_1^\nu$ and for $k\in\mathbb{N}^*$,

$$\mathrm{Obs}_k(x) := D(x_k)+\frac{1}{2}\sum_{\substack{i+j=k\\ i,j\geqslant 1}}[x_i,x_j]\,. \tag{13.33}$$

Then (13.32) can be written out as the infinite list of equations $\mathrm{Obs}_k(x)=0$, where $k\in\mathbb{N}^*$. If x satisfies $\mathrm{Obs}_k(x)=0$ for $k=1,\ldots,\ell$ we say that x is a solution of the Maurer–Cartan equation up to order ℓ and we denote by $\mathrm{MC}_\ell(\mathfrak{g})$ the set of all such elements. Clearly, $x\in\mathrm{MC}_\ell(\mathfrak{g})$ if and only if x satisfies (13.32) modulo

$\nu^{\ell+1}$ and $x \in \mathrm{MC}(\mathfrak{g})$ if and only if $x \in \mathrm{MC}_\ell(\mathfrak{g})$ for all $\ell \in \mathbb{N}^*$. Moreover, we have the following proposition.

Proposition 13.24. *Let $(\mathfrak{g}, [\cdot,\cdot], D)$ be a differential graded Lie algebra. Suppose that $x \in \mathrm{MC}_\ell(\mathfrak{g})$ for some $\ell \in \mathbb{N}^*$. Then $\mathrm{Obs}_{\ell+1}(x)$ is a cocycle.*

Proof. Since D is a graded derivation, satisfying $D^2 = 0$, and since the Lie bracket is graded skew-symmetric, we have that

$$D\left(\mathrm{Obs}_{\ell+1}(x)\right) = \frac{1}{2} \sum_{\substack{j+k=\ell+1 \\ j,k\geqslant 1}} \left([D(x_j), x_k] - [x_j, D(x_k)]\right) = \sum_{\substack{j+k=\ell+1 \\ j,k\geqslant 1}} [D(x_j), x_k] \ .$$

Since x satisfies the Maurer–Cartan equation up to order ℓ, we obtain

$$D\left(\mathrm{Obs}_{\ell+1}(x)\right) = -\frac{1}{2} \sum_{\substack{i+j+k=\ell+1 \\ i,j,k\geqslant 1}} [[x_i, x_j], x_k] = 0 \ ,$$

where we used the graded Jacobi identity in the last step. □

For the second characterization, we use the *exponential map*. For $x \in \nu\mathfrak{g}^\nu$, we define e^x to be the following formal power series

$$e^x := 1 + x + \frac{x^2}{2} + \cdots + \frac{x^k}{k!} + \cdots \in (S^\bullet \mathfrak{g})^\nu \ . \tag{13.34}$$

Notice that this infinite sum is a well-defined formal power series in ν with coefficients in $S^\bullet \bar{\mathfrak{g}}$ since at most the first $k+1$ terms of the infinite sum in (13.34) contribute to the coefficient of ν^k in e^x (recall that x is a formal power series in ν without constant term). The total differential Λ, applied to e^x, can be explicitly computed, as is shown in the following lemma.

Lemma 13.25. *Let $(\mathfrak{g}, [\cdot,\cdot], D)$ be a differential graded Lie algebra with total differential $\Lambda : S^\bullet \bar{\mathfrak{g}} \to S^\bullet \bar{\mathfrak{g}}$. For $x \in \nu\mathfrak{g}_1^\nu$, one has*

$$\Lambda(e^x) = \left(D(x) + \frac{1}{2}[x,x]\right) e^x \ . \tag{13.35}$$

Proof. One computes directly that

$$\begin{aligned} \Lambda(e^x) &= \sum_{k\geqslant 1} \frac{k D(x) x^{k-1}}{k!} + \sum_{k\geqslant 2} \frac{k(k-1)}{2} \frac{[x,x]}{k!} x^{k-2} \\ &= \left(D(x) + \frac{1}{2}[x,x]\right) e^x \ . \end{aligned}$$

This completes the proof. □

The lemma and the remarks made about Obs_k lead at once to the proof of the following proposition.

Proposition 13.26. *Let $(\mathfrak{g}, [\cdot,\cdot], D)$ be a differential graded Lie algebra with total differential $\Lambda : S^\bullet \bar{\mathfrak{g}} \to S^\bullet \bar{\mathfrak{g}}$. For $x \in \nu\mathfrak{g}_1^\nu$, the following statements are equivalent.*

(i) *x satisfies the Maurer–Cartan equation associated to $\mathfrak{g}$;*
(ii) $\Lambda(e^x) = 0$;
(iii) $\mathrm{Obs}_k(x) = 0$ *for every* $k \in \mathbb{N}^*$.

In the case of a pointed differential graded Lie algebra $(\mathfrak{g}, [\cdot,\cdot], z)$, these conditions are also equivalent to

(iv) $[z+x, z+x] = 0$.

13.3.4 Gauge Equivalence

We have introduced in Sections 13.1.1 and 13.2.1 the notion of equivalence of deformations in the case of associative products and of Poisson brackets. We now introduce the corresponding notion for formal power series (without constant term) with coefficients in the degree one component of an arbitrary pointed differential graded Lie algebra.

Definition 13.27. Let $(\mathfrak{g}, [\cdot,\cdot], z)$ be a pointed differential graded Lie algebra. Two formal power series without constant term $x, y \in \nu\mathfrak{g}^\nu$ are said to be *gauge equivalent* denoted by $x \sim y$, if there exists $\xi \in \nu\mathfrak{g}_0^\nu$ such that

$$z + y = e^{\mathrm{ad}_\xi}(z+x) .$$

The definition requires some explanation. As the notation suggests,[2] e^{ad_ξ} stands for the endomorphism of $\mathfrak{g}^\nu$, defined for $x \in \mathfrak{g}^\nu$ by

$$e^{\mathrm{ad}_\xi}(x) := \sum_{k=0}^{\infty} \frac{\mathrm{ad}_\xi^k x}{k!} = x + [\xi, x] + \frac{1}{2}[\xi, [\xi, x]] + \cdots . \tag{13.36}$$

The latter infinite sum is a well-defined formal power series in ν with coefficients in $\mathfrak{g}$ since at most the first $k+1$ terms of the sum contribute to the coefficient of ν^k. Notice that it follows easily from (13.36) that

$$\frac{\mathrm{d}}{\mathrm{d}t} e^{\mathrm{ad}_{t\xi}}(x) = \left[\xi, e^{\mathrm{ad}_{t\xi}}(x)\right] , \tag{13.37}$$

for $\xi \in \nu\mathfrak{g}_0^\nu$ and $x \in \mathfrak{g}^\nu$. When we work in a fixed pointed differential graded Lie algebra $(\mathfrak{g}, [\cdot,\cdot], z)$, we will make extensive use of the notation

$$\xi \odot x := e^{\mathrm{ad}_\xi}(z+x) - z = x + \sum_{k=1}^{\infty} \frac{1}{k!} \mathrm{ad}_\xi^k (z+x) ,$$

[2] The endomorphism e^{ad_ξ} of $\mathfrak{g}^\nu$, defined here, should not be confused with the exponential e^ξ, which is the element of $(S^\bullet \bar{\mathfrak{g}})^\nu$, defined in (13.34).

where x and ξ are as above. In terms of this notation, $x, y \in \nu\mathfrak{g}^\nu$ are gauge equivalent if and only if there exists $\xi \in \nu\mathfrak{g}_0^\nu$ such that $y = \xi \odot x$.

In order to show that the relation $\sim$ is an equivalence relation, we use the *Campbell–Hausdorff formula*

$$\exp(u)\exp(v) = \exp(\mathrm{CH}(u,v)), \tag{13.38}$$

where $\mathrm{CH}(u,v)$ is an infinite sum, each of whose terms consists only of repeated brackets of u and v. The first few terms of $\mathrm{CH}(u,v)$ are given by

$$\mathrm{CH}(u,v) = u + v + \frac{1}{2}[u,v] + \frac{1}{12}\left([u,[u,v]] + [v,[v,u]]\right) + \cdots. \tag{13.39}$$

In particular, $\mathrm{CH}(u,-u) = 0$ for all u. This formula is usually given for u and v in a finite-dimensional Lie algebra, with exp being the exponential map between the Lie algebra and its Lie group (see for example [29]), but the Campbell–Hausdorff formula is essentially a formal identity, in particular it is equally valid when for example u and v are formal power series (without constant term) with coefficients in an arbitrary Lie algebra. In the present case, we apply it for u and v of the form ad_ξ, where ξ is a formal power series without constant term and the Lie bracket is the graded commutator (of graded linear maps, such as ad_ξ). For $\xi, \eta \in \nu\mathfrak{g}_0^\nu$ and $x \in \nu\mathfrak{g}^\nu$, the Campbell–Hausdorff formula and the fact that ad is a representation imply that

$$e^{\mathrm{ad}_\eta} e^{\mathrm{ad}_\xi}(x) = e^{\mathrm{CH}(\mathrm{ad}_\eta, \mathrm{ad}_\xi)}(x) = e^{\mathrm{ad}_{\mathrm{CH}(\eta,\xi)}}(x),$$

so that

$$\eta \odot (\xi \odot x) = \mathrm{CH}(\eta, \xi) \odot x. \tag{13.40}$$

The fact that $\sim$ is an equivalence relation follows at once from it.

To finish this section, we show that the equivalence relation $\sim$ is compatible with the Maurer–Cartan equation, i.e., that if $x \sim y$, then x is a solution of the Maurer–Cartan equation if and only if y is a solution of the Maurer–Cartan equation. To do this, let $x, y \in \nu\mathfrak{g}_1^\nu$ be two equivalent solutions of the Maurer–Cartan equation and let $\xi \in \nu\mathfrak{g}_0^\nu$ be such that $y = \xi \odot x$. Since x is a solution of the Maurer–Cartan equation if and only if $[z+x, z+x] = 0$, we need to show that

$$\left[e^{\mathrm{ad}_\xi}(z+x), e^{\mathrm{ad}_\xi}(z+x)\right] = 0.$$

For $s, t \in \mathfrak{g}^\nu$, the fact that ad_ξ is a derivation of the Lie bracket implies that

$$\begin{aligned} e^{\mathrm{ad}_\xi}[s,t] &= \sum_{j=0}^{\infty} \frac{1}{j!} \mathrm{ad}_\xi^j [s,t] = \sum_{j=0}^{\infty}\sum_{i=0}^{j} \frac{1}{j!}\binom{j}{i}\left[\mathrm{ad}_\xi^i s, \mathrm{ad}_\xi^{j-i} t\right] \\ &= \sum_{j=0}^{\infty}\sum_{i=0}^{j}\left[\frac{1}{i!}\mathrm{ad}_\xi^i s, \frac{1}{(j-i)!}\mathrm{ad}_\xi^{j-i} t\right] = \sum_{k,\ell=0}^{\infty}\left[\frac{1}{k!}\mathrm{ad}_\xi^k s, \frac{1}{\ell!}\mathrm{ad}_\xi^\ell t\right] \end{aligned}$$

$$= \left[e^{\mathrm{ad}_\xi} s, e^{\mathrm{ad}_\xi} t \right] .$$

We may conclude that $\left[e^{\mathrm{ad}_\xi}(z+x), e^{\mathrm{ad}_\xi}(z+x) \right] = e^{\mathrm{ad}_\xi}[z+x, z+x] = 0$, since x is a solution of the Maurer–Cartan equation.

The upshot is that gauge equivalence defines an equivalence relation on the set of solutions of the Maurer–Cartan equation associated to $\mathfrak{g}$. For $x \in \mathrm{MC}(\mathfrak{g})$, we denote by $\mathrm{cl}(x)$ its equivalence class modulo gauge equivalence.

13.3.5 Path Equivalence

We introduce in this section another notion of equivalence of solutions of the Maurer–Cartan equation. This notion, which turns out to be equivalent to the notion of gauge equivalence (see Proposition 13.31 below) seems both less intuitive and technically more complicated, but we will see that it is easier to work with the latter notion when dealing with L_∞-morphisms, introduced in the next section.

We first define the notion of a polynomial path in $V^\nu := V[[\nu]]$, where V is an arbitrary $\mathbb{F}$-vector space. By definition, a *polynomial path* in V^ν is a map from $\mathbb{F}$ to V^ν whose component in $V\nu^k$ is for every $k \in \mathbb{N}$ a polynomial in t; it means that a polynomial path $x(t)$ in V^ν can be written as

$$x(t) = \sum_{k=0}^{\infty} x_k(t) \nu^k ,$$

where each one of the functions $x_k(t)$ is a polynomial function, with values in V. We will often consider such paths satisfying $x_0(t) = 0$; they will be referred to as polynomial paths in νV^ν. Polynomial paths in V^ν admit well-defined derivatives with respect to t, which are themselves also polynomial paths in V^ν. Also, polynomial paths in a (graded) algebra, such as $(S^\bullet V)^\nu$, form an algebra and, for every polynomial path $\gamma(t)$ in νV^ν, its exponential $e^{\gamma(t)}$ (defined as in (13.34)) is a polynomial path in $(S^\bullet V)^\nu$.

Definition 13.28. Let $(\mathfrak{g}, [\cdot,\cdot], z)$ be a pointed differential graded Lie algebra, and let Λ denote the total differential on $(S^\bullet \bar{\mathfrak{g}})^\nu$, as in (13.30). We say that two solutions $x, y \in \nu\mathfrak{g}_1^\nu$ of the Maurer–Cartan equation are *path equivalent* if there exists two polynomial paths $\gamma(t), \eta(t)$ in $\nu\mathfrak{g}_1^\nu$ and $\nu\mathfrak{g}_0^\nu$ respectively, such that $\gamma(0) = x, \gamma(1) = y$ and

$$\frac{\mathrm{d}}{\mathrm{d}t} e^{\gamma(t)} = \Lambda\left(\eta(t) e^{\gamma(t)} \right) . \qquad (13.41)$$

We will show in Proposition 13.31 below that two solutions of the Maurer–Cartan equation are gauge equivalent if and only if they are path equivalent. We will do this by using the following two lemmas, the first one of which can be viewed as giving yet another (equivalent!) notion of equivalence of solutions of the Maurer–Cartan equation.

Lemma 13.29. *Let $(\mathfrak{g},[\cdot,\cdot],z)$ be a pointed differential graded Lie algebra and suppose that $x,y \in \nu\mathfrak{g}_1^\nu$ are two solutions of the Maurer–Cartan equation associated to $\mathfrak{g}$. Let $\gamma(t)$ and $\eta(t)$ be polynomial paths in $\nu\mathfrak{g}_1^\nu$ and $\nu\mathfrak{g}_0^\nu$ respectively, with $\gamma(0)=x$ and $\gamma(1)=y$. Then condition (13.41) is equivalent to the following condition:*

$$\frac{\mathrm{d}}{\mathrm{d}t}\gamma(t) = [z+\gamma(t),\eta(t)] \ . \tag{13.42}$$

Proof. If condition (13.41) holds, then the component in $\bar{\mathfrak{g}}^\nu = S^1\bar{\mathfrak{g}}^\nu \subset (S^\bullet\bar{\mathfrak{g}})^\nu$ of $\frac{\mathrm{d}}{\mathrm{d}t}e^{\gamma(t)} - \Lambda\left(\eta(t)e^{\gamma(t)}\right)$ vanishes. Since the components of $\gamma(t)\in\nu\mathfrak{g}_1^\nu$ commute, this component is given by

$$\frac{\mathrm{d}}{\mathrm{d}t}\gamma(t) - \Lambda_{D_z}(\eta(t)) - \Lambda_{[\cdot,\cdot]}(\eta(t)\gamma(t)) = \frac{\mathrm{d}}{\mathrm{d}t}\gamma(t) - [z+\gamma(t),\eta(t)] \ ,$$

condition (13.41) implies condition (13.42).

Conversely, suppose that $\gamma(t)$ and $\eta(t)$ satisfy condition (13.42), where $\gamma(0)=x$ is a solution of the Maurer–Cartan equation and $\gamma(1)=y$. We first show that $\gamma(t)$ is a solution of the Maurer–Cartan equation for all t, which we do by showing that $[z+\gamma(t),z+\gamma(t)]=0$ for all t. This is in turn done by showing that the polynomial path $F(t) := [z+\gamma(t),z+\gamma(t)]$ is the unique solution of a linear differential equation, with initial condition $F(0)=0$. Taking the derivative of F and using (13.42) and the graded Jacobi identity we find

$$\begin{aligned}
\frac{\mathrm{d}}{\mathrm{d}t}F(t) &= \left[\frac{\mathrm{d}}{\mathrm{d}t}\gamma(t), z+\gamma(t)\right] + \left[z+\gamma(t),\frac{\mathrm{d}}{\mathrm{d}t}\gamma(t)\right] \\
&= [[z+\gamma(t),\eta(t)],z+\gamma(t)] + [z+\gamma(t),[z+\gamma(t),\eta(t)]] \\
&= [[z+\gamma(t),z+\gamma(t)],\eta(t)] \\
&= [F(t),\eta(t)] \ ,
\end{aligned}$$

which yields the required linear differential equation. We write $F(t) = \sum_{k\geqslant 1}F_k(t)\nu^k$ and $\eta(t) = \sum_{k\geqslant 1}\eta_k(t)\nu^k$, where all coefficients $F_k(t)$ and $\eta_k(t)$ depend polynomially on t. Then the differential equation can be written out as an infinite list of equations

$$\frac{\mathrm{d}}{\mathrm{d}t}F_k(t) = \sum_{\ell=1}^{k-1}[F_{k-\ell}(t),\eta_\ell(t)] \quad \text{and} \quad F_k(0)=0 \ ,$$

where $k\in\mathbb{N}^*$. An easy recursion on k implies that $F_k(t)=0$ for all $k\in\mathbb{N}^*$. In conclusion, $\gamma(t)$ is a solution of the Maurer–Cartan equation for all $t\in\mathbb{F}$. Next, we sum up, for $k\in\mathbb{N}$, each one of the following equalities,

$$\begin{aligned}
\Lambda_{D_z}\left(\eta(t)\gamma^k(t)\right) &= \Lambda_{D_z}(\eta(t))\gamma^k(t) + \Lambda_{D_z}(\gamma^k(t))\eta(t), \\
\Lambda_{[\cdot,\cdot]}\left(\eta(t)\gamma^k(t)\right) &= -k\,[\eta(t),\gamma(t)]\,\gamma^{k-1}(t) + \Lambda_{[\cdot,\cdot]}\left(\gamma^k(t)\right)\eta(t) \ ,
\end{aligned}$$

to find that

$$\Lambda\left(\eta(t)e^{\gamma(t)}\right) = [z+\gamma(t),\eta(t)]\,e^{\gamma(t)} + \Lambda\left(e^{\gamma(t)}\right)\eta(t)$$
$$= [z+\gamma(t),\eta(t)]\,e^{\gamma(t)}\ ,$$

where we have used in the last step that $\gamma(t) \in \mathrm{MC}(\mathfrak{g})$ for all t, combined with item *(ii)* of Proposition 13.26. It follows that

$$\frac{\mathrm{d}}{\mathrm{d}t}e^{\gamma(t)} - \Lambda\left(\eta(t)e^{\gamma(t)}\right) = \left(\frac{\mathrm{d}}{\mathrm{d}t}\gamma(t) - [z+\gamma(t),\eta(t)]\right)e^{\gamma(t)}\ ,$$

showing that condition (13.42) implies condition (13.41). □

The second lemma which we will use to prove Proposition 13.31 below can be seen as a generalization of the classical formula, giving the derivative of $e^{f(t)}$, when $f(t)$ takes values in a non-commutative algebra.

Lemma 13.30. *Let $(\mathfrak{g},[\cdot\,,\cdot])$ be a graded Lie algebra and let $\xi(t)$ be a polynomial path in $\nu\mathfrak{g}_0^\nu$. Then*

$$e^{\mathrm{ad}_{\xi(t)}} \circ \frac{\mathrm{d}e^{\mathrm{ad}_{-\xi(t)}}}{\mathrm{d}t} = \mathrm{ad}_{\sum_{k\in\mathbb{N}} \frac{1}{(k+1)!}\mathrm{ad}^k_{\xi(t)}\left(\frac{\mathrm{d}\xi(t)}{\mathrm{d}t}\right)}\ . \tag{13.43}$$

Proof. Given a polynomial path $\xi(t)$ in $\nu\mathfrak{g}_0^\nu$, we consider for fixed t the polynomial path in $\mathrm{End}(\mathfrak{g}_1)^\nu$, defined for $s \in \mathbb{F}$ by

$$G_s := e^{\mathrm{ad}_{s\xi(t)}} \circ \frac{\mathrm{d}e^{\mathrm{ad}_{-s\xi(t)}}}{\mathrm{d}t} + \mathrm{ad}_{\sum_{k\in\mathbb{N}} \frac{s^{k+1}}{(k+1)!}\mathrm{ad}^k_{\xi(t)}\frac{\mathrm{d}\xi(t)}{\mathrm{d}t}}\ . \tag{13.44}$$

It is clear that $G_0 = 0$. We will show that

$$\frac{\mathrm{d}}{\mathrm{d}s}G_s = \mathrm{ad}_{\xi(t)} \circ G_s - G_s \circ \mathrm{ad}_{\xi(t)}\ . \tag{13.45}$$

It follows, as in the proof of Lemma 13.29, by writing G_s and ξ as formal series in ν that all coefficients of G as a series in ν vanish, so that $G_s = 0$ for all $s \in \mathbb{F}$. In particular, $G_1 = 0$, which is tantamount to (13.43).

In order to show (13.45), we write the two terms in the right-hand side of (13.44) as A_s and B_s, so that $G_s = A_s + B_s$. Explicitly, $A_s(x)$ and $B_s(x)$ are given, for $x \in \mathfrak{g}_1^\nu$, by

$$\begin{aligned} A_s(x) &:= e^{\mathrm{ad}_{s\xi(t)}}\left(\frac{\mathrm{d}}{\mathrm{d}t}e^{\mathrm{ad}_{-s\xi(t)}}(x)\right), \\ B_s(x) &:= \left[\sum_{k\in\mathbb{N}} \frac{s^{k+1}}{(k+1)!}\mathrm{ad}^k_{\xi(t)}\dot{\xi}(t), x\right], \end{aligned} \tag{13.46}$$

where $\dot{\xi}(t)$ is an abbreviation for $\frac{\mathrm{d}}{\mathrm{d}t}\xi(t)$. Using (13.37) twice, we compute

$$\begin{aligned}\frac{\mathrm{d}}{\mathrm{d}s}A_s(x) &= [\xi(t),A_s(x)] + e^{\mathrm{ad}_{s\xi(t)}}\left(\frac{\mathrm{d}}{\mathrm{d}t}\left[-\xi(t),e^{\mathrm{ad}_{-s\xi(t)}}(x)\right]\right)\\ &= [\xi(t),A_s(x)] - e^{\mathrm{ad}_{s\xi(t)}}\frac{\mathrm{d}}{\mathrm{d}t}\left(e^{\mathrm{ad}_{-s\xi(t)}}([\xi(t),x])\right)\\ &= [\xi(t),A_s(x)] - A_s([\xi(t),x]) - \left[\dot{\xi}(t),x\right] . \end{aligned} \tag{13.47}$$

For the derivative of B_s, it follows at once from the second equation in (13.46) that

$$\frac{\mathrm{d}}{\mathrm{d}s}B_s(x) = \left[e^{\mathrm{ad}_{s\xi(t)}}\dot{\xi}(t),x\right] ,$$

so that, in view of the graded Jacobi identity,

$$\begin{aligned}[\xi(t),B_s(x)] &= \left[\sum_{k\in\mathbb{N}}\frac{s^{k+1}}{(k+1)!}\,\mathrm{ad}^{k+1}_{\xi(t)}\dot{\xi}(t),x\right] + B_s([\xi(t),x])\\ &= \frac{\mathrm{d}}{\mathrm{d}s}B_s(x) - \left[\dot{\xi}(t),x\right] + B_s([\xi(t),x]) ,\end{aligned}$$

giving the following formula for the derivative of B_s:

$$\frac{\mathrm{d}}{\mathrm{d}s}B_s(x) = [\xi(t),B_s(x)] - B_s([\xi(t),x]) + \left[\dot{\xi}(t),x\right] . \tag{13.48}$$

Equations (13.47) and (13.48), combined, yield the proof of (13.45), and hence of (13.43). □

Using the two lemmas, we show that gauge equivalence and path equivalence are the same (for solutions of the Maurer–Cartan equation).

Proposition 13.31. *Let $(\mathfrak{g},[\cdot,\cdot],z)$ be a pointed differential graded Lie algebra. Two solutions $x,y\in \nu\mathfrak{g}_1^\nu$ of the Maurer–Cartan equation associated to $\mathfrak{g}$ are gauge equivalent if and only if they are path equivalent.*

Proof. Let us first assume that x and y are gauge equivalent. Then there exists $\xi\in \nu\mathfrak{g}_0^\nu$ such that $y=\xi\odot x$. For $t\in\mathbb{F}$, define

$$\gamma(t) := (t\xi)\odot x = e^{\mathrm{ad}_{t\xi}}(z+x) - z ,$$

and let $\eta(t) := -\xi$, which is a constant polynomial path. Then $\gamma(0)=x$, $\gamma(1)=y$ and it follows from (13.37) that $\gamma(t)$ and $\eta(t)$ satisfy condition (13.42). In view of Lemma 13.29, x and y are path equivalent.

Conversely, let us assume that x and y are path equivalent, and let $\gamma(t),\eta(t)$ be polynomial paths in $\nu\mathfrak{g}_1^\nu$ and $\nu\mathfrak{g}_0^\nu$ respectively, such that $\gamma(0)=x$, $\gamma(1)=y$ and such that the equivalent conditions (13.41) and (13.42) are satisfied. In order to establish that x,y are gauge equivalent, we will construct a polynomial path $\xi(t)$ in $\nu\mathfrak{g}_0^\nu$ such that $\gamma(t)=\xi(t)\odot x$, i.e.,

$$e^{-\mathrm{ad}_{\xi(t)}}(z+\gamma(t)) = z+x , \tag{13.49}$$

for all $t \in \mathbb{F}$. It implies that $y = \gamma(1) = \xi(1) \odot x$, so that x and y are gauge equivalent. In order to satisfy (13.49) for $t = 0$, we require $\xi(0) = 0$. With this additional assumption, the condition which we wish to impose on $\xi(t)$ is $\frac{\mathrm{d}}{\mathrm{d}t} e^{-\mathrm{ad}_{\xi(t)}}(z + \gamma(t)) = 0$, or equivalently:

$$e^{\mathrm{ad}_{\xi(t)}} \frac{\mathrm{d}}{\mathrm{d}t}\left(e^{-\mathrm{ad}_{\xi(t)}}(z + \gamma(t))\right) = 0 .$$

We can compute the left-hand side of the previous equation with the help of (13.43) and (13.42) as follows:

$$\begin{aligned}
e^{\mathrm{ad}_{\xi(t)}} \frac{\mathrm{d}}{\mathrm{d}t}\left(e^{-\mathrm{ad}_{\xi(t)}}(z + \gamma(t))\right) &= e^{\mathrm{ad}_{\xi(t)}} \frac{\mathrm{d}e^{-\mathrm{ad}_{\xi(t)}}}{\mathrm{d}t}(z + \gamma(t)) + \frac{\mathrm{d}}{\mathrm{d}t}\gamma(t) \\
&= -\mathrm{ad}_{\sum_{k \in \mathbb{N}} \frac{1}{(k+1)!} \mathrm{ad}^k_{\xi(t)} \frac{\mathrm{d}\xi(t)}{\mathrm{d}t}}(z + \gamma(t)) + \frac{\mathrm{d}}{\mathrm{d}t}\gamma(t) \\
&= \left[z + \gamma(t), \sum_{k \in \mathbb{N}} \frac{1}{(k+1)!} \mathrm{ad}^k_{\xi(t)} \frac{\mathrm{d}\xi(t)}{\mathrm{d}t} + \eta(t) \right] .
\end{aligned}$$

It suffices therefore to find a polynomial path $\xi(t)$ in $\nu\mathfrak{g}_0^\nu$ which satisfies

$$\sum_{k \in \mathbb{N}} \frac{1}{(k+1)!} \mathrm{ad}^k_{\xi(t)} \frac{\mathrm{d}\xi(t)}{\mathrm{d}t} = -\eta(t) \tag{13.50}$$

and $\xi(0) = 0$. To see that such a polynomial path $\xi(t)$ exists, substitute $\xi(t) = \sum_{\ell \in \mathbb{N}^*} \xi_\ell(t)\nu^\ell$ and $\eta(t) = \sum_{\ell \in \mathbb{N}^*} \eta_\ell(t)\nu^\ell$ in (13.50) and observe that the equation decomposes into an infinite number of equations

$$\frac{\mathrm{d}\xi_\ell(t)}{\mathrm{d}t} = -\eta_\ell(t) - \sum_{k=1}^{\ell-1} \frac{1}{(k+1)!} \sum_{\substack{i_1 + \cdots + i_k + j = \ell \\ i_1, \ldots, i_k, j \geqslant 1}} \mathrm{ad}_{\xi_{i_1}(t)} \circ \cdots \circ \mathrm{ad}_{\xi_{i_k}(t)} \frac{\mathrm{d}\xi_j(t)}{\mathrm{d}t} , \tag{13.51}$$

indexed by $\ell \in \mathbb{N}^*$, which express $\frac{\mathrm{d}\xi_\ell(t)}{\mathrm{d}t}$ as a multi-linear expression in $\eta_\ell(t)$ and $\xi_1(t), \ldots, \xi_{\ell-1}(t)$. By integration, we find a unique polynomial $\xi_\ell(t)$, satisfying the initial condition $\xi_\ell(0) = 0$. Thus, the coefficients of $\xi(t)$ can be constructed recursively. It leads to a polynomial path $\xi(t)$ in $\nu\mathfrak{g}_0^\nu$ which at $t = 1$ realizes the gauge equivalence between x and y. □

Remark 13.32. Let $x, y \in \nu\mathfrak{g}_1^\nu$ be solutions of the Maurer–Cartan equation, as in Proposition 13.31. Suppose that x and y are gauge equivalent, $y = \xi \odot x$, where $\xi \in \nu^\ell\mathfrak{g}_0^\nu$. Then it is clear from the first part of the proof of the proposition that the constant polynomial path $\eta(t)$ belongs to $\nu^\ell\mathfrak{g}_0^\nu$. Alternatively, suppose that x and y are path equivalent, where the polynomial path $\eta(t)$ belongs to $\nu^\ell\mathfrak{g}_0^\nu$. Then the integration of (13.51) gives a polynomial path $\xi(t)$ in $\nu^\ell\mathfrak{g}_0^\nu$, since $\eta_k(t) = 0$ for $k < \ell$. Then $\xi(1) \in \nu^\ell\mathfrak{g}_0^\nu$, so that x and y are gauge equivalent via an element of $\nu^\ell\mathfrak{g}_0^\nu$.

13.3.6 Applications to Deformation Theory

We pointed out in Example 13.17 that for every Poisson algebra $(\mathscr{A},\cdot,\pi)$, the triple $(\overline{\mathfrak{X}}^\bullet(\mathscr{A}),[\cdot,\cdot]_S,\pi)$ is a pointed differential graded Lie algebra. Similarly, according to Example 13.19, for every commutative associative algebra $(\mathscr{A},\mu)$, the triple $(\overline{\mathrm{HC}}^\bullet(\mathscr{A}),[\cdot,\cdot]_G,\mu)$ is a pointed differential graded Lie algebra. Up to a sign, the differentials D_π and D_μ coincide with the Poisson and Hochschild coboundary operators, namely

$$\delta_\pi(P) = (-1)^{\bar{p}} D_\pi(P)\,, \qquad \delta_\mu(\psi) = (-1)^{\bar{q}} D_\mu(\psi)\,,$$

for $P \in \overline{\mathfrak{X}}^{\bar{p}}(\mathscr{A})$ and $\psi \in \overline{\mathrm{HC}}^{\bar{q}}(\mathscr{A})$. Thus, the cohomology of $(\overline{\mathfrak{X}}^\bullet(\mathscr{A}),[\cdot,\cdot]_S,\pi)$ is the Poisson cohomology of $(\mathscr{A},\cdot,\pi)$ and the cohomology of $(\overline{\mathrm{HC}}^\bullet(\mathscr{A}),[\cdot,\cdot]_G,\mu)$ is the Hochschild cohomology of $(\mathscr{A},\mu)$. Since for elements of shifted degree 1, δ_π and D_π have opposite sign, the Maurer–Cartan equation (13.32) of $(\overline{\mathfrak{X}}^\bullet(\mathscr{A}),[\cdot,\cdot]_S,\pi)$ is precisely the Maurer–Cartan equation (13.25) of $(\mathscr{A},\cdot,\pi)$, and similarly for the Maurer–Cartan equation (13.11) of $(\mathscr{A},\mu)$.

The fact that the notion of gauge equivalence of solutions of the Maurer–Cartan equation for pointed differential graded Lie algebras coincides in the cases of $(\overline{\mathrm{HC}}^\bullet(\mathscr{A}),[\cdot,\cdot]_G,\mu)$ and $(\overline{\mathfrak{X}}^\bullet(\mathscr{A}),[\cdot,\cdot]_S,\pi)$ with the notion of equivalence of formal deformations of associative products and of Poisson brackets is also true, but it is less obvious. We prove it in the case of formal deformations of associative products, the proof in the case of Poisson brackets being completely analogous. Thus, we consider the pointed differential graded Lie algebra $(\overline{\mathrm{HC}}^\bullet(\mathscr{A}),[\cdot,\cdot]_G,\mu)$. Let x and y be solutions of the Maurer–Cartan equation associated to this differential graded Lie algebra, which is equivalent to saying that $\mu+x$ and $\mu+y$ are formal deformations of μ. Let us suppose that x and y are gauge equivalent, which means that there exists an element $\xi \in \nu\overline{\mathrm{HC}}^0(\mathscr{A})^\nu = \nu\,\mathrm{Hom}(\mathscr{A},\mathscr{A})[[\nu]]$, such that $\mu+y = e^{\mathrm{ad}_\xi}(\mu+x)$, where $\mathrm{ad}_\xi = [\xi,\cdot]_G$. Let $\Phi:\mathscr{A}^\nu \to \mathscr{A}^\nu$ be the $\mathbb{F}^\nu$-linear map, defined by $\Phi := e^\xi$. We show that Φ realizes an equivalence between the formal deformations $\mu+x$ and $\mu+y$ (see Definition 13.2). Since $\Phi(F) = F \pmod{\nu}$ for all $F \in \mathscr{A}$, we only need to show that

$$\Phi((\mu+x)(\Phi^{-1}(F),\Phi^{-1}(G))) = (\mu+y)(F,G)$$

for all $F,G \in \mathscr{A}$, which amounts to showing that

$$e^\xi((\mu+x)(e^{-\xi}(F),e^{-\xi}(G))) = (e^{\mathrm{ad}_\xi}(\mu+x))(F,G)$$

for all $F,G \in \mathscr{A}$. Writing x' for $\mu+x$, the left-hand side of the above equation is given by

$$\begin{aligned} &e^\xi\left(x'\left(e^{-\xi}(F),e^{-\xi}(G)\right)\right) \\ &= x'(F,G) + \sum_{k\in\mathbb{N}^*} \sum_{\substack{r,s,t\in\mathbb{N} \\ r+s+t=k}} (-1)^{s+t} \frac{1}{r!}\frac{1}{s!}\frac{1}{t!}\,\xi^r\left(x'(\xi^s(F),\xi^t(G))\right) \end{aligned}$$

$$= x'(F,G) + \sum_{k\in\mathbb{N}^*} \frac{1}{k!}\,\mathrm{ad}_\xi^k(x')(F,G) = e^{\mathrm{ad}_\xi}(x')(F,G)$$

where the second equality is easily proved by induction on $k \in \mathbb{N}^*$. This shows that if two solutions of the Maurer–Cartan equation are gauge equivalent, then the corresponding deformations are equivalent. The same computation shows that if two deformations are equivalent, then the corresponding solutions of the Maurer–Cartan equation are gauge equivalent: one constructs in this case $\xi \in \nu\,\mathrm{Hom}(\mathscr{A},\mathscr{A})[[\nu]]$ from Φ by taking the logarithm (using the series expansion of $\log(1+x)$). Summing up, we have shown the following proposition.

Proposition 13.33. *Let $(\mathscr{A},\mu)$ be a commutative associative algebra and let $x,y \in \nu\mathrm{HC}^2(\mathscr{A})^\nu$.*

(1) x is a solution of the Maurer–Cartan equation associated to $\overline{\mathrm{HC}}^\bullet(\mathscr{A})$ if and only if $\mu+x$ is a formal deformation of the associative product μ;
(2) If x and y are solutions of the Maurer–Cartan equation associated to $\overline{\mathrm{HC}}^\bullet(\mathscr{A})$, then x and y are gauge equivalent if and only if the formal deformations $\mu+x$ and $\mu+y$ are equivalent.

The above proposition and its proof are easily transcribed to the case of Poisson brackets. The statement is the following.

Proposition 13.34. *Let $(\mathscr{A},\cdot,\pi)$ be a Poisson algebra and let $x,y \in \nu\mathfrak{X}^2(\mathscr{A})^\nu$.*

(1) x is a solution of the Maurer–Cartan equation associated to $\overline{\mathfrak{X}}^\bullet(\mathscr{A})$ if and only if $\pi+x$ is a formal deformation of π;
(2) If x and y are solutions of the Maurer–Cartan equation associated to $\overline{\mathfrak{X}}^\bullet(\mathscr{A})$, then x and y are gauge equivalent if and only if the formal deformations $\pi+x$ and $\pi+y$ are equivalent.

Both propositions are easily specialized to the case of smooth manifolds; in the case of Proposition 13.33 one can also restrict oneself to star products on the algebra of smooth functions on a manifold M, rather than considering arbitrary deformations of the standard product on $C^\infty(M)$. Also, taking all formulas in the above proof modulo ν^{k+1} shows that two k-th order deformations $\mu+x$ and $\mu+y$ are equivalent if and only if x and y are gauge equivalent up to order k.

13.4 L_∞-Morphisms of Differential Graded Lie Algebras

In this section we introduce the notion of an L_∞-morphism and of an L_∞-quasi-isomorphism of differential graded Lie algebras and we show that an L_∞-quasi-isomorphism between two pointed differential graded Lie algebras $(\mathfrak{g},[\cdot\,,\cdot]_\mathfrak{g},z_\mathfrak{g})$ and $(\mathfrak{h},[\cdot\,,\cdot]_\mathfrak{h},z_\mathfrak{h})$ induces a bijection between the sets $\mathrm{MC}(\mathfrak{g})/\sim$ and $\mathrm{MC}(\mathfrak{h})/\sim$ of gauge equivalence classes of solutions of the Maurer–Cartan equation of $(\mathfrak{g},[\cdot\,,\cdot]_\mathfrak{g},z_\mathfrak{g})$ and of $(\mathfrak{h},[\cdot\,,\cdot]_\mathfrak{h},z_\mathfrak{h})$.

Let $\mathfrak{g} = \bigoplus_{\ell\in\mathbb{Z}} \mathfrak{g}_\ell$ and $\mathfrak{h} = \bigoplus_{\ell\in\mathbb{Z}} \mathfrak{h}_\ell$ be two graded Lie algebras. In what follows we will be mainly interested in graded linear maps $\Phi : S^\bullet\bar{\mathfrak{g}} \to \bar{\mathfrak{h}}$ of degree 0. We recall that $\bar{\mathfrak{g}}$ is the same vector space as $\mathfrak{g}$, but the degree of its homogeneous elements is lowered by one; it is this grading which is extended to $S^\bullet\bar{\mathfrak{g}}$ (see (13.28)). Denoting for $k \in \mathbb{N}$ by Φ_k the restriction of Φ to $S^k\bar{\mathfrak{g}}$, the fact that Φ is graded of degree 0 means that Φ_k maps $\mathfrak{g}_{i_1}\dots\mathfrak{g}_{i_k}$ to $\mathfrak{h}_{i_1+\cdots+i_k-k+1}$ for all $i_1,\dots,i_k \in \mathbb{Z}$. In particular, Φ_k maps $\mathfrak{g}_1\dots\mathfrak{g}_1$ to $\mathfrak{h}_1$, a fact which will be useful when we consider solutions of the Maurer–Cartan equations associated to $\mathfrak{g}$ and $\mathfrak{h}$.

We extend Φ to a graded linear map $\widehat{\Phi} : S^\bullet\bar{\mathfrak{g}} \to S^\bullet\bar{\mathfrak{h}}$, whose component in $\bar{\mathfrak{h}} \subset S^\bullet\bar{\mathfrak{h}}$ is Φ. First, the restriction of $\widehat{\Phi}$ to $S^0\bar{\mathfrak{g}} \simeq \mathbb{F}$ is by definition the identity map onto $S^0\bar{\mathfrak{h}} \simeq \mathbb{F}$. For $k,\ell \in \mathbb{N}^*$ and for a homogeneous monomial $X = x_1\dots x_k \in S^k\bar{\mathfrak{g}}$, we define the component in $S^\ell\bar{\mathfrak{h}}$ of $\widehat{\Phi}(X)$ to be given by

$$\sum_{\substack{i_1,\dots,i_\ell\in\mathbb{N}^* \\ i_1+\cdots+i_\ell=k}} \sum_{\sigma\in S_{i_1,\dots,i_\ell}} \operatorname{sgn}(\sigma;X)\frac{1}{\ell!}\prod_{j=1}^{\ell}\Phi_{i_j}\big(x_{\sigma(i_1+\cdots+i_{j-1}+1)}\cdots x_{\sigma(i_1+\cdots+i_j)}\big)\,, \tag{13.52}$$

where $S_{i_1,\dots,i_\ell}$ is the set of $(i_1,\dots,i_\ell)$-shuffles, i.e.,

$$S_{i_1,\dots,i_\ell} := \left\{\sigma\in S_{i_1+\cdots+i_\ell}\;\middle|\; \begin{array}{c} \sigma(1)<\cdots<\sigma(i_1),\\ \sigma(i_1+1)<\cdots<\sigma(i_1+i_2),\\ \cdots\\ \sigma(i_1+i_2+\cdots+i_{\ell-1}+1)<\cdots<\sigma(i_1+i_2+\cdots+i_\ell)\end{array}\right\}.$$

Remark 13.35.

(1) $\widehat{\Phi}$ is the (unique) extension of Φ as a coderivation of $S^\bullet\bar{\mathfrak{g}}$ with values in $S^\bullet\bar{\mathfrak{h}}$, but this fact will not be used in what follows.

(2) It is clear that, when Φ is identically zero on $S^k\bar{\mathfrak{g}}$ for all $k \geqslant 2$, i.e., when Φ is simply a graded linear map of degree 0 from $\mathfrak{g}$ to $\mathfrak{h}$, then $\widehat{\Phi} = S^\bullet\Phi$ (see (13.31)).

(3) It is also clear that the component in $S^\ell\bar{\mathfrak{h}}$ of $\widehat{\Phi}(x_1\dots x_k)$ is zero for all $\ell > k$.

(4) The order of the terms in the product in (13.52) is important: the meaning of the product is that the terms are ordered as follows: $\Phi_{i_1}\Phi_{i_2}\cdots\Phi_{i_\ell}$.

Definition 13.36. Let $(\mathfrak{g},[\cdot\,,\cdot]_\mathfrak{g},D_\mathfrak{g})$ and $(\mathfrak{h},[\cdot\,,\cdot]_\mathfrak{h},D_\mathfrak{h})$ be two differential graded Lie algebras, whose total differentials are denoted by $\Lambda_\mathfrak{g}$ and by $\Lambda_\mathfrak{h}$ respectively. A graded linear map $\Phi : S^\bullet\bar{\mathfrak{g}} \to \bar{\mathfrak{h}}$ of degree 0 is called an *L_∞-morphism* between $\mathfrak{g}$ and $\mathfrak{h}$ if

$$\widehat{\Phi}\circ\Lambda_\mathfrak{g} = \Lambda_\mathfrak{h}\circ\widehat{\Phi}\,. \tag{13.53}$$

For $x \in \mathfrak{g}$ we have that $\Lambda_\mathfrak{g}(x) = D_\mathfrak{g}(x)$ and $\widehat{\Phi}(x) = \Phi_1(x)$. Therefore, if $\Phi : S^\bullet\bar{\mathfrak{g}} \to \bar{\mathfrak{h}}$ is an L_∞-morphism, then

$$\Phi_1\circ D_\mathfrak{g} = D_\mathfrak{h}\circ\Phi_1\,, \tag{13.54}$$

as maps[3] from $\mathfrak{g}$ to $\mathfrak{h}$, so that $\Phi_1 : \mathfrak{g} \to \mathfrak{h}$ induces a morphism in cohomology,

$$\Phi_1^\bullet : H^\bullet_{D_\mathfrak{g}}(\mathfrak{g}) \to H^\bullet_{D_\mathfrak{h}}(\mathfrak{h}) .$$

Notice that Φ_1 is not a morphism of differential graded Lie algebras, in general. However, if a graded linear map $\phi : \mathfrak{g} \to \mathfrak{h}$ is a morphism of differential graded Lie algebras (see Definition 13.23), then one can extend ϕ to a graded linear map $\Phi : S^\bullet\bar{\mathfrak{g}} \to \bar{\mathfrak{h}}$, with $\Phi_1 = \phi$ and $\Phi_k = 0$ for all $k \geqslant 2$. According to item (2) of Remark 13.35, the map Φ is then an L_∞-morphism. In this sense, morphisms of differential graded Lie algebra are particular cases of L_∞-morphisms.

Definition 13.37. Let $(\mathfrak{g}, [\cdot,\cdot]_\mathfrak{g}, D_\mathfrak{g})$ and $(\mathfrak{h}, [\cdot,\cdot]_\mathfrak{h}, D_\mathfrak{h})$ be two differential graded Lie algebras and let $\Phi : S^\bullet\bar{\mathfrak{g}} \to \bar{\mathfrak{h}}$ be an L_∞-morphism. Then Φ is called an *L_∞-quasi-isomorphism* if the morphism $\Phi_1^\bullet : H^\bullet_{D_\mathfrak{g}}(\mathfrak{g}) \to H^\bullet_{D_\mathfrak{h}}(\mathfrak{h})$, induced by the restriction $\Phi_1 : \mathfrak{g} \to \mathfrak{h}$ of Φ to $\mathfrak{g}$, is an isomorphism.

We have seen in (13.54) that the equation $\widehat{\Phi} \circ \Lambda_\mathfrak{g} = \Lambda_\mathfrak{h} \circ \widehat{\Phi}$, which is an equality of maps $S^\bullet\bar{\mathfrak{g}} \to S^\bullet\bar{\mathfrak{h}}$ yields, when restricted to $S^1\bar{\mathfrak{g}} = \mathfrak{g}$, the equality $\Phi_1 \circ D_\mathfrak{g} = D_\mathfrak{h} \circ \Phi_1$. Similarly, we can consider the restriction to $S^k\bar{\mathfrak{g}}$, which yields an equality of maps, involving $\Phi_1, \ldots, \Phi_k$. In the example which follows, we show how this equality is obtained for $k = 2$.

Example 13.38. Let $(\mathfrak{g}, [\cdot,\cdot]_\mathfrak{g}, D_\mathfrak{g})$ and $(\mathfrak{h}, [\cdot,\cdot]_\mathfrak{h}, D_\mathfrak{h})$ be two differential graded Lie algebras and let $\Phi : S^\bullet\bar{\mathfrak{g}} \to \bar{\mathfrak{h}}$ be an L_∞-morphism. We compute the restriction of the compact equation (13.53) to $S^2\bar{\mathfrak{g}}$. To do this, let $x, y \in \mathfrak{g}$, with x homogeneous, and recall that, by definition

$$\begin{aligned} \Lambda_\mathfrak{g}(xy) &= D_\mathfrak{g}(x)y + (-1)^{d^0(x)} x D_\mathfrak{g}(y) + (-1)^{d^0(x)} [x,y]_\mathfrak{g} , \\ \widehat{\Phi}(xy) &= \Phi_2(xy) + \Phi_1(x)\Phi_1(y) , \end{aligned}$$

while $\Lambda_\mathfrak{g}(x) = D_\mathfrak{g}(x)$ and $\widehat{\Phi}(x) = \Phi_1(x)$. It follows that

$$\begin{aligned} \widehat{\Phi} \circ \Lambda_\mathfrak{g}(xy) &= \Phi_1(D_\mathfrak{g}(x))\Phi_1(y) + \Phi_2(D_\mathfrak{g}(x)y) \\ &\quad + (-1)^{d^0(x)} \Big(\Phi_1(x)\Phi_1(D_\mathfrak{g}(y)) + \Phi_2(xD_\mathfrak{g}(y)) + \Phi_1([x,y]_\mathfrak{g}) \Big) \end{aligned}$$

and

$$\begin{aligned} \Lambda_\mathfrak{h} \circ \widehat{\Phi}(xy) &= D_\mathfrak{h}(\Phi_2(xy)) + D_\mathfrak{h}(\Phi_1(x))\Phi_1(y) \\ &\quad + (-1)^{d^0(x)} \Big([\Phi_1(x), \Phi_1(y)]_\mathfrak{h} + \Phi_1(x) D_\mathfrak{h}(\Phi_1(y)) \Big) . \end{aligned}$$

Equating the right-hand sides of these two equations yields, upon using that $\Phi_1 \circ D_\mathfrak{g} = D_\mathfrak{h} \circ \Phi_1$, the following equation, involving both the maps Φ_1 and Φ_2:

[3] Strictly speaking, Φ_1 is a map from $S^1\bar{\mathfrak{g}} \to S^1\bar{\mathfrak{h}}$, but we rather consider it as a map from $\mathfrak{g}$ to $\mathfrak{h}$, i.e., we consider $\mathfrak{g}$ and $\mathfrak{h}$ with their original grading, as differential graded Lie algebras; with this convention Φ_1 induces maps in cohomology with the following labeling: $\Phi_1^k : H^k_{D_\mathfrak{g}}(\mathfrak{g}) \to H^k_{D_\mathfrak{h}}(\mathfrak{h})$.

$$\begin{aligned}&[\Phi_1(x),\Phi_1(y)]_{\mathfrak{h}} - \Phi_1([x,y]_{\mathfrak{g}})\\ &\quad = \Phi_2(xD_{\mathfrak{g}}(y)) + (-1)^{d^0(x)}\left(\Phi_2(D_{\mathfrak{g}}(x)y) - D_{\mathfrak{h}}(\Phi_2(xy))\right) .\end{aligned} \tag{13.55}$$

Let $\Phi : S^\bullet\bar{\mathfrak{g}} \to \bar{\mathfrak{h}}$ be a graded linear map of degree 0. The $\mathbb{F}^\nu$-linear extension of Φ to $(S^\bullet\bar{\mathfrak{g}})^\nu$ will also be denoted by Φ, and similarly for $\widehat{\Phi}$. To Φ, one can associate a map $\tilde{\Omega}_\Phi : \nu\mathfrak{g}^\nu \to \nu\mathfrak{h}^\nu$, defined for $x \in \nu\mathfrak{g}^\nu$ by:

$$\tilde{\Omega}_\Phi(x) := \sum_{k\in\mathbb{N}^*} \frac{1}{k!}\Phi_k(x^k) . \tag{13.56}$$

The infinite sum makes sense because, for every $k \in \mathbb{N}^*$, at most the first k terms contribute to the coefficient of ν^k in $\tilde{\Omega}_\Phi(x)$. Notice that if $x \in \nu\mathfrak{g}_1^\nu$, then all elements x^k are formal power series whose coefficients are elements of degree 0 in $S^\bullet\bar{\mathfrak{g}}$, since x^k is of degree 0 when $x \in \mathfrak{g}_1$ (or $x \in \mathfrak{g}_1^\nu$). It follows that $\tilde{\Phi}_k(x^k) \in \nu\mathfrak{h}_1^\nu$ for all $k \in \mathbb{N}^*$, hence $\tilde{\Omega}_\Phi(x) \in \nu\mathfrak{h}_1^\nu$.

Sections 13.4.1 to 13.4.3 below are devoted to the proof of the following theorem.

Theorem 13.39. *Let $(\mathfrak{g},[\cdot,\cdot]_{\mathfrak{g}},z_{\mathfrak{g}})$ and $(\mathfrak{h},[\cdot,\cdot]_{\mathfrak{h}},z_{\mathfrak{h}})$ be two pointed differential graded Lie algebras. If $\Phi : S^\bullet\bar{\mathfrak{g}} \to \bar{\mathfrak{h}}$ is an L_∞-morphism, then*

$$\begin{aligned}\Omega_\Phi : \mathrm{MC}(\mathfrak{g})/\sim &\to \mathrm{MC}(\mathfrak{h})/\sim\\ \mathrm{cl}(x) &\mapsto \mathrm{cl}\left(\tilde{\Omega}_\Phi(x)\right)\end{aligned}$$

is a well-defined map, which is a bijection if Φ is an L_∞-quasi-isomorphism.

Theorem 13.39 is true more generally for L_∞-morphism of arbitrary (not necessarily pointed) differential graded Lie algebras, but we will not use it in that generality.

13.4.1 Ω_Φ is Well-Defined

We start with a proposition which will be useful for showing that Ω_Φ is well-defined.

Proposition 13.40. *Let $(\mathfrak{g},[\cdot,\cdot]_{\mathfrak{g}},D_{\mathfrak{g}})$ and $(\mathfrak{h},[\cdot,\cdot]_{\mathfrak{h}},D_{\mathfrak{h}})$ be two differential graded Lie algebras. Let $\Phi : S^\bullet\bar{\mathfrak{g}} \to \bar{\mathfrak{h}}$ be a graded linear map of degree 0 and let $\tilde{\Omega}_\Phi$ be the associated map, defined as in (13.56).*

(1) If $x \in \nu\mathfrak{g}_1^\nu$, then

$$\widehat{\Phi}(e^x) = e^{\tilde{\Omega}_\Phi(x)} \tag{13.57}$$

(2) If $x \in \nu\mathfrak{g}_1^\nu$ and $\omega \in \nu\mathfrak{g}^\nu$, then

$$\widehat{\Phi}(\omega e^x) = \left(\sum_{k\in\mathbb{N}^*} \frac{1}{(k-1)!}\Phi_k(\omega x^{k-1})\right) e^{\tilde{\Omega}_\Phi(x)} . \tag{13.58}$$

Proof. Let $x \in \mathsf{v}\mathfrak{g}_1^\vee$. It follows from the definition of $\tilde{\Omega}_\Phi$, that

$$e^{\tilde{\Omega}_\Phi(x)} = 1 + \sum_{\ell\in\mathbb{N}^*} \frac{1}{\ell!} \sum_{i_1,\ldots,i_\ell\in\mathbb{N}^*} \frac{1}{i_1!\ldots i_\ell!} \Phi_{i_1}(x^{i_1})\ldots\Phi_{i_\ell}(x^{i_\ell}) . \tag{13.59}$$

For every $k \in \mathbb{N}^*$ and for every permutation $\sigma \in S_k$, one has $\operatorname{sgn}(\sigma; x^k) = 1$, because $x \in \mathsf{v}\mathfrak{g}_1^\vee$. Hence, by definition of $\hat{\Phi}$, we have:

$$\begin{aligned}
\widehat{\Phi}(e^x) &= \sum_{k\in\mathbb{N}} \frac{1}{k!}\, \widehat{\Phi}(x^k) \\
&= 1 + \sum_{k\in\mathbb{N}^*} \frac{1}{k!} \sum_{\ell\in\mathbb{N}^*} \sum_{\substack{i_1+\cdots+i_\ell=k\\ i_1,\ldots,i_\ell\geqslant 1}} \frac{1}{\ell!} \sum_{\sigma\in S_{i_1,\ldots,i_\ell}} \Phi_{i_1}(x^{i_1})\ldots\Phi_{i_\ell}(x^{i_\ell}) \\
&= 1 + \sum_{\ell\in\mathbb{N}^*} \frac{1}{\ell!} \sum_{i_1,\ldots,i_\ell\in\mathbb{N}^*} \frac{1}{(i_1+\cdots+i_\ell)!} \sum_{\sigma\in S_{i_1,\ldots,i_\ell}} \Phi_{i_1}(x^{i_1})\ldots\Phi_{i_\ell}(x^{i_\ell})
\end{aligned}$$

which is the same as the right-hand side of (13.59), since $\#S_{i_1,\ldots,i_\ell} = \frac{(i_1+\cdots+i_\ell)!}{i_1!\ldots i_\ell!}$. This proves *(1)*.

In order to prove *(2)*, we introduce a formal parameter t and we compute the coefficient in t of both sides of the equality

$$\widehat{\Phi}(e^{x+t\omega}) = e^{\tilde{\Omega}_\Phi(x+t\omega)} \pmod{t^2} . \tag{13.60}$$

The proof that this equality holds is the same as the proof of (13.57) since everything is computed modulo t^2, so that all products of elements of odd degree in $\bar{\mathfrak{h}}$, which are precisely the ones which do *not* commute, are discarded. For the same reason, the coefficient of t in $\widehat{\Phi}(e^{x+t\omega})$ is given by $\widehat{\Phi}(\omega e^x)$, which is the left-hand side of (13.58), and the coefficient of t in $e^{\tilde{\Omega}_\Phi(x+t\omega)}$ is given by the coefficient of t of $e^{\tilde{\Omega}_\Phi(x)} e^{Z_1(t)}$, where $Z_1(t)$ is the linear term in t of $\tilde{\Omega}_\Phi(x+t\omega)$, which is given by

$$Z_1(t) = \sum_{k\in\mathbb{N}^*} \frac{1}{k!} \sum_{\ell=0}^{k-1} \Phi_k(x^\ell(t\omega)x^{k-\ell-1}) = \sum_{k\in\mathbb{N}^*} \frac{1}{(k-1)!} \Phi_k(\omega x^{k-1})t ,$$

so that the coefficient of t in $e^{\tilde{\Omega}_\Phi(x+t\omega)}$ is given by the right-hand side of (13.58), which establishes *(2)*. □

Proposition 13.41. *Let* $(\mathfrak{g}, [\cdot,\cdot]_\mathfrak{g}, z_\mathfrak{g})$ *and* $(\mathfrak{h}, [\cdot,\cdot]_\mathfrak{h}, z_\mathfrak{h})$ *be two pointed differential graded Lie algebras. Let* $\Phi : S^\bullet\bar{\mathfrak{g}} \to \bar{\mathfrak{h}}$ *be an* L_∞*-morphism and let* $\tilde{\Omega}_\Phi : \mathsf{v}\mathfrak{g}^\vee \to \mathsf{v}\mathfrak{h}^\vee$ *be the associated map, defined as in (13.56).*

(1) *If* $x \in \mathsf{v}\mathfrak{g}_1^\vee$ *is a solution of the Maurer–Cartan equation associated to* $\mathfrak{g}$*, then* $\tilde{\Omega}_\Phi(x)$ *is a solution of the Maurer–Cartan equation associated to* $\mathfrak{h}$*;*

(2) *If* $x, y \in \mathsf{v}\mathfrak{g}_1^\vee$ *are gauge equivalent solutions of the Maurer–Cartan equation associated to* $\mathfrak{g}$*, then* $\tilde{\Omega}_\Phi(x)$ *and* $\tilde{\Omega}_\Phi(y)$ *are gauge equivalent;*

(3) *If* $x \in \nu\mathfrak{g}_1^\vee$ *and* $\xi \in \nu^k\mathfrak{g}_0^\vee$ *for some* $k \in \mathbb{N}$, *then there exists* $s \in \nu^k\mathfrak{h}_0^\vee$, *such that* $\tilde{\Omega}_\Phi(\xi \odot x) = s \odot \tilde{\Omega}_\Phi(x)$. *Moreover,* $s = \Phi_1(\xi) \pmod{\nu^{k+1}}$.

Proof. We denote by $\Lambda_\mathfrak{g}$ and $\Lambda_\mathfrak{h}$ the total differential of $\mathfrak{g}$, respectively of $\mathfrak{h}$. According to Proposition 13.26, if $x \in \nu\mathfrak{g}_1^\vee$, then $x \in \mathrm{MC}(\mathfrak{g})$ if and only if $\Lambda_\mathfrak{g}(e^x) = 0$, and $\tilde{\Omega}_\Phi(x) \in \mathrm{MC}(\mathfrak{h})$ if and only if $\Lambda_\mathfrak{h}(e^{\tilde{\Omega}_\Phi(x)}) = 0$. The fact that Φ is an L_∞-morphism and (13.57) imply that

$$\Lambda_\mathfrak{h}(e^{\tilde{\Omega}_\Phi(x)}) = \Lambda_\mathfrak{h} \circ \hat{\Phi}(e^x) = \hat{\Phi} \circ \Lambda_\mathfrak{g}(e^x)\,, \tag{13.61}$$

and *(1)* follows.

Let $x, y \in \nu\mathfrak{g}_1^\vee$ be two gauge equivalent solutions of the Maurer–Cartan equation. As we saw in Proposition 13.31, x and y are also path equivalent, which means, by definition, that there exist a polynomial path $\gamma(t)$ in $\nu\mathfrak{g}_1^\vee$ and a polynomial path $\eta(t)$ in $\nu\mathfrak{g}_0^\vee$, such that $\gamma(0) = x, \gamma(1) = y$ and

$$\frac{\mathrm{d}}{\mathrm{d}t} e^{\gamma(t)} = \Lambda_\mathfrak{g}\left(\eta(t) e^{\gamma(t)}\right).$$

Applying $\hat{\Phi}$ to both sides of this equation, and using the definition of an L_∞-morphism, we obtain

$$\frac{\mathrm{d}}{\mathrm{d}t} \hat{\Phi}\left(e^{\gamma(t)}\right) = \Lambda_\mathfrak{h} \circ \hat{\Phi}\left(\eta(t) e^{\gamma(t)}\right),$$

which gives, in view of both items of Proposition 13.40,

$$\frac{\mathrm{d}}{\mathrm{d}t} e^{\tilde{\Omega}_\Phi(\gamma(t))} = \Lambda_\mathfrak{h}\left(\chi(t) e^{\tilde{\Omega}_\Phi(\gamma(t))}\right), \tag{13.62}$$

where

$$\chi(t) := \sum_{\ell \in \mathbb{N}^*} \frac{1}{(\ell-1)!} \Phi_\ell(\eta(t)\gamma^{\ell-1}(t))\,. \tag{13.63}$$

Now $\tilde{\Omega}_\Phi(\gamma(t))$ is a polynomial path in $\nu\mathfrak{h}_1^\vee$ and $\chi(t)$ is a polynomial path in $\nu\mathfrak{h}_0^\vee$. Equation (13.62) says that $\tilde{\Omega}_\Phi(x) = \tilde{\Omega}_\Phi(\gamma(0))$ and $\tilde{\Omega}_\Phi(y) = \tilde{\Omega}_\Phi(\gamma(1))$ are path equivalent, hence gauge equivalent (by Proposition 13.31), which is the content of *(2)*.

According to Remark 13.32, if $x \in \nu\mathfrak{g}_1^\vee$ and $\xi \in \nu^k\mathfrak{g}_0^\vee$, then the polynomial path $\eta(t)$ which is used in establishing the path equivalence between x and $\xi \odot x$ can be chosen in $\nu^k\mathfrak{g}_0^\vee$, namely we can choose $\eta(t) := -\xi$. Then the polynomial path $\chi(t)$, given by (13.63), belongs to $\nu^k\mathfrak{h}_0^\vee$ so that, again by Remark 13.32, the gauge equivalence between $\tilde{\Omega}_\Phi(\xi \odot x)$ and $\tilde{\Omega}_\Phi(x)$ can be realized by an element $s \in \nu^k\mathfrak{h}_0^\vee$. Since $\chi(t) = \Phi_1(\eta(t)) \pmod{\nu^{k+1}}$, we have $s = \Phi_1(\xi) \pmod{\nu^{k+1}}$, which proves *(3)*. □

13.4.2 Surjectivity of Ω_Φ

We show in this section that if Φ is an L_∞-quasi-isomorphism, then Ω_Φ is surjective, which is part of the statement of Theorem 13.39. We will use the following proposition, which describes how the operators Obs_ℓ behave under an L_∞-morphism (see Section 13.3.3 for the definition of Obs_ℓ and for its relation to the Maurer–Cartan equation).

Proposition 13.42. *Suppose that Φ is an L_∞-morphism between two differential graded Lie algebras $(\mathfrak{g},[\cdot\,,\cdot]_\mathfrak{g},D_\mathfrak{g})$ and $(\mathfrak{h},[\cdot\,,\cdot]_\mathfrak{h},D_\mathfrak{h})$. Let $x \in \nu\mathfrak{g}_1^\nu$ and let $\ell \in \mathbb{N}^*$. If $x \in \mathrm{MC}_\ell(\mathfrak{g})$, then $\tilde{\Omega}_\Phi(x) \in \mathrm{MC}_\ell(\mathfrak{h})$. Moreover, the following equality holds:*

$$\Phi_1\left(\mathrm{Obs}_{\ell+1}(x)\right) = \mathrm{Obs}_{\ell+1}\left(\tilde{\Omega}_\Phi(x)\right). \tag{13.64}$$

Proof. As before, we denote by $\Lambda_\mathfrak{g}$ and $\Lambda_\mathfrak{h}$ the total differentials on $S^\bullet\bar{\mathfrak{g}}$ and $S^\bullet\bar{\mathfrak{h}}$ respectively. For $x \in \nu\mathfrak{g}_1^\nu$, we have according to (13.61) that

$$\Lambda_\mathfrak{h}\left(e^{\tilde{\Omega}_\Phi(x)}\right) = \widehat{\Phi}\left(\Lambda_\mathfrak{g}(e^x)\right). \tag{13.65}$$

Both sides of this equation can be computed from (13.35), namely

$$\Lambda_\mathfrak{h}\left(e^{\tilde{\Omega}_\Phi(x)}\right) = \left(D_\mathfrak{h}(\tilde{\Omega}_\Phi(x)) + \frac{1}{2}\left[\tilde{\Omega}_\Phi(x),\tilde{\Omega}_\Phi(x)\right]_\mathfrak{h}\right) e^{\tilde{\Omega}_\Phi(x)}, \tag{13.66}$$

$$\widehat{\Phi}\left(\Lambda_\mathfrak{g}(e^x)\right) = \widehat{\Phi}\left(\left(D_\mathfrak{g}(x) + \frac{1}{2}[x,x]_\mathfrak{g}\right) e^x\right). \tag{13.67}$$

Let $\ell \in \mathbb{N}^*$ and suppose that $x \in \mathrm{MC}_\ell(\mathfrak{g})$. Then $D_\mathfrak{g}(x) + \frac{1}{2}[x,x]_\mathfrak{g} = 0 \pmod{\nu^{\ell+1}}$, hence the right-hand side of equation (13.67) is zero, modulo $\nu^{\ell+1}$. In view of (13.65), the right-hand side of (13.66) is also zero, modulo $\nu^{\ell+1}$, so that

$$D_\mathfrak{h}(\tilde{\Omega}_\Phi(x)) + \frac{1}{2}\left[\tilde{\Omega}_\Phi(x),\tilde{\Omega}_\Phi(x)\right]_\mathfrak{h} = 0 \pmod{\nu^{\ell+1}}.$$

This shows that $\tilde{\Omega}_\Phi(x) \in \mathrm{MC}_\ell(\mathfrak{h})$, which proves the first part of the proposition.

For $x \in \mathrm{MC}_\ell(\mathfrak{g})$, the first part of the proof and (13.66) imply that

$$\left(\Lambda_\mathfrak{h}\left(\widehat{\Phi}(e^x)\right)\right)_{\ell+1} = \left(\Lambda_\mathfrak{h}(e^{\tilde{\Omega}_\Phi(x)})\right)_{\ell+1} = \mathrm{Obs}_{\ell+1}(\tilde{\Omega}_\Phi(x)). \tag{13.68}$$

Moreover, letting $\omega := D_\mathfrak{g}(x) + \frac{1}{2}[x,x]_\mathfrak{g} \in \nu^{\ell+1}\mathfrak{g}_2^\nu$ we obtain, using (13.67) and Proposition 13.40,

$$\Lambda_\mathfrak{h}\left(\widehat{\Phi}(e^x)\right) = \widehat{\Phi}(\omega e^x) = \left(\sum_{k\in\mathbb{N}^*} \frac{1}{(k-1)!}\Phi_k(\omega x^{k-1})\right) e^{\tilde{\Omega}_\Phi(x)}. \tag{13.69}$$

Therefore,

$$\left(\Lambda_{\mathfrak{h}}\left(\widehat{\Phi}(e^x)\right)\right)_{\ell+1} = (\Phi_1(\omega))_{\ell+1} = \Phi_1(\omega_{\ell+1}) = \Phi_1(\mathrm{Obs}_{\ell+1}(x)),$$

which leads, combined with (13.68), to Eq. (13.64). □

We now prove that if Φ is an L_∞-quasi-isomorphism between two pointed differential graded Lie algebras $(\mathfrak{g}, [\cdot,\cdot]_{\mathfrak{g}}, z_{\mathfrak{g}})$ and $(\mathfrak{h}, [\cdot,\cdot]_{\mathfrak{h}}, z_{\mathfrak{h}})$, then Ω_Φ is surjective, which is part of the statement of Theorem 13.39.

Proof. First, let us point out a fact which will be used several times in this proof. Let $(\mathfrak{g}, [\cdot,\cdot], D)$ be a differential graded Lie algebra, let $\ell \in \mathbb{N}^*$ and let $x = \sum_{k\in\mathbb{N}^*} x_k \nu^k$ and $x' = \sum_{k\in\mathbb{N}^*} x'_k \nu^k$. If $x = x' \pmod{\nu^k}$, then

$$\mathrm{Obs}_k(x) - \mathrm{Obs}_k(x') = D(x_k - x'_k), \tag{13.70}$$

in particular $\mathrm{Obs}_k(x) - \mathrm{Obs}_k(x')$ is a coboundary. This fact follows immediately from the definition (13.33) of the operator Obs_k.

Suppose that $y \in \mathrm{MC}(\mathfrak{h})$ (in particular, all non-zero coefficients y_k of y belong to $\mathfrak{h}_1$). We show that there exist formal power series $x \in \nu\mathfrak{g}_1^\nu$ and $\xi \in \nu\mathfrak{h}_0^\nu$, such that $y = \xi \odot \tilde{\Omega}_\Phi(x)$. These series will be constructed term by term. Thus, we assume that we have constructed the first k terms $x_1, \ldots, x_k$ of x and the first k terms $\xi_1, \ldots, \xi_k$ of ξ, such that

(a$_k$) $\bar{x}_k := x_1\nu + \cdots + x_k\nu^k \in \mathrm{MC}_k(\mathfrak{g})$;
(b$_k$) $\bar{\xi}_k := \xi_1\nu + \cdots + \xi_k\nu^k$ and $\bar{x}_k$ satisfy the following equation:

$$y = \bar{\xi}_k \odot \tilde{\Omega}_\Phi(\bar{x}_k) \pmod{\nu^{k+1}}.$$

We show how to construct an extra term x_{k+1} for x and ξ_{k+1} for ξ, such that (a$_{k+1}$) and (b$_{k+1}$) are satisfied. Notice that (a$_0$) and (b$_0$) are trivially satisfied, since by construction $\bar{x}_0 = 0$ and $\bar{\xi}_0 = 0$, while the constant term of y is zero.

First we construct $x'_{k+1} \in \mathfrak{g}_1$ such that $\bar{x}'_{k+1} := \bar{x}_k + x'_{k+1}\nu^{k+1} \in \mathrm{MC}_{k+1}(\mathfrak{g})$. In view of Proposition 13.42, (b$_k$) and (13.70), there exists $u \in \mathfrak{h}_1$ such that

$$\Phi_1(\mathrm{Obs}_{k+1}(\bar{x}_k)) = \mathrm{Obs}_{k+1}(\tilde{\Omega}_\Phi(\bar{x}_k)) = \mathrm{Obs}_{k+1}\left((-\bar{\xi}_k)\odot y\right) + D(u) = D(u),$$

where we used in the last step that y, and hence $(-\bar{\xi}_k)\odot y$, belongs to $\mathrm{MC}(\mathfrak{h})$. Since $\bar{x}_k \in \mathrm{MC}_k(\mathfrak{g})$, Proposition 13.24 implies that $\mathrm{Obs}_{k+1}(\bar{x}_k)$ is a cocycle, so it defines a cohomology class in $H^2_{D_{z_{\mathfrak{g}}}}(\mathfrak{g})$; by the above computation, the image under Φ_1 of this cohomology class is trivial. Now $\Phi_1^2 : H^2_{D_{z_{\mathfrak{g}}}}(\mathfrak{g}) \to H^2_{D_{z_{\mathfrak{h}}}}(\mathfrak{h})$ is injective, since Φ is an L_∞-quasi-isomorphism, hence the cohomology class of $\mathrm{Obs}_{k+1}(\bar{x}_k)$ is trivial, i.e., there exists $x'_{k+1} \in \mathfrak{g}_1$ satisfying $\mathrm{Obs}_{k+1}(\bar{x}_k) = \left[x'_{k+1}, z_{\mathfrak{g}}\right]_{\mathfrak{g}}$. Letting $\bar{x}'_{k+1} := \bar{x}_k + x'_{k+1}\nu^{k+1}$, we have that

$$\mathrm{Obs}_i(\bar{x}'_{k+1}) = \mathrm{Obs}_i(\bar{x}_k) = 0, \text{ for all } 1 \leqslant i \leqslant k,$$

while (13.70) implies that

$$\mathrm{Obs}_{k+1}(\bar{x}'_{k+1}) = \mathrm{Obs}_{k+1}(\bar{x}_k) + D_{z_{\mathfrak{g}}}(x'_{k+1}) = \mathrm{Obs}_{k+1}(\bar{x}_k) + \left[z_{\mathfrak{g}}, x'_{k+1}\right]_{\mathfrak{g}} = 0\,,$$

so that $\bar{x}'_{k+1} \in \mathrm{MC}_{k+1}(\mathfrak{g})$, i.e., $\bar{x}'_{k+1}$ satisfies property (a_{k+1}).

We now prove that there exist a cocycle $x''_{k+1} \in \mathfrak{g}_1$ and an element $\xi_{k+1} \in \mathfrak{h}_0$, such that $\bar{x}_{k+1} := \bar{x}'_{k+1} + x''_{k+1}\nu^{k+1} = \bar{x}_k + (x'_{k+1} + x''_{k+1})\nu^{k+1}$ and $\bar{\xi}_{k+1} := \bar{\xi}_k + \xi_{k+1}\nu^{k+1}$ satisfy property (b_{k+1}); property (a_{k+1}) remains satisfied when we replace $\bar{x}'_{k+1}$ by $\bar{x}_{k+1}$ because x''_{k+1} is a cocycle.

Property (b_k) and the fact that $\bar{x}'_{k+1} = \bar{x}_k \pmod{\nu^{k+1}}$ imply that

$$y = \bar{\xi}_k \odot \tilde{\Omega}_\Phi(\bar{x}'_{k+1}) \pmod{\nu^{k+1}}\,.$$

Writing $\bar{\xi}_k \odot \tilde{\Omega}_\Phi(\bar{x}'_{k+1}) = \sum_{k\in\mathbb{N}^*} X_k \nu^k$, with $X_k \in \mathfrak{h}_1$ for all k, we have according to (13.70) that

$$\mathrm{Obs}_{k+1}(y) - \mathrm{Obs}_{k+1}\left(\bar{\xi}_k \odot \tilde{\Omega}_\Phi(\bar{x}'_{k+1})\right) = D_{z_{\mathfrak{h}}}(y_{k+1} - X_{k+1})\,. \tag{13.71}$$

As $\bar{x}'_{k+1} \in \mathrm{MC}_{k+1}(\mathfrak{g})$, Proposition 13.42 implies that $\tilde{\Omega}_\Phi(\bar{x}'_{k+1}) \in \mathrm{MC}_{k+1}(\mathfrak{h})$, hence $\bar{\xi}_k \odot \tilde{\Omega}_\Phi(\bar{x}'_{k+1}) \in \mathrm{MC}_{k+1}(\mathfrak{h})$, so that $\mathrm{Obs}_{k+1}\left(\bar{\xi}_k \odot \tilde{\Omega}_\Phi(\bar{x}'_{k+1})\right) = 0$. Moreover, since $y \in \mathrm{MC}(\mathfrak{h})$, we also have $\mathrm{Obs}_{k+1}(y) = 0$, so that (13.71) gives $D_{z_{\mathfrak{h}}}(y_{k+1} - X_{k+1}) = 0$. Since $\Phi_1^1 : H^1_{D_{z_{\mathfrak{g}}}}(\mathfrak{g}) \to H^1_{D_{z_{\mathfrak{h}}}}(\mathfrak{h})$ is surjective, there exists a cocycle $x''_{k+1} \in \mathfrak{g}_1$ and an element $\xi_{k+1} \in \mathfrak{h}_0$, such that

$$y_{k+1} - X_{k+1} = \Phi_1(x''_{k+1}) - D_{z_{\mathfrak{h}}}(\xi_{k+1}) = \Phi_1(x''_{k+1}) + \left[\xi_{k+1}, z_{\mathfrak{h}}\right]_{\mathfrak{h}}\,,$$

in other words, we obtain

$$y - \left(\bar{\xi}_k \odot \tilde{\Omega}_\Phi(\bar{x}'_{k+1})\right) = \left(\Phi_1(x''_{k+1}) + \left[\xi_{k+1}, z_{\mathfrak{h}}\right]_{\mathfrak{h}}\right)\nu^{k+1} \pmod{\nu^{k+2}}\,. \tag{13.72}$$

With the help of (13.56) and (13.72), we compute modulo ν^{k+2}:

$$\begin{aligned}\bar{\xi}_{k+1} \odot \tilde{\Omega}_\Phi(\bar{x}_{k+1}) &= (\bar{\xi}_k + \xi_{k+1}\nu^{k+1}) \odot \tilde{\Omega}_\Phi(\bar{x}'_{k+1} + x''_{k+1}\nu^{k+1}) \\ &= \bar{\xi}_k \odot \tilde{\Omega}_\Phi(\bar{x}'_{k+1}) + \left(\Phi_1(x''_{k+1}) + \left[\xi_{k+1}, z_{\mathfrak{h}}\right]_{\mathfrak{h}}\right)\nu^{k+1} = y\,.\end{aligned}$$

Thus, $\bar{x}_{k+1}$ and $\bar{\xi}_{k+1}$ satisfy (a_{k+1}) and (b_{k+1}). □

Remark 13.43. Notice that the fact that $\Phi_1^\bullet$ is an isomorphism is not fully used in the proof. In fact, we only used that $\Phi_1^2 : H^2_{D_{z_{\mathfrak{g}}}}(\mathfrak{g}) \to H^2_{D_{z_{\mathfrak{h}}}}(\mathfrak{h})$ is injective and that $\Phi_1^1 : H^1_{D_{z_{\mathfrak{g}}}}(\mathfrak{g}) \to H^1_{D_{z_{\mathfrak{h}}}}(\mathfrak{h})$ is surjective.

13.4.3 Injectivity of Ω_Φ

In this section we prove that Ω_Φ is injective, which is also part of the statement of Theorem 13.39. To do this, we use the following lemma. Recall that, for a formal power series x (in ν), we denote by x_k its coefficient of ν^k.

Lemma 13.44. *Let $(\mathfrak{g}, [\cdot,\cdot], z)$ be a pointed differential graded Lie algebra. Suppose that $x \in \nu\mathfrak{g}_1^\nu$ and that $\xi \in \nu^k \mathfrak{g}_0^\nu$.*

(1) $(\xi \odot x - x)_k = -D_z(\xi_k)$;
(2) *If $s \in \mathfrak{g}_0$ is a cocycle, then* $(\xi + s\nu^k) \odot x = \xi \odot x \pmod{\nu^{k+1}}$;
(3) *If* $\xi \odot x = x \pmod{\nu^{k+1}}$, *then ξ_k is a cocycle;*
(4) *If ξ_k is a coboundary and x is solution of the Maurer–Cartan equation, then there exists $\xi' \in \nu^{k+1}\mathfrak{g}_0^\nu$, such that $\xi \odot x = \xi' \odot x$.*

Proof. Items *(1)–(3)* follow from the fact that, since $\xi \in \nu^k \mathfrak{g}_0^\nu$, the term in ν^ℓ in $\xi \odot x$ is x_ℓ when $1 \leqslant \ell < k$ and is $x_k + [\xi_k, z] = x_k - D_z(\xi_k)$ when $\ell = k$.

For item *(4)*, assume that ξ_k is a coboundary, $\xi_k = D_z(\eta)$ for some $\eta \in \mathfrak{g}_{-1}$, and that $x \in \mathrm{MC}(\mathfrak{g})$, so that $[z+x, z+x] = 0$. For $w \in \nu\mathfrak{g}^\nu{}_{-1}$, the graded Jacobi identity implies that $\mathrm{ad}_{[w,z+x]}(z+x) = -\frac{1}{2}[[z+x,z+x],w] = 0$, so that

$$[w, z+x] \odot x = x + \sum_{k\in\mathbb{N}^*} \frac{1}{k!} \mathrm{ad}^k_{[w,z+x]}(z+x) = x\,.$$

In particular, if $w \in \nu^k \mathfrak{g}^\nu_{-1}$, then

$$\xi \odot x = \xi \odot ([w,z+x] \odot x) = \mathrm{CH}(\xi, [w,z+x]) \odot x\,, \tag{13.73}$$

and it follows from (13.39) that $(\mathrm{CH}(\xi,[w,z+x]))_k = \xi_k + [w_k, z] = D_z(\eta - w_k)$. Therefore, taking $w := \eta\nu^k$, we have that $\xi' := \mathrm{CH}(\xi,[w,z+x]) \in \nu^{k+1}\mathfrak{g}_0^\nu$, while $\xi' \odot x = \xi \odot x$, according to (13.73). This proves *(4)*. □

We can now prove that if Φ is an L_∞-quasi-isomorphism between two pointed differential graded Lie algebras $(\mathfrak{g}, [\cdot,\cdot]_\mathfrak{g}, z_\mathfrak{g})$ and $(\mathfrak{h}, [\cdot,\cdot]_\mathfrak{h}, z_\mathfrak{h})$, then Ω_Φ is injective, which is part of the statement of Theorem 13.39.

Proof. Suppose that $x, y \in \mathrm{MC}(\mathfrak{g})$ have the same image under Ω_Φ, which means that $\tilde{\Omega}_\Phi(x)$ and $\tilde{\Omega}_\Phi(y)$ are gauge equivalent. We need to show that x and y are gauge equivalent. To do this, we construct three sequences: a sequence $(x^{(k)})_{k\in\mathbb{N}}$ in $\nu\mathfrak{g}_1^\nu$ with $x^{(0)} = x$, a sequence $(\xi^{(k)})_{k\in\mathbb{N}}$ in $\mathfrak{g}_0$ and a sequence $(u^{(k)})_{k\in\mathbb{N}}$ in $\mathfrak{h}_0^\nu$, with $u^{(k)} \in \nu^{k+1}\mathfrak{h}_0^\nu$ for all k. These sequences will be constructed such that their elements satisfy, for all $k \in \mathbb{N}^*$, the following four properties:

(A_k) $x^{(k)} = (\xi^{(k)}\nu^k) \odot x^{(k-1)}$;
(B_k) $x^{(k)} = y \pmod{\nu^{k+1}}$;
(C_k) $u_k^{(k-1)} + \Phi_1(\xi^{(k)})$ is a coboundary;
(D_k) $\tilde{\Omega}_\Phi(x^{(k)}) = u^{(k)} \odot \tilde{\Omega}_\Phi(y)$.

We first show that the existence of these sequences satisfying the properties (A_k) and (B_k) proves that x and y are gauge equivalent; properties (C_k) and (D_k) will turn out to be useful for constructing the sequences satisfying the former properties. Consider the formal power series $\Xi := \lim_{k\to\infty}\Xi^{(k)}$, where the polynomials $\Xi^{(k)}$ are recursively defined by $\Xi^{(k)} := \mathrm{CH}(\xi^{(k)}\nu^k, \Xi_{k-1})$ for $k \in \mathbb{N}^*$ and $\Xi_0 := 0$. Notice that the limit is well-defined, because $\Xi^{(k)} = \Xi^{(k-1)} \pmod{\nu^k}$ for $k \in \mathbb{N}^*$, and that it belongs to $\nu\mathfrak{g}_0^\nu$. In view of (A_1),...,(A_k), (B_k), and (13.40),

$$\Xi^{(k)} \odot x = x^{(k)} = y \pmod{\nu^{k+1}},$$

for all $k \in \mathbb{N}$, so that $\Xi \odot x = y$.

We proceed to the construction of the three sequences. First, we set $\xi^{(0)} := 0$ and we choose for $u^{(0)}$ an element in $\nu\mathfrak{h}_0^\nu$ for which $u^{(0)} \odot \tilde{\Omega}_\Phi(y) = \tilde{\Omega}_\Phi(x)$, which can be done because, by assumption, $\tilde{\Omega}_\Phi(x)$ and $\tilde{\Omega}_\Phi(y)$ are gauge equivalent. In order to satisfy (A_0), (B_0), (C_0) and (D_0), we define $x^{(-1)} := x^{(0)} = x$ and $u_0^{(-1)} := 0$. Let us assume that $x^{(0)},\ldots,x^{(k)}$ and $u^{(0)},\ldots,u^{(k)}$ and $\xi^{(0)},\ldots,\xi^{(k)}$ are constructed, and satisfy the above conditions (A_ℓ), (B_ℓ), (C_ℓ) and (D_ℓ), for $\ell = 0,\ldots,k$.

We first construct $x^{(k+1)}$ and $\xi^{(k+1)}$ satisfying (A_{k+1}) and (B_{k+1}). In view of (B_k), $x^{(k)}$ and y agree modulo ν^{k+1}. As a first consequence, since both are solutions of the Maurer–Cartan equation, (13.70) implies that $D(x^{(k)}_{k+1} - y_{k+1}) = 0$, i.e., $(x^{(k)} - y)_{k+1}$ is a cocycle. As a second consequence, $\Phi_\ell(y^\ell) = \Phi_\ell((x^{(k)})^\ell) \pmod{\nu^{k+2}}$ for $\ell \geqslant 2$, so that

$$\left(\tilde{\Omega}_\Phi(x^{(k)}) - \tilde{\Omega}_\Phi(y)\right)_{k+1} = \Phi_1\left(x^{(k)}_{k+1} - y_{k+1}\right). \tag{13.74}$$

The left-hand side of (13.74) is a coboundary, in view of (D_k) and item *(1)* of Lemma 13.44. Since $\Phi_1^1 : H^1_{D_{z_\mathfrak{g}}}(\mathfrak{g}) \to H^1_{D_{z_\mathfrak{h}}}(\mathfrak{h})$ is injective, $(x^{(k)} - y)_{k+1}$ is also a coboundary, i.e., there exists $\xi^{(k+1)} \in \mathfrak{g}_0$ such that $D_{z_\mathfrak{g}}(\xi^{(k+1)}) = x^{(k)}_{k+1} - y_{k+1}$. Now $((\xi^{(k+1)}\nu^{k+1}) \odot x^{(k)})_{k+1} = x^{(k)}_{k+1} + \left[\xi^{(k+1)}, z_\mathfrak{g}\right]$, so that $(\xi^{(k+1)}\nu^{k+1}) \odot x^{(k)} = y \pmod{\nu^{k+2}}$. Defining $x^{(k+1)} := (\xi^{(k+1)}\nu^{k+1}) \odot x^{(k)}$, it follows that $x^{(k+1)}$ and $\xi^{(k+1)}$ satisfy (A_{k+1}) and (B_{k+1}).

The constructed element $\xi^{(k+1)}$ does not satisfy (C_{k+1}), namely $u^{(k)}_{k+1} + \Phi_1(\xi^{(k+1)})$ need not be a coboundary, but it is a cocycle. In fact, in view of (A_{k+1}) and (B_{k+1}),

$$(\xi^{(k+1)}\nu^{k+1}) \odot x^{(k)} = y \pmod{\nu^{k+2}},$$

so that, in view of (D_k), we have the following equalities, modulo ν^{k+2}:

$$u^{(k)} \odot \tilde{\Omega}_\Phi\left((\xi^{(k+1)}\nu^{k+1}) \odot x^{(k)}\right) = u^{(k)} \odot \tilde{\Omega}_\Phi(y) = \tilde{\Omega}_\Phi(x^{(k)}). \tag{13.75}$$

Since Φ is an L_∞-morphism, item *(3)* of Proposition 13.41 implies that

$$\tilde{\Omega}_\Phi(\xi^{(k+1)}\nu^{k+1} \odot x^{(k)}) = \Phi_1(\xi^{(k+1)}\nu^{k+1}) \odot \tilde{\Omega}_\Phi(x^{(k)}) \pmod{\nu^{k+2}},$$

so that, still modulo ν^{k+2}, the left-hand side of (13.75) is $\mathrm{CH}(u^{(k)}, \Phi_1(\xi^{(k+1)}\nu^{k+1}))\odot \tilde{\Omega}_\Phi(x^{(k)})$. According to (13.39),

$$(\mathrm{CH}(u^{(k)}, \Phi_1(\xi^{(k+1)}\nu^{k+1})))_{k+1} = u^{(k)}_{k+1} + \Phi_1(\xi^{(k+1)}),$$

so that we can apply item in *(3)* of Lemma 13.44 to (13.75), to conclude that $u^{(k)}_{k+1} + \Phi_1(\xi^{(k+1)})$ is a cocycle.

We now show how $\xi^{(k+1)}$ can be modified, so that (C_{k+1}) is satisfied, i.e., so that $u^{(k)}_{k+1} + \Phi_1(\xi^{(k+1)})$ is a coboundary. We will do this without much altering $x^{(k+1)}$. Since $u^{(k)}_{k+1} + \Phi_1(\xi^{(k+1)})$ is a cocycle it defines a cohomology class in $H^0_{D_{z_\mathfrak{h}}}(\mathfrak{h})$; since $\Phi^0_1 : H^0_{D_{z_\mathfrak{g}}}(\mathfrak{g}) \to H^0_{D_{z_\mathfrak{h}}}(\mathfrak{h})$ is surjective, this cohomology class is the image of a cocycle $\kappa \in \mathfrak{g}_0$. Upon replacing $\xi^{(k+1)}$ by $\xi^{(k+1)} - \kappa$, the new $x^{(k+1)}$, defined by (A_{k+1}), still satisfies (B_{k+1}), in view of item *(2)* of Lemma 13.44, i.e., removing a cocycle from $\xi^{(k+1)}$ does not alter $x^{(k+1)}$ modulo ν^{k+1}. However, the cohomology class of $u^{(k)}_{k+1} + \Phi_1(\xi^{(k+1)})$ is now trivial, i.e., the latter element is a coboundary.

The last step of the construction is the construction of u_{k+1}, such that (D_{k+1}) is satisfied. Item *(3)* in Proposition 13.41 yields the existence of $s \in \nu^{k+1}\mathfrak{h}^\nu_0$ such that $\tilde{\Omega}_\Phi(\xi^{(k+1)}\nu^{k+1} \odot x^{(k)}) = s \odot \tilde{\Omega}_\Phi(x^{(k)})$; by the same proposition, $s = \Phi_1(\xi^{(k+1)})$ (mod ν^{k+2}). Using (A_{k+1}) and (D_k), it follows that

$$\tilde{\Omega}_\Phi(x^{(k+1)}) = s \odot \tilde{\Omega}_\Phi(x^{(k)}) = s \odot \left(u^{(k)} \odot \tilde{\Omega}_\Phi(y)\right) = \mathrm{CH}(s, u^{(k)}) \odot \tilde{\Omega}_\Phi(y).$$

In view of (13.39), $(\mathrm{CH}(s, u^{(k)}))_{k+1} = \Phi_1(\xi^{(k+1)}) + u^{(k)}_{k+1}$, which is a coboundary, according to (C_{k+1}). Item *(4)* in Lemma 13.44 implies that there exists $u^{(k+1)}$ in $\nu^{k+2}\mathfrak{h}^\nu_0$ such that

$$u^{(k+1)} \odot \tilde{\Omega}_\Phi(y) = \mathrm{CH}(s, u^{(k)}) \odot \tilde{\Omega}_\Phi(y) = \tilde{\Omega}_\Phi(x^{(k+1)}),$$

which is (D_{k+1}). □

Remark 13.45. The fact that $\Phi^\bullet_1$ is an isomorphism is not fully used in the proof of the injectivity of Ω_Φ. Indeed, we only used that $\Phi^1_1 : H^1_{D_{z_\mathfrak{g}}}(\mathfrak{g}) \to H^1_{D_{z_\mathfrak{h}}}(\mathfrak{h})$ is injective and that $\Phi^0_1 : H^0_{D_{z_\mathfrak{g}}}(\mathfrak{g}) \to H^0_{D_{z_\mathfrak{h}}}(\mathfrak{h})$ is surjective (see Remark 13.43 for a similar remark concerning the proof of the surjectivity of Ω_Φ).

13.5 Kontsevich's Formality Theorem and Its Consequences

The present section is devoted to a fundamental result by Kontsevich, whose statement and implications are accessible for the readers who have become sufficiently familiar with the concepts developed in the previous sections of this chapter. Instead of giving a full proof of Kontsevich's theorem, which is beyond the scope of this book, we present some of the main ingredients, and we concentrate on a few of its consequences, which have deep implications to both Poisson geometry and Lie theory. The most popular consequence can be stated in a few words: the algebra of smooth functions on a real Poisson manifold admits a formal deformation quantization. This result will be proved in Theorem 13.51, as a consequence of Kontsevich's formality theorem (Theorem 13.46).

13.5.1 Kontsevich's Formality Theorem

Let M be a real manifold and denote by μ the usual product on $C^\infty(M)$. Recall that there are two pointed differential graded Lie algebras, which are naturally associated to M:

- $(\overline{\mathfrak{X}}^\bullet(M),[\cdot\,,\cdot]_S,0)$ is the graded vector space of multivector fields on M, with shifted degree, equipped with the Schouten bracket and with the zero differential (see Example 13.18);
- $(\overline{\mathrm{HC}}^\bullet_{\mathrm{diff}}(M),[\cdot\,,\cdot]_G,\mu)$ is the graded vector space of differential Hochschild cochains of $C^\infty(M)$, with shifted degree, equipped with the Gerstenhaber bracket and with the Hochschild differential associated to μ (see Example 13.20).

First, we show how these differential graded Lie algebras are related. Let P be a p-vector field on M, where $p \geqslant 1$. A p-linear map $\phi(P):(C^\infty(M))^p \to C^\infty(M)$ is naturally associated to P by setting, for all $F_1,\ldots,F_p \in C^\infty(M)$,

$$\phi(P)(F_1,\ldots,F_p) := \frac{1}{p!}P[F_1,\ldots,F_p]\,. \tag{13.76}$$

It leads for every p to a graded linear map of degree 0,

$$\phi : \overline{\mathfrak{X}}^{\overline{p}}(M) \to \overline{\mathrm{HC}}^{\overline{p}}_{\mathrm{diff}}(M)\,, \tag{13.77}$$

which is injective, because P can be recovered from $\phi(P)$. We have, according to (13.7), for all $F_0,\ldots,F_p \in C^\infty(M)$,

$$\begin{aligned}\delta^p_\mu(p!\,\phi(P))(F_0,\ldots,F_p) &= F_0P[F_1,\ldots,F_p]+(-1)^{p+1}P[F_0,\ldots,F_{p-1}]F_p\\ &\quad+\sum_{i=1}^{p}(-1)^iP[F_0,\ldots,F_{i-1}F_i,\ldots,F_p]\\ &=0\,,\end{aligned}$$

where we used in the last equality that P is a derivation in each of its arguments. It follows that the map ϕ takes values in the Hochschild cocycles of $(C^\infty(M),\mu)$. In particular, ϕ induces a linear map $H^\bullet\phi$ between the cohomologies of the pointed differential graded Lie algebras $(\overline{\mathfrak{X}}^\bullet(M),[\cdot\,,\cdot]_S,0)$ and $(\overline{\mathrm{HC}}^\bullet_{\mathrm{diff}}(M),[\cdot\,,\cdot]_G,\mu)$. According to the Hochschild–Kostant–Rosenberg theorem (see [131]), $H^\bullet\phi$ is an isomorphism of differential graded Lie algebras. The map ϕ itself is however not a morphism of (differential) graded Lie algebras, since $[P,Q]_S \neq [P,Q]_G$ for general multivector fields P,Q as one easily checks from the definitions.

The following result, due to Kontsevich, states that, even if ϕ is not a morphism of differential graded Lie algebras, it extends to an L_∞-morphism Φ between $(\overline{\mathfrak{X}}^\bullet(M),[\cdot,\cdot]_S,0)$ and $(\overline{\mathrm{HC}}^\bullet_{\mathrm{diff}}(M),[\cdot,\cdot]_G,\mu)$; since the induced map in cohomology $H^\bullet\Phi_1$ is $H^\bullet\phi$, which is according to the Hochschild–Kostant–Rosenberg theorem an isomorphism, Φ is an L_∞-quasi-isomorphism (see Definition 13.37).

Theorem 13.46 (Kontsevich's formality theorem). *Let M be a real manifold and denote by μ the usual product on $C^\infty(M)$. There exists an L_∞-quasi-isomorphism Φ from $(\overline{\mathfrak{X}}^\bullet(M),[\cdot,\cdot]_S,0)$ to $(\overline{\mathrm{HC}}^\bullet_{\mathrm{diff}}(M),[\cdot,\cdot]_G,\mu)$, such that the map $\Phi_1 : \overline{\mathfrak{X}}^\bullet(M) \to \overline{\mathrm{HC}}^\bullet_{\mathrm{diff}}(M)$ is the natural inclusion map ϕ, defined in (13.77).*

For a proof of this theorem, which is quite involved and is beyond the scope of this book, we refer to [38, 39, 107, 189] and [14]. We will give in the following section some more details about the construction of the L_∞-quasi-isomorphism in the case of $M = \mathbb{R}^d$. But first, we say a few words about the origin of the name of the theorem. Recall from Example 13.21 that the cohomology of a differential graded Lie algebra is itself also a (pointed) differential graded Lie algebra. In the case of $(\overline{\mathfrak{X}}^\bullet(M),[\cdot,\cdot]_S,0)$, the differential is trivial hence the latter differential graded Lie algebra and its cohomology are isomorphic differential graded Lie algebras. According to the Hochschild–Kostant–Rosenberg theorem, cited above, this cohomology and the cohomology of $(\overline{\mathrm{HC}}^\bullet_{\mathrm{diff}}(M),[\cdot,\cdot]_G,\mu)$ are isomorphic differential graded Lie algebras. Thus, Kontsevich's theorem says that there exists an L_∞-quasi-isomorphism between $(\overline{\mathrm{HC}}^\bullet_{\mathrm{diff}}(M),[\cdot,\cdot]_G,\mu)$ and its cohomology. A differential graded Lie algebra which admits an L_∞-quasi-isomorphism to its cohomology is called *formal*, a terminology which is borrowed from topology. Thus, Kontsevich's theorem says that $(\overline{\mathrm{HC}}^\bullet_{\mathrm{diff}}(M),[\cdot,\cdot]_G,\mu)$ is formal, which explains the name "formality theorem".

13.5.2 Kontsevich's Formality Theorem for $\mathbb{R}^d$

Without going into technical details, we present in this section Kontsevich's explicit construction of the L_∞-quasi-isomorphism Φ between $(\overline{\mathfrak{X}}^\bullet(M),[\cdot,\cdot]_S,0)$ and $(\overline{\mathrm{HC}}^\bullet_{\mathrm{diff}}(M),[\cdot,\cdot]_G,\mu)$ in the particular case where $M = \mathbb{R}^d$. In order to do this, we first need to explain how multi-differential operators are naturally coded by graphs, how weights are associated to these graphs and how they all sum up to yield a formula for the L_∞-quasi-isomorphism Φ.

13.5.2.1 A – Kontsevich's Graphs

We describe in this subsection the graphs which are used in Kontsevich's construction. Recall that an oriented graph is a pair (V,A), where A is a subset of $V \times V$. The elements of V are called *vertices* and the elements of A *arrows*. For an arrow $a = (x,y) \in A$, we call y its *head*, denoted $h(a)$ and x its *tail*, denoted $t(a)$. We will assume that Γ contains no loops, which means that if $(x,y) \in A$, then $x \neq y$. For given integers $k, \ell \in \mathbb{N}$, the vertex set of the graphs which are considered is the set $V_{k,\ell}$, defined by

$$V_{k,\ell} := \{1,\ldots,k\} \cup \{\bar{1},\ldots,\bar{\ell}\} .$$

For reasons which will be clear later, it is demanded that these integers satisfy $2k + \ell - 2 \geqslant 0$, that is the cases $(k,\ell) = (0,0)$ and $(k,\ell) = (1,0)$ are excluded. The k vertices $1,\ldots,k$ of $V_{k,\ell}$ are called *vertices of the first type*, while the ℓ vertices $\bar{1},\ldots,\bar{\ell}$ of $V_{k,\ell}$ are called *vertices of the second type*. Another piece of data added to the graphs which are considered here is an ordering of each of the sets of arrows with a common tail, i.e., of each of the sets

$$\mathrm{Star}(s) := \{a \in A \mid t(a) = s\} ,$$

for $s \in \{1,\ldots,k\}$. Denoting the number of elements in $\mathrm{Star}(s)$ by p_s, such an ordering amounts to a numbering σ_s of the elements of $\mathrm{Star}(s)$ by the integers $1,\ldots,p_s$; using this numbering we write $\mathrm{Star}(s)$ as $\mathrm{Star}(s) = \{\sigma_s^1,\ldots,\sigma_s^{p_s}\}$.

Definition 13.47. Let $k,\ell \in \mathbb{N}$, such that $2k + \ell - 2 \geqslant 0$. A triplet $\Gamma = (V_{k,\ell}, A, \sigma)$ is said to be a *Kontsevich graph* if

(1) The pair $(V_{k,\ell}, A)$ is an oriented graph without loops, called the *underlying graph* of Γ;
(2) For every arrow $a \in A$, its tail $t(a)$ is a vertex of the first type;
(3) For every $s \in \{1,\ldots,k\}$, σ_s is an ordering of $\mathrm{Star}(s)$.

The set of Kontsevich graphs $\Gamma = (V_{k,\ell}, A, \sigma)$ is denoted by $G_{k,\ell}$. The underlying graph of some element $G_{3,4}$ is depicted in Fig. 13.1.

13.5.2.2 B – The Weight of a Kontsevich Graph

Let $k,\ell \in \mathbb{N}$, such that $2k + \ell - 2 \geqslant 0$. To a Kontsevich graph $\Gamma \in G_{k,\ell}$ we associate a number ϖ_Γ, called its *weight*. The definition involves angles in the Poincaré half-plane and integration over a configuration space. We recall first that the Poincaré half-plane is the open subset

$$\mathscr{H} := \left\{(x,y) \in \mathbb{R}^2 \mid y > 0\right\} ,$$

equipped with the metric

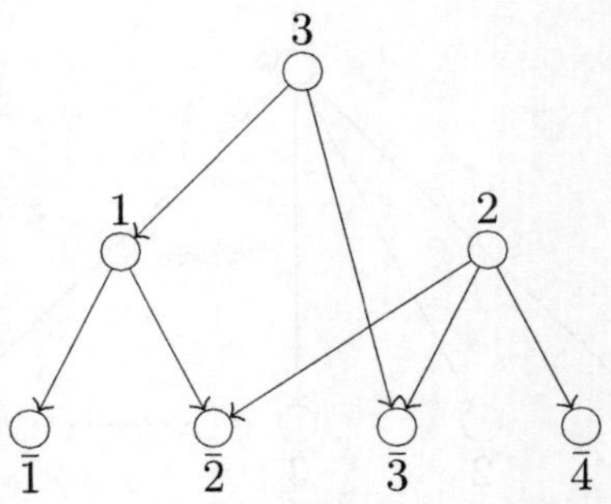

Fig. 13.1 The elements in $G_{k,\ell}$ are triplets (V,A,σ), where the set of vertices consists of k vertices of the first type, denoted $1,\dots,k$ and ℓ vertices of the second type, denoted $\bar{1},\dots,\bar{\ell}$. All arrows start from vertices of the first type and are ordered by σ. Depicted here is the graph (V,A) that underlies some element of $G_{3,4}$. For reasons that will become clear later, the graph is drawn in the upper half-plane, with the vertices of the second type on its boundary.

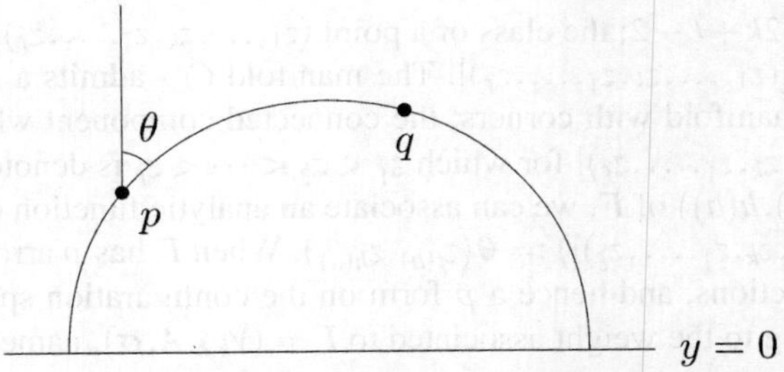

Fig. 13.2 For p and q, points in the Poincaré half-plane $\mathscr{H}$, $\theta(p,q)$ denotes the angle θ between the vertical geodesic through p and the geodesic through p and q.

$$(\mathrm{d}s)^2 = \frac{(\mathrm{d}x)^2 + (\mathrm{d}y)^2}{y^2},$$

whose geodesics are the vertical half-lines, starting from the horizontal axis (the line $y = 0$) and half-circles, with center on the horizontal axis. For $p,q \in \overline{\mathscr{H}} := \mathscr{H} \cup \{y = 0\}$, with $p \neq q$, we denote by $\theta(p,q)$ the angle (at p) between the vertical geodesic through p and the geodesic through p and q (see Fig. 13.2). The angle function θ leads, for each arrow a of Γ, to an analytic function θ_a on a configuration space of points, whose construction we briefly recall (see the appendix by A. Bruguières in [39] for details). With k,ℓ as above, consider $\mathrm{Conf}_{k,\ell}$, the space of $(k+\ell)$-tuples of distinct points, the first k points belonging to $\mathscr{H}$ and the ℓ remaining points belonging to the boundary of $\mathscr{H}$. We write a point of $\mathrm{Conf}_{k,\ell}$ as a $(k+\ell)$-tuple $(z_1,\dots,z_k,z_{\bar{1}},\dots,z_{\bar{\ell}})$, since we think of such a point as corresponding to the vertices of a graph of $G_{k,\ell}$, drawn in $\overline{\mathscr{H}}$. The group $\mathbf{G}$ of affine transformations of $\mathbb{C}$ of the form $z \mapsto \lambda z + \mu$, with $\lambda \in \mathbb{R}_+^*$ and $\mu \in \mathbb{R}$, acts in a natural way on $\mathrm{Conf}_{k,\ell}$. This action is free and the quotient manifold

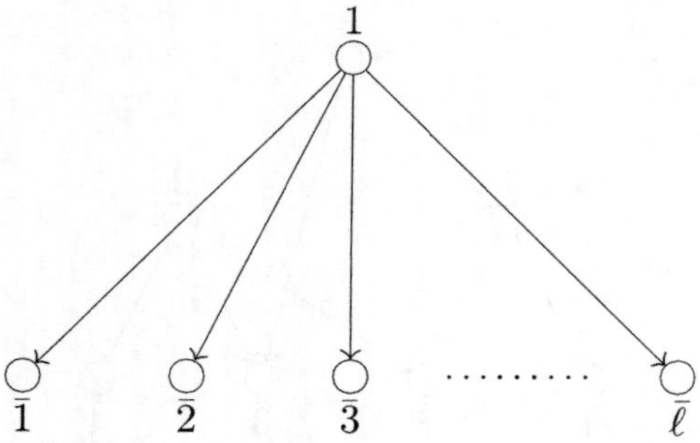

Fig. 13.3 The graph Γ_ℓ belongs to $G_{1,\ell}$. It has one vertex of the first type and ℓ vertices of the second type. Its weight is $1/(\ell!)^2$.

$$C_{k,\ell} := \mathrm{Conf}_{k,\ell}/\mathbf{G}$$

has dimension $2k+\ell-2$; the class of a point $(z_1,\dots,z_k,z_{\bar{1}},\dots,z_{\bar{\ell}}) \in \mathrm{Conf}_{k,\ell}$ in $C_{k,\ell}$ is denoted by $[(z_1,\dots,z_k,z_{\bar{1}},\dots,z_{\bar{\ell}})]$. The manifold $C_{k,\ell}$ admits a natural compactification as a manifold with corners; the connected component which contains the points $[(z_1,\dots,z_k,z_{\bar{1}},\dots,z_{\bar{\ell}})]$ for which $z_{\bar{1}} < z_{\bar{2}} < \cdots < z_{\bar{\ell}}$ is denoted by $\bar{C}_{k,\ell}$. To an arrow $a = (t(a),h(a))$ of Γ, we can associate an analytic function on $C_{k,\ell}$, by defining $\theta_a([(z_1,\dots,z_k,z_{\bar{1}},\dots,z_{\bar{\ell}})]) := \theta(z_{t(a)},z_{h(a)})$. When Γ has p arrows, we obtain in this way p functions, and hence a p-form on the configuration space, which leads upon integrating to the weight associated to $\Gamma = (V_{k,\ell},A,\sigma)$, namely

$$\varpi_\Gamma := \prod_{s=1}^{k} \frac{1}{(\#\mathrm{Star}(s))!} \int_{\bar{C}_{k,\ell}} \bigwedge_{a\in A} \frac{\mathrm{d}\theta_a}{2\pi}.$$

We do not discuss here the technical issues, related to the fact that $\bar{C}_{k,\ell}$ has corners and concerning the choice of ordering in the above wedge product. We simply point out that, for dimensional reasons, $\varpi_\Gamma = 0$ whenever the number of arrows in Γ is different from the dimension $2k+\ell-2$ of $\bar{C}_{k,\ell}$.

Example 13.48. Up to the ordering of its arrows, there is a single graph Γ_ℓ in $G_{1,\ell}$ which has ℓ arrows. It is depicted in Fig. 13.3. The Kontsevich weight of Γ_ℓ is given by

$$\varpi_{\Gamma_\ell} = \frac{1}{\ell!\,(2\pi)^\ell} \int_{\bar{C}_{1,\ell}} \bigwedge_{t=1}^{\ell} \mathrm{d}\theta_{(1,\bar{t})} = \frac{1}{\ell!\,(2\pi)^\ell} \int_{0\leqslant\theta_1<\cdots<\theta_\ell\leqslant 2\pi} \mathrm{d}\theta_1\wedge\cdots\wedge\mathrm{d}\theta_\ell = \frac{1}{(\ell!)^2}. \tag{13.78}$$

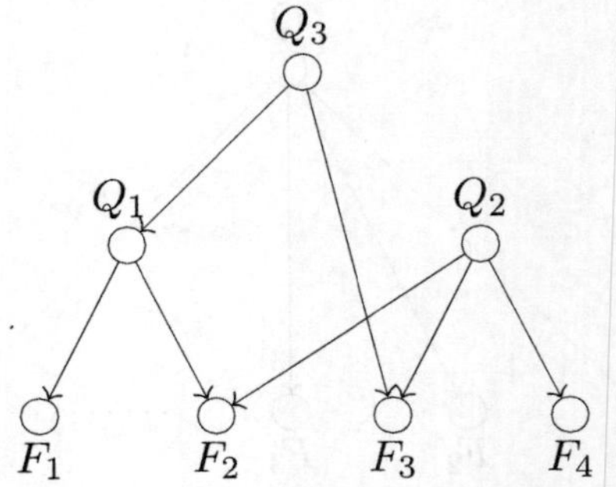

Fig. 13.4 To each vertex of a graph one associates a multivector field on $\mathbb{R}^d$ (a 0-vector field, i.e., a function, when the vertex is of the second type). Depicted is the graph of Fig. 13.1, with the extra data.

13.5.2.3 C – The Differential Operator Associated to a Kontsevich Graph

Let $\Gamma = (V_{k,\ell}, A, \sigma) \in G_{k,\ell}$ be a Kontsevich graph, with $2k + \ell - 2 \geqslant 0$, as before. We show how there is naturally associated to Γ (equipped with some extra data) a multi-differential operator on $\mathbb{R}^d$; the natural coordinates on $\mathbb{R}^d$ are denoted by $x_1, \ldots, x_d$. We add the following data to the graph:

(1) To each vertex s of the first type, a p_s-vector field Q_s on $\mathbb{R}^d$;
(2) To each vertex $\bar{t}$ of the second type, a smooth function $F_{\bar{t}}$ on $\mathbb{R}^d$.

See Fig. 13.4, which is Fig. 13.1 with the extra data. Let $\gamma : A \to \{1, \ldots, d\}$ be any map. Using the above data and γ, we can associate to each vertex $\bar{t}$ of the second type a function $D^\Gamma_{\gamma,\bar{t}}(F_{\bar{t}})$, namely

$$D^\Gamma_{\gamma,\bar{t}}(F_{\bar{t}}) := \left(\prod_{\substack{a\in A \\ h(a)=\bar{t}}} \frac{\partial}{\partial x_{\gamma(a)}} \right) F_{\bar{t}} .$$

Similarly, we associate a function $D^\Gamma_{\gamma,s}(Q_s)$ to each vertex s of the first type, by setting

$$D^\Gamma_{\gamma,s}(Q_s) := \left(\prod_{\substack{a\in A \\ h(a)=s}} \frac{\partial}{\partial x_{\gamma(a)}} \right) \left(Q_s \left[x_{\gamma(\sigma_s^1)}, \ldots, x_{\gamma(\sigma_s^{p_s})} \right] \right) ,$$

where we recall that $\{\sigma_s^1, \ldots, \sigma_s^{p_s}\} = \mathrm{Star}(s)$ is the set of arrows whose tail is s, ordered by σ_s. When the above products are taken over the empty set, i.e., if $\bar{t}$, respectively s, is not the head of an arrow in A, then the derivatives reduce to a zero-th order derivative, i.e., no derivatives are taken. Taking the product of the thus

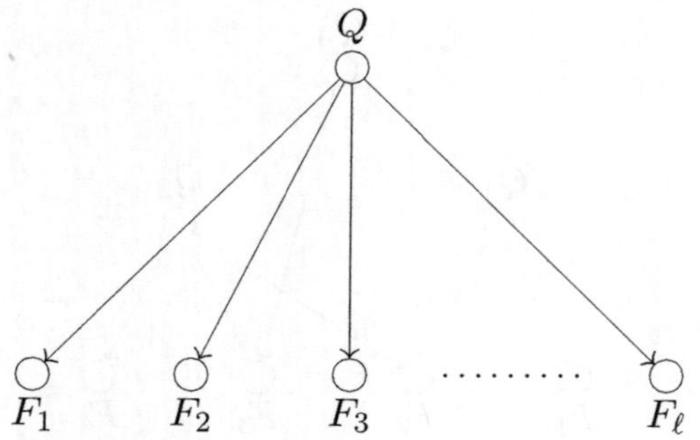

Fig. 13.5 The graph Γ_ℓ with some extra data.

constructed $k+\ell$ functions and summing over all maps $\gamma : A \to \{1,\dots,d\}$, leads to the function

$$B_\Gamma(Q_1,\dots,Q_k)(F_1,\dots,F_\ell) := \sum_{\gamma:A\to\{1,\dots,d\}} \prod_{s=1}^{k} D^\Gamma_{\gamma,s}(Q_s) \prod_{t=1}^{\ell} D^\Gamma_{\gamma,\bar{t}}(F_t)\,.$$

The right-hand side of this equation is clearly a multi-differential operator in the functions $F_1,\dots,F_\ell$. In summary, we have associated to each Kontsevich graph $\Gamma = (V_{k,\ell},A)$ and to each choice of multivector fields (of the right degree) an ℓ-differential operator $B_\Gamma(Q_1,\dots,Q_k)$, whose action on functions is defined as above. If at least one of the multivector fields Q_s is not of the right degree (i.e., its degree is different from p_s, the number of arrows of Γ whose tail is s), we define $B_\Gamma(Q_1,\dots,Q_k)$ to be equal to the zero ℓ-differential operator. Extending these definitions by linearity, we get for each Kontsevich graph $\Gamma = (V_{k,\ell},A) \in G_{k,\ell}$ a linear map $B_\Gamma : S^k\bar{\mathfrak{g}} \to \mathfrak{h}$, where we recall that $\mathfrak{g}$ and $\mathfrak{h}$ are respectively the pointed differential graded Lie algebras $(\overline{\mathfrak{X}}^\bullet(\mathbb{R}^d),[\cdot,\cdot]_S,0)$ and $(\overline{\mathrm{HC}}^\bullet_{\mathrm{diff}}(\mathbb{R}^d),[\cdot,\cdot]_G,\mu)$.

Example 13.49. We write down B_{Γ_ℓ}, where Γ_ℓ is the graph in $G_{1,\ell}$ which we introduced in Example 13.48. Let Q be an ℓ-vector field on $\mathbb{R}^d$ and let $F_1,\dots,F_\ell$ be functions on $\mathbb{R}^d$. See Fig. 13.5, where these data have been added to the graph. We label the arrows of the graph from 1 to ℓ, so that the arrow labeled i connects 1 to $\bar{i}$, for all $i = 1,\dots,\ell$. For a map $\gamma : A = \{1,\dots,\ell\} \to \{1,\dots,d\}$ we write $\gamma(i)$ as γ_i; then, summing over all such maps γ simply amounts to summing over all $\gamma_1,\dots,\gamma_\ell$, ranging from 1 to d. With these notations,

$$\begin{aligned} B_{\Gamma_\ell}(Q)(F_1,\dots,F_\ell) &= \sum_{1\leqslant\gamma_1,\dots,\gamma_\ell\leqslant d} Q[x_{\gamma_1},\dots,x_{\gamma_\ell}]\,\frac{\partial F_1}{\partial x_{\gamma_1}}\frac{\partial F_2}{\partial x_{\gamma_2}}\cdots\frac{\partial F_\ell}{\partial x_{\gamma_\ell}} \\ &= Q[F_1,\dots,F_\ell], \end{aligned}$$

so that $B_{\Gamma_\ell}(Q) = Q$.

Example 13.50. Similarly, we write down B_Γ, where Γ is the graph of Fig. 13.4. The arrows of the graph are labeled $1,\dots,7$, starting with the arrows whose tail is $Q_1 \in \mathfrak{X}^2(\mathbb{R}^d)$, then those whose tail is $Q_2 \in \mathfrak{X}^3(\mathbb{R}^d)$, finally the ones whose tail is $Q_3 \in \mathfrak{X}^2(\mathbb{R}^d)$. It leads to the following formula:

$$B_\Gamma(Q_1,Q_2,Q_3)(F_1,\dots,F_4) = \sum_{1\leqslant \gamma_1,\dots,\gamma_7\leqslant d} \frac{\partial \tilde{Q}_1}{\partial x_{\gamma_6}} \tilde{Q}_2 \tilde{Q}_3 \frac{\partial F_1}{\partial x_{\gamma_1}} \frac{\partial^2 F_2}{\partial x_{\gamma_2}\partial x_{\gamma_3}} \frac{\partial^2 F_3}{\partial x_{\gamma_4}\partial x_{\gamma_7}} \frac{\partial F_4}{\partial x_{\gamma_5}},$$

where

$$\tilde{Q}_1 := Q_1[x_{\gamma_1},x_{\gamma_2}],\ \tilde{Q}_2 := Q_2[x_{\gamma_3},x_{\gamma_4},x_{\gamma_5}],\ \tilde{Q}_3 := Q_3[x_{\gamma_6},x_{\gamma_7}].$$

13.5.2.4 D – Kontsevich's Formula for the L_∞-Quasi-Isomorphism

We have introduced in the previous subsections all the ingredients which are used in Kontsevich's explicit construction of the L_∞-quasi-isomorphism Φ between $(\overline{\mathfrak{X}}^\bullet(\mathbb{R}^d),[\cdot,\cdot]_S,0)$ and $(\overline{\mathrm{HC}}^\bullet_{\mathrm{diff}}(\mathbb{R}^d),[\cdot,\cdot]_G,\mu)$. As above, we denote these pointed differential graded Lie algebras by $\mathfrak{g}$ and $\mathfrak{h}$ respectively, and we denote the natural coordinates on $\mathbb{R}^d$ by $x_1,\dots,x_d$. Recall that $\Phi : S^\bullet\bar{\mathfrak{g}} \to \bar{\mathfrak{h}}$ is completely determined by the family of maps $(\Phi_k)_{k\in\mathbb{N}}$, where for each $k \in \mathbb{N}$, Φ_k is the restriction of Φ to $S^k\bar{\mathfrak{g}}$. They are defined as follows: for all $p_1,\dots,p_k \in \mathbb{N}$ and for all $Q_1Q_2\dots Q_k \in \mathfrak{g}_{p_1-1}\cdots\mathfrak{g}_{p_k-1}$, the ℓ-differential operator $\Phi_k(Q_1Q_2\dots Q_k) \in \mathfrak{h}_{\ell-1}$, where $\ell := \sum_{i=1}^k p_i - 2k + 2$, is defined by:

$$\Phi_k(Q_1Q_2\dots Q_k) := \sum_{\Gamma\in G_{k,\ell}} \varpi_\Gamma B_\Gamma(Q_1,\dots,Q_k). \tag{13.79}$$

In this definition, the sum is taken over all Kontsevich graphs $\Gamma = (V_{k,\ell},A) \in G_{k,\ell}$ and for each such graph Γ, ϖ_Γ and B_Γ are the weight and the ℓ-differential operator associated to Γ, as defined in Subsections 13.5.2.2 and 13.5.2.3.

A few comments about this definition are in order:

(1) As we have seen, $\varpi_\Gamma = 0$ when Γ is a graph in $G_{k,\ell}$ whose number of arrows is different from $2k+\ell-2$. This explain why the sum in (13.79) is only taken over $G_{k,\ell}$, where $\ell := \sum_{i=1}^k p_i - 2k + 2$. Recall also that B_Γ is zero when Γ has a vertex i for which the number of arrows which start from i is *not* equal to p_i.

(2) With respect to the grading which we introduced on $S^\bullet\bar{\mathfrak{g}}$, the monomial $Q_1Q_2\cdots Q_k$ in (13.79) is of degree $\sum_{i=1}^k p_i - 2k$, while $B_\Gamma(Q_1Q_2\dots Q_k)$ is of weight $\ell-2$, with ℓ as above. It follows that Φ is a graded map (of degree 0).

(3) Each Φ_k is well-defined: the order of $Q_1,\dots,Q_k$ in the right-hand side of (13.79) is irrelevant, since both ϖ_Γ and B_Γ are independent of the ordering of the vertices of the first type.

(4) The above formula for Φ_k is particularly simple for $k = 1$. In fact, there are $\ell!$ graphs $\Gamma_\ell = (V_{1,\ell},A,\sigma)$ in $G_{1,\ell}$ with ℓ arrows, differing only by their ordering (as given by σ). As we have seen in Example 13.48, the weight of these graphs is $1/(\ell!)^2$; according to Example 13.49, $B_{\Gamma_\ell}(Q) = Q$. It follows that

$$\Phi_1(Q) = \sum_{\Gamma \in G_{1,\ell}} \varpi_\Gamma B_\Gamma(Q) = \frac{1}{\ell!} Q \,.$$

In view of (13.76), it follows that $\Phi_1 = \phi$, as stated in Kontsevich's formality theorem (Theorem 13.46).

The main result is that Φ is an L_∞-quasi-isomorphism between $\mathfrak{g}$ and $\mathfrak{h}$. As a consequence, Kontsevich's formality theorem (Theorem 13.46) holds for $\mathbb{R}^d$ and the L_∞-quasi-isomorphism which proves this theorem can be constructed quite explicitly. For a proof that Φ is an L_∞-quasi-isomorphism we refer to [39, 107].

13.5.3 A Few Consequences of Kontsevich's Formality Theorem

The following theorem is a first important corollary of Kontsevich's formality theorem.

Theorem 13.51. *Let M be a real manifold and denote by μ the usual product on $C^\infty(M)$.*

(1) Every Poisson structure π on M admits a star product, i.e., there exists a star product $\mu_\star = \mu + \sum_{i\in\mathbb{N}^*} \mu_i \nu^i$ on $C^\infty(M)$, with $\mu_1 = \frac{\pi}{2}$;

(2) There is a one-to-one correspondence between the set of equivalence classes of formal deformations of the trivial Poisson structure on M and the set of equivalence classes of star products on $C^\infty(M)$;

(3) The correspondence in (2) is natural in the following sense: if the equivalence class of a formal deformation $\pi_\star = \sum_{i\in\mathbb{N}^*} \pi_i \nu^i$ of the zero Poisson structure on M and the equivalence class of a star product $\mu_\star = \mu + \sum_{i\in\mathbb{N}^*} \mu_i \nu^i$ on M correspond under (2), then π_1 is the skew-symmetric part of $2\mu_1$.

Proof. Let π be a Poisson structure on M. It is an element of degree one of $(\overline{\mathfrak{X}}^\bullet(M), [\cdot\,,\cdot]_S, 0)$ and since it satisfies the Jacobi identity, $\nu\pi$ is a solution of the Maurer–Cartan equation associated to $(\overline{\mathfrak{X}}^\bullet(M), [\cdot\,,\cdot]_S, 0)$. According to Theorem 13.46, there exists an L_∞-quasi-isomorphism Φ between the differential graded Lie algebras $(\overline{\mathfrak{X}}^\bullet(M), [\cdot\,,\cdot]_S, 0)$ and $(\overline{\mathrm{HC}}^\bullet_{\mathrm{diff}}(M), [\cdot\,,\cdot]_G, \mu)$, with $\Phi_1 = \phi$. According to the first item of Proposition 13.41, $\tilde{\Omega}_\Phi(\nu\pi)$ is a solution of the Maurer–Cartan equation associated to $(\overline{\mathrm{HC}}^\bullet_{\mathrm{diff}}(M), [\cdot\,,\cdot]_G, \mu)$, so that $\mu + \tilde{\Omega}_\Phi(\nu\pi)$ is a star product on $C^\infty(M)$ (Proposition 13.33). Moreover, by definition of $\tilde{\Omega}_\Phi$ (see (13.56)), we can write

$$\tilde{\Omega}_\Phi(\nu\pi) = \sum_{k\geqslant 1} \frac{1}{k!} \Phi_k((\nu\pi)^k) = \sum_{k\geqslant 1} \frac{\nu^k}{k!} \Phi_k(\pi^k) \,, \tag{13.80}$$

so that the coefficient in ν of the star product $\mu + \tilde{\Omega}_\Phi(\nu\pi)$ is $\Phi_1(\pi) = \phi(\pi) = \frac{\pi}{2}$, according to Theorem 13.46. This shows *(1)*.

Theorems 13.46 and 13.39, combined, show that there is a one-to-one correspondence between the set of gauge equivalence classes of solutions of the Maurer–Cartan equation associated to the differential graded Lie algebra $(\overline{\mathfrak{X}}^\bullet(M), [\cdot,\cdot]_S, 0)$ and the set of gauge equivalence classes of solutions of the Maurer–Cartan equation associated to $(\overline{\mathrm{HC}}^\bullet_{\mathrm{diff}}(M), [\cdot,\cdot]_G, \mu)$. According to Proposition 13.33, the first set is the set of equivalence classes of deformations of the trivial Poisson structure, while the second set, after a translation by μ, is the set of equivalence classes of star products on $C^\infty(M)$. This yields the one-to-one correspondence in *(2)*.

Let $\pi_\star = \sum_{k\in\mathbb{N}^*} \pi_k \nu^k$ be a formal deformation of the trivial Poisson structure on M and consider $\mu_\star := \mu + \tilde{\Omega}_\Phi(\pi_\star)$, which is a star product on $C^\infty(M)$. Recall that the equivalence class of $\pi_\star$ and the equivalence class of $\mu_\star$ are related by the one-to-one correspondence in *(2)*. We have, as in (13.80), that $\mu_\star = \mu + \frac{\pi_1}{2}\nu \pmod{\nu^2}$. Let $\mu'_\star = \mu + \sum_{k\in\mathbb{N}^*} \mu'_k \nu^k$ be an arbitrary star product on $C^\infty(M)$ which is equivalent to $\mu_\star$, which means that $\mu'_\star - \mu$ and $\mu_\star - \mu$ are gauge equivalent. Then there exists $\xi = \sum_{k\in\mathbb{N}^*} \xi_k \nu^k$ in $\nu\overline{\mathrm{HC}}^0_{\mathrm{diff}}(M)^\nu$ such that $\mu'_\star = e^{\mathrm{ad}_\xi}(\mu_\star)$, so that

$$\mu'_\star = \mu + \left(\frac{\pi_1}{2} + [\xi_1, \mu]_G\right)\nu \pmod{\nu^2}.$$

The bilinear map $[\xi_1, \mu]_G \in \overline{\mathrm{HC}}^1_{\mathrm{diff}}(M)$ is symmetric, because

$$[\xi_1, \mu]_G(F, G) = \xi_1(\mu(F, G)) - \mu(\xi_1(F), G) - \mu(F, \xi_1(G)),$$

for all $F, G \in C^\infty(M)$. This shows that π_1 is the skew-symmetric part of $2\mu'_1$, which is the content of *(3)*. □

If (M, π) is a real Poisson manifold, one can also consider the formal deformations of π rather than of the trivial Poisson structure on M. We show in the following theorem how these deformations are related to star products on $C^\infty(M)M$.

Theorem 13.52. *Let (M, π) be a real Poisson manifold and let μ denote the usual product on $C^\infty(M)$. There is a one-to-one correspondence between the set of equivalence classes of formal deformations of the Poisson structure π and the set of equivalence classes of star products $\mu_\star = \mu + \sum_{k\in\mathbb{N}^*} \mu_k \nu^k$ on $C^\infty(M)$, for which π is the skew-symmetric part of $2\mu_1$.*

Proof. Let us denote by $\mathscr{D}_0$ the set of equivalence classes of formal deformations of the trivial Poisson structure on M and by $\mathscr{D}_\pi$ the set of equivalence classes of formal deformations of the Poisson structure π on M. Let us also denote by $\mathscr{S}_\mu$ the set of equivalence classes of star products on $C^\infty(M)$ and finally by $\mathscr{S}_{\mu,\pi}$ the set of equivalence classes of star products $\mu_\star = \mu + \sum_{k\in\mathbb{N}^*} \mu_k \nu^k$ on $C^\infty(M)$, such that the skew-symmetric part of $2\mu_1$ is π. Of course, one has $\mathscr{S}_{\mu,\pi} \subseteq \mathscr{S}_\mu$ and according to item *(2)* of Theorem 13.51, there exists a bijection $\psi : \mathscr{D}_0 \to \mathscr{S}_\mu$.

An element $\pi_\star = \pi + \sum_{k\in\mathbb{N}^*} \pi_k \nu^k \in \overline{\mathfrak{X}}^1(M)^\nu$ is a formal deformation of π if and only if $[\pi_\star, \pi_\star]_S = 0$, which is also equivalent to $[\pi_\star\nu, \pi_\star\nu]_S = 0$, meaning that $\pi_\star\nu = \pi\nu + \sum_{k\in\mathbb{N}^*} \pi_k \nu^{k+1}$ is a formal deformation of the trivial Poisson structure. Moreover, saying that two formal deformations $\pi_\star$ and $\pi'_\star$ of π are equivalent means that $\pi_\star - \pi$ and $\pi'_\star - \pi$ are two gauge equivalent solutions of the Maurer–Cartan equation associated to the pointed differential graded Lie algebra $(\overline{\mathfrak{X}}^\bullet(M), [\cdot,\cdot]_S, \pi)$, i.e., that there exists $\xi \in \nu\overline{\mathfrak{X}}^0(M)^\nu$ satisfying

$$\pi'_\star = e^{\mathrm{ad}_\xi}(\pi_\star)\;.$$

Since this equation is equivalent to $\pi'_\star\nu = \nu e^{\mathrm{ad}_\xi}(\pi_\star) = e^{\mathrm{ad}_\xi}(\pi_\star\nu)$, two formal deformations $\pi_\star$ and $\pi'_\star$ of π are equivalent if and only if $\pi_\star\nu$ and $\pi'_\star\nu$ are equivalent as formal deformations of the trivial Poisson structure. This fact permits us to identify $\nu\mathscr{D}_\pi$ with a subset of $\mathscr{D}_0$. Now, according to item *(3)* of Theorem 13.51, we have $\psi(\nu\mathscr{D}_\pi) \subseteq \mathscr{S}_{\mu,\pi}$. Moreover, if $\mu_\star = \mu + \sum_{k\in\mathbb{N}^*}\mu_k\nu^k$ is an element of $\mathscr{S}_{\mu,\pi}$, one has $2\mu_1^- = \pi$ and, once more according to item *(3)* of Theorem 13.51, there exists a formal deformation $\pi_\star = \sum_{k\in\mathbb{N}^*}\pi_k\nu^k$ of the trivial Poisson structure, whose equivalence class is sent by ψ to $\mu_\star$ and such that $\pi = 2\mu_1^- = \pi_1$. This implies in particular that $\frac{1}{\nu}\pi_\star$ is a formal deformation of the Poisson structure π and that $\mu_\star$ lies in $\psi(\nu\mathscr{D}_\pi)$. This shows that $\psi(\nu\mathscr{D}_\pi) = \mathscr{S}_{\mu,\pi}$, so that ψ induces a bijection between $\nu\mathscr{D}_\pi$ and $\mathscr{S}_{\mu,\pi}$, hence between $\mathscr{D}_\pi$ and $\mathscr{S}_{\mu,\pi}$. □

Example 13.53. In the case of $\mathbb{R}^d$, the explicit construction of the L_∞-quasi-isomorphism which we have given in Subsection 13.5.2.4 leads to an explicit formula for the star product on $C^\infty(\mathbb{R}^d)$, equipped with an arbitrary Poisson structure π. Specializing (13.80) to (13.79), gives the following formula for the star product on $C^\infty(\mathbb{R}^d)$, for $F,G \in C^\infty(\mathbb{R}^d)$,

$$\begin{aligned}\mu_\star(F,G) &= \mu(F,G) + \sum_{k\geqslant 1}\frac{\nu^k}{k!}\Phi_k(\pi^k)(F,G)\\ &= \mu(F,G) + \sum_{k\geqslant 1}\frac{\nu^k}{k!}\sum_{\Gamma\in G_{k,2}}\varpi_\Gamma B_\Gamma(\pi,\ldots,\pi)(F,G)\;.\end{aligned}$$

In the above sum, only those graphs for which every vertex of the first type is the tail of exactly two arrows contribute, since $B_\Gamma(\pi,\ldots,\pi)$ is equal to zero for all other graphs.

13.6 Notes

Deformation quantization was proposed by Bayen, Flato, Fronsdal, Lichnerowicz and Sternheimer in [19] as an approach to reverse the standard operation of taking the classical limit of a quantum system, which is loosely speaking done by letting Planck's constant $\hbar$ tend to zero. They raise the question whether a deformation quantization of every Poisson manifold exists. For symplectic manifolds, an affirmative answer to this question was given by De Wilde–Lecomte [57], who use the Moyal–Weyl product, which answers the question locally (for symplectic manifolds). A geometric proof of the latter result was given by Fedosov [73].

The fact that every Poisson manifold admits a deformation quantization has been shown by Kontsevich [107]. An alternative proof of Kontsevich's result was given by Tamarkin [189].

In a purely algebraic context, the problem of (formally) deforming commutative associative algebras was considered by Gerstenhaber in [82], who establishes the link with Hochschild cohomology. Deformations of Poisson structures were introduced in [126]. Deformations of Poisson manifolds, in which both the associative product and the Poisson bracket are deformed, are considered in Ginzburg–Kaledin [85].

Appendix A
Multilinear Algebra

In this appendix, we recall some basic definitions and properties related to multilinear algebra, including (graded) algebra and coalgebra structures, derivations and coderivations. We need them in two particular cases, namely for vector spaces over a field $\mathbb{F}$, and for $\mathscr{A}$-modules, where $\mathscr{A}$ is a commutative associative algebra over a field $\mathbb{F}$. Notice that the latter modules can also be thought of as vector spaces over $\mathbb{F}$, a fact which we often use, since many operations which we consider are $\mathbb{F}$-linear, rather than $\mathscr{A}$-linear. In order to cover both cases, we consider in this appendix the structures which we need on modules over an arbitrary commutative ring R with unit; the reader may find it useful to keep in mind the example of the $C^\infty(M)$-module of vector fields or differential forms over a manifold M.

Throughout the appendix, $\mathbb{F}$ denotes a field of characteristic zero and R denotes an arbitrary commutative ring with unit, denoted by 1.

A.1 Tensor Algebra

For R-modules V and W, the set of linear[1] maps $V \to W$ is denoted by $\operatorname{Hom}_R(V,W)$, or by $\operatorname{Hom}(V,W)$ when it is clear that V and W are considered as R-modules. $\operatorname{Hom}(V,W)$ is itself an R-module in a natural way. When V is a free R-module (for example when $R=\mathbb{F}$, so that V is an $\mathbb{F}$-vector space) and $\mathscr{B}$ is a basis of V, every map $\mathscr{B} \to W$ extends to a unique element of $\operatorname{Hom}(V,W)$ by linearity; when both V and W are finite-dimensional $\mathbb{F}$-vector spaces, and bases for V and W have been fixed, we often think of elements of $\operatorname{Hom}(V,W)$ as matrices (with $\dim W$ rows and $\dim V$ columns).

The *dual* of an R-module V is the R-module $V^* := \operatorname{Hom}(V,R)$. For a finite-dimensional $\mathbb{F}$-vector space V, the dual V^* is an $\mathbb{F}$-vector space, isomorphic to V; the isomorphism is however not canonical, since it depends on the choice of a basis

[1] We rarely use the word R-linear, to avoid confusion with the terminology for multilinear maps, i.e., k-linear maps, with $k \in \mathbb{N}^*$.

C. Laurent-Gengoux et al., *Poisson Structures*,
Grundlehren der mathematischen Wissenschaften 347,
DOI 10.1007/978-3-642-31090-4, © Springer-Verlag Berlin Heidelberg 2013

for V. The natural pairing (evaluation map)

$$\begin{aligned} \langle\cdot,\cdot\rangle : V^* \times V &\to R \\ (\xi, v) &\mapsto \langle \xi, v\rangle := \xi(v)\,, \end{aligned} \tag{A.1}$$

leads for fixed $v \in V$ to a linear map $\langle\cdot, v\rangle : V^* \to R$, i.e., to an element of $(V^*)^* = \mathrm{Hom}(V^*, R)$. The R-module $(V^*)^*$ is called the *bidual* of V. The resulting linear map $V \to (V^*)^*$ is, in general, neither injective nor surjective, but there are two important particular cases in which it is an isomorphism:

- When V is a finite-dimensional vector space over $R = \mathbb{F}$;
- When $R = C^\infty(M)$, the algebra of smooth functions on a smooth manifold M, with V the space of smooth differential k-forms on M; in this case, V^* can be identified with the R-module of smooth vector fields on M.

We will often deal with bilinear maps, sometimes with more general multilinear maps between R-modules, where multilinear means (R-)linear in each of its arguments, keeping the other arguments fixed. The language of tensor products, which we introduce now, is very useful for this. Let V_1, V_2, V and W be R-modules and let $p : V_1 \times V_2 \to V$ be a bilinear map. If we compose p with a linear map $V \to W$, then we obtain a bilinear map $V_1 \times V_2 \to W$; one may wonder if, given V_1 and V_2, there exists an R-module V, such that, for every R-module W, every bilinear map $V_1 \times V_2 \to W$ can be obtained in this way from a linear map on $V \to W$. In fact, there is a unique (up to isomorphism) such R-module, called the *tensor product* of V_1 and V_2, denoted by $V_1 \otimes_R V_2$, or $V_1 \otimes V_2$, and it comes with a natural bilinear map $p : V_1 \times V_2 \to V_1 \otimes V_2$. Formally, the stated property means that every bilinear map $\phi : V_1 \times V_2 \to W$ factors uniquely via p, meaning that there exists a unique linear map $\overline{\phi}$, such that $\overline{\phi} \circ p = \phi$. This property is displayed in the following diagram:

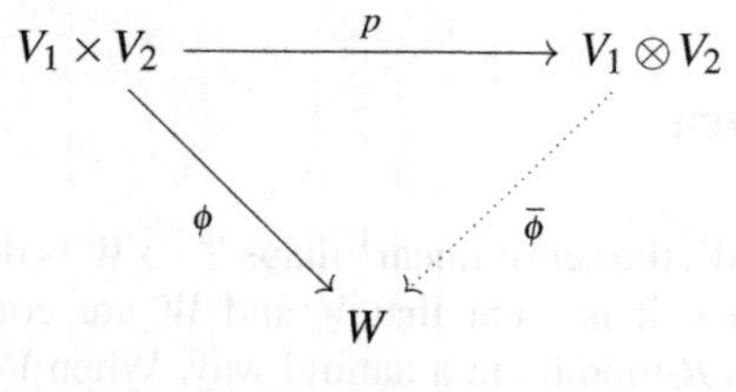

In practice we do not make a distinction between the bilinear map ϕ and the linear map $\overline{\phi}$, so we simply write $\phi \in \mathrm{Hom}(V_1 \otimes V_2, W)$. A natural construction of the tensor product $V_1 \otimes V_2$ is as the quotient of the free R-module which is generated by all formal expressions $v_1 \otimes v_2$, with $v_1 \in V_1$ and $v_2 \in V_2$, divided by the equivalence relation defined by

$$\begin{aligned} &(v_1 + v_1') \otimes v_2 = v_1 \otimes v_2 + v_1' \otimes v_2, \\ &v_1 \otimes (v_2 + v_2') = v_1 \otimes v_2 + v_1 \otimes v_2', \\ &a(v_1 \otimes v_2) = (av_1) \otimes v_2 = v_1 \otimes (av_2), \end{aligned}$$

where $v_1, v_1' \in V_1$ and $v_2, v_2' \in V_2$ and $a \in R$. One says that $v_1 \otimes v_2$ is the *tensor product* of v_1 and v_2. The maps p and $\overline{\phi}$ are in this notation simply given by

$p(v_1, v_2) = v_1 \otimes v_2$ and $\overline{\phi}(v_1 \otimes v_2) = \phi(v_1, v_2)$. In view of the above three properties, p is a bilinear map.

As an application, consider the bilinear map $V_1 \times V_2 \to V_2 \otimes V_1$, which is given by $(v_1, v_2) \mapsto v_2 \otimes v_1$. It factors via p to yield an R-module isomorphism $S : V_1 \otimes V_2 \to V_2 \otimes V_1$, given by $S(v_1 \otimes v_2) = v_2 \otimes v_1$, and called the *twist map*.

One easily shows that for arbitrary R-modules V_1, V_2 and V_3, one has a natural isomorphism $(V_1 \otimes V_2) \otimes V_3 \simeq V_1 \otimes (V_2 \otimes V_3)$. The latter is also denoted as $V_1 \otimes V_2 \otimes V_3$ as it can also be described, like in the case of bilinear maps, as a universal object for trilinear maps, defined on $V_1 \times V_2 \times V_3$. Under the natural isomorphisms, $(V_1 \otimes V_2) \otimes V_3 \simeq V_1 \otimes (V_2 \otimes V_3) \simeq V_1 \otimes V_2 \otimes V_3$, one has that

$$(v_1 \otimes v_2) \otimes v_3 \leftrightarrow v_1 \otimes (v_2 \otimes v_3) \leftrightarrow v_1 \otimes v_2 \otimes v_3 \ , \tag{A.2}$$

so that, in the sequel, we will not make a notational distinction between the elements in (A.2). The extension to several R-modules is clear. For a given R-module V, we obtain a natural sequence of R-modules $T^k V := V^{\otimes k}$, where $k = 0, 1, 2, \ldots$, which is defined, as the notation suggests, by $V^{\otimes k} := V \otimes \cdots \otimes V$ (k factors), when $k \geqslant 1$ and $V^{\otimes 0} := R$. Equipped with the product $(X, Y) \mapsto X \otimes Y$, where $X \in V^{\otimes k}$ and $Y \in V^{\otimes \ell}$, the R-module

$$T^{\bullet} V := \bigoplus_{k=0}^{\infty} T^k V = \bigoplus_{k=0}^{\infty} V^{\otimes k}$$

becomes a graded R-algebra, called the *tensor algebra* of V. See Section A.3 below for the basic definitions on graded algebras.

Fixing one R-module Z, it is useful to think of taking the tensor product with Z as a functor, which means on the one hand that a linear map $\phi \in \mathrm{Hom}(V, W)$ yields, in a natural way, a linear map

$$\tilde{\phi} \in \mathrm{Hom}(V \otimes Z, W \otimes Z) \ , \tag{A.3}$$

simply by putting $\tilde{\phi}(v \otimes z) := \phi(v) \otimes z$, for $v \in V$ and $z \in Z$, which is well-defined; the map $\tilde{\phi}$ is usually denoted by $\phi \otimes \mathbb{1}_Z$, where $\mathbb{1}_Z$ stands for the identity map on Z. On the other hand, it means that taking the tensor product with Z has the usual functorial properties, which make it into a covariant functor. In formulas, this is written as follows:

$$\mathbb{1}_V \otimes \mathbb{1}_Z = \mathbb{1}_{V \otimes Z}, \qquad (\phi \circ \psi) \otimes \mathbb{1}_Z = (\phi \otimes \mathbb{1}_Z) \circ (\psi \otimes \mathbb{1}_Z) \ ,$$

where $\phi \in \mathrm{Hom}(V, W)$ and $\psi \in \mathrm{Hom}(U, V)$. Above, we tensored with Z on the right, but we could also have tensored with Z on the left.

For R-modules V_i and W_i, with $i = 1, 2$, there is also a natural injective morphism (which is not surjective, in general)

$$\mathrm{Hom}(V_1, W_1) \otimes \mathrm{Hom}(V_2, W_2) \to \mathrm{Hom}(V_1 \otimes V_2, W_1 \otimes W_2) \ ,$$

where $\phi_1 \otimes \phi_2 \in \mathrm{Hom}(V_1, W_1) \otimes \mathrm{Hom}(V_2, W_2)$ is, as a linear map from $V_1 \otimes V_2$ to $W_1 \otimes W_2$, given by

$$(\phi_1 \otimes \phi_2)(v_1 \otimes v_2) := \phi_1(v_1) \otimes \phi_2(v_2)\,,$$

for all $v_1 \in V_1$ and $v_2 \in V_2$. This justifies the notation $\phi \otimes \mathbb{1}_Z$ introduced above.

For R-modules V_1, V_2 and W, bilinear maps $V_1 \times V_2 \to W$ are also in natural correspondence with linear maps $V_1 \to \mathrm{Hom}(V_2, W)$, or with linear maps $V_2 \to \mathrm{Hom}(V_1, W)$. We use this usually in the form of the two natural isomorphisms

$$\mathrm{Hom}(V_1 \otimes V_2, W) \simeq \mathrm{Hom}(V_1, \mathrm{Hom}(V_2, W)) \simeq \mathrm{Hom}(V_2, \mathrm{Hom}(V_1, W))\,,$$

which allows one to use a single bilinear map $V_1 \times V_2 \to W$ to associate to each element of V_1 (respectively V_2) a linear map $V_2 \to W$ (respectively a linear map $V_1 \to W$). For example, for arbitrary R-modules V and W,

$$(V \otimes W)^* \simeq \mathrm{Hom}(V, W^*) \simeq \mathrm{Hom}(W, V^*)\,. \tag{A.4}$$

For given R-modules V and W, there is also a natural linear map $\Psi : V^* \otimes W \to \mathrm{Hom}(V, W)$, which associates to an element $\xi \otimes w$, with $\xi \in V^*$ and $w \in W$, the linear map

$$\begin{aligned} \Psi(\xi \otimes w) : V &\to W \\ v &\mapsto \xi(v)\,w = \langle \xi, v\rangle\, w\,. \end{aligned}$$

The map Ψ is always injective, but is in general not surjective. When $R = \mathbb{F}$, then Ψ is an isomorphism if and only if V or W is finite-dimensional. Combining the injection $V^* \otimes W^* \to \mathrm{Hom}(V, W^*)$ with the first isomorphism in (A.4), we obtain a natural inclusion

$$V^* \otimes W^* \to (V \otimes W)^*\,,$$

which is an isomorphism when $R = \mathbb{F}$ and V or W is finite-dimensional.

When $\phi_1 \in \mathrm{Hom}(V_1, \mathscr{A})$ and $\phi_2 \in \mathrm{Hom}(V_2, \mathscr{A})$, where $\mathscr{A}$ is an R-algebra, then $\phi_1 \otimes \phi_2$ is often implicitly combined with the multiplication map $\mathscr{A} \otimes \mathscr{A} \to \mathscr{A}$, yielding $\phi_1 \otimes \phi_2 \in \mathrm{Hom}(V_1 \otimes V_2, \mathscr{A})$, given for $v_1 \in V_1$ and $v_2 \in V_2$, by

$$(\phi_1 \otimes \phi_2)(v_1 \otimes v_2) := \phi_1(v_1)\phi_2(v_2)\,, \tag{A.5}$$

where the latter product is the multiplication in $\mathscr{A}$. When this multiplication has extra properties, they yield similar properties for the product of maps: for example, if $\mathscr{A}$ is commutative, then

$$\phi_1 \otimes \phi_2 = (\phi_2 \otimes \phi_1) \circ S\,,$$

as elements of $\mathrm{Hom}(V_1 \otimes V_2, \mathscr{A})$. Associativity of $\mathscr{A}$ implies that

$$(\phi_1 \otimes \phi_2) \otimes \phi_3 = \phi_1 \otimes (\phi_2 \otimes \phi_3)\,,$$

as elements of $\mathrm{Hom}((V_1 \otimes V_2) \otimes V_3, \mathscr{A}) \simeq \mathrm{Hom}(V_1 \otimes (V_2 \otimes V_3), \mathscr{A})$.

A.2 Exterior and Symmetric Algebra

Our multilinear maps are usually skew-symmetric k-linear maps $V^k \to V$, where V is an R-module, so we recall here the corresponding tensorial notions. In $T^\bullet V$, consider the submodule N which is generated by all ℓ-tensors ($\ell \in \mathbb{N}^*$) of the form $v_{i_1} \otimes \cdots \otimes v_{i_\ell}$, where $v_{i_s} = v_{i_t}$ for some $1 \leqslant s < t \leqslant \ell$, and let $N_k := N \cap T^k V$ for $k \in \mathbb{N}^*$ and $N_0 := \{0\}$. Then we may consider

$$\wedge^\bullet V := T^\bullet V / N = \bigoplus_{k=0}^{\infty} T^k V / N_k = \bigoplus_{k=0}^{\infty} \wedge^k V ,$$

where $\wedge^k V := T^k V / N_k$, for $k \in \mathbb{N}$. Notice that $\wedge^k V = \{0\}$ as soon as k is bigger than the (minimal) number of generators of V. We denote the quotient maps by $p : T^\bullet V \to \wedge^\bullet V$ and $p_k : T^k V \to \wedge^k V$. Since N is a two-sided ideal of $T^\bullet V$, we have that the associative product $\otimes$ on $T^\bullet V$ induces an associative product on $\wedge^\bullet V$, which is denoted by $\wedge$. Thus $p(v_1 \otimes \cdots \otimes v_k) = p(v_1) \wedge \cdots \wedge p(v_k)$, which we also write as $v_1 \wedge \cdots \wedge v_k$, because p_1 (i.e., the restriction of p to V) is injective. One easily verifies that, if $X \in \wedge^i V$ and $Y \in \wedge^j V$, then $X \wedge Y \in \wedge^{i+j} V$ and

$$X \wedge Y = (-1)^{ij} Y \wedge X . \tag{A.6}$$

In the language of the next section, this property is called graded commutativity. The associative, graded commutative R-algebra $(\wedge^\bullet V, \wedge)$ is called the *exterior algebra* of V and elements of $\wedge^\bullet V$ are called *multivectors*. One similarly constructs the *symmetric algebra* $(S^\bullet V, \cdot)$ which is the associative commutative graded R-algebra obtained as $T^\bullet V / N'$, where N' is the two-sided ideal of $T^\bullet V$, generated by all $v \otimes w - w \otimes v$, where $v, w \in V$.

It is clear that every skew-symmetric k-linear map $\phi \in \mathrm{Hom}(V^{\otimes k}, W)$ vanishes on N_k. Therefore, for R-modules V and W, we have that every skew-symmetric k-linear map $\phi \in \mathrm{Hom}(V^{\otimes k}, W)$ corresponds in a canonical way to a linear map $\overline{\phi} : \wedge^k V \to W$, as in the following commutative diagram:

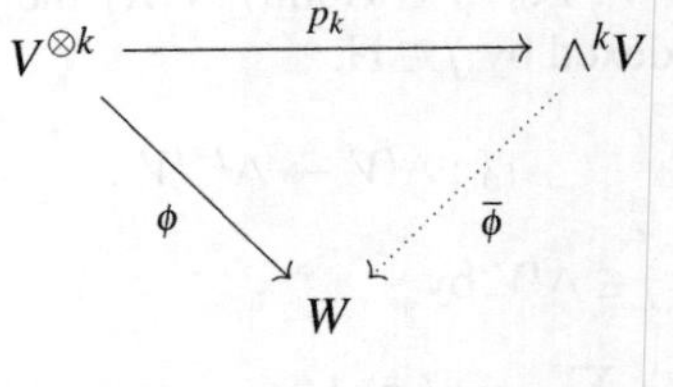

In formulas, $\overline{\phi}(v_1 \wedge \cdots \wedge v_k) = \phi(v_1, \ldots, v_k)$. From now on, we do not distinguish notationally between the maps $\overline{\phi}$ and ϕ: we write $\phi \in \mathrm{Hom}(\wedge^k V, W)$ and we simply say that ϕ is a skew-symmetric k-linear map.

One usually thinks of elements of $\wedge^k V$ as skew-symmetric tensors: the permutation group S_k defines a natural linear action on $V^{\otimes k}$ which is defined for $\sigma \in S_k$ and $v_1 \otimes \cdots \otimes v_k \in V^{\otimes k}$ by

$$\sigma(v_1 \otimes \cdots \otimes v_k) := v_{\sigma(1)} \otimes \cdots \otimes v_{\sigma(k)} \,.$$

An element $X \in V^{\otimes k}$ is called a *symmetric tensor* when $\sigma(X) = X$ for all $\sigma \in S_k$, while it is called a *skew-symmetric tensor* when $\sigma(X) = \operatorname{sgn}(\sigma)X$ for all $\sigma \in S_k$, where $\operatorname{sgn}(\sigma)$ denotes the signature of σ. In order to identify the skew-symmetric k-tensors with $\wedge^k V$, one defines a linear map ρ_k^-, the *skew-symmetrization map*, by

$$\begin{aligned} \rho_k^- : \quad & \wedge^k V \to T^k V \\ & v_1 \wedge \cdots \wedge v_k \mapsto \frac{1}{k!} \sum_{\sigma \in S_k} \operatorname{sgn}(\sigma) v_{\sigma(1)} \otimes \cdots \otimes v_{\sigma(k)} \,. \end{aligned} \tag{A.7}$$

It is easy to see that this map is well-defined and injective, and that its image consists precisely of all skew-symmetric k-tensors. One similarly identifies the symmetric k-tensors with $S^k V$ by using the *symmetrization map*

$$\rho_k^+ : S^k V \to T^k V \,,$$

whose definition is formally the same as the above definition (A.7) of the skew-symmetrization map, except that one leaves out the factor $\operatorname{sgn}(\sigma)$.

There are two types of *internal products* related to the exterior algebra. Let V and W be arbitrary R-modules. For $X \in \wedge^j V$, the *internal product* ι_X yields, for every $i \in \mathbb{N}$, a linear map

$$\iota_X : \operatorname{Hom}(\wedge^i V, W) \to \operatorname{Hom}(\wedge^{i-j} V, W) \,,$$

which is given by

$$\iota_X \phi\,(Z) := \phi(X \wedge Z)$$

where $\phi : \wedge^i V \to W$ and $Z \in \wedge^{i-j} V$, assuming $i \geqslant j$; otherwise, i.e., when $i < j$, then $\iota_X \phi := 0$. It is easily verified that

$$\iota_{X \wedge Y} = \iota_Y \circ \iota_X \,,$$

for $X \in \wedge^j V$ and $Y \in \wedge^k V$. For $\phi \in \operatorname{Hom}(\wedge^i V, R)$ the *internal product* ι_ϕ, is the family of linear maps, indexed by $j \in \mathbb{N}$,

$$\iota_\phi : \wedge^j V \to \wedge^{j-i} V \,,$$

defined for all $v_1 \wedge \cdots \wedge v_j \in \wedge^j V$ by

$$\iota_\phi(v_1 \wedge \cdots \wedge v_j) := \sum_{\sigma \in S_{i,j-i}} \operatorname{sgn}(\sigma)\, \phi(v_{\sigma(1)}, \ldots, v_{\sigma(i)})\, v_{\sigma(i+1)} \wedge \cdots \wedge v_{\sigma(j)}$$

when $i \leqslant j$, and ι_ϕ is the zero map otherwise. In this formula, $S_{i,k}$ denotes the set of all (i,k)-*shuffles*, i.e., all permutations $\sigma \in S_{i+k}$ for which $\sigma(1) < \cdots < \sigma(i)$ and $\sigma(i+1) < \cdots < \sigma(i+k)$; $\operatorname{sgn}(\sigma)$ is the signature of σ as a permutation.

To finish this section, we define a product on skew-symmetric multilinear maps which take values in a commutative associative R-algebra $\mathscr{A}$. We write FG for the product of elements F and G in $\mathscr{A}$. As above, V denotes an arbitrary R-module; often, in the body of the book, this product will be considered for $V = \mathscr{A}$, viewed as an $\mathbb{F}$-vector space ($R = \mathbb{F}$). For $\phi \in \mathrm{Hom}(\wedge^i V, \mathscr{A})$ and $\psi \in \mathrm{Hom}(\wedge^j V, \mathscr{A})$, where $i, j > 0$, we define $\phi \wedge \psi \in \mathrm{Hom}(\wedge^{i+j} V, \mathscr{A})$ by

$$(\phi \wedge \psi)(v_1, \ldots, v_{i+j}) := \sum_{\sigma \in S_{i,j}} \mathrm{sgn}(\sigma)\, \phi(v_{\sigma(1)}, \ldots, v_{\sigma(i)})\, \psi(v_{\sigma(i+1)}, \ldots, v_{\sigma(i+j)}),$$

for all $v_1, \ldots, v_{i+j} \in V$. We also write

$$\langle \phi \wedge \psi, v_1 \wedge \cdots \wedge v_{i+j} \rangle := (\phi \wedge \psi)(v_1, \ldots, v_{i+j}),$$

thereby extending the notation of (A.1).

Notice that, if $\mathscr{A} = R$, so that $\phi \in (\wedge^i V)^*$ and $\psi \in (\wedge^j V)^*$, then we can alternatively define $\phi \wedge \psi \in (\wedge^{i+j} V)^*$ as being the linear map $\psi \circ \iota_\phi$. In this particular case, the definition implies the following relation between ι and $\wedge$:

$$\iota_{\phi \wedge \psi} = \iota_\psi \circ \iota_\phi,$$

for $\phi, \psi \in \oplus_{k \in \mathbb{N}} (\wedge^k V)^*$. Indeed, let $\phi \in \mathrm{Hom}(\wedge^i V, \mathscr{A})$ and $\psi \in \mathrm{Hom}(\wedge^j V, \mathscr{A})$, where $i, j > 0$, and let $v_1, \ldots, v_k \in V$, with $k \geqslant i + j$. Then,

$$\begin{aligned}
&\iota_{\phi \wedge \psi}(v_1, \ldots, v_k) \\
&= \sum_{\sigma \in S_{i+j,k-i-j}} \mathrm{sgn}(\sigma) (\phi \wedge \psi)(v_{\sigma(1)}, \ldots, v_{\sigma(i+j)})\, v_{\sigma(i+j+1)} \wedge \cdots \wedge v_{\sigma(k)} \\
&= \sum_{\tau \in S_{i,j,k-i-j}} \mathrm{sgn}(\tau)\, \phi(v_{\tau(1)}, \ldots, v_{\tau(i)})\, \psi(v_{\tau(i+1)}, \ldots, v_{\tau(i+j)}) \\
&\qquad\qquad v_{\tau(i+j+1)} \wedge \cdots \wedge v_{\tau(k)} \qquad\qquad \text{(A.8)} \\
&= \iota_\psi \left(\sum_{\rho \in S_{i,k-i}} \mathrm{sgn}(\rho) \phi(v_{\rho(1)}, \ldots, v_{\rho(i)})\, v_{\rho(i+1)} \wedge \cdots \wedge v_{\rho(k)} \right) \\
&= (\iota_\psi \circ \iota_\phi)(v_1, \ldots, v_k),
\end{aligned}$$

where we have used the notation $S_{i,j,k-i-j}$ for the set of all permutations of $\{1, \ldots, k\}$ that are increasing on the intervals $1, \ldots, i$ and $i+1, \ldots, i+j$ and $i+j+1, \ldots, k$.

We return to the general case for which $\mathscr{A}$ is not necessarily R, but an arbitrary commutative associative R-algebra. We point out that, if $\phi \in \mathrm{Hom}(\wedge^i V, \mathscr{A})$, $\psi \in \mathrm{Hom}(\wedge^j V, \mathscr{A})$ and $\chi \in \mathrm{Hom}(\wedge^k V, \mathscr{A})$, where $i, j, k \geqslant 0$, then

$$(\phi \wedge \psi) \wedge \chi = \phi \wedge (\psi \wedge \chi),$$

i.e., $\wedge$ is an associative product on $\oplus_{\ell \in \mathbb{N}} \mathrm{Hom}(\wedge^\ell, \mathscr{A})$. This is obtained by exactly the same computation as in (A.8), with $k = \ell - i - j$ and with χ acting on the last part (i.e., on the last v_s that appear out of the arguments of ϕ and ψ). Moreover, for $\phi \in \mathrm{Hom}(\wedge^i V, \mathscr{A})$ and $\psi \in \mathrm{Hom}(\wedge^j V, \mathscr{A})$, we have

$$\phi \wedge \psi = (-1)^{ij} \psi \wedge \phi .$$

In the language of the next section, the product $\wedge$ makes $\oplus_{k\in\mathbb{N}} \mathrm{Hom}(\wedge^k V, \mathscr{A})$ into a graded R-algebra which is associative and graded commutative. Notice that, in the notation which we use, if $\phi_1, \dots, \phi_k \in \mathrm{Hom}(V, \mathscr{A})$ and $v_1, \dots, v_k \in V$, then

$$\langle \phi_1 \wedge \cdots \wedge \phi_k, v_1 \wedge \cdots \wedge v_k \rangle = \det\left(\langle \phi_i, v_j \rangle \right)_{1 \leqslant i,j \leqslant k} . \tag{A.9}$$

A.3 Algebras and Graded Algebras

In this section we recall the basic definitions of algebras and graded algebras. These definitions will be dualized in the next section, to obtain the notions of a coalgebra and of a graded coalgebra.

Let V be an R-module. An algebra structure on V is a bilinear map $\mu : V \times V \to V$, called a *product*. We also view μ as an element of $\mathrm{Hom}(V \otimes V, V)$, and we say that (V, μ) is an R-algebra. Usually, μ is assumed to have additional properties; the typical extra properties that μ may be supposed to have are summarized in Table A.1.

Table A.1 A product μ on an R-module, which makes it into an R-algebra, is usually assumed to have one or two additional properties, taken from the list which appears in this table. We write the properties in their usual form (with $u, v, w \in V$) and in their functional form; the latter is useful for obtaining the "co"-version (see Section A.4). S is the twist map $u \otimes v \mapsto v \otimes u$ and $\mathscr{S}$ is the cycle map $u \otimes v \otimes w \mapsto v \otimes w \otimes u$.

Property	Usual / functional form
commutative	$\mu(v,u) = \mu(u,v)$ $\mu \circ S = \mu$
skew-symmetric	$\mu(v,u) = -\mu(u,v)$ $\mu \circ S = -\mu$
associative	$\mu(u,\mu(v,w)) = \mu(\mu(u,v),w)$ $\mu \circ (\mathbb{1}_V \otimes \mu) = \mu \circ (\mu \otimes \mathbb{1}_V)$
Jacobi identity	$\mu(u,\mu(v,w)) + \circlearrowleft (u,v,w) = 0$ $\sum_{\ell=0}^{2} \mu \circ (\mathbb{1}_V \otimes \mu) \circ \mathscr{S}^\ell = 0$

The usual combinations of adjectives are the following: (1) If μ is skew-symmetric and satisfies the Jacobi identity, then (V, μ) is called a *Lie algebra*, and μ is called a *Lie bracket* on V. We employ the standard custom of using brackets, such as $[\cdot,\cdot]$ and $\{\cdot,\cdot\}$, for the product. (2) An equally important combination of properties that μ

may have are commutativity and associativity; some authors call (V,μ) in this case simply an *algebra* (or R-algebra), but we will not use this convention here since our modules will usually have two algebra structures, one of which is a Lie algebra structure, and the other one is associative and commutative.

Example A.1. A simple example of a Lie algebra structure is given by the vector space $\mathrm{Hom}(W,W)$, where W is an arbitrary vector space, equipped with the *commutator*

$$[\phi_1,\phi_2] := \phi_1\circ\phi_2 - \phi_2\circ\phi_1 ,$$

where $\phi_1,\phi_2\in\mathrm{Hom}(W,W)$. The Jacobi identity for $[\cdot,\cdot]$ is a direct consequence of the associativity of the composition of (linear) maps.

A linear map $\phi : V\to W$ between R-algebras (V,μ) and (W,μ') is called an *algebra homomorphism* if $\phi(\mu(v_1,v_2)) = \mu'(\phi(v_1),\phi(v_2))$, for all $v_1, v_2\in V$, which is written in functional form as $\phi\circ\mu = \mu'\circ(\phi\otimes\phi)$, and which corresponds to the commutativity of the following diagram.

$$\begin{array}{ccc} V\otimes V & \xrightarrow{\mu} & V \\ \Big\downarrow{\scriptstyle \phi\otimes\phi} & & \Big\downarrow{\scriptstyle \phi} \\ W\otimes W & \xrightarrow{\mu'} & W \end{array} \tag{A.10}$$

In the case of Lie algebras $(V,[\cdot,\cdot])$ and $(W,[\cdot,\cdot]')$, such a linear map ϕ is called a *Lie algebra homomorphism*. The homomorphism property then takes the form $\phi([v_1,v_2]) = [\phi(v_1),\phi(v_2)]'$ for all $v_1,v_2\in V$.

We now turn to the graded version of these definitions. For this, it is assumed that V is a *graded R-module*,

$$V=\bigoplus_{i\in\mathbb{Z}} V_i , \tag{A.11}$$

where each of the subspaces V_i is invariant under the action of R. The notation $V_\bullet$ (or $V^\bullet$, when the subspaces are indexed by superscripts) is also used for V. In many cases, one has $V=\oplus_{i\in\mathbb{N}}V_i$, or even $V=\oplus_{i=0}^{k}V_i$, the other V_i being undefined, but one easily arrives at the form (A.11) by defining $V_i := \{0\}$, for those values of i where V_i was undefined. An element of V_i is called a *homogeneous element* of V of *degree i*. A linear map $\phi : V\to W$ between graded R-modules is said to be *graded* of *degree r* if $\phi(V_i)\subset W_{i+r}$ for every $i\in\mathbb{Z}$. When V and W are written as $V_\bullet$ and $W_\bullet$, the suggestive notation $\phi : V_\bullet \to W_{\bullet+r}$ is also used. We denote the R-module of all graded linear maps from V to W of degree r by $\mathrm{Hom}_r(V,W)$. A *graded product* on V is a product μ on V such that

$$\mu(V_i\otimes V_j)\subset V_{i+j} , \text{ for all } i,j\in\mathbb{Z} .$$

Then V, equipped with μ, becomes a *graded R-algebra* and we have induced maps $\mu_{i,j} : V_i\otimes V_j\to V_{i+j}$ for all i,j. A graded linear map of degree zero $\phi : V\to W$ which is a (Lie) algebra homomorphism is called a *graded (Lie) algebra homomorphism*. The graded analog of Table A.1 is given by Table A.2.

Table A.2 The graded analog of Table A.1 is displayed. The only difference between the graded and ungraded notions lies in the signs; in fact, as there are no signs in the case of graded associativity, the notion of associativity and graded associativity coincide. For the graded Jacobi identity, it is understood that when one sums over the three cyclic permutations of (i,j,k), the exponent ℓ takes the consecutive values 0, 1 and 2. The elements u,v and w are assumed to be homogeneous of respective degrees i, j and k.

Property	Usual / functional form
graded commutative	$\mu(v,u) = (-1)^{ij}\mu(u,v)$ $\mu_{j,i}\circ S = (-1)^{ij}\mu_{i,j}$
graded skew-symmetric	$\mu(v,u) = -(-1)^{ij}\mu(u,v)$ $\mu_{j,i}\circ S = -(-1)^{ij}\mu_{i,j}$
(graded) associative	$\mu(u,\mu(v,w)) = \mu(\mu(u,v),w)$ $\mu_{i,j+k}\circ(\mathbb{1}_{V_i}\otimes\mu_{j,k}) = \mu_{i+j,k}\circ(\mu_{i,j}\otimes\mathbb{1}_{V_k})$
graded Jacobi identity	$(-1)^{ik}\mu(u,\mu(v,w)) + \circlearrowleft(u,v,w) = 0$ $(-1)^{ik}\mu_{i,j+k}\circ(\mathbb{1}_{V_i}\otimes\mu_{j,k})\circ\mathscr{S}^{\ell} + \circlearrowleft(i,j,k) = 0$

As in the ungraded case, the combination of graded skew-symmetric and the graded Jacobi identity leads to the notion of a *graded Lie bracket* and of a *graded Lie algebra*. The combination of graded commutativity and associativity leads to the notion of an *associative, graded commutative algebra*.

Example A.2. To give a simple example of a graded Lie algebra, we define for graded linear maps $\phi_i \in \operatorname{Hom}_{r_i}(V,V)$, where $i = 1,2$, their *graded commutator* $[\phi_1,\phi_2]$ as the graded linear map of degree $r_1 + r_2$, given by

$$[\phi_1,\phi_2] := \phi_1\circ\phi_2 - (-1)^{r_1r_2}\phi_2\circ\phi_1 \,. \tag{A.12}$$

The graded R-module $\oplus_{r\in\mathbb{Z}}\operatorname{Hom}_r(V,V)$, equipped with this bracket, is a graded Lie algebra. For graded linear maps ϕ_1,ϕ_2,ϕ_3 of degree r_1,r_2,r_3, the graded Jacobi identity takes the following form

$$(-1)^{r_1r_3}[\phi_1,[\phi_2,\phi_3]] + (-1)^{r_2r_1}[\phi_2,[\phi_3,\phi_1]] + (-1)^{r_3r_2}[\phi_3,[\phi_1,\phi_2]] = 0\,, \tag{A.13}$$

which can also be written, in view of the graded skew-symmetry of $[\cdot\,,\cdot]$, as

$$[\phi_1,[\phi_2,\phi_3]] = [[\phi_1,\phi_2],\phi_3] + (-1)^{r_1r_2}[\phi_2,[\phi_1,\phi_3]] \,.$$

In the language of Section A.5, this means that $[\phi_1,\cdot]$, which is a graded linear map of degree r_1, is a graded derivation of $[\cdot\,,\cdot]$.

We have in this appendix already met the following three examples of graded algebras, associated to an R-module V.

$(T^\bullet V, \otimes)$ is a graded R-algebra, which is associative;
$(\wedge^\bullet V, \wedge)$ is a graded R-algebra, which is associative and graded commutative;
$(S^\bullet V, \cdot)$ is a graded R-algebra, which is associative and commutative.

The grading on $T^\bullet V$ comes from the natural decomposition $T^\bullet V = \oplus_{i\in\mathbb{N}} V^{\otimes i}$; for $\wedge^\bullet V$ and $S^\bullet V$, the induced grading is used. The commutativity of the graded algebra $(S^\bullet V, \cdot)$ should not be confused with the *graded* commutativity of the graded algebra $(\wedge^\bullet V, \wedge)$: in the former the arguments commute, but in the latter they only commute up to a sign, see (A.6). We think of $T^\bullet$ as a functor: given a linear map $\phi \in \mathrm{Hom}(V,W)$, we obtain a homomorphism of graded algebras $T^\bullet\phi : T^\bullet V \to T^\bullet W$, whose restriction to $T^k V$ is the linear map $T^k V \to T^k W$, defined by $T^k\phi := \phi \otimes \phi \otimes \cdots \otimes \phi$, i.e.,

$$T^k\phi\,(v_1 \otimes v_2 \otimes \cdots \otimes v_k) := \phi(v_1) \otimes \phi(v_2) \otimes \cdots \otimes \phi(v_k)\,.$$

$T^\bullet$ is a covariant functor: for $\phi \in \mathrm{Hom}(V,W)$ and $\psi \in \mathrm{Hom}(W,Z)$ one obtains

$$T^\bullet(\mathbb{1}_V) = \mathbb{1}_{T^\bullet V}\,, \qquad T^\bullet(\psi\circ\phi) = T^\bullet(\psi)\circ T^\bullet(\phi)\,.$$

Similarly, $\wedge^\bullet$ and $S^\bullet$ are covariant functors.

Associated to a graded vector space $V = \bigoplus_{i\in\mathbb{Z}} V_i$ there is a graded exterior algebra which takes into account the grading on V. This algebra, denoted by $\wedge^\bullet V$ is the associative, graded commutative algebra, obtained by dividing the tensor algebra $T^\bullet V = \bigoplus_{k\in\mathbb{N}} T^k V$ of V by the ideal generated by the elements of the form $x \otimes y + (-1)^{ij} y \otimes x$, with $x \in V_i$ and $y \in V_j$. Denoting by $\wedge$ the product in $\wedge^\bullet V$, one then has

$$x \wedge y = -(-1)^{ij} y \wedge x\,,$$

for all $x \in V_i$ and $y \in V_j$. For integers $i_1, \ldots, i_k$ and for $\sigma \in \mathscr{S}_k$ an arbitrary permutation of $\{1, \ldots, k\}$, let $\mathrm{sgn}(\sigma; i_1, \ldots, i_k) \in \{1, -1\}$ denote the sign, defined by the equality

$$x_1 \wedge \cdots \wedge x_k = \mathrm{sgn}(\sigma; i_1, \ldots, i_k)\, x_{\sigma(1)} \wedge \cdots \wedge x_{\sigma(k)}\,,$$

where x_ℓ is an arbitrary homogeneous element of V of degree i_ℓ, for $\ell = 1, \ldots, k$. As in the ungraded case, if $\phi : V^{\otimes k} \to W$ is a linear map, satisfying

$$\phi(x_1 \otimes \cdots \otimes x_k) = \mathrm{sgn}(\sigma; i_1, \ldots, i_k)\, \phi(x_{\sigma(1)} \otimes \cdots \otimes x_{\sigma(k)})\,,$$

for all homogeneous elements $x_1, \ldots, x_k$, with i_ℓ denoting the degree of x_ℓ for $\ell = 1, \ldots, k$, then ϕ descends to a linear map $\wedge^\bullet V \to W$, which we also denote by ϕ.

A.4 Coalgebras and Graded Coalgebras

We now consider the "co"-versions of the concepts which were introduced in the previous section. This is done in the customary way, namely we obtain the "co"-versions by dualizing the definitions of the previous section, written in their functional forms (see Tables A.1 and A.2), which is done by reversing all arrows and switching the order of their composition. Let us first dualize the definition of an algebra: a *coalgebra* structure on an R-module V is an element Δ of $\mathrm{Hom}(V, V\otimes V)$, called a *coproduct*. The most important additional properties which Δ may have, are summarized in Table A.3.

Table A.3 Dualizing the usual properties of the product of an algebra, written in functional form, as in Table A.1, we obtain the usual properties which the coproduct Δ, defining a coalgebra structure on an R-module, can have. As before, S is the twist map and $\mathscr{S}$ is the cycle map.

Property	Functional form
cocommutative	$S\circ\Delta=\Delta$
co-skew-symmetric	$S\circ\Delta=-\Delta$
coassociative	$(\mathbb{1}_V\otimes\Delta)\circ\Delta=(\Delta\otimes\mathbb{1}_V)\circ\Delta$
co-Jacobi identity	$\Sigma_{\ell=0}^{2}\mathscr{S}^{\ell}\circ(\mathbb{1}_V\otimes\Delta)\circ\Delta=0$

In order to define a *homomorphism of coalgebras*, we just reverse the arrows in (A.10): for given coalgebras (V,Δ) and (W,Δ') we call a linear map $\phi: W\to V$ a homomorphism if the following diagram is commutative.

$$\begin{array}{ccc} V\otimes V & \xleftarrow{\;\Delta\;} & V \\ \Big\uparrow{\scriptstyle \phi\otimes\phi} & & \Big\uparrow{\scriptstyle \phi} \\ W\otimes W & \xleftarrow[\;\Delta'\;]{} & W \end{array}$$

We now consider the graded versions of the above "co"-concepts. Let V be a graded R-module, $V=\oplus_{i\in\mathbb{Z}}V_i$. A *graded coproduct* is a coproduct Δ on V such that for every $k\in\mathbb{Z}$,

$$\Delta(V_k)\subset\bigoplus_{i+j=k}V_i\otimes V_j\,,$$

is finitely supported, i.e., $\Delta(V_k)$ has a non-trivial intersection with only a finite number of $V_i\otimes V_j$. Notice that $\Delta(V_k)$ is automatically finitely supported when $V_i=\{0\}$ for all $i<0$. A graded R-module V, equipped with a graded coproduct Δ, is called a *graded R-coalgebra*. For fixed $i,j\in\mathbb{Z}$, we will need the linear map $\Delta_{i,j}: V_{i+j}\to V_i\otimes V_j$, which is obtained by composing Δ, restricted to V_{i+j}, with the natural projection $V \otimes V \to V_i \otimes V_j$. The usual properties of a graded coproduct are displayed in Table A.4.

Table A.4 This last table deals with the case of graded coalgebras. We recall that S is the twist map and $\mathscr{S}$ is the cycle map (see Table A.2).

Property	Functional form
graded cocommutative	$S \circ \Delta_{j,i} = (-1)^{ij} \Delta_{i,j}$
graded co-skew-symmetric	$S \circ \Delta_{j,i} = -(-1)^{ij} \Delta_{i,j}$
graded coassociative	$(\mathbb{1}_{V_i} \otimes \Delta_{j,k}) \circ \Delta_{i,j+k} = (\Delta_{i,j} \otimes \mathbb{1}_{V_k}) \circ \Delta_{i+j,k}$
graded co-Jacobi identity	$(-1)^{ik} \mathscr{S}^{\ell} \circ (\mathbb{1}_{V_i} \otimes \Delta_{j,k}) \circ \Delta_{i,j+k} + \circlearrowleft (i,j,k) = 0$

We have seen in the previous section that the graded R-modules $T^\bullet V$, $\wedge^\bullet V$ and $S^\bullet V$ have a natural (graded) algebra structure. We now show that they also have a graded coalgebra structure; the latter structure is important at a few places in this book. The coalgebra structure Δ on $T^\bullet V$ is called *de-concatenation* and is defined, for $k \in \mathbb{N}$ and for $v_1, \ldots, v_k \in V$, by

$$\Delta(v_1 \otimes \cdots \otimes v_k) := \sum_{i=0}^{k} (v_1 \otimes \cdots \otimes v_i) \otimes (v_{i+1} \otimes \cdots \otimes v_k), \tag{A.14}$$

which means that the linear maps $\Delta_{i,j}$ are given by

$$\begin{array}{rrl} \Delta_{i,j}: & V^{\otimes(i+j)} & \to V^{\otimes i} \otimes V^{\otimes j} \\ & v_1 \otimes \cdots \otimes v_{i+j} & \mapsto (v_1 \otimes \cdots \otimes v_i) \otimes (v_{i+1} \otimes \cdots \otimes v_{i+j}). \end{array} \tag{A.15}$$

Graded coassociativity of Δ is an immediate consequence of the associativity of $\otimes$:

$$\begin{aligned} &((\Delta_{i,j} \otimes \mathbb{1}_{V_{k-i-j}}) \circ \Delta_{i+j,k-i-j})(v_1 \otimes \cdots \otimes v_k) \\ &\quad = ((v_1 \otimes \cdots \otimes v_i) \otimes (v_{i+1} \otimes \cdots \otimes v_j)) \otimes (v_{j+1} \otimes \cdots \otimes v_k) \\ &\quad = (v_1 \otimes \cdots \otimes v_i) \otimes ((v_{i+1} \otimes \cdots \otimes v_j) \otimes (v_{j+1} \otimes \cdots \otimes v_k)) \\ &\quad = ((\mathbb{1}_{V_i} \otimes \Delta_{j,k-i-j}) \circ \Delta_{i,k-i})(v_1 \otimes \cdots \otimes v_k). \end{aligned}$$

We now turn to the natural coalgebra structure of the exterior algebra $\wedge^\bullet V$. Let us denote by δ the diagonal map $V \to V \times V : v \mapsto (v,v)$. By functoriality of $\wedge^\bullet$, it induces a linear map

$$\wedge^\bullet \delta : \wedge^\bullet V \to \wedge^\bullet (V \times V),$$

which we view as a linear map

$$\Delta := \rho \circ \wedge^\bullet \delta : \wedge^\bullet V \to \wedge^\bullet V \otimes \wedge^\bullet V,$$

where $\rho : \wedge^\bullet(V \times V) \to \wedge^\bullet V \otimes \wedge^\bullet V$ is the natural isomorphism, given by

$$\rho\big((v_1,0)\wedge\cdots\wedge(v_i,0)\wedge(0,v_{i+1})\wedge\cdots\wedge(0,v_k)\big) := (v_1\wedge\cdots\wedge v_i)\otimes(v_{i+1}\wedge\cdots\wedge v_k)\,,$$

where $v_1,\ldots,v_k \in V$. We denote the natural product on $\wedge^\bullet V \otimes \wedge^\bullet V$ which makes ρ into a homomorphism of graded algebras by $\triangle$ (not to be confused with Δ). It follows easily from the graded commutativity of $\wedge$ that, if $v_i \in \wedge^{r_i} V$ for $i = 1,\ldots,4$, then

$$(v_1\otimes v_2)\triangle(v_3\otimes v_4) = (-1)^{r_2 r_3}(v_1\wedge v_3)\otimes(v_2\wedge v_4)\,. \tag{A.16}$$

The commutative diagram

$$\begin{array}{ccc}
V & \xrightarrow{\ \delta\ } & V\times V \\
\Big\downarrow{\scriptstyle \delta} & & \Big\downarrow{\scriptstyle \delta\times\mathbb{1}_V} \\
V\times V & \xrightarrow[\ \mathbb{1}_V\times\delta\ ]{} & V\times V\times V
\end{array}$$

leads to a commutative diagram

$$\begin{array}{ccc}
\wedge^\bullet V & \xrightarrow{\ \wedge^\bullet\delta\ } & \wedge^\bullet(V\times V) \\
\Big\downarrow{\scriptstyle \wedge^\bullet\delta} & & \Big\downarrow{\scriptstyle \wedge^\bullet(\delta\times\mathbb{1}_V)} \\
\wedge^\bullet(V\times V) & \xrightarrow[\ \wedge^\bullet(\mathbb{1}_V\times\delta)\ ]{} & \wedge^\bullet(V\times V\times V)
\end{array}$$

which, in terms of Δ, becomes the coassociativity property of the coproduct Δ.

$$\begin{array}{ccc}
\wedge^\bullet V & \xrightarrow{\ \Delta\ } & \wedge^\bullet V\otimes\wedge^\bullet V \\
\Big\downarrow{\scriptstyle \Delta} & & \Big\downarrow{\scriptstyle \Delta\otimes\mathbb{1}_{\wedge^\bullet V}} \\
\wedge^\bullet V\otimes\wedge^\bullet V & \xrightarrow[\ \mathbb{1}_{\wedge^\bullet V}\otimes\Delta\ ]{} & \wedge^\bullet V\otimes\wedge^\bullet V\otimes\wedge^\bullet V
\end{array}$$

The graded algebra structure on $\wedge^\bullet V\otimes\wedge^\bullet V\otimes\wedge^\bullet V$ is defined as in (A.16). An explicit formula for Δ is given by

$$\begin{aligned}
\Delta(v_1\wedge v_2\wedge\cdots\wedge v_k) &= \rho(\wedge^\bullet\delta(v_1\wedge\cdots\wedge v_k)) = \rho(\delta(v_1)\wedge\cdots\wedge\delta(v_k)) \\
&= \rho\big(((v_1,0)+(0,v_1))\wedge\cdots\wedge((v_k,0)+(0,v_k))\big) \\
&= (v_1\otimes 1 + 1\otimes v_1)\triangle\cdots\triangle(v_k\otimes\mathbb{1}_V + \mathbb{1}_V\otimes v_k)
\end{aligned}$$

$$= \sum_{i+j=k} \sum_{\sigma \in S_{i,j}} \operatorname{sgn}(\sigma)(v_{\sigma(1)} \wedge \cdots \wedge v_{\sigma(i)}) \otimes (v_{\sigma(i+1)} \wedge \cdots \wedge v_{\sigma(k)}) .$$

In particular, $\Delta_{i,j}$ is given by

$$\Delta_{i,j}(v_1 \wedge \cdots \wedge v_{i+j}) = \sum_{\sigma \in S_{i,j}} \operatorname{sgn}(\sigma)(v_{\sigma(1)} \wedge \cdots \wedge v_{\sigma(i)}) \otimes (v_{\sigma(i+1)} \wedge \cdots \wedge v_{\sigma(i+j)}) , \tag{A.17}$$

when $i, j \in \mathbb{N}$ and $\Delta_{i,j} = 0$ when $i < 0$ or $j < 0$. The above formula is similar to the de-concatenation formula (A.14) which we have introduced in the case of the tensor algebra.

A.5 Graded Derivations and Coderivations

Let (V, μ) be a (not necessarily associative) graded algebra, where $V = \oplus_{i\in\mathbb{Z}} V_i$ is a graded R-module. A graded linear map $\phi \in \operatorname{Hom}_r(V,V)$ of degree r is said to be a *graded derivation* of degree r if

$$\phi(vw) = \phi(v)w + (-1)^{rp} v\phi(w) ,$$

for all $v \in V_p$ and $w \in V_q$, where vw is a shorthand for $\mu(v,w)$. In functional notation this means that for every $p, q \in \mathbb{Z}$,

$$\phi \circ \mu_{p,q} = \mu_{p+r,q} \circ (\phi \otimes \mathbb{1}_{V_q}) + (-1)^{rp} \mu_{p,r+q} \circ (\mathbb{1}_{V_p} \otimes \phi) , \tag{A.18}$$

as linear maps $V_p \otimes V_q \to V_{p+r+q}$. We denote the R-module of all graded derivations of degree r of V by $\operatorname{Der}_r(V)$. It is easily verified by direct computation that the graded commutator of two graded derivations of degrees r_1 and r_2 is a graded derivation of degree $r_1 + r_2$. This implies that $\oplus_{r\in\mathbb{Z}} \operatorname{Der}_r(V)$ is a graded Lie algebra, with the graded commutator as Lie bracket. It is a Lie subalgebra of the graded Lie algebra $\oplus_{r\in\mathbb{Z}} \operatorname{Hom}_r(V,V)$, equipped with the graded commutator, which we considered in Section A.3. Clearly, a derivation is completely determined by its values on an arbitrary set of elements of V, which generates V as a (graded) algebra.

Example A.3. Let V be an R-module and consider the graded algebra $\wedge^\bullet V$. Equipped with the graded commutator, the R-module $\oplus_{r\in\mathbb{Z}} \operatorname{Der}_r(\wedge^\bullet V)$ is a graded Lie algebra. Particular elements of $\operatorname{Der}_r(\wedge^\bullet V)$ can be constructed from linear maps $V \to \wedge^{r+1} V$: every linear map

$$\phi : V \to \wedge^{r+1} V$$

extends to a derivation of degree r of the graded algebra $(\wedge^\bullet V, \wedge)$ by putting

$$\begin{aligned} \tilde{\phi} : \quad \wedge^\bullet V &\to \wedge^{\bullet + r} V \\ v_1 \wedge \cdots \wedge v_k &\mapsto \sum_{i=1}^{k} (-1)^{i-1} \phi(v_i) \wedge v_1 \wedge \cdots \wedge \widehat{v_i} \wedge \cdots \wedge v_k . \end{aligned}$$

Indeed, for $X \in \wedge^p V$ and $Y \in \wedge^q V$,

$$\tilde{\phi}(X \wedge Y) = \tilde{\phi}(X) \wedge Y + (-1)^{pr} X \wedge \tilde{\phi}(Y),$$

which is an easy consequence of the fact that ϕ will either be applied to a factor which appears in X, leaving Y untouched, or vice versa.

We now formulate the notion of a derivation for the case of a graded coalgebra. As before, (V, Δ) is a graded coalgebra, where $V = \oplus_{i \in \mathbb{Z}} V_i$ is a graded R-module. Dualizing (A.18), a linear map $\phi : V \to V$ of degree $-r$ is called a *graded coderivation* of degree r if for every $p, q \in \mathbb{Z}$,

$$\Delta_{p,q} \circ \phi = (\phi \otimes \mathbb{1}_{V_q}) \circ \Delta_{p+r,q} + (-1)^{rp} (\mathbb{1}_{V_p} \otimes \phi) \circ \Delta_{p,r+q} \, ,$$

as maps from V_{p+r+q} to $V_p \otimes V_q$. We denote the R-module of all graded coderivations of degree r of V by $\mathrm{CoDer}_r(V)$. Again, it follows by direct computation that the graded commutator of two graded coderivations of degrees r_1 and r_2 is a graded coderivation of degree $r_1 + r_2$. This implies that $\oplus_{r \in \mathbb{Z}} \mathrm{CoDer}_r(V)$ is also a graded Lie algebra, with the graded commutator as Lie bracket. Like $\oplus_{r \in \mathbb{Z}} \mathrm{Der}_r(V)$, it is a Lie subalgebra of the graded Lie algebra $\oplus_{r \in \mathbb{Z}} \mathrm{Hom}_r(V, V)$, equipped with the graded commutator. The following example is the "co"-version of Example A.3.

Example A.4. Let V be an R-module and consider the graded coalgebra $\wedge^\bullet V$. Equipped with the graded commutator, the R-module $\oplus_{r \in \mathbb{Z}} \mathrm{CoDer}_r(\wedge^\bullet V)$ is a graded Lie algebra. Particular elements of $\mathrm{CoDer}_r(\wedge^\bullet V)$ can be constructed from linear maps $\wedge^{r+1} V \to V$: every linear map

$$\phi : \wedge^{r+1} V \to V$$

extends to a coderivation of degree r of the graded R-coalgebra $(\wedge^\bullet V, \Delta)$ by putting

$$\begin{aligned} \tilde{\phi} : \quad \wedge^\bullet V \quad &\to \wedge^{\bullet - r} V \\ v_1 \wedge \cdots \wedge v_k &\mapsto \sum_{\tau \in S_{r+1,k-1-r}} \mathrm{sgn}(\tau)\, \phi(v_{\tau(1)}, \ldots, v_{\tau(r+1)}) \wedge v_{\tau(r+2)} \wedge \cdots \wedge v_{\tau(k)} \, . \end{aligned}$$

It is understood that the above definition means that $\tilde{\phi} = 0$ on $\wedge^k V$ for $k \leqslant r$. It follows that, for $p, q \in \mathbb{N}$ we have that

$$\Delta_{p,q} \circ \tilde{\phi} = (\tilde{\phi} \otimes \mathbb{1}_{\wedge^q V}) \circ \Delta_{p+r,q} + (-1)^{rp} (\mathbb{1}_{\wedge^p V} \otimes \tilde{\phi}) \circ \Delta_{p,r+q} \, . \tag{A.19}$$

All coderivations of $\wedge^\bullet V$ are obtained in this way: if $\Phi : \wedge^\bullet V \to \wedge^{\bullet - r} V$ is a coderivation of degree r of $(\wedge^\bullet V, \Delta)$, then $\Phi = \tilde{\phi}$, where $\phi : \wedge^{r+1} V \to V$ is the restriction of Φ to $\wedge^{r+1} V$. Indeed, since Φ and $\tilde{\phi}$ agree on $\wedge^k V$, for $k \leqslant r+1$ (they are both zero when $k \leqslant r$, for degree reasons), they also agree, in view of (A.19), on $\wedge^{r+2} V$ (take $p = q = 1$ in (A.19)), and similarly for the higher exterior powers of V.

Appendix B
Real and Complex Differential Geometry

In this appendix we recall the basic notions of differential geometry: the definition of a real manifold, of a complex manifold and of a vector field on such a manifold. We also recall briefly the main properties of vector fields on manifolds: the existence of integral curves of a vector field, the flow of a vector field, the bracket of vector fields and the straightening theorem, which says that a vector field takes, in well-chosen coordinates, a simple form. Our definition of vector fields on a manifold is based on the concept of a pointwise derivation. This approach easily generalizes to the introduction of the concept of a bivector field on a manifold, a crucial element in the (geometrical!) definition of the notion of a Poisson structure on a (real or complex) manifold (see Section 1.3).

B.1 Real and Complex Manifolds

We adopt the following geometric point of view: a differentiable manifold is a (second countable, Hausdorff) topological space which is covered by a family of *coordinate charts* (U,x), where U is an open subset of M, called the *domain* of the chart, and $x = (x_1,\ldots,x_d)$ is a homeomorphism from U to an open subset of $\mathbb{R}^d$. The functions $x_1,\ldots,x_d$ are called *local coordinates*; they are said to be *centered* at m if $x(m) = o$, where o stands for the origin of $\mathbb{R}^d$. The coordinate charts are demanded to be compatible in the sense that, if (U,x) and (V,y) are coordinate charts, with $U \cap V \neq \emptyset$, then the homeomorphism

$$y \circ x^{-1}\big|_{x(U\cap V)} : x(U \cap V) \to y(U \cap V)$$

is a smooth map, see Fig. B.1. This homeomorphism is called a *transition map*. A collection of compatible coordinate charts of M, whose domains cover M, is called an *atlas* of M. For connected differentiable manifolds, the integer d is independent of the coordinate chart; it is called the *dimension* of M, a terminology which we also use in the non-connected case, when d is independent of the coordinate chart; it is

C. Laurent-Gengoux et al., *Poisson Structures*,
Grundlehren der mathematischen Wissenschaften 347,
DOI 10.1007/978-3-642-31090-4, © Springer-Verlag Berlin Heidelberg 2013

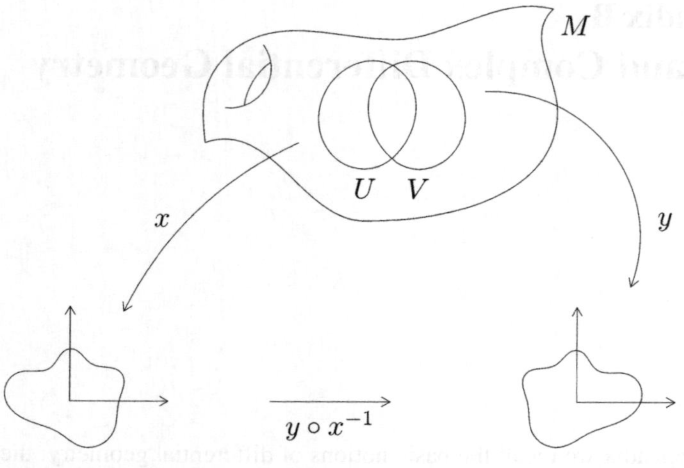

Fig. B.1 A manifold comes equipped with an atlas, a collection of coordinate charts (U,x), where the transition maps $y \circ x^{-1}$ between coordinate charts (U,x) and (V,y) (with $U \cap V \neq \emptyset$) are demanded to be smooth.

denoted by $\dim M$. When the integer d is even, we may interpret the homeomorphisms x as taking values in $\mathbb{C}^{d/2}$; in this case, if all transition maps are complex analytic (holomorphic), then M is called a *complex manifold* and $d/2$ is called the (complex) *dimension* of M. It is a trivial, but important, fact that every non-empty open subset of a real or complex manifold is itself, in a natural way, a real or complex manifold. In particular, every open subset of a (real or complex) vector space is a (real or complex) manifold.

The main virtue of manifolds is that we can do calculus on them, hence also analytic geometry. Roughly speaking, the coordinate charts allow us to identify objects on the manifold, locally, with standard objects on open subsets of $\mathbb{F}^d$ ($\mathbb{F} = \mathbb{R}$ or $\mathbb{F} = \mathbb{C}$) and the transition maps allow us to compare these standard objects. The first object one thinks of is that of a smooth function: a function $F : M \to \mathbb{F}$ on a real (respectively complex) manifold M is called a *smooth function* (respectively a *holomorphic function*) if for every coordinate chart (U,x) of M the function

$$\tilde{F} = F \circ x^{-1} : x(U) \to \mathbb{F}$$

is smooth (respectively holomorphic). The function $\tilde{F}$ is called the *coordinate expression* of F in the coordinate chart (U,x). See Fig. B.2. The (commutative associative) algebra of all such functions F on M will be denoted by $\mathscr{F}(M)$. For an open subset U, viewing U itself as a manifold, we have an algebra of functions $\mathscr{F}(U)$ and there are obvious restriction maps $\mathscr{F}(M) \to \mathscr{F}(U)$, which are in general neither injective nor surjective; the restriction of $F \in \mathscr{F}(M)$ to U will be denoted by $F_{|U}$. Notice that $\mathscr{F}(M)$ may consist only of the constant functions, for example when M is a compact complex manifold.

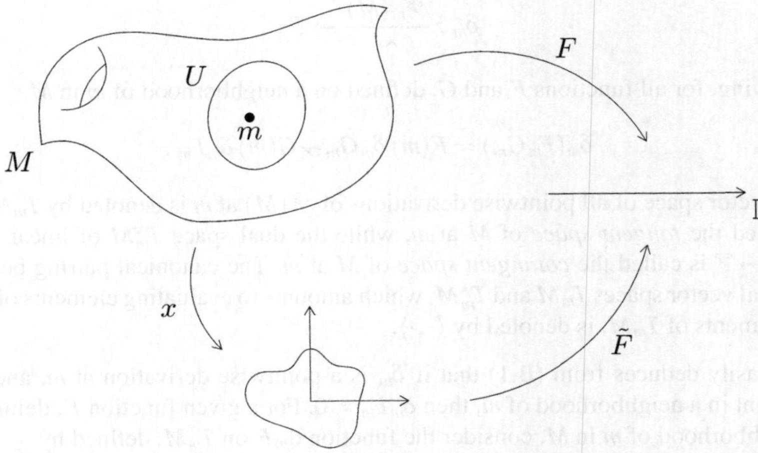

Fig. B.2 Smooth functions on a manifold M are functions on M which are smooth in terms of local coordinates.

One similarly defines the notion of a *smooth map* between real manifolds and a *holomorphic map* between complex manifolds. It leads to two categories: the category of real manifolds, whose objects are real manifolds with smooth maps as morphisms, and the category of complex manifolds, whose objects are complex manifolds and whose morphisms are holomorphic maps. Since complex manifolds are in a natural way also real manifolds, and since holomorphic maps are smooth, there is a natural forgetful functor from the latter category to the former.

B.2 The Tangent Space

Let M be a manifold and let $m \in M$ be an arbitrary point. We define the tangent space $T_m M$ of M at m. To do this, we consider the set $\mathscr{F}_m(M)$ of all pairs (F,U), where U is an open subset of M which contains m, and F is an element of $\mathscr{F}(U)$. Two elements (F,U) and (G,V) of $\mathscr{F}_m(M)$ are defined to be equivalent, denoted $(F,U) \sim (G,V)$, if there exists a pair $(H,W) \in \mathscr{F}_m(M)$, such that $W \subset U \cap V$ and $H = F|_W = G|_W$. For $(F,U) \in \mathscr{F}_m(M)$, we denote its equivalence class by F_m and we call it the *germ* of F (or of (F,U)) at m. It is clear that the quotient set $\mathscr{F}_m(M)/\sim$ of all function germs at m inherits from $\mathscr{F}(M)$ the structure of an associative $\mathbb{F}$-algebra. For example, in the complex case, the algebra $\mathscr{F}_m(M)/\sim$ is isomorphic to the algebra of power series in d variables, whose radius of convergence is positive. Notice that a function germ F_m has a well-defined value at m, which is simply $F(m)$.

Definition B.1. Let M be a manifold and let $m \in M$. A *pointwise derivation* δ_m of $\mathscr{F}(M)$ at m is a linear function

$$\delta_m : \frac{\mathscr{F}_m(M)}{\sim} \to \mathbb{F} \,,$$

satisfying, for all functions F and G, defined on a neighborhood of m in M,

$$\delta_m(F_m G_m) = F(m)\,\delta_m G_m + G(m)\,\delta_m F_m \,. \tag{B.1}$$

The vector space of all pointwise derivations of $\mathscr{F}(M)$ at m is denoted by $T_m M$, and is called the *tangent space* of M at m, while the dual space $T_m^* M$ of linear forms $T_m M \to \mathbb{F}$ is called the *cotangent space* of M at m. The canonical pairing between the dual vector spaces $T_m M$ and $T_m^* M$, which amounts to evaluating elements of $T_m^* M$ on elements of $T_m M$, is denoted by $\langle \cdot , \cdot \rangle$.

One easily deduces from (B.1) that if δ_m is a pointwise derivation at m, and F is constant in a neighborhood of m, then $\delta_m F_m = 0$. For a given function F, defined on a neighborhood of m in M, consider the function $\mathrm{d}_m F$ on $T_m M$, defined by

$$\begin{aligned} \mathrm{d}_m F \,:\, T_m M &\to \mathbb{F} \\ \delta_m &\mapsto \delta_m F_m. \end{aligned} \tag{B.2}$$

Clearly, $\mathrm{d}_m F$ is a linear function, hence it is an element of the cotangent space $T_m^* M$. It is called the *differential* of F at m. The differential of a function admits a natural generalization to the case of maps between manifolds. Let M and N be two manifolds and let Ψ be a map, defined on a neighborhood of a point $m \in M$, with values in N. The linear map $T_m\Psi : T_m M \to T_{\Psi(m)} N$, called the *tangent map* of Ψ at m, associates to a pointwise derivation δ_m of $\mathscr{F}(M)$ at m, the pointwise derivation $T_m\Psi(\delta_m)$ of $\mathscr{F}(N)$ at $\Psi(m)$, defined for every germ $G_{\Psi(m)}$ at $\Psi(m)$ by

$$(T_m\Psi)(\delta_m)\,G_{\Psi(m)} := \delta_m (G\circ\Psi)_m \,,$$

see Fig. B.3. This is well-defined, because the germ $(G\circ\Psi)_m$ is independent of the function G which represents the germ $G_{\Psi(m)}$. When F is a function, defined on a neighborhood of m, the tangent map at m is a linear map $T_m F : T_m M \to T_{F(m)}\mathbb{F}$, and we can recover the differential $\mathrm{d}_m F$, upon composing $T_m F$ with the canonical isomorphism $T_{F(m)}\mathbb{F} \simeq \mathbb{F}$, which will be explained below.

The tangent map obeys the usual rules of calculus: for example, if M, N and P are manifolds, $\Psi : M \to N$ and $\Xi : N \to P$ are maps and $m \in M$, then

$$T_m(\Xi\circ\Psi) = (T_{\Psi(m)}\Xi)\circ T_m\Psi \,.$$

In particular, if Ψ is a diffeomorphism (at least in the neighborhood of m), then $T_m\Psi$ is invertible and

$$(T_m\Psi)^{-1} = T_{\Psi(m)}\Psi^{-1} \,.$$

Consider the vector space $\mathbb{F}^d$, viewed as a d-dimensional manifold, and let m be a point of $\mathbb{F}^d$. There is a natural isomorphism between the vector spaces $T_m\mathbb{F}^d$ and $\mathbb{F}^d$. Namely to a vector $v \in \mathbb{F}^d$ we can associate a pointwise derivation v_m at m by setting,

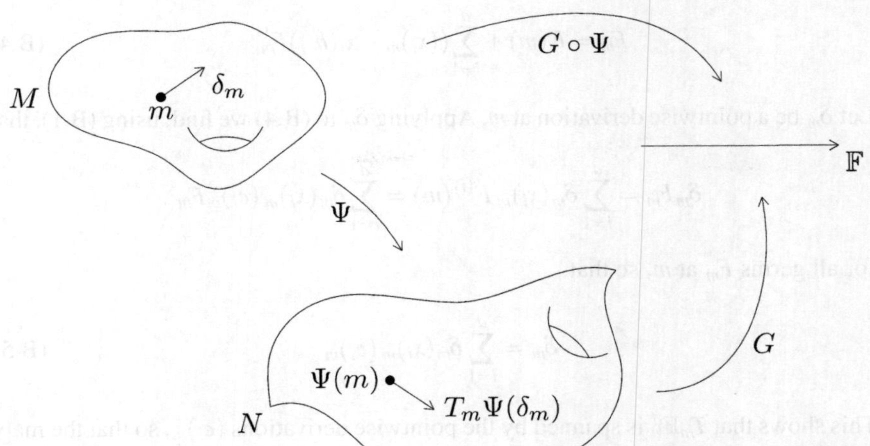

Fig. B.3 A map Ψ between two manifolds M and N leads for every point $m \in M$ to a tangent map $T_m\Psi$, which is a linear map between the tangent space T_mM and $T_{\Psi(m)}N$.

for every germ F_m at m,

$$v_m F_m := \frac{\mathrm{d}}{\mathrm{d}t}_{|t=0} F(m+tv) = \lim_{t\to 0} \frac{F(m+tv)-F(m)}{t} .$$

It is clear that $v \mapsto v_m$ defines an injective linear map from $\mathbb{F}^d$ to $T_m\mathbb{F}^d$. The surjectivity of this map follows from the Hadamard lemma.

Lemma B.2 (Hadamard's lemma). *Let $F \in \mathscr{F}(U)$, where U is an open subset of $\mathbb{F}^d$ and let $m \in U$. On a small neighborhood $V \subset U$ of m in $\mathbb{F}^d$,*

$$F = F(m) + \sum_{i=1}^{d} (x_i - x_i(m))\, F^{(i)} , \tag{B.3}$$

where each of the functions $F^{(i)}$ belongs to $\mathscr{F}(V)$. In particular,

$$F^{(i)}(m) = (e_i)_m F_m ,$$

where $(e_1, \ldots, e_d)$ denotes the natural basis of $\mathbb{F}^d$.

For a proof of this lemma, which is essentially a first-order form of Taylor's theorem, see [155, p. 17].

Let $F \in \mathscr{F}(U)$ and consider (B.3), germified at $m \in U$,

$$F_m = F(m) + \sum_{i=1}^{d} ((x_i)_m - x_i(m)) F_m^{(i)} . \tag{B.4}$$

Let δ_m be a pointwise derivation at m. Applying δ_m to (B.4) we find, using (B.1), that

$$\delta_m F_m = \sum_{i=1}^{d} \delta_m(x_i)_m F^{(i)}(m) = \sum_{i=1}^{d} \delta_m(x_i)_m (e_i)_m F_m ,$$

for all germs F_m at m, so that

$$\delta_m = \sum_{i=1}^{d} \delta_m(x_i)_m (e_i)_m . \tag{B.5}$$

This shows that $T_m\mathbb{F}^d$ is spanned by the pointwise derivations $(e_i)_m$, so that the map, defined by $v \mapsto v_m$, is an isomorphism between $\mathbb{F}^d$ and $T_m\mathbb{F}^d$.

If (U,x) is a coordinate chart of a manifold M, centered at m, then each of the d vectors of the natural basis $(e_1,\dots,e_d)$ of $\mathbb{F}^d$ leads to a pointwise derivation of $\mathscr{F}(M)$ at m, defined by

$$\left(\frac{\partial}{\partial x_i}\right)_m := (T_o x^{-1})\,(e_i)_o .$$

For a function F, defined on an open neighborhood of m in M, this means that

$$\left(\frac{\partial}{\partial x_i}\right)_m F_m = \partial_i \tilde{F}(x(m)) ,$$

where $\tilde{F} : x(U) \to \mathbb{F}$ is the coordinate expression of a representative $F \in \mathscr{F}(U)$ of the germ F_m (see Fig. B.2), and $1 \leqslant i \leqslant d$; also, $\partial_i \tilde{F}$ denotes the derivative of $\tilde{F}$ with respect to its i-th variable (real or complex). Since $T_o\mathbb{F}^d$ is spanned by the pointwise derivations $(e_1)_o,\dots,(e_d)_o$ and since $T_o x^{-1}$ is an isomorphism, the pointwise derivations $\left(\frac{\partial}{\partial x_1}\right)_m,\dots,\left(\frac{\partial}{\partial x_d}\right)_m$ span T_mM.

Remark B.3. When the tangent space of M at m is viewed as an equivalence class of curves, passing through m, as is done in a more analytic approach to elementary differential geometry, then the differential $\mathrm{d}_m F$ of a function $F : M \to \mathbb{F}$ at m, is

$$\mathrm{d}_m F(\tilde{\gamma}_m) := \frac{\mathrm{d}}{\mathrm{d}t}_{|t=0} F(\gamma(t))$$

where $\gamma : I \to M$ is a curve, defined on a neighborhood I of 0 in $\mathbb{F}$, with $\gamma(0) = m$, whose equivalence class has been denoted by $\tilde{\gamma}_m$. As we have seen, in our approach to defining the tangent space, the definition of the differential takes the more algebraic form

$$\langle \mathrm{d}_m F, \delta_m \rangle := \delta_m F_m ,$$

for all $\delta_m \in T_mM$. Thus, in our setup, we view tangent vectors as objects which act on equivalence classes of functions, rather than viewing functions as objects which define linear forms on equivalence classes of curves, although both points of view are equivalent (see [198, Ch. 1]).

B.3 Vector Fields

Let us consider a manifold M and a map $\mathscr{V}$, which assigns to every $m \in M$ an element $\mathscr{V}_m$ of T_mM. To each function $F \in \mathscr{F}(U)$, where U is an open subset of M, we can associate a function $\mathscr{V}[F]$ on U by defining, for all $m \in U$,

$$\mathscr{V}[F](m) := \mathscr{V}_m F_m \in \mathbb{F} . \tag{B.6}$$

We also write $\mathscr{V}_m[F]$ for $\mathscr{V}_m F_m$, so that $\mathscr{V}_m[F] = \mathscr{V}[F](m)$. We say that $\mathscr{V}$ is a *smooth vector field* (respectively *holomorphic vector field*) on M if for every open subset $U \subset M$ and for every function $F \in \mathscr{F}(U)$, the function $\mathscr{V}[F]$, defined by (B.6), belongs to $\mathscr{F}(U)$ (i.e., it is a smooth, respectively holomorphic function on U). When the type of manifold which is considered is irrelevant or is clear from the context, we simply say *vector field* for smooth or holomorphic vector field. Notice that we use square brackets to denote the action of a vector field on a function.

With respect to pointwise multiplication, the vector fields on M form an $\mathscr{F}(M)$-module, which is denoted by $\mathfrak{X}^1(M)$. Viewed as a vector space, $\mathfrak{X}^1(M)$ is a Lie algebra, where the Lie bracket is the commutator of vector fields, defined as follows. Let $\mathscr{V}$ and $\mathscr{W}$ be vector fields on M, and let $m \in M$. For every function F, defined in a neighborhood of m, letting

$$[\mathscr{V},\mathscr{W}]_m(F_m) := \mathscr{V}_m(\mathscr{W}[F])_m - \mathscr{W}_m(\mathscr{V}[F])_m ,$$

leads to a well-defined linear map $[\mathscr{V},\mathscr{W}]_m : \mathscr{F}_m(M)/\sim \to \mathbb{F}$, which is easily shown to be a pointwise derivation at m. For given vector fields $\mathscr{V}$ and $\mathscr{W}$ on M, the map which assigns to every $m \in M$ the element $[\mathscr{V},\mathscr{W}]_m$ of the tangent space T_mM is a (smooth or holomorphic) vector field, hence we have a map $[\cdot,\cdot] : \mathfrak{X}^1(M) \times \mathfrak{X}^1(M) \to \mathfrak{X}^1(M)$. Clearly, $[\cdot,\cdot]$ is a skew-symmetric bilinear map, which satisfies the Jacobi identity, hence it defines a Lie algebra structure on $\mathfrak{X}^1(M)$; it is called the *Lie bracket* on vector fields.

We have seen that a map $\Psi : M \to N$ leads for every $m \in M$ to a linear map $T_m\Psi : T_mM \to T_{\Psi(m)}N$. However, since Ψ is in general neither injective nor surjective, this collection of linear maps cannot be used to associate to a vector field $\mathscr{V}$ on M, a vector field on N. Nevertheless, when Ψ is bijective, so that Ψ is a diffeomorphism (or biholomorphism), we get a vector field $\Psi_*\mathscr{V}$ on N by setting

$$(\Psi_*\mathscr{V})_{\Psi(m)} := (T_m\Psi)\mathscr{V}_m ,$$

for all $m \in M$. The vector field $\Psi_*\mathscr{V}$ is called the *pushforward* of $\mathscr{V}$ by Ψ.

It is clear that vector fields can be restricted to open subsets; we usually do not make a notational distinction between a vector field on M and its restriction to some open subset of M. It is also clear from (B.6) that $\mathscr{V}$ defines, for every open subset U of M, a derivation of $\mathscr{F}(U)$, i.e., we have

$$\mathscr{V}[FG] = F\mathscr{V}[G] + G\mathscr{V}[F] ,$$

for all $F,G \in \mathscr{F}(U)$. In particular, a vector field on M defines a derivation of $\mathscr{F}(M)$.

Remark B.4. It is shown in standard books on differential geometry that for a *real manifold* M, the above natural correspondence between (smooth) vector fields on M and derivations of $\mathscr{F}(M)$ is bijective. For complex manifolds however, this is not true in general: think of a compact complex torus $\mathbb{C}^d/\mathbb{Z}^{2d}$, which has non-trivial holomorphic vector fields, but whose algebra of holomorphic functions consists of constant functions only, so that all its derivations are trivial. The same phenomenon occurs for skew-symmetric biderivations and bivector fields (e.g., Poisson structures), introduced in Chapter 1.

Remark B.5. The set of all tangent vectors at m, for m ranging through M, has a natural vector bundle structure over M, denoted $TM \to M$. The fiber over m is the vector space T_mM and the vector fields on M can be defined as the (smooth, holomorphic) sections of $TM \to M$. In abstract geometrical constructions, it is the latter point of view on vector fields which is often the most appropriate.

On a coordinate chart (U,x) of M, there are d distinguished vector fields $\partial/\partial x_i$, $i = 1,\ldots,d$, which are defined by $m \mapsto \left(\frac{\partial}{\partial x_i}\right)_m$, for all $m \in U$, which amounts to defining

$$\frac{\partial}{\partial x_i}[F](m) := \partial_i \tilde{F}(x(m)) , \tag{B.7}$$

for all $F \in \mathscr{F}(M)$, where $\tilde{F} : x(U) \to \mathbb{F}$ is the coordinate expression of F, as in Fig. B.2. By a slight abuse of notation, we usually write $\frac{\partial F}{\partial x_i}(m)$ instead of either expression in (B.7). If $\mathscr{V}$ is a map which assigns to every $m \in U$ an element $\mathscr{V}_m \in T_mM$, then $\mathscr{V}$ can be written in a unique way as $\mathscr{V} = \sum_{i=1}^d \mathscr{V}^{(i)}\,\partial/\partial x_i$, since the pointwise derivations $\left(\frac{\partial}{\partial x_i}\right)_m$ form a basis of T_mM, for every $m \in U$. Then $\mathscr{V}$ is a (smooth) vector field on U if and only if the coefficients $\mathscr{V}^{(i)}$ in this expression belong to $\mathscr{F}(U)$. It is clear that these coefficients $\mathscr{V}^{(i)}$ are given by $\mathscr{V}^{(i)} = \mathscr{V}[x_i]$. By a slight abuse of language, we often refer to the expression

$$\mathscr{V} = \sum_{i=1}^{d} \mathscr{V}[x_i] \frac{\partial}{\partial x_i} \tag{B.8}$$

as a *coordinate expression* of $\mathscr{V}$ in the coordinate chart (U,x). It is very useful for explicit computations, as it allows us to compute with vector fields on a manifold, locally, in the same way as on $\mathbb{F}^d$.

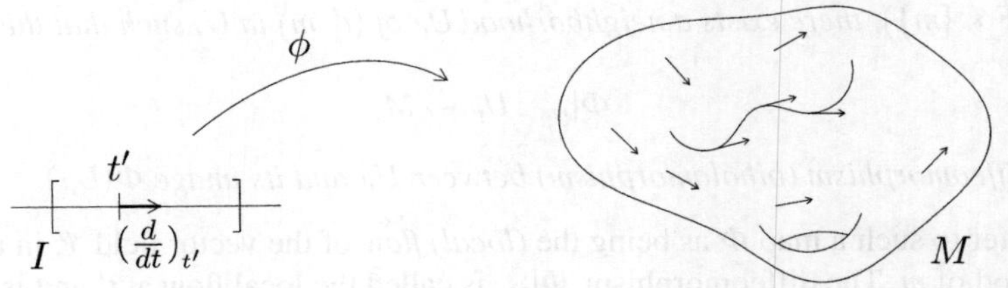

Fig. B.4 For a given vector field on a manifold, there passes through every point of the manifold a unique integral curve.

B.4 The Flow of a Vector Field

Let $\mathscr{V}$ be a vector field on a manifold M and let $m \in M$. The fundamental theorem on the existence and uniqueness of solutions of first order ordinary differential equations with initial conditions, tells us that there exists a connected open neighborhood I of 0 in $\mathbb{F}$, and there exists a map $\phi : I \to M$, such that $\phi(0) = m$ and such that

$$(T_{t'}\phi)\left(\frac{\mathrm{d}}{\mathrm{d}t}\right)_{t'} = \mathscr{V}_{\phi(t')}, \tag{B.9}$$

for all $t' \in I$; the map is unique in the sense that if $\phi_1 : I_1 \to M$ and $\phi_2 : I_2 \to M$ are two such maps, then they coincide on $(I_1 \cap I_2)^0$, the connected component of their common intersection, which contains 0,

$$\phi_1|_{(I_1\cap I_2)^0} = \phi_2|_{(I_1\cap I_2)^0}.$$

The map ϕ or the pair (I, ϕ) is called an *integral curve* of $\mathscr{V}$, passing through m, see Fig. B.4. By a slight abuse of notation, the left-hand side in (B.9) is often denoted by $\frac{\mathrm{d}\phi}{\mathrm{d}t}(t')$; using this notation (B.9) takes the more familiar form

$$\frac{\mathrm{d}\phi}{\mathrm{d}t}(t') = \mathscr{V}_{\phi(t')}$$

The integral curves of a vector field depend smoothly on the initial data; this is stated in a precise way in the following theorem (see [187, Ch. 5] for a proof).

Theorem B.6. *Let $\mathscr{V}$ be a vector field on a manifold M and let $m \in M$. There exists a neighborhood U of $(0, m)$ in $\mathbb{F} \times M$ and there exists a map $\Phi : U \to M$ such that, for every $(0, m') \in U$, the restriction*

$$\Phi|_{I_{m'}} : I_{m'} \to M, \tag{B.10}$$

is an integral curve of $\mathscr{V}$, passing through m'; the subset $I_{m'}$ in (B.10) is the connected component of $U \cap (\mathbb{F} \times \{m'\})$ which contains $(0, m')$. Moreover, Φ has the following property: for every (t', m) in the connected component of $(0, m)$ in

$U \cap (\mathbb{F} \times \{m\})$, there exists a neighborhood $U_{t'}$ of (t', m) in U, such that the restriction

$$\Phi|_{U_{t'}} : U_{t'} \to M ,$$

is a diffeomorphism (biholomorphism) between $U_{t'}$ and its image $\Phi(U_{t'})$.

We refer to such a map Φ as being the *(local) flow* of the vector field $\mathscr{V}$ in a neighborhood of m. The diffeomorphism $\Phi|_{U_{t'}}$ is called the local flow at t' and is usually denoted by $\Phi_{t'}$ (omitting its domain of definition, which is all of M in good cases, for example when M is a compact real manifold).

An important and useful consequence of the (local) existence of the flow of a vector field, is the straightening theorem, which says that the coordinate expression of a vector field on a manifold takes a particularly simple form, at points where the vector field does not vanish.

Theorem B.7 (Straightening theorem). *Let M be a manifold and let $\mathscr{V}$ be a vector field on M. If $m \in M$ is such that $\mathscr{V}(m) \neq 0$, then there exist local coordinates $x_1, \ldots, x_d$ on a neighborhood U of m, such that $\mathscr{V} = \partial/\partial x_1$ on U.*

B.5 The Frobenius Theorem

Instead of having a vector at every point of a manifold M, as is the case of a vector field on M, one may have a one-dimensional subspace of the tangent space to M, at every point of M. This is what is called a 1-dimensional *distribution* on M; a k-dimensional distribution $\mathscr{D}$ on M is then the datum of a k-dimensional subspace $\mathscr{D}(m)$ of $T_m M$ for every $m \in M$. One says that $\mathscr{D}$ is *smooth* (or *holomorphic*) if there exist for every $m \in M$ smooth (or holomorphic) vector fields $\mathscr{V}_1, \ldots, \mathscr{V}_k$, on a neighborhood U of m, such that

$$\mathscr{D}(m) = \operatorname{span}\{(\mathscr{V}_1)_m, \ldots, (\mathscr{V}_k)_m\} ,$$

for every $m \in U$. When it is clear from the context, we often simply say *distribution* for smooth (or holomorphic) distribution. A vector field $\mathscr{V}$, defined on an open subset U of M, is said to be *adapted* to $\mathscr{D}$ on U if $\mathscr{V}_m \in \mathscr{D}(m)$ for every $m \in U$. A distribution $\mathscr{D}$ on M is said to be *involutive* if for every open subset U of M, and for every pair of vector fields on U, which are adapted to $\mathscr{D}$ on U, their Lie bracket is also adapted to $\mathscr{D}$ on U. Involutivity of a distribution is a very strong condition, as is plain from Frobenius' theorem.

Theorem B.8 (Frobenius' theorem). *Let M be a d-dimensional manifold and suppose that $\mathscr{D}$ is a k-dimensional distribution on M. If $\mathscr{D}$ is involutive, then every point $m \in M$ admits a coordinate chart (U, x), with $m \in U$, such that*

$$\mathscr{D}(m') = \operatorname{span}\left\{\left(\frac{\partial}{\partial x_1}\right)_{m'}, \ldots, \left(\frac{\partial}{\partial x_k}\right)_{m'}\right\}, \quad \textit{for every } m' \in U .$$

For a short and elementary proof, which is immediately adapted to the holomorphic case, we refer to [41] or [138]. It is clear that Frobenius' theorem can be seen as a generalization of the straightening theorem (Theorem B.7).

References

1. R. Abraham and J. Marsden. *Foundations of mechanics*. Benjamin/Cummings Publishing Co. Inc. Advanced Book Program, Reading, MA, 1978. Second edition, revised and enlarged, With the assistance of T. Ratiu and R. Cushman.
2. M. Adler. On a trace functional for formal pseudo differential operators and the symplectic structure of the Korteweg-de Vries type equations. *Invent. Math.*, 50(3):219–248, 1978/79.
3. M. Adler, P. van Moerbeke, and P. Vanhaecke. *Algebraic integrability, Painlevé geometry and Lie algebras*, volume 47 of *Ergebnisse der Mathematik und ihrer Grenzgebiete. 3. Folge. A Series of Modern Surveys in Mathematics [Results in Mathematics and Related Areas. 3rd Series. A Series of Modern Surveys in Mathematics]*. Springer-Verlag, Berlin, 2004.
4. A. Alekseev. A new proof of the convexity theorem for the Poisson-Lie moment map. In *Geometry and physics (Aarhus, 1995)*, volume 184 of *Lecture Notes in Pure and Appl. Math.*, pages 231–236. Dekker, New York, 1997.
5. A. Alekseev. On Poisson actions of compact Lie groups on symplectic manifolds. *J. Differential Geom.*, 45(2):241–256, 1997.
6. A. Alekseev and Y. Kosmann-Schwarzbach. Manin pairs and moment maps. *J. Differential Geom.*, 56(1):133–165, 2000.
7. A. Alekseev, Y. Kosmann-Schwarzbach, and E. Meinrenken. Quasi-Poisson manifolds. *Canad. J. Math.*, 54(1):3–29, 2002.
8. A. Alekseev and A. Malkin. Symplectic structure of the moduli space of flat connection[s] on a Riemann surface. *Comm. Math. Phys.*, 169(1):99–119, 1995.
9. A. Alekseev, A. Malkin, and E. Meinrenken. Lie group valued moment maps. *J. Differential Geom.*, 48(3):445–495, 1998.
10. A. Alekseev and E. Meinrenken. Clifford algebras and the classical dynamical Yang-Baxter equation. *Math. Res. Lett.*, 10(2–3):253–268, 2003.
11. A. Alekseev and E. Meinrenken. On the Kashiwara-Vergne conjecture. *Invent. Math.*, 164(3):615–634, 2006.
12. A. Alekseev, E. Meinrenken, and C. Woodward. Linearization of Poisson actions and singular values of matrix products. *Ann. Inst. Fourier (Grenoble)*, 51(6):1691–1717, 2001.
13. J. Andersen, J. Mattes, and N. Reshetikhin. The Poisson structure on the moduli space of flat connections and chord diagrams. *Topology*, 35(4):1069–1083, 1996.
14. D. Arnal, D. Manchon, and M. Masmoudi. Choix des signes pour la formalité de M. Kontsevich. *Pacific J. Math.*, 203(1):23–66, 2002.
15. V. Arnol'd. *Mathematical methods of classical mechanics*. Springer-Verlag, New York, 1978. Translated from the Russian by K. Vogtmann and A. Weinstein, Graduate Texts in Mathematics, 60.
16. V. Arnol'd. Remarks on Poisson structures on a plane and on other powers of volume elements. *J. Soviet Math.*, 47(3):2509–2516, 1989.

C. Laurent-Gengoux et al., *Poisson Structures*,
Grundlehren der mathematischen Wissenschaften 347,
DOI 10.1007/978-3-642-31090-4, © Springer-Verlag Berlin Heidelberg 2013

17. V. I. Arnol'd, S. M. Guseĭn-Zade, and A. N. Varchenko. *Singularities of differentiable maps. Vol. I*, volume 82 of *Monographs in Mathematics*. Birkhäuser Boston Inc., Boston, MA, 1985. The classification of critical points, caustics and wave fronts, Translated from the Russian by Ian Porteous and Mark Reynolds.
18. M. Audin. *Torus actions on symplectic manifolds*, volume 93 of Progress in Mathematics. Birkhäuser Verlag, Basel, revised edition, 2004.
19. F. Bayen, M. Flato, C. Fronsdal, A. Lichnerowicz, and D. Sternheimer. Deformation theory and quantization. I. Deformations of symplectic structures. *Ann. Physics*, 111(1):61–110, 1978.
20. F. Bayen, M. Flato, C. Fronsdal, A. Lichnerowicz, and D. Sternheimer. Deformation theory and quantization. II. Physical applications. *Ann. Physics*, 111(1):111–151, 1978.
21. F. A. Berezin. Quantization. *Izv. Akad. Nauk SSSR Ser. Mat.*, 38:1116–1175, 1974.
22. R. Bezrukavnikov and V. Ginzburg. On deformations of associative algebras. *Ann. of Math. (2)*, 166(2):533–548, 2007.
23. K. H. Bhaskara and K. Viswanath. *Poisson algebras and Poisson manifolds*, volume 174 of *Pitman Research Notes in Mathematics Series*. Longman Scientific & Technical, Harlow, 1988.
24. F. Bottacin. Poisson structures on moduli spaces of sheaves over Poisson surfaces. *Invent. Math.*, 121(2):421–436, 1995.
25. F. Bottacin. Poisson structures on Hilbert schemes of points of a surface and integrable systems. *Manuscripta Math.*, 97(4):517–527, 1998.
26. F. Bottacin. Poisson structures on moduli spaces of parabolic bundles on surfaces. *Manuscripta Math.*, 103(1):31–46, 2000.
27. N. Bourbaki. *Lie groups and Lie algebras. Chapters 1–3*. Elements of Mathematics (Berlin). Springer-Verlag, Berlin, 1998. Translated from the French, Reprint of the 1989 English translation.
28. J.-L. Brylinski. A differential complex for Poisson manifolds. *J. Differential Geom.*, 28(1):93–114, 1988.
29. D. Bump. *Lie groups*, volume 225 of *Graduate Texts in Mathematics*. Springer-Verlag, New York, 2004.
30. F. Butin. Poisson homology in degree 0 for some rings of symplectic invariants. *J. Algebra*, 322(10):3580–3613, 2009.
31. M. Cahen, S. Gutt, and J. Rawnsley. Nonlinearizability of the Iwasawa Poisson Lie structure. *Lett. Math. Phys.*, 24(1):79–83, 1992.
32. M. Cahen, S. Gutt, and J. Rawnsley. Some remarks on the classification of Poisson Lie groups. In *Symplectic geometry and quantization (Sanda and Yokohama, 1993)*, volume 179 of *Contemp. Math.*, pages 1–16. Amer. Math. Soc., Providence, RI, 1994.
33. A. Cannas da Silva. *Lectures on symplectic geometry*, volume 1764 of *Lecture Notes in Mathematics*. Springer-Verlag, Berlin, 2001.
34. A. Cannas da Silva and A. Weinstein. *Geometric models for noncommutative algebras*, volume 10 of *Berkeley Mathematics Lecture Notes*. American Mathematical Society, Providence, RI, 1999.
35. A. Cattaneo. On the integration of Poisson manifolds, Lie algebroids, and coisotropic submanifolds. *Lett. Math. Phys.*, 67(1):33–48, 2004.
36. A. Cattaneo. Deformation quantization from functional integrals. In *Déformation, quantification, théorie de Lie*, volume 20 of *Panoramas et Synthèses [Panoramas and Syntheses]*, pages 123–164. Société Mathématique de France, Paris, 2005.
37. A. Cattaneo and G. Felder. A path integral approach to the Kontsevich quantization formula. *Comm. Math. Phys.*, 212(3):591–611, 2000.
38. A. Cattaneo, G. Felder, and L. Tomassini. From local to global deformation quantization of Poisson manifolds. *Duke Math. J.*, 115(2):329–352, 2002.
39. A. Cattaneo, B. Keller, Ch. Torossian, and A. Bruguières. *Déformation, quantification, théorie de Lie*, volume 20 of *Panoramas et Synthèses [Panoramas and Syntheses]*. Société Mathématique de France, Paris, 2005.

40. V. Chari and A. Pressley. *A guide to quantum groups*. Cambridge University Press, Cambridge, 1995. Corrected reprint of the 1994 original.
41. S. Chern and J. Wolfson. A simple proof of Frobenius theorem. In *Manifolds and Lie groups (Notre Dame, IN, 1980)*, volume 14 of *Progr. Math.*, pages 67–69. Birkhäuser, Boston, MA, 1981.
42. C. Chevalley. Invariants of finite groups generated by reflections. *Amer. J. Math.*, 77:778–782, 1955.
43. N. Chriss and V. Ginzburg. *Representation theory and complex geometry*. Birkhäuser Boston Inc., Boston, MA, 1997.
44. D. Collingwood and W. McGovern. *Nilpotent orbits in semisimple Lie algebras*. Van Nostrand Reinhold Mathematics Series. Van Nostrand Reinhold Co., New York, 1993.
45. J. Conn. Normal forms for analytic Poisson structures. *Ann. of Math. (2)*, 119(3):577–601, 1984.
46. J. Conn. Normal forms for smooth Poisson structures. *Ann. of Math. (2)*, 121(3):565–593, 1985.
47. A. Coste, P. Dazord, and A. Weinstein. Groupoïdes symplectiques. In *Publications du Département de Mathématiques. Nouvelle Série. A, Vol. 2*, volume 87 of *Publ. Dép. Math. Nouvelle Sér. A*, pages i–ii, 1–62. Univ. Claude-Bernard, Lyon, 1987.
48. Th. Courant. Dirac manifolds. *Trans. Amer. Math. Soc.*, 319(2):631–661, 1990.
49. M. Crainic and R. Fernandes. Integrability of Lie brackets. *Ann. of Math. (2)*, 157(2):575–620, 2003.
50. M. Crainic and R. Fernandes. Integrability of Poisson brackets. *J. Differential Geom.*, 66(1):71–137, 2004.
51. M. Crainic and R. Fernandes. Stability of symplectic leaves. *Invent. Math.*, 180(3):481–533, 2010.
52. M. Crainic and R. Fernandes. A geometric approach to Conn's linearization theorem. *Ann. of Math. (2)*, 173(2):1121–1139, 2011.
53. M. Crainic and R. Fernandes. Lectures on integrability of Lie brackets. In *Lectures on Poisson geometry*, volume 17 of *Geom. Topol. Monogr.*, pages 1–107. Geom. Topol. Publ., Coventry, 2011.
54. P. Damianou and R. Fernandes. From the Toda lattice to the Volterra lattice and back. *Rep. Math. Phys.*, 50(3):361–378, 2002.
55. P. Damianou and R. Fernandes. Integrable hierarchies and the modular class. *Ann. Inst. Fourier (Grenoble)*, 58(1):107–137, 2008.
56. P. Damianou, H. Sabourin, and P. Vanhaecke. Transverse Poisson structures to adjoint orbits in semisimple Lie algebras. *Pacific J. Math.*, 232(1):111–138, 2007.
57. M. De Wilde and P. Lecomte. Existence of star-products and of formal deformations of the Poisson Lie algebra of arbitrary symplectic manifolds. *Lett. Math. Phys.*, 7(6):487–496, 1983.
58. V. Drinfel'd. Hamiltonian structures on Lie groups, Lie bialgebras and the geometric meaning of classical Yang-Baxter equations. *Dokl. Akad. Nauk SSSR*, 268(2):285–287, 1983.
59. V. Drinfel'd. Quantum groups. In *Proceedings of the International Congress of Mathematicians, Vol. 1, 2 (Berkeley, CA, 1986)*, pages 798–820, Providence, RI, 1987. Amer. Math. Soc.
60. J.-P. Dufour and A. Haraki. Rotationnels et structures de Poisson quadratiques. *C. R. Acad. Sci. Paris Sér. I Math.*, 312(1):137–140, 1991.
61. J.-P. Dufour and M. Zhitomirskii. Classification of nonresonant Poisson structures. *J. London Math. Soc. (2)*, 60(3):935–950, 1999.
62. J.-P. Dufour and M. Zhitomirskii. Singularities and bifurcations of 3-dimensional Poisson structures. *Israel J. Math.*, 121:199–220, 2001.
63. J.-P. Dufour and N. T. Zung. *Poisson structures and their normal forms*, volume 242 of *Progress in Mathematics*. Birkhäuser Verlag, Basel, 2005.
64. J.-P. Dufour and N.T. Zung. Linearization of Nambu structures. *Compositio Math.*, 117(1):77–98, 1999.

65. J. Duistermaat. On global action-angle coordinates. *Comm. Pure Appl. Math.*, 33(6):687–706, 1980.
66. J. Duistermaat and J. Kolk. *Lie groups*. Universitext. Springer-Verlag, Berlin, 2000.
67. Ch. Ehresmann. Les connexions infinitésimales dans un espace fibré différentiable. In *Colloque de topologie (espaces fibrés), Bruxelles, 1950*, pages 29–55. Georges Thone, Liège, 1951.
68. P. Etingof and V. Ginzburg. Noncommutative del Pezzo surfaces and Calabi-Yau algebras. *J. Eur. Math. Soc. (JEMS)*, 12(6):1371–1416, 2010.
69. P. Etingof and A. Varchenko. Geometry and classification of solutions of the classical dynamical Yang-Baxter equation. *Comm. Math. Phys.*, 192(1):77–120, 1998.
70. P. Etingof and A. Varchenko. Solutions of the quantum dynamical Yang-Baxter equation and dynamical quantum groups. *Comm. Math. Phys.*, 196(3):591–640, 1998.
71. S. Evens, J.-H. Lu, and A. Weinstein. Transverse measures, the modular class and a cohomology pairing for Lie algebroids. *Quart. J. Math. Oxford Ser. (2)*, 50(200):417–436, 1999.
72. L. Faddeev and L. Takhtajan. *Hamiltonian methods in the theory of solitons*. Classics in Mathematics. Springer, Berlin, English edition, 2007. Translated from the 1986 Russian original by A. Reyman.
73. B. Fedosov. A simple geometrical construction of deformation quantization. *J. Differential Geom.*, 40(2):213–238, 1994.
74. R. Fernandes. Connections in Poisson geometry. I. Holonomy and invariants. *J. Differential Geom.*, 54(2):303–365, 2000.
75. R. Fernandes. Lie algebroids, holonomy and characteristic classes. *Adv. Math.*, 170(1):119–179, 2002.
76. R. Fernandes and Ph. Monnier. Linearization of Poisson brackets. *Lett. Math. Phys.*, 69:89–114, 2004.
77. V. V. Fock and A. A. Rosly. Poisson structure on moduli of flat connections on Riemann surfaces and the r-matrix. In *Moscow Seminar in Mathematical Physics*, volume 191 of *Amer. Math. Soc. Transl. Ser. 2*, pages 67–86. Amer. Math. Soc., Providence, RI, 1999.
78. A. Fomenko. *Integrability and nonintegrability in geometry and mechanics*, volume 31 of *Mathematics and its Applications (Soviet Series)*. Kluwer Academic Publishers Group, Dordrecht, 1988. Translated from the Russian by M. V. Tsaplina.
79. J.-P. Françoise, G. L. Naber, and T. S. Tsun, editors. *Encyclopedia of mathematical physics. Vol. 1, 2, 3, 4, 5*. Academic Press/Elsevier Science, Oxford, 2006.
80. W. Fulton and J. Harris. *Representation theory*, volume 129 of *Graduate Texts in Mathematics*. Springer-Verlag, New York, 1991. A first course, Readings in Mathematics.
81. I. Gel'fand and I. Dorfman. Hamiltonian operators and the classical Yang-Baxter equation. *Funktsional. Anal. i Prilozhen.*, 16(4):1–9, 96, 1982.
82. M. Gerstenhaber. On the deformation of rings and algebras. *Ann. of Math. (2)*, 79:59–103, 1964.
83. V. Ginzburg. Momentum mappings and Poisson cohomology. *Internat. J. Math.*, 7(3):329–358, 1996.
84. V. Ginzburg. Coisotropic intersections. *Duke Math. J.*, 140(1):111–163, 2007.
85. V. Ginzburg and D. Kaledin. Poisson deformations of symplectic quotient singularities. *Adv. Math.*, 186(1):1–57, 2004.
86. V. Ginzburg and A. Weinstein. Lie-Poisson structure on some Poisson Lie groups. *J. Amer. Math. Soc.*, 5(2):445–453, 1992.
87. J. Grabowski, G. Marmo, and P. W. Michor. Construction of completely integrable systems by Poisson mappings. *Modern Phys. Lett. A*, 14(30):2109–2118, 1999.
88. J. Grabowski, G. Marmo, and A. M. Perelomov. Poisson structures: towards a classification. *Modern Phys. Lett. A*, 8(18):1719–1733, 1993.
89. M. Gualtieri. Generalized complex geometry. *PhD. Thesis, Oxford*, 2004.
90. L. Guieu and C. Roger. *L'algèbre et le groupe de Virasoro*. Les Publications CRM, Montreal, QC, 2007. Aspects géométriques et algébriques, généralisations. [Geometric and algebraic aspects, generalizations], With an appendix by V. Sergiescu.

91. V. Guillemin. *Moment maps and combinatorial invariants of Hamiltonian T^n-spaces*, volume 122 of *Progress in Mathematics*. Birkhäuser Boston Inc., Boston, MA, 1994.
92. V. Guillemin and S. Sternberg. *Symplectic techniques in physics*. Cambridge University Press, Cambridge, second edition, 1990.
93. R. Hartshorne. *Algebraic geometry*. Springer-Verlag, New York, 1977. Graduate Texts in Mathematics, No. 52.
94. N. Hitchin. Instantons, Poisson structures and generalized Kähler geometry. *Comm. Math. Phys.*, 265(1):131–164, 2006.
95. G. Hochschild and J.-P. Serre. Cohomology of Lie algebras. *Ann. of Math. (2)*, 57:591–603, 1953.
96. J. Huebschmann. Poisson cohomology and quantization. *J. Reine Angew. Math.*, 408:57–113, 1990.
97. J. Huebschmann. Poisson structures on certain moduli spaces for bundles on a surface. *Ann. Inst. Fourier (Grenoble)*, 45(1):65–91, 1995.
98. J. Huebschmann. Symplectic and Poisson structures of certain moduli spaces. I. *Duke Math. J.*, 80(3):737–756, 1995.
99. J. Humphreys. *Introduction to Lie algebras and representation theory*, volume 9 of *Graduate Texts in Mathematics*. Springer-Verlag, New York, 1978. Second printing, revised.
100. C. Jacobi. Sur le mouvement d'un point et sur un cas particulier du problème des trois corps. *Compt. Rend.*, 3:59–61, 1836.
101. C. Jacobi. Note von der geodätischen linie auf einem ellipsoid und den verschiedenen anwendungen einer merkwürdigen analytischen substitution. *J. Reine Angew. Math.*, 19:309–313, 1839.
102. N. Jacobson. *Lie algebras*. Dover Publications Inc., New York, 1979. Republication of the 1962 original.
103. Ch. Kassel. *Quantum groups*, volume 155 of *Graduate Texts in Mathematics*. Springer-Verlag, New York, 1995.
104. S. Khoroshkin, A. Radul, and V. Rubtsov. A family of Poisson structures on Hermitian symmetric spaces. *Comm. Math. Phys.*, 152(2):299–315, 1993.
105. A. A. Kirillov. Unitary representations of nilpotent Lie groups. *Uspehi Mat. Nauk*, 17(4(106)):57–110, 1962.
106. A. A. Kirillov. Local Lie algebras. *Uspehi Mat. Nauk*, 31(4(190)):57–76, 1976.
107. M. Kontsevich. Deformation quantization of Poisson manifolds. *Lett. Math. Phys.*, 66(3):157–216, 2003 (q-alg/9709040, 1997).
108. Y. Kosmann-Schwarzbach. Quantum and classical Yang-Baxter equations. *Modern Phys. Lett. A*, 5(13):981–990, 1990.
109. Y. Kosmann-Schwarzbach. From Poisson algebras to Gerstenhaber algebras. *Ann. Inst. Fourier (Grenoble)*, 46(5):1243–1274, 1996.
110. Y. Kosmann-Schwarzbach. Lie bialgebras, Poisson Lie groups and dressing transformations. In *Integrability of nonlinear systems (Pondicherry, 1996)*, volume 495 of *Lecture Notes in Phys.*, pages 104–170. Springer, Berlin, 1997.
111. Y. Kosmann-Schwarzbach. Odd and even Poisson brackets. In *Quantum theory and symmetries (Goslar, 1999)*, pages 565–571. World Sci. Publ., River Edge, NJ, 2000.
112. Y. Kosmann-Schwarzbach. Quasi, twisted, and all that...in Poisson geometry and Lie algebroid theory. In *The breadth of symplectic and Poisson geometry*, volume 232 of *Progr. Math.*, pages 363–389. Birkhäuser Boston, Boston, MA, 2005.
113. Y. Kosmann-Schwarzbach. Poisson manifolds, Lie algebroids, modular classes: a survey. *SIGMA Symmetry Integrability Geom. Methods Appl.*, 4:Paper 005, 30, 2008.
114. Y. Kosmann-Schwarzbach. *The Noether theorems*. Sources and Studies in the History of Mathematics and Physical Sciences. Springer, New York, 2011. Invariance and conservation laws in the twentieth century, Translated, revised and augmented from the 2006 French edition by Bertram E. Schwarzbach.
115. Y. Kosmann-Schwarzbach and F. Magri. Poisson-Nijenhuis structures. *Ann. Inst. H. Poincaré Phys. Théor.*, 53(1):35–81, 1990.

116. Y. Kosmann-Schwarzbach and A. Weinstein. Relative modular classes of Lie algebroids. *C. R. Math. Acad. Sci. Paris*, 341(8):509–514, 2005.
117. B. Kostant. Orbits, symplectic structures and representation theory. In *Proc. U.S.-Japan Seminar in Differential Geometry (Kyoto, 1965)*, page 71. Nippon Hyoronsha, Tokyo, 1966.
118. J.-L. Koszul. Crochet de Schouten-Nijenhuis et cohomologie. *Astérisque*, (Numero Hors Serie):257–271, 1985. The mathematical heritage of Élie Cartan (Lyon, 1984).
119. A. Kotov, P. Schaller, and Th. Strobl. Dirac sigma models. *Comm. Math. Phys.*, 260(2):455–480, 2005.
120. C. Laurent-Gengoux, E. Miranda, and P. Vanhaecke. Action-angle coordinates for integrable systems on Poisson manifolds. *Int. Math. Res. Not. IMRN*, (8):1839–1869, 2011.
121. C. Laurent-Gengoux, M. Stiénon, and P. Xu. Holomorphic Poisson manifolds and holomorphic Lie algebroids. *Int. Math. Res. Not. IMRN*, pages Art. ID rnn 130, 30, 2008.
122. A. Le Blanc. Quasi-Poisson structures and integrable systems related to the moduli space of flat connections on a punctured Riemann sphere. *J. Geom. Phys.*, 57(8):1631–1652, 2007.
123. P. Lecomte and C. Roger. Modules et cohomologies des bigèbres de Lie. *C. R. Acad. Sci. Paris Sér. I Math.*, 310(6):405–410, 1990.
124. L.-C. Li and S. Parmentier. On dynamical Poisson groupoids. I. *Mem. Amer. Math. Soc.*, 174(824):vi+72, 2005.
125. P. Libermann and C.-M. Marle. *Symplectic geometry and analytical mechanics*, volume 35 of *Mathematics and its Applications*. D. Reidel Publishing Co., Dordrecht, 1987. Translated from the French by B. E. Schwarzbach.
126. A. Lichnerowicz. Les variétés de Poisson et leurs algèbres de Lie associées. *J. Differential Geometry*, 12(2):253–300, 1977.
127. A. Lichnerowicz. Variétés de Poisson et feuilletages. *Ann. Fac. Sci. Toulouse Math. (5)*, 4(3–4):195–262 (1983), 1982.
128. S. Lie and F. Engel. Theorie der transformationsgruppen. *Teubner Verlag, Leipzig*, 2, 1890.
129. Z.-J. Liu and P. Xu. On quadratic Poisson structures. *Lett. Math. Phys.*, 26(1):33–42, 1992.
130. Z.-J. Liu and P. Xu. Dirac structures and dynamical r-matrices. *Ann. Inst. Fourier (Grenoble)*, 51(3):835–859, 2001.
131. J.-L. Loday. *Cyclic homology*, volume 301 of *Grundlehren der Mathematischen Wissenschaften [Fundamental Principles of Mathematical Sciences]*. Springer-Verlag, Berlin, 1992. Appendix E by M. O. Ronco.
132. J.-H. Lu. Multiplicative and Affine Poisson structures on Lie groups. *PhD. Thesis, University of California at Berkeley*, 1990.
133. J.-H. Lu. Momentum mappings and reduction of Poisson actions. In *Symplectic geometry, groupoids, and integrable systems (Berkeley, CA, 1989)*, volume 20 of *Math. Sci. Res. Inst. Publ.*, pages 209–226. Springer, New York, 1991.
134. J.-H. Lu. Coordinates on Schubert cells, Kostant's harmonic forms, and the Bruhat Poisson structure on G/B. *Transform. Groups*, 4(4):355–374, 1999.
135. J.-H. Lu. Classical dynamical r-matrices and homogeneous Poisson structures on G/H and K/T. *Comm. Math. Phys.*, 212(2):337–370, 2000.
136. J.-H. Lu and A. Weinstein. Poisson Lie groups, dressing transformations, and Bruhat decompositions. *J. Differential Geom.*, 31(2):501–526, 1990.
137. J.-H. Lu and M. Yakimov. Group orbits and regular partitions of Poisson manifolds. *Comm. Math. Phys.*, 283(3):729–748, 2008.
138. A. Lundell. A short proof of the Frobenius theorem. *Proc. Amer. Math. Soc.*, 116(4):1131–1133, 1992.
139. K. Mackenzie. *General theory of Lie groupoids and Lie algebroids*, volume 213 of *London Mathematical Society Lecture Note Series*. Cambridge University Press, Cambridge, 2005.
140. K. Mackenzie and P. Xu. Lie bialgebroids and Poisson groupoids. *Duke Math. J.*, 73(2):415–452, 1994.
141. F. Magri. A simple model of the integrable Hamiltonian equation. *J. Math. Phys.*, 19(5):1156–1162, 1978.

142. S. Majid. *Foundations of quantum group theory*. Cambridge University Press, Cambridge, 1995.
143. Ch.-M. Marle. The Schouten-Nijenhuis bracket and interior products. *J. Geom. Phys.*, 23(3–4):350–359, 1997.
144. J. Marsden and T. Ratiu. Reduction of Poisson manifolds. *Lett. Math. Phys.*, 11(2):161–169, 1986.
145. J. Marsden and T. Ratiu. *Introduction to mechanics and symmetry*, volume 17 of *Texts in Applied Mathematics*. Springer-Verlag, New York, second edition, 1999. A basic exposition of classical mechanical systems.
146. J. Marsden and A. Weinstein. Reduction of symplectic manifolds with symmetry. *Rep. Mathematical Phys.*, 5(1):121–130, 1974.
147. J. McKay. Graphs, singularities, and finite groups. In *The Santa Cruz Conference on Finite Groups (Univ. California, Santa Cruz, CA, 1979)*, volume 37 of *Proc. Sympos. Pure Math.*, pages 183–186. Amer. Math. Soc., Providence, RI, 1980.
148. Kenneth R. Meyer. Symmetries and integrals in mechanics. In *Dynamical systems (Proc. Sympos., Univ. Bahia, Salvador, 1971)*, pages 259–272. Academic Press, New York, 1973.
149. E. Miranda and N.T. Zung. A note on equivariant normal forms of Poisson structures. *Math. Res. Lett.*, 13(5-6):1001–1012, 2006.
150. Ph. Monnier. Une cohomologie associée à une fonction. Applications aux cohomologies de Poisson et de Nambu-poisson. *PhD. Thesis, Université de Montpellier*, 2001.
151. Ph. Monnier. Formal Poisson cohomology of quadratic Poisson structures. *Lett. Math. Phys.*, 59(3):253–267, 2002.
152. Ph. Monnier. Poisson cohomology in dimension two. *Israel J. Math.*, 129:189–207, 2002.
153. Ph. Monnier and N. T. Zung. Levi decomposition for smooth Poisson structures. *J. Differential Geom.*, 68(2):347–395, 2004.
154. D. Mumford, J. Fogarty, and F. Kirwan. *Geometric invariant theory*, volume 34 of *Ergebnisse der Mathematik und ihrer Grenzgebiete (2) [Results in Mathematics and Related Areas (2)]*. Springer-Verlag, Berlin, third edition, 1994.
155. J. Nestruev. *Smooth manifolds and observables*, volume 220 of *Graduate Texts in Mathematics*. Springer-Verlag, New York, 2003. Joint work of A. M. Astashov, A. B. Bocharov, S. V. Duzhin, A. B. Sossinsky, A. M. Vinogradov and M. M. Vinogradov, Translated from the 2000 Russian edition by Sossinsky, I. S. Krasil'schik and Duzhin.
156. A. Nijenhuis and R. Richardson. Deformations of Lie algebra structures. *J. Math. Mech.*, 17:89–105, 1967.
157. A. Odesskiĭ and V. Rubtsov. Polynomial Poisson algebras with a regular structure of symplectic leaves. *Teoret. Mat. Fiz.*, 133(1):1321–1337, 2002.
158. P. Olver. *Applications of Lie groups to differential equations*, volume 107 of *Graduate Texts in Mathematics*. Springer-Verlag, New York, second edition, 1993.
159. J.-P. Ortega and T. Ratiu. The optimal momentum map. In *Geometry, mechanics, and dynamics*, pages 329–362. Springer, New York, 2002.
160. J.-P. Ortega and T. Ratiu. *Momentum maps and Hamiltonian reduction*, volume 222 of *Progress in Mathematics*. Birkhäuser Boston Inc., Boston, MA, 2004.
161. V. Ovsienko and K. Rozhe. Deformations of Poisson brackets and extensions of Lie algebras of contact vector fields. *Russian Math. Surveys*, 47(6):135–191, 1992.
162. J.-S. Park. Topological open p-branes. In *Symplectic geometry and mirror symmetry (Seoul, 2000)*, pages 311–384. World Sci. Publ., River Edge, NJ, 2001.
163. S. Pelap. Poisson (co)homology of polynomial Poisson algebras in dimension four: Sklyanin's case. *J. Algebra*, 322(4):1151–1169, 2009.
164. A. M. Perelomov. *Integrable systems of classical mechanics and Lie algebras. Vol. I.* Birkhäuser Verlag, Basel, 1990. Translated from the Russian by A. G. Reyman.
165. D. Perrin. *Algebraic geometry*. Universitext. Springer-Verlag London Ltd., London, 2008. An introduction, Translated from the 1995 French original by C. Maclean.
166. A. Pichereau. Poisson (co)homology and isolated singularities. *J. Algebra*, 299(2):747–777, 2006.

167. A. Pichereau. Formal deformations of Poisson structures in low dimensions. *Pacific J. Math.*, 239(1):105–133, 2009.
168. A. Pichereau and G. Van de Weyer. Double Poisson cohomology of path algebras of quivers. *J. Algebra*, 319(5):2166–2208, 2008.
169. S. Poisson. Mémoire sur la variation des constantes arbitraires dans les questions de mécanique. *J. Ecole Polytec.*, 8:266–344, 1809.
170. A. Polishchuk. Algebraic geometry of Poisson brackets. *J. Math. Sci. (New York)*, 84(5):1413–1444, 1997. Algebraic geometry, 7.
171. J. Pradines. Théorie de Lie pour les groupoïdes différentiables. Calcul différentiel dans la catégorie des groupoïdes infinitésimaux. *C. R. Acad. Sci. Paris Sér. A-B*, 264:A245–A248, 1967.
172. B. Przybylski. Complex Poisson manifolds. In *Differential geometry and its applications (Opava, 1992)*, volume 1 of *Math. Publ.*, pages 227–241. Silesian Univ. Opava, Opava, 1993.
173. O. Radko. A classification of topologically stable Poisson structures on a compact oriented surface. *J. Symplectic Geom.*, 1(3):523–542, 2002.
174. C. Roger and P. Vanhaecke. Poisson cohomology of the affine plane. *J. Algebra*, 251(1):448–460, 2002.
175. V. Roubtsov and T. Skrypnyk. Compatible Poisson brackets, quadratic Poisson algebras and classical r-matrices. In *Differential equations: geometry, symmetries and integrability*, volume 5 of *Abel Symp.*, pages 311–333. Springer, Berlin, 2009.
176. P. Saksida. Lattices of Neumann oscillators and Maxwell-Bloch equations. *Nonlinearity*, 19(3):747–768, 2006.
177. J. A. Schouten. Ueber Differentialkomitanten zweier kontravarianter Grössen. *Nederl. Akad. Wetensch., Proc.*, 43:449–452, 1940.
178. J. A. Schouten. *On the differential operators of first order in tensor calculus*. Rapport ZA 1953-012. Math. Centrum Amsterdam, 1953.
179. M. Semenov-Tian-Shansky. What a classical r-matrix is. *Functional Anal. Appl.*, 17(4):259–272, 1983.
180. M. Semenov-Tian-Shansky. Dressing transformations and Poisson group actions. *Publ. Res. Inst. Math. Sci.*, 21(6):1237–1260, 1985.
181. J.-P. Serre. *Local fields*, volume 67 of *Graduate Texts in Mathematics*. Springer-Verlag, New York, 1979. Translated from the French by M. J. Greenberg.
182. I. Shafarevich. *Basic algebraic geometry. 1*. Springer-Verlag, Berlin, second edition, 1994. Varieties in projective space, Translated from the 1988 Russian edition and with notes by M. Reid.
183. P. Slodowy. *Simple singularities and simple algebraic groups*, volume 815 of *Lecture Notes in Mathematics*. Springer, Berlin, 1980.
184. J. Śniatycki and W. M. Tulczyjew. Generating forms of Lagrangian submanifolds. *Indiana Univ. Math. J.*, 22:267–275, 1972/73.
185. J.-M. Souriau. Quantification géométrique. *Comm. Math. Phys.*, 1:374–398, 1966.
186. M. Spivak. *Calculus on manifolds. A modern approach to classical theorems of advanced calculus*. W. A. Benjamin, Inc., New York-Amsterdam, 1965.
187. M. Spivak. *A comprehensive introduction to differential geometry. Vol. I*. Publish or Perish Inc., Wilmington, DE, second edition, 1979.
188. T. A. Springer. *Invariant theory*. Lecture Notes in Mathematics, Vol. 585. Springer-Verlag, Berlin, 1977.
189. D. E. Tamarkin. *Operadic proof of M. Kontsevich's formality theorem*. ProQuest LLC, Ann Arbor, MI, 1999. Thesis (Ph.D.)–The Pennsylvania State University.
190. P. Tauvel and R. Yu. *Lie algebras and algebraic groups*. Springer Monographs in Mathematics. Springer-Verlag, Berlin, 2005.
191. A. Thimm. Integrable geodesic flows on homogeneous spaces. *Ergodic Theory Dynamical Systems*, 1(4):495–517 (1982), 1981.

192. W. M. Tulczyjew. *Geometric formulations of physical theories*, volume 11 of *Monographs and Textbooks in Physical Science. Lecture Notes*. Bibliopolis, Naples, 1989. Statics and dynamics of mechanical systems.
193. I. Vaisman. Remarks on the Lichnerowicz-Poisson cohomology. *Ann. Inst. Fourier (Grenoble)*, 40(4):951–963 (1991), 1990.
194. I. Vaisman. *Lectures on the geometry of Poisson manifolds*, volume 118 of *Progress in Mathematics*. Birkhäuser Verlag, Basel, 1994.
195. I. Vaisman. *Lectures on symplectic and Poisson geometry*. Textos de Matemática. Série B [Texts in Mathematics. Series B], 23. Universidade de Coimbra Departamento de Matemática, Coimbra, 2000. Notes edited by Fani Petalidou.
196. P. Vanhaecke. *Integrable systems in the realm of algebraic geometry*, volume 1638 of *Lecture Notes in Mathematics*. Springer-Verlag, Berlin, second edition, 2001.
197. A. M. Vinogradov. The union of the Schouten and Nijenhuis brackets, cohomology, and superdifferential operators. *Mat. Zametki*, 47(6):138–140, 1990.
198. F. Warner. *Foundations of differentiable manifolds and Lie groups*, volume 94 of *Graduate Texts in Mathematics*. Springer-Verlag, New York, 1983. Corrected reprint of the 1971 edition.
199. A. Weinstein. The local structure of Poisson manifolds. *J. Differential Geom.*, 18(3):523–557, 1983.
200. A. Weinstein. Affine Poisson structures. *Internat. J. Math.*, 1(3):343–360, 1990.
201. A. Weinstein. The modular automorphism group of a Poisson manifold. *J. Geom. Phys.*, 23(3-4):379–394, 1997.
202. A. Weinstein. Poisson geometry. *Differential Geom. Appl.*, 9(1-2):213–238, 1998.
203. R. Wells. *Differential analysis on complex manifolds*, volume 65 of *Graduate Texts in Mathematics*. Springer, New York, third edition, 2008. With a new appendix by O. Garcia-Prada.
204. E. Whittaker. *A treatise on the analytical dynamics of particles and rigid bodies*. Cambridge Mathematical Library. Cambridge University Press, Cambridge, 1988. With an introduction to the problem of three bodies, Reprint of the 1937 edition, With a foreword by W. McCrea.
205. P. Xu. Poisson cohomology of regular Poisson manifolds. *Ann. Inst. Fourier (Grenoble)*, 42(4):967–988, 1992.
206. P. Xu. Gerstenhaber algebras and BV-algebras in Poisson geometry. *Comm. Math. Phys.*, 200(3):545–560, 1999.
207. P. Xu. Dirac submanifolds and Poisson involutions. *Ann. Sci. École Norm. Sup. (4)*, 36(3):403–430, 2003.
208. P. Xu. Momentum maps and Morita equivalence. *J. Differential Geom.*, 67(2):289–333, 2004.
209. M. Yakimov. Symplectic leaves of complex reductive Poisson-Lie groups. *Duke Math. J.*, 112(3):453–509, 2002.

Index

C. Laurent-Gengoux et al., *Poisson Structures*,
Grundlehren der mathematischen Wissenschaften 347,
DOI 10.1007/978-3-642-31090-4, © Springer-Verlag Berlin Heidelberg 2013

List of Notations

C. Laurent-Gengoux et al., *Poisson Structures*,
Grundlehren der mathematischen Wissenschaften 347,
DOI 10.1007/978-3-642-31090-4, © Springer-Verlag Berlin Heidelberg 2013